LINEAR PROGRAMMING AND NETWORK FLOWS

LINEAR PROGRAMMING AND NETWORK FLOWS

SECOND EDITION

MOKHTAR S. BAZARAA
Senior Vice President
Burnham Service Corporation

JOHN J. JARVIS
School of Industrial and Systems Engineering
Georgia Institute of Technology

HANIF D. SHERALI
Department of Industrial Engineering and Operations Research
Virginia Polytechnic Institute and State University

WILEY

LINEAR PROGRAMMING AND NETWORK FLOWS
Second Edition

Authorized reprint by Wiley India (P.) Ltd., 4435/7, Ansari Road, Daryaganj, New Delhi 110 002.

Reprint : 2014

Printed at : *Ar Emm International, New Delhi*

ISBN : 978-81-265-1892-0

Dedicated to Our Parents

PREFACE

Linear Programming deals with the problem of minimizing or maximizing a linear function in the presence of linear equality and/or inequality constraints. Since the development of the simplex method by George B. Dantzig in 1947, linear programming has been extensively used in the military, industrial, governmental, and urban planning fields, among others. The popularity of linear programming can be attributed to many factors including its ability to model large and complex problems, and the ability of the users to solve large problems in a reasonable amount of time by the use of effective algorithms and modern computers.

During and after World War II it became evident that planning and coordination among various projects and the efficient utilization of scarce resources were essential. Intensive work by the United States Air Force team SCOOP (Scientific Computation of Optimum Programs) began in June 1947. As a result, the simplex method was developed by George B. Dantzig by the end of the summer of 1947. Interest in linear programming spread quickly among economists, mathematicians, statisticians, and government institutions. In the summer of 1949 a conference on linear programming was held under the sponsorship of the Cowles Commission for Research in Economics. The papers presented at that conference were later collected in 1951 by T. C. Koopmans into the book *Activity Analysis of Production and Allocation*.

Since the development of the simplex method many people have contributed to the growth of linear programming by developing its mathematical theory, devising efficient computational methods and codes, exploring new algorithms and new applications, and by their use of linear programming as an aiding tool for solving more complex problems, for instance, discrete programs, nonlinear programs, combinatorial problems, stochastic programming problems, and problems of optimal control.

This book addresses linear programming and network flows. Both the general theory and characteristics of these optimization problems as well as effective solution algorithms are presented. The simplex algorithm provides considerable insight into the theory of linear programming and yields an efficient algorithm in practice. Hence, in this text we study this method in detail. Whenever possible, the simplex algorithm is specialized to take advantage of the problem structure, such as in network flow problems. We also present Khachian's and Karmarkar's polynomial-time algorithms for linear programming problems. The latter algorithm compares favorably with the simplex method, particularly for general large-scale problems, and is therefore discussed in some detail. Throughout, we first present the concepts or the techniques, illustrate these by numerical examples, and then provide further insights along with a detailed mathematical analysis and justification. Rigorous proofs of the results are given without the theorem–proof format. Even though this may bother some readers, we believe that the format and mathematical level adopted in this book will provide an adequate and smooth study for those who wish to learn the techniques and the know-how to use them, as well as for those who wish to study the theory and the algorithms at a more rigorous level.

The book can be used both as a reference and as a textbook for advanced undergraduate students and first-year graduate students in the fields of industrial engineering, management, operations research, computer science, mathematics, and other engineering disciplines that deal with the subjects of linear programming and network flows. Even though the book's material requires some mathematical maturity, the only prerequisite is linear algebra and elementary calculus. For the convenience of the reader, pertinent results from linear algebra and convex analysis are summarized in Chapter 2.

This book can be used in several ways. It can be used in a two-course sequence on linear programming and network flows, in which case all of its material could be easily covered. The book can also be utilized in a one-semester course on linear programming and network flows. The instructor may have to omit some topics at his or her discretion. The book can also be used as a text for a course on either linear programming or network flows.

Following the introductory first chapter, the second chapter presents basic results on linear algebra and convex analysis, along with an insightful, geometrically motivated study of the structure of polyhedral sets. Following this, the book is organized into two parts: linear programming and networks flows. The linear programming part consists of Chapters 3 to 8. In Chapter 3 the simplex method is developed in detail, and in Chapter 4 the initiation of the simplex method by the use of artificial variables and the problem of degeneracy are discussed. Chapter 5 deals with some specializations of the simplex method and the development of optimality criteria in linear programming. In Chapter 6 we consider the dual problem, develop several computational procedures, based on duality, and discuss sensitivity (including the tolerance approach) and parametric analyses. Chapter 7 introduces the reader to the decomposition principle and large-scale programming. The equivalence of several decomposition techniques for linear programming problems is exhibited in this chapter. Chapter 8 discusses some basic computational complexity issues, exhibits the worst-case exponential behavior of the simplex algorithm, and presents Karmarkar's polynomial-time algorithm. Khachian's polynomial ellipsoid algorithm and the affine scaling variant of Karmarkar's algorithm are presented in the Exercises.

The part on network flows consists of Chapters 9 to 12. In Chapter 9 we study the principal characteristics of network structured linear programming problems and discuss the specialization of the simplex algorithm to solve these problems. A detailed discussion of list structures, useful from a terminology as well as from an implementation viewpoint, is also presented. Chapter 10 deals with the popular transportation and assignment network flow problems. Although the algorithmic justifications and some special techniques rely on the material in Chapter 9, it is possible to study Chapter 10 separately if one is simply interested in the fundamental properties and algorithms for transportation and assignment problems. Chapter 11 presents the out-of-kilter algorithm along with some basic ingredients of primal–dual types of algorithms for network flow problems. Finally, Chapter 12 covers the special topics of the maximal flow problem, the shortest path problem (including several efficient polynomial-time algorithms for this popular problem), the multicommodity minimal cost flow problem, and a network synthesis or design problem. The last of these topics compliments as well as relies on the techniques developed for the problems of *analysis*, which are the types of problems considered in the rest of this book.

In revising the first edition, we followed two principal objectives. The first objective was to offer further concepts and insights into linear programming theory and algorithmic techniques. Toward this end, we have included detailed geometrically motivated discussions dealing with the structure of polyhedral sets, optimality conditions, and the nature of solution algorithms. We have also added examples and remarks throughout the book that provide insights, and improve the understanding of and correlate the various topics discussed in the book. The second objective was to update the book to the state-of-the-art while keeping the exposition transparent and easy to follow. In keeping with this spirit, several new topics have been introduced, such as cycling and stalling phenomena and their prevention (including special approaches for network flow problems), numerically stable implementation techniques and empirical studies dealing with the simplex algorithm, the tolerance approach to sensitivity analysis, the equivalence of the Dantzig–Wolfe decomposition, Benders' partitioning method and Lagrangian relaxation techniques for linear programming problems, computational complexity issues, the worst-case behavior of the simplex method, Khachian's and Karmarkar's polynomial-time algorithms for linear programming problems, list structures for network simplex implementations, a successive shortest path algorithm for linear assignment problems, polynomial-time partitioned shortest path algorithms, and the network synthesis or design problem, among others. The writing style enables the instructor to skip several of these advanced topics in an undergraduate or introductory level graduate course without any loss of continuity. Also, most of the old exercises have been revised and new exercises have been added. The notes and references sections and the bibliography have been considerably extended.

We re-express our gratitude to Dr. Jeff Kennington, Dr. Gene Ramsay, Dr. Ron Rardin, and Dr. Michael Todd for their many fine suggestions during the preparation of the first edition; to Dr. Robert N. Lehrer, former director of the School of Industrial and Systems Engineering at the Georgia Institute of Technology, for his support during the preparation of this edition; to Mr. Carl H. Wohlers for preparing its bibliography; and to Mrs. Alice Jarvis, Mrs. Carolyn Piersma, Miss Kaye Watkins, and Mrs. Amelia Williams for their typing assistance. We also thank Dr. Faiz Al-Khayyal, Dr. Dick Cottle, Dr. Joanna Leleno, Dr. Craig Tovey, and Dr. Hossam Zaki among many others for their helpful comments in preparing this manuscript. We are grateful to Dr. Robert D. Dryden, chairman of the Department of Industrial Engineering and Operations Research at the Virginia Polytechnic Institute and State University for his support during the preparation of this second edition, and to Mrs. Semeen

Sherali for her assistance with typing the revisions. Special thanks are due to Dr. Suleyman Tufekci, and to Dr. Joanna Leleno and to Dr. Zhuangyi Liu for respectively preparing the first and second versions of the solutions' manual.

Mokhtar S. Bazaraa
John J. Jarvis
Hanif D. Sherali

CONTENTS

ONE: INTRODUCTION

Linear programming is concerned with the optimization (minimization or maximization) of a linear function while satisfying a set of linear equality and/or inequality constraints or restrictions. The linear programming problem was first conceived by George B. Dantzig around 1947 while he was working as a Mathematical Advisor to the United States Air Force Comptroller on developing a mechanized planning tool for a time-staged deployment, training, and logistical supply *program*. Although the Soviet mathematician and economist L. V. Kantorovich formulated and solved a problem of this type dealing with organization and planning in 1939, his work remained unknown until 1959. Hence the conception of the *general class* of linear programming problems is usually credited to Dantzig. Because the Air Force refers to its various plans and schedules to be implemented as "programs," Dantzig's first published paper addressed this problem as "Programming in a Linear Structure." The term *linear programming* was actually coined by the economist and mathematician T. C. Koopmans in the summer of 1948 while he and Dantzig strolled near the Santa Monica beach in California.

In 1949 George B. Dantzig published the "simplex method" for solving linear programs. Since that time a number of individuals have contributed to the field of linear programming in many different ways, including theoretical development, computational aspects, and exploration of new applications of the subject. The simplex method of linear programming enjoys wide acceptance because of (1) its ability to model important and complex management decision problems and (2) its capability for producing solutions in a reasonable amount of time. In subsequent chapters of this text we shall consider the simplex method and its variants, with emphasis on the understanding of the methods.

In this chapter the linear programming problem is introduced. The following topics are discussed: basic definitions in linear programming, assumptions leading to linear models, manipulation of the problem, examples of linear problems, and geometric solution in the feasible region space and the requirement space.

This chapter is elementary and may be skipped if the reader has previous knowledge of linear programming.

1.1 THE LINEAR PROGRAMMING PROBLEM

We begin our discussion by formulating a particular type of linear programming problem. As will be seen subsequently, any general linear programming problem may be manipulated into this form.

Basic Definitions

Consider the following linear programming problem.

$$
\begin{array}{ll}
\text{Minimize} & c_1x_1 + c_2x_2 + \cdots + c_nx_n \\
\text{Subject to} & a_{11}x_1 + a_{12}x_2 + \cdots + a_{1n}x_n \geq b_1 \\
& a_{21}x_1 + a_{22}x_2 + \cdots + a_{2n}x_n \geq b_2 \\
& \quad \vdots \qquad\quad \vdots \qquad\qquad\quad \vdots \qquad\;\; \vdots \\
& a_{m1}x_1 + a_{m2}x_2 + \cdots + a_{mn}x_n \geq b_m \\
& \quad x_1, \qquad x_2, \quad \ldots, \qquad x_n \geq 0
\end{array}
$$

Here $c_1x_1 + c_1x_2 + , \ldots, + c_nx_n$ is the *objective function* (or criterion function) to be minimized and will be denoted by z. The coefficients $c_1, c_2, \ldots, c_n$ are the (known) *cost coefficients* and $x_1, x_2, \ldots, x_n$ are the *decision variables* (variables, structural variables, or activity levels) to be determined. The inequality $\Sigma_{j=1}^{n} a_{ij}x_j \geq b_i$ denotes the ith *constraint* (or restriction or functional, structural, or technological constraint). The coefficients a_{ij} for $i = 1, 2, \ldots, m$, $j = 1, 2, \ldots, n$ are called the *technological coefficients*. These technological coefficients form the *constraint matrix* $\mathbf{A}$.

$$
\mathbf{A} = \begin{bmatrix} a_{11} & a_{12} & \cdots & a_{1n} \\ a_{21} & a_{22} & \cdots & a_{2n} \\ \vdots & \vdots & & \vdots \\ a_{m1} & a_{m2} & \cdots & a_{mn} \end{bmatrix}
$$

The column vector whose ith component is b_i, which is referred to as the *right-hand-side vector*, represents the minimal requirements to be satisfied. The constraints $x_1, x_2, \ldots, x_n \geq 0$ are the *nonnegativity constraints*. A set of variables $x_1, \ldots, x_n$ satisfying all the constraints is called a *feasible point* or a *feasible vector*. The set of all such points constitutes the *feasible region* or the *feasible space*.

Using the foregoing terminology, the linear programming problem can be stated as follows: Among all feasible vectors, find one that minimizes (or maximizes) the objective function.

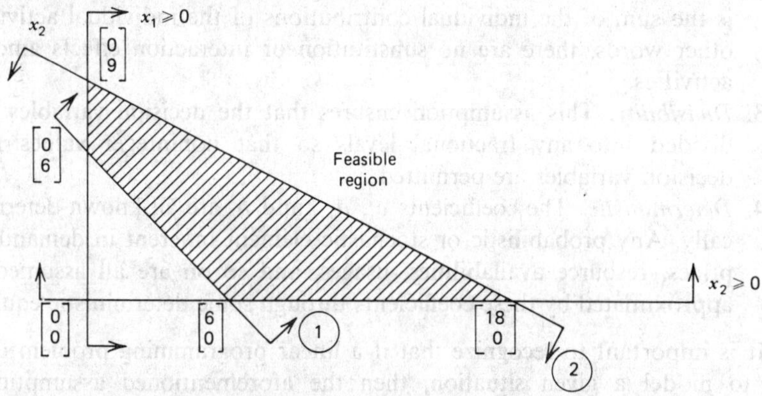

Figure 1.1. **Illustration of the feasible region.**

Example 1.1

Consider the following linear problem.

$$\text{Minimize} \quad 2x_1 + 5x_2$$

$$\begin{aligned} \text{Subject to} \quad x_1 + x_2 &\geq 6 \\ -x_1 - 2x_2 &\geq -18 \\ x_1, x_2 &\geq 0 \end{aligned}$$

In this case we have two decision variables x_1 and x_2. The objective function to be minimized is $2x_1 + 5x_2$. The constraints and the feasible region are illustrated in Figure 1.1. The optimization problem is thus to find a point in the feasible region with the smallest possible objective.

Assumptions of Linear Programming

To represent an optimization problem as a linear program, several assumptions that are implicit in the linear programming formulation discussed previously are needed. A brief discussion of these assumptions is given.

1. *Proportionality*. Given a variable x_j, its contribution to cost is c_jx_j and its contribution to the ith constraint is $a_{ij}x_j$. This means that if x_j is doubled, say, so is its contribution to cost and to each of the constraints. To illustrate, suppose that x_j is the amount of activity j used. For instance, if $x_j = 10$, then the cost of this activity is $10c_j$. If $x_j = 20$, then the cost is $20c_j$, and so on. This means that no savings (or extra costs) are realized by using more of activity j, that is, there are no economies or returns to scale or discounts. Also no setup cost for starting the activity is realized.
2. *Additivity*. This assumption guarantees that the total cost is the sum of the individual costs, and that the total contribution to the ith restriction

is the sum of the individual contributions of the individual activities. In other words, there are no substitution or interaction effects among the activities.

3. *Divisibility*. This assumption ensures that the decision variables can be divided into any fractional levels so that noninteger values for the decision variables are permitted.
4. *Deterministic*. The coefficients c_j, a_{ij}, and b_i are all known deterministically. Any probabilistic or stochastic element inherent in demand, costs, prices, resource availabilities, usages, and so on are all assumed to be approximated by these coefficients through some deterministic equivalent.

It is important to recognize that if a linear programming problem is being used to model a given situation, then the aforementioned assumptions are implied to hold, at least over some anticipated operating range for the activities. When Dantzig first presented his linear programming model to a meeting of the econometric society in Wisconsin, the famous economist H. Hotelling critically remarked that in reality, the world is indeed nonlinear. As Dantzig recounts, the well-known mathematician John von Neumann came to his rescue by countering that the talk was about "Linear" Programming and was based on a set of postulated axioms. Quite simply, a user may apply this technique if and only if the application fits the stated axioms.

Despite the seemingly restrictive assumptions, linear programs are among the most widely used models today. They represent several systems quite satisfactorily, and they are capable of providing a large amount of information besides simply a solution, as we shall see later, particularly in Chapter 6. Moreover, they are also often used to solve certain types of nonlinear optimization problems via (successive) linear approximations and constitute an important tool in solution methods for linear discrete optimization problems having integer restricted variables.

Problem Manipulation

Recall that a linear program is a problem of minimizing or maximizing a linear function in the presence of linear inequality and/or equality constraints. By simple manipulations the problem can be transformed from one form to another equivalent form. These manipulations are most useful in linear programming, as will be seen throughout the text.

INEQUALITIES AND EQUATIONS

An inequality can be easily transformed into an equation. To illustrate, consider the constraint given by $\sum_{j=1}^{n} a_{ij}x_j \geq b_i$. This constraint can be put in an equation form by subtracting the nonnegative *surplus* or *slack variable* x_{n+i} (sometimes denoted by s_i) leading to $\sum_{j=1}^{n} a_{ij}x_j - x_{n+i} = b_i$ and $x_{n+i} \geq 0$. Similarly the constraint $\sum_{j=1}^{n} a_{ij}x_j \leq b_i$ is equivalent to $\sum_{j=1}^{n} a_{ij}x_j + x_{n+i} = b_i$ and $x_{n+i} \geq 0$. Also an equation of the form $\sum_{j=1}^{n} a_{ij}x_j = b_i$ can be transformed into the two inequalities $\sum_{j=1}^{n} a_{ij}x_j \leq b_i$ and $\sum_{j=1}^{n} a_{ij}x_j \geq b_i$, although this is not the practice.

NONNEGATIVITY OF THE VARIABLES

For most practical problems the variables represent physical quantities and hence must be nonnegative. The simplex method is designed to solve linear programs where the variables are nonnegative. If a variable x_j is unrestricted in sign, then it can be replaced by $x_j' - x_j''$ where $x_j' \geq 0$ and $x_j'' \geq 0$. If $x_1, \ldots, x_k$ are some k variables that are all unrestricted in sign, then only one additional variable x'' is needed in the equivalent transformation $x_j = x_j' - x''$ for $j = 1, \ldots, k$, where $x_j' \geq 0$ and $x'' \geq 0$. (Here $-x''$ plays the role of representing the most negative variable, while all the other variables x_j are x_j' above this value.) Alternatively, one could solve for each unrestricted variable in terms of the other variables using any equation in which it appears, eliminate this variable from the problem by substitution using this equation, and then discard this equation from the problem. However this strategy is seldom used from a data management and numerical implementation viewpoint. Continuing, if $x_j \geq l_j$, then the new variable $x_j' = x_j - l_j$ is automatically nonnegative. Also if a variable x_j is restricted such that $x_j \leq u_j$ where $u_j \leq 0$, then the substitution $x_j' = u_j - x_j$ produces a nonnegative variable x_j'.

MINIMIZATION AND MAXIMIZATION PROBLEMS

Another problem manipulation is to convert a maximization problem into a minimization problem and conversely. Note that over any region

$$\text{Maximum} \sum_{j=1}^{n} c_j x_j = -\text{minimum} \sum_{j=1}^{n} -c_j x_j$$

So a maximization (minimization) problem can be converted into a minimization (maximization) problem by multiplying the coefficients of the objective function by -1. After the optimization of the new problem is completed, the objective of the old problem is -1 times the optimal objective of the new problem.

Standard and Canonical Formats

For the foregoing discussion we see that a given linear program can be put in different equivalent forms by suitable manipulations. Two forms in particular will be useful. These are the standard and the canonical forms. A linear program is said to be in *standard format* if all restrictions are equalities and all variables are nonnegative. The simplex method is designed to be applied only after the problem is put in the standard form. The canonical form is also useful especially in exploiting duality relationships. A minimization problem is in *canonical form* if all variables are nonnegative and all the constraints are of the $\geq$ type. A maximization problem is in canonical format if all the variables are nonnegative and all the constraints are of the $\leq$ type. The standard and canonical forms are summarized in Table 1.1.

Table 1.1 Standard and Canonical Forms

	MINIMIZATION PROBLEM	MAXIMIZATION PROBLEM
STANDARD FORM	Minimize $\sum_{j=1}^{n} c_j x_j$ Subject to $\sum_{j=1}^{n} a_{ij} x_j = b_i \quad i = 1, \ldots, m$ $x_j \geq 0 \quad j = 1, \ldots, n$	Maximize $\sum_{j=1}^{n} c_j x_j$ Subject to $\sum_{j=1}^{n} a_{ij} x_j = b_i \quad i = 1, \ldots, m$ $x_j \geq 0 \quad j = 1, \ldots, n$
CANONICAL FORM	Minimize $\sum_{j=1}^{n} c_j x_j$ Subject to $\sum_{j=1}^{n} a_{ij} x_j \geq b_i, \quad i = 1, \ldots, m$ $x_j \geq 0 \quad j = 1, \ldots, n$	Maximize $\sum_{j=1}^{n} c_j x_j$ Subject to $\sum_{j=1}^{n} a_{ij} x_j \leq b_i \quad i = 1, \ldots, m$ $x_j \geq 0 \quad j = 1, \ldots, n$

Linear Programming in Matrix Notation

A linear programming problem can be stated in a more convenient form using matrix notation. To illustrate, consider the following problem.

$$\text{Minimize} \quad \sum_{j=1}^{n} c_j x_j$$

$$\begin{aligned} \text{Subject to} \quad & \sum_{j=1}^{n} a_{ij} x_j = b_i \qquad i = 1, 2, \ldots, m \\ & x_j \geq 0 \qquad j = 1, 2, \ldots, n \end{aligned}$$

Denote the row vector $(c_1, c_2, \ldots, c_n)$ by **c**, and consider the following column vectors **x** and **b**, and the $m \times n$ matrix **A**.

$$\mathbf{x} = \begin{bmatrix} x_1 \\ x_2 \\ \vdots \\ x_n \end{bmatrix} \qquad \mathbf{b} = \begin{bmatrix} b_1 \\ b_2 \\ \vdots \\ b_m \end{bmatrix} \qquad \mathbf{A} = \begin{bmatrix} a_{11} & a_{12} & \cdots & a_{1n} \\ a_{21} & a_{22} & \cdots & a_{2n} \\ \vdots & \vdots & & \vdots \\ a_{m1} & a_{m2} & \cdots & a_{mn} \end{bmatrix}$$

Then the problem can be written as follows.

$$\text{Minimize} \quad \mathbf{cx}$$

$$\begin{aligned} \text{Subject to} \quad & \mathbf{Ax} = \mathbf{b} \\ & \mathbf{x} \geq \mathbf{0} \end{aligned}$$

The problem can also be conveniently represented via the columns of **A**. Denoting **A** by $[\mathbf{a}_1, \mathbf{a}_2, \ldots, \mathbf{a}_n]$ where $\mathbf{a}_j$ is the jth column of **A**, the problem can be formulated as follows.

$$\text{Minimize} \quad \sum_{j=1}^{n} c_j x_j$$

$$\begin{aligned} \text{Subject to} \quad & \sum_{j=1}^{n} \mathbf{a}_j x_j = \mathbf{b} \\ & x_j \geq 0 \qquad j = 1, 2, \ldots, n \end{aligned}$$

1.2 LINEAR PROGRAMMING MODELING AND EXAMPLES

The modeling and analysis of an operations research problem in general and a linear programming problem in particular evolves through several stages. The *problem formulation phase* stage involves a detailed study of the system, data collection, and the identification of the specific problem that needs to be analyzed, along with the system constraints, restrictions, or limitations and the objective function(s). Often the problem analyzed may only be part of an overall system problem. The next stage involves the construction of an abstraction or an idealization of the problem through a *mathematical model*. Care must be taken to ensure that the model satisfactorily represents the system being analyzed, but keeps the model mathematically tractable. This compromise must be made judiciously, and the underlying assumptions inherent in the model must be properly considered. It must be borne in mind that from this point onward, the solutions obtained will be solutions to the model and not necessarily solutions to the actual system unless the model adequately represents the true situation.

The third step is to *derive a solution*. A proper technique that exploits any special structures (if present) must be chosen or designed. One or more optimal solutions may be sought, or only a heuristic or an approximate solution may be determined along with some assessment of its quality. In the case of multiple objective functions, one may seek *efficient* or *Pareto–optimal* solutions, that is, solutions that are such that a further improvement in any objective function value is necessarily accompanied by a detriment in some other objective function value.

The fourth stage is *model testing, analysis, and (possibly) restructuring*. One examines the model solution and its sensitivity to various system parameters, and studies its predictions to various what-if types of scenarios. This analysis provides insights into the system. One can also use this analysis to ascertain the reliability of the model by comparing the predicted outcomes with the expected outcomes, using either past experience or conducting the test retroactively using historical data. At this stage, one may wish to *enrich* the model further by incorporating other important features of the system that have not been modeled as yet, or, on the other hand, one may choose to *simplify* the model.

The final stage is *implementation*. The model is set up to interactively aid in the decision-making process. The model should never *replace* the decision maker. Often a "frank-factor" based on judgement and experience needs to be applied to the model solution before making policy decisions.

We describe several problems that can be formulated as linear programs. The purpose is to show the varieties of problems that can be recognized and expressed in precise mathematical terms as linear programs.

Feed Mix Problem

An agricultural mill manufactures feed for chickens. This is done by mixing several ingredients, such as corn, limestone, or alfalfa. The mixing is to be done in such a way that the feed meets certain levels for different types of nutrients, such as protein, calcium, carbohydrates, and vitamins. To be more specific, suppose that n ingredients $j = 1, 2, \ldots, n$ and m nutrients $i = 1, 2, \ldots, m$ are considered. Let the unit cost of ingredient j be c_j and let the amount of ingredient j to be used be x_j. The cost is therefore $\Sigma_{j=1}^{n} c_j x_j$. If the amount of the final product needed is b, then we must have $\Sigma_{j=1}^{n} x_j = b$. Further suppose that a_{ij} is the amount of nutrient i present in a unit of ingredient j, and that the acceptable lower and upper limits of nutrient i in a unit of the chicken feed are l_i' and u_i' respectively. Therefore we must have the constraints $l_i' b \leq \Sigma_{j=1}^{n} a_{ij} x_j \leq u_i' b$ for $i = 1, 2, \ldots, m$. Finally, because of shortages, suppose that the mill cannot acquire more than u_j units of ingredient j. The problem of mixing the ingredients such that the cost is minimized and the restrictions are met can be formulated as follows.

$$\begin{array}{lllllll}
\text{Minimize} & & c_1x_1 + & c_2x_2 \cdots + & c_nx_n & & \\
\text{Subject to} & & x_1 + & x_2 \cdots + & x_n = b & & \\
 & bl'_1 \le & a_{11}x_1 + & a_{12}x_2 \cdots + & a_{1n}x_n & \le bu'_1 & \\
 & bl'_2 \le & a_{21}x_1 + & a_{22}x_2 \cdots + & a_{2n}x_n & \le bu'_2 & \\
 & \vdots & \vdots & \vdots & \vdots & \vdots & \\
 & bl'_m \le & a_{m1}x_1 + & a_{m2}x_2 \cdots + & a_{mn}x_n & \le bu'_m & \\
 & & & & 0 \le x_1 \le u_1 & & \\
 & & & & 0 \le x_2 \le u_2 & & \\
 & & & & \vdots & & \\
 & & & & 0 \le x_n \le u_n & &
\end{array}$$

Production Scheduling: An Optimal Control Problem

A company wishes to determine the production rate over the planning horizon of the next T weeks such that the known demand is satisfied and the total production and inventory cost is minimized. Let the known demand rate at time t be $g(t)$, and similarly denote the production rate and inventory at t by $x(t)$ and $y(t)$. Further suppose that the initial inventory at time 0 is y_0 and that the desired inventory at the end of the planning horizon is y_T. Suppose that the inventory cost is proportional to the units in storage, so that the inventory cost is given by $c_1\int_0^T y(t)\,dt$ where $c_1 > 0$ is known. Also suppose that the production cost is proportional to the rate of production, and so is given by $c_2\int_0^T x(t)\,dt$. Then the total cost is $\int_0^T [c_1y(t) + c_2x(t)]\,dt$. Also note that the inventory at any time is given according to the relationship

$$y(t) = y_0 + \int_0^t [x(\tau) - g(\tau)]\,d\tau \qquad t \in [0, T]$$

Suppose that no backlogs are allowed; that is, all demand must be satisfied. Further suppose that the present manufacturing capacity restricts the production rate so that it does not exceed b_1 at any time. Also the available storage restricts the maximum inventory to be less than or equal to b_2. Hence the production scheduling problem can be stated as follows.

$$\begin{array}{lll}
\text{Minimize} & \displaystyle\int_0^T [c_1y(t) + c_2x(t)]\,dt & \\
\text{Subject to} & \displaystyle y(t) = y_0 + \int_0^t [x(\tau) - g(\tau)]\,d\tau & t \in [0, T] \\
 & y(T) = y_T & \\
 & 0 \le x(t) \le b_1 & t \in [0, T] \\
 & 0 \le y(t) \le b_2 & t \in [0, T]
\end{array}$$

The foregoing model is a linear control problem, where the *control variable* is the production rate $x(t)$ and the *state variable* is the inventory level $y(t)$. The problem can be approximated by a linear program by discretizing the continuous variables x and y. First the planning horizon $[0, T]$ is divided into n smaller periods $[0, \Delta], [\Delta, 2\Delta], \ldots, [(n-1)\Delta, n\Delta]$ where $n\Delta = T$. The production rate, the inventory, and the demand rate are assumed constant over each period. In particular let the production rate, the inventory, and the demand rate on period j be x_j, y_j, and g_j respectively. Then the production scheduling problem can be approximated by the following linear program (why?).

$$\begin{array}{lll} \text{Minimize} & \sum_{j=1}^{n} (c_1\Delta)y_j + \sum_{j=1}^{n} (c_2\Delta)x_j & \\ \text{Subject to} & y_j = y_{j-1} + (x_j - g_j)\Delta & j = 1, 2, \ldots, n \\ & y_n = y_T & \\ & 0 \le x_j \le b_1 & j = 1, 2, \ldots, n \\ & 0 \le y_j \le b_2 & j = 1, 2, \ldots, n \end{array}$$

Cutting Stock Problem

A manufacturer of metal sheets produces rolls of standard fixed width w and of standard length l. A large order is placed by a customer who needs sheets of width w and varying lengths. In particular, b_i sheets with length l_i and width w for $i = 1, 2, \ldots, m$ are ordered. The manufacturer would like to cut the standard rolls in such a way as to satisfy the order and to minimize the waste. Since scrap pieces are useless to the manufacturer, the objective is to minimize the number of rolls needed to satisfy the order. Given a standard sheet of length l, there are many ways of cutting it. Each such way is called a *cutting pattern*. The jth cutting pattern is characterized by the column vector $\mathbf{a}_j$ where the ith component $\mathbf{a}_j$, namely a_{ij}, is a nonnegative integer denoting the number of sheets of length l_i in the jth pattern. For instance, suppose that the standard sheets have length $l = 10$ meters and that sheets of lengths 1.5, 2.5, 3.0, and 4.0 meters are needed. The following are typical cutting patterns:

$$\mathbf{a}_1 = \begin{bmatrix} 3 \\ 2 \\ 0 \\ 0 \end{bmatrix}, \quad \mathbf{a}_2 = \begin{bmatrix} 0 \\ 4 \\ 0 \\ 0 \end{bmatrix}, \quad \mathbf{a}_3 = \begin{bmatrix} 0 \\ 0 \\ 3 \\ 0 \end{bmatrix}, \ldots$$

Note that the vector $\mathbf{a}_j$ represents a cutting pattern if and only if $\Sigma_{i=1}^{n} a_{ij} l_i \le l$ and each a_{ij} is a nonnegative integer. The number of cutting patterns n is finite. If we let x_j be the number of standard rolls cut according to the jth pattern, the problem can be formulated as follows.

$$\begin{array}{lll} \text{Minimize} & \sum_{j=1}^{n} x_j & \\ \text{Subject to} & \sum_{j=1}^{n} a_{ij} x_j \ge b_i & i = 1, 2, \ldots, m \\ & x_j \ge 0 & j = 1, 2, \ldots, n \\ & x_j \text{ integer} & j = 1, 2, \ldots, n \end{array}$$

If the integrality requirement on the x_j's is dropped, the problem is a linear program. Of course the difficulty of this problem is that the number of possible cutting patterns n is very large, and also it is not computationally feasible to enumerate each cutting pattern and its column $\mathbf{a}_j$ beforehand. The decomposition algorithm of Chapter 7 is particularly suited to solve the preceding problem where a new cutting pattern is generated at each iteration (see also Exercise 7.28). In Section 6.7 we suggest a method for handling the integrality requirements.

The Transportation Problem

The Brazilian coffee company processes coffee beans into coffee at m plants. The coffee is then shipped every week to n warehouses in major cities for retail, distribution, and exporting. Suppose that the unit shipping cost from plant i to warehouse j is c_{ij}. Further suppose that the production capacity at plant i is a_i and that the demand at warehouse j is b_j. It is desired to find the production–shipping pattern x_{ij} from plant i to warehouse j, $i = 1, \ldots, m$, $j = 1, \ldots, n$, that minimizes the overall shipping cost. This is the well-known transportation problem. The essential elements of the problem are shown in the network of Figure 1.2. The transportation problem can be formulated as the following linear program.

$$\text{Minimize} \quad \sum_{i=1}^{m} \sum_{j=1}^{n} c_{ij} x_{ij}$$

$$\begin{aligned}
\text{Subject to} \quad & \sum_{j=1}^{n} x_{ij} \le a_i && i = 1, 2, \ldots, m \\
& \sum_{i=1}^{m} x_{ij} = b_j && j = 1, 2, \ldots, n \\
& x_{ij} \ge 0 && i = 1, 2, \ldots, m \quad j = 1, 2, \ldots, n
\end{aligned}$$

Capital Budgeting Problem

A municipal construction project has funding requirements over the next four years of \$2 million, \$4 million, \$8 million, and \$5 million respectively. Assume

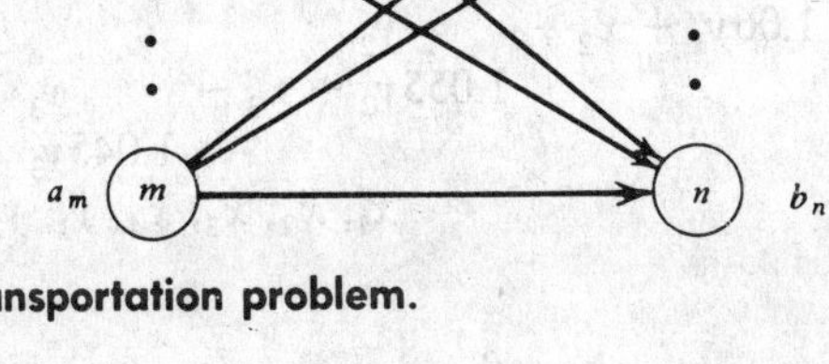

Figure 1.2. The transportation problem.

that all of the money for a given year is required at the beginning of the year. The city intends to sell exactly enough long-term bonds to cover the project funding requirements, and all of these bonds, regardless of when they are sold, will be paid off (*mature*) on the same date in a distant future year. The long-term bond market interest rates (that is, the costs of selling bonds) for the next four years are projected to be 7 percent, 6 percent, 6.5 percent, and 7.5 percent respectively. Bond interest paid will commence one year after the project is completed and will continue for 20 years, after which the bonds will be paid off. During the same period the short-term interest rates on time deposits (that is, what the city can earn on deposits) are projected to be 6 percent, 5.5 percent, and 4.5 percent respectively (the city will clearly not invest money in short-term deposits during the fourth year). What is the city's optimal strategy for selling bonds and depositing funds in time accounts in order to complete the construction project?

To formulate this problem as a linear program, let x_j, $j = 1, \ldots, 4$ be the amount of bonds sold at the beginning of each year j. When bonds are sold, some of the money will immediately be used for construction and some money will be placed in short-term deposits to be used in later years. Let y_j, $j = 1, \ldots, 3$ be the money placed in time deposits at the beginning of year j. Consider the beginning of the first year. The amount of bonds sold minus the amount of time deposits made will be used for the funding requirement at that year. Thus we may write

$$x_1 - y_1 = 2$$

We could have expressed the constraint as $\geq$. However, it is clear in this case that any excess funds will be deposited so that $=$ is also acceptable.

Consider the beginning of the second year. In addition to bonds sold and time deposits made, we also have time deposits plus interest becoming available from the previous year. Thus we have

$$1.06y_1 + x_2 - y_2 = 4$$

The third and fourth constraints are constructed in a similar manner.

Ignoring the fact that the amounts occur in different years (that is, the time value of money), the unit cost of selling bonds is 20 times the interest rate. Thus for bonds sold at the beginning of the first year we have $c_1 = 20(0.07)$. The other cost coefficients are computed similarly.

Finally, the linear programming model becomes as follows.

$$\begin{array}{llr}
\text{Minimize} & 20(0.07)x_1 + 20(0.06)x_2 + 20(0.065)x_3 + 20(0.075)x_4 & \\
\text{Subject to} & x_1 - y_1 & = 2 \\
& 1.06y_1 + x_2 - y_2 & = 4 \\
& 1.055y_2 + x_3 - y_3 & = 8 \\
& 1.045y_3 + x_4 & = 5 \\
& x_1, x_2, x_3, x_4, y_1, y_2, y_3 & \geq 0
\end{array}$$

Tanker Scheduling Problem

A shipline company requires a fleet of ships to service requirements for carrying cargo between six cities. There are four specific routes that must be served daily. These routes and the number of ships required for each route are as follows.

ROUTE #	ORIGIN	DESTINATION	NUMBER OF SHIPS PER DAY NEEDED
1	Dhahran	New York	3
2	Marseilles	Istanbul	2
3	Naples	Bombay	1
4	New York	Marseilles	1

All cargo are compatible, and therefore only one type of ship is needed. The travel time matrix between the various cities is shown.

	NAPLES	MARSEILLES	ISTANBUL	NEW YORK	DHAHRAN	BOMBAY
Naples	0	1	2	14	7	7
Marseilles	1	0	3	13	8	8
Istanbul	2	3	0	15	5	5
New York	14	13	15	0	17	20
Dhahran	7	8	5	17	0	3
Bombay	7	8	5	20	3	0

t_{ij} matrix (days)

It takes one day to off-load and one day to on-load each ship. How many ships must the shipline company purchase?

In addition to nonnegativity there are two types of constraints that must be maintained in this problem. First, we must ensure that ships coming off of some route get assigned to some (other) route. Second, we must ensure that each route gets its required number of ships per day. Let x_{ij} be the number of ships per day coming off of route i and assigned to route j. Let b_i represent the number of ships per day required on route i.

To ensure that ships from a given route get assigned to other routes, we write the constraint

$$\sum_{j=1}^{4} x_{ij} = b_i \qquad i = 1, \ldots, 4$$

To ensure that a given route gets its required number of ships, we write the constraint

$$\sum_{k=1}^{4} x_{ki} = b_i \qquad i = 1,\ldots,4$$

Computing the cost coefficients is a bit more involved. Since the objective is to minimize the total number of ships, let c_{ij} be the number of ships required to ensure a continuous daily flow of one ship coming off of route i and assigned to route j. To illustrate the computation of the c_{ij}'s consider c_{23}. It takes one day to load a ship at Marseilles, three days to travel from Marseilles to Istanbul, one day to unload cargo at Istanbul, and two days to head from Istanbul to Naples—a total of seven days. This implies that seven ships are needed to ensure that one ship will be assigned daily from route 2 to route 3 (why?). In particular, one ship will be on-loading at Marseilles, three ships en route from Marseilles to Istanbul, one ship off-loading at Istanbul, and two ships en route from Istanbul to Naples.

In general c_{ij} is given as follows:

$$\begin{aligned} c_{ij} = {} & \text{one day for on-loading} + \text{number of days for transit on route } i \\ & + \text{one day for off-loading} \\ & + \text{number of days for travel from the destination of route } i \\ & \quad \text{to the origin of route } j \end{aligned}$$

Therefore the tanker scheduling problem becomes as follows.

$$\begin{aligned} \text{Minimize} \quad & 36x_{11} + 32x_{12} + 33x_{13} + 19x_{14} + 10x_{21} + 8x_{22} + 7x_{23} \\ & + 20x_{24} + 12x_{31} + 17x_{32} + 16x_{33} + 29x_{34} + 23x_{41} \\ & + 15x_{42} + 16x_{43} + 28x_{44} \end{aligned}$$

$$\begin{aligned} \text{Subject to} \quad & \sum_{j=1}^{4} x_{ij} = b_i \qquad i = 1, 2, 3, 4 \\ & \sum_{k=1}^{4} x_{ki} = b_i \qquad i = 1, 2, 3, 4 \\ & x_{ij} \geq 0 \qquad i, j = 1, 2, 3, 4 \end{aligned}$$

where $b_1 = 3$, $b_2 = 2$, $b_3 = 1$, and $b_4 = 1$.

It can be easily seen that this is another application of the transportation problem (it will be instructive for the reader to form the origins and destinations of the corresponding transportation problem).

1.3 GEOMETRIC SOLUTION

We describe here a geometric procedure for solving a linear programming problem. Even though this method is only suitable for very small problems, it

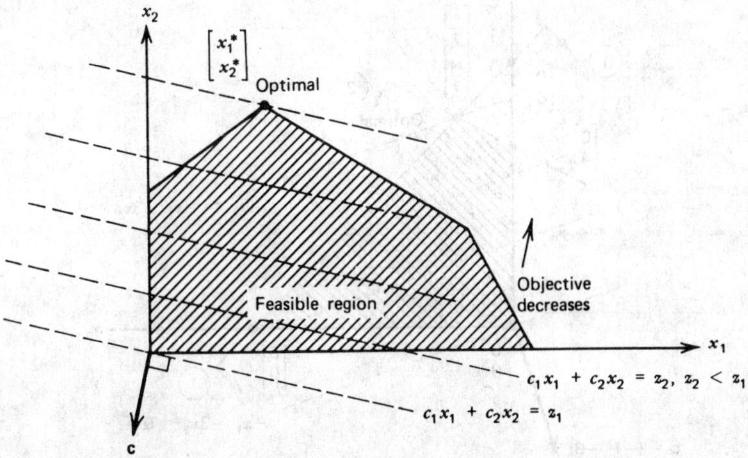

Figure 1.3. Geometric solution.

provides a great deal of insight into the linear programming problem. To be more specific, consider the following problem.

$$\begin{aligned} &\text{Minimize} && \mathbf{cx} \\ &\text{Subject to} && \mathbf{Ax} \geq \mathbf{b} \\ & && \mathbf{x} \geq \mathbf{0} \end{aligned}$$

Note that the feasible region consists of all vectors $\mathbf{x}$ satisfying $\mathbf{Ax} \geq \mathbf{b}$ and $\mathbf{x} \geq \mathbf{0}$. Among all such points we wish to find a point with minimal $\mathbf{cx}$ value. Note that points with the same objective z satisfy the equation $\mathbf{cx} = z$, that is, $\sum_{j=1}^{n} c_j x_j = z$. Since z is to be minimized, then the plane (line in a two-dimensional space) $\sum_{j=1}^{n} c_j x_j = z$ must be moved parallel to itself in the direction that minimizes the objective most. This direction is $-\mathbf{c}$, and hence the plane is moved in the direction $-\mathbf{c}$ as much as possible. This process is illustrated in Figure 1.3. Note that as the optimal point $\mathbf{x}^*$ is reached, the line $c_1x_1 + c_2x_2 = z^*$, where $z^* = c_1x_1^* + c_2x_2^*$, cannot be moved farther in the direction $-\mathbf{c} = (-c_1, -c_2)$ because this will lead to only points outside the feasible region. In other words, one cannot move from $\mathbf{x}^*$ in a direction that makes an acute angle with $-\mathbf{c}$ and hence reduces the objective function value while remaining feasible. We therefore conclude that $\mathbf{x}^*$ is indeed the optimal solution. Needless to say, for a maximization problem, the plane $\mathbf{cx} = z$ must be moved as much as possible in the direction $\mathbf{c}$.

The foregoing process is convenient for problems with two variables and is obviously impractical for problems with more than three variables. It is worth noting that the optimal point $\mathbf{x}^*$ in Figure 1.3 is one of the five corner points that are called *extreme points*. We shall show in Section 3.1 that if a linear program in standard or canonical form has a finite optimal solution, then it has an optimal corner (or extreme) point solution.

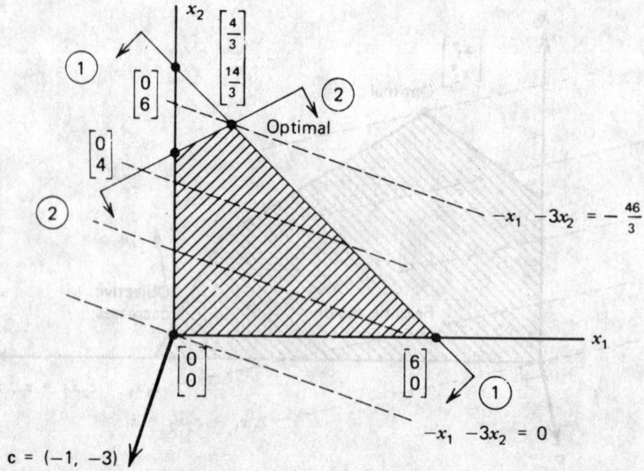

Figure 1.4. **Numerical example.**

Example 1.2

$$\begin{array}{ll} \text{Minimize} & -x_1 - 3x_2 \\ \\ \text{Subject to} & x_1 + \ \ x_2 \le 6 \\ & -x_1 + 2x_2 \le 8 \\ & x_1, \quad\ x_2 \ge 0 \end{array}$$

The feasible region is illustrated in Figure 1.4. For example, consider the second constraint. The equation associated with this constraint is $-x_1 + 2x_2 = 8$. The gradient or the partial derivative vector of the linear function $-x_1 + 2x_2$ is $\begin{bmatrix} -1 \\ 2 \end{bmatrix}$. Hence, $-x_1 + 2x_2$ increases (steepest) in the direction $\begin{bmatrix} -1 \\ 2 \end{bmatrix}$ and decreases steepest in the direction $\begin{bmatrix} 1 \\ -2 \end{bmatrix}$. Consequently, the region feasible to $-x_1 + 2x_2 \le 8$ relative to the equation $-x_1 + 2x_2 = 8$ is as shown in Figure 1.4 and encompasses points having decreasing values of $-x_1 + 2x_2$. (Alternatively, this region may be determined relative to the equation $-x_1 + 2x_2 = 8$ by testing the feasibility of a point, for example, the origin.) Similarly, the region feasible to the first constraint is as shown. (Try adding the constraint $-2x_1 + 3x_2 \ge 0$ to this figure.) The nonnegativity constraints restrict the points to be in the first quadrant. The equations $-x_1 - 3x_2 = z$ are called the *objective contours* and are represented by dotted lines in Figure 1.4. In particular the contour $-x_1 - 3x_2 = z = 0$ passes through the origin. The contours are moved in the direction $-\mathbf{c} = (1, 3)$ as much as possible until the optimal point $(4/3, 14/3)$ is reached.

In this example we had a unique optimal solution. Other cases may occur depending on the problem structure. All possible cases that may arise are summarized (for a minimization problem).

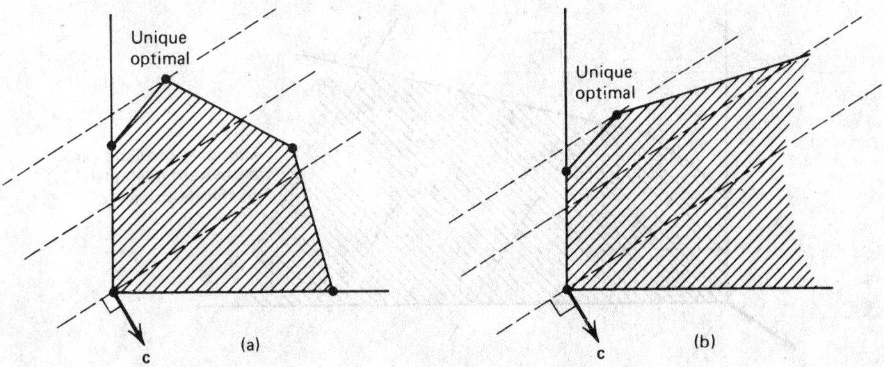

Figure 1.5. **Unique finite optimal solution: (*a*) Bounded region. (*b*) Unbounded region.**

1. *Unique Finite Optimal Solution*. If the optimal finite solution is unique, then it occurs at an extreme point. Figures 1.5*a* and *b* show a unique optimal solution. In Figure 1.5*a* the feasible region is bounded; that is, there is a ball around the origin that contains the feasible region. In Figure 1.5*b* the feasible region is not bounded. In each case, however, the unique optimal solution is finite.
2. *Alternative Finite Optimal Solutions*. This case is illustrated in Figure 1.6. Note that in Figure 1.6*a* the feasible region is bounded. The two corner points $\mathbf{x}_1^*$ and $\mathbf{x}_2^*$ are optimal, as well as any point on the line segment joining them. In Figure 1.6*b* the feasible region is unbounded but the optimal objective is finite. Any point on the "ray" with vertex $\mathbf{x}^*$ in Figure 1.6*b* is optimal. Hence the *optimal solution set* is unbounded.

In both cases (1) and (2), it is instructive to make the following observation. Pick an optimal solution $\mathbf{x}^*$ in Figure 1.5 or 1.6, corner point

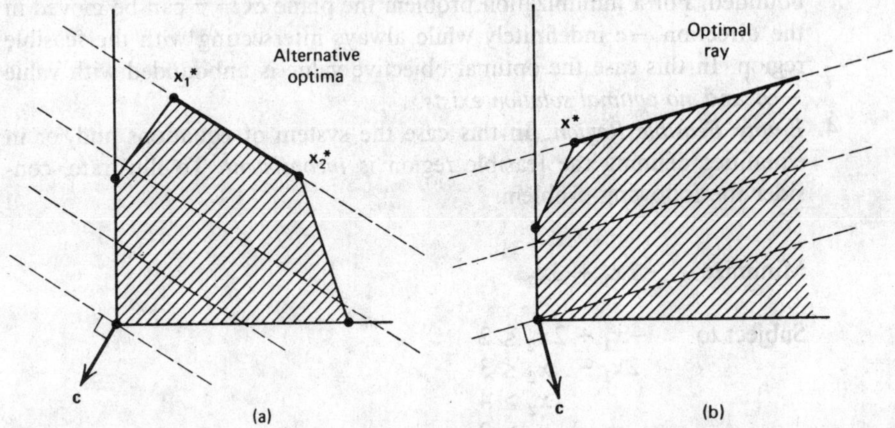

Figure 1.6. **Alternate finite optima: (*a*) Bounded region. (*b*) Unbounded region.**

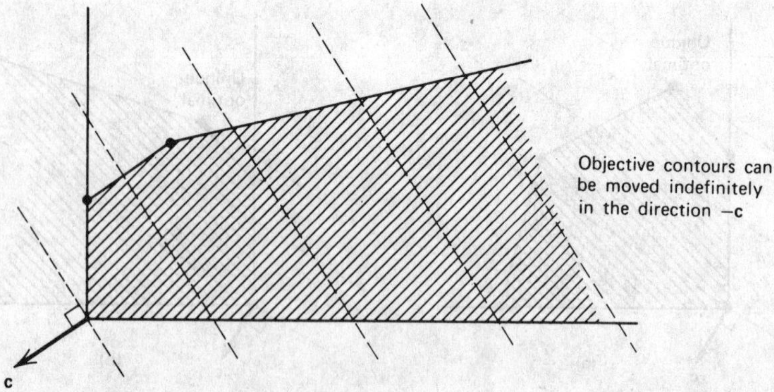

Figure 1.7. **Unbounded optimal solution.**

or not. Draw the normal vectors to the constraints passing through $\mathbf{x}^*$ pointing in the outward direction with respect to the feasible region. Also construct the vector $-\mathbf{c}$ at $\mathbf{x}^*$. Note that the "cone" spanned by the normals to the constraints passing through $\mathbf{x}^*$ contains the vector $-\mathbf{c}$. This is in fact the necessary and sufficient condition for $\mathbf{x}^*$ to be optimal, and will be formally established later. Intuitively, when this condition occurs, one can see that there is no direction along which one can move and improve the objective function while remaining feasible. Such a direction would have to make an acute angle with $-\mathbf{c}$ to improve the objective value and simultaneously make an angle $\geq 90°$ with respect to each of the normals to the constraints passing through $\mathbf{x}^*$ in order to maintain feasibility for some step length along this direction. This is impossible at any optimal solution, although it is possible at any nonoptimal solution.

3. *Unbounded Optimal Solution Value*. This case is illustrated in Figure 1.7 where both the feasible region and the optimal solution value are unbounded. For a minimization problem the plane $\mathbf{cx} = z$ can be moved in the direction $-\mathbf{c}$ indefinitely while always intersecting with the feasible region. In this case the optimal objective value is unbounded with value $-\infty$ and *no optimal solution exists*.
4. *Empty Feasible Region*. In this case the system of equations and/or inequalities defining the feasible region is *inconsistent*. To illustrate, consider the following problem.

$$\begin{array}{ll} \text{Minimize} & -2x_1 + 3x_2 \\ \\ \text{Subject to} & -x_1 + 2x_2 \leq 2 \\ & 2x_1 - x_2 \leq 3 \\ & x_2 \geq 4 \\ & x_1, \quad x_2 \geq 0 \end{array}$$

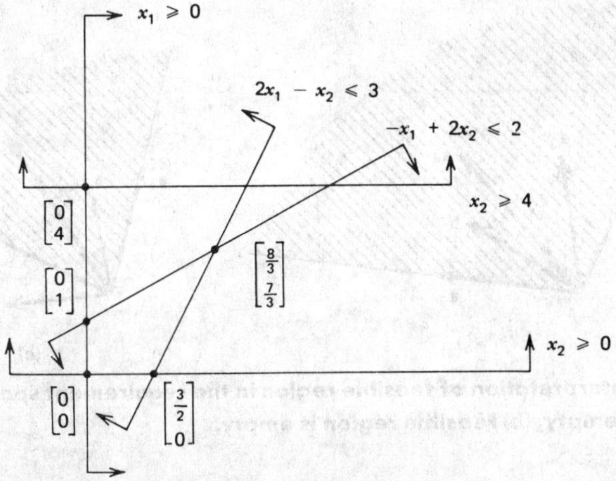

Figure 1.8. An example of an empty feasible region.

Examining Figure 1.8 it is clear that there exists no point (x_1, x_2) satisfying these inequalities. The problem is said to be *infeasible*, *inconsistent*, or with an *empty feasible region*. Again we say that *no optimal solution exists* in this case.

1.4 THE REQUIREMENT SPACE

The linear programming problem can be interpreted and solved geometrically in another space usually referred to as the requirement space.

Interpretation of Feasibility

Consider the following linear programming problem in standard form.

$$\begin{array}{ll} \text{Minimize} & \mathbf{cx} \\ \text{Subject to} & \mathbf{Ax} = \mathbf{b} \\ & \mathbf{x} \geq \mathbf{0} \end{array}$$

where $\mathbf{A}$ is an $m \times n$ matrix whose jth column is denoted by $\mathbf{a}_j$. The problem can be rewritten as follows.

$$\begin{array}{ll} \text{Minimize} & \sum_{j=1}^{n} c_j x_j \\ \text{Subject to} & \sum_{j=1}^{n} \mathbf{a}_j x_j = \mathbf{b} \\ & x_j \geq 0 \qquad j = 1, 2, \ldots, n \end{array}$$

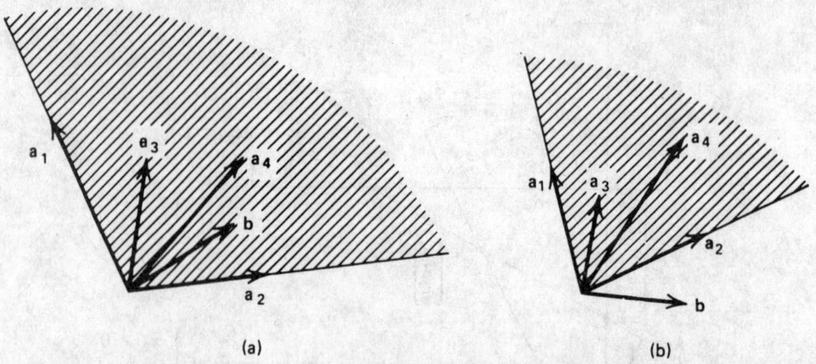

Figure 1.9. **Interpretation of feasible region in the requirement space: (a) Feasible region is not empty. (b) Feasible region is empty.**

Given the vectors $\mathbf{a}_1, \mathbf{a}_2, \ldots, \mathbf{a}_n$, we wish to find nonnegative scalars $x_1, x_2, \ldots, x_n$ such that $\sum_{j=1}^{n} \mathbf{a}_j x_j = \mathbf{b}$ and such that $\sum_{j=1}^{n} c_j x_j$ is minimized. Note, however, that the collection of vectors of the form $\sum_{j=1}^{n} \mathbf{a}_j x_j$, where $x_1, x_2, \ldots, x_n \geq 0$, is the cone generated by $\mathbf{a}_1, \mathbf{a}_2, \ldots, \mathbf{a}_n$ (see Figure 1.9). Thus the problem has a feasible solution if and only if the vector $\mathbf{b}$ belongs to this cone. Since the vector $\mathbf{b}$ usually reflects requirements to be satisfied, Figure 1.9 is usually referred to as the requirement space.

Example 1.3

Consider the following two systems.

System 1

$$
\begin{aligned}
2x_1 + x_2 + x_3 &= 2 \\
-x_1 + 3x_2 + x_4 &= 3 \\
x_1, \quad x_2, \quad x_3, \quad x_4 &\geq 0
\end{aligned}
$$

System 2

$$
\begin{aligned}
2x_1 + x_2 + x_3 &= -1 \\
-x_1 + 3x_2 + x_4 &= 2 \\
x_1, \quad x_2, \quad x_3, \quad x_4 &\geq 0
\end{aligned}
$$

Figure 1.10 shows the requirement space of both systems. For System 1 the vector $\mathbf{b}$ belongs to the cone generated by the vectors $\begin{bmatrix} 2 \\ -1 \end{bmatrix}$, $\begin{bmatrix} 1 \\ 3 \end{bmatrix}$, $\begin{bmatrix} 1 \\ 0 \end{bmatrix}$, and $\begin{bmatrix} 0 \\ 1 \end{bmatrix}$ and hence admits feasible solutions. For the second system, $\mathbf{b}$ does not belong to the cone and the system is hence inconsistent.

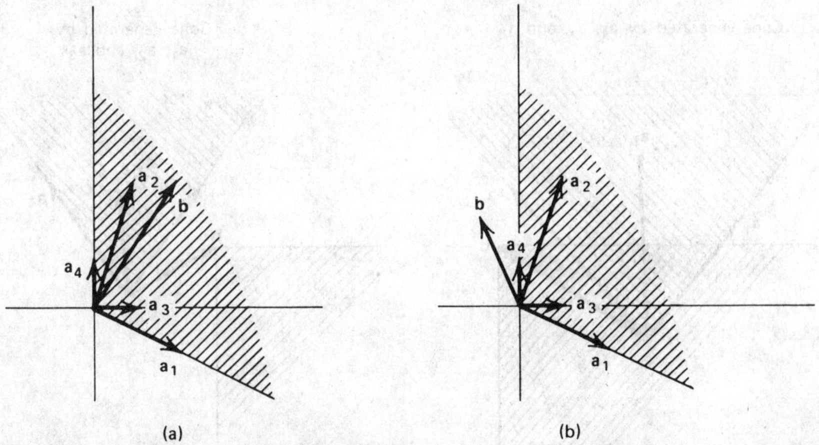

Figure 1.10. **Illustration of the requirement space: (*a*) System 1 is feasible. (*b*) System 2 is inconsistent.**

The Requirement Space and Inequality Constraints

We now illustrate the interpretation of feasibility for the inequality case. Consider the following inequality system:

$$\sum_{j=1}^{n} \mathbf{a}_j x_j \leq \mathbf{b}$$

$$x_j \geq 0 \qquad j = 1, 2, \ldots, n$$

Note that the collection of vectors $\Sigma_{j=1}^{n}\mathbf{a}_j x_j$ where $x_j \geq 0$ for $j = 1, 2, \ldots, n$ is the cone generated by $\mathbf{a}_1, \mathbf{a}_2, \ldots, \mathbf{a}_n$. If a feasible solution exists, then this cone must intersect the collection of vectors that are less than or equal to the requirement vector $\mathbf{b}$. Figure 1.11 shows both a feasible system and an infeasible system.

Optimality

We have seen that the system $\Sigma_{j=1}^{n}\mathbf{a}_j x_j = \mathbf{b}$ and $x_j \geq 0$ for $j = 1, 2, \ldots, n$ is feasible if and only if $\mathbf{b}$ belongs to the cone generated by $\mathbf{a}_1, \mathbf{a}_2, \ldots, \mathbf{a}_n$. The variables $x_1, x_2, \ldots, x_n$ must be chosen so that feasibility is satisfied and $\Sigma_{j=1}^{n} c_j x_j$ is minimized. Therefore the linear programming problem can be stated as follows. Find nonnegative $x_1, x_2, \ldots, x_n$ such that

$$\begin{bmatrix} c_1 \\ \mathbf{a}_1 \end{bmatrix} x_1 + \begin{bmatrix} c_2 \\ \mathbf{a}_2 \end{bmatrix} x_2 + \cdots + \begin{bmatrix} c_n \\ \mathbf{a}_n \end{bmatrix} x_n = \begin{bmatrix} z \\ \mathbf{b} \end{bmatrix}$$

where the objective z is to be minimized. In other words we seek to represent the vector $\begin{bmatrix} z \\ \mathbf{b} \end{bmatrix}$, for the smallest possible z, in the cone spanned by the vectors $\begin{bmatrix} c_1 \\ \mathbf{a}_1 \end{bmatrix}$,

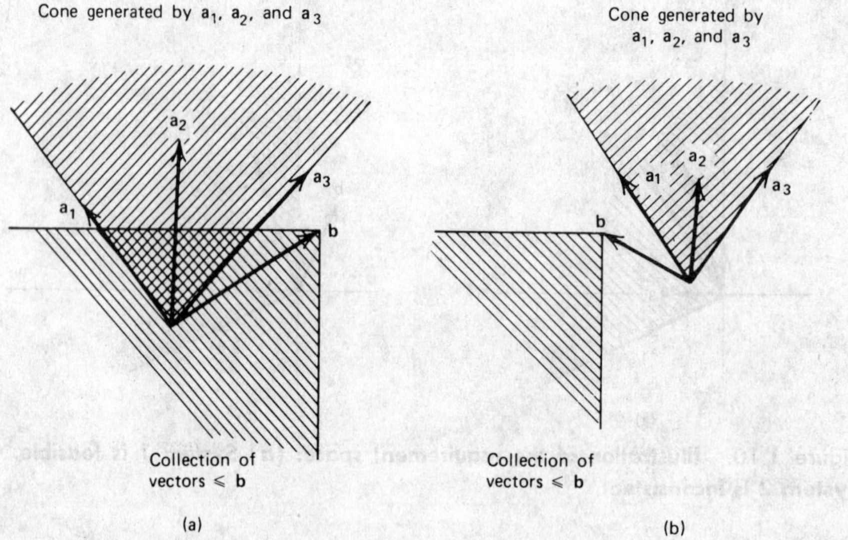

Figure 1.11. Requirement space and inequality constraints: (*a*) System is feasible. (*b*) System is infeasible.

$\begin{bmatrix} c_2 \\ \mathbf{a}_2 \end{bmatrix}, \ldots,$ and $\begin{bmatrix} c_n \\ \mathbf{a}_n \end{bmatrix}$. The reader should note that the price we must pay for including the objective function explicitly in the requirement space is to increase the dimensionality from m to $m + 1$.

Example 1.4

$$\begin{aligned} &\text{Minimize} && -2x_1 - 3x_2 \\ &\text{Subject to} && x_1 + 2x_2 \leq 2 \\ & && x_1, \quad x_2 \geq 0 \end{aligned}$$

Add the slack variable $x_3 \geq 0$. The problem is then to choose $x_1, x_2, x_2 \geq 0$ such that

$$\begin{bmatrix} -2 \\ 1 \end{bmatrix} x_1 + \begin{bmatrix} -3 \\ 2 \end{bmatrix} x_2 + \begin{bmatrix} 0 \\ 1 \end{bmatrix} x_3 = \begin{bmatrix} z \\ 2 \end{bmatrix}$$

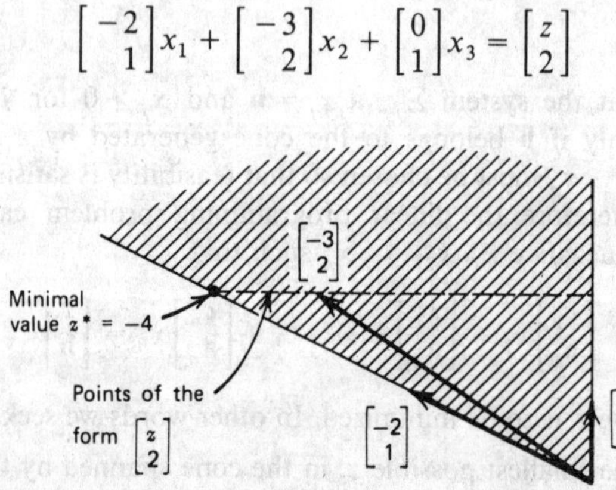

Figure 1.12. Optimal bounded objective in the requirement space.

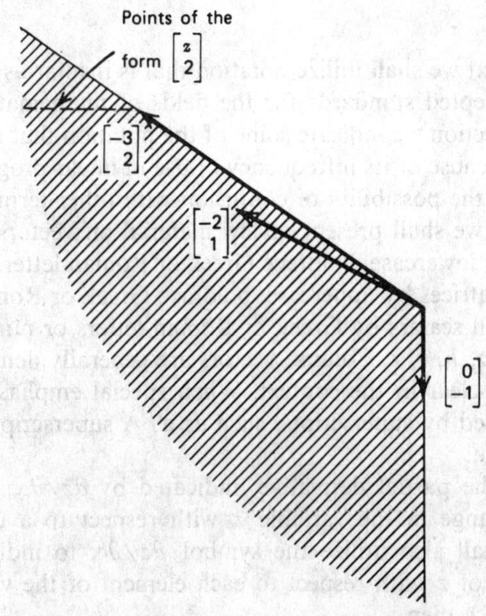

Figure 1.13. Unbounded optimal objective in the requirement space.

where z is to be minimized. The cone generated by the vectors $\begin{bmatrix} -2 \\ 1 \end{bmatrix}$, $\begin{bmatrix} -3 \\ 2 \end{bmatrix}$, and $\begin{bmatrix} 0 \\ 1 \end{bmatrix}$ is shown in Figure 1.12. We want to choose a vector $\begin{bmatrix} z \\ 2 \end{bmatrix}$ in this cone with minimal z. This gives the optimal solution $z^* = -4$ with $x_1^* = 2$ (the multiplier associated with $\begin{bmatrix} -2 \\ 1 \end{bmatrix}$) and $x_2^* = x_3^* = 0$.

Example 1.5

$$\begin{array}{ll} \text{Minimize} & -2x_1 - 3x_2 \\ \text{Subject to} & x_1 + 2x_2 \geq 2 \\ & x_1, \quad x_2 \geq 0 \end{array}$$

Obviously the optimal solution value is unbounded. We illustrate this fact in the requirement space. Subtracting the slack variable $x_3 \geq 0$, the problem can be restated as follows. Find $x_1, x_2, x_3 \geq 0$ such that

$$\begin{bmatrix} -2 \\ 1 \end{bmatrix} x_1 + \begin{bmatrix} -3 \\ 2 \end{bmatrix} x_2 + \begin{bmatrix} 0 \\ -1 \end{bmatrix} x_3 = \begin{bmatrix} z \\ 2 \end{bmatrix}$$

such that z is minimized. The cone generated by $\begin{bmatrix} -2 \\ 1 \end{bmatrix}$, $\begin{bmatrix} -3 \\ 2 \end{bmatrix}$, and $\begin{bmatrix} 0 \\ -1 \end{bmatrix}$ is shown in Figure 1.13. We want to choose $\begin{bmatrix} z \\ 2 \end{bmatrix}$ in this cone with smallest possible z. Note that we can find points of the form $\begin{bmatrix} z \\ 2 \end{bmatrix}$ in the cone with arbitrarily small z. Therefore the optimal solution value is $-\infty$ or it is unbounded.

1.5 NOTATION

Throughout the text we shall utilize notation that is insofar as possible consistent with generally accepted standards for the fields of mathematics and operations research. In this section we indicate some of the notation that may require special attention either because of its infrequency of use in linear programming literature or else because of the possibility of confusion with other terms.

In Chapter 2 we shall present review material on vectors and matrices. We indicate vectors by lowercase, boldface Greek or Roman letters or numerals, such as $\mathbf{a}$, $\mathbf{b}$, $\mathbf{x}$, $\mathbf{1}$, $\boldsymbol{\lambda}$; matrices by uppercase, boldface Greek or Roman letters, such as $\mathbf{A}$, $\mathbf{B}$, $\mathbf{N}$, $\boldsymbol{\Phi}$; and all scalars by Greek or Roman letters or numerals that are not boldface, such as a, b, 1, ϵ. Column vectors are generally denoted by subscripts, such as $\mathbf{a}_j$, unless clear in the context. When special emphasis is required, row vectors are indicated by superscripts, such as $\mathbf{a}^i$. A superscript t will denote the transpose operation.

In calculus, the partial derivative, indicated by $\partial z/\partial x$, represents the expected rate of change in the variable z with respect to a unit change in the variable x. We shall also utilize the symbol $\partial z/\partial \mathbf{x}$ to indicate the vector of partial derivatives of z with respect to each element of the vector $\mathbf{x}$. That is, if $\mathbf{x} = (x_1, x_2, \ldots, x_n)$, then

$$\frac{\partial z}{\partial \mathbf{x}} = \left(\frac{\partial z}{\partial x_1}, \frac{\partial z}{\partial x_2}, \ldots, \frac{\partial z}{\partial x_n} \right)$$

Also, we shall sometimes consider the partial derivative of one vector with respect to another vector, such as $\partial \mathbf{y}/\partial \mathbf{x}$. If $\mathbf{y} = (y_1, y_2, \ldots, y_m)$ and $\mathbf{x} = (x_1, x_2, \ldots, x_n)$ then

$$\frac{\partial \mathbf{y}}{\partial \mathbf{x}} = \begin{bmatrix} \frac{\partial y_1}{\partial x_1} & \frac{\partial y_1}{\partial x_2} & \cdots & \frac{\partial y_1}{\partial x_n} \\ \frac{\partial y_2}{\partial x_1} & \frac{\partial y_2}{\partial x_2} & \cdots & \frac{\partial y_2}{\partial x_n} \\ \vdots & \vdots & & \vdots \\ \frac{\partial y_m}{\partial x_1} & \frac{\partial y_m}{\partial x_2} & \cdots & \frac{\partial y_m}{\partial x_n} \end{bmatrix}$$

Note that if z is a function of the vector $\mathbf{x} = (x_1, x_2, \ldots, x_n)$, then $\partial z/\partial \mathbf{x}$ is called the *gradient*.

We shall, when necessary, use (a, b) to refer to the *open interval* $a < x < b$, and $[a, b]$ to refer to the *closed interval* $a \le x \le b$. Finally we shall utilize the standard set operators $\cup$, $\cap$, $\subset$, and $\in$ to refer to union, intersection, set inclusion, and set membership respectively.

EXERCISES

1.1 An agricultural mill manufactures feed for cattle, sheep, and chickens. This is done by mixing the following main ingredients: corn, limestone, soybeans, and fish meal. These ingredients contain the following nutrients: vitamins, protein, calcium, and crude fat. The contents of the nutrients in each kilogram of the ingredients are summarized in the table.

	NUTRIENT			
INGREDIENT	VITAMINS	PROTEIN	CALCIUM	CRUDE FAT
Corn	8	10	6	8
Limestone	6	5	10	6
Soybeans	10	12	6	6
Fish meal	4	8	6	9

The mill is contracted to produce 10, 6, and 8 (metric) tons of cattle feed, sheep feed, and chicken feed. Because of shortages, a limited amount of the ingredients is available—namely, 6 tons of corn, 10 tons of limestone, 4 tons of soybeans, and 5 tons of fish meal. The price per kilogram of these ingredients is respectively \$0.20, \$0.12, \$0.24, and \$0.12. The minimal and maximal units of the various nutrients that are permitted is summarized for a kilogram of the cattle feed, the sheep feed, and the chicken feed.

	NUTRIENT							
	VITAMINS		PROTEIN		CALCIUM		CRUDE FAT	
PRODUCT	MIN	MAX	MIN	MAX	MIN	MAX	MIN	MAX
Cattle feed	6	∞	6	∞	7	∞	4	8
Sheep feed	6	∞	6	∞	6	∞	4	6
Chicken feed	4	6	6	∞	6	∞	4	6

Formulate this feed-mix problem so that the total cost is minimized.

1.2 The technical staff of a hospital wishes to develop a computerized menu-planning system. To start with, a lunch menu is sought. The menu is divided into three major categories: vegetables, meat, and dessert. At least one equivalent serving of each category is desired. The cost per serving of some suggested items as well as their content of carbohydrates, vitamins, protein, and fats is summarized.

	CARBOHYDRATES	VITAMINS	PROTEIN	FATS	COST IN \$/SERVING
Vegetables					
Peas	1	3	1	0	0.10
Green beans	1	5	2	0	0.12

	CARBOHYDRATES	VITAMINS	PROTEIN	FATS	COST IN $/SERVING
Vegetables					
Okra	1	5	1	0	0.13
Corn	2	6	1	2	0.09
Macaroni	4	2	1	1	0.10
Rice	5	1	1	1	0.07
Meat					
Chicken	2	1	3	1	0.70
Beef	3	8	5	2	1.20
Fish	3	6	6	1	0.63
Dessert					
Orange	1	3	1	0	0.28
Apple	1	2	0	0	0.42
Pudding	1	0	0	0	0.15
Jello	1	0	0	0	0.12

Suppose that the minimal requirements of carbohydrates, vitamins, protein, and fats per meal are respectively 5, 10, 10, and 2.

a. Formulate the menu-planing problem as a linear program.

b. Many practical aspects have been left out in the foregoing model. These include planning the breakfast, lunch, and supper menus together, weekly planning so that different varieties of food are used, and special menus for patients on particular diets. Discuss in detail how can these aspects be incorporated in a comprehensive menu-planning system.

1.3 Consider the problem of locating a new machine to an existing layout consisting of four machines. These machines are located at the following x_1 and x_2 coordinates: $\begin{pmatrix} 3 \\ 0 \end{pmatrix}$, $\begin{pmatrix} 0 \\ -3 \end{pmatrix}$, $\begin{pmatrix} -2 \\ 1 \end{pmatrix}$, and $\begin{pmatrix} 1 \\ 4 \end{pmatrix}$. Let the coordinates of the new machine be $\begin{pmatrix} x_1 \\ x_2 \end{pmatrix}$. Formulate the problem of finding an optimal location as a *linear program* for each of the following cases.

a. The sum of the distances from the new machine to the four machines is minimized. Use the street distance; for example, the distance from $\begin{pmatrix} x_1 \\ x_2 \end{pmatrix}$ to the first machine located at $\begin{pmatrix} 3 \\ 0 \end{pmatrix}$ is $|x_1 - 3| + |x_2|$.

b. Because of various amounts of flow between the new machine and the existing machines, reformulate the problem where the sum of the weighted distances is minimized, where the weights corresponding to the four machines are 5, 7, 3, and 1 respectively.

c. In order to avoid congestion, suppose that the new machine must be located in the square $\{(x_1, x_2): -1 \le x_1 \le 2, 0 \le x_2 \le 1\}$. Formulate (a) and (b) with this added restriction.

d. Suppose that the new machine must be located so that its distance from the first machine does not exceed 3/2. Formulate the problem with this added restriction.

1.4 Consider the following problem of launching a rocket to a fixed altitude b in a given time T while expending a minimum amount of fuel. Let $u(t)$ be the acceleration force exerted at time t and let $y(t)$ be the rocket altitude at time t. The problem can be formulated as follows.

$$\text{Minimize} \quad \int_0^T |u(t)|\,dt$$

$$\begin{aligned}\text{Subject to} \quad & \ddot{y}(t) = u(t) - g \\ & y(T) = b \\ & y(t) \ge 0 \qquad t \in [0, T]\end{aligned}$$

where g is the gravitational force and $\ddot{y}$ is the second derivative of the altitude y. Discretize the problem and reformulate it as a linear programming problem. In particular formulate the problem where $T = 10$, $b = 15$, and $g = 32$. (*Hint*. Replace the integration by proper summation and the differentiation by difference equations. Make the change of variable $|u_j| = x_j$ and note that $x_j \ge u_j$ and $x_j \ge -u_j$.)

1.5 A company wishes to plan its production of two items with seasonal demands over a 12-month period. The monthly demand of item 1 is 100,000 units during the months of October, November, and December; 10,000 units during the months of January, February, March, and April; and 30,000 units during the remaining months. The demand of item 2 is 50,000 during the months of October through February and 15,000 during the remaining months. Suppose that the unit product cost of items 1 and 2 is \$5.00 and \$8.00, respectively, provided that these were manufactured prior to June. After June, the unit costs are reduced to \$4.50 and \$7.00 because of the installation of an improved manufacturing system. The total units of items 1 and 2 that can be manufactured during any particular month cannot exceed 120,000 for Jan–Sept, and 150,000 for Oct–Dec. Furthermore, each unit of item 1 occupies 2 cubic feet and each unit of item 2 occupies 4 cubic feet of inventory. Suppose that the maximum inventory space allocated to these items is 150,000 cubic feet and that the holding cost per cubic foot during any month is \$0.10. Formulate the production scheduling problem so that the total production and inventory costs are minimized.

1.6 A textile mill produces five types of fabrics. The demand (in thousand yards) over a quarter-year time horizon for these fabrics is 16, 48, 35, 21, and 82 respectively. These five fabrics are woven, finished, and sold in the market at prices 0.9, 0.8, 0.8, 1.2, and 0.6 \$/per yard respectively. Besides weaving and finishing the fabrics at the mill itself, the fabrics are also purchased woven from outside sources and are then finished at the mill before being sold. If the unfinished fabrics are purchased outside, the costs in $/per yard for the five fabrics are 0.8, 0.7, 0.75, 0.9, and 0.7 respectively. If produced at the mill itself, the respective costs are 0.6, 0.5, 0.6, 0.7, and 0.3 $/per yard. There are two types of looms that can produce the fabrics at the mill, that is, there are 10 Dobbie looms and 80 regular looms. The production rate of each Dobbie loom is 4.6, 4.6, 5.2, 3.8, and 4.2 yards per hour for the five fabrics. The regular looms have the same production rates as the Dobbie looms, but they can only produce fabric types 3, 4, and 5. Assuming that the mill operates seven days a week and 24 hours a day, formulate the problem of optimally planning to meet the demand over a quarter-year horizon as a linear program. Is your formulation a transportation problem? If not, reformulate the problem as a transportation problem.

1.7 Fred has $2200 to invest over the next five years. At the beginning of each year he can invest money in one- or two-year time deposits. The bank pays 8 percent interest on one-year time deposits and 17 percent (total) on two-year time deposits. In addition, West World Limited will offer three-year certificates starting at the beginning of the second year. These certificates will return 27 percent (total). If Fred reinvests his money available every year, formulate a linear program to show him how to maximize his total cash on hand at the end of the fifth year.

1.8 A steel manufacturer produces four sizes of I beams: small, medium, large, and extra large. These beams can be produced on any one of three machine types: A, B, and C. The lengths in feet of the I beams that can be produced on the machines per hour are summarized.

	MACHINE		
BEAM	A	B	C
Small	300	600	800
Medium	250	400	700
Large	200	350	600
Extra large	100	200	300

Assume that each machine can be used up to 50 hours per week and that the hourly operating costs of these machines are respectively $30.00, $50.00, and $80.00. Further suppose that 10,000, 8,000, 6,000, and 6,000 feet of the different size I beams are required weekly. Formulate the machine scheduling problem as a linear program.

1.9 A cheese firm produces two types of cheese: Swiss cheese and sharp cheese. The firm has 60 experienced workers and would like to increase its working force to 90 workers during the next eight weeks. Each experienced worker can train 3 new employees in a period of two weeks during which the workers involved virtually produce nothing. It takes one hour to produce 10 pounds of Swiss cheese and one hour to produce 6 pounds of sharp cheese. A work week is 40 hours. The weekly demands (in 1000 pounds) are summarized.

	WEEK							
CHEESE TYPE	1	2	3	4	5	6	7	8
Swiss cheese	12	12	12	16	16	20	20	20
Sharp cheese	8	8	10	10	12	12	12	12

Suppose that a trainee receives full salary as an experienced worker. Further suppose that overaging destroys the flavor of the cheese so that inventory is limited to one week. How should the company hire and train its new employees so that the labor cost is minimized? Formulate the problem as a linear program.

1.10 A lathe is used to reduce the diameter of a steel shaft whose length is 36 in. from 14 in. to 12 in. The speed x_1 (in revolutions per minute), the depth feed x_2 (in inches per minute), and the length feed x_3 (in inches per minute) must be determined. The duration of the cut is given by $36/x_2x_3$. The compression and side stresses exerted on the cutting tool are given by $30x_1 + 4000x_2$ and $40x_1 + 6000x_2 + 6000x_3$ pounds per square inch respectively. The temperature (in degrees Fahrenheit) at the tip of the cutting tool is $200 + 0.5x_1 + 150(x_2 + x_3)$. The maximum compression stress, side stress, and temperature allowed are 150,000 psi, 100,000 psi, and 800°F. It is desired to determine the speed (which must be in the range from 600 to 800 rpm), the depth feed, and the length feed such that the duration of the cut is minimized. In order to use a linear model the following approximation is made. Since $36/x_2x_3$ is minimized if and only if x_2x_3 is maximized, it was decided to replace the objective by the maximization of the minimum of x_2 and x_3. Formulate the problem as a linear model and comment on the validity of the approximation used in the objective function. (We ask the reader to solve this problem in Exercise 3.26.)

1.11 An oil refinery can buy two types of oil: light crude oil and heavy crude oil. The cost per barrel of these types is respectively \$11 and \$9. The following quantities of gasoline, kerosene, and jet fuel are produced per barrel of each type of oil.

	GASOLINE	KEROSENE	JET FUEL
Light crude oil	0.4	0.2	0.35
Heavy crude oil	0.32	0.4	0.2

Note that 5 percent and 8 percent of the crude are lost respectively during the refining process. The refinery has contracted to deliver 1 million barrels of gasoline, 400,000 barrels of kerosene, and 250,000 barrels of jet fuel. Formulate the problem of finding the number of barrels of each crude oil that satisfies the demand and minimizes the total cost as a linear program. (We ask the reader to solve this problem in Exercise 3.27.)

1.12 A company manufactures an assembly consisting of a frame, a shaft, and a ball bearing. The company manufactures the shafts and frames but purchases the ball bearings from a ball bearing manufacturer. Each shaft must be processed on a forging machine, a lathe, and a grinder. These operations require 0.5 hours, 0.2 hours, and 0.3 hours per shaft respectively. Each frame requires 0.8 hours on a forging machine, 0.1 hours on a drilling machine, 0.3 hours on a milling machine, and 0.5 hours on a grinder. The company has 5 lathes, 10 grinders, 20 forging machines, 3 drillers, and 6 millers. Assume that each machine operates a maximum of 2400 hours per year. Formulate the problem of finding the maximum number of assembled components that can be produced as a linear program. (We ask the reader to solve this problem as Exercise 3.44.)

1.13 A television set manufacturing firm has to decide on the mix of color and black-and-white TV's to be produced. A market research indicates that at most 1000 units and 4000 units of color and black-and-white TV's can be sold per month. The maximum number of man-hours available is 50,000 per month. A color TV requires 20 man-hours and a black-and-white TV requires 15 man-hours. The unit profits of the color and black-and-white TV's are \$60 and \$30 respectively. It is desired to find the number of units of each TV type that the firm must produce in order to maximize its profit. Formulate the problem. (We ask the reader to solve this problem in Exercise 3.45.)

1.14 A manufacturer of plastics is planning to blend a new product from four chemical compounds. These compounds are mainly composed of three elements: A, B, and C. The composition and unit cost of these chemicals are shown in the following table:

CHEMICAL COMPOUND	1	2	3	4
Percentage A	30	20	40	20
Percentage B	20	60	30	40
Percentage C	40	15	25	30
Cost/kilogram	20	30	20	15

The new product consists of 20 percent element A, at least 30 percent element B, and at least 20 percent element C. Owing to side effects of compounds 1 and 2, they must not exceed 30 percent and 40 percent of the content of the new product. Formulate the problem of finding the least costly way of blending as a linear program. (We ask the reader to solve this problem in Exercise 5.30.)

1.15 A production manager is planning the scheduling of three products on four machines. Each product can be manufactured on each of the machines. The unit production costs (in $) are summarized.

PRODUCT	MACHINE 1	2	3	4
1	4	4	5	7
2	6	7	5	6
3	12	10	8	11

The time (in hours) required to produce each unit product on each of the machines is summarized.

PRODUCT	MACHINE 1	2	3	4
1	0.3	0.25	0.2	0.2
2	0.2	0.3	0.2	0.25
3	0.8	0.6	0.6	0.5

Suppose that 4000, 5000, and 3000 units of the products are required, and that the available machine-hours are 1500, 1200, 1500, and 2000 respectively. Formulate the scheduling problem as a linear program. (We ask the reader to solve this problem in Exercise 10.21.)

1.16 A furniture manufacturer has three plants that need 500, 700, and 600 tons of lumber weekly. The manufacturer may purchase the lumber from three lumber companies. The first two lumber manufacturers virtually have unlimited supply, and because of other commitments the third manufacturer cannot ship more than 500 tons weekly. The first lumber manufacturer uses rail for transportation and there is no limit on the tonnage that can be shipped to the furniture facilities. On the other hand, the last two lumber companies use trucks that limit the maximum tonnage that can be shipped to any of the furniture companies to 200 tons. The following table gives the transportation cost from the lumber companies to the furniture manufacturers ($ per ton).

LUMBER COMPANY	FURNITURE FACILITY 1	2	3
1	2	3	5
2	2.5	4	4.8
3	3	3.6	3.2

Formulate the problem as a linear program. (We ask the reader to solve this problem in Exercise 10.34.)

1.17 A corporation has $30 million available for the coming year to allocate to its three subsidiaries. Because of commitments to stability of personnel employment and for other reasons, the corporation has established a minimal level of funding for each subsidiary. These funding levels are $3 million, $5 million, and $8 million respectively. Owing to the nature of its operation, subsidiary 2 cannot utilize more than $17 million without major new capital expansion. The corporation is unwilling to undertake such expansion at this time. Each subsidiary has the opportunity to conduct various projects with the funds it receives. A rate of return (as a percent of investment) has been established for each project. In addition, certain of the projects permit only limited investment. The data of each project are given.

SUBSIDIARY	PROJECT	RATE OF RETURN	UPPER LIMIT OF INVESTMENT
1	1	8%	$6 million
	2	6%	$5 million
	3	7%	$9 million
2	4	5%	$7 million
	5	8%	$10 million
	6	9%	$4 million
3	7	10%	$6 million
	8	6%	$3 million

Formulate this problem as a linear program. (We ask the reader to solve this problem in Exercise 9.50.)

1.18 A ten-acre slum in New York City is to be cleared. The officials of the city must decide on the redevelopment plan. Two housing plans are to be considered: low-income housing and middle-income housing. The types of housing can be developed at 20 and 15 units per acre respectively. The unit costs of the low- and middle-income housing are $13,000 and $18,000. The lower and upper limits set by the officials on the number of low-income housing units are 60 and 100. Similarly, the number of middle-income housing units must lie between 30 and 70. The combined maximum housing market potential is estimated to be 150 (which is less than the sum of the individual market limits due to the overlap between the two markets). The total mortgage committed to the renewal plan is not to exceed $2 million. Finally, it was suggested by the architectural adviser that the number of low-income housing units be at least 50 units greater than one-half the number of the middle-income housing units.

a. Formulate the minimum cost renewal planning problem as a linear program and solve it graphically.
b. Resolve the problem if the objective is to maximize the number of houses being constructed.

1.19 A region is divided into m residential and central business districts. Each district is represented by a node and the nodes are interconnected by links representing major routes. People living in the various districts go to their business in the same and/or at other districts so that each node attracts and/or generates a number of trips. In particular, let a_{ij} be the number of trips generated at node i with final destination at node j and let b_{ij} be the time to travel from node i to node j. It is desired to determine the routes to be taken by the people living in the region.

a. Illustrate the problem by a suitable network.
b. Develop some measures of effectiveness for this *traffic assignment problem* and for each measure devise a suitable model.

1.20 Consider the problem of scheduling court hearings over a planning horizon consisting of n periods. Let b_j be the available judge-hours at period j, h_{ij} be the number of hearings of class i arriving at period j, and a_i be the number of judge-hours required to process a hearing of class i. It is desired to determine the number of hearings x_{ij} of class i processed at period j

a. Formulate the problem as a linear program.
b. Modify the model in part a so that hearings would not be delayed for too long.

1.21 Suppose that there are m sources that generate waste and n disposal sites. The amount of waste generated at source i is a_i and the capacity of site j is b_j. It is desired to select appropriate transfer facilities from among K candidate facilities. Potential transfer facility k has fixed cost f_k, capacity q_k and unit processing cost α_k per ton of waste. Let c_{ik} and $\bar{c}_{kj}$ be the unit shipping costs from source i to transfer station k and from transfer station k to disposal site j respectively. The problem is to choose the transfer facilities and the shipping pattern that minimize the total capital and operating costs of the transfer stations plus the transportation costs. Formulate this *distribution* problem.

(*Hint*. Let y_k be 1 if transfer station k is selected and 0 otherwise.)

1.22 A governmental planning agency wishes to determine the sources of purchase of fuel for use by n depots from among m bidders. Suppose that the maximum quantity offered by bidder i is a_i gallons and that the demand of depot j is b_j gallons. Let c_{ij} be the unit delivery cost of bidder i to the jth depot.

a. Formulate the problem of minimizing the total purchasing cost as a linear program.

b. Suppose that a discount in the unit delivery cost is offered by bidder i if the ordered quantity exceeds the level α_i. How would you incorporate this modification in the model developed in part (a)?

1.23 The quality of air in an industrial region largely depends on the effluent emission from n plants. Each plant can use m different types of fuel. Suppose that the total energy needed at plant j is b_j british thermal units per day and that c_{ij} is the effluent emission per ton of fuel type i at plant j. Further suppose that fuel type i costs c_i dollars per ton and that each ton of this fuel type generates α_{ij} british thermal units at plant j. The level of air pollution in the region is not to exceed b micrograms per cubic meter. Finally, let γ_j be a meteorological parameter relating emissions at plant j to air quality at the region.

a. Formulate the problem of determining the mix of fuels to be used at each plant.
b. How would you incorporate technology constraints that prohibit the use of certain mixes of fuel at certain plants?
c. How could you ensure equity among the plants?

1.24 For some industry, let $p(Q)$, $Q \geq 0$ be an inverse demand curve, that is, $p(Q)$ is the price at which a quantity Q will be demanded. Let $f(Q)$, $Q \geq 0$ by a supply or marginal cost curve, that is, $f(Q)$ is the price at which the industry is willing to supply Q units of the product. Consider the problem of determining a (perfect competition) equilibrium price and production quantity p^* and Q^* for $p(\cdot)$ and $f(\cdot)$ of the type shown in Figure 1.14 via the intersection of these supply and demand curves. Note that Q^* is obtainable by finding that Q that maximizes the area under the demand curve minus the area under the supply curve. Now suppose that $f(\cdot)$ is not available directly, but being a marginal cost curve, is given implicitly by

$$\int_0^Q f(x)\,dx = \min\{\mathbf{cy} : \mathbf{Ay} = \mathbf{b},\ \boldsymbol{\alpha}\mathbf{y} = Q,\ \mathbf{y} \geq \mathbf{0}\}$$

where $\mathbf{A}$ is $m \times n$, $\boldsymbol{\alpha}$ and $\mathbf{c}$ are $1 \times n$, and $\mathbf{y}$ is an n-vector representing a set

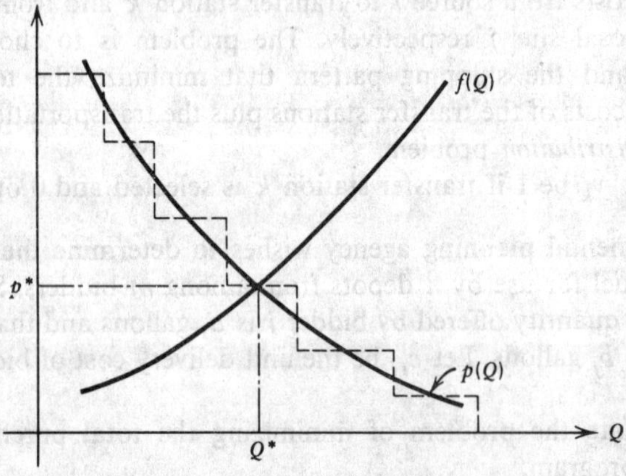

Figure 1.14. Supply demand equilibrium for Problem 1.24.

of production activities. Further suppose that the demand curve $p(\cdot)$ is approximated using the steps shown in the figure. (Let there be s steps used, each of height H_i and width W_i for $i = 1, \ldots, s$.) With this approximation, formulate the problem of determining the price–quantity equilibrium as a linear program.

1.25 Consider the following linear programming problem.

$$\begin{array}{ll} \text{Minimize} & x_1 - 2x_2 - 3x_3 \\ \text{Subject to} & x_1 + 2x_2 + x_3 \leq 14 \\ & x_1 + 2x_2 + 4x_3 \geq 12 \\ & x_1 - x_2 + x_3 = 2 \\ & x_1, \quad x_2 \quad \text{unrestricted} \\ & x_3 \leq -3 \end{array}$$

a. Reformulate the problem so that it is in standard format.
b. Reformulate the problem so that it is in canonical format.
c. Convert the problem into a maximization problem.

1.26 Consider the feasible region sketched in Figure 1.5*b*. Geometrically, identify conditions on the objective gradient vector **c** for which the different points in the feasible region will be optimal and identify the vectors **c** for which no optimum exists. (Assume a minimization problem.)

1.27 Consider the following problem.

$$\begin{array}{ll} \text{Maximize} & x_1 - x_2 \\ \text{Subject to} & -x_1 + 2x_2 \leq 0 \\ & -3x_1 + x_2 \geq -3 \\ & x_1, \quad x_2 \geq 0 \end{array}$$

a. Sketch the feasible region in the (x_1, x_2) space.
b. Identify the regions in the (x_1, x_2) space where the slack variables x_3 and x_4, say, are equal to zero.
c. Solve the problem geometrically.
d. Draw the requirement space and interpret feasibility.

1.28 Sketch the feasible region of the set $\{\mathbf{x}: \mathbf{Ax} \leq \mathbf{b}\}$ where **A** and **b** are as given. In each case state whether the feasible region is empty or not and whether it is bounded or not.

a. $\mathbf{A} = \begin{bmatrix} 1 & 1 \\ 2 & -1 \\ 0 & 1 \end{bmatrix} \quad \mathbf{b} = \begin{bmatrix} 6 \\ 6 \\ 2 \end{bmatrix}$

b. $\mathbf{A} = \begin{bmatrix} -1 & 0 \\ 0 & -1 \\ 2 & 3 \\ 1 & -1 \end{bmatrix} \quad \mathbf{b} = \begin{bmatrix} 0 \\ 0 \\ 12 \\ 5 \end{bmatrix}$

c. $\mathbf{A} = \begin{bmatrix} 1 & 1 \\ -1 & -2 \\ -1 & 0 \end{bmatrix} \quad \mathbf{b} = \begin{bmatrix} 4 \\ -12 \\ 0 \end{bmatrix}$

1.29 Consider the following problem.

$$\text{Maximize} \quad 2x_1 + 3x_2$$

$$\begin{aligned} \text{Subject to} \quad x_1 + x_2 &\le 2 \\ 4x_1 + 6x_2 &\le 9 \\ x_1, \quad x_2 &\ge 0 \end{aligned}$$

a. Sketch the feasible region.
b. Find two alternative optimal extreme (corner) points.
c. Find an infinite class of optimal solutions.

1.30 Consider the following problem.

$$\text{Maximize} \quad 3x_1 + x_2$$

$$\begin{aligned} \text{Subject to} \quad -x_1 + 2x_2 &\le 6 \\ x_2 &\le 4 \end{aligned}$$

a. Sketch the feasible region.
b. Verify that the problem has an unbounded optimal solution value.

1.31 Consider the following problem.

$$\text{Maximize} \quad -x_1 - x_2 + 2x_3 + x_4$$

$$\begin{aligned} \text{Subject to} \quad 2x_1 + x_2 + x_3 + x_4 &\ge 6 \\ x_1 + x_2 - 2x_3 + x_4 &\le 4 \\ x_1, \quad x_2, \quad x_3, \quad x_4 &\ge 0 \end{aligned}$$

b. Introduce slack variables and draw the requirement space.
b. Interpret feasibility in the requirement space.
c. You are told that an optimal solution can be obtained by having at most two positive variables while all other variables are set at zero. Utilize this statement and the requirement space to find an optimal solution.

1.32 Consider the problem: Minimize **cx** subject to **Ax** ≥ **b**, **x** ≥ **0**. Suppose that one component of the vector **b**, say b_i, is increased by one unit to $b_i + 1$.
a. What happens to the feasible region?
b. What happens to the optimal objective?

1.33 From the results of the previous problem, assuming $\partial z^*/\partial b_i$ exists, is it ≤ 0, $= 0$, or ≥ 0?

1.34 Solve Exercises 1.32 and 1.33 if the restrictions **Ax** ≥ **b** are replaced by **Ax** ≤ **b**.

1.35 Consider the problem: Minimize **cx** subject to **Ax** ≥ **b**, **x** ≥ **0**. Suppose that a new constraint, $m + 1$, is added to the problem.
a. What happens to the feasible region?
b. What happens to the optimal objective z^*?

1.36 Consider the problem: Minimize $\mathbf{cx}$ subject to $\mathbf{Ax} \geq \mathbf{b}$, $\mathbf{x} \geq \mathbf{0}$. Suppose that a new variable, $n + 1$, is added to the problem.
   a. What happens to the feasible region?
   b. What happens to the optimal objective z^*?

1.37 Consider the problem: Minimize $\mathbf{cx}$ subject to $\mathbf{Ax} \geq \mathbf{b}$, $\mathbf{x} \geq \mathbf{0}$. Suppose that a constraint, say constraint i, is deleted from the problem.
   a. What happens to the feasible region?
   b. What happens to the optimal objective z^*?

1.38 Consider the problem: Minimize $\mathbf{cx}$ subject to $\mathbf{Ax} \geq \mathbf{b}$, $\mathbf{x} \geq \mathbf{0}$. Suppose that a variable, say, x_k, is deleted from the problem.
   a. What happens to the feasible region?
   b. What happens to the optimal objective z^*?

NOTES AND REFERENCES

1. Linear programming and the simplex method were developed by Dantzig in 1947 in connection with military planning. A great deal of work has influenced the development of linear programming, including World War II developments and the need for scheduling supply and maintenance operations as well as training of flying personnel (Leontief's input–output model [1951], von Neumann's Equilibrium Model [1937], Koopman's Model of Transportation [1949], the Hitchcock transportation problem [1941], the work of Kantorovich [1960], von Neumann–Morgenstern game theory [1944]) and the rapid progress in electronic computing machines. The papers by Dantzig [1982] and by Albers and Reid [1986] provide good historical developments.
2. Linear programming has found numerous applications in the military, the government, industry, and urban engineering. See Swanson [1980], for example.
3. Linear programming is also frequently used as a part of general computational schemes for solving nonlinear programming problems, discrete programs, combinatorial problems, problems of optimal control, and programming under uncertainty.
4. Exercise 1.6 is derived from Camm *et al.* [1987] who discuss some modeling issues. For more detailed discussions, see Woolsey and Swanson [1975] and Swanson [1980]. For a discussion on structured modeling and on computer-assisted/Artificial Intelligence approaches to linear programming modeling, see Geoffrion [1987], Greenberg [1983], and Murphy and Stohr [1986], for example.

TWO: LINEAR ALGEBRA, CONVEX ANALYSIS, AND POLYHEDRAL SETS

In this chapter we review some basic results from linear algebra and convex analysis. These results will be used throughout the book. The reader may skip any sections of this chapter, according to his or her familiarity with the subject material. Sections 2.1 and 2.2 review some elementary results from vector and matrix algebra. In Section 2.3 we discuss solvability of a system of linear equations and introduce the important notion of basic solutions. The remaining sections discuss results from convex analysis, including the notions of convex sets, convex and concave functions, convex cones, hyperplanes, and polyhedral sets. The sections on polyhedral sets, which are sets over which linear programs are solved, and their representation in terms of extreme points and extreme directions are very important in linear programming, and hence they deserve thorough study. In particular, Section 2.6 provides geometric insights into the structure of polyhedral sets and develops its various characterizations using only fundamental definitions.

2.1 VECTORS

An n vector is a row or a column array of n numbers. For example, $\mathbf{a} = (1, 2, 3, -1, 4)$ is a row vector of size $n = 5$, and $\mathbf{a} = \begin{pmatrix} 1 \\ -1 \end{pmatrix}$ is a column vector of size $n = 2$. Row and column vectors are denoted by lowercase boldface letters, such as **a**, **b**, **c**. Whether a vector is a row or a column vector will be clear from the context. Figure 2.1 shows some vectors in a two-dimensional space. Each vector can be represented by a point or by a line from the origin to the point, with an arrowhead at the end point of the line.

Special Vectors

ZERO VECTOR

The *zero vector*, denoted by $\mathbf{0}$, is a vector with all components equal to zero. This vector is also referred to as the origin.

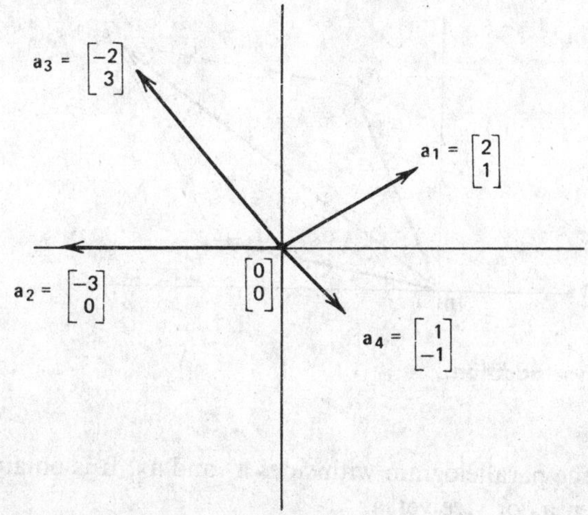

Figure 2.1. **Some examples of vectors.**

i TH UNIT VECTOR

This is a vector with zero components, except for a 1 in the ith position. This vector is denoted by $\mathbf{e}_i$ and is sometimes called the ith *coordinate vector*.

$$\mathbf{e}_i = (0, 0, \ldots, \underset{\substack{\uparrow \\ i\text{th position}}}{1}, \ldots, 0, 0)$$

SUM VECTOR

This is a vector with each component equal to one. The sum vector is denoted by $\mathbf{1}$.

Addition and Multiplication of Vectors

ADDITION

Two vectors of the same size can be added, where addition is performed componentwise. To illustrate, let $\mathbf{a}_1$ and $\mathbf{a}_2$ be the following two n vectors:

$$\mathbf{a}_1 = (a_{11}, a_{21}, \ldots, a_{n1})$$
$$\mathbf{a}_2 = (a_{12}, a_{22}, \ldots, a_{n2})$$

Then the addition of $\mathbf{a}_1$ and $\mathbf{a}_2$, denoted by $\mathbf{a}_1 + \mathbf{a}_2$, is the following vector:

$$\mathbf{a}_1 + \mathbf{a}_2 = (a_{11} + a_{12}, a_{21} + a_{22}, \ldots, a_{n1} + a_{n2})$$

The operation of vector addition is illustrated in Figure 2.2. Note that $\mathbf{a}_1 + \mathbf{a}_2$ is

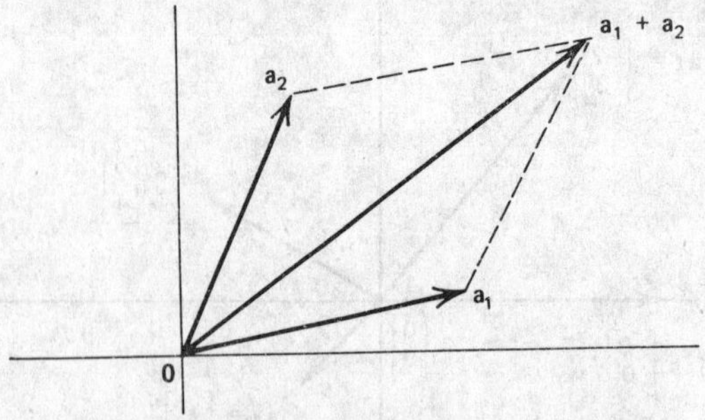

Figure 2.2. **Vector addition.**

the diagonal of the parallelogram with sides $\mathbf{a}_1$ and $\mathbf{a}_2$. It is obtained by stepping along $\mathbf{a}_1$ and then $\mathbf{a}_2$ or vice versa.

SCALAR MULTIPLICATION

The operation of multiplying a vector $\mathbf{a}$ with a scalar k is performed componentwise. If $\mathbf{a} = (a_1, a_2, \ldots, a_n)$, then the vector $k\mathbf{a} = (ka_1, ka_2, \ldots, ka_n)$. This operation is shown in Figure 2.3. If $k > 0$, then $k\mathbf{a}$ points in the same direction as $\mathbf{a}$. On the other hand, if $k < 0$, then $k\mathbf{a}$ points in the opposite direction.

Inner Product

Any two n vectors $\mathbf{a}$ and $\mathbf{b}$ can be multiplied. The result of this multiplication is a real number called the *inner* or *dot product* of the two vectors. It is defined here:

$$\mathbf{ab} = a_1b_1 + a_2b_2 + \cdots + a_nb_n = \sum_{j=1}^{n} a_jb_j$$

For example, if $\mathbf{a} = (1, -1)$ and $\mathbf{b} = \begin{pmatrix} -2 \\ 1 \end{pmatrix}$, then $\mathbf{ab} = -3$.

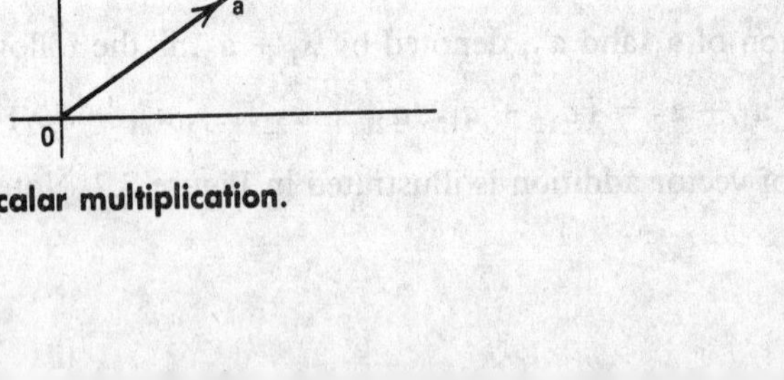

Figure 2.3. **Scalar multiplication.**

Norm of a Vector

Various norms (measures of size) of a vector can be used. We shall use here the *Euclidean norm*. This norm is the square root of the inner product of the vector and itself. In other words, the norm of a vector **a**, denoted by $\|\mathbf{a}\|$, is given by $\sqrt{\sum_{j=1}^{n} a_j^2}$. Note that

$$\|\mathbf{a} + \mathbf{b}\|^2 = \|\mathbf{a}\|^2 + \|\mathbf{b}\|^2 + 2\mathbf{ab}$$

for any two vectors **a** and **b** of the same size.

Schwartz Inequality

Given two vectors **a** and **b** of the same size, the following inequality, called the *Schwartz inequality*, holds:

$$|\mathbf{ab}| \le \|\mathbf{a}\|\,\|\mathbf{b}\|$$

To illustrate, let $\mathbf{a} = (0, 2)$ and $\mathbf{b} = \begin{pmatrix} 3 \\ 4 \end{pmatrix}$. Then $\mathbf{ab} = 8$, whereas $\|\mathbf{a}\| = 2$ and $\|\mathbf{b}\| = 5$. Clearly $8 \le 2 \times 5$. In fact, the "discrepancy" between the inner product **ab** and $\|\mathbf{a}\|\,\|\mathbf{b}\|$ measures the angle θ between the two vectors. In particular, $\cos\theta = \mathbf{ab}/\|\mathbf{a}\|\,\|\mathbf{b}\|$. Of course if $\mathbf{ab} = 0$, then $\cos\theta = 0$; that is, the two vectors **a** and **b** are *orthogonal* or *normal* or *perpendicular* to each other. Figure 2.4 shows two orthogonal vectors **a** and **b**.

Euclidean Space

An n-dimensional *Euclidean* space, denoted by E^n, is the collection of all vectors of dimension n. Addition and scalar multiplication of vectors in E^n are defined previously. Also, associated with any vector in E^n is its norm, and associated with any two vectors in E^n is their inner product, defined previously.

Linear and Affine Combinations

A vector **b** in E^n is said to be a *linear combination* of $\mathbf{a}_1, \mathbf{a}_2, \ldots, \mathbf{a}_k$ in E^n, if $\mathbf{b} = \sum_{j=1}^{k} \lambda_j \mathbf{a}_j$, where $\lambda_1, \lambda_2, \ldots, \lambda_k$ are real numbers. If, in addition, $\sum_{j=1}^{k} \lambda_j = 1$, then **b** is said to be an *affine combination* of $\mathbf{a}_1, \mathbf{a}_2, \ldots, \mathbf{a}_k$.

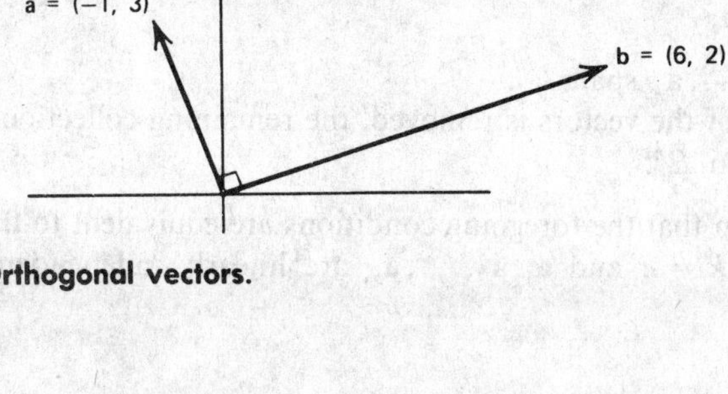

Figure 2.4. **Orthogonal vectors.**

Linear and Affine Subspaces

A *linear subspace* S_L of E^n is a subset of E^n such that if $\mathbf{a}_1$ and $\mathbf{a}_2 \in S_L$, then every linear combination of $\mathbf{a}_1$ and $\mathbf{a}_2$ belongs to S_L. Similarly, an *affine subspace* S_A of E^n is a subset of E^n such that if $\mathbf{a}_1$ and $\mathbf{a}_2 \in S_A$, then every affine combination of $\mathbf{a}_1$ and $\mathbf{a}_2$ belongs to S_A.

Linear Independence

A collection of vectors $\mathbf{a}_1, \mathbf{a}_2, \ldots, \mathbf{a}_k$ of dimension n is called *linearly independent* if

$$\sum_{j=1}^{k} \lambda_j \mathbf{a}_j = \mathbf{0} \text{ implies that } \lambda_j = 0 \quad \text{for } j = 1, 2, \ldots, k$$

For example, let $\mathbf{a}_1 = (1, 2)$ and $\mathbf{a}_2 = (-1, 1)$. These two vectors are linearly independent because $\lambda_1(1, 2) + \lambda_2(-1, 1) = (0, 0)$ implies that $\lambda_1 = \lambda_2 = 0$.

A collection of vectors is called *linearly dependent* if they are not linearly independent. Therefore $\mathbf{a}_1, \mathbf{a}_2, \ldots, \mathbf{a}_k$ are linearly dependent if there exist $\lambda_1, \lambda_2, \ldots, \lambda_k$, *not all zero*, such that $\sum_{j=1}^{k} \lambda_j \mathbf{a}_j = \mathbf{0}$. For example, let $\mathbf{a}_1 = (1, 2, 3)$, $\mathbf{a}_2 = (-1, 1, -1)$, and $\mathbf{a}_3 = (0, 3, 2)$. These three vectors are linearly dependent because $\lambda_1\mathbf{a}_1 + \lambda_2\mathbf{a}_2 + \lambda_3\mathbf{a}_3 = \mathbf{0}$ for $\lambda_1 = \lambda_2 = 1$ and $\lambda_3 = -1$.

Spanning Set

A collection of vectors $\mathbf{a}_1, \mathbf{a}_2, \ldots, \mathbf{a}_k$ in E^n is said to *span* E^n if any vector in E^n can be represented as a linear combination of $\mathbf{a}_1, \mathbf{a}_2, \ldots, \mathbf{a}_k$. In other words, given any vector $\mathbf{b}$ in E^n, we must be able to find scalars $\lambda_1, \lambda_2, \ldots, \lambda_k$ such that $\mathbf{b} = \sum_{j=1}^{k} \lambda_j \mathbf{a}_j$.

To illustrate, let $n = 2$, and consider $\mathbf{a}_1 = (1, 0)$, $\mathbf{a}_2 = (-1, 3)$, and $\mathbf{a}_3 = (2, 1)$. The vectors $\mathbf{a}_1, \mathbf{a}_2, \mathbf{a}_3$ span E^2 since any vector $\mathbf{b}$ in E^2 can be represented as a linear combination of these vectors. For example, $\mathbf{b} = (b_1, b_2)$ can be represented as $\lambda_1\mathbf{a}_1 + \lambda_2\mathbf{a}_2 + \lambda_3\mathbf{a}_3$ where $\lambda_1 = b_1 + \frac{1}{3}b_2$, $\lambda_2 = \frac{1}{3}b_2$, and $\lambda_3 = 0$. In this case the representation is not unique. Another representation is given by letting $\lambda_1 = b_1 - 2b_2$, $\lambda_2 = 0$, and $\lambda_3 = b_2$.

Basis

A collection of vectors $\mathbf{a}_1, \mathbf{a}_2, \ldots, \mathbf{a}_k$ forms a *basis* of E^n if the following conditions hold:

1. $\mathbf{a}_1, \mathbf{a}_2, \ldots, \mathbf{a}_k$ span E^n.
2. If any of the vectors is removed, the remaining collection of vectors does not span E^n.

It can be shown that the foregoing conditions are equivalent to the following two requirements: $k = n$ and $\mathbf{a}_1, \mathbf{a}_2, \ldots, \mathbf{a}_n$ are linearly independent. To illustrate, consider the two vectors $\mathbf{a}_1 = \begin{pmatrix} 1 \\ 1 \end{pmatrix}$ and $\mathbf{a}_2 = \begin{pmatrix} 0 \\ 1 \end{pmatrix}$ in E^2. These two vectors form a basis of E^2 since $k = n = 2$, and $\mathbf{a}_1$ and $\mathbf{a}_2$ are linearly independent.

Given a basis of E^n, say $\mathbf{a}_1, \mathbf{a}_2, \ldots, \mathbf{a}_n$, any vector $\mathbf{b}$ in E^n is uniquely represented in terms of this basis. If $\mathbf{b} = \sum_{j=1}^n \lambda_j \mathbf{a}_j$ and also $\mathbf{b} = \sum_{j=1}^n \lambda'_j \mathbf{a}_j$, then, $\sum_{j=1}^n (\lambda_j - \lambda'_j)\mathbf{a}_j = \mathbf{0}$, which implies that $\lambda_j = \lambda'_j$ for each j, since otherwise we would violate the linear independence of $\mathbf{a}_1, \mathbf{a}_2, \ldots, \mathbf{a}_n$.

Since a basis in E^n must always have n vectors, then the dimension of a basis is unique, namely n. But a basis itself is not unique, since any set of n vectors that are linearly independent will form a basis.

Replacing a Vector in the Basis by Another Vector

In the simplex method, different bases will be generated, where one vector from the last basis is replaced by another vector. We have to be careful in choosing the vectors entering and leaving the basis, because otherwise the new vectors may not be linearly independent, and hence will not form a basis. To illustrate, the vectors $\mathbf{a}_1 = (1, 2, 1)$, $\mathbf{a}_2 = (3, 0, 1)$, and $\mathbf{a}_3 = (2, -2, 1)$ are linearly independent, and hence form a basis of E^3. We cannot replace $\mathbf{a}_3$ by $(2, -2, 0)$, because $\mathbf{a}_1$, $\mathbf{a}_2$, and $(2, -2, 0)$ are linearly dependent and do not form a basis.

This leads to the following natural question: If we have a basis of E^n, what is the condition that will guarantee that if a vector of the basis, say $\mathbf{a}_j$, is replaced by another vector, say $\mathbf{a}$, then the new set of vectors still forms a basis?

Let $\mathbf{a}_1, \mathbf{a}_2, \ldots, \mathbf{a}_n$ form a basis of E^n. We want to replace $\mathbf{a}_j$ by $\mathbf{a}$. Since $\mathbf{a}_1, \ldots, \mathbf{a}_n$ form a basis, then $\mathbf{a}$ can be represented as a linear combination of these vectors, that is,

$$\mathbf{a} = \sum_{i=1}^{n} \lambda_i \mathbf{a}_i$$

Suppose that $\lambda_j \neq 0$. We shall show that the vectors $\mathbf{a}_1, \mathbf{a}_2, \ldots, \mathbf{a}_{j-1}, \mathbf{a}, \mathbf{a}_{j+1}, \ldots, \mathbf{a}_n$ are linearly independent, and hence form a basis. Suppose that there exist μ and $\mu_i (i \neq j)$, such that

$$\sum_{i \neq j} \mu_i \mathbf{a}_i + \mu \mathbf{a} = \mathbf{0}$$

Substituting $\mathbf{a} = \sum_{i=1}^n \lambda_i \mathbf{a}_i$, we get

$$\sum_{i \neq j} \mu_i \mathbf{a}_i + \mu \sum_{i=1}^{n} \lambda_i \mathbf{a}_i = \mathbf{0}$$

$$\sum_{i \neq j} (\mu_i + \mu\lambda_i)\mathbf{a}_i + \mu\lambda_j \mathbf{a}_j = \mathbf{0}$$

But since $\mathbf{a}_1, \mathbf{a}_2, \ldots, \mathbf{a}_j, \ldots, \mathbf{a}_n$ are linearly independent, then $\mu\lambda_j = 0$, and $\mu_i + \mu\lambda_i = 0$ for $i \neq j$. Since $\lambda_j \neq 0$ by assumption, then $\mu = 0$. But this implies that $\mu_i = 0$ for $i \neq j$. In other words, $\sum_{i \neq j} \mu_i \mathbf{a}_i + \mu \mathbf{a} = \mathbf{0}$ is only possible if $\mu = 0$ and $\mu_i = 0$ for $i \neq j$, and hence $\mathbf{a}$ and $\mathbf{a}_i (i \neq j)$ are linearly independent and must form a basis. From this discussion it is obvious that the condition $\lambda_j \neq 0$ is sufficient for the new set of vectors to be linearly independent. Obviously the condition is also necessary, because if λ_j were zero, then $\mathbf{a} - \sum_{i \neq j} \lambda_i \mathbf{a}_i = \mathbf{0}$, and hence $\mathbf{a}$ and $\mathbf{a}_i (i \neq j)$ would be linearly dependent.

2.2 MATRICES

A *matrix* is a rectangular array of numbers. If the matrix has m rows and n columns, it is called an $m \times n$ matrix (reads "m by n"). Matrices will be denoted by capital boldface letters, such as **A**, **B**, **C**. An example of a 3×2 matrix is given

$$\mathbf{A} = \begin{bmatrix} 1 & -1 \\ 2 & 2 \\ 3 & 1 \end{bmatrix}$$

The entry in row i and column j is denoted by a_{ij}; for example, $a_{12} = -1$ and $a_{31} = 3$. An $m \times n$ matrix **A** can be represented by its columns or by its rows. If we denote the columns of **A** by $\mathbf{a}_1, \mathbf{a}_2, \ldots, \mathbf{a}_n$, then $\mathbf{A} = [\mathbf{a}_1, \mathbf{a}_2, \ldots, \mathbf{a}_n]$. Similarly, **A** can be represented as

$$\begin{bmatrix} \mathbf{a}^1 \\ \mathbf{a}^2 \\ \vdots \\ \mathbf{a}^m \end{bmatrix}$$

where $\mathbf{a}^1, \mathbf{a}^2, \ldots, \mathbf{a}^m$ are the rows of **A**. Note that every vector is a matrix, but not every matrix is a vector.

Addition of Matrices

The addition of two matrices of the same dimension is defined componentwise; that is, if **A** and **B** are $m \times n$ matrices, then $\mathbf{C} = \mathbf{A} + \mathbf{B}$ is defined by letting $c_{ij} = a_{ij} + b_{ij}$ for $i = 1, 2, \ldots, m$ and $j = 1, 2, \ldots, n$.

Multiplication by a Scalar

Let **A** be an $m \times n$ matrix and let k be a scalar. Then $k\mathbf{A}$ is an $m \times n$ matrix whose ij entry is ka_{ij}.

Matrix Multiplication

Let **A** be an $m \times n$ matrix and **B** be an $n \times p$ matrix. Then the product **AB** is defined to be the $m \times p$ matrix **C** with

$$c_{ij} = \sum_{k=1}^{n} a_{ik} b_{kj} \qquad \text{for } i = 1, \ldots, m \quad \text{and } j = 1, \ldots, p$$

In other words, the ij entry of **C** is determined as the inner product of the ith row of **A** and the jth column of **B**. Let

$$\mathbf{A} = \begin{bmatrix} 1 & -1 & 1 \\ 4 & -2 & 5 \\ 2 & 0 & 1 \end{bmatrix} \quad \text{and} \quad \mathbf{B} = \begin{bmatrix} 5 & 0 \\ 3 & 0 \\ 1 & 1 \end{bmatrix}$$

Then

$$\mathbf{C} = \mathbf{AB} = \begin{bmatrix} 1 & -1 & 1 \\ 4 & -2 & 5 \\ 2 & 0 & 1 \end{bmatrix} \begin{bmatrix} 5 & 0 \\ 3 & 0 \\ 1 & 1 \end{bmatrix} = \begin{bmatrix} 3 & 1 \\ 19 & 5 \\ 11 & 1 \end{bmatrix}$$

The following points need to be emphasized. If **A** is an $m \times n$ matrix and **B** is a $p \times q$ matrix, then

1. **AB** is defined only if $n = p$. **AB** is then an $m \times q$ matrix.
2. **BA** is defined only if $q = m$. **BA** is then an $p \times n$ matrix.
3. Even if **AB** and **BA** are both defined (if $m = n = p = q$), then **AB** is not necessarily equal to **BA**. Note that **AB** in the foregoing example is defined, but **BA** is not defined.

Special Matrices

ZERO MATRIX

An $m \times n$ matrix is called the *zero* matrix if each entry in the matrix is zero.

IDENTITY MATRIX

A square $n \times n$ matrix is called the *identity* matrix, denoted by **I** (sometimes the notation $\mathbf{I}_n$ is used to denote the size), if it has entries equal to one on the diagonal and zero entries everywhere else. Note that $\mathbf{AI}_n = \mathbf{A}$ and $\mathbf{I}_m\mathbf{A} = \mathbf{A}$ for any $m \times n$ matrix **A**.

TRIANGULAR MATRIX

A square $n \times n$ matrix is called an *upper triangular* matrix if all the entries below the diagonal are zeros. Similarly, an $n \times n$ matrix is called a *lower triangular*, matrix if all elements above the diagonal are zeros.

Transposition

Given an $m \times n$ matrix **A** with a_{ij} as its *ij* entry, the *transpose* of **A**, denoted by $\mathbf{A}^t$, is an $n \times m$ matrix whose *ij* entry is a_{ji}. In other words, $\mathbf{A}^t$ is formed by letting the jth column of **A** be the jth row of $\mathbf{A}^t$ (similarly by letting the jth row of **A** be the jth column of $\mathbf{A}^t$). A square matrix **A** is called *symmetric* if $\mathbf{A} = \mathbf{A}^t$ and *skew-symmetric* if $\mathbf{A} = -\mathbf{A}^t$. The following results are obvious.

1. $(\mathbf{A}^t)^t = \mathbf{A}$.
2. If **A** and **B** have the same dimension, then $(\mathbf{A} + \mathbf{B})^t = \mathbf{A}^t + \mathbf{B}^t$.
3. If **AB** is defined, then $(\mathbf{AB})^t = \mathbf{B}^t\mathbf{A}^t$.

Partitioned Matrices

Given an $m \times n$ matrix $\mathbf{A}$, we can obtain a submatrix of $\mathbf{A}$ by deleting certain rows and/or columns of $\mathbf{A}$. Hence we can think of a matrix $\mathbf{A}$ as being *partitioned* into submatrices. For example, consider

$$\mathbf{A} = \left[\begin{array}{cc|cc} a_{11} & a_{12} & a_{13} & a_{14} \\ a_{21} & a_{22} & a_{23} & a_{24} \\ a_{31} & a_{32} & a_{33} & a_{34} \\ \hline a_{41} & a_{42} & a_{43} & a_{44} \end{array}\right]$$

Here $\mathbf{A}$ has been partitioned into four submatrices: $\mathbf{A}_{11}$, $\mathbf{A}_{12}$, $\mathbf{A}_{21}$, and $\mathbf{A}_{22}$; therefore

$$\mathbf{A} = \left[\begin{array}{c|c} \mathbf{A}_{11} & \mathbf{A}_{12} \\ \hline \mathbf{A}_{21} & \mathbf{A}_{22} \end{array}\right]$$

where

$$\mathbf{A}_{11} = \begin{bmatrix} a_{11} & a_{12} \\ a_{21} & a_{22} \\ a_{31} & a_{32} \end{bmatrix} \qquad \mathbf{A}_{12} = \begin{bmatrix} a_{13} & a_{14} \\ a_{23} & a_{24} \\ a_{33} & a_{34} \end{bmatrix}$$

$$\mathbf{A}_{21} = \begin{bmatrix} a_{41} & a_{42} \end{bmatrix} \qquad \mathbf{A}_{22} = \begin{bmatrix} a_{43} & a_{44} \end{bmatrix}$$

Now suppose that $\mathbf{A}$ and $\mathbf{B}$ are partitioned as follows:

$$\mathbf{A} = \overset{\quad n_1 \qquad n_2}{\left[\begin{array}{c|c} \mathbf{A}_{11} & \mathbf{A}_{12} \\ \hline \mathbf{A}_{21} & \mathbf{A}_{22} \end{array}\right]} \begin{array}{c} m_1 \\ m_2 \end{array} \qquad \mathbf{B} = \overset{\quad q_1 \qquad q_2 \qquad q_3}{\left[\begin{array}{c|c|c} \mathbf{B}_{11} & \mathbf{B}_{12} & \mathbf{B}_{13} \\ \hline \mathbf{B}_{21} & \mathbf{B}_{22} & \mathbf{B}_{23} \end{array}\right]} \begin{array}{c} p_1 \\ p_2 \end{array}$$

Then $\mathbf{AB}$ is defined by

$$\begin{aligned} \mathbf{AB} &= \left[\begin{array}{c|c} \mathbf{A}_{11} & \mathbf{A}_{12} \\ \hline \mathbf{A}_{21} & \mathbf{A}_{22} \end{array}\right] \left[\begin{array}{c|c|c} \mathbf{B}_{11} & \mathbf{B}_{12} & \mathbf{B}_{13} \\ \hline \mathbf{B}_{21} & \mathbf{B}_{22} & \mathbf{B}_{23} \end{array}\right] \\ &= \left[\begin{array}{c|c|c} \mathbf{A}_{11}\mathbf{B}_{11} + \mathbf{A}_{12}\mathbf{B}_{21} & \mathbf{A}_{11}\mathbf{B}_{12} + \mathbf{A}_{12}\mathbf{B}_{22} & \mathbf{A}_{11}\mathbf{B}_{13} + \mathbf{A}_{12}\mathbf{B}_{23} \\ \hline \mathbf{A}_{21}\mathbf{B}_{11} + \mathbf{A}_{22}\mathbf{B}_{21} & \mathbf{A}_{21}\mathbf{B}_{12} + \mathbf{A}_{22}\mathbf{B}_{22} & \mathbf{A}_{21}\mathbf{B}_{13} + \mathbf{A}_{22}\mathbf{B}_{23} \end{array}\right] \end{aligned}$$

Note that we must have $n_1 = p_1$ and $n_2 = p_2$, so that the product of the submatrices is well defined.

Elementary Matrix Operations

Given an $m \times n$ matrix $\mathbf{A}$, we can perform some elementary row and column operations. These operations are most helpful in solving a system of linear equations and in finding the inverse of a matrix (to be defined later).

An *elementary row operation* on a matrix $\mathbf{A}$ is one of the following operations:

1. Row i and row j of $\mathbf{A}$ are interchanged.
2. Row i is multiplied by a nonzero scalar k.
3. Row i is replaced by row i plus k times row j.

Elementary row operations on a matrix **A** are equivalent to premultiplying **A** by a specific matrix. *Elementary column operations* are defined similarly. Elementary column operations on **A** are equivalent to postmultiplying **A** by a specific matrix.

Example 2.1

Let

$$\mathbf{A} = \begin{bmatrix} 2 & 1 & 1 & 10 \\ -1 & 2 & 1 & 8 \\ 1 & -1 & 2 & 2 \end{bmatrix}.$$

We shall perform the following elementary operations on **A**. Divide row 1 by 2, add the new row 1 to row 2, and subtract it from row 3. This gives

$$\begin{bmatrix} 1 & \frac{1}{2} & \frac{1}{2} & 5 \\ 0 & \frac{5}{2} & \frac{3}{2} & 13 \\ 0 & -\frac{3}{2} & \frac{3}{2} & -3 \end{bmatrix}$$

Now multiply row 2 by $\frac{2}{5}$, multiply the new row 2 by $\frac{3}{2}$, and add to row 3. This gives

$$\begin{bmatrix} 1 & \frac{1}{2} & \frac{1}{2} & 5 \\ 0 & 1 & \frac{3}{5} & \frac{26}{5} \\ 0 & 0 & \frac{24}{10} & \frac{24}{5} \end{bmatrix}$$

Divide row 3 by $\frac{24}{10}$. This gives

$$\begin{bmatrix} 1 & \frac{1}{2} & \frac{1}{2} & 5 \\ 0 & 1 & \frac{3}{5} & \frac{26}{5} \\ 0 & 0 & 1 & 2 \end{bmatrix}$$

Note that the matrix **A** is reduced to the foregoing matrix through elementary row operations. In particular, the first operation is performed by premultiplying **A** by the following matrix:

$$\begin{bmatrix} \frac{1}{2} & 0 & 0 \\ \frac{1}{2} & 1 & 0 \\ -\frac{1}{2} & 0 & 1 \end{bmatrix}$$

Solving a System of Linear Equations by Elementary Matrix Operations

Consider the system $\mathbf{Ax} = \mathbf{b}$ of m equations of n unknown, where **A** is an $m \times n$ matrix, **b** is an m vector, and **x** is an n vector of variables. The following fact is helpful in solving this system: $\mathbf{Ax} = \mathbf{b}$ if and only if $\mathbf{A'x} = \mathbf{b'}$, where $(\mathbf{A'}, \mathbf{b'})$ is obtained from $(\mathbf{A}, \mathbf{b})$ by a finite number of elementary row operations. To illustrate, consider the following system:

$$\begin{aligned} 2x_1 + \ x_2 + \ x_3 &= 10 \\ -x_1 + 2x_2 + \ x_3 &= \ 8 \\ x_1 - \ x_2 + 2x_3 &= \ 2 \end{aligned}$$

$$(\mathbf{A}, \mathbf{b}) = \begin{bmatrix} 2 & 1 & 1 & 10 \\ -1 & 2 & 1 & 8 \\ 1 & -1 & 2 & 2 \end{bmatrix}$$

This matrix was reduced in Example 2.1 through elementary row operations to

$$\begin{bmatrix} 1 & \frac{1}{2} & \frac{1}{2} & 5 \\ 0 & 1 & \frac{3}{5} & \frac{26}{5} \\ 0 & 0 & 1 & 2 \end{bmatrix}$$

Therefore $\mathbf{x}$ solves the original system if and only if it solves the following system.

$$\begin{aligned} x_1 + \tfrac{1}{2}x_2 + \tfrac{1}{2}x_3 &= 5 \\ x_2 + \tfrac{3}{5}x_3 &= \tfrac{26}{5} \\ x_3 &= 2 \end{aligned}$$

Note that $\mathbf{A}'$ is upper triangular, and we can solve the system by *back substitution*. From the third equation $x_3 = 2$, from the second equation $x_2 = 4$, and from the first equation $x_1 = 2$. The process of reducing $\mathbf{A}$ into an upper triangular matrix with ones on the diagonal is called *Gaussian reduction* of the system. By performing further row operations, we could have reduced $\mathbf{A}$ into an identity matrix, from which the vector $\mathbf{b}$ would have been transformed to $(2, 4, 2)'$. This process is called a *Gauss–Jordan* reduction of the system.

Matrix Inversion

Let $\mathbf{A}$ be a square $n \times n$ matrix. If $\mathbf{B}$ is an $n \times n$ matrix such that $\mathbf{AB} = \mathbf{I}$ and $\mathbf{BA} = \mathbf{I}$, then $\mathbf{B}$ is called the *inverse* of $\mathbf{A}$. The inverse matrix, if it exists, is unique and is denoted by $\mathbf{A}^{-1}$. If $\mathbf{A}$ has an inverse, $\mathbf{A}$ is called *nonsingular*; otherwise $\mathbf{A}$ is called *singular*.

CONDITION FOR EXISTENCE OF THE INVERSE

Given an $n \times n$ matrix $\mathbf{A}$, it has an inverse if and only if the rows of $\mathbf{A}$ are linearly independent or, equivalently, if the columns of $\mathbf{A}$ are linearly independent.

CALCULATION OF THE INVERSE

The inverse matrix, if it exists, can be obtained through a finite number of elementary row operations. This can be done by noting that if a sequence of elementary row operations reduce $\mathbf{A}$ to the identity, then the same sequence of operations will reduce $(\mathbf{A}, \mathbf{I})$ to $(\mathbf{I}, \mathbf{A}^{-1})$. In fact, this is equivalent to premultiplying the system by $\mathbf{A}^{-1}$. Further, if $(\mathbf{A}, \mathbf{B})$ is reduced to $(\mathbf{I}, \mathbf{F})$ by elementary row operations, then $\mathbf{F} = \mathbf{A}^{-1}\mathbf{B}$.

In order to calculate the inverse, we adjoin the identity to $\mathbf{A}$. The matrix $\mathbf{A}$ is reduced to the identity by elementary row operations. This will result in reducing the identity to $\mathbf{A}^{-1}$. Of course, if $\mathbf{A}^{-1}$ does not exist, then the elementary row operations will fail to produce the identity. This discussion is made clear by the following two examples.

Example 2.2

(A^{-1} exists)

Consider the matrix $\mathbf{A}$:

$$\mathbf{A} = \begin{bmatrix} 2 & 1 & 1 \\ -1 & 2 & 1 \\ 1 & -1 & 2 \end{bmatrix}$$

To find the inverse, form the augmented matrix $(\mathbf{A}, \mathbf{I})$. Reduce $\mathbf{A}$ by elementary row operations to the identity. The matrix in place of $\mathbf{I}$ is $\mathbf{A}^{-1}$.

$$\left[\begin{array}{rrr|rrr} 2 & 1 & 1 & 1 & 0 & 0 \\ -1 & 2 & 1 & 0 & 1 & 0 \\ 1 & -1 & 2 & 0 & 0 & 1 \end{array}\right]$$

Divide the first row by 2. Add the new first row to the second row and subtract it from the third row.

$$\left[\begin{array}{rrr|rrr} 1 & \frac{1}{2} & \frac{1}{2} & \frac{1}{2} & 0 & 0 \\ 0 & \frac{5}{2} & \frac{3}{2} & \frac{1}{2} & 1 & 0 \\ 0 & -\frac{3}{2} & \frac{3}{2} & -\frac{1}{2} & 0 & 1 \end{array}\right]$$

Multiply the second row by $\frac{2}{5}$. Multiply the new second row by $-\frac{1}{2}$ and add to the first row. Multiply the new second row by $\frac{3}{2}$ and add to the third row.

$$\left[\begin{array}{rrr|rrr} 1 & 0 & \frac{1}{5} & \frac{2}{5} & -\frac{1}{5} & 0 \\ 0 & 1 & \frac{3}{5} & \frac{1}{5} & \frac{2}{5} & 0 \\ 0 & 0 & \frac{12}{5} & -\frac{1}{5} & \frac{3}{5} & 1 \end{array}\right]$$

Multiply the third row by $\frac{5}{12}$. Multiply the new third row by $-\frac{3}{5}$ and add to the second row. Multiply the new third row by $-\frac{1}{5}$ and add to the first row.

$$\left[\begin{array}{rrr|rrr} 1 & 0 & 0 & \frac{5}{12} & -\frac{3}{12} & -\frac{1}{12} \\ 0 & 1 & 0 & \frac{3}{12} & \frac{3}{12} & -\frac{3}{12} \\ 0 & 0 & 1 & -\frac{1}{12} & \frac{3}{12} & \frac{5}{12} \end{array}\right]$$

Therefore the inverse of **A** exists and is given by

$$\mathbf{A}^{-1} = \frac{1}{12}\begin{bmatrix} 5 & -3 & -1 \\ 3 & 3 & -3 \\ -1 & 3 & 5 \end{bmatrix}$$

Example 2.3

(A^{-1} does not exist)

Consider the matrix **A**:

$$\mathbf{A} = \begin{bmatrix} 1 & 1 & 2 \\ 2 & -1 & 1 \\ 1 & 2 & 3 \end{bmatrix}$$

The inverse does not exist since $\mathbf{a}_3 = \mathbf{a}_1 + \mathbf{a}_2$. If we use the foregoing procedure, the elementary matrix operations will fail to produce the identity.

$$\left[\begin{array}{ccc|ccc} 1 & 1 & 2 & 1 & 0 & 0 \\ 2 & -1 & 1 & 0 & 1 & 0 \\ 1 & 2 & 3 & 0 & 0 & 1 \end{array}\right]$$

Multiply the first row by -2 and add to the second row. Multiply the first row by -1 and add to the third row.

$$\left[\begin{array}{ccc|ccc} 1 & 1 & 2 & 1 & 0 & 0 \\ 0 & -3 & -3 & -2 & 1 & 0 \\ 0 & 1 & 1 & -1 & 0 & 1 \end{array}\right]$$

Multiply the second row by $-\frac{1}{3}$. Then multiply the new second row by -1 and add to the first row. Multiply the new second row by -1 and add to the third row.

$$\left[\begin{array}{ccc|ccc} 1 & 0 & 1 & \frac{1}{3} & \frac{1}{3} & 0 \\ 0 & 1 & 1 & \frac{2}{3} & -\frac{1}{3} & 0 \\ 0 & 0 & 0 & -\frac{5}{3} & \frac{1}{3} & 1 \end{array}\right]$$

There is no way that the left-hand-side matrix can be transformed into the identity matrix by elementary row operations, and hence the matrix **A** has no inverse.

The following facts about matrix inversion are useful.

1. If **A** is nonsingular, then $\mathbf{A}'$ is also nonsingular and $(\mathbf{A}')^{-1} = (\mathbf{A}^{-1})'$.
2. If **A** and **B** are both $n \times n$ nonsingular matrices, then **AB** is nonsingular and $(\mathbf{AB})^{-1} = \mathbf{B}^{-1}\mathbf{A}^{-1}$.
3. A triangular matrix (either lower or upper triangular) with nonzero diagonal elements has an inverse. This can be easily established by noting that such a matrix can be reduced to the identity by a finite number of elementary row operations. In particular, let $D = \text{diag}\{d_1, \ldots, d_n\}$ be a *diagonal matrix* with diagonal elements $d_1, \ldots, d_n$ and all other elements being zero. If $d_1, \ldots, d_n$ are all nonzero, then $D^{-1} = \text{diag}\{1/d_1, \ldots, 1/d_n\}$.

4. Let $\mathbf{A}$ be partitioned as follows, where $\mathbf{D}$ is nonsingular.

$$\mathbf{A} = \begin{array}{c} \begin{matrix} n_1 & n_2 \end{matrix} \\ \left[\begin{array}{c|c} \mathbf{I} & \mathbf{C} \\ \hline \mathbf{0} & \mathbf{D} \end{array}\right] \end{array} \begin{matrix} n_1 \\ n_2 \end{matrix}$$

Then $\mathbf{A}$ is nonsingular and

$$\mathbf{A}^{-1} = \left[\begin{array}{c|c} \mathbf{I} & -\mathbf{C}\mathbf{D}^{-1} \\ \hline \mathbf{0} & \mathbf{D}^{-1} \end{array}\right]$$

Determinant of a Matrix

Associated with each square $n \times n$ matrix is a real number called the *determinant* of the matrix. Let $\mathbf{A}$ be an $n \times n$ matrix whose ij element is a_{ij}. The determinant of $\mathbf{A}$, denoted by $\det \mathbf{A}$, is defined as follows:

$$\det \mathbf{A} = \sum_{i=1}^{n} a_{i1} A_{i1}$$

where A_{i1} is the *cofactor* of a_{i1} defined as $(-1)^{i+1}$ times the determinant of the submatrix of $\mathbf{A}$ obtained by deleting the ith row and first column. The determinant of a 1×1 matrix is just the element itself. To illustrate, consider the following example.

$$\det \begin{bmatrix} 1 & 0 & 1 \\ 2 & 1 & -3 \\ -3 & 2 & 1 \end{bmatrix} = 1A_{11} + 2A_{21} - 3A_{31}$$

$$= \det \begin{bmatrix} 1 & -3 \\ 2 & 1 \end{bmatrix} - 2\det \begin{bmatrix} 0 & 1 \\ 2 & 1 \end{bmatrix} - 3\det \begin{bmatrix} 0 & 1 \\ 1 & -3 \end{bmatrix}$$

Note that the foregoing definition reduces the calculation of a determinant of an $n \times n$ matrix to n determinants of $(n-1) \times (n-1)$ matrices. The same definition can be used to reduce a determinant of an $(n-1) \times (n-1)$ matrix to determinants of $(n-2) \times (n-2)$ matrices, and so forth. Obviously, by this definition, the determinant of a 2×2 matrix, say $\mathbf{A}' = \begin{bmatrix} a_{11} & a_{12} \\ a_{21} & a_{22} \end{bmatrix}$, is simply $a_{11}a_{22} - a_{21}a_{12}$.

To summarize, the determinant of an $n \times n$ matrix can be calculated by successively applying the foregoing definition. The determinant of $\mathbf{A}$ is therefore given by $1(1 + 6) - 2(0 - 2) - 3(0 - 1) = 14$. We summarize some important facts about determinants of square matrices.

1. In the definition, the first column was used as a reference in calculating $\det \mathbf{A}$. Any column or row can be used as a reference in the calculations; that is,

$$\det \mathbf{A} = \sum_{i=1}^{n} a_{ij} A_{ij} \qquad \text{for any } j = 1, 2, \ldots, n$$

and similarly

$$\det \mathbf{A} = \sum_{j=1}^{n} a_{ij}A_{ij} \qquad \text{for any } i = 1,2,\ldots,n$$

where A_{ij} is the cofactor of a_{ij} given as $(-1)^{i+j}$ times the determinant of the submatrix obtained from **A** by deleting the ith row and jth column.

2. $\det \mathbf{A} = \det \mathbf{A}'$.
3. Let **B** be obtained from **A** by interchanging two rows (or columns). Then $\det \mathbf{B} = -\det \mathbf{A}$.
4. Let **B** be obtained from **A** by adding to one row (column) a constant times another row (column). Then $\det \mathbf{B} = \det \mathbf{A}$.
5. Let **B** be obtained from **A** by multiplying a row (or column) by a scalar k. Then $\det \mathbf{B} = k \det \mathbf{A}$.
6. Let **A** be partitioned as follows, where **B** and **C** are square.

$$\mathbf{A} = \left[\begin{array}{c|c} \mathbf{B} & \mathbf{0} \\ \hline \mathbf{D} & \mathbf{C} \end{array}\right]$$

Then $\det \mathbf{A} = \det \mathbf{B} \cdot \det \mathbf{C}$.

7. By a repeated application of fact 6, the determinant of a triangular matrix is simply the product of the diagonal entries.
8. Let **A** and **B** be $n \times n$ matrices. Then $\det(\mathbf{AB}) = \det \mathbf{A} \cdot \det \mathbf{B}$.
9. $\det \mathbf{A} \neq 0$ if and only if the columns (and rows) of **A** are linearly independent. Equivalently, $\det \mathbf{A} = 0$ if and only if the rows (columns) of **A** are linearly dependent. Therefore a square matrix **A** has an inverse if and only if its determinant is not zero.
10. Let **A** be an $n \times n$ matrix whose determinant is not zero. Then $\mathbf{A}^{-1}$ exists and is given by

$$\mathbf{A}^{-1} = \frac{\mathbf{B}}{\det \mathbf{A}}$$

where **B** is the transpose of the matrix whose ij entry is A_{ij}, the cofactor of a_{ij}. Here **B** is called the *adjoint matrix* of **A**.

11. Consider the system $\mathbf{Ax} = \mathbf{b}$ where **A** is $n \times n$, **b** is an n vector, and **x** is an n vector of unknowns. If **A** has an inverse (that is, if $\det \mathbf{A} \neq 0$), then the unique solution to this system is given by

$$x_j = \frac{\det \mathbf{A}_j}{\det \mathbf{A}} \qquad \text{for } j = 1,2,\ldots,n$$

where $\mathbf{A}_j$ is obtained from **A** by replacing the jth column of **A** by **b**. This method for solving the system is called *Cramer's rule*.

The Rank of a Matrix

Let **A** be an $m \times n$ matrix. The *row rank* of the matrix is equal to the maximum number of linearly independent rows of **A**. The *column rank* of **A** is the maximum number of linearly independent column of **A**.

It can be shown that the row rank of a matrix is always equal to its column rank, and hence the *rank* of the matrix is equal to the maximum number of linearly independent rows (or columns) of **A**. Thus it is clear that rank(**A**) $\leq$ minimum (m, n). If rank(**A**) = minimum (m, n), **A** is said to be of *full rank*. It can be shown that the rank of **A** is k if and only if **A** can be reduced to $\left[\begin{array}{c|c} \mathbf{I}_k & \mathbf{Q} \\ \hline \mathbf{0} & \mathbf{0} \end{array}\right]$ through a finite sequence of elementary matrix operations.

2.3 SIMULTANEOUS LINEAR EQUATIONS

Consider the system $\mathbf{Ax} = \mathbf{b}$ and the augmented matrix $(\mathbf{A}, \mathbf{b})$ with m rows and $n + 1$ columns. If the rank of $(\mathbf{A}, \mathbf{b})$ is greater than the rank of **A**, then **b** cannot be represented as a linear combination of $\mathbf{a}_1, \mathbf{a}_2, \ldots, \mathbf{a}_n$, and hence there is no solution to the system $\mathbf{Ax} = \mathbf{b}$ (and in particular there is no solution to the system $\mathbf{Ax} = \mathbf{b}$, $\mathbf{x} \geq \mathbf{0}$).

Now let us suppose that rank(**A**) = rank(**A**, **b**) = k. Possibly after rearranging the rows of (**A**, **b**), let

$$(\mathbf{A}, \mathbf{b}) = \begin{bmatrix} \mathbf{A}_1 & \mathbf{b}_1 \\ \mathbf{A}_2 & \mathbf{b}_2 \end{bmatrix}$$

where $\mathbf{A}_1$ is $k \times n$, $\mathbf{b}_1$ is a k vector, $\mathbf{A}_2$ is an $(m - k) \times n$ matrix, $\mathbf{b}_2$ is an $m - k$ vector, and rank$(\mathbf{A}_1)$ = rank$(\mathbf{A}_1, \mathbf{b}_1) = k$.

Note that if a vector **x** satisfies $\mathbf{A}_1\mathbf{x} = \mathbf{b}_1$, then it satisfies $\mathbf{A}_2\mathbf{x} = \mathbf{b}_2$ automatically. Thus we can throw away the "redundant" or "dependent" constraints $\mathbf{A}_2\mathbf{x} = \mathbf{b}_2$ and keep the independent constraints $\mathbf{A}_1\mathbf{x} = \mathbf{b}_1$. Since rank$(\mathbf{A}_1) = k$, we can pick k linearly independent columns of $\mathbf{A}_1$. Possibly after rearranging the columns of $\mathbf{A}_1$, let $\mathbf{A}_1 = (\mathbf{B}, \mathbf{N})$, where **B** is a $k \times k$ nonsingular matrix and **N** is $k \times (n - k)$. Note that such a matrix **B** exists since $\mathbf{A}_1$ has rank k. Here **B** is called the *basic matrix* (since the columns of **B** form a basis of E^k) and **N** is called the *nonbasic matrix*. Let us decompose **x** accordingly into $\mathbf{x}_B$ and $\mathbf{x}_N$, where $\mathbf{x}_B$ is composed of $x_1, x_2, \ldots, x_k$ and $\mathbf{x}_N$ is composed of $x_{k+1}, \ldots, x_n$. Now $\mathbf{A}_1\mathbf{x} = \mathbf{b}_1$ means that $(\mathbf{B}, \mathbf{N})\begin{bmatrix} \mathbf{x}_B \\ \mathbf{x}_N \end{bmatrix} = \mathbf{b}_1$; that is, $\mathbf{Bx}_B + \mathbf{Nx}_N = \mathbf{b}_1$. Since **B** has an inverse, then we can solve $\mathbf{x}_B$ in terms of $\mathbf{x}_N$ by premultiplying by $\mathbf{B}^{-1}$, and we get

$$\mathbf{x}_B = \mathbf{B}^{-1}\mathbf{b}_1 - \mathbf{B}^{-1}\mathbf{Nx}_N$$

In the case $k = n$, **N** is vacuous, and we have a unique solution to the system $\mathbf{A}_1\mathbf{x} = \mathbf{b}_1$, namely $\mathbf{x}_B = \mathbf{B}^{-1}\mathbf{b}_1 = \mathbf{A}_1^{-1}\mathbf{b}_1$. On the other hand, if $n > k$, then by assigning arbitrary values to the vector $\mathbf{x}_N$, we can uniquely solve for $\mathbf{x}_B$, using the equation $\mathbf{x}_B = \mathbf{B}^{-1}\mathbf{b}_1 - \mathbf{B}^{-1}\mathbf{Nx}_N$, to obtain a solution $\begin{bmatrix} \mathbf{x}_B \\ \mathbf{x}_N \end{bmatrix}$ to the system $\mathbf{A}_1\mathbf{x} = \mathbf{b}_1$. In this case we have an infinite number of solutions to the system $\mathbf{A}_1\mathbf{x} = \mathbf{b}_1$ (and hence to the system $\mathbf{Ax} = \mathbf{b}$). Note that the notion of decomposing $\mathbf{A}_1$ into **B** and **N** and solving $\mathbf{x}_B = \mathbf{B}^{-1}\mathbf{b}_1 - \mathbf{B}^{-1}\mathbf{Nx}_N$ can be interpreted as follows. We have a system of k equations in n unknowns. Assign arbitrary values to $n - k$ of the variables, corresponding to $\mathbf{x}_N$, and then solve for the remaining system of k equations in k unknowns. This is done such that the k equations in

k unknowns have a *unique solution*, and that is why we require $\mathbf{B}$ to have an inverse. Such a solution obtained by letting $\mathbf{x}_N = \mathbf{0}$ and $\mathbf{x}_B = \mathbf{B}^{-1}\mathbf{b}_1$ is called a *basic solution* of the system $\mathbf{A}_1\mathbf{x} = \mathbf{b}_1$. Let us now summarize the different possible cases that may arise:

1. Rank $(\mathbf{A}, \mathbf{b}) >$ rank $(\mathbf{A})$ and hence $\mathbf{Ax} = \mathbf{b}$ has no solution.
2. Rank $(\mathbf{A}, \mathbf{b}) =$ rank $(\mathbf{A}) = k = n$, and hence there exists a unique solution to the system $\mathbf{Ax} = \mathbf{b}$.
3. Rank $(\mathbf{A}, \mathbf{b}) =$ rank $(\mathbf{A}) = k < n$, and hence we have an infinite number of solutions to the system $\mathbf{Ax} = \mathbf{b}$.

In general these cases may be determined through a Gauss–Jordan reduction of the system $\mathbf{Ax} = \mathbf{b}$ to the form $\mathbf{A}'\mathbf{x} = \mathbf{b}'$ via elementary row operations.

$$\begin{bmatrix} \mathbf{I}_k & \mathbf{Q} \\ 0 & 0 \end{bmatrix} \mathbf{x} = \begin{bmatrix} \mathbf{b}'_1 \\ \mathbf{b}'_2 \end{bmatrix}$$

Hence rank $(\mathbf{A}) = k \leq \min\{m, n\}$. If $k < m$ and $\mathbf{b}'_2 \neq 0$, we have case 1. If $k = n \leq m$ and $\mathbf{b}'_2 = 0$ when $m > n$, we have case 2. If $k < n$ and $b'_2 = 0$ when $k < m$, we have case 3.

Example 2.4

Consider the following system:

$$\begin{aligned} x_1 + 2x_2 + x_3 - 2x_4 &= 10 \\ -x_1 + 2x_2 - x_3 + x_4 &= 6 \\ x_2 + x_3 \qquad &= 2 \end{aligned}$$

We shall solve this system by matrix inversion (Gauss–Jordan reduction) and by only using a Gaussian reduction.

1. *Matrix Inversion*. Reduce three column of $\mathbf{A}$ to the identity [this is possible since rank $(\mathbf{A}) = 3$]

$$\left[\begin{array}{rrrr|r} 1 & 2 & 1 & -2 & 10 \\ -1 & 2 & -1 & 1 & 6 \\ 0 & 1 & 1 & 0 & 2 \end{array}\right]$$

Add the first row to the second row.

$$\left[\begin{array}{rrrr|r} 1 & 2 & 1 & -2 & 10 \\ 0 & 4 & 0 & -1 & 16 \\ 0 & 1 & 1 & 0 & 2 \end{array}\right]$$

Divide the second row by 4. Multiply the new second row by -2 and add to the first row. Multiply the new second row by -1 and add to the third row.

$$\left[\begin{array}{rrrr|r} 1 & 0 & 1 & -\frac{3}{2} & 2 \\ 0 & 1 & 0 & -\frac{1}{4} & 4 \\ 0 & 0 & 1 & \frac{1}{4} & -2 \end{array}\right]$$

Multiply the third row by -1 and add to the first row.

$$\begin{array}{cccc|c} x_1 & x_2 & x_3 & x_4 & \\ \hline 1 & 0 & 0 & -\frac{7}{4} & 4 \\ 0 & 1 & 0 & -\frac{1}{4} & 4 \\ 0 & 0 & 1 & \frac{1}{4} & -2 \end{array}$$

The original system has been reduced to the system shown. Equivalence of the two systems is assured since the new system is obtained from the original system after performing a finite number of elementary row operations. The solution to the system is as follows. Assign x_4 arbitrarily, say $x_4 = \lambda$. Then $x_1 = 4 + \frac{7}{4}\lambda$, $x_2 = 4 + \frac{1}{4}\lambda$, and $x_3 = -2 - \frac{1}{4}\lambda$.

2. *Gaussian Reduction*.

$$\left[\begin{array}{cccc|c} 1 & 2 & 1 & -2 & 10 \\ -1 & 2 & -1 & 1 & 16 \\ 0 & 1 & 1 & 0 & 2 \end{array}\right]$$

Add the first row to the second row.

$$\left[\begin{array}{cccc|c} 1 & 2 & 1 & -2 & 10 \\ 0 & 4 & 0 & -1 & 16 \\ 0 & 1 & 1 & 0 & 2 \end{array}\right]$$

Divide the second row by 4. Subtract the new second row from the third row.

$$\begin{array}{cccc|c} x_1 & x_2 & x_3 & x_4 & \\ \hline 1 & 2 & 1 & -2 & 10 \\ 0 & 1 & 0 & -\frac{1}{4} & 4 \\ 0 & 0 & 1 & \frac{1}{4} & -2 \end{array}$$

The foregoing matrix has an upper triangular submatrix. Let x_4 be equal to an arbitrary value λ. Then $x_3 = -2 - \frac{1}{4}\lambda$, $x_2 = 4 + \frac{1}{4}\lambda$, and $x_1 = 10 + 2\lambda - 2x_2 - x_3 = 4 + \frac{7}{4}\lambda$. This gives the same general solution obtained earlier.

2.4 CONVEX SETS AND CONVEX FUNCTIONS

In this section we consider some basic properties of convex sets, convex functions, and concave functions.

Convex Sets

A set X in E^n is called a *convex set* if given any two points $\mathbf{x}_1$ and $\mathbf{x}_2$ in X, then $\lambda\mathbf{x}_1 + (1 - \lambda)\mathbf{x}_2 \in X$ for each $\lambda \in [0, 1]$.

Note that $\lambda\mathbf{x}_1 + (1 - \lambda)\mathbf{x}_2$ for λ in the interval $[0, 1]$ represents a point on the line segment joining $\mathbf{x}_1$ and $\mathbf{x}_2$. Any point of the form $\lambda\mathbf{x}_1 + (1 - \lambda)\mathbf{x}_2$ where $0 \leq \lambda \leq 1$ is called a *convex combination* (or weighted average) of $\mathbf{x}_1$ and $\mathbf{x}_2$. If

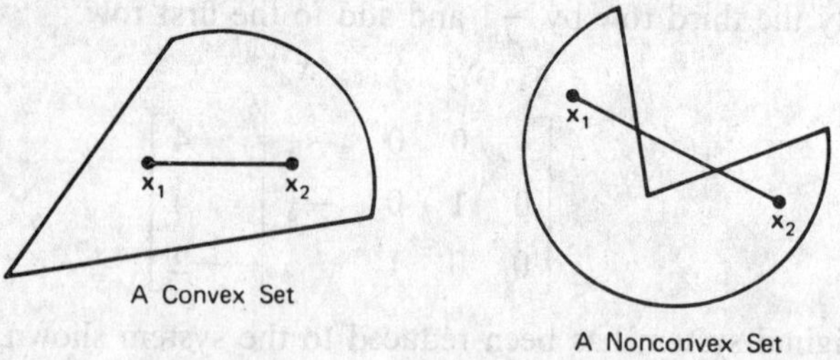

Figure 2.5. Example of convex and nonconvex sets.

$\lambda \in (0,1)$, then the convex combination is called *strict*. Hence convexity of X can be interpreted geometrically as follows. For each pair of points $\mathbf{x}_1$ and $\mathbf{x}_2$ in X, the line segment joining them, or the convex combinations of the two points, must belong to X.

Figure 2.5 shows an example of a convex set and an example of a nonconvex set. In the latter case, we see that not all convex combinations of $\mathbf{x}_1$ and $\mathbf{x}_2$ belong to X. The following are some examples of convex sets.

1. $\{(x_1, x_2) : x_1^2 + x_2^2 \le 1\}$.
2. $\{\mathbf{x} : \mathbf{A}\mathbf{x} = \mathbf{b}\}$, where $\mathbf{A}$ is an $m \times n$ matrix and $\mathbf{b}$ is an m vector.
3. $\{\mathbf{x} : \mathbf{A}\mathbf{x} = \mathbf{b}, \mathbf{x} \ge \mathbf{0}\}$, where $\mathbf{A}$ is an $m \times n$ matrix and $\mathbf{b}$ is an m vector.
4. $\{\mathbf{x} : \mathbf{A}\mathbf{x} \le \mathbf{b}, \mathbf{x} \ge \mathbf{0}\}$ where $\mathbf{A}$ is an $m \times n$ matrix and $\mathbf{b}$ is an m vector.
5. $\left\{\mathbf{x} : \mathbf{x} = \lambda_1 \begin{bmatrix} 1 \\ 0 \\ 0 \end{bmatrix} + \lambda_2 \begin{bmatrix} 1 \\ 2 \\ 1 \end{bmatrix} + \lambda_3 \begin{bmatrix} -1 \\ 2 \\ -3 \end{bmatrix}, \lambda_1 + \lambda_2 + \lambda_3 = 1, \lambda_1, \lambda_2, \lambda_3 \ge 0 \right\}$.

Extreme Points

The notion of extreme points play an especially important role in the theory of linear programming. A point $\mathbf{x}$ in a convex set X is called an *extreme point* of X, if $\mathbf{x}$ cannot be represented as a strict convex combination of two distinct points in X. In other words, if $\mathbf{x} = \lambda\mathbf{x}_1 + (1-\lambda)\mathbf{x}_2$ with $\lambda \in (0,1)$ and $\mathbf{x}_1, \mathbf{x}_2 \in X$, then $\mathbf{x} = \mathbf{x}_1 = \mathbf{x}_2$.

Figure 2.6 shows some examples of extreme and nonextreme points of convex sets. Note that $\mathbf{x}_1$ is an extreme point of X whereas $\mathbf{x}_2$ and $\mathbf{x}_3$ are not. More insights into extreme points are given in Section 2.6.

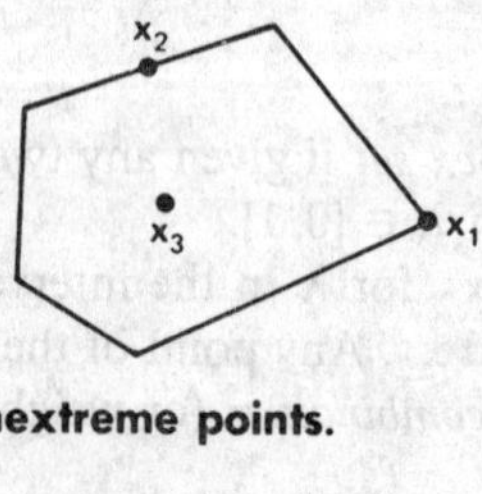

Figure 2.6. Extreme and nonextreme points.

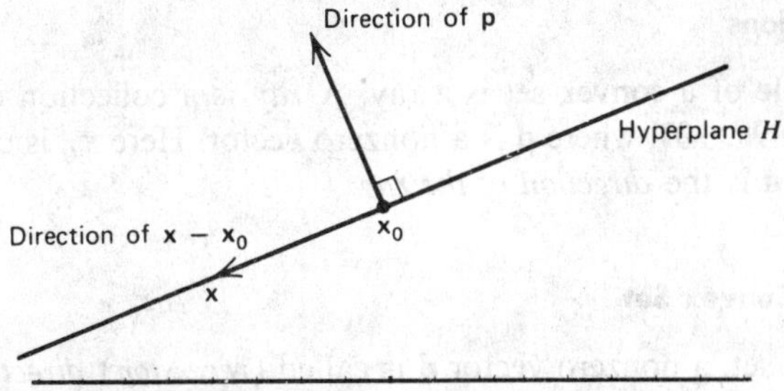

Figure 2.7. Hyperplane.

Hyperplanes and Halfspaces

A hyperplane in E^n generalizes the notion of a straight line in E^2 and the notion of a plane in E^3. A *hyperplane* H in E^n is a set of the form $\{\mathbf{x} : \mathbf{px} = k\}$ where $\mathbf{p}$ is a nonzero vector in E^n and k is a scalar. Here $\mathbf{p}$ is usually called the *normal* or the *gradient* to the hyperplane.

Equivalently, a hyperplane consists of all points $\mathbf{x} = (x_1, x_2, \ldots, x_n)$ satisfying the equation $\Sigma_{j=1}^n p_j x_j = k$. The constant k can be eliminated by referring to a fixed point $\mathbf{x}_0$ on the hyperplane. If $\mathbf{x}_0 \in H$, then $\mathbf{px}_0 = k$, and for any $\mathbf{x} \in H$, we have $\mathbf{px} = k$. Upon subtraction we get $\mathbf{p}(\mathbf{x} - \mathbf{x}_0) = 0$. In other words, H can be represented as the collection of points satisfying $\mathbf{p}(\mathbf{x} - \mathbf{x}_0) = 0$, where $\mathbf{x}_0$ is any fixed point in H. A hyperplane is a convex set.

Figure 2.7 shows a hyperplane and its normal vector $\mathbf{p}$. Note that $\mathbf{p}$ is orthogonal to $\mathbf{x} - \mathbf{x}_0$ for each $\mathbf{x}$ in the hyperplane H.

A hyperplane divides E^n into two regions, called halfspaces. Hence a *halfspace* is a collection of points of the form $\{\mathbf{x} : \mathbf{px} \geq k\}$, where $\mathbf{p}$ is a nonzero vector in E^n and k is a scalar. A halfspace can also be represented as a set of points of the form $\{\mathbf{x} : \mathbf{px} \leq k\}$. The union of the two halfspaces $\{\mathbf{x} : \mathbf{px} \geq k\}$ and $\{\mathbf{x} : \mathbf{px} \leq k\}$ is E^n. Referring to a fixed point $\mathbf{x}_0$ in the hyperplane defining the two halfspaces the latter can be represented as $\{\mathbf{x} : \mathbf{p}(\mathbf{x} - \mathbf{x}_0) \geq 0\}$ or as $\{\mathbf{x} : \mathbf{p}(\mathbf{x} - \mathbf{x}_0) \leq 0\}$, as shown in Figure 2.8. Note that the first halfspace consists of points $\mathbf{x}$ for which $(\mathbf{x} - \mathbf{x}_0)$ makes an acute angle ($\leq 90°$) with $\mathbf{p}$, whereas the second halfspace is comprised of points $\mathbf{x}$ such that $(\mathbf{x} - \mathbf{x}_0)$ makes an obtuse angle ($\geq 90°$) with $\mathbf{p}$.

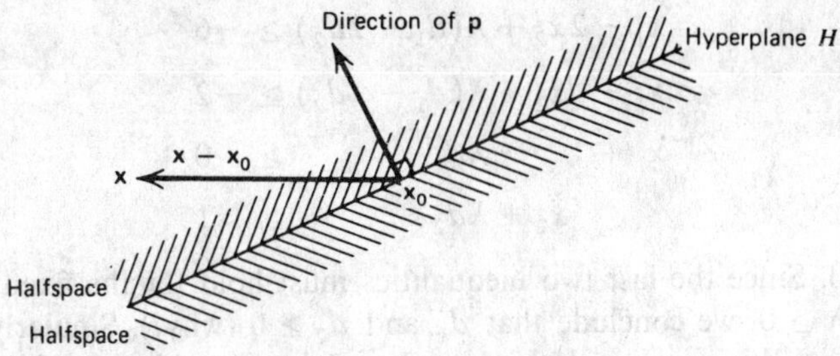

Figure 2.8. Halfspaces.

Rays and Directions

Another example of a convex set is a ray. A *ray* is a collection of points of the form $\{\mathbf{x}_0 + \lambda\mathbf{d} : \lambda \geq 0\}$, where $\mathbf{d}$ is a nonzero vector. Here $\mathbf{x}_0$ is called the *vertex* of the ray, and $\mathbf{d}$ is the *direction of the ray*.

Directions of a Convex Set

Given a convex set, a nonzero vector $\mathbf{d}$ is called (*recession*) *direction of the set*, if for each $\mathbf{x}_0$ in the set, the ray $\{\mathbf{x}_0 + \lambda\mathbf{d} : \lambda \geq 0\}$ also belongs to the set. Hence starting at any point $\mathbf{x}_0$ in the set, one can *recede* along $\mathbf{d}$ for any step length $\lambda \geq 0$ and remain within the set. Clearly if the set is bounded, then it has no directions.

Consider the nonempty polyhedral set $X = \{\mathbf{x} : \mathbf{Ax} \leq \mathbf{b}, \mathbf{x} \geq \mathbf{0}\}$. Then a nonzero $\mathbf{d}$ is a direction of X if and only if

$$\mathbf{A}(\mathbf{x} + \lambda\mathbf{d}) \leq \mathbf{b}$$

$$\mathbf{x} + \lambda\mathbf{d} \geq \mathbf{0}$$

for each $\lambda \geq 0$ and each $\mathbf{x} \in X$. Since $\mathbf{x} \in X$, then $\mathbf{Ax} \leq \mathbf{b}$ and the first inequality holds for $\lambda \geq 0$ arbitrarily large if and only if $\mathbf{Ad} \leq \mathbf{0}$. Similarly, $\mathbf{x} + \lambda\mathbf{d}$ is nonnegative for λ arbitrarily large, if and only if $\mathbf{d}$ is nonnegative. To summarize, $\mathbf{d}$ is a (recession) direction of X if and only if

$$\mathbf{d} \geq \mathbf{0}, \mathbf{d} \neq \mathbf{0}, \quad \text{and} \quad \mathbf{Ad} \leq \mathbf{0}$$

Similarly, if $X = \{\mathbf{x} : \mathbf{Ax} = \mathbf{b}, \mathbf{x} \geq \mathbf{0}\} \neq \varnothing$, then by replacing the equality by two inequalities, it follows that $\mathbf{d}$ is a direction of X if and only if $\mathbf{d} \neq \mathbf{0}$, $\mathbf{d} \geq \mathbf{0}$, and $\mathbf{Ad} = \mathbf{0}$. The set of directions in either case forms a convex set.

Example 2.5

Consider the set $X = \{(x_1, x_2) : x_1 - 2x_2 \geq -6, x_1 - x_2 \geq -2, x_1 \geq 0, x_2 \geq 1\}$ depicted in Figure 2.9. Let $\mathbf{x}_0 = \begin{pmatrix} x_1 \\ x_2 \end{pmatrix}$ be an arbitrary fixed feasible point. Then $\mathbf{d} = \begin{pmatrix} d_1 \\ d_2 \end{pmatrix}$ is a direction of X if and only if $\begin{pmatrix} d_1 \\ d_2 \end{pmatrix} \neq \begin{pmatrix} 0 \\ 0 \end{pmatrix}$ and $\mathbf{x}_0 + \lambda\mathbf{d} = \begin{pmatrix} x_1 + \lambda d_1 \\ x_2 + \lambda d_2 \end{pmatrix}$ belongs to X for all $\lambda \geq 0$. Therefore

$$\begin{aligned} x_1 - 2x_2 + \lambda(d_1 - 2d_2) &\geq -6 \\ x_1 - x_2 + \lambda(d_1 - d_2) &\geq -2 \\ x_1 + \lambda d_1 &\geq 0 \\ x_2 + \lambda d_2 &\geq 1 \end{aligned}$$

for all $\lambda \geq 0$. Since the last two inequalities must hold for the fixed x_1 and x_2 and for all $\lambda \geq 0$, we conclude that d_1 and $d_2 \geq 0$ (why?). Similarly, from the first two inequalities we conclude that $d_1 - 2d_2 \geq 0$ and $d_1 - d_2 \geq 0$ (why?).

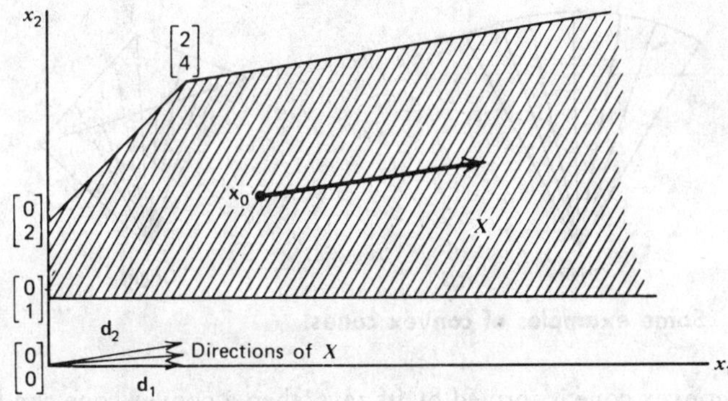

Figure 2.9. **Directions of convex sets.**

Since d_1 and $d_2 \geq 0$, then $d_1 \geq 2d_2$ implies that $d_1 \geq d_2$. Therefore $\begin{pmatrix} d_1 \\ d_2 \end{pmatrix}$ is a direction of X if and only if

$$\begin{pmatrix} d_1 \\ d_2 \end{pmatrix} \neq \begin{pmatrix} 0 \\ 0 \end{pmatrix}$$
$$d_1 \geq 0, \qquad d_2 \geq 0$$
$$d_1 \geq 2d_2$$

This collection of vectors is shown in Figure 2.9 and can be *normalized* such that each direction has norm (or length) equal to 1.

EXTREME DIRECTIONS OF A CONVEX SET

The notion of *extreme directions* is similar to the notion of extreme points. An extreme direction of a convex set is a direction of the set that cannot be represented as a positive combination of two distinct directions of the set. Two vectors, $\mathbf{d}_1$ and $\mathbf{d}_2$, are said to be *distinct* or not equivalent if $\mathbf{d}_1$ cannot be represented as a positive multiple of $\mathbf{d}_2$. In the foregoing example, after normalization, we have two extreme directions $\mathbf{d}_1 = (1, 0)$ and $\mathbf{d}_2 = (2/\sqrt{5}, 1/\sqrt{5})$. Any other direction of the set, which is not a multiple of $\mathbf{d}_1$ or $\mathbf{d}_2$, can be represented as $\lambda_1 \mathbf{d}_1 + \lambda_2 \mathbf{d}_2$ where $\lambda_1, \lambda_2 > 0$. Any ray that is contained in the convex set and whose direction is an extreme direction is called an *extreme ray*. Section 2.6 provides further insights into recession directions and extreme directions.

Convex Cones

A special important class of convex sets is convex cones. A *convex cone* C is a convex set with the additional property that $\lambda \mathbf{x} \in C$ for each $\mathbf{x} \in C$ and for each $\lambda \geq 0$. Note that a convex cone always contains the origin by letting $\lambda = 0$, and also that given any point $\mathbf{x} \in C$, the ray or *halfline* $\{\lambda \mathbf{x} : \lambda \geq 0\}$ belongs to C. Hence a convex cone is a convex set that consists entirely of rays emanating from the origin. Figure 2.10 shows some examples of convex cones.

Figure 2.10. **Some examples of convex cones.**

Since a convex cone is formed by its rays, then a convex cone can be entirely characterized by its directions. In fact, not all directions are needed, since a nonextreme direction can be represented as a positive combination of extreme directions. In other words, a convex cone is fully characterized by its extreme directions.

As an example, consider the convex cone whose extreme directions are (1, 1) and (0, 1). From Figure 2.11 it is clear that the convex cone must be the set $\{(x_1, x_2): x_1 \geq 0, x_1 \leq x_2\}$.

Given a set of vectors $\mathbf{a}_1, \mathbf{a}_2, \ldots, \mathbf{a}_k$, we can form the convex cone C generated by these vectors. This cone consists of all nonnegative combinations of $\mathbf{a}_1, \mathbf{a}_2, \ldots, \mathbf{a}_k$, that is,

$$C = \left\{ \sum_{j=1}^{k} \lambda_j \mathbf{a}_j : \lambda_j \geq 0 \quad \text{for } j = 1, 2, \ldots, k \right\}$$

Figure 2.11 shows the convex cone generated by the points (0, 1) and (1, 1).

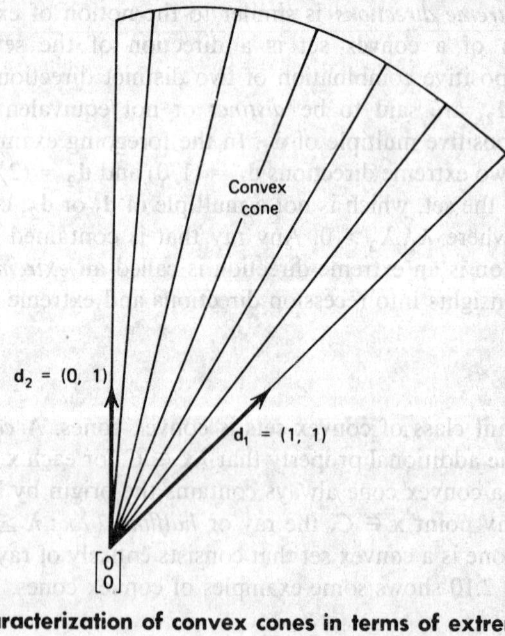

Figure 2.11. **Characterization of convex cones in terms of extreme directions.**

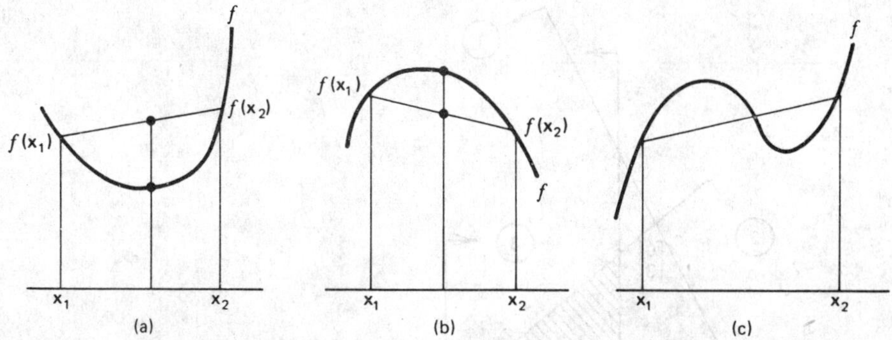

Figure 2.12. **Examples of convex and concave functions: (*a*) Convex function. (*b*) Concave function. (*c*) Neither convex nor concave.**

Convex and Concave Functions

Convex and concave functions play an important role in optimization problems. These functions naturally arise in linear optimization problems when dealing with parametric analysis.

A function f of the vector $(x_1, x_2, \ldots, x_n)$ is said to be *convex* if the following inequality holds for any two vectors $\mathbf{x}_1$ and $\mathbf{x}_2$:

$$f(\lambda \mathbf{x}_1 + (1-\lambda)\mathbf{x}_2) \leq \lambda f(\mathbf{x}_1) + (1-\lambda) f(\mathbf{x}_2) \qquad \text{for all} \qquad \lambda \in [0,1].$$

Figure 2.12a shows an example of a convex function. Note that the foregoing inequality can be interpreted as follows: $\lambda f(\mathbf{x}_1) + (1-\lambda) f(\mathbf{x}_2)$ where $\lambda \in [0,1]$ represents the height of the *chord* joining $(\mathbf{x}_1, f(\mathbf{x}_1))$ and $(\mathbf{x}_2, f(\mathbf{x}_2))$ at the point $\lambda \mathbf{x}_1 + (1-\lambda)\mathbf{x}_2$. Since $\lambda f(\mathbf{x}_1) + (1-\lambda) f(\mathbf{x}_2) \geq f(\lambda \mathbf{x}_1 + (1-\lambda)\mathbf{x}_2)$, then the height of the chord is at least as large as the height of the function itself.

A function f is *concave* if and only if $-f$ is convex. This can be restated as follows:

$$f(\lambda \mathbf{x}_1 + (1-\lambda)\mathbf{x}_2) \geq \lambda f(\mathbf{x}_1) + (1-\lambda) f(\mathbf{x}_2) \qquad \text{for all} \qquad \lambda \in [0,1]$$

for any given $\mathbf{x}_1$ and $\mathbf{x}_2$. Figure 2.12b shows an example of a concave function. An example of a function that is neither convex nor concave is depicted in Figure 2.12c.

2.5 POLYHEDRAL SETS AND POLYHEDRAL CONES

Polyhedral sets and polyhedral cones represent important special cases of convex sets and convex cones. A *polyhedral set* or a *polyhedron* is the intersection of a finite number of halfspaces. A bounded polyhedral set is called a *polytope*. Since a halfspace can be represented by an inequality of the type $\mathbf{a}^i\mathbf{x} \leq b_i$, then a polyhedral set can be represented by the system $\mathbf{a}^i\mathbf{x} \leq b_i$ for $i = 1, \ldots, m$. Hence a polyhedral set can be represented by $\{\mathbf{x} : \mathbf{A}\mathbf{x} \leq \mathbf{b}\}$ where $\mathbf{A}$ is an $m \times n$ matrix whose ith row is $\mathbf{a}^i$ and $\mathbf{b}$ an m vector. Since an equation can be written as two

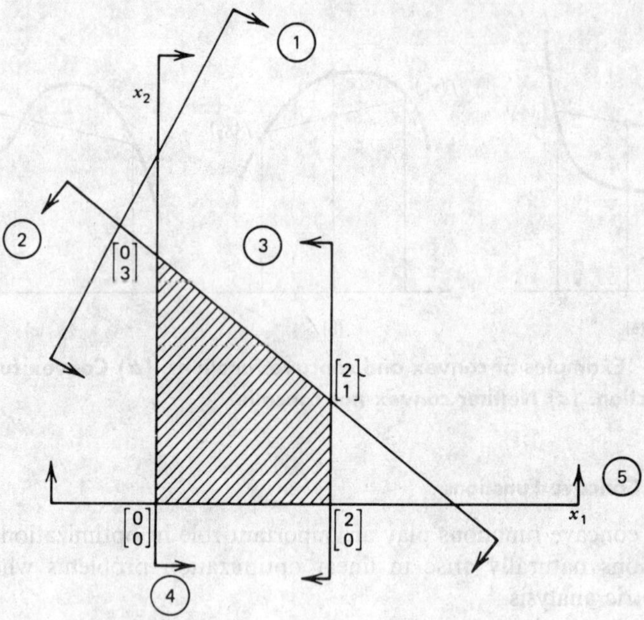

Figure 2.13. **Polyhedral set.**

inequalities, a polyhedral set can be represented by a finite number of linear inequalities and/or equations. As an example, consider the polyhedral set defined by the following inequalities:

$$\begin{aligned} -2x_1 + x_2 &\leq 4 \\ x_1 + x_2 &\leq 3 \\ x_1 \qquad &\leq 2 \\ x_1 \qquad &\geq 0 \\ x_2 &\geq 0 \end{aligned}$$

The intersection of these five halfspaces gives the shaded set of Figure 2.13. Clearly the set is a convex set. We can see a distinct difference between the first inequality and the remaining inequalities. If the first inequality is disregarded, the polyhedral set is not affected. Such an inequality is called (*geometrically*) *redundant* or *irrelevant* to the polyhedral set.

A special class of polyhedral sets is *polyhedral cones*. A polyhedral cone is the intersection of a finite number of halfspaces, whose hyperplanes pass through the origin. In other words, C is a polyhedral cone if it can be represented as $\{\mathbf{x} : \mathbf{Ax} \leq \mathbf{0}\}$, where $\mathbf{A}$ is an $m \times n$ matrix. Note that the ith row of the matrix $\mathbf{A}$ is the normal vector to the hyperplane defining the ith halfspace. Figure 2.11 shows an example of a polyhedral cone.

2.6 EXTREME POINTS, FACES, DIRECTIONS, AND EXTREME DIRECTIONS OF POLYHEDRAL SETS: GEOMETRIC INSIGHTS

In Section 2.4 we discussed extreme points and directions of general convex sets, although we often used polyhedral sets for illustration. In this section, we will provide some geometric insights and equivalent definitions of extreme points and directions for polyhedral sets. We will assume that the polyhedral set under discussion in this section is of the form

$$X = \{\mathbf{x} : \mathbf{A}\mathbf{x} \leq \mathbf{b}, \mathbf{x} \geq \mathbf{0}\} \tag{2.1}$$

where $\mathbf{A}$ is $m \times n$ and $\mathbf{b}$ is an m vector. As shown in Chapter 1, through suitable linear transformations, the feasible region of any linear programming problem can be put into canonical form.

Extreme Points

We have already seen that an extreme point of X is a point in X that cannot be made to lie in the interior of a line segment contained within X. We will now provide an equivalent geometric definition. Toward this end, consider the following simple but important observation. Let $\bar{\mathbf{x}} \in X$ and suppose that some constraint $\boldsymbol{\alpha}\mathbf{x} \leq \beta$ of X is *binding, or tight, or active*, at $\bar{\mathbf{x}}$ that is, $\boldsymbol{\alpha}\bar{\mathbf{x}} = \beta$. Suppose further that we can write $\bar{\mathbf{x}} = \lambda\mathbf{x}' + (1 - \lambda)\mathbf{x}''$, where $0 < \lambda < 1$ and where $\mathbf{x}'$ and $\mathbf{x}'' \in X$. Then it must be also true that $\boldsymbol{\alpha}\mathbf{x}' = \beta$ and $\boldsymbol{\alpha}\mathbf{x}'' = \beta$, that is, $\boldsymbol{\alpha}\mathbf{x} \leq \beta$ must be binding at $\mathbf{x}'$ and at $\mathbf{x}''$ as well. This clearly follows from the fact that $\boldsymbol{\alpha}\mathbf{x}' \leq \beta$ and $\boldsymbol{\alpha}\mathbf{x}'' \leq \beta$ (since $\mathbf{x}', \mathbf{x}'' \in X$), but $\lambda(\boldsymbol{\alpha}\mathbf{x}') + (1 - \lambda)(\boldsymbol{\alpha}\mathbf{x}'') = \beta$ (since $\boldsymbol{\alpha}\bar{\mathbf{x}} = \beta$), where $0 < \lambda < 1$.

An extreme point now can be defined as follows. Let the hyperplanes associated with the $(m + n)$ defining halfspaces of X be referred to as *defining hyperplanes* of X. Furthermore, note that a set of defining hyperplanes are linearly independent if the coefficient matrix associated with this set of equations has full row rank. Then a point $\bar{\mathbf{x}} \in X$ is said to be an *extreme point* or a *corner point*, or a *vertex* of X if $\bar{\mathbf{x}}$ lies on some n linearly independent defining hyperplanes of X. If more than n defining hyperplanes pass through an extreme point, then such an extreme point is called a *degenerate extreme point*. The excess number of planes over n is called its *order of degeneracy*. A polyhedron with at least one degenerate extreme point is said to be a *degenerate polyhedral set*.

This definition of extreme points is readily seen to be equivalent to saying that $\bar{\mathbf{x}}$ cannot be written as a strict convex combination of two distinct points in X. When $\bar{\mathbf{x}} \in X$ lies on some n linearly independent defining hyperplanes, then if we can write $\bar{\mathbf{x}} = \lambda\mathbf{x}' + (1 - \lambda)\mathbf{x}''$, where $0 < \lambda < 1$ and $\mathbf{x}', \mathbf{x}'' \in X$, we have that both $\mathbf{x}'$ and $\mathbf{x}''$ also lie on these n hyperplanes. The solution to these n hyperplane equations, however, is unique. Hence $\mathbf{x}' = \mathbf{x}''$. Conversely, if the maximum number of linearly independent defining hyperplanes binding at $\bar{\mathbf{x}} \in X$ are $r < n$ and are given by $\mathbf{G}\mathbf{x} = \mathbf{g}$ where $\mathbf{G}$ is $r \times n$, let $\mathbf{d} \neq \mathbf{0}$ be a solution to $\mathbf{G}\mathbf{d} = \mathbf{0}$. Note that such a $\mathbf{d}$ exists (why?). Then there exists an $\epsilon > 0$ such that $\mathbf{x}' = (\bar{\mathbf{x}} + \epsilon\mathbf{d}) \in X$ and $\mathbf{x}'' = (\bar{\mathbf{x}} - \epsilon\mathbf{d}) \in X$, since $\mathbf{G}\mathbf{x}' = \mathbf{g}$, $\mathbf{G}\mathbf{x}'' = \mathbf{g}$ and the constraints of X that are nonbinding at $\bar{\mathbf{x}}$ remain satisfied provided $\epsilon > 0$ is small enough. Consequently, since $\bar{\mathbf{x}} = 0.5\mathbf{x}' + 0.5\mathbf{x}''$ and $\mathbf{x}', \mathbf{x}'' \in X$, $\bar{\mathbf{x}}$ is not an extreme point of X. Referring to the polyhedral set in Figure 2.14, note how every extreme point is formed by some $n = 3$ linearly independent defining hyperplanes, and at every other point, fewer than $n = 3$ linearly independent defining hyperplanes are binding. Note also that at the indicated degenerate extreme point, four defining hyperplanes are binding. The order of degeneracy is one. In particular, there is more than one choice of selecting $n = 3$ linearly independent hyperplanes that give this vertex as their unique intersection point. However, no defining hyperplane or halfspace can be omitted without changing the structure of this polyhedron. That is, there are no *redundant* constraints.

Faces, Edges, and Adjacent Extreme Points

We have seen that each extreme point $\bar{\mathbf{x}}$ of X is a unique solution to some n linearly independent defining hyperplanes binding at $\bar{\mathbf{x}}$. In general the set of points in X that correspond to some nonempty set of binding defining hyperplanes of X are called *proper faces* of X. Now given any face F of X if $r(F)$ is the maximum number of linearly independent defining hyperplanes binding at *all* points feasible to F, then the *dimension of F*, denoted $\dim(F)$, is equal to $n - r(F)$. In other words, each linearly independent binding hyperplane results in the loss of one "degree of freedom." Consequently, extreme points are zero-dimensional faces of X. Similarly, an *edge* of X is a one-dimensional face of X, that is, it is the set of points in X formed by the intersection of some $(n - 1)$ linearly independent defining hyperplanes (with some points in this set not having more than $(n - 1)$ binding linearly independent hyperplanes). That is, an edge has one "degree of freedom" imparted by having one less than n linearly independent hyperplanes that are binding at all points. Similarly, if $r(X)$ is defined with respect to X itself, then although X is in E^n, it is actually of dimension $\dim(X) = n - r(X)$. The set X is said to be *full dimensional* if $r(X) = 0$, that is, $\dim(X) = n$. Furthermore, the set X and the empty set are also sometimes called *improper faces* of X itself. The other faces are called *proper*

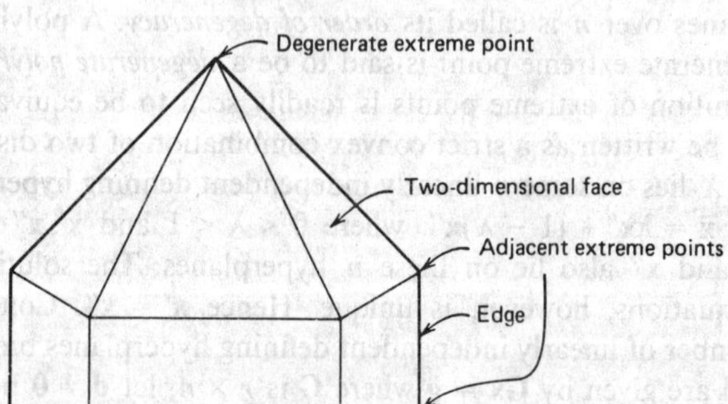

Figure 2.14. Extreme points and edges of X.

faces. The highest dimensional proper face of X is of dimension $\dim(X) - 1$ (when $\dim(X) \geq 1$) and is called a *facet* of X.

Finally, two extreme points of X are said to be *adjacent* if the line segment joining them is an edge of X. Hence adjacent extreme points have some $(n-1)$ common binding linearly independent defining hyperplanes. Figure 2.14 illustrates these definitions.

Recession Directions and Extreme Directions

In Section 2.4, we characterized the directions of X as vectors $\mathbf{d}$ satisfying the conditions

$$\mathbf{d} \geq \mathbf{0}, \mathbf{d} \neq \mathbf{0}, \quad \text{and} \quad \mathbf{A}\mathbf{d} \leq \mathbf{0}.$$

Geometrically, this so-called *homogeneous* system defines a polyhedral cone (excepting the origin), also known as a *recession cone*, and is obtained by translating the defining hyperplanes of X parallel to themselves until they pass through the origin. To eliminate duplication, these directions may be normalized. To maintain linearity, it is convenient to normalize the directions using the (rectilinear) norm $|d_1| + \cdots + |d_n| = 1$, which leads to the normalization constraint $d_1 + \cdots + d_n = 1$, since $\mathbf{d} \geq \mathbf{0}$. Hence the set

$$D = \{\mathbf{d} : \mathbf{A}\mathbf{d} \leq \mathbf{0}, \mathbf{1}\mathbf{d} = 1, \mathbf{d} \geq \mathbf{0}\} \tag{2.2}$$

characterizes the *set of recession directions* of X, where $\mathbf{1}$ is a vector of ones. Figure 2.15 illustrates this set. In fact, as seen in the figure, the *extreme directions*

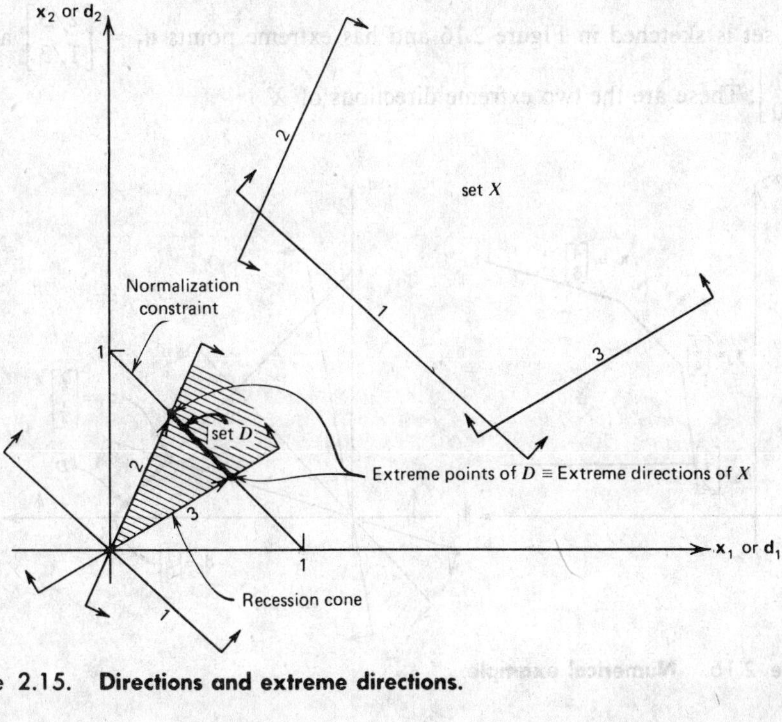

Figure 2.15. **Directions and extreme directions.**

of X are precisely the extreme points of D. Clearly, because of the normalization constraint, a direction $\mathbf{d} \in D$ cannot be written as a positive linear combination of two distinct directions if and only if it cannot be written as a strict convex combination of two distinct directions in D.

Example 2.6.

Consider the polyhedral set X given by the inequalities

$$\begin{aligned} -3x_1 + x_2 &\leq -2 \\ -x_1 + x_2 &\leq 2 \\ -x_1 + 2x_2 &\leq 8 \\ -x_2 &\leq -2 \\ x_1, \quad x_2 &\geq 0 \end{aligned}$$

The set is illustrated in Figure 2.16. Its extreme points are given as:

$$\mathbf{x}_1 = \begin{bmatrix} 4 \\ 6 \end{bmatrix}, \mathbf{x}_2 = \begin{bmatrix} 2 \\ 4 \end{bmatrix}, \text{and } \mathbf{x}_3 = \begin{bmatrix} 4/3 \\ 2 \end{bmatrix}$$

The set D of (2.2) is given by

$$D = \{(d_1, d_2) : -3d_1 + d_2 \leq 0, -d_1 + d_2 \leq 0, -d_1 + 2d_2 \leq 0, -d_2 \leq 0, \\ d_1 + d_2 = 1, d_1 \geq 0, d_2 \geq 0\}$$

This set is sketched in Figure 2.16 and has extreme points $\mathbf{d}_1 = \begin{bmatrix} 2/3 \\ 1/3 \end{bmatrix}$ and $\mathbf{d}_2 = \begin{bmatrix} 1 \\ 0 \end{bmatrix}$. These are the two extreme directions of X.

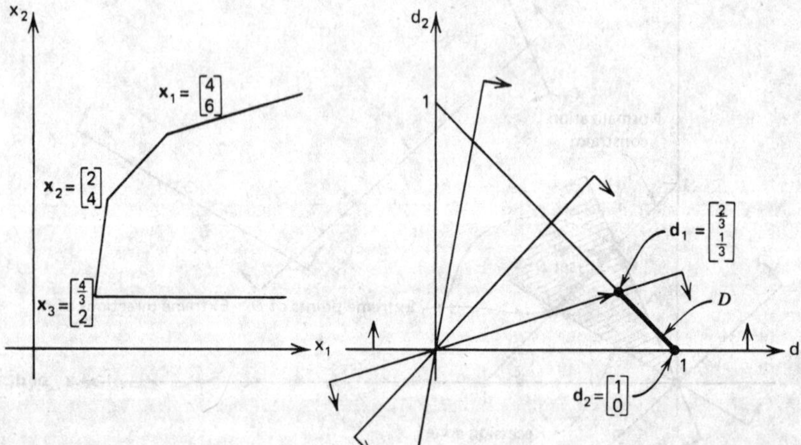

Figure 2.16. **Numerical example.**

2.7 REPRESENTATION OF POLYHEDRAL SETS

In this section we discuss the representation of a polyhedral set in terms of extreme points and extreme directions. This alternative representation will prove very useful throughout the book. The proof of the representation theorem given here is insightful and instructive. The reader may want to study this proof in order to develop more confidence with the notions of extreme points and extreme directions.

Basic Ideas

1. Bounded Polyhedral Sets (Polytopes)

Consider the bounded polyhedral set (recall that a set is bounded if there is a number k such that $\|\mathbf{x}\| < k$ for each point $\mathbf{x}$ in the set) of Figure 2.17, which is formed as the intersection of five halfspaces. We have five extreme points, namely $\mathbf{x}_1$, $\mathbf{x}_2$, $\mathbf{x}_3$, $\mathbf{x}_4$, and $\mathbf{x}_5$. Note that any point in the set can be represented as a convex combination, or a weighted average, of these five extreme points. To illustrate, choose the point $\mathbf{x}$ shown in Figure 2.17. Note that $\mathbf{x}$ can be represented as a convex combination of $\mathbf{y}$ and $\mathbf{x}_4$, that is,

$$\mathbf{x} = \lambda\mathbf{y} + (1-\lambda)\mathbf{x}_4 \quad \text{where} \quad \lambda \in (0,1)$$

But $\mathbf{y}$ can itself be represented as a convex combination of $\mathbf{x}_1$ and $\mathbf{x}_2$, that is,

$$\mathbf{y} = \mu\mathbf{x}_1 + (1-\mu)\mathbf{x}_2 \quad \text{where} \quad \mu \in (0,1)$$

Substituting, we get

$$\mathbf{x} = \lambda\mu\mathbf{x}_1 + \lambda(1-\mu)\mathbf{x}_2 + (1-\lambda)\mathbf{x}_4$$

Since $\lambda \in (0,1)$ and $\mu \in (0,1)$, then $\lambda\mu$, $\lambda(1-\mu)$, and $(1-\lambda) \in (0,1)$. Also $\lambda\mu + \lambda(1-\mu) + (1-\lambda) = 1$. In other words, $\mathbf{x}$ has been represented as a convex combination of the extreme points $\mathbf{x}_1$, $\mathbf{x}_2$, and $\mathbf{x}_4$. In general any point in a bounded polyhedral set can be represented as a convex combination of its extreme points.

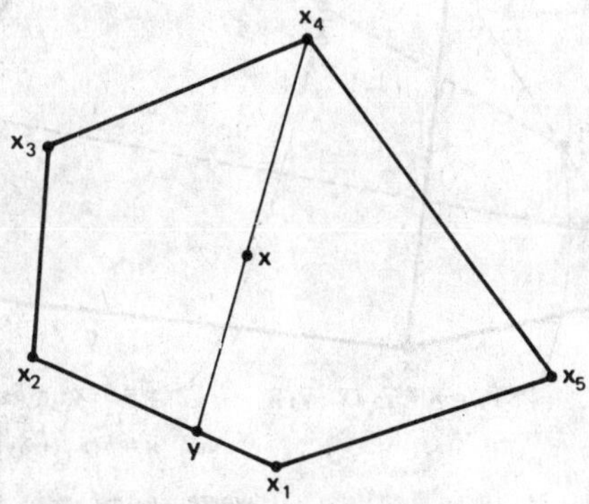

Figure 2.17. **Representation in terms of extreme points.**

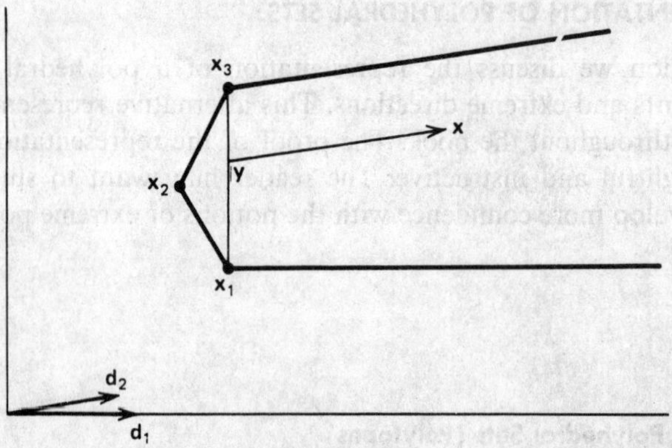

Figure 2.18. Representation of polyhedral sets in terms of extreme points and extreme directions.

2. Unbounded Polyhedral Sets

Let us now consider the case of an unbounded polyhedral set. An example is shown in Figure 2.18. We see that the set has three extreme points $\mathbf{x}_1$, $\mathbf{x}_2$, and $\mathbf{x}_3$, as well as two extreme directions $\mathbf{d}_1$ and $\mathbf{d}_2$. From Figure 2.18 it is clear that in general we can represent every point in the set as a convex combination of the extreme points, plus a nonnegative linear combination of the extreme directions. To illustrate, consider the point $\mathbf{x}$ in Figure 2.18. The point $\mathbf{x}$ can be represented as $\mathbf{y}$ plus a positive multiple of the extreme direction $\mathbf{d}_2$. Note that $\mathbf{x} - \mathbf{y}$ points

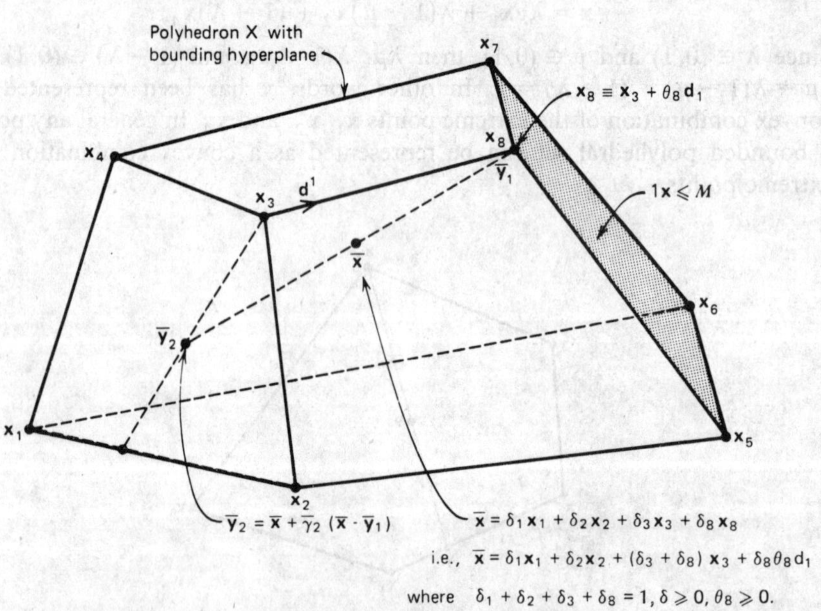

Figure 2.19. Illustration for the proof of Theorem 2.1.

in the direction $\mathbf{d}_2$. But $\mathbf{y}$ itself is a convex combination of the extreme points $\mathbf{x}_1$ and $\mathbf{x}_3$, and hence

$$\begin{aligned}\mathbf{x} &= \mathbf{y} + \mu\mathbf{d}_2 \\ &= \lambda\mathbf{x}_1 + (1-\lambda)\mathbf{x}_3 + \mu\mathbf{d}_2\end{aligned}$$

where $\lambda \in (0,1)$ and $\mu > 0$.

This discussion is made more precise later.

We now state and prove the Representation/Resolution/Caratheodory Theorem for the general case of bounded or unbounded sets X. Figure 2.19 is provided to help conceptualize the ideas of the proof.

***Theorem* 2.1** Representation Theorem for the General Case

Let $X = \{\mathbf{x} : \mathbf{A}\mathbf{x} \leq \mathbf{b}, \mathbf{x} \geq \mathbf{0}\}$ be a nonempty (polyhedral) set. Then the set of extreme points is not empty and has a finite number of points, say $\mathbf{x}_1, \mathbf{x}_2, \ldots, \mathbf{x}_k$. Furthermore, the set of extreme directions is empty if and only if X is bounded. If X is not bounded, then the set of extreme directions is nonempty and has a finite number of vectors, say $\mathbf{d}_1, \mathbf{d}_2, \ldots, \mathbf{d}_l$. Moreover, $\bar{\mathbf{x}} \in X$ if and only if it can be represented as a convex combination of $\mathbf{x}_1, \ldots, \mathbf{x}_k$ plus a nonnegative linear combination of $\mathbf{d}_1, \ldots, \mathbf{d}_l$, that is,

$$\begin{aligned}\bar{\mathbf{x}} &= \sum_{j=1}^{k} \lambda_j \mathbf{x}_j + \sum_{j=1}^{l} \mu_j \mathbf{d}_j \\ \sum_{j=1}^{k} \lambda_j &= 1 \\ \lambda_j &\geq 0 \qquad j = 1, 2, \ldots, k \\ \mu_j &\geq 0 \qquad j = 1, 2, \ldots, l\end{aligned} \tag{2.3}$$

Proof. For convenience, denote by S_p and S_d the sets of extreme points and extreme directions of X. Let us first show that $1 \leq k < \infty$, where $|S_p| = k$. Toward this end, let $\bar{\mathbf{x}} \in X$. If $\bar{\mathbf{x}} \in S_p$, then $k \geq 1$. Otherwise, let r be the maximum number of linearly independent defining hyperplanes binding at $\bar{\mathbf{x}}$ and let $\bar{\mathbf{x}} = \lambda\mathbf{y}_1 + (1-\lambda)\mathbf{y}_2$, where $0 < \lambda < 1$ and $\mathbf{y}_1, \mathbf{y}_2 \in X$, $\mathbf{y}_1 \neq \mathbf{y}_2$. Note that $0 \leq r < n$. Let $\mathbf{d} = \mathbf{y}_2 - \mathbf{y}_1 \neq \mathbf{0}$, so that $\mathbf{y}_1 = \bar{\mathbf{x}} - (1-\lambda)\mathbf{d}$ and $\mathbf{y}_2 = \bar{\mathbf{x}} + \lambda\mathbf{d}$. Consider moving from $\bar{\mathbf{x}}$ along $\mathbf{d}$ and $-\mathbf{d}$. Both directions permit positive step lengths from the foregoing statement, but both cannot permit infinite step lengths since $X \subseteq \{\mathbf{x} : \mathbf{x} \geq \mathbf{0}\}$. Hence without loss of generality, let us say that $\bar{\gamma} = \max\{\gamma : \bar{\mathbf{x}} - \gamma\mathbf{d} \in X\} < \infty$ and put $\bar{\mathbf{y}}_1 = \bar{\mathbf{x}} - \bar{\gamma}\mathbf{d}$. Note that the maximum number of hyperplanes binding at $\bar{\mathbf{y}}_1$ must be $\bar{r} \geq r + 1$. This follows since the defining hyperplanes binding at $\bar{\mathbf{x}}$ are also binding at $\bar{\mathbf{y}}_1$ (because $\bar{\mathbf{x}}$ is a strict convex combination of $\bar{\mathbf{y}}_1$ and $\mathbf{y}_2$). Moreover, at least one additional linearly independent hyperplane must be binding at $\bar{\mathbf{y}}_1$ to block any further motion along $-\mathbf{d}$ and hence determine $\bar{\gamma}$. If $\bar{r} = n$, then $\bar{\mathbf{y}}_1 \in S_p$, and so $k \geq 1$. Otherwise, one can replace $\bar{\mathbf{x}}$ by $\bar{\mathbf{y}}_1$ and repeat this process until one obtains $\bar{r} = n$, which, clearly, must occur finitely. Hence $k \geq 1$. Since the number of ways n linearly independent hyperplanes can be chosen from $(m + n)$ hyperplanes is finite, we also have $k < \infty$.

Next, by definition, if X is bounded, then $D = \varnothing$, and conversely, if X is unbounded, then $D \neq \varnothing$ since X is a convex set. Noting that D is of the same form as X (where the equality in Equation (2.2) can be equivalently written as two inequalities) and that the extreme points of D correspond to the extreme directions of X, we have that $1 \leq l < \infty$.

Now suppose that $\bar{x}$ can be written as Equation (2.3). Then it is easily verified using Equations (2.1) and (2.2) that $\bar{\mathbf{x}} \in X$. Conversely, consider any $\bar{\mathbf{x}} \in X$, and let us show that $\bar{\mathbf{x}}$ can be written as Equation (2.3). Define the set

$$\bar{X} = X \cap \{\mathbf{x} : \mathbf{1x} \leq M\}$$

where M is large enough so that $\mathbf{1x}_j < M$ for each $j = 1, \ldots, k$, and $\mathbf{1}\bar{\mathbf{x}} < M$. Note that $\bar{X}$ is bounded, and moreover, extreme points of X are also extreme points of $\bar{X}$. Let $\bar{S}_p \equiv \{\mathbf{x}_1, \ldots, \mathbf{x}_k, \ldots, \mathbf{x}_{k+u}\}$ be the extreme points of $\bar{X}$, where $0 \leq u < \infty$. Let us first show that we can write $\bar{\mathbf{x}}$ as a convex combination of points in $\bar{S}_p$. If $\bar{\mathbf{x}} \in \bar{S}_p$, this is obviously true. Otherwise, let the system $\mathbf{Gx} = \mathbf{g}$ represent the hyperplanes of $\bar{X}$ that are binding at $\bar{\mathbf{x}}$, and note that rank $(\mathbf{G}) \leq (n - 1)$. Find a solution $\mathbf{d} \neq \mathbf{0}$ to the system $\mathbf{Gd} = \mathbf{0}$, and compute $\bar{\gamma}_1 = \max\{\gamma : \bar{\mathbf{x}} + \gamma\mathbf{d} \in \bar{X}\}$. Note that $0 < \bar{\gamma}_1 < \infty$ (why?). Denote $\bar{\mathbf{y}}_1 = \bar{\mathbf{x}} + \bar{\gamma}_1\mathbf{d}$. Hence at $\bar{\mathbf{y}}_1 \in \bar{X}$, we have at least one additional linearly independent hyperplane of $\bar{X}$ binding. If the new binding hyperplane(s) along with $\mathbf{Gx} = \mathbf{g}$ produce a system of rank n, then $\bar{\mathbf{y}}_1$ is a vertex of $\bar{X}$. Otherwise, we may repeat this step at $\bar{\mathbf{y}}_1$ until after at most $[n - \text{rank}(\mathbf{G})]$ such steps, we obtain a vertex $\bar{\mathbf{y}}_1$ of $\bar{X}$ satisfying $\mathbf{G}\bar{\mathbf{y}}_1 = \mathbf{g}$. Now define

$$\bar{\gamma}_2 = \max\left\{\gamma : \bar{\mathbf{x}} + \gamma(\bar{\mathbf{x}} - \bar{\mathbf{y}}_1) \in \bar{X}\right\}$$

and put

$$\bar{\mathbf{y}}_2 = \bar{\mathbf{x}} + \bar{\gamma}_2(\bar{\mathbf{x}} - \bar{\mathbf{y}}_1).$$

Observe that $\bar{\gamma}_2 < \infty$ since $\bar{X}$ is bounded. Also, $\bar{\gamma}_2 > 0$ since $\mathbf{G}[\bar{\mathbf{x}} + \gamma(\bar{\mathbf{x}} - \bar{\mathbf{y}}_1)] = \mathbf{g}$ for all $\gamma \geq 0$ implies that $\bar{\gamma}_2$ is determined by some constraint not binding at $\bar{\mathbf{x}}$. Hence, in particular, $\mathbf{G}\bar{\mathbf{y}}_2 = \mathbf{g}$, and at least one additional linearly independent hyperplane is binding at $\bar{\mathbf{y}}_2$. Furthermore, $\bar{\mathbf{x}}$ is a convex combination $\delta\bar{\mathbf{y}}_1 + (1 - \delta)\bar{\mathbf{y}}_2$ of $\bar{\mathbf{y}}_1$ and $\bar{\mathbf{y}}_2$, where $\delta = \bar{\gamma}_2/(1 + \bar{\gamma}_2)$. Now since $\bar{\mathbf{y}}_1 \in \bar{S}_p$, if we also have $\bar{\mathbf{y}}_2 \in \bar{S}_p$, then we have written $\bar{\mathbf{x}}$ as a convex combination of points in $\bar{S}_p$. Otherwise, we can write $\bar{\mathbf{y}}_2$ itself as a strict convex combination of some $\bar{\mathbf{y}}_3$ and $\bar{\mathbf{y}}_4$, where $\bar{\mathbf{y}}_3 \in \bar{S}_p$ and $\bar{\mathbf{y}}_4$ has at least one additional linearly independent hyperplane binding. Continuing this way, we can ultimately write $\bar{\mathbf{x}}$ as a convex combination of vertices of $\bar{S}_p$, in fact, using no more than n-rank $(\mathbf{G}) + 1$ vertices of $\bar{S}_p$. Let this representation be given by

$$\bar{\mathbf{x}} = \sum_{j=1}^{k+u} \delta_j \mathbf{x}_j, \quad \text{where } \sum_{j=1}^{k+u} \delta_j = 1, \text{ and } \delta \geq 0. \tag{2.4}$$

Now if $\delta_j = 0$ for $j > k$, then Equation (2.4) is of the form Equation (2.3). Otherwise, consider any $\mathbf{x}_v$ with $v > k$ and $\delta_v > 0$. Note that $\mathbf{x}_v$ is a new extreme point created by the constraint $\mathbf{1x} \leq M$, that is, $\mathbf{1x} = M$ is one of any n linearly independent hyperplanes defining $\mathbf{x}_v$ in $\overline{X}$. The other $(n-1)$ linearly independent hyperplanes come from the defining hyperplanes of X, and therefore define an edge of X. Consequently, there exists some extreme point $\mathbf{x}_{i(v)}$ of X, $1 \leq i(v) \leq k$, which is the adjacent extreme point of $\mathbf{x}_v$ in $\overline{X}$ along this edge. Moreover, $(\mathbf{x}_v - \mathbf{x}_{i(v)})$ is a recession direction of X since no other hyperplane of X stops the progress along this direction from $x_{i(v)}$. More importantly, let $\bar{\mathbf{d}} = (\mathbf{x}_v - \mathbf{x}_{i(v)})/\theta_v$, where $\theta_v = \mathbf{1}(\mathbf{x}_v - \mathbf{x}_{i(v)}) > 0$. Observe that $\bar{\mathbf{d}} \in D$. Furthermore, the $(n-1)$ linearly independent hyperplanes of the homogeneous system $\mathbf{Ad} \leq \mathbf{0}$, $\mathbf{d} \geq \mathbf{0}$, which correspond to the $(n-1)$ linearly independent hyperplanes of X defining $\mathbf{x}_v$, are binding at $\bar{\mathbf{d}}$. Also, from the hyperplanes defining $\mathbf{x}_v$, these $(n-1)$ hyperplanes along with $\mathbf{1d} = 1$ produce n linearly independent binding hyperplanes of D at $\bar{\mathbf{d}}$. Hence the direction $\bar{\mathbf{d}}$ must be an extreme point $\mathbf{d}_{j(v)}$, say, of D and therefore an extreme direction of X. Consequently, we have $\mathbf{x}_v \equiv \mathbf{x}_{i(v)} + \theta_v \mathbf{d}_{j(v)}$. Substituting this into Equation (2.4) for each such v, and arbitrarily letting $i(v) = j(v) = 1$ if $\delta_v = 0$, we get

$$\bar{\mathbf{x}} = \sum_{j=1}^{k} \delta_j \mathbf{x}_j + \sum_{v=k+1}^{k+u} \delta_v \mathbf{x}_{i(v)} + \sum_{v=k+1}^{k+u} \delta_v \theta_v \mathbf{d}_{j(v)}$$

which is of the form Equation (2.3). This completes the proof.

***Corollary* 2.1**

Any $\bar{\mathbf{x}} \in X$ can be represented as in Equation (2.3) using no more than $\min\{(n+1), (k+l)\}$ positive λ_j and μ_j variables.

Proof. Given $\bar{\mathbf{x}} \in X$, Theorem 2.1 asserts that there exists a solution to Equation (2.3). Let $r =$ rank of the coefficient matrix associated with the equality system in Equation (2.3). Hence $r = \min\{(n+1), (k+l)\}$. Furthermore, note that the set of (λ, μ) satisfying Equation (2.3) forms a polyhedral set in E^{k+l}, and by Theorem 2.1, this set has an extreme point. Since there must be $(k+l)$ linearly independent hyperplanes binding at any such extreme point and the equality system provides r such hyperplanes, we must have at least $(k+l-r)$ additional λ_j and μ_j variables equal to zero at an extreme point of Equation (2.3). Consequently, there exists a representation of $\bar{\mathbf{x}}$ in Equation (2.3) in which at most r λ_j and μ_j, variables are positive. This completes the proof.

Theorem 2.1 suggests a constructive algorithm for actually representing a given point $\bar{\mathbf{x}} \in X$ in terms of no more than $(n+1)$ extreme points and n extreme directions, using a number of elementary computational operations such as additions, multiplications, comparisons, and so on bounded above by a polynomial in n of degree ≤ 4. That is, it suggests a *polynomial-time algorithm of complexity bounded by* $O(n^4)$ to obtain such a representation. (See Exercise 2.57.) Note that if X is bounded, then the algorithm uses no more than $\min\{(n+1), k\}$ of its extreme points in the representation

Example 2.7

Consider the polyhedral set of Example 2.6 that is illustrated in Figure 2.18. Let us represent the point $\bar{\mathbf{x}} = \begin{pmatrix} 4 \\ 3 \end{pmatrix} \in X$ as in Equation (2.3). Bound X by constructing the plane $x_1 + x_2 \leq M$ as in Figure 2.18 to obtain the set $\bar{X}$. Proceeding as in the proof of Theorem 2.1, since no constraints of $\bar{X}$ are binding at $\bar{\mathbf{x}}$, pick $\mathbf{d} = (0, 1)$ arbitrarily. Thus $\bar{\gamma}_1 = \max\{\gamma : (4, 3) + \gamma(0, 1) \in \bar{X}\} = 3$ and hence we get $\bar{\mathbf{y}}_1 = (4, 6)$. Since two linearly independent equations are binding at $\bar{\mathbf{y}}_1$ we have that $\bar{\mathbf{y}}_1$ is an extreme point $\mathbf{x}_1$ of $\bar{X}$. Next, we compute $\bar{\gamma}_2 = \max\{\gamma : (4, 3) + \gamma(0, -3) \in \bar{X}\} = 1/3$, and hence we get $\bar{\mathbf{y}}_2 = (4, 2)$. Therefore, we have $\bar{\mathbf{x}} = \delta\bar{\mathbf{y}}_1 + (1 - \delta)\bar{\mathbf{y}}_2$ where $\delta = (1/3)/(4/3) = 1/4$. Hence

$$\bar{\mathbf{x}} = (1/4)\mathbf{x}_1 + (3/4)\bar{\mathbf{y}}_2.$$

Next, observe that one constraint $x_2 = 2$ is binding at $\bar{\mathbf{y}}_2$. Repeating the process at $\bar{\mathbf{y}}_2$, and using $\mathbf{d} = (1, 0)$ as a solution to the homogeneous system $x_2 = 0$, we get

$$\bar{\mathbf{y}}_2 = \frac{(M - 6)}{(M - 10/3)}\mathbf{x}_3 + \frac{8/3}{(M - 10/3)}\mathbf{x}_4$$

where $\mathbf{x}_3 = (4/3, 2)$ and $\mathbf{x}_4 = (M - 2, 2)$. Substituting, we obtain

$$\bar{\mathbf{x}} = \frac{1}{4}\mathbf{x}_1 + \frac{3(M - 6)}{4(M - 10/3)}\mathbf{x}_3 + \frac{2}{(M - 10/3)}\mathbf{x}_4$$

Now at $\mathbf{x}_4 = (M - 2, 2)$, which is of the form Equation (2.4), only one constraint $(x_2 = 2)$ of X is binding. This gives via its homogeneous solution the extreme direction $\mathbf{d}_1 = (1, 0)$ of X, and so we get the adjacent extreme point of $\mathbf{x}_4$ in $\bar{X}$ by searching along $-\mathbf{d}_1$ as $\mathbf{x}_3 = (4/3, 2)$. This gives $\mathbf{x}_4 = \mathbf{x}_3 + \theta\mathbf{d}_1$, where $\theta = \mathbf{1}(\mathbf{x}_4 - \mathbf{x}_3) = (M - 10/3)$. Substituting for $\mathbf{x}_4$, we finally obtain

$$\bar{\mathbf{x}} = \frac{1}{4}\mathbf{x}_1 + \frac{3(M - 6)}{4(M - 10/3)}\mathbf{x}_3 + \frac{2}{(M - 10/3)}[\mathbf{x}_3 + (M - 10/3)\mathbf{d}_1]$$

or

$$\bar{\mathbf{x}} = \tfrac{1}{4}\mathbf{x}_1 + \tfrac{3}{4}\mathbf{x}_3 + 2\mathbf{d}_1$$

which is of the required form Equation (2.3). Note that this representation is not unique. Another representation for $\bar{\mathbf{x}}$ is

$$\bar{\mathbf{x}} = \tfrac{1}{2}\mathbf{x}_2 + \tfrac{1}{2}\mathbf{x}_3 + \tfrac{7}{3}\mathbf{d}_1$$

Note from Figure 2.18 that $n = 2$, $k = 3$, and $l = 2$ in this problem. Hence $\min\{(n + 1), (k + l)\} = 3$, and hence both representations use the upper bound number of terms prescribed by Corollary 2.1. In this case, 3 also happens to be a lower bound on the number of terms required in any representation of $\bar{\mathbf{x}} = (4, 3)$.

Some Insights into the Equality Constrained Polyhedron

Consider the nonempty polyhedral set $X = \{\mathbf{x} : \mathbf{A}\mathbf{x} = \mathbf{b}, \mathbf{x} \geq \mathbf{0}\}$ where $\mathbf{A}$ is $m \times n$ and $\text{rank}(\mathbf{A}) = m$. Note by equivalently replacing each equality constraint by two inequalities, all the discussion in this section continues to hold for this case as well. In particular, consider the extreme points of X. As before, these are points in X formed by the intersection of n linearly independent hyperplanes. However, from the equality constraints, we know that m linearly independent hyperplanes are always binding at any feasible solution. Hence at any extreme point $\bar{\mathbf{x}}$ of X, there must be some $(n - m)$ additional hyperplanes binding from the nonnegativity constraints, say $\mathbf{x}_N = 0$, which together with $\mathbf{A}\mathbf{x} = \mathbf{b}$ produce n linearly independent equations in n unknowns, thereby yielding $\bar{\mathbf{x}}$ as the unique solution. Similarly, one can characterize edges and adjacent extreme points of X following Section 2.6, and note as before that extreme directions of X correspond to extreme points of

$$D = \{\mathbf{d} \cdot \mathbf{A}\mathbf{d} = \mathbf{0}, \mathbf{1}\mathbf{d} = 1, \mathbf{d} \geq \mathbf{0}\}.$$

Finally, in light of the remark concerning extreme points of X, the reader may find it instructive to read through the proof of Theorem 2.1, thinking of X as the equality constrained set.

EXERCISES

2.1 Which of the following collection of vectors form a basis of E^3, span E^3, or neither?

a. $\mathbf{a}_1 = \begin{pmatrix} 1 \\ 2 \\ 1 \end{pmatrix}, \quad \mathbf{a}_2 = \begin{pmatrix} -1 \\ 0 \\ -1 \end{pmatrix}, \quad \mathbf{a}_3 = \begin{pmatrix} 0 \\ 0 \\ 1 \end{pmatrix}$

b. $\mathbf{a}_1 = \begin{pmatrix} 1 \\ 3 \\ 2 \end{pmatrix}, \quad \mathbf{a}_2 = \begin{pmatrix} 1 \\ 0 \\ 5 \end{pmatrix}$

c. $\mathbf{a}_1 = \begin{pmatrix} -1 \\ 2 \\ 3 \end{pmatrix} \quad \mathbf{a}_2 = \begin{pmatrix} 0 \\ 1 \\ 0 \end{pmatrix}, \quad \mathbf{a}_3 = \begin{pmatrix} 1 \\ 2 \\ 3 \end{pmatrix}, \quad \mathbf{a}_4 = \begin{pmatrix} -3 \\ 2 \\ 4 \end{pmatrix}$

d. $\mathbf{a}_1 = \begin{pmatrix} 1 \\ 2 \\ 1 \end{pmatrix}, \quad \mathbf{a}_2 = \begin{pmatrix} -3 \\ -1 \\ 2 \end{pmatrix}, \quad \mathbf{a}_3 = \begin{pmatrix} -5 \\ 0 \\ 5 \end{pmatrix}$

e. $\mathbf{a}_1 = \begin{pmatrix} 1 \\ 4 \\ 2 \end{pmatrix}, \quad \mathbf{a}_2 = \begin{pmatrix} -1 \\ -4 \\ -1 \end{pmatrix}, \quad \mathbf{a}_3 = \begin{pmatrix} 0 \\ 1 \\ 0 \end{pmatrix}$

2.2 Let

$$\mathbf{a}_1 = \begin{pmatrix} -1 \\ 2 \\ 0 \end{pmatrix}, \quad \mathbf{a}_2 = \begin{pmatrix} 3 \\ 2 \\ 5 \end{pmatrix}, \quad \text{and} \quad \mathbf{a}_3 = \begin{pmatrix} \frac{5}{2} \\ 3 \\ 5 \end{pmatrix}$$

Are these vectors linearly independent? Do they span E^3?

2.3 Let $\mathbf{a}_1, \mathbf{a}_2, \ldots, \mathbf{a}_k$ form a basis for E^n. Show that $\mathbf{a}_1, \mathbf{a}_2, \ldots, \mathbf{a}_k$ are linearly independent. Also show that $k = n$.

2.4 Show that the vectors

$$\mathbf{a}_1 = \begin{pmatrix} 1 \\ 0 \\ 0 \end{pmatrix}, \quad \mathbf{a}_2 = \begin{pmatrix} 0 \\ 1 \\ 0 \end{pmatrix}, \quad \text{and} \quad \mathbf{a}_3 = \begin{pmatrix} 1 \\ 5 \\ 3 \end{pmatrix}$$

form a basis for E^3. Supposing that $\mathbf{a}_2$ is replaced by $\begin{bmatrix} 0 \\ 1 \\ 1 \end{bmatrix}$, indicate whether the new set of vectors still forms a basis of E^3.

2.5 Suppose that $\mathbf{a}_1, \mathbf{a}_2, \ldots, \mathbf{a}_n$ form a basis of E^n and $\mathbf{y} = \lambda_1\mathbf{a}_1 + \lambda_2\mathbf{a}_2 + \cdots + \lambda_n\mathbf{a}_n$ with $\lambda_j = 0$. Prove that $\mathbf{a}_1, \ldots, \mathbf{a}_{j-1}, \mathbf{y}, \mathbf{a}_{j+1}, \ldots, \mathbf{a}_n$ do not form a basis of E^n.

2.6 Let $\mathbf{B}$ be an invertible matrix. Show that $\mathbf{B}^{-1}$ is unique.

2.7 Let

$$\mathbf{A} = \left[\begin{array}{c|c} \mathbf{B} & \mathbf{0} \\ \hline \mathbf{T} & \mathbf{I} \end{array}\right]$$

where $\mathbf{B}$ is an $m \times m$ invertible matrix, $\mathbf{I}$ is a $k \times k$ identity matrix, $\mathbf{0}$ is an $m \times k$ zero matrix, and $\mathbf{T}$ is an arbitrary $k \times m$ matrix. Show that $\mathbf{A}$ has an inverse and that

$$\mathbf{A}^{-1} = \left[\begin{array}{c|c} \mathbf{B}^{-1} & \mathbf{0} \\ \hline -\mathbf{T}\mathbf{B}^{-1} & \mathbf{I} \end{array}\right]$$

2.8 Find the inverse of the following triangular matrix.

$$\mathbf{A} = \begin{bmatrix} 1 & 4 & -3 & -4 \\ 0 & 4 & 1 & 2 \\ 0 & 0 & 2 & 5 \\ 0 & 0 & 0 & -1 \end{bmatrix}$$

2.9 Does the following matrix have an inverse? If the answer is yes, find $\mathbf{A}^{-1}$.

$$\mathbf{A} = \begin{bmatrix} 1 & -1 & 2 & 0 \\ 2 & 5 & 0 & 1 \\ 0 & 2 & 0 & 1 \\ 1 & 3 & 1 & 2 \end{bmatrix}$$

2.10 Let $\mathbf{A}$ be a 3×5 matrix. Consider the matrix $\mathbf{C}$ obtained by weighting the columns of $\mathbf{A}$ with scalars 2, $\frac{3}{2}$, -2, 3, and -1 respectively and adding them up. Write $\mathbf{C}$ as a matrix product. Similarly, let $\mathbf{D}$ be the weighted sum of the rows of $\mathbf{A}$, using weights 2, -1, and $\frac{3}{2}$ respectively. Write $\mathbf{D}$ as a matrix product. Likewise, let $\mathbf{E}$ be a matrix whose first row is the sum of the rows of $\mathbf{A}$, whose second row is the third row of $\mathbf{A}$, and whose third row is obtained by summing twice the first row of $\mathbf{A}$, $-\frac{3}{2}$ times the second row of $\mathbf{A}$, and thrice the third row of $\mathbf{A}$. Write $\mathbf{E}$ as a matrix product.

2.11 Let $\mathbf{A}$ be an $n \times n$ invertible matrix. Show that $\mathbf{A}'$ has an inverse and that $(\mathbf{A}')^{-1} = (\mathbf{A}^{-1})'$.

2.12 Show that if $\mathbf{A}$ and $\mathbf{B}$ are $n \times n$ matrices that are both invertible, then $(\mathbf{AB})^{-1} = \mathbf{B}^{-1}\mathbf{A}^{-1}$.

2.13 Let $\mathbf{A} = (\mathbf{a}_1, \mathbf{a}_2, \ldots, \mathbf{a}_j, \ldots, \mathbf{a}_m)$ be an invertible $m \times m$ matrix. Show that $\mathbf{A}^{-1}\mathbf{a}_j = \mathbf{e}_j$, where $\mathbf{e}_j$ is a vector of zeros except for a 1 at position j.

2.14 If the ith row of a square nonsingular matrix $\mathbf{B}$ is multiplied by a scalar $\lambda \neq 0$, what changes would result in $\mathbf{B}^{-1}$?

2.15 If the ith column of a square nonsingular matrix $\mathbf{B}$ is multiplied by a scalar $\lambda \neq 0$, what changes would result in $\mathbf{B}^{-1}$?

2.16 Let $\mathbf{B}$ be an invertible matrix with nonnegative entries. Show that every row of $\mathbf{B}^{-1}$ has at least one positive entry.

2.17 Let $\mathbf{A}$ be an $n \times n$ matrix. Suppose that $\mathbf{B}$ is an $n \times n$ matrix such that $\mathbf{AB} = \mathbf{I}$. Is it necessarily true that $\mathbf{A}$ has an inverse? Is it necessarily true that $\mathbf{B} = \mathbf{A}^{-1}$?

2.18 Find the determinants of the following matrices.

a. $\mathbf{A} = \begin{bmatrix} 1 & 0 & 1 \\ 2 & 1 & -1 \\ 0 & 1 & 2 \end{bmatrix}$

b. $\mathbf{A} = \begin{bmatrix} 1 & 0 & -2 & 1 \\ 3 & 1 & -1 & 1 \\ -2 & 2 & 1 & 2 \\ 1 & 2 & 1 & 5 \end{bmatrix}$

c. $\mathbf{A} = \begin{bmatrix} 1 & -2 & 1 \\ 1 & 1 & -2 \\ 3 & -2 & 3 \end{bmatrix}$

2.19 Find the rank of the following matrices.

$$\mathbf{A} = \begin{bmatrix} 1 & 0 & 1 & 1 \\ 2 & 2 & 4 & -1 \\ 1 & 0 & 5 & 3 \end{bmatrix}$$

$$\mathbf{A} = \begin{bmatrix} -1 & 1 & 3 \\ 1 & 4 & 2 \\ 2 & 3 & -1 \end{bmatrix}$$

2.20 Show that the determinant of a square triangular matrix is the product of the diagonal entries.

2.21 Solve the following system by Cramer's rule.

$$\begin{aligned} 2x_1 + x_2 &= 6 \\ 5x_1 - 2x_2 &= 4 \end{aligned}$$

2.22 Demonstrate by enumeration that every basis matrix of the following system is triangular.

$$\begin{aligned} x_1 + x_3 + x_4 &= 4 \\ -x_1 + x_2 + x_5 &= 2 \\ -x_2 - x_4 &= -3 \\ -x_3 &= -2 \end{aligned}$$

2.23 Solve the following system of equations.

$$\begin{aligned} x_1 + 2x_2 + x_3 &= 1 \\ -x_1 + x_2 - x_3 &= 5 \\ 2x_1 + 3x_2 + x_3 &= -4 \end{aligned}$$

Without resolving the system, what is the solution if the right-hand side of the first equation is changed from 1 to 2?

2.24 Construct a general solution of the system $\mathbf{Ax} = \mathbf{b}$ where $\mathbf{A}$ is an $m \times n$ matrix with rank m.

2.25 What is the general solution of the following system?

$$\begin{aligned} x_1 + 2x_2 + x_3 &= 3 \\ -x_1 + 3x_2 + x_3 &= 6 \end{aligned}$$

2.26 Find all basic solutions of the following system.

$$\begin{aligned} -x_1 + 2x_2 + x_3 + x_4 - 2x_5 &= 4 \\ x_1 - 2x_2 \qquad + 2x_4 - x_5 &= 3 \end{aligned}$$

2.27 Determine whether the following system possesses: (a) no solution, (b) a unique solution, or (c) many (how many?) solutions.

$$\begin{aligned} x_1 + 3x_2 + x_3 - x_4 &= 1 \\ 5x_2 - 6x_3 + x_4 &= 0 \\ x_1 - 2x_2 + 4x_3 \qquad &= 1 \end{aligned}$$

2.28 Consider the system $\mathbf{Ax} = \mathbf{b}$ where $\mathbf{A} = [\mathbf{a}_1, \mathbf{a}_2, \ldots, \mathbf{a}_n]$ is an $m \times n$ matrix of rank m. Let $\mathbf{x}$ be any solution of this system. Starting with $\mathbf{x}$, construct a basic solution. (*Hint*. Suppose that $x_1, \ldots, x_p \neq 0$ and $x_{p+1}, \ldots, x_n = 0$. If $p > m$, represent one of the columns $\mathbf{a}_j$ for $j = 1, 2, \ldots, p$ as a linear combination of the remaining vectors. This results in a new solution with a smaller number of nonzero variables. Repeat the process.)

2.29 Which of the following sets are convex and which are not?

a. $\{(x_1, x_2) : x_1^2 + x_2^2 \geq 1\}$
b. $\{(x_1, x_2, x_3) : x_1 + 2x_2 \leq 1,\ x_1 - x_3 \leq 2\}$
c. $\{(x_1, x_2) : x_2 - x_1^2 = 0\}$
d. $\{(x_1, x_2, x_3) : x_2 \geq x_1^2,\ x_1 + x_2 + x_3 \leq 6\}$
e. $\{(x_1, x_2) : x_1 = 1,\ |x_2| \leq 4\}$
f. $\{(x_1, x_2, x_3) : x_3 = |x_2|,\ x_1 \leq 4\}$

2.30 Show that a hyperplane $H = \{\mathbf{x} : \mathbf{px} = k\}$ and a halfspace $H^+ = \{\mathbf{x} : \mathbf{px} \geq k\}$ are convex sets.

2.31 Consider the set $\{(x_1, x_2) : -x_1 + x_2 \leq 2,\ x_1 + 2x_2 \leq 8,\ x_1 \geq 0,\ x_2 \geq 0\}$. What is the minimum distance from $(-2, 3)$ to the set? What is the point in the set closest to $(-2, 3)$?

2.32 Consider the set $X = \{(x_1, x_2) : x_1, x_2 \geq 0,\ x_1 + x_2 \geq 2,\ x_2 \leq 4\}$. Find a hyperplane H such that X and the point $(3, -2)$ are on different sides of the hyperplane. Write the equation of the hyperplane.

2.33 Let $\mathbf{a}_1 = \begin{pmatrix} 1 \\ 0 \end{pmatrix}$, $\mathbf{a}_2 = \begin{pmatrix} 2 \\ 3 \end{pmatrix}$, $\mathbf{a}_3 = \begin{pmatrix} -1 \\ 4 \end{pmatrix}$, $\mathbf{a}_4 = \begin{pmatrix} 5 \\ -3 \end{pmatrix}$, and $\mathbf{a}_5 = \begin{pmatrix} -4 \\ 3 \end{pmatrix}$. Illustrate geometrically the collection of all convex combinations of these five points.

2.34 Show that the set of feasible solutions to the following linear program forms a convex set.

$$\begin{array}{ll} \text{Minimize} & \mathbf{cx} \\ \text{Subject to} & \mathbf{Ax} = \mathbf{b} \\ & \mathbf{x} \geq \mathbf{0} \end{array}$$

2.35 Which of the following functions are convex, concave, or neither?

a. $f(x) = x^2$

b. $f(x_1, x_2) = e^{-x_1-x_2} + x_1^2 - 2x_1$

c. $f(x_1, x_2) = \text{Maximum}\,(f_1(x_1, x_2),\ f_2(x_1, x_2))$ where $f_1(x_1, x_2) = x_1^2 + x_2^2$ and $f_2(x_1, x_2) = 2x_1^2 - x_2$

d. $f(x_1, x_2) = \text{Minimum}\,(f_1(x_1, x_2),\ f_2(x_1, x_2))$ where $f_1(\cdot)$ and $f_2(\cdot)$ are defined in part (c).

e. $f(x_1, x_2, x_3) = -x_1^2 - 2x_2^2 - x_3^2 + 2x_1x_2 - x_2x_3 + 2x_1 + 5x_3$

f. $f(x_1, x_2) = x_1^2 + 2x_2^2 - 2x_1x_2 + x_2$

2.36 Show that f is convex if and only if its *epigraph* $= \{(\mathbf{x}, y) : \mathbf{x} \in E^n, y \in E^1, y \geq f(\mathbf{x})\}$ is a convex set. Similarly, show that f is concave if and only if its *hypograph* $= \{(\mathbf{x}, y) : \mathbf{x} \in E^n, y \in E^1, y \leq f(\mathbf{x})\}$ is a convex set.

2.37 Show that a differentiable function f is convex if and only if the following inequality holds for each fixed point $\mathbf{x}_0$ in E^n: $f(\mathbf{x}) \geq f(\mathbf{x}_0) + \nabla f(\mathbf{x}_0)(\mathbf{x} - \mathbf{x}_0)$ for all $\mathbf{x} \in E^n$, where $\nabla f(\mathbf{x}_0)$ is the gradient vector of f at $\mathbf{x}_0$ given by

$$\left(\frac{\partial f(\mathbf{x}_0)}{\partial x_1}, \frac{\partial f(\mathbf{x}_0)}{\partial x_2}, \ldots, \frac{\partial f(\mathbf{x}_0)}{\partial x_n} \right).$$

2.38 If S is an *open set*, show that the problem

$$\begin{array}{ll} \text{Maximize} & \mathbf{cx} \\ \text{Subject to} & \mathbf{x} \in S \end{array}$$

where $\mathbf{c} \neq \mathbf{0}$ possesses no optimal point. (*Note*. S is *open* if for each $\mathbf{x}_0 \in S$, there is an $\epsilon > 0$ such that $\|\mathbf{x} - \mathbf{x}_0\| < \epsilon$ implies that $\mathbf{x} \in S$.)

2.39 Show that if C is a convex cone, then C has at most one extreme point, namely the origin.

2.40 Show that C is a convex cone if and only if $\mathbf{x}$ and $\mathbf{y} \in C$ imply that $\lambda\mathbf{x} + \mu\mathbf{y} \in C$ for all $\lambda \geq 0$ and $\mu \geq 0$.

2.41 Find all extreme points of the following polyhedral set.

$$X = \{(x_1, x_2, x_3) : x_1 - x_2 + x_3 \leq 1, -x_1 + 2x_2 \leq 4, x_1, x_2, x_3 \geq 0\}$$

Does X have any recession directions? Why?

2.42 Find the extreme points of the region defined by the following inequalities.

$$\begin{aligned} x_1 + x_2 + x_3 &\le 5 \\ -x_1 + x_2 + 2x_3 &\le 6 \\ x_1, \quad x_2, \quad x_3 &\ge 0 \end{aligned}$$

(*Hint*. Consider $n = 3$ intersecting defining hyperplanes at a time).

2.43 Given a nonempty polyhedron X, a face F of X is also defined as the intersection of a supporting hyperplane of X with X itself. (A hyperplane *supports* X if X lies completely in one halfspace defined by the hyperplane and there is at least one point in common between X and the hyperplane). Show the equivalence between this definition and the one given in Section 2.6.

2.44 Consider the nonempty polyhedral set $X = \{\mathbf{x} : \mathbf{A}\mathbf{x} = \mathbf{b},\ \mathbf{x} \ge \mathbf{0}\}$. Show directly by definition that $\mathbf{d}$ is a direction of the set if and only if $\mathbf{d} \ne \mathbf{0}$, $\mathbf{A}\mathbf{d} = \mathbf{0}$, and $\mathbf{d} \ge \mathbf{0}$. Obtain analogous results if $\mathbf{A}\mathbf{x} = \mathbf{b}$ is replaced by $\mathbf{A}_1\mathbf{x} = \mathbf{b}_1$, $\mathbf{A}_2\mathbf{x} \ge \mathbf{b}_2$.

2.45 Let $X = \{\mathbf{x} : \mathbf{A}\mathbf{x} \le \mathbf{b}\}$ and let $\mathbf{x}_0$ be such at $\mathbf{A}\mathbf{x}_0 < \mathbf{b}$. Show that $\mathbf{x}_0$ cannot be an extreme point of X.

2.46 Prove in detail that a polyhedral set x is bounded if and only if it has no directions.

2.47 Why does the following set have directions? Find all its extreme directions.

$$\begin{aligned} -x_1 + x_2 \qquad &= 4 \\ x_1 - 2x_2 + x_3 &\le 6 \\ x_3 &\ge 1 \\ x_1, \quad x_2, \quad x_3 &\ge 0 \end{aligned}$$

(*Hint*: Enumerate the extreme points of its normalized set of directions.)

2.48 Let $X = \{\mathbf{x} : \mathbf{A}\mathbf{x} = \mathbf{b},\ \mathbf{x} \ge \mathbf{0}\}$ where $\mathbf{A}$ is an $m \times n$ matrix with rank m. Show that $\mathbf{d}$ is an extreme direction of X if and only if $\mathbf{d}$ is a positive multiple of the vector $(-\mathbf{y}_j', 0, 0, \ldots, 1, 0, \ldots, 0)'$ where

$\mathbf{y}_j = \mathbf{B}^{-1}\mathbf{a}_j \le \mathbf{0}$

$\mathbf{A} = [\mathbf{B}, \mathbf{N}]$ where $\mathbf{B}$ is an *$m \times m$ invertible matrix*

$\mathbf{a}_j =$ *a* column of $\mathbf{N}$

the 1 appears in position j

(*Hint*: Consider extreme points of the normalized direction set.) Illustrate by the following system.

$$\begin{aligned} x_1 - 3x_2 + x_3 \qquad &= 0 \\ -3x_1 + x_2 \qquad + x_4 &= 3 \\ x_1, \quad x_2, \quad x_3, \quad x_4 &\ge 0 \end{aligned}$$

2.49 Show that an unbounded polyhedral set of the form $\{\mathbf{x} : \mathbf{Ax} = \mathbf{b}, \mathbf{x} \geq \mathbf{0}\}$ has at least one extreme direction. In particular, using the normalized direction set D in lieu of X in the proof of Theorem 2.1, show how you can start with any direction $\mathbf{d}$ in D and reduce it to an extreme direction.

2.50 You are given the following polyhedral set. Identify the faces, extreme points, extreme directions, and extreme rays of the set.

$$\begin{aligned} x_1 - x_2 + x_3 &\leq 10 \\ 2x_1 - x_2 + 2x_3 &\leq 40 \\ 3x_1 - 2x_2 + 3x_3 &\leq 50 \\ x_1, \quad x_2, \quad x_3 &\geq 0 \end{aligned}$$

2.51 Consider the polyhedral set $X = \{\mathbf{x} : \mathbf{px} = k\}$ where $\mathbf{p}$ is a nonzero vector and k is a scalar. Show that X has neither extreme points nor extreme rays. How do you explain this in terms of the general Representation Theorem?

2.52 Let $X = \{(x_1, x_2) : x_1 - x_2 \leq 3,\ 2x_1 + x_2 \leq 4,\ x_1 \geq -3\}$. Find all extreme points of X and represent $\mathbf{x} = (0, 1)$ as a convex combination of the extreme points.

2.53 Find all extreme points and extreme directions of the following polyhedral set:

$$X = \{(x_1, x_2, x_3, x_4) : x_1, x_2, x_3, x_4 \geq 0, -x_1 + x_2 - 2x_3 \leq 1, \\ -2x_1 - x_3 + 2x_4 \leq 2\}$$

Represent $x = (1, 1, 1, 2)$ as a convex combination of the extreme points of X plus a nonnegative combination of the extreme directions of X.

2.54 Answer the following questions and provide a brief explanation or illustration.

a. Is it possible for X in Equation (2.1) to be empty but D in Equation (2.2) to be nonempty?

b. Is there a relationship between redundancy and degeneracy of a polyhedral set?

c. Does degeneracy imply redundancy in two dimensions?

d. If the intersection of a finite number of halfspaces is nonempty, then this set has at least one extreme point. True or false?

e. An unbounded n-dimensional polyhedral set can have at most n extreme directions. True or false?

f. What is the maximum (actual) dimension of $X = \{\mathbf{x} : \mathbf{Ax} = \mathbf{b}, \mathbf{x} \geq \mathbf{0}\}$, where $\mathbf{A}$ is $m \times n$ of rank t, $t \leq m \leq n$?

2.55 Consider the polyhedral set $X = \{\mathbf{x} : \mathbf{Qx} \leq \mathbf{q}\}$, where $\mathbf{Q}$ is $m \times n$. Mathematically characterize the statement that an *entire line belongs to X*. What does this imply about the rank of $\mathbf{Q}$? Show that a nonempty polyhedron has extreme points if and only if it contains no lines.

2.56 In the proof of Theorem 2.1, consider the direction $\bar{\mathbf{d}}$ given by $(\mathbf{x}_v - \mathbf{x}_{i(v)})/\theta_v$. Show how you would algebraically obtain $\bar{\mathbf{d}}$, given the extreme point $\mathbf{x}_v$ of $\bar{X}$, $v > k$.

2.57 Based on Theorem 2.1 and its proof, construct a polynomial-time algorithm of complexity bounded by $O(n^4)$ in order to obtain a representation of $\bar{\mathbf{x}} \in X$ in terms of no more than $(n + 1)$ extreme points and n extreme directions of X. Show that the number of extreme points needed in the representation by this algorithm is no more than the bound provided by Corollary 2.1 when X is a bounded set.

NOTES AND REFERENCES

1. Sections 2.1 through 2.3 present a quick review of some relevant results of vector and matrix algebra.
2. Sections 2.4 and 2.5 give some basic definitions and properties of convex sets, convex cones, and convex functions. For more details the reader may refer to Eggleston [1958], Mangasarian [1969], and the more advanced text of Rockafellar [1970].
3. Correspondence between bases and extreme points is established in Section 3.2, and an algebraic characterization of extreme directions is presented in Exercise 2.48.
4. The representation theorem for polyhedral sets evolved from the work of Minkowski [1910] and Goldman and Tucker [1956]. The result is also true for (nonpolyhedral) convex sets that contain no lines. See Rockafellar [1970] and Bazaraa and Shetty [1976]. The geometric proof of Theorem 2.1 in Section 2.6 is taken from Sherali [1987]. (Parts of the proof, parts of Example 2.7, and Figure 2.17 are reproduced from Sherali [1987]—Copyright © (1987) by the American Sciences Press, Inc., Columbus, Ohio 43221-0161. Reprinted by permission.) For further geometric insights, see Akgul [1986], Grunbaum [1967], Murty [1983, 1985], and Sommerville [1958].

THREE: THE SIMPLEX METHOD

In this chapter we begin to discuss Dantzig's Simplex Method, which was conceived in the summer of 1947 for solving linear programming problems. The first significant application of this method occurred soon after in the fall of 1947. J. Laderman solved a diet-planning linear program with nine equality constraints in 27 nonnegative variables at the National Bureau of Standards. Using desk calculators, this problem took 120 man-days to solve and the worksheets were laboriously glued together and spread out like a "table cloth." Today, using modern-age computer facilities and a sophisticated implementation of the simplex method, linear programs with more than 5000 constraints and 10,000 variables are readily solvable. Although several variants of the simplex method have evolved and other new competing algorithms have been proposed (see Chapter 8), the simplex method remains a viable and popular tool for solving linear programming problems. It also provides further insights into the facial structure of polyhedral sets that define the underlying feasible region for linear programs.

We begin our discussion of the simplex method by showing that if an optimal solution exists, then an optimal extreme point also exists. Extreme points are then characterized in terms of basic feasible solutions. We then describe the simplex method for improving these solutions until optimality is reached, or else until we conclude that the optimal value is unbounded. Both geometric and algebraic descriptions are provided. The well-known tableau format of the simplex method is also discussed. This is a key chapter, fundamental to the development of many other chapters in the book.

3.1 EXTREME POINTS AND OPTIMALITY

We observed from Figure 1.3 that when an optimal solution of a linear programming problem exists, an optimal extreme point also exists. This observation is always true, as will be shown shortly.

Consider the following linear programming problem.

$$\text{Minimize} \quad \mathbf{cx}$$

$$\text{Subject to} \quad \mathbf{Ax} = \mathbf{b}$$
$$\mathbf{x} \geq \mathbf{0}$$

Let $\mathbf{x}_1, \mathbf{x}_2, \ldots, \mathbf{x}_k$ be the extreme points of the constraint set, and let $\mathbf{d}_1, \mathbf{d}_2, \ldots, \mathbf{d}_l$ be the extreme directions of the constraint set. Recall that any point $\mathbf{x}$ such that $\mathbf{Ax} = \mathbf{b}$ and $\mathbf{x} \geq \mathbf{0}$ can be represented as

$$\mathbf{x} = \sum_{j=1}^{k} \lambda_j \mathbf{x}_j + \sum_{j=1}^{l} \mu_j \mathbf{d}_j$$

where

$$\sum_{j=1}^{k} \lambda_j = 1$$

$$\lambda_j \geq 0 \quad j = 1, 2, \ldots, k$$
$$\mu_j \geq 0 \quad j = 1, 2, \ldots, l$$

Therefore the linear programming problem can be transformed into a problem in the variables $\lambda_1, \lambda_2, \ldots, \lambda_k, \mu_1, \mu_2, \ldots, \mu_l$, resulting in the following linear program.

$$\text{Minimize} \quad \sum_{j=1}^{k} (\mathbf{cx}_j)\lambda_j + \sum_{j=1}^{l} (\mathbf{cd}_j)\mu_j$$

$$\text{Subject to} \quad \sum_{j=1}^{k} \lambda_j = 1$$

$$\lambda_j \geq 0 \qquad j = 1, 2, \ldots, k$$
$$\mu_j \geq 0 \qquad j = 1, 2, \ldots, l$$

Since the μ_j's can be made arbitrarily large, the minimum is $-\infty$ if $\mathbf{cd}_j < 0$ for some $j = 1, 2, \ldots, l$. If $\mathbf{cd}_j \geq 0$ for all $j = 1, 2, \ldots, l$, then the corresponding μ_j can be chosen as zero. Now in order to minimize $\Sigma_{j=1}^{k}(\mathbf{cx}_j)\lambda_j$ over $\lambda_1, \lambda_2, \ldots, \lambda_k$ satisfying $\lambda_j \geq 0$ for $j = 1, 2, \ldots, k$, and $\Sigma_{j=1}^{k}\lambda_j = 1$, we simply find the Minimum $\mathbf{cx}_j$, say $\mathbf{cx}_p$, let $\lambda_p = 1$, and all other λ_j's equal to zero.

To summarize, the optimal solution of the linear problem is finite if and only if $\mathbf{cd}_j \geq 0$ for all extreme directions. Furthermore, if this is the case, then we find the minimum point by picking the minimum objective value among all extreme points. This shows that if an optimal solution exists, we must be able to find an optimal extreme point. Of course, if the Minimum $\mathbf{cx}_j$ occurs at more than one index, then each corresponding extreme point is an optimal point and each convex combination of these points is an optimal solution (why?). In fact, the entire set of alternative optimal solutions is given by the set of convex combinations of such points plus a nonnegative linear combination of the extreme directions $\mathbf{d}_j$ that satisfy $\mathbf{cd}_j = 0$ (why?).

Example 3.1

Consider the region defined by the following constraints:

$$\begin{aligned} -x_1 + x_2 &\le 2 \\ -x_1 + 2x_2 &\le 6 \\ x_1, \quad x_2 &\ge 0 \end{aligned}$$

Note that this region has three extreme points $\mathbf{x}_1$, $\mathbf{x}_2$, and $\mathbf{x}_3$, and two extreme directions $\mathbf{d}_1$ and $\mathbf{d}_2$ (see Figure 3.1). These are (without normalizing $\mathbf{d}_2$)

$$\mathbf{x}_1 = \begin{bmatrix} 0 \\ 0 \end{bmatrix} \quad \mathbf{x}_2 = \begin{bmatrix} 0 \\ 2 \end{bmatrix} \quad \mathbf{x}_3 = \begin{bmatrix} 2 \\ 4 \end{bmatrix}$$

$$\mathbf{d}_1 = \begin{bmatrix} 1 \\ 0 \end{bmatrix} \quad \mathbf{d}_2 = \begin{bmatrix} 2 \\ 1 \end{bmatrix}$$

Now suppose that we are minimizing $x_1 - 3x_2$ over the foregoing region. We see from Figure 3.1*a* that the problem is unbounded and has value $-\infty$. In this case we have

$$\mathbf{cx}_1 = (1, -3)\begin{bmatrix} 0 \\ 0 \end{bmatrix} = 0$$

$$\mathbf{cx}_2 = (1, -3)\begin{bmatrix} 0 \\ 2 \end{bmatrix} = -6$$

$$\mathbf{cx}_3 = (1, -3)\begin{bmatrix} 2 \\ 4 \end{bmatrix} = -10$$

$$\mathbf{cd}_1 = (1, -3)\begin{bmatrix} 1 \\ 0 \end{bmatrix} = 1$$

$$\mathbf{cd}_2 = (1, -3)\begin{bmatrix} 2 \\ 1 \end{bmatrix} = -1$$

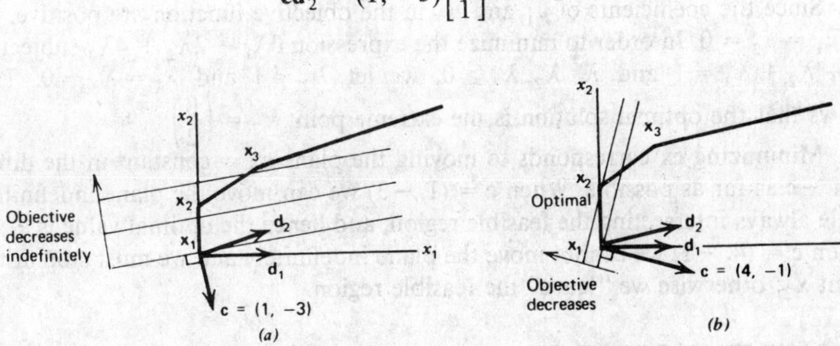

Figure 3.1. **Extreme directions and optimality: (*a*) Unbounded optimal solution. (*b*) Bounded optimal solution.**

The problem is equivalent to the following.

$$\begin{array}{ll} \text{Minimize} & 0\lambda_1 - 6\lambda_2 - 10\lambda_3 + \mu_1 - \mu_2 \\ \text{Subject to} & \lambda_1 + \lambda_2 + \lambda_3 = 1 \\ & \lambda_1, \quad \lambda_2, \quad \lambda_3, \quad \mu_1, \quad \mu_2 \geq 0 \end{array}$$

Since $\mathbf{cd}_2 = -1 < 0$ and μ_2 can be made arbitrarily large without violating the foregoing constraints, the objective value can be made $-\infty$ by letting $\mu_2 = \infty$. Then μ_1 can be chosen equal to zero. Any set of nonnegative $\lambda_1, \lambda_2, \lambda_3$ adding to 1 satisfies the foregoing constraints, for example, $\lambda_1 = 1$, $\lambda_2 = \lambda_3 = 0$. This illustrates the necessary and sufficient condition for unboundedness for a feasible linear program, namely $\mathbf{cd} < 0$ for some extreme direction $\mathbf{d}$.

Now consider the problem of minimizing $4x_1 - x_2$ over the same region. From Figure 3.1*b* the optimal solution is the extreme point $\mathbf{x}_2 = \begin{bmatrix} 0 \\ 2 \end{bmatrix}$. In this case we have

$$\mathbf{cx}_1 = (4, -1)\begin{bmatrix} 0 \\ 0 \end{bmatrix} = 0$$

$$\mathbf{cx}_2 = (4, -1)\begin{bmatrix} 0 \\ 2 \end{bmatrix} = -2$$

$$\mathbf{cx}_3 = (4, -1)\begin{bmatrix} 2 \\ 4 \end{bmatrix} = 4$$

$$\mathbf{cd}_1 = (4, -1)\begin{bmatrix} 1 \\ 0 \end{bmatrix} = 4$$

$$\mathbf{cd}_2 = (4, -1)\begin{bmatrix} 2 \\ 1 \end{bmatrix} = 7$$

The problem is therefore equivalent to the following.

$$\begin{array}{ll} \text{Minimize} & 0\lambda_1 - 2\lambda_2 + 4\lambda_3 + 4\mu_1 + 7\mu_2 \\ \text{Subject to} & \lambda_1 + \lambda_2 + \lambda_3 = 1 \\ & \lambda_1, \quad \lambda_2, \quad \lambda_3, \quad \mu_1, \quad \mu_2 \geq 0 \end{array}$$

Since the coefficients of μ_1 and μ_2 in the objective function are positive, we let $\mu_1 = \mu_2 = 0$. In order to minimize the expression $0\lambda_1 - 2\lambda_2 + 4\lambda_3$ subject to $\lambda_1 + \lambda_2 + \lambda_3 = 1$ and $\lambda_1, \lambda_2, \lambda_3 \geq 0$, we let $\lambda_2 = 1$ and $\lambda_1 = \lambda_3 = 0$. This shows that the optimal solution is the extreme point $\mathbf{x}_2 = \begin{pmatrix} 0 \\ 2 \end{pmatrix}$.

Minimizing $\mathbf{cx}$ corresponds to moving the plane $\mathbf{cx}$ = constant in the direction $-\mathbf{c}$ as far as possible. When $\mathbf{c} = (1, -3)$ we can move the plane indefinitely while always intersecting the feasible region, and hence the optimal value is $-\infty$. When $\mathbf{c} = (4, -1)$ we cannot move the plane indefinitely and we must stop at the point $\mathbf{x}_2$; otherwise we "leave" the feasible region.

3.2 BASIC FEASIBLE SOLUTIONS

We have developed, in the previous section, a necessary and sufficient condition for an unbounded solution. We also showed that if an optimal solution exists, then an optimal extreme point also exists. The notion of an extreme point is a geometric notion, and an algebraic characterization of extreme points is needed before they can be utilized from a computational point of view.

In this section we introduce basic feasible solutions and show that they correspond to extreme points. This characterization will enable us to algebraically describe the simplex method.

Definition **(Basic Feasible Solutions)**

Consider the system $\mathbf{Ax} = \mathbf{b}$ and $\mathbf{x} \geq \mathbf{0}$, where $\mathbf{A}$ is an $m \times n$ matrix and $\mathbf{b}$ is an m vector. Suppose that rank $(\mathbf{A}, \mathbf{b})$ = rank $(\mathbf{A}) = m$. After possibly rearranging the columns of $\mathbf{A}$, let $\mathbf{A} = [\mathbf{B}, \mathbf{N}]$ where $\mathbf{B}$ is an $m \times m$ invertible matrix and $\mathbf{N}$ is an $m \times (n - m)$ matrix. The solution $\mathbf{x} = \begin{bmatrix} \mathbf{x}_B \\ \mathbf{x}_N \end{bmatrix}$ to the equations $\mathbf{Ax} = \mathbf{b}$, where

$$\mathbf{x}_B = \mathbf{B}^{-1}\mathbf{b}$$

and

$$\mathbf{x}_N = \mathbf{0}$$

is called a *basic solution* of the system. If $\mathbf{x}_B \geq \mathbf{0}$, then $\mathbf{x}$ is called a *basic feasible solution* of the system. Here $\mathbf{B}$ is called the *basic matrix* (or simply the basis) and $\mathbf{N}$ is called the *nonbasic matrix*. The components of $\mathbf{x}_B$ are called *basic variables* (or dependent variables) and the components of $\mathbf{x}_N$ are called *nonbasic variables* (or independent variables). If $\mathbf{x}_B > \mathbf{0}$, then $\mathbf{x}$ is called a *nondegenerate basic feasible solution*, and if at least one component of $\mathbf{x}_B$ is zero, then $\mathbf{x}$ is called a *degenerate basic feasible solution*.

The notion of a basic feasible solution is illustrated by the following two examples.

Example 3.2

(Basic Feasible Solutions)

Consider the polyhedral set defined by the following inequalities (and illustrated in Figure 3.2):

$$\begin{aligned} x_1 + x_2 &\leq 6 \\ x_2 &\leq 3 \\ x_1, x_2 &\geq 0 \end{aligned}$$

By introducing the slack variables x_3 and x_4, the problem is put in the following standard format:

$$\begin{aligned} x_1 + x_2 + x_3 \qquad &= 6 \\ x_2 \qquad + x_4 &= 3 \\ x_1, \quad x_2, \quad x_3, \quad x_4 &\geq 0 \end{aligned}$$

Note that the constraint matrix $\mathbf{A} = [\mathbf{a}_1, \mathbf{a}_2, \mathbf{a}_3, \mathbf{a}_4] = \begin{bmatrix} 1 & 1 & 1 & 0 \\ 0 & 1 & 0 & 1 \end{bmatrix}$. From the foregoing definition, basic feasible solutions correspond to finding a 2×2 basis

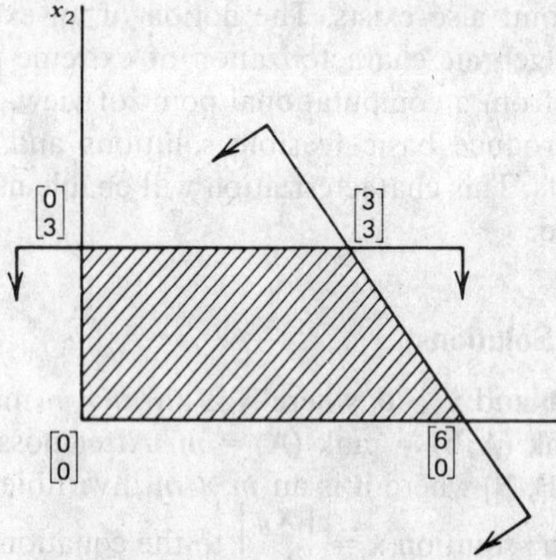

Figure 3.2. **Basic feasible solutions.**

B with nonnegative $\mathbf{B}^{-1}\mathbf{b}$. The following are the possible ways of extracting **B** out of **A**.

1. $\mathbf{B} = [\mathbf{a}_1, \mathbf{a}_2] = \begin{bmatrix} 1 & 1 \\ 0 & 1 \end{bmatrix}$

$$\mathbf{x}_B = \begin{bmatrix} x_1 \\ x_2 \end{bmatrix} = \mathbf{B}^{-1}\mathbf{b} = \begin{bmatrix} 1 & -1 \\ 0 & 1 \end{bmatrix} \begin{bmatrix} 6 \\ 3 \end{bmatrix} = \begin{bmatrix} 3 \\ 3 \end{bmatrix}, \qquad \mathbf{x}_N = \begin{bmatrix} x_3 \\ x_4 \end{bmatrix} = \begin{bmatrix} 0 \\ 0 \end{bmatrix}$$

2. $\mathbf{B} = [\mathbf{a}_1, \mathbf{a}_4] = \begin{bmatrix} 1 & 0 \\ 0 & 1 \end{bmatrix}$

$$\mathbf{x}_B = \begin{bmatrix} x_1 \\ x_4 \end{bmatrix} = \mathbf{B}^{-1}\mathbf{b} = \begin{bmatrix} 1 & 0 \\ 0 & 1 \end{bmatrix} \begin{bmatrix} 6 \\ 3 \end{bmatrix} = \begin{bmatrix} 6 \\ 3 \end{bmatrix}, \qquad \mathbf{x}_N = \begin{bmatrix} x_2 \\ x_3 \end{bmatrix} = \begin{bmatrix} 0 \\ 0 \end{bmatrix}$$

3. $\mathbf{B} = [\mathbf{a}_2, \mathbf{a}_3] = \begin{bmatrix} 1 & 1 \\ 1 & 0 \end{bmatrix}$

$$\mathbf{x}_B = \begin{bmatrix} x_2 \\ x_3 \end{bmatrix} = \mathbf{B}^{-1}\mathbf{b} = \begin{bmatrix} 0 & 1 \\ 1 & -1 \end{bmatrix} \begin{bmatrix} 6 \\ 3 \end{bmatrix} = \begin{bmatrix} 3 \\ 3 \end{bmatrix}, \qquad \mathbf{x}_N = \begin{bmatrix} x_1 \\ x_4 \end{bmatrix} = \begin{bmatrix} 0 \\ 0 \end{bmatrix}$$

4. $\mathbf{B} = [\mathbf{a}_2, \mathbf{a}_4] = \begin{bmatrix} 1 & 0 \\ 1 & 1 \end{bmatrix}$

$$\mathbf{x}_B = \begin{bmatrix} x_2 \\ x_4 \end{bmatrix} = \mathbf{B}^{-1}\mathbf{b} = \begin{bmatrix} 1 & 0 \\ -1 & 1 \end{bmatrix} \begin{bmatrix} 6 \\ 3 \end{bmatrix} = \begin{bmatrix} 6 \\ -3 \end{bmatrix}, \qquad \mathbf{x}_N = \begin{bmatrix} x_1 \\ x_3 \end{bmatrix} = \begin{bmatrix} 0 \\ 0 \end{bmatrix}$$

5. $\mathbf{B} = [\mathbf{a}_3, \mathbf{a}_4] = \begin{bmatrix} 1 & 0 \\ 0 & 1 \end{bmatrix}$

$$\mathbf{x}_B = \begin{bmatrix} x_3 \\ x_4 \end{bmatrix} = \mathbf{B}^{-1}\mathbf{b} = \begin{bmatrix} 1 & 0 \\ 0 & 1 \end{bmatrix} \begin{bmatrix} 6 \\ 3 \end{bmatrix} = \begin{bmatrix} 6 \\ 3 \end{bmatrix}, \qquad \mathbf{x}_N = \begin{bmatrix} x_1 \\ x_2 \end{bmatrix} = \begin{bmatrix} 0 \\ 0 \end{bmatrix}$$

Note that the points corresponding to 1, 2, 3, and 5 are basic feasible solutions. The point obtained in 4 is a basic solution, but is not feasible because it violates the nonnegativity restrictions. In other words, we have four basic feasible solutions, namely

$$\mathbf{x}_1 = \begin{bmatrix} 3 \\ 3 \\ 0 \\ 0 \end{bmatrix}, \quad \mathbf{x}_2 = \begin{bmatrix} 6 \\ 0 \\ 0 \\ 3 \end{bmatrix}, \quad \mathbf{x}_3 = \begin{bmatrix} 0 \\ 3 \\ 3 \\ 0 \end{bmatrix}, \quad \mathbf{x}_4 = \begin{bmatrix} 0 \\ 0 \\ 6 \\ 3 \end{bmatrix}$$

These points belong to E^4 since after introducing the slack variables we have four variables. These basic feasible solutions, projected in E^2—that is, in the (x_1, x_2) space—give rise to the following four points:

$$\begin{bmatrix} 3 \\ 3 \end{bmatrix}, \quad \begin{bmatrix} 6 \\ 0 \end{bmatrix}, \quad \begin{bmatrix} 0 \\ 3 \end{bmatrix}, \quad \begin{bmatrix} 0 \\ 0 \end{bmatrix}$$

These four points are illustrated in Figure 3.2. Note that these points are precisely the extreme points of the feasible region.

In this example, the possible number of basic feasible solutions is bounded by the number of ways of extracting two columns out of four columns to form the basis. Therefore the number of basic feasible solutions is less than or equal to

$$\binom{4}{2} = \frac{4!}{2!2!} = 6.$$

Out of these six possibilities, one point violates the nonnegativity of $\mathbf{B}^{-1}\mathbf{b}$. Furthermore, $\mathbf{a}_1$ and $\mathbf{a}_3$ could not have been used to form a basis since $\mathbf{a}_1 = \mathbf{a}_3 = \begin{bmatrix} 1 \\ 0 \end{bmatrix}$ are linearly dependent, and hence the matrix $\begin{bmatrix} 1 & 1 \\ 0 & 0 \end{bmatrix}$ does not qualify as a basis. This leaves four basic feasible solutions. In general, the number of basic feasible solutions is less than or equal to

$$\binom{n}{m} = \frac{n!}{m!(n-m)!}$$

There is another intuitive way of viewing basic solutions and basic feasible solutions. (This is made more precise in the sequel.) Each constraint, including the nonnegativity constraints, can be associated uniquely with a certain variable. Thus $x_1 \geq 0$ can be associated with the variable x_1, and the line $x_1 = 0$ is the boundary of the halfspace corresponding to $x_1 \geq 0$. Also, $x_1 + x_2 \leq 6$ can be associated with the variable x_3, and $x_3 = 0$ is the boundary of the halfspace

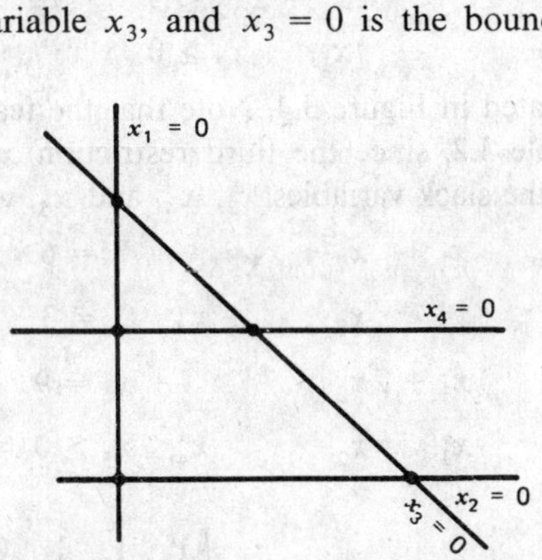

Figure 3.3. Associating basic solutions with nonbasic variables.

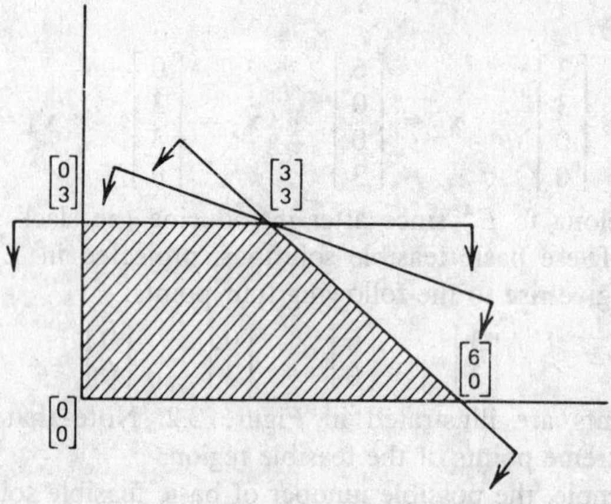

Figure 3.4. **Degenerate basic feasible solutions.**

corresponding to $x_1 + x_2 \leq 6$. Graphically portraying the boundary of the various constraints, we get the graph of Figure 3.3. Now, basic solutions correspond to the intersection of two lines in this graph. The lines correspond to the nonbasic variables. In the graph there are five intersections corresponding to five basic solutions. Note that there is no intersection of the lines $x_2 = 0$ and $x_4 = 0$ and thus no basic solution corresponding to these two variables being nonbasic. As soon as the feasible region is identified. we can distinguish the basic solutions from those that are also basic feasible solutions.

Example 3.3

(Degenerate Basic Feasible Solutions)

Consider the following system of inequalities:

$$\begin{aligned} x_1 + x_2 &\leq 6 \\ x_2 &\leq 3 \\ x_1 + 2x_2 &\leq 9 \\ x_1, \quad x_2 &\geq 0 \end{aligned}$$

This system is illustrated in Figure 3.4. Note that the feasible region is precisely the region of Example 3.2, since the third restriction $x_1 + 2x_2 \leq 9$ is "redundant." After adding the slack variables x_3, x_4, and x_5, we get

$$\begin{aligned} x_1 + x_2 + x_3 \qquad\qquad &= 6 \\ x_2 \qquad + x_4 \qquad &= 3 \\ x_1 + 2x_2 \qquad\qquad + x_5 &= 9 \\ x_1, \quad x_2, \quad x_3, \quad x_4, \quad x_5 &\geq 0 \end{aligned}$$

Note that

$$\mathbf{A} = [\mathbf{a}_1, \mathbf{a}_2, \mathbf{a}_3, \mathbf{a}_4, \mathbf{a}_5] = \begin{bmatrix} 1 & 1 & 1 & 0 & 0 \\ 0 & 1 & 0 & 1 & 0 \\ 1 & 2 & 0 & 0 & 1 \end{bmatrix}.$$

Let us consider the basic feasible solution with $\mathbf{B} = [\mathbf{a}_1, \mathbf{a}_2, \mathbf{a}_3]$.

$$\mathbf{x}_B = \begin{bmatrix} x_1 \\ x_2 \\ x_3 \end{bmatrix} = \begin{bmatrix} 1 & 1 & 1 \\ 0 & 1 & 0 \\ 1 & 2 & 0 \end{bmatrix}^{-1} \begin{bmatrix} 6 \\ 3 \\ 9 \end{bmatrix} = \begin{bmatrix} 0 & -2 & 1 \\ 0 & 1 & 0 \\ 1 & 1 & -1 \end{bmatrix} \begin{bmatrix} 6 \\ 3 \\ 9 \end{bmatrix} = \begin{bmatrix} 3 \\ 3 \\ 0 \end{bmatrix}$$

$$\mathbf{x}_N = \begin{bmatrix} x_4 \\ x_5 \end{bmatrix} = \begin{bmatrix} 0 \\ 0 \end{bmatrix}$$

Note that this basic feasible solution is degenerate since the basic variable $x_3 = 0$. Now consider the basic feasible solution with $\mathbf{B} = [\mathbf{a}_1, \mathbf{a}_2, \mathbf{a}_4]$.

$$\mathbf{x}_B = \begin{bmatrix} x_1 \\ x_2 \\ x_4 \end{bmatrix} = \begin{bmatrix} 1 & 1 & 0 \\ 0 & 1 & 1 \\ 1 & 2 & 0 \end{bmatrix}^{-1} \begin{bmatrix} 6 \\ 3 \\ 9 \end{bmatrix} = \begin{bmatrix} 2 & 0 & -1 \\ -1 & 0 & 1 \\ 1 & 1 & -1 \end{bmatrix} \begin{bmatrix} 6 \\ 3 \\ 9 \end{bmatrix} = \begin{bmatrix} 3 \\ 3 \\ 0 \end{bmatrix}$$

$$\mathbf{x}_N = \begin{bmatrix} x_3 \\ x_5 \end{bmatrix} = \begin{bmatrix} 0 \\ 0 \end{bmatrix}$$

Note that this basis gives rise to the same basic feasible solution obtained by $\mathbf{B} = [\mathbf{a}_1, \mathbf{a}_2, \mathbf{a}_3]$. It can be also checked that the basic feasible solution with basis $\mathbf{B} = [\mathbf{a}_1, \mathbf{a}_2, \mathbf{a}_5]$ is given by

$$\mathbf{x}_B = \begin{bmatrix} x_1 \\ x_2 \\ x_5 \end{bmatrix} = \begin{bmatrix} 3 \\ 3 \\ 0 \end{bmatrix} \qquad \mathbf{x}_N = \begin{bmatrix} x_3 \\ x_4 \end{bmatrix} = \begin{bmatrix} 0 \\ 0 \end{bmatrix}$$

Note that all three of the foregoing *bases* represent the single extreme point or basic feasible solution $(x_1, x_2, x_3, x_4, x_5) = (3, 3, 0, 0, 0)$. This basic feasible solution is degenerate since each associated basis involves a basic variable at level zero. The remaining extreme points of Figure 3.4 correspond to nondegenerate basic feasible solutions (why?). The reader should also note that degeneracy is not always simply the result of redundant constraints. (See Figure 2.14, for example.)

Correspondence Between Basic Feasible Solutions and Extreme Points

We shall now show that the collection of basic feasible solutions and the collection of extreme points are equivalent. In other words, a point is a basic feasible solution if and only if it is an extreme point. Since a linear programming problem with a finite optimal value has an optimal solution at an extreme point, an optimal basic feasible solution can always be found for such problems.

Consider the following problem.

$$\begin{aligned} &\text{Minimize} \quad \mathbf{cx} \\ &\text{Subject to} \quad \mathbf{Ax} = \mathbf{b} \\ &\qquad\qquad\quad\ \mathbf{x} \geq \mathbf{0} \end{aligned}$$

where $\mathbf{A}$ is an $m \times n$ matrix with rank m. Let $\mathbf{x}$ be an extreme point of the feasible region. We shall show that $\mathbf{x}$ is also a basic feasible solution of the system $\mathbf{Ax} = \mathbf{b}$, $\mathbf{x} \geq \mathbf{0}$.

By the definition given in Section 2.6, there are some n linearly independent defining hyperplanes binding at $\mathbf{x}$. Since $\mathbf{Ax} = \mathbf{b}$ provides m linear independent binding hyperplanes, there must be some $p = n - m$ additional binding defining hyperplanes from the nonnegativity constraints that together with $\mathbf{Ax} = \mathbf{b}$ provide n linearly independent defining hyperplanes binding at $\mathbf{x}$. Denoting these p additional hyperplanes by $\mathbf{x}_N = \mathbf{0}$, we therefore conclude that the system $\mathbf{Ax} = \mathbf{b}$, $\mathbf{x}_N = \mathbf{0}$ has $\mathbf{x}$ as the unique solution. Now let $\mathbf{N}$ represent the columns of the variables $\mathbf{x}_N$ in $\mathbf{A}$ and let $\mathbf{B}$ be the remaining columns of $\mathbf{A}$ with $\mathbf{x}_B$ as the associated variables. Since $\mathbf{Ax} = \mathbf{b}$ can be written as $\mathbf{Bx}_B + \mathbf{Nx}_N = \mathbf{b}$, this means that $\mathbf{B}$ is $m \times m$ and invertible, and moreover, $\mathbf{x}_B = \mathbf{B}^{-1}\mathbf{b} \geq \mathbf{0}$ since $\mathbf{x} = (\mathbf{x}_B, \mathbf{x}_N)$ is a feasible solution. Therefore $\mathbf{x}$ is a basic feasible solution.

Conversely, suppose that $\mathbf{x}$ is a basic feasible solution of the system $\mathbf{Ax} = \mathbf{b}$, $\mathbf{x} \geq \mathbf{0}$. We want to show that $\mathbf{x}$ is an extreme point. By definition, $\mathbf{x} = (\mathbf{x}_B, \mathbf{x}_N)$ where correspondingly $\mathbf{A} = (\mathbf{B}, \mathbf{N})$ such that $\mathbf{x}_B = \mathbf{B}^{-1}\mathbf{b} \geq \mathbf{0}$ and $\mathbf{x}_N = \mathbf{0}$. But this means that the n hyperplanes $\mathbf{Ax} = \mathbf{b}$, $\mathbf{x}_N = \mathbf{0}$ are binding at $\mathbf{x}$ and are moreover linearly independent since $\mathbf{B}^{-1}$ exists. Hence by the definition in Section 2.6, $\mathbf{x}$ is an extreme point. Therefore we have shown that every basis feasible solution is an extreme point and vice versa. Exercise 3.9 asks the reader to construct an alternative proof for this characterization.

Every basic feasible solution is equivalent to an extreme point. However, there may exist more than one basis corresponding to the same basic feasible solution or extreme point. A case of this type can occur in the presence of degeneracy, as illustrated in Example 3.3. Referring to the preceding proof, this case corresponds to that of an extreme point at which some $r > p \equiv n - m$ defining hyperplanes from $\mathbf{x} \geq \mathbf{0}$ are binding. Hence for any associated basis, $(r - p)$ of the $\mathbf{x}_B$ variables are also zero. Consequently, the number of positive variables is $q = m - (r - p) < m$. In this case, each possible choice of a basis $\mathbf{B}$ that includes the columns of these q positive variables represents this point (why?).

Clearly, if there exists more than one basis representing an extreme point, then this extreme point is degenerate. However, the converse is *not* necessarily true. Consider the following example of a polyhedral set:

$$\begin{aligned} x_1 + x_2 + x_3 &= 1 \\ -x_1 + x_2 + x_3 &= 1 \\ x_1, \quad x_2, \quad x_3 &\geq 0 \end{aligned}$$

Consider the solution $\bar{\mathbf{x}} = (0, 1, 0)$. Observe that this is an extreme point or a basic feasible solution with a corresponding basis with x_1 and x_2 as basic variables. Moreover, this is a degenerate extreme point. There are four defining hyperplanes binding at $\bar{\mathbf{x}}$. Moreover, there are three ways of choosing three linearly independent hyperplanes from this set that yield $\bar{\mathbf{x}}$ as the (unique) solution (how?). However, the basis associated with $\bar{\mathbf{x}}$ is unique.

In this example, the constraints $\mathbf{Ax} = \mathbf{b}$ imply that $x_1 = 0$. However, after fixing $x_1 = 0$, we no longer have a full rank system. If this were not the case, then for any degenerate extreme point, there would be more than one corresponding basis. To see this, consider the system $\mathbf{Ax} = \mathbf{b}$, $\mathbf{x} \geq \mathbf{0}$, where $\mathbf{A}$ is $m \times n$ of rank m. Let $\mathbf{x}$ be a degenerate extreme point solution, and let $\mathbf{x}_B$ and $\mathbf{x}_N$ be a set of basic and nonbasic variables respectively corresponding to a particular basis. As usual, let $\mathbf{B}$ and $\mathbf{N}$ be the columns of $\mathbf{A}$ associated with the variables $\mathbf{x}_B$ and $\mathbf{x}_N$ respectively. Then the given system is equivalent to

$$\mathbf{B}^{-1}\mathbf{N}\mathbf{x}_N + \mathbf{x}_B = \mathbf{B}^{-1}\mathbf{b} \equiv \bar{\mathbf{b}}$$

$$\mathbf{x}_N, \mathbf{x}_B \geq 0$$

Consider a degenerate basic variable x_{B_r} (with $\bar{b}_r = 0$) that is such that $\mathbf{Ax} = \mathbf{b}$ does not necessarily imply that $x_{B_r} = 0$. Given that such a variable exists, we will construct another basis representing this point. Let x_k be some component of $\mathbf{x}_N$ that has a nonzero component θ_r in the row corresponding to x_{B_r}. Note that x_k exists. Then consider a new choice of $(n - m)$ nonbasic variables given by x_{B_r} and $\mathbf{x}_{N-k}$, where $\mathbf{x}_{N-k}$ represents the components of $\mathbf{x}_N$ other than x_k. Putting $x_{B_r} = 0$ and $\mathbf{x}_{N-k} = \mathbf{0}$ above *uniquely* gives $x_k = \bar{b}_r/\theta_r = 0$ from row r, and so $x_{B_i} = \bar{b}_i$ is obtained as before from the other rows. Hence this corresponds to an alternative basis that represents the same extreme point. Finally note that if no degenerate basic variable x_{B_r} of this type exists, then there is only one basis that represents this extreme point.

Existence of Basic Feasible Solutions

Since a nonempty set $X = \{\mathbf{x} : \mathbf{Ax} = \mathbf{b}, \mathbf{x} \geq \mathbf{0}\}$ has extreme points (from Theorem 2.1), we know that the system $\mathbf{Ax} = \mathbf{b}$, $\mathbf{x} \geq \mathbf{0}$ has at least one basic feasible solution. In fact, starting from any feasible point $\mathbf{x} \in X$, the proof of Theorem 2.1 shows how to *constructively* obtain an extreme point and hence a basic feasible solution of X. Assuming without loss of generality, that rank $(\mathbf{A}) = m$, we can examine the binding nonnegativity constraints at this extreme point, and select $p = (n - m)$ of them to correspond to the variables $\mathbf{x}_N$, such that the system $\mathbf{Ax} = \mathbf{b}$, $\mathbf{x}_N = \mathbf{0}$ has rank n (how?). Calling the remaining m variables $\mathbf{x}_B$ as basic variables, we obtain a corresponding basis.

Summary of Results

Let us summarize some of the important facts about the following linear programming problem, where $\mathbf{A}$ is an $m \times n$ matrix with rank m.

$$\begin{aligned} \text{Minimize} \quad & \mathbf{cx} \\ \text{Subject to} \quad & \mathbf{Ax} = \mathbf{b} \\ & \mathbf{x} \geq \mathbf{0} \end{aligned}$$

***Theorem* 3.1**

The collection of extreme points corresponds to the collection of basic feasible solutions, and both are nonempty, provided that the feasible region is not empty.

***Theorem* 3.2**

Assume that the feasible region is nonempty. Then a finite optimal solution exists if and only if $\mathbf{cd}_j \geq 0$ for $j = 1, 2, \ldots, l$, where $\mathbf{d}_1, \ldots, \mathbf{d}_l$ are the extreme directions of the feasible region. Otherwise, the optimal solution value is unbounded.

***Theorem* 3.3**

If an optimal solution exists, then an optimal extreme point (or equivalently an optimal basic feasible solution) exists.

***Theorem* 3.4**

For every extreme point (basic feasible solution) there corresponds a basis (not necessarily unique), and, conversely, for every basis there corresponds a (unique) extreme point. Moreover, if an extreme point has more than one basis representing it, then it is degenerate. Conversely, a degenerate extreme point has more than one basis representing it if and only if the system $\mathbf{Ax} = \mathbf{b}$ itself does not imply that the degenerate basic variables corresponding to an associated basis are identically zero.

Now since the extreme points may be enumerated by algebraically enumerating all basic feasible solutions that are bounded by $\binom{n}{m}$, one may think of simply listing all basic feasible solutions and picking the one with the minimal objective value. This is unsatisfactory, however, for a number of reasons. First, the number of basic feasible solutions is bounded by $\binom{n}{m}$, which is large, even for moderate values of m and n. Second, this simple approach does not tell us if the problem has an unbounded solution that may occur if the feasible region is unbounded. Last, if the feasible region is empty and if we apply the foregoing "simple-minded procedure," we shall discover that the feasible region is empty, only after all possible ways of extracting m columns out of n columns of the matrix $\mathbf{A}$ fail to produce a basic feasible solution, on the grounds that $\mathbf{B}$ does not have an inverse, or else $\mathbf{B}^{-1}\mathbf{b} \ngeq \mathbf{0}$.

The simplex method is a clever procedure that moves from an extreme point to another extreme point, with a better (at least not worse) objective. It also discovers whether the feasible region is empty and whether the optimal solution is unbounded. In practice, the method only enumerates a small portion of the extreme points of the feasible region. The key to this method is exposed next.

3.3 KEY TO THE SIMPLEX METHOD

The key to the simplex method lies in recognizing the optimality of a given extreme point solution based on local considerations without having to (globally) enumerate all extreme points or basic feasible solutions.

Consider the following linear programming problem

LP: Minimize $\mathbf{cx}$

Subject to $\mathbf{Ax} = \mathbf{b}$
$\mathbf{x} \geq \mathbf{0}$

where $\mathbf{A}$ is an $m \times n$ matrix with rank m. Suppose that we have a basic feasible solution $\begin{pmatrix} \mathbf{B}^{-1}\mathbf{b} \\ \mathbf{0} \end{pmatrix}$ whose objective value z_0 is given by

$$z_0 = \mathbf{c}\begin{pmatrix} \mathbf{B}^{-1}\mathbf{b} \\ \mathbf{0} \end{pmatrix} = (\mathbf{c}_B, \mathbf{c}_N)\begin{pmatrix} \mathbf{B}^{-1}\mathbf{b} \\ \mathbf{0} \end{pmatrix} = \mathbf{c}_B\mathbf{B}^{-1}\mathbf{b} \tag{3.1}$$

Now let $\mathbf{x}_B$ and $\mathbf{x}_N$ denote the set of basic and nonbasic variables for the given basis. Then feasibility requires that $\mathbf{x}_B \geq \mathbf{0}$, $\mathbf{x}_N \geq \mathbf{0}$, and that $\mathbf{b} = \mathbf{Ax} = \mathbf{Bx}_B + \mathbf{Nx}_N$. Multiplying the last equation by $\mathbf{B}^{-1}$ and rearranging the terms, we get

$$\begin{aligned} \mathbf{x}_B &= \mathbf{B}^{-1}\mathbf{b} - \mathbf{B}^{-1}\mathbf{N}\mathbf{x}_N \\ &= \mathbf{B}^{-1}\mathbf{b} - \sum_{j \in R} \mathbf{B}^{-1}\mathbf{a}_j x_j \\ &= \bar{\mathbf{b}} - \sum_{j \in R} (\mathbf{y}_j) x_j, \quad \text{say,} \end{aligned} \tag{3.2}$$

where R is the current set of the indices of the nonbasic variables. Noting Equations (3.2) and (3.1) and letting z denote the objective function value, we get

$$\begin{aligned} z &= \mathbf{cx} \\ &= \mathbf{c}_B\mathbf{x}_B + \mathbf{c}_N\mathbf{x}_N \\ &= \mathbf{c}_B\left(\mathbf{B}^{-1}\mathbf{b} - \sum_{j \in R} \mathbf{B}^{-1}\mathbf{a}_j x_j\right) + \sum_{j \in R} c_j x_j \\ &= z_0 - \sum_{j \in R} (z_j - c_j) x_j \end{aligned} \tag{3.3}$$

where $z_j = \mathbf{c}_B\mathbf{B}^{-1}\mathbf{a}_j$ for each nonbasic variable.

Using the foregoing transformations, the linear programming problem LP may be rewritten as

$$\begin{aligned} &\text{Minimize} \quad z = z_0 - \sum_{j \in R} (z_j - c_j) x_j \\ &\text{Subject to} \quad \sum_{j \in R} (\mathbf{y}_j) x_j + \mathbf{x}_B = \bar{\mathbf{b}} \\ &\qquad x_j \geq 0, \ j \in R, \quad \text{and} \quad \mathbf{x}_B \geq \mathbf{0} \end{aligned} \tag{3.4}$$

Without loss of generality, let us assume that no row in Equation (3.4) has all zeros in the columns of the nonbasic variables x_j, $j \in R$. Otherwise the basic variable in such a row is known in value, and this row can be deleted from the problem. Now observe that the variables $\mathbf{x}_B$ simply play the role of slack variables in Equation (3.4). Hence we can equivalently write LP in the *nonbasic variable space*, that is, in terms of the nonbasic variables as follows.

$$\text{Minimize} \quad z = z_0 - \sum_{j \in R} (z_j - c_j)x_j$$

$$\begin{aligned} \text{Subject to} \quad & \sum_{j \in R} (\mathbf{y}_j)x_j \leq \bar{\mathbf{b}} \\ & x_j \geq 0, \qquad j \in R \end{aligned} \tag{3.5}$$

Note that the number of nonbasic variables is $p = (n - m)$ and so we have represented LP in some p-dimensional space. This is to be expected since there are p independent variables or p *degrees of freedom* in our constraint system. The values $(c_j - z_j)$ are sometimes referred to as *reduced cost coefficients* since they are the coefficients of the (nonbasic) variables in this reduced space. The representation (3.4) in which the objective function z and the basic variables $\mathbf{x}_B$ have been solved for in terms of the nonbasic variables is referred to as a representation of the basic solution in *canonical form*. The key result now simply says the following:

$$\text{If } (z_j - c_j) \leq 0 \text{ for all } j \in R, \text{ then the current basic feasible solution is optimal.} \tag{3.6}$$

This should be clear by noting that since $z_j - c_j \leq 0$ for all $j \in R$, we have $z \geq z_0$ for any feasible solution, and for the current (basic) feasible solution, we know that $z = z_0$ since $x_j = 0$ for all $j \in R$.

3.4 GEOMETRIC MOTIVATION OF THE SIMPLEX METHOD

It is instructive to examine the foregoing operations geometrically. Observe that in the nonbasic variable space representation, the feasible region of LP is defined in terms of n intersecting halfspaces, m associated with the inequality constraints in Equation (3.5), and $p = n - m$ associated with the nonnegativity constraints. Associated with each of these halfspaces, there is a certain *defining variable* that takes on a value of zero on the corresponding defining hyperplane, and takes on nonnegative values over the corresponding halfspace. For the inequalities in Equation (3.5), these defining variables are the basic variables $\mathbf{x}_B$, and for the nonnegativity constraints, these defining variables are the variables x_j, $j \in R$ themselves.

Figure 3.5 illustrates a situation in $p = 2$ dimensions, where $n = 6$, $m = 4$, $R = \{1, 2\}$ and where $\mathbf{x}_B = \{x_3, x_4, x_5, x_6\}$. The defining variables associated with the $n = 6$ defining hyperplanes are shown against the corresponding hyperplanes in this figure. The extreme points or vertices of the feasible region are labeled as $\mathbf{v}_1, \ldots, \mathbf{v}_5$ and $\bar{\mathbf{c}}$ is the reduced cost vector. Note that the feasible region is defined by the restrictions $x_j \geq 0$, $j = 1, \ldots, n$ in this space. The current basic feasible solution corresponds to the vertex $\mathbf{v}_1$, that is, the origin. Notice also that

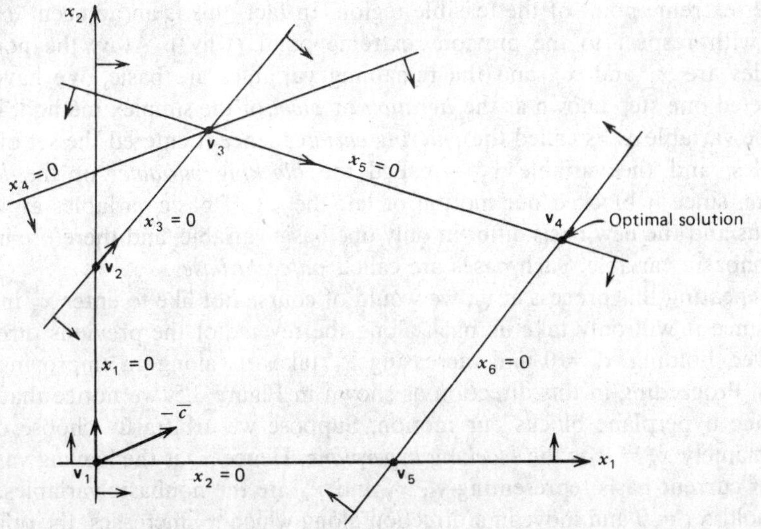

Figure 3.5. Geometric motivation.

every other extreme point is also defined by the intersection of some $p = 2$ linearly independent hyperplanes, with the corresponding p defining variables being the nonbasic variables. Hence for the vertex $\mathbf{v}_2$ defined by $x_1 = 0$ and $x_3 = 0$, the nonbasic variables are x_1 and x_3, and the basic variables are x_2, x_4, x_5, and x_6. On the other hand, note that the degenerate vertex $\mathbf{v}_3$ has three defining hyperplanes passing through it, any two of which uniquely define this extreme point as their intersection. Hence there are three bases representing $\mathbf{v}_3$. (Enumerate them.)

Now consider the origin $\mathbf{v}_1$ and examine the p defining hyperplanes associated with the corresponding nonbasic variables. Holding $(p - 1)$ of these hyperplanes binding and moving in a direction feasible to the remaining hyperplane takes us along a one-dimensional ray with vertex at $\mathbf{v}_1$. There are p such rays. Hence, at $\mathbf{v}_1$, since $(p - 1) = 1$, holding $x_2 = 0$ and increasing x_1 takes us along the x_1 axis, with the objective function changing at a rate of $\partial z/\partial x_1 = -(z_1 - c_1) = \bar{c}_1 < 0$. Similarly, holding $x_1 = 0$ and increasing x_2 takes us along the x_2 axis, with the objective function changing at a rate of $\partial z/\partial x_2 = -(z_2 - c_2) = \bar{c}_2 < 0$. Hence either ray describes an attractive direction of motion. Suppose we choose the latter direction of motion. Since this is a feasible direction, it takes us along an edge of the feasible region (why?). Naturally we would like to move as far as possible along this edge since $\partial z/\partial x_2 = \bar{c}_2$ is a negative constant. However, our motion is *blocked* by the hyperplane $x_3 = 0$, since x_3 has been driven to zero and moving any further would drive x_3 negative. (Of course, if no such *blocking hyperplane* existed, then the optimal solution value would have been unbounded.) Furthermore, since $(p - 1)$ linearly independent hyperplanes were binding all along and we were blocked by a new hyperplane, we now have p linearly independent hyperplanes binding, and so we are at another extreme point of the feasible region. In fact, this is an adjacent extreme point with respect to the previous extreme point (why?). At $\mathbf{v}_2$ the nonbasic variables are x_1 and x_3, and the remaining variables are basic. We have now completed one step known as the *iteration* or *pivot* of the simplex method. In this step the variable x_2 is called the *entering variable* since it entered the set of basic variables, and the variable x_3 is called the *blocking variable*, or the *leaving variable*, since it blocked our motion or left the set of basic variables at $\mathbf{v}_1$. The previous and the new basis differ in only one basic variable, and therefore in only one nonbasic variable. Such bases are called *adjacent bases*.

Repeating this process at $\mathbf{v}_2$, we would of course not like to enter x_3 into the basis since it will only take us back along the reverse of the previous direction. However, holding $x_3 = 0$ and increasing x_1 takes us along an improving edge (why?). Proceeding in this direction as shown in Figure 3.5, we notice that more than one hyperplane blocks our motion. Suppose we arbitrarily choose one of these, namely $x_4 = 0$ as the *blocking hyperplane*. Hence x_4 is the leaving variable, and for current basis representing $\mathbf{v}_3$, x_3 and x_4 are the nonbasic variables. Now if we hold $x_4 = 0$ and move in a direction along which x_3 increases, the objective function value decreases since this direction is making an acute angle with $-\bar{\mathbf{c}}$. However, this direction is not feasible. We are blocked by the hyperplane $x_5 = 0$ even before we begin to move! That is, x_5 leaves the basis giving x_4 and x_5 as our new nonbasic variables, while we are still at the same vertex $\mathbf{v}_3$. Such a step in which one changes from one basis to an adjacent basis, both representing the *same* extreme point solution, is called a *degenerate iteration* or a *degenerate pivot*. While such a pivot is undesirable, it may not always be avoidable, unlike as in Figure 3.5 (how?). Certainly, one would not like to be stuck in an infinite loop at an extreme point, performing a sequence of degenerate pivots that takes us in a closed loop of bases, all representing the same extreme point, with the objective function therefore remaining a constant. Such a phenomenon is called *cycling* and as shown in Chapter 4, it can indeed occur. Special precautions need to be taken for its prevention.

With x_4 and x_5 as the nonbasic variables at $\mathbf{v}_3$, holding $x_5 = 0$ and increasing x_4 takes us along an improving, feasible edge direction and therefore results in a *nondegenerate pivot*. The blocking or leaving variable that gets driven to zero is x_6, and the new nonbasic variables are x_5 and x_6. Observe that corresponding to this basis, none of the p rays, defined by holding some $(p - 1)$ nonbasic variables equal to zero and increasing the remaining nonbasic variable, lead to an improvement in the objective function, and so our "key result" declares $\mathbf{v}_4$ to be optimal. In fact, $\mathbf{v}_4$ is optimal in the polyhedral cone formed by the halfspaces having the current nonbasic variables as defining variables, which indeed contains the feasible region (why?). A path followed by the simplex algorithm along edges of the polyhedron (from $\mathbf{v}_1$ to $\mathbf{v}_2$ to $\mathbf{v}_3$ to $\mathbf{v}_4$ in the previous example) is known as a *simplex path*.

The reader may note that when we implement the simplex method algebraically, we will be representing the linear program in the nonbasic variable space at every iteration, just as we did at the extreme point $\mathbf{v}_1$. This will conveniently give us the rate of change in objective function value along each of the p incident rays directly from the corresponding reduced cost vector, and so at

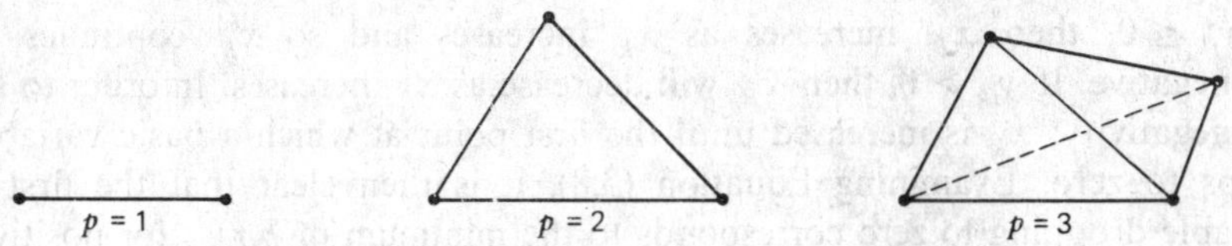

Figure 3.6. **Examples of simplices in E_p, $p = 1, 2, 3$.**

each iteration, we will be inquiring whether the current origin is optimal or whether we ought to consider moving along one of the current "axis" directions.

Finally let us point out the origin of the term "simplex" in this algorithm (which is not an antonym of "complex"!). A *simplex* in p dimensions is the convex hull of a set of $(p + 1)$ noncoplanar points in E^p, that is, points not all lying on the same hyperplane in E^p. Hence for $p = 1$ this is a line segment, for $p = 2$ it is a triangle, and for $p = 3$ it is a tetrahedron, as shown in Figure 3.6. In particular, given a basis corresponding to an extreme point, examine for convenience the feasible region represented in the space of the nonbasic variables. Note that the convex hull of the current vertex which is the origin, and some p points with one point chosen along each axis defines a simplex in E^p (why?). In fact, by examining the axis directions, the simplex method is verifying whether the origin is optimal for this simplex or whether any of the other vertices of the simplex have a better objective value. Consequently, the simplex method proceeds by examining one such simplex after another, each time inquiring whether the current vertex of the present simplex is optimal for this simplex or not.

3.5 ALGEBRA OF THE SIMPLEX METHOD

We will now translate the geometric process described in the foregoing section into algebra. Toward this end, consider the representation of the linear program LP in the nonbasic variable space, written in equality form as in Equation (3.4). If $(z_j - c_j) \leq 0$ for all $j \in R$, then $x_j = 0$, $j \in R$ and $\mathbf{x}_B = \bar{\mathbf{b}}$ is optimal for LP as noted in Equation (3.6). Otherwise while holding $(p - 1)$ nonbasic variables fixed at zero, the simplex method considers increasing the remaining variable, say, x_k. Naturally we would like $z_k - c_k$ to be positive, and perhaps the most positive of all the $z_j - c_j$, $j \in R$. Now fixing $x_j = 0$ for $j \in R - \{k\}$, we obtain from Equation (3.4) that

$$z = z_0 - (z_k - c_k)x_k \tag{3.7}$$

and

$$\begin{bmatrix} x_{B_1} \\ x_{B_2} \\ \vdots \\ x_{B_r} \\ \vdots \\ x_{B_m} \end{bmatrix} = \begin{bmatrix} \bar{b}_1 \\ \bar{b}_2 \\ \vdots \\ \bar{b}_r \\ \vdots \\ \bar{b}_m \end{bmatrix} - \begin{bmatrix} y_{1k} \\ y_{2k} \\ \vdots \\ y_{rk} \\ \vdots \\ y_{mk} \end{bmatrix} x_k \tag{3.8}$$

If $y_{ik} \leq 0$, then x_{B_i} increases as x_k increases and so x_{B_i} continues to be nonnegative. If $y_{ik} > 0$, then x_{B_i} will decrease as x_k increases. In order to satisfy nonnegativity, x_k is increased until the first point at which a basic variable x_{B_r} drops to zero. Examining Equation (3.8), it is then clear that the first basic variable dropping to zero corresponds to the minimum of $\bar{b}_i/y_{ik}$ for positive y_{ik}. More precisely, we can increase x_k until

$$x_k = \frac{\bar{b}_r}{y_{rk}} \equiv \underset{1 \leq i \leq m}{\text{Minimum}} \left\{ \frac{\bar{b}_i}{y_{ik}} : y_{ik} > 0 \right\} \tag{3.9}$$

In the absence of degeneracy, $\bar{b}_r > 0$, and hence $x_k = \bar{b}_r/y_{rk} > 0$. From Equation (3.7) and the fact that $z_k - c_k > 0$, it then follows that $z < z_0$ and the objective function strictly improves. As x_k increases from level 0 to $\bar{b}_r/y_{rk}$, a new feasible solution is obtained. Substituting $x_k = \bar{b}_r/y_{rk}$ in Equation (3.8) gives the following point:

$$x_{B_i} = \bar{b}_i - \frac{y_{ik}}{y_{rk}}\bar{b}_r \qquad i = 1, 2, \ldots, m$$

$$x_k = \frac{\bar{b}_r}{y_{rk}} \tag{3.10}$$

$$\text{All other } x_j\text{'s are zero}$$

From Equation (3.10), $x_{B_r} = 0$ and hence at most m variables are positive. The corresponding columns in $\mathbf{A}$ are $\mathbf{a}_{B_1}, \mathbf{a}_{B_2}, \ldots, \mathbf{a}_{B_{r-1}}, \mathbf{a}_k, \mathbf{a}_{B_{r+1}}, \ldots, \mathbf{a}_{B_m}$. Note that these columns are linearly independent since $y_{rk} \neq 0$. (Recall that if $\mathbf{a}_{B_1}, \ldots, \mathbf{a}_{B_r}, \ldots, \mathbf{a}_{B_m}$ are linearly independent, and if $\mathbf{a}_k$ replaces $\mathbf{a}_{B_r}$, then the new columns are linearly independent if and only if $y_{rk} \neq 0$; see Section 2.1). Therefore the point given by Equation (3.10) is a basic feasible solution.

To summarize, we have algebraically described an *iteration*, that is, the process of transforming from one basis to an adjacent basis. This is done by increasing the value of a nonbasic variable x_k with positive $z_k - c_k$ and adjusting the current basic variables. In the process, the variable x_{B_r} drops to zero. The variable x_k hence enters the basis and x_{B_r} leaves the basis. In the absence of degeneracy the objective function value strictly decreases and hence the basic feasible solutions generated are distinct. Since there is only a finite number of basic feasible solutions, the procedure would terminate in a finite number of steps.

Example 3.4

$$\begin{array}{ll} \text{Minimize} & x_1 + x_2 \\ \text{Subject to} & x_1 + 2x_2 \leq 4 \\ & x_2 \leq 1 \\ & x_1, \quad x_2 \geq 0 \end{array}$$

Introduce the slack variables x_3 and x_4 to put the problem in a standard form. This leads to the following constraint matrix **A**:

$$\mathbf{A} = [\mathbf{a}_1, \mathbf{a}_2, \mathbf{a}_3, \mathbf{a}_4] = \begin{bmatrix} 1 & 2 & 1 & 0 \\ 0 & 1 & 0 & 1 \end{bmatrix}$$

Consider the basic feasible solution corresponding to $\mathbf{B} = [\mathbf{a}_1, \mathbf{a}_2]$. In other words, x_1 and x_2 are the basic variables while x_3 and x_4 are the nonbasic variables. The representation of the problem in this nonbasic variable space as in Equation (3.4) with $R = \{3, 4\}$ may be obtained as follows. First, compute:

$$\mathbf{B}^{-1} = \begin{bmatrix} 1 & 2 \\ 0 & 1 \end{bmatrix}^{-1} = \begin{bmatrix} 1 & -2 \\ 0 & 1 \end{bmatrix}, \qquad \mathbf{c}_B\mathbf{B}^{-1} = (1, 1)\begin{bmatrix} 1 & -2 \\ 0 & 1 \end{bmatrix} = (1, -1)$$

Hence

$$\mathbf{y}_3 = \mathbf{B}^{-1}\mathbf{a}_3 = \begin{bmatrix} 1 & -2 \\ 0 & 1 \end{bmatrix}\begin{pmatrix} 1 \\ 0 \end{pmatrix} = \begin{bmatrix} 1 \\ 0 \end{bmatrix},$$

$$\mathbf{y}_4 = \mathbf{B}^{-1}\mathbf{a}_4 = \begin{bmatrix} 1 & -2 \\ 0 & 1 \end{bmatrix}\begin{pmatrix} 0 \\ 1 \end{pmatrix} = \begin{bmatrix} -2 \\ 1 \end{bmatrix},$$

and

$$\bar{\mathbf{b}} = \mathbf{B}^{-1}\mathbf{b} = \begin{bmatrix} 1 & -2 \\ 0 & 1 \end{bmatrix}\begin{pmatrix} 4 \\ 1 \end{pmatrix} = \begin{bmatrix} 2 \\ 1 \end{bmatrix}.$$

Also

$$z_0 = \mathbf{c}_B\mathbf{B}^{-1}\mathbf{b} = (1, -1)\begin{pmatrix} 4 \\ 1 \end{pmatrix} = 3$$

$$z_3 - c_3 = \mathbf{c}_B\mathbf{B}^{-1}\mathbf{a}_3 - c_3 = (1, -1)\begin{bmatrix} 1 \\ 0 \end{bmatrix} - 0 = 1$$

$$z_4 - c_4 = \mathbf{c}_B\mathbf{B}^{-1}\mathbf{a}_4 - c_4 = (1, -1)\begin{bmatrix} 0 \\ 1 \end{bmatrix} - 0 = -1$$

Hence the required representation of the problem is

$$\begin{array}{lrcl} \text{Minimize} & 3 - x_3 + x_4 & & \\ \text{Subject to} & x_3 - 2x_4 + x_1 & = & 2 \\ & x_4 \qquad + x_2 & = & 1 \\ & x_1, \quad x_2, \quad x_3, \quad x_4 & \geq & 0 \end{array}$$

The feasible region of this problem is shown in Figure 3.7 in both the original (x_1, x_2) space as well as in the current (x_3, x_4) space. Since $z_3 - c_3 > 0$, then the

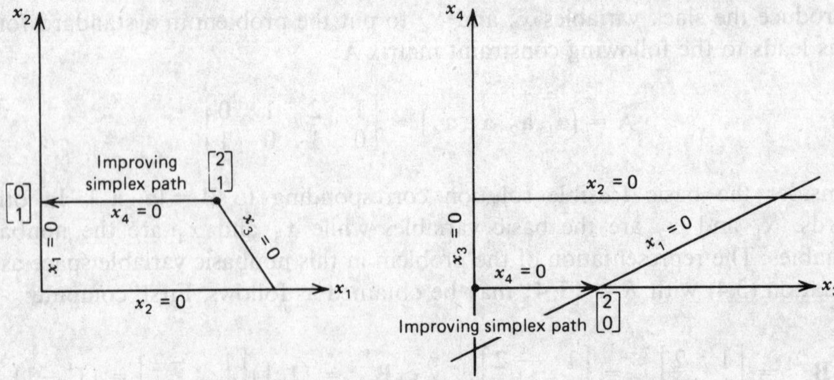

Figure 3.7. **Improving a basic feasible solution.**

objective improves by increasing x_3. The modified solution is given by

$$\mathbf{x}_B = \mathbf{B}^{-1}\mathbf{b} - \mathbf{B}^{-1}\mathbf{a}_3x_3$$

$$\begin{bmatrix}x_1\\x_2\end{bmatrix} = \begin{bmatrix}2\\1\end{bmatrix} - \begin{bmatrix}1\\0\end{bmatrix}x_3$$

The maximum value of x_3 is 2 (any larger value of x_3 will force x_1 to be negative). Therefore the new basic feasible solution is

$$(x_1, x_2, x_3, x_4) = (0, 1, 2, 0)$$

Here x_3 enters the basis and x_1 leaves the basis. Note that the new point has an objective value equal to 1, which is an improvement over the previous objective value of 3. The improvement is precisely $(z_3 - c_3)x_3 = 2$. The reader is asked to continue the improvement process starting from the new point.

Interpretation of Entering and Leaving the Basis

We now look more closely at the process of entering and leaving the basis and their interpretation.

INTERPRETATION OF $z_k - c_k$

The criterion $z_k - c_k > 0$ for a nonbasic variable x_k to enter the basis will be used over and over again throughout the text. It will be helpful at this stage to review the definition of z_k and make a few comments on the meaning of the entry criterion $z_k - c_k > 0$. Recall that $z = \mathbf{c}_B\bar{\mathbf{b}} - (z_k - c_k)x_k$, where

$$z_k = \mathbf{c}_B\mathbf{B}^{-1}\mathbf{a}_k = \mathbf{c}_B\mathbf{y}_k = \sum_{i=1}^{m} c_{B_i}y_{ik} \tag{3.11}$$

and c_{B_i} is the cost of the ith basic variable. Note that if x_k is raised from zero level, while the other nonbasic variables are kept at zero level, then the basic variables $x_{B_1}, x_{B_2}, \ldots, x_{B_m}$ must be modified according to Equation (3.8). In other words, if x_k is increased by one unit, then $x_{B_1}, x_{B_2}, \ldots,$ and x_{B_m} will be decreased respectively by $y_{1k}, y_{2k}, \ldots, y_{mk}$ units (if $y_{ik} < 0$, then x_{B_i} will be increased). The saving (a negative saving means more cost) that results from the modification of the basic variables, as a result of increasing x_k by one unit, is therefore, $\Sigma_{i=1}^m c_{B_i} y_{ik}$, which is z_k (see Equation 3.11). However, the cost of increasing x_k itself by one unit is c_k. Hence $z_k - c_k$ is the saving minus the cost of increasing x_k by one unit. Naturally, if $z_k - c_k$ is positive, it will be to our advantage to increase x_k. For each unit of x_k, the cost will be reduced by an amount $z_k - c_k$ and hence it will be to our advantage to increase x_k as much as possible. On the other hand, if $z_k - c_k < 0$, then by increasing x_k, the net saving is negative, and this action will result in a larger cost. So this action is prohibited. Finally if $z_k - c_k = 0$, then increasing x_k will lead to a different solution with the same cost. So whether x_k is kept at zero level, or increased, no change in cost takes place.

Now suppose that x_k is a basic variable. In particular, suppose that x_k is the tth basic variable, that is, $x_k = x_{B_t}$, $c_k = c_{B_t}$, and $\mathbf{a}_k = \mathbf{a}_{B_t}$. Recall that $z_k = \mathbf{c}_B\mathbf{B}^{-1}\mathbf{a}_k = \mathbf{c}_B\mathbf{B}^{-1}\mathbf{a}_{B_t}$. But $\mathbf{B}^{-1}\mathbf{a}_{B_t}$ is a vector of zeros except for one at the tth position (see Exercise 2.13). Therefore, $z_k = c_{B_t}$, and hence $z_k - c_k = c_{B_t} - c_{B_t} = 0$.

Leaving the Basis and the Blocking Variable

Suppose that we decided to increase a nonbasic variable x_k with a positive $z_k - c_k$. From Equation (3.7), the larger the value of x_k, the smaller is the objective z. As x_k is increased, the basic variables are modified according to Equation (3.8). If the vector $\mathbf{y}_k$ has any positive component(s), then the corresponding basic variable(s) is decreased as x_k is increased. Therefore the nonbasic variable x_k cannot be indefinitely increased, because otherwise the nonnegativity of the basic variables will be violated. Recall that a first basic variable x_{B_r} that drops to zero is called the *blocking variable* because it blocks further increase of x_k. Thus x_k enters the basis and x_{B_r} leaves the basis.

Example 3.5

$$\begin{aligned} \text{Minimize} \quad & 2x_1 - x_2 \\ \text{Subject to} \quad & -x_1 + x_2 \leq 2 \\ & 2x_1 + x_2 \leq 6 \\ & x_1, \quad x_2 \geq 0 \end{aligned}$$

Introduce the slack variables x_3 and x_4. This leads to the following constraints:

$$\begin{aligned} -\,x_1 + x_2 + x_3 \qquad &= 2 \\ 2x_1 + x_2 \qquad + x_4 &= 6 \\ x_1, \quad x_2, \quad x_3, \quad x_4 &\geq 0 \end{aligned}$$

Consider the basic feasible solution with basis $\mathbf{B} = [\mathbf{a}_1, \mathbf{a}_2] = \begin{bmatrix} -1 & 1 \\ 2 & 1 \end{bmatrix}$ and

$$\mathbf{B}^{-1} = \begin{bmatrix} -\frac{1}{3} & \frac{1}{3} \\ \frac{2}{3} & \frac{1}{3} \end{bmatrix}$$

$$\mathbf{x}_B = \mathbf{B}^{-1}\mathbf{b} - \mathbf{B}^{-1}\mathbf{N}\mathbf{x}_N$$

$$\begin{bmatrix} x_{B_1} \\ x_{B_2} \end{bmatrix} = \begin{bmatrix} x_1 \\ x_2 \end{bmatrix} = \begin{bmatrix} -\frac{1}{3} & \frac{1}{3} \\ \frac{2}{3} & \frac{1}{3} \end{bmatrix}\begin{bmatrix} 2 \\ 6 \end{bmatrix} - \begin{bmatrix} -\frac{1}{3} & \frac{1}{3} \\ \frac{2}{3} & \frac{1}{3} \end{bmatrix}\begin{bmatrix} 1 & 0 \\ 0 & 1 \end{bmatrix}\begin{bmatrix} x_3 \\ x_4 \end{bmatrix}$$

$$= \begin{bmatrix} \frac{4}{3} \\ \frac{10}{3} \end{bmatrix} - \begin{bmatrix} -\frac{1}{3} \\ \frac{2}{3} \end{bmatrix} x_3 - \begin{bmatrix} \frac{1}{3} \\ \frac{1}{3} \end{bmatrix} x_4 \tag{3.12}$$

Currently $x_3 = x_4 = 0$, $x_1 = \frac{4}{3}$ and $x_2 = \frac{10}{3}$. Note that

$$z_4 - c_4 = \mathbf{c}_B\mathbf{B}^{-1}\mathbf{a}_4 - c_4 = (2, -1)\begin{bmatrix} -\frac{1}{3} & \frac{1}{3} \\ \frac{2}{3} & \frac{1}{3} \end{bmatrix}\begin{bmatrix} 0 \\ 1 \end{bmatrix} - 0 = \frac{1}{3} > 0$$

Hence the objective improves by holding x_3 nonbasic and introducing x_4 in the basis. Then x_3 is kept at zero level, x_4 is increased, and x_1 and x_2 are modified according to Equation (3.12). We see that x_4 can be increased to 4, at which instant x_1 drops to zero. Any further increase of x_4 results in violating the nonnegativity of x_1, and so x_1 is the *blocking variable*. With $x_4 = 4$ and $x_3 = 0$, the modified values of x_1 and x_2 are 0 and 2 respectively. The new basic feasible solution is

$$(x_1, x_2, x_3, x_4) = (0, 2, 0, 4)$$

Note that $\mathbf{a}_4$ replaces $\mathbf{a}_1$; that is, x_1 drops from the basis and x_4 enters the basis. The new set of basic and nonbasic variables and their values are given as:

$$\mathbf{x}_B = \begin{bmatrix} x_{B_1} \\ x_{B_2} \end{bmatrix} = \begin{bmatrix} x_4 \\ x_2 \end{bmatrix} = \begin{bmatrix} 4 \\ 2 \end{bmatrix}, \qquad \mathbf{x}_N = \begin{bmatrix} x_3 \\ x_1 \end{bmatrix} = \begin{bmatrix} 0 \\ 0 \end{bmatrix}$$

Moving from the old to the new basic feasible solution is illustrated in Figure 3.8 in the original (x_1, x_2) space. Note that as x_4 increased by one unit, x_1 decreases by $\frac{1}{3}$ unit and x_2 decreases by $\frac{1}{3}$ unit; that is, we move in the direction $(-\frac{1}{3}, -\frac{1}{3})$ in the (x_1, x_2) space. This continues until we are blocked by the nonnegativity restriction $x_1 \geq 0$. At this point x_1 drops to zero and leaves the basis.

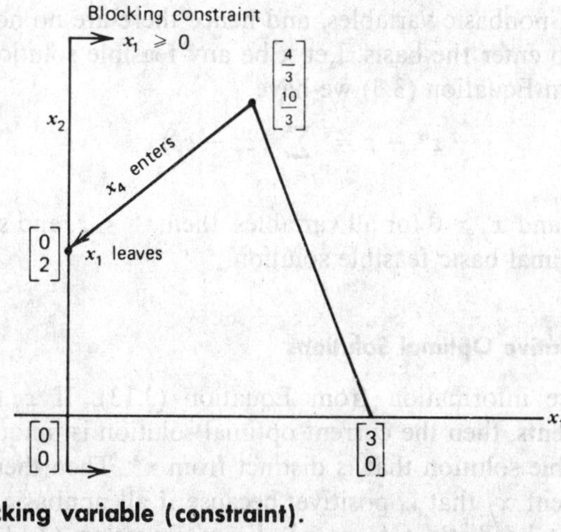

Figure 3.8. **Blocking variable (constraint).**

3.6 TERMINATION: OPTIMALITY AND UNBOUNDEDNESS

We have discussed a procedure that moves from one basis to an adjacent basis, by introducing one variable into the basis, and removing another variable from the basis. The criteria for entering and leaving are summarized below.

1. *Entering*: x_k may enter if $z_k - c_k > 0$
2. *Leaving*: x_{B_r} may leave if

$$\frac{\bar{b}_r}{y_{rk}} = \underset{1 \le i \le m}{\text{Minimum}} \left\{ \frac{\bar{b}_i}{y_{ik}} : y_{ik} > 0 \right\}$$

Two logical questions immediately arise. First, what happens if each nonbasic variable x_j has $z_j - c_j \le 0$? In this case, as seen by Equation (3.6), we have obtained an optimal basic feasible solution. Second, suppose that $z_k - c_k > 0$, and hence x_k is eligible to enter the basis, but we cannot find any positive component y_{ik}, that is, $\mathbf{y}_k \le \mathbf{0}$. In this case, the optimal solution value is unbounded. These cases will be discussed in more detail in this section.

Termination with an Optimal Solution

Consider the following problem, where $\mathbf{A}$ is an $m \times n$ matrix with rank m.

$$\begin{aligned} &\text{Minimize} && \mathbf{cx} \\ &\text{Subject to} && \mathbf{Ax} = \mathbf{b} \\ &&& \mathbf{x} \ge \mathbf{0} \end{aligned}$$

Suppose that $\mathbf{x}^*$ is a basic feasible solution with basis $\mathbf{B}$; that is, $\mathbf{x}^* = \begin{pmatrix} \mathbf{B}^{-1}\mathbf{b} \\ \mathbf{0} \end{pmatrix}$. Let z^* denote the objective of $\mathbf{x}^*$, that is, $z^* = \mathbf{c}_B\mathbf{B}^{-1}\mathbf{b}$. Suppose further that

$z_j - c_j \leq 0$ for all nonbasic variables, and hence there are no nonbasic variables that are eligible to enter the basis. Let $\mathbf{x}$ be any feasible solution with objective value z. Then from Equation (3.3) we have

$$z^* - z = \sum_{j \in R} (z_j - c_j) x_j \tag{3.13}$$

Since $z_j - c_j \leq 0$ and $x_j \geq 0$ for all variables, then $z^* \leq z$, and so as in Equation (3.6), $\mathbf{x}^*$ is an optimal basic feasible solution.

Unique and Alternative Optimal Solutions

We can get more information from Equation (3.13). If $z_j - c_j < 0$ for all nonbasic components, then the current optimal solution is unique. To show this, let $\mathbf{x}$ be any feasible solution that is distinct from $\mathbf{x}^*$. Then there is at least one nonbasic component x_j that is positive, because if all nonbasic components are zero, $\mathbf{x}$ would not be distinct from $\mathbf{x}^*$. From Equation (3.13) it follows that $z > z^*$, and hence $\mathbf{x}^*$ is the unique optimal solution.

Now consider the case where $z_j - c_j \leq 0$ for all nonbasic components, but $z_k - c_k = 0$ for at least one nonbasic variable x_k. As x_k is increased, we get (in the absence of degeneracy) points that are distinct from $\mathbf{x}^*$ but have the same objective value (why?). If x_k is increased until it is blocked by a basic variable, we get an alternative optimal basic feasible solution. The process of increasing x_k from level zero until it is blocked generates an infinite number of alternative optimal solutions.

Example 3.6

$$\begin{array}{lrcrcrcrl}
\text{Minimize} & -3x_1 & + & x_2 & & & & & \\
\text{Subject to} & x_1 & + & 2x_2 & + & x_3 & & & = 4 \\
& -x_1 & + & x_2 & & & + & x_4 & = 1 \\
& x_1, & & x_2, & & x_3, & & x_4 & \geq 0
\end{array}$$

Consider the basic feasible solution with basis $\mathbf{B} = [\mathbf{a}_1, \mathbf{a}_4] = \begin{bmatrix} 1 & 0 \\ -1 & 1 \end{bmatrix}$ and $\mathbf{B}^{-1} = \begin{bmatrix} 1 & 0 \\ 1 & 1 \end{bmatrix}$. The corresponding point is given by

$$\mathbf{x}_B = \begin{bmatrix} x_1 \\ x_4 \end{bmatrix} = \mathbf{B}^{-1}\mathbf{b} = \begin{bmatrix} 1 & 0 \\ 1 & 1 \end{bmatrix}\begin{bmatrix} 4 \\ 1 \end{bmatrix} = \begin{bmatrix} 4 \\ 5 \end{bmatrix}, \qquad \mathbf{x}_N = \begin{bmatrix} x_2 \\ x_3 \end{bmatrix} = \begin{bmatrix} 0 \\ 0 \end{bmatrix}$$

and the objective value is -12. To see if we can improve the solution, calculate $z_2 - c_2$ and $z_3 - c_3$ as follows, noting that $\mathbf{c}_B\mathbf{B}^{-1} = (-3, 0)$, where $\mathbf{c}_B = (-3, 0)$.

$$z_2 - c_2 = \mathbf{c}_B\mathbf{B}^{-1}\mathbf{a}_2 - c_2 = (-3, 0)\begin{bmatrix} 2 \\ 1 \end{bmatrix} - 1 = -7$$

$$z_3 - c_3 = \mathbf{c}_B\mathbf{B}^{-1}\mathbf{a}_3 - c_3 = (-3, 0)\begin{bmatrix} 1 \\ 0 \end{bmatrix} - 0 = -3$$

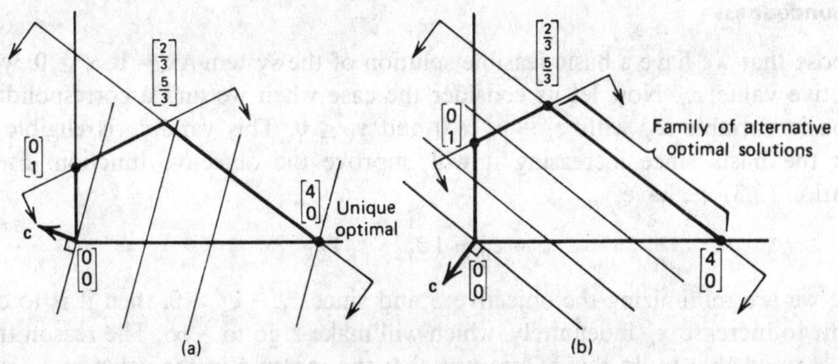

Figure 3.9. Termination criterion: (*a*) Unique optimal. (*b*) Alternative optima.

Since both $z_2 - c_2 < 0$ and $z_3 - c_3 < 0$, then the basic feasible solution $(x_1, x_2, x_3, x_4) = (4, 0, 0, 5)$ is the unique optimal point. This unique optimal solution is illustrated in Figure 3.9*a*. Now consider a new problem where the objective function $-2x_1 - 4x_2$ is to be minimized over the same region. Again, consider the same point $(4, 0, 0, 5)$. The objective value is -8, and the new $\mathbf{c}_B = (-2, 0)$, $\mathbf{c}_B\mathbf{B}^{-1} = (-2, 0)$. Calculate $z_2 - c_2$ and $z_3 - c_3$ as follows:

$$z_2 - c_2 = \mathbf{c}_B\mathbf{B}^{-1}\mathbf{a}_2 - c_2 = (-2, 0)\begin{bmatrix} 2 \\ 1 \end{bmatrix} + 4 = 0$$

$$z_3 - c_3 = \mathbf{c}_B\mathbf{B}^{-1}\mathbf{a}_3 - c_3 = (-2, 0)\begin{bmatrix} 1 \\ 0 \end{bmatrix} - 0 = -2$$

In this case, the given basic feasible solution is optimal, but it is no longer a unique optimal solution. We see that by increasing x_2, a family of optimal solutions is obtained. Actually, if we increase x_2, keep $x_3 = 0$, and modify x_1 and x_4, we get

$$\begin{bmatrix} x_1 \\ x_4 \end{bmatrix} = \mathbf{B}^{-1}\mathbf{b} - \mathbf{B}^{-1}\mathbf{a}_2 x_2$$

$$= \begin{bmatrix} 4 \\ 5 \end{bmatrix} - \begin{bmatrix} 2 \\ 3 \end{bmatrix} x_2$$

For any $x_2 \le \frac{5}{3}$, the solution

$$\begin{bmatrix} x_1 \\ x_2 \\ x_3 \\ x_4 \end{bmatrix} = \begin{bmatrix} 4 - 2x_2 \\ x_2 \\ 0 \\ 5 - 3x_2 \end{bmatrix}$$

is an optimal solution with objective -8. In particular, if $x_2 = \frac{5}{3}$, we get an alternative basic feasible optimal solution, where x_4 drops from the basis. This is illustrated in Figure 3.9*b*. Note that the new objective function contours are parallel to the hyperplane $x_1 + 2x_2 = 4$ corresponding to the first constraint. That is why we obtain alternative optimal solutions in this example. In general whenever the optimal objective function contour contains a face of dimension > 0, we will have alternative optimal solutions.

Unboundedness

Suppose that we have a basic feasible solution of the system $\mathbf{Ax} = \mathbf{b}$, $\mathbf{x} \geq \mathbf{0}$, with objective value z_0. Now let us consider the case when we find a corresponding nonbasic variable x_k with $z_k - c_k > 0$ and $\mathbf{y}_k \leq \mathbf{0}$. This variable is eligible to enter the basis since increasing it will improve the objective function. From Equation (3.3) we have

$$z = z_0 - (z_k - c_k)x_k$$

Since we are minimizing the objective z and since $z_k - c_k > 0$, then it is to our benefit to increase x_k indefinitely, which will make z go to $-\infty$. The reason that we were not able to do this before was that the increase in the value of x_k was blocked by a basic variable. This puts a "ceiling" on x_k beyond which a basic variable will be negative. But if blocking is not encountered, there is no reason why we should stop increasing x_k. This is precisely the case when $\mathbf{y}_k \leq \mathbf{0}$. Recall that from Equation (3.8) we have

$$\mathbf{x}_B = \mathbf{B}^{-1}\mathbf{b} - \mathbf{y}_k x_k$$

and so if $\mathbf{y}_k \leq \mathbf{0}$, then x_k can be increased indefinitely without any of the basic variables becoming negative. Therefore the solution $\mathbf{x}$ (where $\mathbf{x}_B = \mathbf{B}^{-1}\mathbf{b} - \mathbf{y}_k x_k$, x_k is arbitrarily large and other nonbasic components are zero) is feasible and its objective value $z = z_0 - (z_k - c_k)x_k$, which approaches $-\infty$ as x_k approaches $+\infty$.

To summarize, if we have a basic feasible solution with $z_k - c_k > 0$ for some nonbasic variable x_k, and meanwhile $\mathbf{y}_k \leq \mathbf{0}$, then the optimal is unbounded with objective $-\infty$. This is obtained by increasing x_k indefinitely and adjusting the values of the current basic variables, and is equivalent to moving along the ray:

$$\left\{ \begin{bmatrix} \mathbf{B}^{-1}\mathbf{b} \\ \hline 0 \\ \vdots \\ 0 \\ \vdots \\ 0 \end{bmatrix} + x_k \begin{bmatrix} -\mathbf{y}_k \\ \hline 0 \\ \vdots \\ 1 \\ \vdots \\ 0 \end{bmatrix} : x_k \geq 0 \right\}$$

Note that the vertex of the ray is the current basic feasible solution $\begin{pmatrix} \mathbf{B}^{-1}\mathbf{b} \\ \mathbf{0} \end{pmatrix}$ and the direction of the ray is

$$\mathbf{d} = \begin{bmatrix} -\mathbf{y}_k \\ \hline 0 \\ \vdots \\ 1 \\ \vdots \\ 0 \end{bmatrix}$$

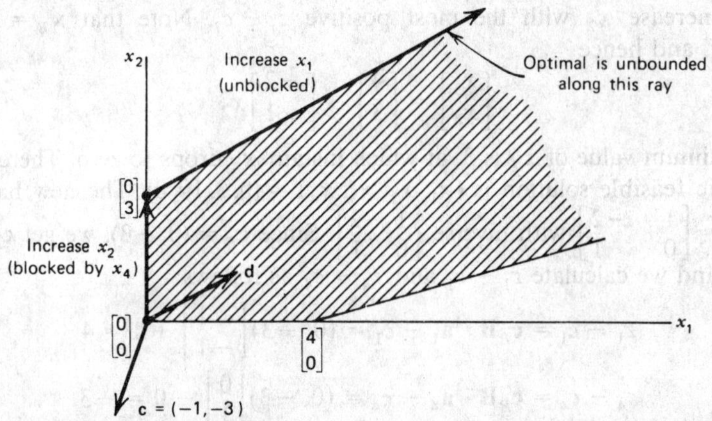

Figure 3.10. Unbounded optimal.

where the 1 appears in the kth position. It may be noted that

$$\mathbf{cd} = (\mathbf{c}_B, \mathbf{c}_N)\mathbf{d} = -\mathbf{c}_B\mathbf{y}_k + c_k = -z_k + c_k$$

But since $c_k - z_k < 0$ (because x_k was eligible to enter the basis), then $\mathbf{cd} < 0$, which is the necessary and sufficient condition for unboundedness. In Exercise 3.32 we ask the reader to verify that $\mathbf{d}$ is indeed an (extreme) direction of the feasible region.

Example 3.7

(Unboundedness)

$$\begin{array}{ll} \text{Minimize} & -x_1 - 3x_2 \\ \text{Subject to} & x_1 - 2x_2 \leq 4 \\ & -x_1 + x_2 \leq 3 \\ & x_1, \quad x_2 \geq 0 \end{array}$$

The problem, illustrated in Figure 3.10, clearly has an unbounded optimal solution. After introducing the slack variables x_3 and x_4, we get the constraint matrix $\mathbf{A} = \begin{bmatrix} 1 & -2 & 1 & 0 \\ -1 & 1 & 0 & 1 \end{bmatrix}$. Now consider the basic feasible solution whose basis $\mathbf{B} = [\mathbf{a}_3, \mathbf{a}_4] = \begin{bmatrix} 1 & 0 \\ 0 & 1 \end{bmatrix}$.

$$\mathbf{x}_B = \begin{bmatrix} x_3 \\ x_4 \end{bmatrix} = \begin{bmatrix} 1 & 0 \\ 0 & 1 \end{bmatrix} \begin{bmatrix} 4 \\ 3 \end{bmatrix} = \begin{bmatrix} 4 \\ 3 \end{bmatrix}, \qquad \mathbf{x}_N = \begin{bmatrix} x_1 \\ x_2 \end{bmatrix} = \begin{bmatrix} 0 \\ 0 \end{bmatrix}$$

Calculate $z_1 - c_1$ and $z_2 - c_2$ as follows, noting $\mathbf{c}_B\mathbf{B}^{-1} = (0, 0)$:

$$z_1 - c_1 = \mathbf{c}_B\mathbf{B}^{-1}\mathbf{a}_1 - c_1 = -c_1 = 1$$

$$z_2 - c_2 = \mathbf{c}_B\mathbf{B}^{-1}\mathbf{a}_2 - c_2 = -c_2 = 3$$

So we increase x_2 with the most positive $z_j - c_j$. Note that $\mathbf{x}_B = \mathbf{B}^{-1}\mathbf{b} - \mathbf{B}^{-1}\mathbf{a}_2 x_2$, and hence

$$\begin{bmatrix} x_3 \\ x_4 \end{bmatrix} = \begin{bmatrix} 4 \\ 3 \end{bmatrix} - \begin{bmatrix} -2 \\ 1 \end{bmatrix} x_2$$

The maximum value of x_2 is 3, at which instant x_4 drops to zero. Therefore the new basic feasible solution is $(x_1, x_2, x_3, x_4) = (0, 3, 10, 0)$. The new basis $\mathbf{B}$ is $[\mathbf{a}_3, \mathbf{a}_2] = \begin{bmatrix} 1 & -2 \\ 0 & 1 \end{bmatrix}$ with inverse $\begin{bmatrix} 1 & 2 \\ 0 & 1 \end{bmatrix}$. Since $\mathbf{c}_B = (0, -3)$, we get $\mathbf{c}_B\mathbf{B}^{-1} = (0, -3)$ and we calculate $z_1 - c_1$ and $z_4 - c_4$ as follows:

$$z_1 - c_1 = \mathbf{c}_B\mathbf{B}^{-1}\mathbf{a}_1 - c_1 = (0, -3)\begin{bmatrix} 1 \\ -1 \end{bmatrix} + 1 = 4$$

$$z_4 - c_4 = \mathbf{c}_B\mathbf{B}^{-1}\mathbf{a}_4 - c_4 = (0, -3)\begin{bmatrix} 0 \\ 1 \end{bmatrix} - 0 = -3$$

Note that $z_1 - c_1 > 0$ and $\mathbf{y}_1 = \mathbf{B}^{-1}\mathbf{a}_1 = \begin{bmatrix} -1 \\ -1 \end{bmatrix} \le \begin{bmatrix} 0 \\ 0 \end{bmatrix}$. Therefore the optimal solution value is unbounded. In this case, if x_1 is increased and x_4 is kept zero, we get the following solution:

$$\mathbf{x}_B = \mathbf{B}^{-1}\mathbf{b} - \mathbf{B}^{-1}\mathbf{a}_1 x_1$$

$$\begin{bmatrix} x_3 \\ x_2 \end{bmatrix} = \begin{bmatrix} 10 \\ 3 \end{bmatrix} - \begin{bmatrix} -1 \\ -1 \end{bmatrix} x_1 = \begin{bmatrix} 10 + x_1 \\ 3 + x_1 \end{bmatrix}$$

$$x_4 = 0$$

Note that this solution is feasible for all $x_1 \ge 0$. In particular,

$$x_1 - 2x_2 + x_3 = x_1 - 2(3 + x_1) + (10 + x_1) = 4,$$

and

$$-x_1 + x_2 + x_4 = -x_1 + (3 + x_1) + 0 = 3$$

Furthermore, $z = -9 - 4x_1$, which approaches $-\infty$ as x_1 approaches ∞. Therefore the optimal solution value is unbounded by moving along the ray

$$\{(0, 3, 10, 0) + x_1(1, 1, 1, 0) : x_1 \ge 0\}$$

Again note that the necessary and sufficient condition for unboundedness holds, namely

$$\mathbf{cd} = (-1, -3, 0, 0)\begin{bmatrix} 1 \\ 1 \\ 1 \\ 0 \end{bmatrix} = -4 < 0$$

3.7 THE SIMPLEX METHOD

All the machinery that is needed to describe the simplex algorithm and to prove its convergence in a finite number of iterations (in the absence of degeneracy) has been generated. Given a basic feasible solution and a corresponding basis, we can either improve it if $z_k - c_k > 0$ for some nonbasic variable x_k, or stop with an optimal point if $z_j - c_j \le 0$ for all nonbasic variables. If $z_k - c_k > 0$ and the vector $\mathbf{y}_k$ contains at least one positive component, then the increase in x_k will be blocked by one of the current basic variables, which drops to zero and leaves the basis. On the other hand, if $z_k - c_k > 0$ and $\mathbf{y}_k \leq \mathbf{0}$, then x_k can be increased indefinitely, and the optimal solution value is unbounded ($-\infty$). This discussion is exactly what the simplex method does.

We now give a summary of the simplex method for solving the following linear programming problem.

$$\begin{array}{ll} \text{Minimize} & \mathbf{cx} \\ \text{Subject to} & \mathbf{Ax} = \mathbf{b} \\ & \mathbf{x} \geq \mathbf{0} \end{array}$$

where $\mathbf{A}$ is an $m \times n$ matrix with rank m (the requirement that rank $(\mathbf{A}) = m$ will be relaxed in Chapter 4).

The Simplex Algorithm (Minimization Problem)

INITIALIZATION STEP

Choose a starting basic feasible solution with basis $\mathbf{B}$. (Several procedures for finding an initial basis will be described in Chapter 4.)

MAIN STEP

1. Solve the system $\mathbf{Bx}_B = \mathbf{b}$ (with unique solution $\mathbf{x}_B = \mathbf{B}^{-1}\mathbf{b} = \bar{\mathbf{b}}$). Let $\mathbf{x}_B = \bar{\mathbf{b}}$, $\mathbf{x}_N = \mathbf{0}$, and $z = \mathbf{c}_B\mathbf{x}_B$.
2. Solve the system $\mathbf{wB} = \mathbf{c}_B$ (with unique solution $\mathbf{w} = \mathbf{c}_B\mathbf{B}^{-1}$). (The vector $\mathbf{w}$ is referred to as the vector of *simplex multipliers* because its components are the multipliers of the rows of $\mathbf{A}$ that are added to the objective function in order to bring it into canonical form.) Calculate $z_j - c_j = \mathbf{wa}_j - c_j$ for all nonbasic variables. (This is known as the *pricing* operation.) Let

$$z_k - c_k = \underset{j \in R}{\text{Maximum}}\, z_j - c_j$$

   where R is the current set of indices associated with the nonbasic variables. If $z_k - c_k \leq 0$, then stop with the current basic feasible solution as an optimal solution. Otherwise go to step 3 with x_k as the entering variable. (This strategy for selecting an entering variable is known as *Dantzig's Rule*.)
3. Solve the system $\mathbf{By}_k = \mathbf{a}_k$ (with unique solution $\mathbf{y}_k = \mathbf{B}^{-1}\mathbf{a}_k$). If $\mathbf{y}_k \leq \mathbf{0}$, then stop with the conclusion that the optimal solution is unbounded along the ray

$$\left\{ \begin{bmatrix} \bar{\mathbf{b}} \\ \mathbf{0} \end{bmatrix} + x_k \begin{bmatrix} -\mathbf{y}_k \\ \mathbf{e}_k \end{bmatrix} : x_k \geq 0 \right\}$$

   where $\mathbf{e}_k$ is an $n - m$ vector of zeros except for a 1 at the kth position. If $\mathbf{y}_k \nleq \mathbf{0}$, go to step 4.

4. Let x_k enter the basis. The index r of the blocking variable x_{B_r} which leaves the basis is determined by the following *minimum ratio test*:

$$\frac{\bar{b}_r}{y_{rk}} = \underset{1 \le i \le m}{\text{Minimum}} \left\{ \frac{\bar{b}_i}{y_{ik}} : y_{ik} > 0 \right\}$$

Update the basis $\mathbf{B}$ where $\mathbf{a}_k$ replaces $\mathbf{a}_{B_r}$, update the index set R, and repeat step 1.

Modification for a Maximization Problem

A maximization problem can be transformed into a minimization problem by multiplying the objective coefficients by -1. A maximization problem can also be handled directly as follows. Let $z_k - c_k$ instead be the minimum $z_j - c_j$ for j nonbasic; the stopping criterion is that $z_k - c_k \ge 0$. Otherwise, the steps are as previously.

Finite Convergence of the Simplex Method in the Absence of Degeneracy

Note that at each iteration, one of the following three actions is executed. We may stop with an optimal extreme point if $z_k - c_k \le 0$; we may stop with an unbounded solution if $z_k - c_k > 0$ and $\mathbf{y}_k \le \mathbf{0}$; or else we generate a new basic feasible solution if $z_k - c_k > 0$ and $\mathbf{y}_k \nleq \mathbf{0}$. In the absence of degeneracy, $\bar{b}_r > 0$ and hence $x_k = \bar{b}_r / y_{rk} > 0$. Therefore the difference between the objective values at the previous iteration and the current iteration is $x_k\ (z_k - c_k) > 0$. In other words, the objective function strictly decreases at each iteration and hence the basic feasible solutions generated by the simplex method are distinct. Since there is only a finite number of basic feasible solutions, the method would stop in a finite number of steps with a finite optimal solution or with an unbounded optimal solution. From this discussion the following theorem is obvious.

***Theorem* 3.5** (Finite Convergence)

In the absence of degeneracy (and assuming feasibility), the simplex method stops in a finite number of iterations, either with an optimal basic feasible solution or with the conclusion that the optimal value is unbounded.

In the presence of degeneracy as we have seen, there is the possibility of cycling in an infinite loop. This issue is discussed at length in Chapter 4.

Example 3.8

$$\begin{array}{ll} \text{Minimize} & -x_1 - 3x_2 \\ \text{Subject to} & 2x_1 + 3x_2 \le 6 \\ & -x_1 + x_2 \le 1 \\ & x_1, x_2 \ge 0 \end{array}$$

The problem is illustrated in Figure 3.11. After introducing the nonnegative slack

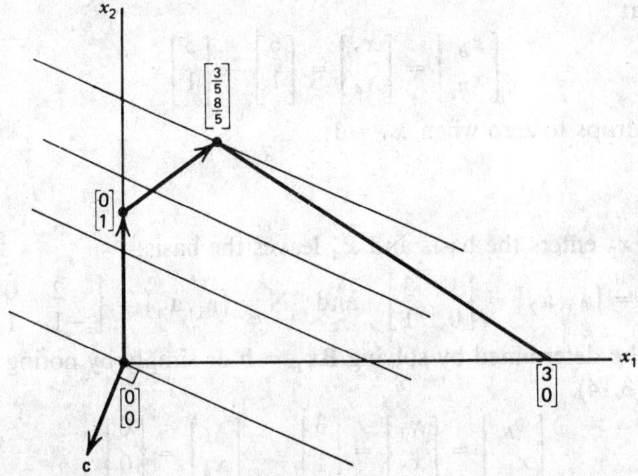

Figure 3.11. Example of the simplex method.

variables x_3 and x_4, we get the following constraints:

$$\begin{aligned} 2x_1 + 3x_2 + x_3 \quad &= 6 \\ -x_1 + \ \ x_2 + \quad x_4 &= 1 \\ x_1, \quad x_2, \quad x_3, x_4 &\geq 0 \end{aligned}$$

Iteration 1

Let $\mathbf{B} = [\mathbf{a}_3, \mathbf{a}_4] = \begin{bmatrix} 1 & 0 \\ 0 & 1 \end{bmatrix}$ and $\mathbf{N} = [\mathbf{a}_1, \mathbf{a}_2] = \begin{bmatrix} 2 & 3 \\ -1 & 1 \end{bmatrix}$. Solving the system $\mathbf{B}\mathbf{x}_B = \mathbf{b}$ leads to $x_{B_1} = x_3 = 6$ and $x_{B_2} = x_4 = 1$. The nonbasic variables are x_1 and x_2 and the objective $z = \mathbf{c}_B\mathbf{x}_B = (0,0)\begin{bmatrix} 6 \\ 1 \end{bmatrix} = 0$. In order to determine which variable enters the basis, calculate $z_j - c_j = \mathbf{c}_B\mathbf{B}^{-1}\mathbf{a}_j - c_j = \mathbf{w}\mathbf{a}_j - c_j$. First we find $\mathbf{w}$ by solving the system $\mathbf{w}\mathbf{B} = \mathbf{c}_B$:

$$(w_1, w_2)\begin{bmatrix} 1 & 0 \\ 0 & 1 \end{bmatrix} = (0,0) \Rightarrow w_1 = w_2 = 0$$

$$z_1 - c_1 = \mathbf{w}\mathbf{a}_1 - c_1 = 1$$
$$z_2 - c_2 = \mathbf{w}\mathbf{a}_2 - c_2 = 3$$

Therefore x_2 is increased. In order to determine x_2 we need to calculate $\mathbf{y}_2$ by solving the system $\mathbf{B}\mathbf{y}_2 = \mathbf{a}_2$:

$$\begin{bmatrix} 1 & 0 \\ 0 & 1 \end{bmatrix}\begin{bmatrix} y_{12} \\ y_{22} \end{bmatrix} = \begin{bmatrix} 3 \\ 1 \end{bmatrix} \Rightarrow y_{12} = 3 \quad \text{and} \quad y_{22} = 1$$

The variable x_{B_r} leaving the basis is determined by the following minimum ratio test:

$$\text{Minimum}\left\{\frac{\bar{b}_1}{y_{12}}, \frac{\bar{b}_2}{y_{22}}\right\} = \text{Minimum}\left\{\frac{6}{3}, \frac{1}{1}\right\} = 1$$

Therefore the index $r = 2$; that is, $x_{B_2} = x_4$ leaves the basis. This is also obvious by noting that

$$\begin{bmatrix} x_{B_1} \\ x_{B_2} \end{bmatrix} = \begin{bmatrix} x_3 \\ x_4 \end{bmatrix} = \begin{bmatrix} 6 \\ 1 \end{bmatrix} - \begin{bmatrix} 3 \\ 1 \end{bmatrix} x_2 \tag{3.14}$$

and x_4 first drops to zero when $x_2 = 1$.

Iteration 2

The variable x_2 enters the basis and x_4 leaves the basis:

$$\mathbf{B} = [\mathbf{a}_3, \mathbf{a}_2] = \begin{bmatrix} 1 & 3 \\ 0 & 1 \end{bmatrix} \quad \text{and} \quad \mathbf{N} = [\mathbf{a}_1, \mathbf{a}_4] = \begin{bmatrix} 2 & 0 \\ -1 & 1 \end{bmatrix}$$

Now $\mathbf{x}_B$ can be determined by solving $\mathbf{Bx}_B = \mathbf{b}$ or simply by noting that $x_2 = 1$ in Equation (3.14).

$$\begin{bmatrix} x_{B_1} \\ x_{B_2} \end{bmatrix} = \begin{bmatrix} x_3 \\ x_2 \end{bmatrix} = \begin{bmatrix} 3 \\ 1 \end{bmatrix} \qquad \begin{bmatrix} x_1 \\ x_4 \end{bmatrix} = \begin{bmatrix} 0 \\ 0 \end{bmatrix}$$

The objective value is $z = -3$. Calculate $\mathbf{w}$ by $\mathbf{wB} = \mathbf{c}_B$:

$$(w_1, w_2)\begin{bmatrix} 1 & 3 \\ 0 & 1 \end{bmatrix} = (0, -3) \Rightarrow w_1 = 0 \text{ and } w_2 = -3$$

$$\begin{aligned} z_1 - c_1 &= \mathbf{wa}_1 - c_1 \\ &= (0, -3)\begin{bmatrix} 2 \\ -1 \end{bmatrix} + 1 = 4 \end{aligned}$$

The variable x_4 left the basis in the previous iteration and cannot enter the basis in this iteration since $z_4 - c_4 < 0$ (see Exercise 3.37). Therefore x_1 is increased. Solve the system $\mathbf{By}_1 = \mathbf{a}_1$:

$$\begin{bmatrix} 1 & 3 \\ 0 & 1 \end{bmatrix}\begin{bmatrix} y_{11} \\ y_{21} \end{bmatrix} = \begin{bmatrix} 2 \\ -1 \end{bmatrix} \Rightarrow y_{11} = 5 \text{ and } y_{21} = -1$$

Since $y_{21} < 0$, then $x_{B_1} = x_3$ leaves the basis as x_1 is increased. This is also clear by noting that

$$\begin{bmatrix} x_{B_1} \\ x_{B_2} \end{bmatrix} = \begin{bmatrix} x_3 \\ x_2 \end{bmatrix} = \begin{bmatrix} 3 \\ 1 \end{bmatrix} - \begin{bmatrix} 5 \\ -1 \end{bmatrix} x_1$$

and x_3 drops to zero when $x_1 = \frac{3}{5}$.

Iteration 3

Here x_1 enters the basis and x_3 leaves the basis:

$$\mathbf{B} = [\mathbf{a}_1, \mathbf{a}_2] = \begin{bmatrix} 2 & 3 \\ -1 & 1 \end{bmatrix} \quad \text{and} \quad \mathbf{N} = [\mathbf{a}_3, \mathbf{a}_4] = \begin{bmatrix} 1 & 0 \\ 0 & 1 \end{bmatrix}$$

$$\begin{bmatrix} x_{B_1} \\ x_{B_2} \end{bmatrix} = \begin{bmatrix} x_1 \\ x_2 \end{bmatrix} = \begin{bmatrix} \frac{3}{5} \\ \frac{8}{5} \end{bmatrix} \qquad \mathbf{x}_N = \begin{bmatrix} x_3 \\ x_4 \end{bmatrix} = \begin{bmatrix} 0 \\ 0 \end{bmatrix}$$

The objective value $z = \frac{-27}{5}$. Calculate $\mathbf{w}$ by $\mathbf{wB} = \mathbf{c}_B$.

$$(w_1, w_2)\begin{bmatrix} 2 & 3 \\ -1 & 1 \end{bmatrix} = (-1, -3) \Rightarrow w_1 = \tfrac{-4}{5} \text{ and } w_2 = \tfrac{-3}{5}$$

The variable x_3 left the basis in the last iteration and cannot enter the basis in this iteration (because $z_3 - c_3 < 0$).

$$z_4 - c_4 = \mathbf{wa}_4 - c_4$$
$$= \left(\tfrac{-4}{5}, \tfrac{-3}{5}\right)\begin{bmatrix}0\\1\end{bmatrix} - 0 = \tfrac{-3}{5}$$

Therefore $z_j - c_j \leq 0$ for all nonbasic variables, and the current point is optimal. The optimal solution is therefore

$$(x_1, x_2, x_3, x_4) = \left(\tfrac{3}{5}, \tfrac{8}{5}, 0, 0\right)$$

with objective value $\frac{-27}{5}$. Figure 3.11 displays the simplex path, that is, the sequence of steps that the simplex method took to reach the optimal point.

3.8 THE SIMPLEX METHOD IN TABLEAU FORMAT

At each iteration the following linear systems of equations need be solved: $\mathbf{Bx}_B = \mathbf{b}$, $\mathbf{wB} = \mathbf{c}_B$, and $\mathbf{By}_k = \mathbf{a}_k$. Various procedures for solving and updating these systems will lead to different algorithms that all lie under the general framework of the simplex method described previously. In this section we describe the simplex method in tableau format. In subsequent chapters we shall describe several procedures for solving the preceding systems for general linear programs as well as for problems with special structure such as network flow problems.

Suppose that we have a starting basic feasible solution $\mathbf{x}$ with basis $\mathbf{B}$. The linear programming problem can be represented as follows.

$$\begin{aligned}
&\text{Minimize} \quad z \\
&\text{Subject to} \quad z - \mathbf{c}_B\mathbf{x}_B - \mathbf{c}_N\mathbf{x}_N = 0 && (3.15)\\
&\qquad\qquad\quad \mathbf{Bx}_B + \mathbf{Nx}_N = \mathbf{b} && (3.16)\\
&\qquad\qquad\quad \mathbf{x}_B, \quad \mathbf{x}_N \geq \mathbf{0}
\end{aligned}$$

From Equation (3.16) we have

$$\mathbf{x}_B + \mathbf{B}^{-1}\mathbf{Nx}_N = \mathbf{B}^{-1}\mathbf{b} \tag{3.17}$$

Multiplying (3.17) by $\mathbf{c}_B$ and adding to Equation (3.15), we get

$$z + \mathbf{0x}_B + \left(\mathbf{c}_B\mathbf{B}^{-1}\mathbf{N} - \mathbf{c}_N\right)\mathbf{x}_N = \mathbf{c}_B\mathbf{B}^{-1}\mathbf{b} \tag{3.18}$$

Currently $\mathbf{x}_N = \mathbf{0}$, and from Equations (3.17) and (3.18) we get $\mathbf{x}_B = \mathbf{B}^{-1}\mathbf{b}$ and $z = \mathbf{c}_B\mathbf{B}^{-1}\mathbf{b}$. Also, from (3.17) and (3.18) we can conveniently represent the current basic feasible solution with basis $\mathbf{B}$ in the following tableau. Here we think of z as a (basic) variable to be minimized. The objective row will be referred to as row 0 and the remaining rows are rows 1 through m. The right-hand-side column (RHS) will denote the values of the basic variables (including the objective function). The basic variables are identified on the far left column.

	z	$\mathbf{x}_B$	$\mathbf{x}_N$	RHS	
z	1	$\mathbf{0}$	$\mathbf{c}_B\mathbf{B}^{-1}\mathbf{N} - \mathbf{c}_N$	$\mathbf{c}_B\mathbf{B}^{-1}\mathbf{b}$	Row 0
$\mathbf{x}_B$	0	$\mathbf{I}$	$\mathbf{B}^{-1}\mathbf{N}$	$\mathbf{B}^{-1}\mathbf{b}$	Rows 1 through m

The *tableau* in which z and $\mathbf{x}_B$ have been solved in terms of $\mathbf{x}_N$ is said to be in *canonical form*. Not only does this tableau give us the value of the objective function $\mathbf{c}_B\mathbf{B}^{-1}\mathbf{b}$ and the basic variables $\mathbf{B}^{-1}\mathbf{b}$ on the right-hand-side, but it also gives us all the information we need to proceed with the simplex method. Actually the cost row gives us $\mathbf{c}_B\mathbf{B}^{-1}\mathbf{N} - \mathbf{c}_N$, which consists of the $z_j - c_j$'s for the nonbasic variables. So row zero will tell us if we are at the optimal solution (if each $z_j - c_j \leq 0$), and which nonbasic variable to increase otherwise. If x_k is increased, then the vector $\mathbf{y}_k = \mathbf{B}^{-1}\mathbf{a}_k$, which is stored in the tableau in rows 1 through m under the variable x_k, will help us determine by how much x_k can be increased. If $\mathbf{y}_k \leq \mathbf{0}$, then x_k can be increased indefinitely without being blocked, and the optimal objective is unbounded. On the other hand, if $\mathbf{y}_k \nleq \mathbf{0}$, that is, if $\mathbf{y}_k$ has at least one positive component, then the increase in x_k will be blocked by one of the current basic variables, which drops to zero. The minimum ratio test (which can be performed since $\mathbf{B}^{-1}\mathbf{b} = \bar{\mathbf{b}}$ and $\mathbf{y}_k$ are both available in the tableau) determines the blocking variable. We would like to have a scheme that will do the following.

1. Update the basic variables and their values.
2. Update the $z_j - c_j$ values of the new nonbasic variables.
3. Update the $\mathbf{y}_j$ columns.

Pivoting

All of the foregoing tasks can be simultaneously accomplished by a simple pivoting operation. If x_k enters the basis and x_{B_r} leaves the basis, then pivoting on y_{rk} can be stated as follows.

1. Divide row r by y_{rk}.
2. For $i = 1, 2, \ldots, m$ and $i \neq r$, update the ith row by adding to it $-y_{ik}$ times the new rth row.
3. Update row zero by adding to it $c_k - z_k$ times the new rth row. The two tableaux of Tables 3.1 and 3.2 represent the situation immediately before and after pivoting.

Table 3.1 Before Pivoting

	z	x_{B_1}		x_{B_r}		x_{B_m}		x_j		x_k		RHS
z	1	0	$\cdots$	0	$\cdots$	0	$\cdots$	$z_j - c_j$	$\cdots$	$z_k - c_k$	$\cdots$	$\mathbf{c}_B\bar{\mathbf{b}}$
x_{B_1}	0	1	$\cdots$	0	$\cdots$	0	$\cdots$	y_{1j}	$\cdots$	y_{1k}	$\cdots$	$\bar{b}_1$
$\vdots$	$\vdots$	$\vdots$		$\vdots$		$\vdots$		$\vdots$		$\vdots$		$\vdots$
x_{B_r}	0	0	$\cdots$	1	$\cdots$	0	$\cdots$	y_{rj}	$\cdots$	$\textcircled{y_{rk}}$	$\cdots$	$\bar{b}_r$
$\vdots$	$\vdots$	$\vdots$		$\vdots$		$\vdots$		$\vdots$		$\vdots$		$\vdots$
x_{B_m}	0	0	$\cdots$	0	$\cdots$	1	$\cdots$	y_{mj}	$\cdots$	y_{mk}	$\cdots$	$\bar{b}_m$

Table 3.2 After Pivoting

	z	x_{B_1}		x_{B_r}		x_{B_m}				x_k		RHS
z	1	0	$\cdots$	$\frac{c_k - z_k}{y_{rk}}$	$\cdots$	0	$\cdots$	$(z_j - c_j) - \frac{y_{rj}}{y_{rk}}(z_k - c_k)$	$\cdots$	0	$\cdots$	$\mathbf{c}_B\bar{\mathbf{b}} - (z_k - c_k)\frac{\bar{b}_r}{y_{rk}}$
x_{B_1}	0	1	$\cdots$	$\frac{-y_{1k}}{y_{rk}}$	$\cdots$	0	$\cdots$	$y_{1j} - \frac{y_{rj}}{y_{rk}} y_{1k}$	$\cdots$	0	$\cdots$	$\bar{b}_1 - \frac{y_{1k}}{y_{rk}}\bar{b}_r$
$\vdots$	$\vdots$	$\vdots$		$\vdots$		$\vdots$		$\vdots$		$\vdots$		$\vdots$
x_k	0	0	$\cdots$	$\frac{1}{y_{rk}}$	$\cdots$	0	$\cdots$	$\frac{y_{rj}}{y_{rk}}$	$\cdots$	1	$\cdots$	$\frac{\bar{b}_r}{y_{rk}}$
$\vdots$	$\vdots$	$\vdots$		$\vdots$		$\vdots$		$\vdots$		$\vdots$		$\vdots$
x_{B_m}	0	0	$\cdots$	$\frac{-y_{mk}}{y_{rk}}$	$\cdots$	1	$\cdots$	$y_{mj} - \frac{y_{rj}}{y_{rk}} y_{mk}$	$\cdots$	0	$\cdots$	$\bar{b}_m - \frac{y_{mk}}{y_{1k}}\bar{b}_r$

Let us examine the implications of the pivoting operation.

1. The variable x_k entered the basis and x_{B_r} left the basis. This is illustrated on the left-hand-side of the tableau by replacing x_{B_r} with x_k. For the purpose of the following iteration, the new x_{B_r} is now x_k.
2. The right-hand-side of the tableau gives the current values of the basic variables (review Equation 3.10). The nonbasic variables are kept zero.
3. Suppose that the original columns of the new basic and nonbasic variables are $\hat{\mathbf{B}}$ and $\hat{\mathbf{N}}$ respectively. Through a sequence of elementary row operations (characterized by pivoting at the intermediate iterations), the original tableau reduces to the current tableau with $\hat{\mathbf{B}}$ replaced by $\mathbf{I}$. From Chapter 2 we know that this is equivalent to premultiplication by $\hat{\mathbf{B}}^{-1}$. Thus pivoting results in a new tableau that gives the new $\hat{\mathbf{B}}^{-1}\hat{\mathbf{N}}$ under the nonbasic variables, an updated set of $z_j - c_j$'s for the new nonbasic variables, and the values of the new basic variables and objective function.

The Simplex Method in Tableau Format (Minimization Problem)

INITIALIZATION STEP

Find an initial basic feasible solution with basis $\mathbf{B}$. Form the following initial tableau.

	z	$\mathbf{x}_B$	$\mathbf{x}_N$	RHS
z	1	$\mathbf{0}$	$\mathbf{c}_B\mathbf{B}^{-1}\mathbf{N} - \mathbf{c}_N$	$\mathbf{c}_B\bar{\mathbf{b}}$
$\mathbf{x}_B$	$\mathbf{0}$	$\mathbf{I}$	$\mathbf{B}^{-1}\mathbf{N}$	$\bar{\mathbf{b}}$

MAIN STEP

Let $z_k - c_k = \text{Maximum}\{z_j - c_j : j \in R\}$. If $z_k - c_k \le 0$, then stop; the current solution is optimal. Otherwise examine $\mathbf{y}_k$. If $\mathbf{y}_k \le \mathbf{0}$, then stop; the optimal solution is unbounded along the ray

$$\left\{\begin{bmatrix}\mathbf{B}^{-1}\mathbf{b}\\ \mathbf{0}\end{bmatrix} + x_k\begin{bmatrix}-\mathbf{y}_k\\ \mathbf{e}_k\end{bmatrix} : x_k \ge 0\right\}$$

where $\mathbf{e}_k$ is a vector of zeros except for a 1 at the kth position. If $\mathbf{y}_k \not\le \mathbf{0}$, determine the index r as follows:

$$\frac{\bar{b}_r}{y_{rk}} = \underset{1 \le i \le m}{\text{Minimum}}\left\{\frac{\bar{b}_i}{y_{ik}} : y_{ik} > 0\right\}$$

Update the tableau by pivoting at y_{rk}. Update the basic and nonbasic variables where x_k enters the basis and x_{B_r} leaves the basis, and repeat the main step.

Example 3.9

$$\begin{array}{ll} \text{Minimize} & x_1 + x_2 - 4x_3 \\ \text{Subject to} & x_1 + x_2 + 2x_3 \le 9 \\ & x_1 + x_2 - x_3 \le 2 \\ & -x_1 + x_2 + x_3 \le 4 \\ & x_1, \quad x_2, \quad x_3 \ge 0 \end{array}$$

Introduce the nonnegative slack variables x_4, x_5, and x_6. The problem becomes the following.

$$\begin{array}{ll} \text{Minimize} & x_1 + x_2 - 4x_3 + 0x_4 + 0x_5 + 0x_6 \\ \text{Subject to} & x_1 + x_2 + 2x_3 + x_4 = 9 \\ & x_1 + x_2 - x_3 + x_5 = 2 \\ & -x_1 + x_2 + x_3 + x_6 = 4 \\ & x_1, \quad x_2, \quad x_3, \quad x_4, \quad x_5, \quad x_6 \ge 0 \end{array}$$

Since $\mathbf{b} \ge \mathbf{0}$, then we can choose our initial basis as $\mathbf{B} = [\mathbf{a}_4, \mathbf{a}_5, \mathbf{a}_6] = \mathbf{I}_3$, and we indeed have $\mathbf{B}^{-1}\mathbf{b} = \bar{\mathbf{b}} \ge \mathbf{0}$. This gives the following initial tableau.

Iteration 1

	z	x_1	x_2	x_3	x_4	x_5	x_6	RHS
z	1	−1	−1	4	0	0	0	0
x_4	0	1	1	2	1	0	0	9
x_5	0	1	1	−1	0	1	0	2
x_6	0	−1	1	(1)	0	0	1	4

Iteration 2

	z	x_1	x_2	x_3	x_4	x_5	x_6	RHS
z	1	3	−5	0	0	0	−4	−16
x_4	0	(3)	−1	0	1	0	−2	1
x_5	0	0	2	0	0	1	1	6
x_3	0	−1	1	1	0	0	1	4

Iteration 3

	z	x_1	x_2	x_3	x_4	x_5	x_6	RHS
z	1	0	-4	0	-1	0	-2	-17
x_1	0	1	$-\frac{1}{3}$	0	$\frac{1}{3}$	0	$-\frac{2}{3}$	$\frac{1}{3}$
x_5	0	0	2	0	0	1	1	6
x_3	0	0	$\frac{2}{3}$	1	$\frac{1}{3}$	0	$\frac{1}{3}$	$\frac{13}{3}$

This is the optimal tableau since $z_j - c_j \leq 0$ for all nonbasic variables. The optimal solution is given by

$$x_1 = \tfrac{1}{3}, \quad x_2 = 0, \quad x_3 = \tfrac{13}{3}$$

$$z = -17$$

Note that the current optimal basis consists of the columns $\mathbf{a}_1$, $\mathbf{a}_5$, and $\mathbf{a}_3$, namely

$$\mathbf{B} = [\mathbf{a}_1, \mathbf{a}_5, \mathbf{a}_3] = \begin{bmatrix} 1 & 0 & 2 \\ 1 & 1 & -1 \\ -1 & 0 & 1 \end{bmatrix}$$

Interpretation of Entries in the Simplex Tableau

Consider the following typical simplex tableau and assume nondegeneracy.

	z	$\mathbf{x}_B$	$\mathbf{x}_N$	RHS
z	1	$\mathbf{0}$	$\mathbf{c}_B\mathbf{B}^{-1}\mathbf{N} - \mathbf{c}_N$	$\mathbf{c}_B\mathbf{B}^{-1}\mathbf{b}$
$\mathbf{x}_B$	$\mathbf{0}$	$\mathbf{I}$	$\mathbf{B}^{-1}\mathbf{N}$	$\mathbf{B}^{-1}\mathbf{b}$

The tableau may be thought of as a representation of both the basic variables $\mathbf{x}_B$ and the cost variable z in terms of the nonbasic variables $\mathbf{x}_N$. The nonbasic variables can therefore be thought of as independent variables, whereas $\mathbf{x}_B$ and z are dependent variables. From row zero we have

$$\begin{aligned} z &= \mathbf{c}_B\mathbf{B}^{-1}\mathbf{b} - \left(\mathbf{c}_B\mathbf{B}^{-1}\mathbf{N} - \mathbf{c}_N\right)\mathbf{x}_N \\ &= \mathbf{c}_B\mathbf{B}^{-1}\mathbf{b} + \sum_{j \in R} (c_j - z_j)x_j \end{aligned}$$

and hence the rate of change of z as a function of a typical nonbasic variable x_j, namely $\partial z/\partial x_j$, is simply $c_j - z_j$. In order to minimize z, we should increase x_j if $\partial z/\partial x_j < 0$, that is, if $z_j - c_j > 0$. (Here and later, each partial derivatives is defined in the usual rate of change sense, holding all the other independent variables constant at their current values.)

Also, the basic variables can be represented in terms of the nonbasic variables as follows:

$$\mathbf{x}_B = \mathbf{B}^{-1}\mathbf{b} - \mathbf{B}^{-1}\mathbf{N}\mathbf{x}_N$$

$$= \mathbf{B}^{-1}\mathbf{b} - \sum_{j \in R} \mathbf{B}^{-1}\mathbf{a}_j x_j$$

$$= \mathbf{B}^{-1}\mathbf{b} - \sum_{j \in R} \mathbf{y}_j x_j$$

Therefore $\partial \mathbf{x}_B/\partial x_j = -\mathbf{y}_j$; that is, $-\mathbf{y}_j$ is the rate of change of the basic variables as a function of the nonbasic variable x_j. In other words, if x_j increases by one unit, then the ith basic variable x_{B_i} decreases by an amount y_{ij}, or simply $\partial x_{B_i}/\partial x_j = -y_{ij}$. A column $\mathbf{y}_j$ can be alternatively interpreted as follows. Recall that $\mathbf{B}\mathbf{y}_j = \mathbf{a}_j$, and hence $\mathbf{y}_j$ represents the linear combination of the basic columns that are needed to represent $\mathbf{a}_j$. More specifically,

$$\mathbf{a}_j = \sum_{i=1}^{m} \mathbf{a}_{B_i} y_{ij}$$

The simplex tableau also gives us a convenient way of predicting the rate of change of the objective function and the value of the basic variables as a function of the right-hand-side vector $\mathbf{b}$. Since the right-hand-side vector usually represents scarce resources, we can predict the rate of change of the objective function as the availability of the resources is varied. In particular,

$$z = \mathbf{c}_B\mathbf{B}^{-1}\mathbf{b} - \sum_{j \in R} (z_j - c_j)x_j$$

and hence $\partial z/\partial \mathbf{b} = \mathbf{c}_B\mathbf{B}^{-1}$. If the original identity consists of slack variables with zero costs, then the elements of row zero at the final tableau under the slacks give $\mathbf{c}_B\mathbf{B}^{-1}\mathbf{I} - \mathbf{0} = \mathbf{c}_B\mathbf{B}^{-1}$, which is $\partial z/\partial \mathbf{b}$. More specifically, if we let $\mathbf{w} = \mathbf{c}_B\mathbf{B}^{-1}$, then $\partial z/\partial b_i = w_i$.

Similarly, the rate of change of the basic variables as a function of the right-hand-side vector $\mathbf{b}$ is given by

$$\frac{\partial \mathbf{x}_B}{\partial \mathbf{b}} = \mathbf{B}^{-1}$$

In particular, $\partial x_{B_i}/\partial \mathbf{b}$ is the ith row of $\mathbf{B}^{-1}$, $\partial \mathbf{x}_B/\partial b_j$ is the jth column of $\mathbf{B}^{-1}$, and $\partial x_{B_i}/\partial b_j$ is the (i, j) entry of $\mathbf{B}^{-1}$.

Note that if the tableau corresponds to a degenerate basic feasible solution, then as a nonbasic variable x_j actually increases, at least one of the basic variables may become immediately negative (see Equation 3.8) destroying feasibility. In this case, the indicated rate of change in objective value with respect to x_j is not realizable, because when x_j increases holding the other nonbasic variables at zero, the resulting points are infeasible. Similar remarks hold for the other partial derivatives. In particular, if $\partial z/\partial b_i = w_i$ at optimality, then although this rate of change is valid in the usual sense of the partial derivative, it may not be the actual realizable change in the optimal objective function value when b_i is increased in the case of degeneracy. As we will be seeing later in Chapter 6, the objective function z as a function of b_i may not be differentiable in such a case, and so w_i may only be the slope of one possible tangential support to this function at the current value of b_i.

Identifying $\mathbf{B}^{-1}$ from the Simplex Tableau

The basis inverse matrix can be identified from the simplex tableau as follows. Assume that the original tableau has an identity matrix. The process of reducing the basis matrix $\mathbf{B}$ of the original tableau to an identity matrix in the current tableau, is equivalent to premultiplying rows 1 through m of the original tableau by $\mathbf{B}^{-1}$ to produce the current tableau (why?). This also converts the identity matrix of the original tableau to $\mathbf{B}^{-1}$. Therefore $\mathbf{B}^{-1}$ can be extracted from the current tableau as the submatrix in rows 1 through m under the original identity columns.

Example 3.10

To illustrate the interpretations of the simplex tableau, consider Example 3.9 at iteration 2. Then

$$\frac{\partial z}{\partial x_1} = -3, \quad \frac{\partial z}{\partial x_2} = 5, \quad \frac{\partial z}{\partial x_6} = 4$$

$$\frac{\partial x_4}{\partial x_1} = -3, \quad \frac{\partial x_5}{\partial x_1} = 0, \quad \frac{\partial x_3}{\partial x_6} = -1$$

$$\frac{\partial \mathbf{x}_B}{\partial x_2} = \begin{bmatrix} 1 \\ 2 \\ -1 \end{bmatrix}$$

$$\frac{\partial z}{\partial b_1} = \frac{\partial z}{\partial b_2} = 0, \quad \frac{\partial z}{\partial b_3} = -4$$

$$\frac{\partial x_5}{\partial b_2} = 1, \quad \frac{\partial x_4}{\partial b_3} = -2$$

$$\mathbf{B}^{-1} = \begin{bmatrix} 1 & 0 & -2 \\ 0 & 1 & 1 \\ 0 & 0 & 1 \end{bmatrix}$$

The vector $\mathbf{a}_2$ can be represented as a linear combination of the basic columns as follows: $\mathbf{a}_2 = -1\mathbf{a}_4 + 2\mathbf{a}_5 + \mathbf{a}_3$. At iteration 3 we have

$$\frac{\partial z}{\partial x_2} = 4, \quad \frac{\partial z}{\partial x_6} = 2$$

$$\frac{\partial x_5}{\partial x_2} = -2, \quad \frac{\partial x_5}{\partial x_6} = -1, \quad \frac{\partial x_3}{\partial x_2} = -\frac{2}{3}$$

$$\frac{\partial \mathbf{x}_B}{\partial x_2} = \begin{bmatrix} \frac{1}{3} \\ -2 \\ \frac{-2}{3} \end{bmatrix}$$

$$\frac{\partial z}{\partial b_1} = -1, \quad \frac{\partial z}{\partial b_2} = 0, \quad \frac{\partial z}{\partial b_3} = -2$$

$$\frac{\partial \mathbf{x}_B}{\partial b_3} = \begin{bmatrix} \frac{-2}{3} \\ 1 \\ \frac{1}{3} \end{bmatrix}$$

$$\frac{\partial x_1}{\partial b_1} = \frac{1}{3}, \quad \frac{\partial x_3}{\partial b_1} = \frac{1}{3}$$

$$\mathbf{B}^{-1} = \begin{bmatrix} \frac{1}{3} & 0 & \frac{-2}{3} \\ 0 & 1 & 1 \\ \frac{1}{3} & 0 & \frac{1}{3} \end{bmatrix}$$

The vector $\mathbf{a}_2$ can be represented as a linear combination of the basic columns as follows: $\mathbf{a}_2 = -\frac{1}{3}\mathbf{a}_1 + 2\mathbf{a}_5 + \frac{2}{3}\mathbf{a}_3$.

3.9 BLOCK PIVOTING

Throughout this chapter we have considered the possibility of entering a single nonbasic variable into the basis at each iteration. Recall that whenever a nonbasic variable enters the basis we must ensure that the new set of variables, the current basic set minus the exit variable plus the entering variable (1) also forms a basis, (2) remains feasible, that is, $x_{B_i} \geq 0$ for all i, and (3) the value of the objective function either remains constant or decreases (for a minimization problem). It is possible to enter sets of nonbasic variables as long as the same three conditions are satisfied. The process of introducing several nonbasic variables simultaneously is called *block pivoting*. However, in the case of multiple entries, the foregoing conditions become harder to ensure.

Suppose that we enter several nonbasic variables into the basis in such a way that condition 2 is maintained. Note that

$$z = \mathbf{c}_B\mathbf{B}^{-1}\mathbf{b} - \sum_{j \in R} (z_j - c_j)x_j$$

If we, for example, use the entry criterion that $z_j - c_j > 0$ for all entering variables, we shall ensure that the value of z will either remain constant or else decrease.

With respect to the question whether the new set still forms a basis, we must extend the rule that the pivot element be nonzero. Consider the basic equations before pivoting:

$$\mathbf{x}_B = \mathbf{B}^{-1}\mathbf{b} - \mathbf{B}^{-1}\mathbf{N}\mathbf{x}_N$$

Let $\bar{\mathbf{b}} = \mathbf{B}^{-1}\mathbf{b}$, $\mathbf{Y}_N = \mathbf{B}^{-1}\mathbf{N}$, $\mathbf{x}_B = \begin{pmatrix} \mathbf{x}_{B_1} \\ \mathbf{x}_{B_2} \end{pmatrix}$, $\mathbf{x}_N = \begin{pmatrix} \mathbf{x}_{N_1} \\ \mathbf{x}_{N_2} \end{pmatrix}$ where the vector $\mathbf{x}_{N_1}$ enters and the vector $\mathbf{x}_{B_2}$ leaves the basis. Here $\mathbf{x}_{B_2}$ and $\mathbf{x}_{N_1}$ each contain the same number of variables (why?). On partitioning the basic system, we get

$$\begin{bmatrix} \mathbf{I}_1 & \mathbf{0} \\ \mathbf{0} & \mathbf{I}_2 \end{bmatrix}\begin{pmatrix} \mathbf{x}_{B_1} \\ \mathbf{x}_{B_2} \end{pmatrix} = \begin{bmatrix} \bar{\mathbf{b}}_1 \\ \bar{\mathbf{b}}_2 \end{bmatrix} - \begin{bmatrix} \mathbf{Y}_{N_{11}} & \mathbf{Y}_{N_{12}} \\ \mathbf{Y}_{N_{21}} & \mathbf{Y}_{N_{22}} \end{bmatrix}\begin{pmatrix} \mathbf{x}_{N_1} \\ \mathbf{x}_{N_2} \end{pmatrix}$$

On rearranging, we get

$$\begin{bmatrix} \mathbf{I}_1 & \mathbf{Y}_{N_{11}} \\ \mathbf{0} & \mathbf{Y}_{N_{21}} \end{bmatrix}\begin{pmatrix} \mathbf{x}_{B_1} \\ \mathbf{x}_{N_1} \end{pmatrix} = \begin{bmatrix} \bar{\mathbf{b}}_1 \\ \bar{\mathbf{b}}_2 \end{bmatrix} - \begin{bmatrix} \mathbf{0} & \mathbf{Y}_{N_{12}} \\ \mathbf{I}_2 & \mathbf{Y}_{N_{22}} \end{bmatrix}\begin{pmatrix} \mathbf{x}_{B_2} \\ \mathbf{x}_{N_2} \end{pmatrix}$$

Now, the new set of variables, $\mathbf{x}_{B_1}$ and $\mathbf{x}_{N_1}$, will form a basis if and only if the matrix $\begin{bmatrix} \mathbf{I}_1 & \mathbf{Y}_{N_{11}} \\ \mathbf{0} & \mathbf{Y}_{N_{12}} \end{bmatrix}$ can be converted into the identity matrix; that is, if this matrix is invertible. From Chapter 2 we know that this matrix is invertible if and only if the determinant of the square matrix $\mathbf{Y}_{N_{12}}$ is nonzero. This is a natural extension of the rule for one variable entry. The new rule for maintaining a basis is as follows. Consider the square submatrix formed by the elements in the leaving rows and entering columns of the current tableau. If the determinant of this submatrix is nonzero, the new set will form a basis.

Rules for checking feasibility of the new basic set are more involved. Except in special circumstances, such as network flows, the rules for feasibility are difficult to check. This is primarily the reason why block pivoting is generally avoided in practice.

EXERCISES

3.1 Consider the linear program: Minimize $\mathbf{cx}$ subject to $\mathbf{Ax} \leq \mathbf{b}$, $\mathbf{x} \geq \mathbf{0}$, where $\mathbf{c}$ is a nonzero vector. Suppose that the point $\mathbf{x}_0$ is such that $\mathbf{Ax}_0 < \mathbf{b}$ and $\mathbf{x}_0 > \mathbf{0}$. Show that $\mathbf{x}_0$ cannot be an optimal solution.

3.2 Consider the following linear programming problem.

$$\begin{array}{lrcl} \text{Maximize} & x_1 + 3x_2 & & \\ \text{Subject to} & x_1 - 3x_2 & \leq & 3 \\ & -2x_1 + x_2 & \leq & 2 \\ & -3x_1 + 4x_2 & \leq & 12 \\ & 3x_1 + x_2 & \leq & 9 \\ & x_1, \quad x_2 & \geq & 0 \end{array}$$

a. Sketch the feasible region in the (x_1, x_2) space and identify the optimal solution.
b. Identify all the extreme points and reformulate the problem in terms of the convex combinations of the extreme points. Solve the resulting problem.
c. Suppose that the fourth constraint is dropped. Identify the extreme points and directions and reformulate the problem in terms of convex combinations of the extreme points and linear combinations of the extreme directions. Solve the resulting problem and interpret the solution.
d. Is the procedure described in parts (b) and (c) practical for solving larger problems? Discuss.

3.3 Consider a maximization linear programming problem with extreme points $\mathbf{x}_1$, $\mathbf{x}_2$, $\mathbf{x}_3$, and $\mathbf{x}_4$, and extreme directions $\mathbf{d}_1$, $\mathbf{d}_2$, and $\mathbf{d}_3$, and with an objective function gradient $\mathbf{c}$ such that $\mathbf{cx}_1 = 5$, $\mathbf{cx}_2 = 7$, $\mathbf{cx}_3 = 4$, $\mathbf{cx}_4 = 7$, $\mathbf{cd}_1 = 0$, $\mathbf{cd}_2 = -3$, and $\mathbf{cd}_3 = 0$. Characterize the set of alternative optimal solutions to this problem.

3.4 Consider the region defined by the constraints $\mathbf{Ax} \geq \mathbf{b}$, where $\mathbf{A}$ is an $m \times n$ matrix where $m > n$. Further suppose that rank $(\mathbf{A}) = n$. Show that $\mathbf{x}_0$ is an extreme point of the region if and only if the following decomposition of $\mathbf{A}$ and $\mathbf{b}$ is possible.

$$\mathbf{A} = \begin{bmatrix} \mathbf{A}_1 \\ \mathbf{A}_2 \end{bmatrix} \begin{matrix} n \text{ rows} \\ m-n \text{ rows} \end{matrix} \qquad \mathbf{b} = \begin{bmatrix} \mathbf{b}_1 \\ \mathbf{b}_2 \end{bmatrix} \begin{matrix} n \text{ rows} \\ m-n \text{ rows} \end{matrix}$$

$$\mathbf{A}_1\mathbf{x}_0 = \mathbf{b}_1, \quad \mathbf{A}_2\mathbf{x}_0 \geq \mathbf{b}_2$$

$$\text{rank}\,(\mathbf{A}_1) = n$$

3.5 Consider the following linear programming problem.

$$\begin{array}{ll} \text{Maximize} & 2x_1 + x_2 - x_3 \\ \text{Subject to} & x_1 + x_2 + 2x_3 \leq 6 \\ & x_1 + 4x_2 - x_3 \leq 4 \\ & x_1, x_2, x_3 \geq 0 \end{array}$$

Find the optimal solution by evaluating the objective function at all extreme points of the constraint set. Show that this approach is valid in this problem. Now suppose that the first constraint is replaced by $x_1 + x_2 - 2x_3 \leq 6$. Can the same approach for finding the optimal point be used? Explain why.

3.6 Consider the following linear programming problem.

$$\begin{array}{ll} \text{Maximize} & 2x_1 + x_2 + 4x_3 + 5x_5 + x_6 \\ \text{Subject to} & 3x_1 + 6x_2 + 3x_3 + 2x_4 + 3x_5 + 4x_6 \leq 60 \\ & x_1, x_2, x_3, x_4, x_5, x_6 \geq 0 \end{array}$$

This problem has one constraint in addition to the nonnegativity constraints, and is called a *knapsack problem*. Find all basic feasible solutions of the problem, and find the optimal by comparing these basic feasible solutions.

3.7 Consider the following constraints.

$$
\begin{aligned}
x_1 + x_2 &\leq 3 \\
-2x_1 + x_2 &\leq 2 \\
x_1 - 2x_2 &\leq 0 \\
x_1, \quad x_2 &\geq 0
\end{aligned}
$$

a. Draw the feasible region.
b. Identify the extreme points, and at each extreme point identify all possible basic and nonbasic variables.
c. Suppose that a move is made from the extreme point $\begin{bmatrix} 2 \\ 1 \end{bmatrix}$ to the extreme point $\begin{bmatrix} 0 \\ 0 \end{bmatrix}$ in the (x_1, x_2) space. Specify the possible entering and leaving variables.

3.8 Consider the polyhedral set consisting of points (x_1, x_2) such that

$$
\begin{aligned}
x_1 + x_2 &\leq 1 \\
x_1, \quad x_2 & \quad \text{unrestricted}
\end{aligned}
$$

Verify geometrically and algebraically that this set has no extreme points. Formulate an equivalent set in a higher dimension where all variables are restricted to be nonnegative. Show that extreme points of the new set indeed exist.

3.9 Let $X = \{\mathbf{x} : \mathbf{A}\mathbf{x} = \mathbf{b}, \ \mathbf{x} \geq \mathbf{0}\}$ where $\mathbf{A}$ is $m \times n$ of rank m. Let $\mathbf{x}$ be a feasible solution such that $x_1, \ldots, x_q$ are positive and $x_{q+1}, \ldots, x_n$ are zero. Assuming that the columns $\mathbf{a}_1, \ldots, \mathbf{a}_q$ of $\mathbf{A}$ corresponding to the positive variables are linearly dependent, construct feasible points $\mathbf{x}'$ and $\mathbf{x}''$ such that $\mathbf{x}$ is a convex combination of these points. Hence argue that if $\mathbf{x}$ is an extreme point of $\mathbf{X}$, it is also a basic feasible solution. Conversely, suppose that $\mathbf{x}$ is a basic feasible solution of X with basis $\mathbf{B}$ and that $\mathbf{x} = \lambda\mathbf{x}' + (1 - \lambda)\mathbf{x}''$ for some $0 < \lambda < 1$ and $\mathbf{x}', \mathbf{x}'' \in X$. Denoting $\mathbf{x}_B$ and $\mathbf{x}_N$ as the corresponding basic and nonbasic variables, show that $\mathbf{x}'_N = \mathbf{x}''_N = \mathbf{0}$ and $\mathbf{x}'_B = \mathbf{x}''_B = \mathbf{B}^{-1}\mathbf{b}$. Hence argue that $\mathbf{x}$ is an extreme point of X.

3.10 Consider the following system.

$$
\begin{aligned}
x_1 + x_2 + x_3 &\leq 2 \\
-x_1 + 2x_2 + 2x_3 &\leq 3 \\
x_1, \quad x_2, \quad x_3 &\geq 0
\end{aligned}
$$

The point $(\frac{1}{2}, \frac{1}{2}, \frac{1}{2})$ is feasible. Verify whether it is basic. If not, use the method described in the text for reducing it to a basic feasible solution.

3.11 Corresponding to the sequence of simplex iterations discussed in Section 3.4 with respect to the problem in Figure 3.5, draw on this figure the sequence of simplices considered by the simplex algorithm.

3.12 Solve the following problem.

$$\begin{array}{ll} \text{Maximize} & 5x_1 + 4x_2 \\ \text{Subject to} & x_1 + 2x_2 \leq 6 \\ & -2x_1 + x_2 \leq 4 \\ & 5x_1 + 3x_2 \leq 15 \\ & x_1, \quad x_2 \geq 0 \end{array}$$

a. Graphically.
b. By the simplex method.

3.13 Solve the following linear programming problem by the simplex method, at each iteration identifying $\mathbf{B}$ and $\mathbf{B}^{-1}$.

$$\begin{array}{ll} \text{Maximize} & 3x_1 + 2x_2 + x_3 \\ \text{Subject to} & 2x_1 - 3x_2 + 2x_3 \leq 3 \\ & -x_1 + x_2 + x_3 \leq 5 \\ & x_1, \quad x_2, \quad x_3 \geq 0 \end{array}$$

3.14 Solve the following problem by the simplex method starting with the basic feasible solution $(x_1, x_2) = (4, 0)$. Sketch the feasible region in the nonbasic variable space.

$$\begin{array}{ll} \text{Maximize} & -x_1 + 2x_2 \\ \text{Subject to} & 3x_1 + 4x_2 = 12 \\ & 2x_1 - x_2 \leq 12 \\ & x_1, \quad x_2 \geq 0 \end{array}$$

(*Hint*: Identify the initial basis and find its inverse.)

3.15 Consider the following problem.

$$\begin{array}{ll} \text{Maximize} & -3x_1 - 2x_2 \\ \text{Subject to} & -x_1 + x_2 \leq 1 \\ & 5x_1 + 3x_2 \leq 15 \\ & x_1 \geq 0 \\ & x_2 \geq \tfrac{3}{2} \end{array}$$

a. Solve the problem graphically.
b. Set up the problem in tableau form for the simplex method, and obtain an initial basic feasible solution.
c. Perform one pivot. After one pivot
   i. indicate the basic vectors
   ii. indicate the values of all variables
   iii. is the solution optimal?
d. Draw the requirements space representation.
   i. give the possible bases
   ii. give the possible feasible bases
e. Relate each basis in part (d)(i) to a point in part (a).

3.16 Consider the constraints $\mathbf{Ax} = \mathbf{b}$, $\mathbf{x} \geq \mathbf{0}$ and assume that they form a bounded region. Consider the following two problems, where x_n is the nth component of $\mathbf{x}$.

$$\begin{array}{ll} \text{Minimize} & x_n \\ \text{Subject to} & \mathbf{Ax} = \mathbf{b} \\ & \mathbf{x} \geq \mathbf{0} \end{array}$$

$$\begin{array}{ll} \text{Maximize} & x_n \\ \text{Subject to} & \mathbf{Ax} = \mathbf{b} \\ & \mathbf{x} \geq \mathbf{0} \end{array}$$

Let the optimal objectives of both problems be x_n' and x_n''. Let x_n be any number in the interval $[x_n', x_n'']$. Show that there exists a feasible point whose nth component is equal to x_n.

3.17 Answer the following questions along with a concise explanation with respect to the linear program to maximize $\mathbf{cx}$ subject to $\mathbf{x} \in X = \{\mathbf{x} : \mathbf{Ax} = \mathbf{b}, \mathbf{x} \geq \mathbf{0}\}$, where $\mathbf{A}$ is $m \times n$ of rank $m < n$.
a. In a simplex tableau, if $z_j - c_j = -10$ for a nonbasic variable x_j, what is the change in objective value when x_j enters the basis given that the minimum ratio is 2 in the pivot?
b. If an extreme point is optimal, then is it possible that not all $z_j - c_j \geq 0$ for an associated basis?
c. If there exists a $\mathbf{d}$ such that $\mathbf{Ad} = \mathbf{0}$, $\mathbf{d} \geq \mathbf{0}$, and $\mathbf{cd} > 0$, then is the optimal value unbounded?
d. Let $\bar{\mathbf{x}}$ be a feasible solution with exactly m positive components. Is $\bar{\mathbf{x}}$ necessarily an extreme point of X?
e. If a nonbasic variable x_k has $z_k - c_k = 0$ at optimality, then can one claim that alternative optimal solutions exist?
f. If $\mathbf{x}_1$ and $\mathbf{x}_2$ are adjacent points and if $\mathbf{B}_1$ and $\mathbf{B}_2$ are respective associated bases, then these bases are also adjacent. True or false?
g. Is it possible for an optimal solution to have more than m positive variables?

h. Suppose $n = m + 1$. What is the least upper bound on the number of extreme points and feasible bases?

i. A p-dimensional polyhedron can have at most p extreme directions. True or false?

j. Let $\overline{\mathbf{x}}$ be an extreme point with $(m-1)$ positive components. Then there are $(p+1)$ bases associated with this extreme point, where $p = n - m$. True or false? (Assume $\mathbf{Ax} = \mathbf{b}$ does not imply any variable to be a constant.)

3.18 a. Characterize the set of alternative optimal solutions given the following optimal tableau:

	z	x_1	x_2	x_3	x_4	x_5	x_6	RHS
z	1	0	0	0	0	2	3	0
x_1	0	1	0	2	−1	−1	1	0
x_2	0	0	1	−2	1	2	3	0

b. Consider the problem to maximize $\mathbf{cx}$ subject to the inequality constraints shown. Give all bases for which $z_j - c_j \geq 0$ for the nonbasic variables.

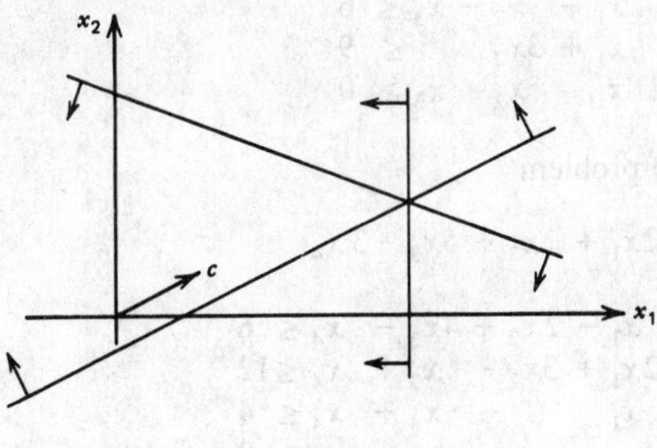

3.19 Suppose we have a basic feasible solution of the system $\mathbf{Ax} = \mathbf{b}$, $\mathbf{x} \geq \mathbf{0}$ with basis $\mathbf{B}$. Suppose that $z_k - c_k > 0$ and x_k is introduced into the basis and x_{B_r} is removed from the basis. Denote the new basis by $\hat{\mathbf{B}}$. Show algebraically that after pivoting:

a. The column under x_j is $(\hat{\mathbf{B}})^{-1}\mathbf{a}_j$.

b. The right-hand-side is $(\hat{\mathbf{B}})^{-1}\mathbf{b}$.

c. The new cost row is composed of $(\mathbf{c}_{\hat{B}})(\hat{\mathbf{B}})^{-1}\mathbf{a}_j - c_j$.

(*Hint*: Suppose

$$\mathbf{B} = (\mathbf{a}_1, \mathbf{a}_2, \ldots, \mathbf{a}_r, \ldots, \mathbf{a}_m)$$

$$\hat{\mathbf{B}} = (\mathbf{a}_1, \mathbf{a}_2, \ldots, \mathbf{a}_k, \ldots, \mathbf{a}_m)$$

First show that $\hat{\mathbf{B}} = \mathbf{BE}$, and $(\hat{\mathbf{B}})^{-1} = \mathbf{E}^{-1}\mathbf{B}^{-1}$, where

$$\mathbf{E} = \begin{bmatrix} 1 & 0 & \cdots & \overset{\overset{r\text{th column}}{\downarrow}}{y_{1k}} & \cdots & 0 \\ 0 & 1 & \cdots & y_{2k} & \cdots & 0 \\ \vdots & \vdots & & \vdots & & \vdots \\ 0 & 0 & \cdots & y_{rk} & \cdots & 0 \\ \vdots & \vdots & & \vdots & & \vdots \\ 0 & 0 & \cdots & y_{mk} & \cdots & 1 \end{bmatrix} \begin{matrix} \\ \\ \\ \leftarrow r\text{th row} \\ \\ \\ \end{matrix}$$

This form is usually called the *product form of the inverse* and is discussed in more detail in Section 5.1).

3.20 Suppose that we have a basic feasible solution that is nondegenerate. Further suppose that an improving nonbasic variable enters the basis. Show that if the minimum ratio criterion for exiting from the basis occurs at a unique index, then the resulting basic feasible solution is nondegenerate. Is this necessarily true if the first solution is degenerate?

3.21 Solve the following problem by the simplex method.

$$\begin{array}{ll} \text{Maximize} & x_1 - 2x_2 + x_3 \\ \text{Subject to} & \begin{array}{rcrcrcl} x_1 &+& 2x_2 &+& x_3 &\le& 12 \\ 2x_1 &+& x_2 &-& x_3 &\le& 6 \\ -x_1 &+& 3x_2 && &\le& 9 \\ x_1, && x_2, && x_3 &\ge& 0 \end{array} \end{array}$$

3.22 Consider the problem

$$\begin{array}{ll} \text{Maximize} & 2x_1 + x_2 + 5x_3 - 3x_4 \\ \text{Subject to} & \begin{array}{rcrcrcrcl} x_1 &+& 2x_2 &+& 4x_3 &-& x_4 &\le& 6 \\ 2x_1 &+& 3x_2 &-& x_3 &+& x_4 &\le& 12 \\ x_1 && &+& x_3 &+& x_4 &\le& 4 \\ x_1, && x_2, && x_3, && x_4 &\ge& 0 \end{array} \end{array}$$

Find a basic feasible solution with the basic variables as x_1, x_2, and x_4. Is this solution optimal? If not, then starting with this solution find the optimal solution.

3.23 Use the simplex method to solve the following problem. Note that the variables are unrestricted in sign.

$$\begin{array}{ll} \text{Minimize} & 3x_1 - x_2 \\ \text{Subject to} & \begin{array}{rcrcl} x_1 &-& 3x_2 &\ge& -3 \\ 2x_1 &+& x_2 &\ge& -2 \\ 2x_1 &+& x_2 &\le& 8 \\ 4x_1 &-& x_2 &\le& 16 \end{array} \end{array}$$

3.24 An agricultural mill produces feed for cattle. The cattle feed consists of three main ingredients: corn, lime, and fish meal. These ingredients contain three nutrients: protein, calcium, and vitamins. The following table gives the nutrient contents per pound of each ingredient.

	INGREDIENT		
NUTRIENT	CORN	LIME	FISH MEAL
Protein	25	15	25
Calcium	15	30	20
Vitamins	5	12	8

The protein, calcium, and vitamins content per pound of the cattle feed must be in the following intervals respectively: $[18, 22]$, $[20, \infty)$, and $[6, 12]$. If the selling prices per pound of corn, lime, and fish meal are respectively \$0.10, \$0.08, and \$0.12, find the least expensive mix.
(*Hint*. First find a basis $\mathbf{B}$ such that $\mathbf{B}^{-1}\mathbf{b} \geq \mathbf{0}$.)

3.25 A firm makes three products 1, 2, and 3. Each product requires production time in three departments as shown.

PRODUCT	DEPARTMENT 1	DEPARTMENT 2	DEPARTMENT 3
1	3 hr/unit	2 hr/unit	1 hr/unit
2	4 hr/unit	1 hr/unit	3 hr/unit
3	2 hr/unit	2 hr/unit	3 hr/unit

There are 600, 400, and 300 hours of production time available in the three departments respectively. If each unit of products 1, 2, and 3 contribute \$2, \$4, and \$2.5 to profit respectively, find the optimal product mix.

3.26 Solve Exercise 1.10 as a linear model by the simplex method. Find the objective $36/x_2x_3$ corresponding to the optimal point obtained from the simplex method. By trial and error see if you can find a feasible point whose objective $36/x_2x_3$ is better than the objective obtained previously.

3.27 Solve Exercise 1.11 to find the number of barrels of each crude oil that satisfy the demand and minimize the total cost.
(*Hint*. First find a basis $\mathbf{B}$ with $\mathbf{B}^{-1}\mathbf{b} \geq \mathbf{0}$.)

3.28 A nut packager has on hand 150 pounds of peanuts, 100 pounds of cashews, and 50 pounds of almonds. The packager can sell three kinds of mixtures of these nuts: a cheap mix consisting of 80 percent peanuts and 20 percent cashews; a party mix with 50 percent peanuts, 30 percent cashews, and 20 percent almonds; and a delux mix with 20 percent peanuts, 50 percent cashews, and 30 percent almonds. If the 12-ounce can of the cheap mix, the party mix, and the deluxe mix can be sold for \$0.90, \$1.10, and \$1.30 respectively, how many cans of each type would the packager produce in order to maximize the return?

3.29 Suppose, *a priori*, we know that a solution cannot be optimal unless it involved a variable at a positive value. Show that this variable can be eliminated and that the reduced system with one less equation and variable can be solved in its place. Illustrate by an example.

3.30 Consider the linear program: Minimize $\mathbf{cx}$ subject to $\mathbf{Ax} \geq \mathbf{b}$, $\mathbf{x} \geq \mathbf{0}$. Converting the inequality constraints to equality constraints, suppose that the optimal basis is $\mathbf{B}$. Show that $\mathbf{w} = \mathbf{c}_B\mathbf{B}^{-1} \geq \mathbf{0}$.

3.31 Suppose that some tableau for a linear programming problem exhibits an indication of unboundedness. Considering only the basic vectors and the nonbasic vector corresponding to the column giving the unboundedness indication, demonstrate which quantities, if any, satisfy the definition of an extreme point. Also demonstrate which quantities, if any, satisfy the definition of an extreme ray. Give an example.

3.32 A necessary and sufficient condition for unboundedness of the objective value of a minimization problem is that there exists a direction of the feasible region such that $\mathbf{cd} < 0$. A condition for unboundedness in the simplex method is the existence of an index j such that $z_j - c_j > 0$ and $\mathbf{y}_j \leq \mathbf{0}$. Discuss in detail the correspondence between the two conditions. (*Hint*. Let

$$\mathbf{d} = \begin{bmatrix} -\mathbf{y}_j \\ \hline 0 \\ \vdots \\ 1 \\ \vdots \\ 0 \end{bmatrix}$$

where the 1 appears at the jth position. Show that $\mathbf{d}$ is a direction of the set and that $\mathbf{cd} = c_j - z_j$. Can you show that $\mathbf{d}$ is an extreme direction?)

3.33 Consider the following problem.

$$\begin{array}{lrl} \text{Maximize} & -3x_1 + 2x_2 - x_3 + x_4 & \\ \text{Subject to} & 2x_1 - 3x_2 - x_3 + x_4 & \leq 8 \\ & -x_1 + 2x_2 + 2x_3 - 3x_4 & \leq 10 \\ & -x_1 + x_2 - 4x_3 + x_4 & \leq 3 \\ & x_1, \quad x_2, \quad x_3, \quad x_4 & \geq 0 \end{array}$$

Use the simplex method to verify that the optimal solution is unbounded. Make use of the final simplex tableau to construct a feasible solution with an objective of at least 3000. Make use of the final tableau to construct a direction $\mathbf{d}$ such that $\mathbf{cd} > 0$.

3.34 Prove or give a counterexample. In order to have a basic variable in a particular row of the simplex tableau, that variable must have a nonzero coefficient in its original column and the particular row.

3.35 We showed in the text that $z_j - c_j = 0$ for a basic variable. Interpret this result.

3.36 Consider the linear programming problem: Maximize **cx** subject to **Ax** = **b**. $\mathbf{x} \geq \mathbf{0}$ where **A** is an $m \times n$ matrix of rank m. Suppose an optimal solution with basis **B** is at hand. Further suppose that **b** is replaced by $\mathbf{b} + \lambda\mathbf{d}$ where λ is a scalar and **d** is a fixed nonzero vector of dimension m. Give a condition such that the basis **B** will be optimal for all $\lambda \geq 0$.

3.37 Show that in the simplex method if a variable x_j leaves the basis, it cannot enter the basis in the next iteration.

3.38 Can a vector that is inserted at one iteration in the simplex method be removed immediately at the next iteration? When can this occur and when is it impossible?

3.39 Find a nonbasic feasible optimal solution of the following problem.

$$\begin{array}{llr}\text{Maximize} & 11x_1 + 2x_2 - x_3 + 3x_4 + 4x_5 + x_6 & \\ \text{Subject to} & 5x_1 + x_2 - x_3 + 2x_4 + x_5 & = 12 \\ & -14x_1 - 3x_2 + 3x_3 - 5x_4 + x_6 & = 2 \\ & 2x_1 + \frac{1}{2}x_2 - \frac{1}{2}x_3 + \frac{1}{2}x_4 & \leq \frac{5}{2} \\ & 3x_1 + \frac{1}{2}x_2 + \frac{1}{2}x_3 + \frac{3}{2}x_4 & \leq 3 \\ & x_1, \quad x_2, \quad x_3, \quad x_4, \quad x_5, \quad x_6 & \geq 0\end{array}$$

(*Hint*. Let the initial basis consist of x_5, x_6 and the slack variables of the last two constraints. Then find alternative optimal solutions of the problem.)

3.40 The following mathematical formulation describes a problem of allocating three resources to the annual production of three commodities by a manufacturing firm. The amounts of the three products to be produced are denoted by x_1, x_2, and x_3. The objective function reflects the dollar contribution to profit of these products.

$$\begin{array}{llr}\text{Maximize} & 10x_1 + 15x_2 + 5x_3 & \\ \text{Subject to} & 2x_1 + x_2 & \leq 6000 \\ & 3x_1 + 3x_2 + x_3 & \leq 9000 \\ & x_1 + 2x_2 + 2x_3 & \leq 4000 \\ & x_1, \quad x_2, \quad x_3 & \geq 0\end{array}$$

a. Without using the simplex method, verify that the optimal basis consists of the slack variable of the first constraint, x_1, and x_2.

b. Make use of the information in (a) to write the optimal tableau.

c. The Research and Development Department proposes a new product whose production coefficients are represented by $[2, 4, 1]^t$. If the contribution to profit is \$12 per unit of this new product, should this product be produced? If so, what is the new optimal program?

d. What is the minimal contribution to profit that should be expected before production of this new product would actually increase the value of the objective function?

3.41 Write a precise argument showing that in the absence of degeneracy and assuming feasibility, the simplex method will provide an optimal solution or show unboundedness of a linear program in a finite number of steps.

3.42 Consider the following linear program.

$$\text{Minimize} \quad -x_1 - 2x_2 + x_3$$

$$\begin{aligned}\text{Subject to} \quad 2x_1 + x_2 + x_3 &\le 6\\ 2x_2 - x_3 &\le 3\\ x_1, \quad x_2, \quad x_3 &\ge 0\end{aligned}$$

a. Find the optimal solution by the simplex method. At each iteration identify $\mathbf{B}, \mathbf{N}, \mathbf{B}^{-1}, \mathbf{B}^{-1}\mathbf{N}, \mathbf{c}_B\mathbf{B}^{-1}$, and the $z_j - c_j$'s.

b. At optimality, find $\partial x_1/\partial x_3$, $\partial x_2/\partial x_4$, $\partial z/\partial x_5$, $\partial \mathbf{x}_B/\partial \mathbf{x}_N$, where x_4 and x_5 are the slack variables. Interpret these values.

c. Suppose that c_1 is changed from -1 to $-1 + \Delta_1$, and c_2 is changed from -2 to $-2 + \Delta_2$. Find the region in the (Δ_1, Δ_2) space that will maintain optimality of the vector you obtained in (a).

d. Suppose a new activity x_6 is considered. Here $c_6 = -3$, $a_{16} = 3$, and $a_{26} = 3$. It is worthwhile to produce the activity? If your answer is yes, find the new optimal solution.

e. Suppose that b_1 is changed from 6 to $6 + \Delta$. Find the range of Δ that will maintain optimality of the basis found in part (a).

f. Make use of the final tableau to represent the column $\mathbf{a}_3$ as a linear combination of $\mathbf{a}_1$ and $\mathbf{a}_2$.

3.43 Consider the following linear programming problem.

$$\text{Maximize} \quad 2x_1 + 12x_2 + 7x_3$$

$$\begin{aligned}\text{Subject to} \quad x_1 + 3x_2 + 2x_3 &\le 10{,}000\\ 2x_1 + 2x_2 + x_3 &\le 4000\\ x_1, \quad x_2, \quad x_3 &\ge 0\end{aligned}$$

The optimal solution is shown, where z is the objective function and x_4 and x_5 are slack variables.

	z	x_1	x_2	x_3	x_4	x_5	RHS
z	1	12	2	0	0	7	28,000
x_4	0	-3	-1	0	1	-2	2,000
x_3	0	2	2	1	0	1	4,000

a. What are the rates of increase of the objective as a function of the right-hand-side of the first and second constraints respectively?
b. Suppose that the right-hand-side of the second constraint is changed to $4000 + \Delta$. What is the range of Δ that will keep the basis of the foregoing tableau optimal?
c. Find explicitly the optimal value z as a function of Δ.
d. Suppose that increasing the right-hand-side of the second constraint involved expanding a manufacturing department. This will involve a fixed cost as well as a variable cost that is a function of Δ. In particular the cost as a function of Δ is

$$h(\Delta) = \begin{cases} 0 & \text{if } \Delta = 0 \\ 3000 + 3\Delta & \text{if } \Delta > 0 \end{cases}$$

What is the break-even point, where the cost and the added profit will balance? What do you recommend for the optimal value of Δ?

3.44 Solve Exercise 1.12 by the simplex method.

3.45 Solve Exercise 1.13 by the simplex method. Suppose that extra man-hours can be obtained by allowing overtime at the average of $12 per hour. Would it be profitable to increase the man-hours? If so, by how much? How would this increase the profit?

3.46 The starting and current tableaux of a given problem are shown. Find the values of the unknowns a through l.

Starting Tableau

z	x_1	x_2	x_3	x_4	x_5	RHS
1	a	1	-3	0	0	0
0	b	c	d	1	0	6
0	-1	2	e	0	1	1

Current Tableau

z	x_1	x_2	x_3	x_4	x_5	RHS
1	0	$-\frac{1}{3}$	j	k	l	-4
0	g	$\frac{2}{3}$	$\frac{2}{3}$	$\frac{1}{3}$	0	f
0	h	i	$-\frac{1}{3}$	$\frac{1}{3}$	1	3

3.47 The following is the current simplex tableau of a given maximization problem. The objective is to maximize $2x_1 - 3x_2$, and the slack variables are x_3 and x_4. The constraints are of the $\leq$ type.

	z	x_1	x_2	x_3	x_4	RHS
z	1	b	1	f	g	6
x_3	0	c	0	1	$\frac{1}{5}$	4
x_1	0	d	e	0	2	a

a. Find the unknowns a through g.
b. Find $\mathbf{B}^{-1}$.
c. Find $\partial x_3/\partial x_2$, $\partial z/\partial b_1$, $\partial z/\partial x_4$, $\partial x_1/\partial b_2$.
d. Is the tableau optimal?

3.48 The following is the current simplex tableau of a linear programming problem. The objective is to minimize $-2x_4 - x_5 - 2x_6$, and x_1, x_2, and x_3 are the slack variables.

	z	x_1	x_2	x_3	x_4	x_5	x_6	RHS
z	1	b	c	0	0	h	g	−14
x_6	0	2	0	$-\frac{14}{3}$	0	1	1	a
x_2	0	3	d	2	0	$\frac{5}{2}$	0	5
x_4	0	0	e	f	1	2	0	0

a. Find the values of the unknowns a through h in the tableau.
b. Find $\mathbf{B}^{-1}$.
c. Find $\partial x_2/\partial x_1$, $\partial z/\partial x_5$, $\partial x_6/\partial b_3$.
d. Without explicitly finding the basic vectors $\mathbf{a}_6, \mathbf{a}_2, \mathbf{a}_4$, give the representation of the vector $\mathbf{a}_5$ in terms of these basic vectors.

3.49 Consider the problem: Maximize $\mathbf{cx}$ subject to $\mathbf{Ax} = \mathbf{b}$, $\mathbf{x}$ unrestricted in sign. Under what conditions does this problem have a bounded optimal solution?

3.50 Consider a linear programming problem in which some of the variables are unrestricted in sign. What are the conditions for a bounded optimal solution? Without introducing additional variables, show how the entry and exit criteria of the simplex method can be modified such that the unrestricted variables are handled directly. How does the simplex method recognize reaching an optimal solution in this case? Illustrate by solving the following problem.

Minimize $-2x_1 + x_2$

Subject to
$$x_1 + x_2 \leq 4$$
$$x_1 - x_2 \leq 6$$
$$x_1 \geq 0$$
$$x_2 \quad \text{unrestricted}$$

3.51 Consider the following simplex tableau for a minimization problem (the constraints are of the $\leq$ type and x_3, x_4, and x_5 are the slacks).

z	x_1	x_2	x_3	x_4	x_5	RHS
1	0	a	0	b	0	f
0	1	−2	0	1	0	c
0	0	−1	1	2	0	d
0	0	0	0	3	1	e

Suppose that $a < 0$, $b \leq 0$, and $c, d, e \geq 0$

a. Find $\mathbf{B}^{-1}$.
b. Find $\mathbf{B}$.
c. Is the tableau optimal?
d. Give the original tableau.
e. From the tableau identify $\mathbf{c}_B\mathbf{B}^{-1}$ and give its interpretation.

Now suppose that $a > 0$, $b \leq 0$; and $c, d, e \geq 0$.

f. Is the new tableau optimal?
g. Give an extreme direction.
h. Let $a = 3$ and $f = -8$. Give a feasible solution with $z = -200$.

3.52 Consider the problem to minimize $z = \mathbf{cx}$, subject to $\mathbf{Ax} \leq \mathbf{b}$ and $\mathbf{x} \geq \mathbf{0}$. Hence we are to find the smallest z for which the inequality system $z \geq \mathbf{cx}$, $\mathbf{Ax} \leq \mathbf{b}$, $\mathbf{x} \geq \mathbf{0}$ has a solution. Consider the variable x_1. Using each inequality with x_1 on one side and all other terms on the other side, show how x_1 can be eliminated from the problem. Hence show how this can be repeated to solve the linear program. Illustrate using the example in Exercise 3.5. (This is known as the Fourier–Motzkin Elimination Method.)

3.53 Consider the linear program to maximize $\mathbf{cx}$ subject to $\mathbf{Ax} = \mathbf{b}$ and $\mathbf{x} \geq \mathbf{0}$, where $\mathbf{A}$ is $m \times n$ of rank m. Let $\mathbf{B}$ be any basis matrix, and let $\mathbf{B}^*$ be the basis associated with any optimal basic feasible solution. Define $z_j = \mathbf{c}_B\mathbf{B}^{-1}\mathbf{a}_j$ and $z_j^* = \mathbf{c}_{B^*}\mathbf{B}^{*-1}\mathbf{a}_j$ for all j. Consider a particular variable x_i for which there exists a set of m columns $\mathbf{a}_{i(j)}$, $j = 1, \ldots, m$ chosen from $(\mathbf{A}, \mathbf{b})$, and a corresponding set of m scalars $\theta_{i(j)}$, $j = 1, \ldots, m$ such that

$$\mathbf{B}^{-1}\left[\mathbf{a}_i - \sum_{j=1}^{m} \theta_{i(j)}\mathbf{a}_{i(j)}\right] \geq 0$$

and

$$(z_i - c_i) - \sum_{j=1}^{m} \theta_{i(j)}[z_{i(j)} - z_{i(j)}^*] > 0$$

Then show that x_i must be nonbasic at optimality. Is this generally implementable? Can you identify an implementable special case? (*Hint*: Since $\mathbf{c}_{B^*}\mathbf{B}^{*-1}\mathbf{B} \geq \mathbf{c}_B$ (why?), obtain an inequality by postmultiplying both sides of this by the expression in the first condition and then use the second condition.)

3.54 Construct a detailed flow diagram of the simplex method. What are the number of operations (additions, subtractions, multiplications, divisions) that are needed at each simplex iteration? Using FORTRAN (or another language), convert the flow diagram into a simplex code.

NOTES AND REFERENCES

1. This chapter describes the simplex algorithm of Dantzig (developed in 1947 and published at a later date in 1949). The material of this chapter is standard and can be found in most linear programming books. The historical information is from Dantzig [1982].

2. In Section 3.1 we proved optimality at an extreme point via the Representation Theorem. The reader may note that the simplex algorithm itself gives a constructive proof of this fact.
3. The material on block pivoting is from Tucker [1960]. For further reading on block pivoting, see Dantzig [1963a], Lasdon [1970], and Cooper and Kennington [1979]. For a discussion on some nonadjacent extreme point methods for linear programs, see Sherali, Soyster, and Baines [1983], Sherali and Baines [1984], Murty and Fathi [1984], Murty [1986], and Sethi and Thompson [1984]. A host of nonlinear programming approaches to linear programming problems have been inspired by Khachian's [1979] and Karmarkar's [1984] polynomial-time algorithms for linear programs (see Chapter 8).
4. Exercise 3.53 relates to Cheng's [1980, 1985] theoretical work (see also Brosius [1981]) on *a priori* recognizing basic or nonbasic variables at optimality.

FOUR: STARTING SOLUTION AND CONVERGENCE

In the previous chapter, we developed the simplex method with the assumption that an initial basic feasible solution is at hand. In many cases, such a solution is not readily available, and some work may be needed to get the simplex method started. In this chapter we describe two procedures (the two-phase method and the big-M method), both involving *artificial variables* to obtain an initial basic feasible solution to a slightly modified set of constraints. The simplex method is used to eliminate the artificial variables and to solve the original problem. We also discuss in more detail the difficulties associated with degeneracy. In particular we show that the simplex method converges in a finite number of steps, even in the presence of degeneracy, provided that a special rule for entering and/or exiting from the basis is adopted.

4.1 THE INITIAL BASIC FEASIBLE SOLUTION

Recall that the simplex method starts with a basic feasible solution and moves to an improved basic feasible solution, until the optimal point is reached or else unboundedness of the objective function is verified. However, in order to initialize the simplex method, a basis $\mathbf{B}$ with $\bar{\mathbf{b}} = \mathbf{B}^{-1}\mathbf{b} \geq \mathbf{0}$ must be available. We shall show that the simplex method can always be initiated with a very simple basis, namely the identity.

Easy Case

Suppose that the constraints are of the form $\mathbf{Ax} \leq \mathbf{b}$, $\mathbf{x} \geq \mathbf{0}$ where $\mathbf{A}$ is an $m \times n$ matrix and $\mathbf{b}$ is an m nonnegative vector. By adding the slack vector $\mathbf{x}_s$, the constraints are put in the following standard form: $\mathbf{Ax} + \mathbf{x}_s = \mathbf{b}$, $\mathbf{x} \geq \mathbf{0}$, $\mathbf{x}_s \geq \mathbf{0}$. Note that the new $m \times (m + n)$ constraint matrix $(\mathbf{A}, \mathbf{I})$ has rank m, and a basic feasible solution of this system is at hand, by letting $\mathbf{x}_s$ be the basic vector, and $\mathbf{x}$ be the nonbasic vector. Hence at this starting basic feasible solution $\mathbf{x}_s = \mathbf{b}$ and $\mathbf{x} = \mathbf{0}$, and the simplex method now can be applied.

Some Bad Cases

In many occasions, finding a starting basic feasible solution is not as straightforward as the case just described. To illustrate, suppose that the constraints are of the form $\mathbf{Ax} \leq \mathbf{b}$, $\mathbf{x} \geq \mathbf{0}$ but the vector $\mathbf{b}$ is not nonnegative. In this case, after introducing the slack vector $\mathbf{x}_s$, we cannot let $\mathbf{x} = \mathbf{0}$, because $\mathbf{x}_s = \mathbf{b}$ violates the nonnegativity requirement.

Another situation occurs when the constraints are of the form $\mathbf{Ax} \geq \mathbf{b}$, $\mathbf{x} \geq \mathbf{0}$, where $\mathbf{b} \nleq \mathbf{0}$. After subtracting the slack vector $\mathbf{x}_s$, we get $\mathbf{Ax} - \mathbf{x}_s = \mathbf{b}$, $\mathbf{x} \geq \mathbf{0}$, and $\mathbf{x}_s \geq \mathbf{0}$. Again, there is no obvious way of picking a basis $\mathbf{B}$ from the matrix $(\mathbf{A}, -\mathbf{I})$ with $\bar{\mathbf{b}} = \mathbf{B}^{-1}\mathbf{b} \geq \mathbf{0}$.

In general any linear programming problem can be transformed into a problem of the following form.

$$\begin{aligned} \text{Minimize} \quad & \mathbf{cx} \\ \text{Subject to} \quad & \mathbf{Ax} = \mathbf{b} \\ & \mathbf{x} \geq \mathbf{0} \end{aligned}$$

where $\mathbf{b} \geq \mathbf{0}$ (if $b_i < 0$, the ith row can be multiplied by -1). This can be accomplished by introducing slack variables and by simple manipulations of the constraints and variables. If $\mathbf{A}$ contains an identity matrix, then an immediate basic feasible solution is at hand, by simply letting $\mathbf{B} = \mathbf{I}$, and since $\mathbf{b} \geq \mathbf{0}$, then $\mathbf{B}^{-1}\mathbf{b} = \mathbf{b} \geq \mathbf{0}$. Otherwise, something else must be done. A trial and error approach may be futile, particularly if the problem is infeasible!

Example 4.1

a. Consider the following constraints:

$$\begin{aligned} x_1 + 2x_2 &\leq 4 \\ -x_1 + x_2 &\leq 1 \\ x_1, \ x_2 &\geq 0 \end{aligned}$$

After adding the slack variables x_3 and x_4, we get

$$\begin{aligned} x_1 + 2x_2 + x_3 &= 4 \\ -x_1 + x_2 + x_4 &= 1 \\ x_1, \quad x_2, \quad x_3, \quad x_4 &\geq 0 \end{aligned}$$

An obvious starting basic feasible solution is given by $\mathbf{x}_B = \begin{bmatrix} x_3 \\ x_4 \end{bmatrix} = \begin{bmatrix} 4 \\ 1 \end{bmatrix}$ and $\mathbf{x}_N = \begin{bmatrix} x_1 \\ x_2 \end{bmatrix} = \begin{bmatrix} 0 \\ 0 \end{bmatrix}$.

b. Consider the following constraints:

$$\begin{aligned} x_1 + x_2 + x_3 &\le 6 \\ -2x_1 + 3x_2 + 2x_3 &\ge 3 \\ x_2, \quad x_3 &\ge 0 \end{aligned}$$

Note that x_1 is unrestricted. So the change of variable $x_1 = x_1^+ - x_1^-$ is made. Also the slack variables x_4 and x_5 are introduced. This leads to the following constraints in standard form:

$$\begin{aligned} x_1^+ - x_1^- + x_2 + x_3 + x_4 \qquad &= 6 \\ -2x_1^+ + 2x_1^- + 3x_2 + 2x_3 \qquad - x_5 &= 3 \\ x_1^+, \quad x_1^-, \quad x_2, \quad x_3, \quad x_4, \quad x_5 &\ge 0 \end{aligned}$$

Note that the constraint matrix does not contain an identity and no obvious feasible basis $\mathbf{B}$ can be extracted.

c. Consider the following constraints:

$$\begin{aligned} x_1 + x_2 - 2x_3 &\le -3 \\ -2x_1 + x_2 + 3x_3 &\le 7 \\ x_1, \quad x_2, \quad x_3 &\ge 0 \end{aligned}$$

Since the right-hand-side of the first constraint is negative, the first inequality is multiplied by -1. Introducing the slack variables x_4 and x_5 leads to the following system:

$$\begin{aligned} -x_1 - x_2 + 2x_3 - x_4 \qquad &= 3 \\ -2x_1 + x_2 + 3x_3 \qquad + x_5 &= 7 \\ x_1, \quad x_2, \quad x_3, \quad x_4, \quad x_5 &\ge 0 \end{aligned}$$

Note again that this constraint matrix contains no identity.

Artificial Variables

After manipulating the constraints and introducing slack variables, suppose that the constraints are put in the format $\mathbf{Ax} = \mathbf{b}$, $\mathbf{x} \ge \mathbf{0}$ where $\mathbf{A}$ is an $m \times n$ matrix and $\mathbf{b} \ge \mathbf{0}$ is an m vector. Further suppose that $\mathbf{A}$ has no identity submatrix (if $\mathbf{A}$ has an identity submatrix then we have an obvious starting basic feasible solution). In this case we shall resort to artificial variables to get a starting basic feasible solution, and then use the simplex method itself and get rid of these artificial variables.

To illustrate, suppose that we change the restrictions by adding an artificial vector $\mathbf{x}_a$ leading to the system $\mathbf{Ax} + \mathbf{x}_a = \mathbf{b}$, $\mathbf{x} \ge \mathbf{0}$, $\mathbf{x}_a \ge \mathbf{0}$. Note that by construction, we forced an identity matrix corresponding to the artificial vector. This gives an immediate basic feasible solution of the new system, namely $\mathbf{x}_a = \mathbf{b}$ and $\mathbf{x} = \mathbf{0}$. Even though we now have a starting basic feasible solution and the simplex method can be applied, we have in effect changed the problem. In order to get back to our original problem, we must force these artificial variables to zero, because $\mathbf{Ax} = \mathbf{b}$ if and only if $\mathbf{Ax} + \mathbf{x}_a = \mathbf{b}$ with $\mathbf{x}_a = \mathbf{0}$. In other words, *artificial variables* are only a tool to get the simplex method started; however, we must guarantee that these variables will eventually drop to zero.

At this stage, it is worthwhile to note the difference between slack and artificial variables. A slack variable is introduced to put the problem in equality form, and the slack variable can very well be positive, which means that the inequality holds as a strict inequality. *Artificial variables*, however, are not legitimate variables, and they may be introduced to facilitate the initiation of the simplex method. These artificial variables, however, must eventually drop to zero in order to attain feasibility in the original problem.

Example 4.2

Consider the following constraints:

$$\begin{aligned} x_1 + 2x_2 &\geq 4 \\ -3x_1 + 4x_2 &\geq 5 \\ 2x_1 + x_2 &\leq 6 \\ x_1, \quad x_2 &\geq 0 \end{aligned}$$

Introducing the slack variables x_3, x_4, and x_5, we get

$$\begin{array}{rrrrrl} x_1 + & 2x_2 & - x_3 & & & = 4 \\ -3x_1 + & 4x_2 & & - x_4 & & = 5 \\ 2x_1 + & x_2 & & & + x_5 & = 6 \\ x_1, & x_2, & x_3, & x_4, & x_5 & \geq 0 \end{array}$$

This constraint matrix has no identity submatrix. We can introduce three artificial variables to obtain a starting basic feasible solution. Note, however, that x_5 appears in the last row only and it has coefficient 1. So we only need to introduce two artificial variables x_6 and x_7, which leads to the following system.

$$\begin{array}{rrrrrrrl} \multicolumn{5}{c}{\overbrace{\hspace{14em}}^{\textit{Legitimate variables}}} & \multicolumn{2}{c}{\overbrace{\hspace{6em}}^{\textit{Artificial variables}}} & \\ x_1 + & 2x_2 & - x_3 & & & + x_6 & & = 4 \\ -3x_1 + & 4x_2 & & - x_4 & & & + x_7 & = 5 \\ 2x_1 + & x_2 & & & + x_5 & & & = 6 \\ x_1, & x_2, & x_3, & x_4, & x_5, & x_6, & x_7 & \geq 0 \end{array}$$

Now we have an immediate starting basic feasible solution of the new system, namely $x_5 = 6$, $x_6 = 4$, and $x_7 = 5$. The rest of the variables are nonbasic and have value zero. Needless to say, we eventually would like for the artificial variables x_6 and x_7 to drop to zero.

4.2 THE TWO-PHASE METHOD

There are various methods that can be used to eliminate the artificial variables. One of these methods is to minimize the sum of the artificial variables, subject to the constraints $\mathbf{Ax} + \mathbf{x}_a = \mathbf{b}$, $\mathbf{x} \geq \mathbf{0}$ and $\mathbf{x}_a \geq \mathbf{0}$. If the original problem has a feasible solution, then the optimal value of this problem is zero, where all the artificial variables drop to zero. More importantly, as the artificial variables drop to zero, they leave the basis, and legitimate variables enter instead. Eventually all artificial variables leave the basis (this is not always the case, because we may have an artificial variable in the basis at level zero; this will be discussed later in greater detail). The basis then consists of legitimate variables. In other words, we get a basic feasible solution of the original system $\mathbf{Ax} = \mathbf{b}$, $\mathbf{x} \geq \mathbf{0}$, and the simplex method can be started with the original objective $\mathbf{cx}$. If, on the other hand, after solving this problem we have a positive artificial variable, then the original problem has no feasible solution (why?). This procedure is called the *two-phase method*. In the first phase we reduce artificial variables to value zero or conclude that the original problem has no feasible solutions. In the former case, the second phase minimizes the original objective function starting with the basic feasible solution obtained at the end of the phase I. The two-phase method is outlined here.

Phase I

Solve the following linear program starting with the basic feasible solution $\mathbf{x} = \mathbf{0}$ and $\mathbf{x}_a = \mathbf{b}$.

$$\begin{aligned} \text{Minimize} \quad & \mathbf{1x}_a \\ \text{Subject to} \quad & \mathbf{Ax} + \mathbf{x}_a = \mathbf{b} \\ & \mathbf{x},\ \mathbf{x}_a \geq \mathbf{0} \end{aligned}$$

If at optimality $\mathbf{x}_a \neq \mathbf{0}$, then stop; the original problem has no feasible solutions. Otherwise let the basic and nonbasic legitimate variables be $\mathbf{x}_B$ and $\mathbf{x}_N$. (We are assuming that all artificial variables left the basis. The case where some artificial variables remain in the basis at zero level will be discussed later.) Go to phase II.

Phase II

Solve the following linear program starting with the basic feasible solution $\mathbf{x}_B = \mathbf{B}^{-1}\mathbf{b}$ and $\mathbf{x}_N = \mathbf{0}$.

$$\begin{aligned} \text{Minimize} \quad & \mathbf{c}_B\mathbf{x}_B + \mathbf{c}_N\mathbf{x}_N \\ \text{Subject to} \quad & \mathbf{x}_B + \mathbf{B}^{-1}\mathbf{N}\mathbf{x}_N = \mathbf{B}^{-1}\mathbf{b} \\ & \mathbf{x}_B, \mathbf{x}_N \geq \mathbf{0} \end{aligned}$$

The foregoing problem is of course equivalent to the original problem.

Example 4.3

$$\begin{array}{ll} \text{Minimize} & x_1 - 2x_2 \\ \text{Subject to} & x_1 + x_2 \geq 2 \\ & -x_1 + x_2 \geq 1 \\ & x_2 \leq 3 \\ & x_1, \quad x_2 \geq 0 \end{array}$$

The feasible region and the path taken by phase I and phase II to reach the optimal point are shown in Figure 4.1. After introducing the slack variables x_3, x_4, and x_5, the following problem is obtained.

$$\begin{array}{ll} \text{Minimize} & x_1 - 2x_2 \\ \text{Subject to} & x_1 + x_2 - x_3 = 2 \\ & -x_1 + x_2 - x_4 = 1 \\ & x_2 + x_5 = 3 \\ & x_1, \quad x_2, \quad x_3, \quad x_4, \quad x_5 \geq 0 \end{array}$$

An initial identity is not available. So introduce the artificial variables x_6 and x_7 (note that the last constraint does not need an artificial variable). Phase I starts by minimizing $x_0 = x_6 + x_7$.

PHASE I

x_0	x_1	x_2	x_3	x_4	x_5	x_6 (ARTIFICIALS)	x_7 (ARTIFICIALS)	RHS
1	0	0	0	0	0	−1	−1	0
0	1	1	−1	0	0	1	0	2
0	−1	1	0	−1	0	0	1	1
0	0	1	0	0	1	0	0	3

Add rows 1 and 2 to row 0 so that $z_6 - c_6 = z_7 - c_7 = 0$ will be displayed.

	x_0	x_1	x_2	x_3	x_4	x_5	x_6	x_7	RHS
x_0	1	0	2	−1	−1	0	0	0	3
x_6	0	1	1	−1	0	0	1	0	2
x_7	0	−1	(1)	0	−1	0	0	1	1
x_5	0	0	1	0	0	1	0	0	3

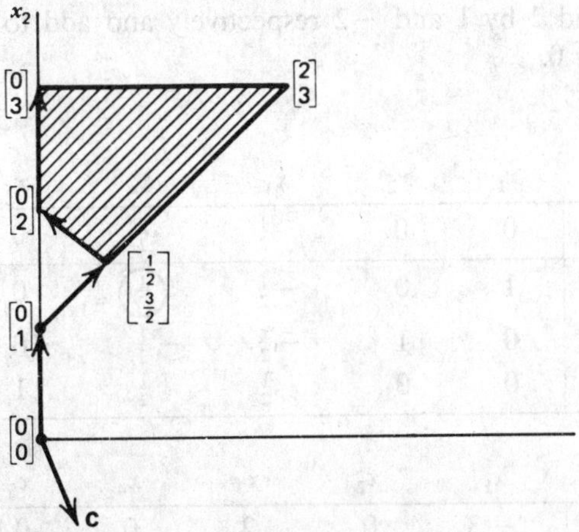

Figure 4.1. Example of the two-phase method.

	x_0	x_1	x_2	x_3	x_4	x_5	x_6	x_7	RHS
x_0	1	2	0	-1	1	0	0	-2	1
x_6	0	(2)	0	-1	1	0	1	-1	1
x_2	0	-1	1	0	-1	0	0	1	1
x_5	0	1	0	0	1	1	0	-1	2

	x_0	x_1	x_2	x_3	x_4	x_5	x_6	x_7	RHS
x_0	1	0	0	0	0	0	-1	-1	0
x_1	0	1	0	$-\frac{1}{2}$	$\frac{1}{2}$	0	$\frac{1}{2}$	$-\frac{1}{2}$	$\frac{1}{2}$
x_2	0	0	1	$-\frac{1}{2}$	$-\frac{1}{2}$	0	$\frac{1}{2}$	$\frac{1}{2}$	$\frac{3}{2}$
x_5	0	0	0	$\frac{1}{2}$	$\frac{1}{2}$	1	$-\frac{1}{2}$	$-\frac{1}{2}$	$\frac{3}{2}$

This is the end of phase I. We have a starting basic feasible solution, $(x_1, x_2) = (\frac{1}{2}, \frac{3}{2})$. Now we are ready to start phase II, where the original objective is minimized starting from the extreme point $(\frac{1}{2}, \frac{3}{2})$ (see Figure 4.1). The artificial variables are disregarded from any further consideration.

PHASE II

z	x_1	x_2	x_3	x_4	x_5	RHS
1	-1	2	0	0	0	0
0	1	0	$-\frac{1}{2}$	$\frac{1}{2}$	0	$\frac{1}{2}$
0	0	1	$-\frac{1}{2}$	$-\frac{1}{2}$	0	$\frac{3}{2}$
0	0	0	$\frac{1}{2}$	$\frac{1}{2}$	1	$\frac{3}{2}$

Multiply rows 1 and 2 by 1 and -2 respectively and add to row 0 producing $z_1 - c_1 = z_2 - c_2 = 0$.

	z	x_1	x_2	x_3	x_4	x_5	RHS
z	1	0	0	$\frac{1}{2}$	$\frac{3}{2}$	0	$-\frac{5}{2}$
x_1	0	1	0	$-\frac{1}{2}$	$\textcircled{\frac{1}{2}}$	0	$\frac{1}{2}$
x_2	0	0	1	$-\frac{1}{2}$	$-\frac{1}{2}$	0	$\frac{3}{2}$
x_5	0	0	0	$\frac{1}{2}$	$\frac{1}{2}$	1	$\frac{3}{2}$

	z	x_1	x_2	x_3	x_4	x_5	RHS
z	1	-3	0	2	0	0	-4
x_4	0	2	0	-1	1	0	1
x_2	0	1	1	-1	0	0	2
x_5	0	-1	0	$\textcircled{1}$	0	1	1

	z	x_1	x_2	x_3	x_4	x_5	RHS
z	1	-1	0	0	0	-2	-6
x_4	0	1	0	0	1	1	2
x_2	0	0	1	0	0	1	3
x_3	0	-1	0	1	0	1	1

Since $z_j - c_j \leq 0$ for all nonbasic variables, the optimal point $(0, 3)$ with objective -6 is reached. Note that phase I moved from the infeasible point $(0, 0)$, to the infeasible point $(0, 1)$, and finally to the feasible point $(\frac{1}{2}, \frac{3}{2})$. From this extreme point, phase II moved to the feasible point $(0, 2)$ and finally to the optimal point $(0, 3)$. This is illustrated in Figure 4.1. The purpose of phase I is to get us to an extreme point of the feasible region, while phase II takes us from this feasible point to an optimal point.

Analysis of the Two-Phase Method

At the end of phase I either $\mathbf{x}_a \neq \mathbf{0}$ or else $\mathbf{x}_a = \mathbf{0}$. These two cases are discussed in detail.

Case A: $\mathbf{x}_a \neq \mathbf{0}$

If $\mathbf{x}_a \neq \mathbf{0}$, then the original problem has no feasible solutions, because if there is an $\mathbf{x} \geq \mathbf{0}$ with $\mathbf{Ax} = \mathbf{b}$, then $\begin{pmatrix} \mathbf{x} \\ \mathbf{0} \end{pmatrix}$ is a feasible solution of the phase I problem and $\mathbf{0}(\mathbf{x}) + \mathbf{1}(\mathbf{0}) = 0 < \mathbf{1x}_a$, violating optimality of $\mathbf{x}_a$.

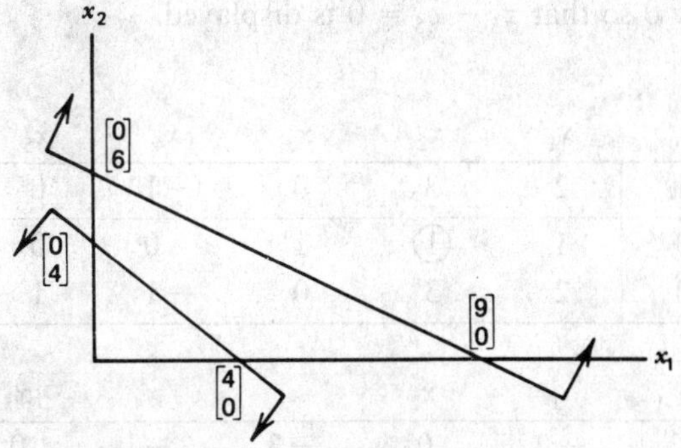

Figure 4.2. **Empty feasible region.**

Example 4.4

(Empty Feasible Region)

$$\text{Minimize} \quad -3x_1 + 4x_2$$

$$\begin{aligned} \text{Subject to} \quad x_1 + x_2 &\leq 4 \\ 2x_1 + 3x_2 &\geq 18 \\ x_1, x_2 &\geq 0 \end{aligned}$$

The constraints admit no feasible points, as shown in Figure 4.2. This will be detected by phase I. Introducing the slack variables x_3 and x_4, we get the following constraints in standard form:

$$\begin{aligned} x_1 + x_2 + x_3 \quad\quad &= 4 \\ 2x_1 + 3x_2 \quad\quad - x_4 &= 18 \\ x_1, \quad x_2, \quad x_3, \quad x_4 &\geq 0 \end{aligned}$$

Since no convenient basis exists, introduce the artificial variable x_5 into the second constraint. Phase I is now used to try to get rid of the artificial variable.

PHASE I

x_0	x_1	x_2	x_3	x_4	x_5	RHS
1	0	0	0	0	−1	0
0	1	1	1	0	0	4
0	2	3	0	−1	1	18

Add row 2 to row 0 so that $z_5 - c_5 = 0$ is displayed.

	x_0	x_1	x_2	x_3	x_4	x_5	RHS
x_0	1	2	3	0	−1	0	18
x_3	0	1	①	1	0	0	4
x_5	0	2	3	0	−1	1	18

	x_0	x_1	x_2	x_3	x_4	x_5	RHS
x_0	1	−1	0	−3	−1	0	6
x_2	0	1	1	1	0	0	4
x_5	0	−1	0	−3	−1	1	6

The optimality criterion of the simplex method, namely $z_j - c_j \le 0$, holds for all variables; but the artificial $x_5 > 0$. We conclude that the original problem has no feasible solutions.

Case B: $\mathbf{x}_a = \mathbf{0}$

This case is further divided into two subcases. In subcase B1 all the artificial variables are out of the basis at the end of phase I. The subcase B2 corresponds to the presence of at least one artificial in the basis at zero level. These cases are discussed now.

Subcase B1 (All Artificials Are Out of the Basis)

Since at the end of phase I we have a basic feasible solution, and since $\mathbf{x}_a$ is out of the basis, then the basis consists entirely of legitimate variables. If the legitimate vector $\mathbf{x}$ is decomposed accordingly into $\mathbf{x}_B$ and $\mathbf{x}_N$, then at the end of phase I we have the following tableau.

	x_0	$\mathbf{x}_B$	$\mathbf{x}_N$	$\mathbf{x}_a$	RHS
x_0	1	$\mathbf{0}$	$\mathbf{0}$	$-\mathbf{1}$	$\mathbf{0}$
$\mathbf{x}_B$	$\mathbf{0}$	$\mathbf{I}$	$\mathbf{B}^{-1}\mathbf{N}$	$\mathbf{B}^{-1}$	$\mathbf{B}^{-1}\mathbf{b}$

Now phase II can be started, with the original objective, after discarding the columns corresponding to $\mathbf{x}_a$. (For academic purposes, these columns may be kept since they would present $\mathbf{B}^{-1}$ at each iteration. Note, however, that an artificial variable should never be permitted to enter the basis again.) The $z_j - c_j$'s for the nonbasic variables are given by the vector $\mathbf{c}_B\mathbf{B}^{-1}\mathbf{N} - \mathbf{c}_N$, which can be easily calculated from the matrix $\mathbf{B}^{-1}\mathbf{N}$ stored in the final tableau of phase I. The following initial tableau of phase II is constructed. Starting with this tableau, the simplex method is used to find the optimal solution.

	z	$\mathbf{x}_B$	$\mathbf{x}_N$	RHS
z	1	$\mathbf{0}$	$\mathbf{c}_B\mathbf{B}^{-1}\mathbf{N}-\mathbf{c}_N$	$\mathbf{c}_B\mathbf{B}^{-1}\mathbf{b}$
$\mathbf{x}_B$	0	$\mathbf{I}$	$\mathbf{B}^{-1}\mathbf{N}$	$\mathbf{B}^{-1}\mathbf{b}$

Subcase B2 (Some Artificials Are in the Basis at Zero Level)

In this case we may proceed directly to phase II, or else eliminate the artificial basic variables and then proceed to phase II. These two actions are discussed in further detail next.

PROCEED DIRECTLY TO PHASE II

First eliminate the columns corresponding to the nonbasic artificial variables of phase I. The starting tableau of phase II consists of some legitimate variables and some artificial variables at zero level. The cost row consisting of $z_j - c_j$'s is constructed for the original objective function so that all legitimate variables that are basic have $z_j - c_j = 0$. The cost coefficients of the artificial variables are given value zero (justify!). While solving the phase II problem by the simplex method, we must be careful that artificial variables never reach a positive level (since this would destroy feasibility). To illustrate, consider the following tableau, where for simplicity we assume that the basis consists of the legitimate variables $x_1, x_2, \ldots, x_k$ and the artificial variables $x_{n+k+1}, \ldots, x_{n+m}$ (the artificial variables $x_{n+1}, \ldots, x_{n+k}$ left the basis during phase I).

	z	x_1	$\cdots$	x_k	$x_{k+1} \cdots x_j \cdots x_n$	x_{n+k+1}	$\cdots$	x_{n+m}	RHS
z	1	0	$\cdots$	0	$z_j - c_j$	0	$\cdots$	0	$\mathbf{c}_B\bar{\mathbf{b}}$
x_1		1			y_{1j}	0	$\cdots$	0	$\bar{b}_1$
x_2			1		y_{2j}	0	$\cdots$	0	$\bar{b}_2$
$\vdots$			$\ddots$		$\vdots$	$\vdots$		$\vdots$	$\vdots$
x_k				1	y_{kj}	0	$\cdots$	0	$\bar{b}_k$
x_{n+k+1}		0	$\cdots$	0	$y_{k+1,j}$	1			0
$\vdots$		$\vdots$		$\vdots$	$\vdots$		$\ddots$		$\vdots$
x_{n+r}		0	$\cdots$	0	y_{rj}		1		0
$\vdots$		$\vdots$		$\vdots$	$\vdots$			$\ddots$	$\vdots$
x_{n+m}		0	$\cdots$	0	y_{mj}			1	0

Suppose that $z_j - c_j > 0$, and so x_j is eligible to enter the basis. If $y_{ij} \geq 0$ for $i = k+1, \ldots, m$, then the artificial variable x_{n+i} will remain zero as x_j

enters the basis and the usual minimum ratio test is performed.[†] If, on the other hand, at least one component $y_{rj} < 0$ ($r = k + 1, \ldots,$ or m), then the artificial variable x_{n+r} becomes positive as x_j is increased. This action must be prohibited since it would destroy feasibility. This can be done by pivoting at y_{rj} rather than using the usual minimum ratio test. Even though $y_{rj} < 0$, pivoting at y_{rj} would maintain feasibility since the right-hand-side at the corresponding row is zero. In this case x_j enters the basis and the artificial x_{n+r} leaves the basis, and the objective value remains constant. With this slight modification the simplex method is used to solve phase II.

FIRST ELIMINATE THE BASIC ARTIFICIAL VARIABLES AT THE END OF PHASE I

Rather than adopting the preceding rule, which guarantees that artificial variables will always be zero during phase II, we can eliminate the artificial variables altogether before proceeding to phase II. The following is a typical tableau (possibly after rearranging) at the end of phase I. The objective row and column do not play any role in the subsequent analysis and are hence omitted.

	BASIC LEGITIMATE VARIABLES				NONBASIC LEGITIMATE VARIABLES	NONBASIC ARTIFICIAL VARIABLES	BASIC ARTIFICIAL VARIABLES				
	x_1	x_2	$\cdots$	x_k	$x_{k+1} \cdots x_n$	$x_{n+1} \cdots x_{n+k}$	x_{n+k+1}		$\cdots$	x_{n+m}	RHS
x_1	1						0	0	$\cdots$	0	$\bar{b}_1$
x_2		1			$\mathbf{R}_1$	$\mathbf{R}_3$	0	0	$\cdots$	0	$\bar{b}_2$
$\vdots$			$\ddots$				$\vdots$	$\vdots$		$\vdots$	$\vdots$
x_k				1			0	0	$\cdots$	0	$\bar{b}_k$
x_{n+k+1}	0	0	$\cdots$	0			1				0
					$\mathbf{R}_2$	$\mathbf{R}_4$		1			0
$\vdots$	$\vdots$	$\vdots$		$\vdots$					$\ddots$		$\vdots$
x_{n+m}	0	0	$\cdots$	0						1	0

We now attempt to drive the artificial variables $x_{n+k+1}, \ldots, x_{n+m}$ out of the basis by placing $m - k$ of the nonbasic legitimate variables $x_{k+1}, \ldots, x_n$ into the basis. For instance, x_{n+k+1} can be driven out of the basis by pivoting at any nonzero element of the first row of $\mathbf{R}_2$. The corresponding nonbasic legitimate variable enters, x_{n+k+1} leaves the basis, and the tableau is updated. This process is continued. If all artificial variables drop from the basis, then the basis would consist entirely of legitimate columns, and we proceed to phase II as described in subcase B1. If, on the other hand, we reach a tableau where the matrix $\mathbf{R}_2 = \mathbf{0}$, none of the artificial variables can leave the basis by introducing x_{k+1}, or $x_{k+2}, \ldots,$ or x_n. Denoting $(x_1, x_2, \ldots, x_k)^t$ and $(x_{k+1}, \ldots, x_n)^t$ by $\mathbf{x}_1$ and $\mathbf{x}_2$

[†]Actually, if any one of the y_{ij}'s is positive for $r = k + 1, \ldots, m$, then we may use the usual minimum ratio test as x_j will enter the basis at zero level.

respectively, and decomposing **A** and **b** accordingly into $\left[\begin{array}{c|c} \mathbf{A}_{11} & \mathbf{A}_{12} \\ \hline \mathbf{A}_{21} & \mathbf{A}_{22} \end{array}\right]$ and $\begin{bmatrix} \mathbf{b}_1 \\ \mathbf{b}_2 \end{bmatrix}$, it is clear that the system

$$\begin{array}{c} \\ k \\ m-k \end{array} \begin{array}{c} \quad k \qquad n-k \\ \left[\begin{array}{c|c} \mathbf{A}_{11} & \mathbf{A}_{12} \\ \hline \mathbf{A}_{21} & \mathbf{A}_{22} \end{array}\right] \end{array} \begin{array}{c} \\ \begin{bmatrix} \mathbf{x}_1 \\ \mathbf{x}_2 \end{bmatrix} = \begin{bmatrix} \mathbf{b}_1 \\ \mathbf{b}_2 \end{bmatrix} \end{array}$$

is transformed into

$$\begin{array}{c} \\ k \\ m-k \end{array} \begin{array}{c} k \; n-k \\ \left[\begin{array}{c|c} \mathbf{I} & \mathbf{R}_1 \\ \hline \mathbf{0} & \mathbf{0} \end{array}\right] \end{array} \begin{array}{c} \\ \begin{bmatrix} \mathbf{x}_1 \\ \mathbf{x}_2 \end{bmatrix} = \begin{bmatrix} \bar{\mathbf{b}}_1 \\ \mathbf{0} \end{bmatrix} \end{array}$$

through a sequence of elementary matrix operations. This shows that rank $(\mathbf{A}, \mathbf{b}) = k < m$; that is, the last $m - k$ equations are *algebraically redundant* and $\mathbf{R}_1 = \mathbf{A}_{11}^{-1}\mathbf{A}_{12}$ and $\bar{\mathbf{b}}_1 = \mathbf{A}_{11}^{-1}\mathbf{b}_1$. The last $m - k$ rows of the last tableau can be thrown away, and we can proceed to phase II without the artificial variables. The basic variables are $x_1, x_2, \ldots, x_k$ and the nonbasic variables are $x_{k+1}, \ldots, x_n$. The starting tableau of phase II is depicted, where $\mathbf{c}_B = (c_1, c_2, \ldots, c_k)$.

z	$x_1 \cdots x_k$	$x_{k+1} \cdots x_n$	RHS
1	$\mathbf{0}$	$\mathbf{c_B}\mathbf{A}_{11}^{-1}\mathbf{A}_{12} - \mathbf{c}_N$	$\mathbf{c}_B\bar{\mathbf{b}}_1$
0	$\mathbf{I}$	$\mathbf{A}_{11}^{-1}\mathbf{A}_{12} = \mathbf{R}_1$	$\bar{\mathbf{b}}_1$

Example 4.5

(Redundancy)

$$\begin{array}{lrcrcrcl} \text{Minimize} & -x_1 & + & 2x_2 & - & 3x_3 & & \\ \text{Subject to} & x_1 & + & x_2 & + & x_3 & = & 6 \\ & -x_1 & + & x_2 & + & 2x_3 & = & 4 \\ & & & 2x_2 & + & 3x_3 & = & 10 \\ & & & & & x_3 & \leq & 2 \\ & x_1, & & x_2, & & x_3 & \geq & 0 \end{array}$$

We need to introduce a slack variable x_4. The constraint matrix **A** is given:

$$\mathbf{A} = \begin{bmatrix} 1 & 1 & 1 & 0 \\ -1 & 1 & 2 & 0 \\ 0 & 2 & 3 & 0 \\ 0 & 0 & 1 & 1 \end{bmatrix}$$

Note that the matrix is not of full rank since if we add the first two rows of **A**, we get the third row; that is, any one of the first three constraints is redundant and can be thrown away. We shall proceed as if this fact were not known, however, and introduce the artificial variables x_5, x_6, and x_7. The phase I objective is: Minimize $x_0 = x_5 + x_6 + x_7$. Phase I proceeds as follows.

PHASE I

x_0	x_1	x_2	x_3	x_4	x_5	x_6	x_7	RHS
1	0	0	0	0	-1	-1	-1	0
1	1	1	1	0	1	0	0	6
0	-1	1	2	0	0	1	0	4
0	0	2	3	0	0	0	1	10
0	0	0	1	1	0	0	0	2

Add rows 1, 2, and 3 to row 0 to display $z_5 - c_5 = z_6 - c_6 = z_7 - c_7 = 0$.

	x_0	x_1	x_2	x_3	x_4	x_5	x_6	x_7	RHS
x_0	1	0	4	6	0	0	0	0	20
x_5	0	1	1	1	0	1	0	0	6
x_6	0	-1	1	2	0	0	1	0	4
x_7	0	0	2	3	0	0	0	1	10
x_4	0	0	0	$\textcircled{1}$	1	0	0	0	2

	x_0	x_1	x_2	x_3	x_4	x_5	x_6	x_7	RHS
x_0	1	0	4	0	-6	0	0	0	8
x_5	0	1	1	0	-1	1	0	0	4
x_6	0	-1	$\textcircled{1}$	0	-2	0	1	0	0
x_7	0	0	2	0	-3	0	0	1	4
x_3	0	0	0	1	1	0	0	0	2

	x_0	x_1	x_2	x_3	x_4	x_5	x_6	x_7	RHS
x_0	1	4	0	0	2	0	-4	0	8
x_5	0	$\textcircled{2}$	0	0	1	1	-1	0	4
x_2	0	-1	1	0	-2	0	1	0	0
x_7	0	2	0	0	1	0	-2	1	4
x_3	0	0	0	1	1	0	0	0	2

	x_0	x_1	x_2	x_3	x_4	x_5	x_6	x_7	RHS
x_0	1	0	0	0	0	-2	-2	0	0
x_1	0	1	0	0	$\frac{1}{2}$	$\frac{1}{2}$	$-\frac{1}{2}$	0	2
x_2	0	0	1	0	$-\frac{3}{2}$	$\frac{1}{2}$	$\frac{1}{2}$	0	2
x_7	0	0	0	0	0	-1	-1	1	0
x_3	0	0	0	1	1	0	0	0	2

Since all the artificial variables are at level zero, we may proceed to phase II with a basic feasible solution of the original problem. We can either proceed directly with the artificial x_7 in the basis at zero level or attempt to eliminate x_7 from the basis. The only legitimate nonbasic variable is x_4, and it has zero coefficient in row 3 corresponding to x_7. This shows that the third row (constraint of the original problem) is redundant and can be thrown away. This will be done as we move to phase II.

PHASE II

Obviously $z_1 - c_1 = z_2 - c_2 = z_3 - c_3 = 0$. Thus x_5 and x_6 are nonbasic artificial variables and will not be introduced in the phase II problem. In order to complete row 0 we need to calculate $z_4 - c_4$:

$$\begin{aligned} z_4 - c_4 &= \mathbf{c}_B \mathbf{B}^{-1} \mathbf{a}_4 - c_4 \\ &= (-1, 2, -3) \begin{pmatrix} \frac{1}{2} \\ -\frac{3}{2} \\ 1 \end{pmatrix} - 0 \\ &= -\tfrac{13}{2} \end{aligned}$$

Since we are minimizing and $z_4 - c_4 \leq 0$ for the only nonbasic variable, then we stop; the solution obtained from phase I is optimal. The tableau displays the optimal solution.

	z	x_1	x_2	x_3	x_4	RHS
z	1	0	0	0	$-\frac{13}{2}$	-4
x_1	0	1	0	0	$\frac{1}{2}$	2
x_2	0	0	1	0	$-\frac{3}{2}$	2
x_3	0	0	0	1	1	2

Organization of the Tableau in the Two-Phase Method

To eliminate the need for recomputing the $z_j - c_j$'s when the phase I objective function is replaced by the phase II (original) objective function, an extra row could be added to the tableau representing the original cost coefficients. The following represents the setup of the initial tableau (not yet in basic form).

						ARTIFICIALS				
	z	x_0	x_1	$\cdots$	x_n	x_{n+1}	$\cdots$	x_{n+m}	RHS	
z	1	0	$-c_1$	$\cdots$	$-c_n$	0	$\cdots$	0	0	← Phase II Objective
x_0	0	1	0	$\cdots$	0	-1	$\cdots$	-1	0	← Phase I Objective
x_{n+1}	0	0	a_{11}	$\cdots$	a_{1n}	1	$\cdots$	0	b_1	
$\vdots$	$\vdots$	$\vdots$	$\vdots$			$\vdots$		$\vdots$	$\vdots$	
x_{n+m}	0	0	a_{m1}	$\cdots$	a_{mn}	0	$\cdots$	1	b_m	

To convert this tableau to canonical form (that is, unit vectors for all basic variables) we must perform preliminary pivoting to obtain zeros for x_{n+1} through x_{n+m} in the x_0 (phase I objective) row. This is done by successively adding every row (except the z row) to the x_0 row.

Once the initial tableau has been converted to canonical form, the simplex method is applied to the resulting tableau using the x_0 row as the (phase I) objective function row until optimality is achieved. During phase I the z row is transformed, just as any other row of the tableau would be, to maintain unit vectors for the basic variables. Note, however, that during phase I the basis entry is solely determined by the entries in row x_0.

At the end of phase I, if x_0 is positive the problem is found to be infeasible and the process is terminated. Otherwise, the x_0 row and x_0 column are deleted and phase II is initiated (after possibly eliminating artificial variables), with the values in the z row being the correct values of $z_j - c_j$ (why?) for the phase II objective.

4.3 THE BIG-*M* METHOD

Recall that artificial variables constitute a tool that can be used to initiate the simplex method. However, the presence of artificial variables at a positive level, by construction, means that the current point is an infeasible solution of the original system. The two-phase method is one way to get rid of the artificials. However, during phase I of the two-phase method the original cost coefficients are essentially ignored. Phase I of the two-phase method seeks any basic feasible solution, not necessarily a good one. Another possibility for eliminating the artificial variables is to assign coefficients for these variables in the original objective function in such a way as to make their presence in the basis at a positive level very unattractive from the objective function point of view. To illustrate, suppose that we want to solve the following linear programming problem, where $\mathbf{b} \geq \mathbf{0}$.

$$\begin{aligned} \text{Minimize} \quad & \mathbf{cx} \\ \text{Subject to} \quad & \mathbf{Ax} = \mathbf{b} \\ & \mathbf{x} \geq \mathbf{0} \end{aligned}$$

If no convenient basis is known, we introduce the artificial vector $\mathbf{x}_a$, which leads to the following system:

$$\mathbf{Ax} + \mathbf{x}_a = \mathbf{b}$$

$$\mathbf{x}, \mathbf{x}_a \geq \mathbf{0}$$

The starting basic feasible solution is given by $\mathbf{x}_a = \mathbf{b}$ and $\mathbf{x} = \mathbf{0}$. In order to reflect the undesirability of a nonzero artificial vector, the objective function is modified such that a large penalty is paid for any such solution. More specifically consider the following problem.

$$\text{Minimize} \quad \mathbf{cx} + M\mathbf{1x}_a$$

$$\text{Subject to} \quad \mathbf{Ax} + \mathbf{x}_a = \mathbf{b}$$
$$\mathbf{x}, \quad \mathbf{x}_a \geq \mathbf{0}$$

where M is a very large positive number (see Section 4.4). The term $M\mathbf{1x}_a$ can be interpreted as a penalty to be paid by any solution with $\mathbf{x}_a \neq \mathbf{0}$. Alternatively, the foregoing strategy can be interpreted as one that minimizes $\mathbf{1x}_a$ with priority one, and among all alternative optimal solutions for this objective, minimizes the secondary objective $\mathbf{cx}$. Hence even though the starting solution $\mathbf{x} = \mathbf{0}$, $\mathbf{x}_a = \mathbf{b}$ is feasible to the new constraints, it has a very unattractive objective value, namely $M\mathbf{1b}$. Therefore the simplex method itself will try to get the artificial variables out of the basis, and then continue to find an optimal solution of the original problem.

The big-M method is illustrated by the following numerical example. Validation of the method and possible complications are discussed subsequently.

Example 4.6

$$\text{Minimize} \quad x_1 - 2x_2$$

$$\begin{aligned} \text{Subject to} \quad x_1 + x_2 &\geq 2 \\ -x_1 + x_2 &\geq 1 \\ x_2 &\leq 3 \\ x_1, \quad x_2 &\geq 0 \end{aligned}$$

This example was solved earlier by the two-phase method (Example 4.3). The slack variables x_3, x_4, and x_5 are introduced and the artificial variables x_6 and x_7 are incorporated in the first two constraints. The modified objective function is $z = x_1 - 2x_2 + Mx_6 + Mx_7$, where M is a large positive number. This leads to the following sequence of tableaux.

z	x_1	x_2	x_3	x_4	x_5	x_6	x_7	RHS
1	−1	2	0	0	0	$-M$	$-M$	0
0	1	1	−1	0	0	1	0	2
0	−1	1	0	−1	0	0	1	1
0	0	1	0	0	1	0	0	3

Multiply rows 1 and 2 by M and add to row 0.

	z	x_1	x_2	x_3	x_4	x_5	x_6	x_7	RHS
z	1	−1	$2 + 2M$	$-M$	$-M$	0	0	0	$3M$
x_6	0	1	1	−1	0	0	1	0	2
x_7	0	−1	①	0	−1	0	0	1	1
x_5	0	0	1	0	0	1	0	0	3

	z	x_1	x_2	x_3	x_4	x_5	x_6	x_7	RHS
z	1	$1+2M$	0	$-M$	$2+M$	0	0	$-2-2M$	$-2+M$
x_6	0	②	0	-1	1	0	1	-1	1
x_2	0	-1	1	0	-1	0	0	1	1
x_5	0	1	0	0	1	1	0	-1	2

	z	x_1	x_2	x_3	x_4	x_5	x_6	x_7	RHS
z	1	0	0	$\frac{1}{2}$	$\frac{3}{2}$	0	$-\frac{1}{2}-M$	$-\frac{3}{2}-M$	$-\frac{5}{2}$
x_1	0	1	0	$-\frac{1}{2}$	$\textcircled{\frac{1}{2}}$	0	$\frac{1}{2}$	$-\frac{1}{2}$	$\frac{1}{2}$
x_2	0	0	1	$-\frac{1}{2}$	$-\frac{1}{2}$	0	$\frac{1}{2}$	$\frac{1}{2}$	$\frac{3}{2}$
x_5	0	0	0	$\frac{1}{2}$	1	1	$-\frac{1}{2}$	$\frac{3}{2}$	$\frac{3}{2}$

	z	x_1	x_2	x_3	x_4	x_5	x_6	x_7	RHS
z	1	-3	0	2	0	0	$-2-M$	$-M$	-4
x_4	0	2	0	-1	1	0	1	-1	1
x_2	0	1	1	-1	0	0	1	0	2
x_5	0	-1	0	①	0	1	-1	0	1

	z	x_1	x_2	x_3	x_4	x_5	x_6	x_7	RHS
z	1	-1	0	0	0	-2	$-M$	$-M$	-6
x_4	0	1	0	0	1	1	0	-1	2
x_2	0	0	1	0	0	1	0	0	3
x_3	0	-1	0	1	0	1	-1	0	1

Since $z_j - c_j \leq 0$ for each nonbasic variable, the last tableau gives the optimal solution. The sequence of points generated in the $(x_1 x_2)$ space is illustrated in Figure 4.3.

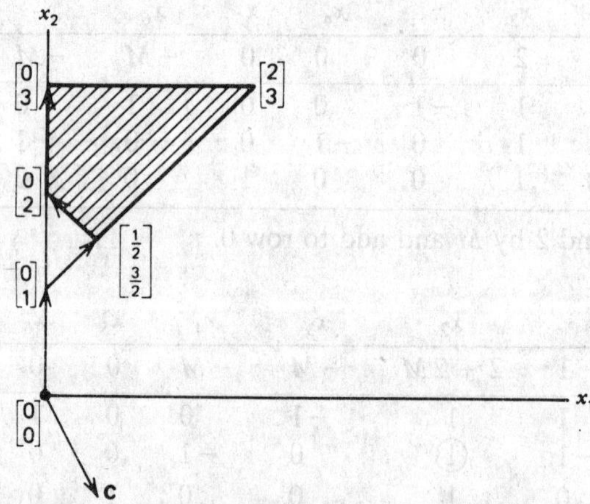

Figure 4.3. **Example of the big-M method.**

Analysis of the Big-*M* Method

We now discuss in more detail the possible cases that may arise while solving the big-M problem. The original problem P and the big-M problem P(M) are stated, where the vector **b** is nonnegative.

Problem P: Minimize $\mathbf{cx}$

Subject to $\mathbf{Ax} = \mathbf{b}$
$\mathbf{x} \geq \mathbf{0}$

Problem P(M): Minimize $\mathbf{cx} + M\mathbf{1x}_a$

Subject to $\mathbf{Ax} + \mathbf{x}_a = \mathbf{b}$
$\mathbf{x}, \quad \mathbf{x}_a \geq \mathbf{0}$

Since problem P(M) has a feasible solution (say $\mathbf{x} = \mathbf{0}$ and $\mathbf{x}_a = \mathbf{b}$), then while solving it by the simplex method one of the following two cases may arise.

1. We arrive at an optimal solution of P(M).
2. We conclude that P(M) has an unbounded optimal solution value, that is, $z \to -\infty$.

Of course, we are interested in conclusions about problem P and not P(M). The following analysis will help us to draw such conclusions.

Case A: Finite Optimal Solution of P(M)

Under this case, we have two possibilities. First, the optimal solution to P(M) has all artificials at value zero, and second, not all artificials are zero. These cases are discussed now.

Subcase A1: $(\mathbf{x}^*, \mathbf{0})$ is an optimal solution of P(M)

In this case $\mathbf{x}^*$ is an optimal solution to problem P. This can be easily verified as follows. Let $\mathbf{x}$ be any feasible solution to problem P, and note that $(\mathbf{x}, \mathbf{0})$ is a feasible solution to problem P(M). Since $(\mathbf{x}^*, \mathbf{0})$ is an optimal solution of problem P(M), then $\mathbf{cx}^* + 0 \leq \mathbf{cx} + 0$, that is, $\mathbf{cx}^* \leq \mathbf{cx}$. Since $\mathbf{x}^*$ is a feasible solution of problem P, then $\mathbf{x}^*$ is indeed an optimal solution of P. This case was illustrated by Example 4.6.

Subcase A2: $(\mathbf{x}^*, \mathbf{x}_a^*)$ is an optimal solution of P(M) and $\mathbf{x}_a^* \neq \mathbf{0}$

If M is a very large positive number, then we conclude that there exists no feasible solution of P. To illustrate this case, suppose on the contrary that $\mathbf{x}$ was a feasible solution of P. Then $(\mathbf{x}, \mathbf{0})$ would be a feasible solution of P(M) and by optimality of $(\mathbf{x}^*, \mathbf{x}_a^*)$ we have

$$\mathbf{cx}^* + M\mathbf{1x}_a^* \leq \mathbf{cx} + 0 = \mathbf{cx}$$

Since M is very large, $\mathbf{x}_a^* \geq \mathbf{0}$ and $\mathbf{x}_a^* \neq \mathbf{0}$, and since $\mathbf{x}_a^*$ corresponds to one of the finite number of basic feasible solutions, the preceding inequality is impossible (why?). Therefore **x** could not have been a feasible solution of P. A formal proof is left as Exercise 4.14.

Example 4.7

(No Feasible Solutions)

$$\begin{array}{lrl} \text{Minimize} & -x_1 - 3x_2 + x_3 & \\ \text{Subject to} & x_1 + x_2 + 2x_3 & \leq 4 \\ & -x_1 + x_3 & \geq 4 \\ & x_3 & \geq 3 \\ & x_1, \quad x_2, \quad x_3 & \geq 0 \end{array}$$

Introduce the slack variables x_4, x_5, and x_6. Also, introduce the two artificial variables x_7 and x_8 for the last two constraints with cost coefficients equal to M. This leads to the following sequence of tableaux.

z	x_1	x_2	x_3	x_4	x_5	x_6	x_7	x_8	RHS
1	1	3	-1	0	0	0	$-M$	$-M$	0
0	1	1	2	1	0	0	0	0	4
0	-1	0	1	0	-1	0	1	0	4
0	0	0	1	0	0	-1	0	1	3

Multiply rows 2 and 3 by M and add to row 0.

	z	x_1	x_2	x_3	x_4	x_5	x_6	x_7	x_8	RHS
z	1	$1-M$	3	$-1+2M$	0	$-M$	$-M$	0	0	$7M$
x_4	0	1	1	②	1	0	0	0	0	4
x_7	0	-1	0	1	0	-1	0	1	0	4
x_8	0	0	0	1	0	0	-1	0	1	3

	z	x_1	x_2	x_3	x_4	x_5	x_6	x_7	x_8	RHS
z	1	$\frac{3}{2}-2M$	$\frac{7}{2}-M$	0	$\frac{1}{2}-M$	$-M$	$-M$	0	0	$2+3M$
x_3	0	$\frac{1}{2}$	$\frac{1}{2}$	1	$\frac{1}{2}$	0	0	0	0	2
x_7	0	$-\frac{3}{2}$	$-\frac{1}{2}$	0	$-\frac{1}{2}$	-1	0	1	0	2
x_8	0	$-\frac{1}{2}$	$-\frac{1}{2}$	0	$-\frac{1}{2}$	0	-1	0	1	2

Since $M > 0$ is very large, then $z_j - c_j \leq 0$ for all nonbasic variables; that is, the simplex optimality criteria hold. Since the artificials x_7 and x_8 still remain in the basis at a positive level, then we conclude that the original problem has no feasible solutions.

Case B: P(M) has an unbounded optimal solution value, that is, $z \to -\infty$.

Suppose that during the solution of the big-M problem, the updated column $\mathbf{y}_k$ is $\leq \mathbf{0}$, where the index k is that of the *most positive* $z_j - c_j$. Then problem P(M) has an unbounded optimal solution. In the meantime, if all artificials are equal to zero, then the original problem has an unbounded optimal value. Otherwise, if at least one artificial variable is positive, then the original problem is infeasible. These two subcases are discussed in more detail.

Subcase B1: $z_k - c_k = \text{Maximum}\,(z_j - c_j) > 0$, $\mathbf{y}_k \leq \mathbf{0}$, and all artificials are equal to zero.

In this case we currently have a feasible solution of the original problem. Furthermore, since problem P(M) is unbounded, then there is a direction, $(\mathbf{d}_1, \mathbf{d}_2) \geq (\mathbf{0}, \mathbf{0})$ of the set $\{(\mathbf{x}, \mathbf{x}_a) : \mathbf{A}\mathbf{x} + \mathbf{x}_a = \mathbf{b},\ \mathbf{x} \geq \mathbf{0},\ \mathbf{x}_a \geq \mathbf{0}\}$ such that $\mathbf{c}\mathbf{d}_1 + M\mathbf{1}\mathbf{d}_2 < 0$ (recall that this is the necessary and sufficient condition for unboundedness of P(M)). Since M is very large and $\mathbf{d}_2 \geq \mathbf{0}$, then $\mathbf{c}\mathbf{d}_1 + M\mathbf{1}\mathbf{d}_2 < 0$ implies that $\mathbf{d}_2 = \mathbf{0}$ and hence $\mathbf{c}\mathbf{d}_1 < 0$. Thus we have found a direction $\mathbf{d}_1$ of the set $\{\mathbf{x} : \mathbf{A}\mathbf{x} = \mathbf{b},\ \mathbf{x} \geq \mathbf{0}\}$ such that $\mathbf{c}\mathbf{d}_1 < 0$ (why?). This implies that problem P has an unbounded optimal value.

Example 4.8

$$\begin{array}{ll}
\text{Minimize} & -x_1 - x_2 \\
\text{Subject to} & x_1 - x_2 - x_3 \qquad\ = 1 \\
& -x_1 + x_2 + 2x_3 - x_4 = 1 \\
& x_1,\quad x_2,\quad x_3,\quad x_4 \geq 0
\end{array}$$

Introduce the artificial variables x_5 and x_6 with cost coefficient M. This leads to the following sequence of tableaux.

z	x_1	x_2	x_3	x_4	x_5	x_6	RHS
1	1	1	0	0	$-M$	$-M$	0
0	1	-1	-1	0	1	0	1
0	-1	1	2	-1	0	1	1

Multiply rows 1 and 2 by M and add to row 0.

	z	x_1	x_2	x_3	x_4	x_5	x_6	RHS
z	1	1	1	M	$-M$	0	0	$2M$
x_5	0	1	-1	-1	0	1	0	1
x_6	0	-1	1	②	-1	0	1	1

	z	x_1	x_2	x_3	x_4	x_5	x_6	RHS
z	1	$1+\frac{1}{2}M$	$1-\frac{1}{2}M$	0	$-\frac{1}{2}M$	0	$-\frac{1}{2}M$	$\frac{3}{2}M$
x_5	0	$\textcircled{\frac{1}{2}}$	$-\frac{1}{2}$	0	$-\frac{1}{2}$	1	$\frac{1}{2}$	$\frac{3}{2}$
x_3	0	$-\frac{1}{2}$	$\frac{1}{2}$	1	$-\frac{1}{2}$	0	$\frac{1}{2}$	$\frac{1}{2}$

	z	x_1	x_2	x_3	x_4	x_5	x_6	RHS
z	1	0	2	0	1	$-M-2$	$-M-1$	-3
x_1	0	1	-1	0	-1	2	1	3
x_3	0	0	0	1	-1	1	1	2

The most positive $z_j - c_j$ corresponds to x_2 and $\mathbf{y}_2 \leq \mathbf{0}$. Therefore the big-$M$ problem is unbounded. Furthermore, the artificial variables x_5 and x_6 are equal to zero, and hence the original program has an unbounded optimal value along the ray $\{(3, 0, 2, 0) + \lambda(1, 1, 0, 0) : \lambda \geq 0\}$.

Subcase B2: $z_k - c_k = \text{Maximum}\,(z_j - c_j) > 0$, $\mathbf{y}_k \leq \mathbf{0}$, and not all artificials are equal to zero.

In this case we show that there could be no feasible solution of the original problem. To illustrate, suppose that the basis consists of the first m columns, where columns 1 through p are formed by real variables and columns $p+1$ through m are formed by artificial variables. The corresponding tableau is depicted here.

	$x_1 \cdots x_p$	$x_{p+1} \cdots x_m$	$x_{m+1} \cdots$	x_j	$\cdots x_n$	RHS
x_1	$1 \cdots 0$	$0 \cdots 0$	$\cdots$	y_{1j}	$\cdots$	$\bar{b}_1$
	$0\ 1 \cdots 0$	$0 \cdots 0$	$\cdots$	y_{2j}	$\cdots$	
$\vdots$				$\vdots$		$\vdots$
x_p	$0 \cdots 1$	$0 \cdots 0$	$\cdots$	y_{pj}	$\cdots$	$\bar{b}_p$
x_{p+1}	$0 \cdots 0$	$1 \cdots 0$	$\cdots$	$y_{p+1,\,j}$	$\cdots$	$\bar{b}_{p+1}$
$\vdots$	$0 \cdots 0$	$0\ 1 \cdots 0$		$\vdots$		$\vdots$
x_m	$0 \cdots 0$	$0 \cdots 1$	$\cdots$	y_{mj}	$\cdots$	$\bar{b}_m$

Noting that $c_i = M$ for $i = p + 1, \ldots, m$, then for $j = m + 1, \ldots,\ n$ we get

$$z_j - c_j = \sum_{i=1}^{p} c_i y_{ij} + M\left(\sum_{i=p+1}^{m} y_{ij}\right) - c_j \tag{4.1}$$

First note that $\Sigma_{i=p+1}^{m} y_{ij} \leq 0$ for all $j = m + 1, \ldots, n$. To show this, first recall that $\mathbf{y}_k \leq \mathbf{0}$ and hence $\Sigma_{i=p+1}^{m} y_{ik} \leq 0$ holds trivially. By contradiction, suppose that $\Sigma_{i=p+1}^{m} y_{ij} > 0$ for some nonbasic variable x_j. From Equation (4.1) and since M is very large, then $z_j - c_j$ is a large positive number, violating the definition of $z_k - c_k$ (recall that $z_k - c_k = \text{Maximum}\,(z_l - c_l)$ and $\mathbf{y}_k \leq \mathbf{0}$). Therefore $\Sigma_{i=p+1}^{m} y_{ij} \leq 0$ for all $j = m + 1, \ldots, n$. Summing the last $m - p$ equations of the previous tableau, we get

$$\sum_{i=p+1}^{m} x_i + \sum_{j=m+1}^{n} x_j\left(\sum_{i=p+1}^{m} y_{ij}\right) = \sum_{i=p+1}^{m} \bar{b}_i \tag{4.2}$$

By contradiction, suppose that problem P has a feasible solution. Then $x_i = 0$ for all artificial variables and hence $x_i = 0$ for $i = p + 1, \ldots, m$. Also $x_j \geq 0$ and $\Sigma_{i=p+1}^{m} y_{ij} \leq 0$ for $j = m + 1, \ldots, n$ as shown previously. Therefore the left-hand-side of Equation (4.2) is ≤ 0. But the right-hand-side is positive since not all artificials are equal to zero. This contradiction shows that there could be no feasible solution of the original problem.

Note that we cannot conclude that the original problem is inconsistent if some $z_j - c_j > 0$, $\mathbf{y}_j \leq \mathbf{0}$ and not all artificials are zero. However, this conclusion is true if we use the *most positive* $z_j - c_j$. See Exercise 4.22.

Example 4.9

Minimize $\quad -x_1 - x_2$

Subject to
$$\begin{aligned} x_1 - x_2 &\geq 1 \\ -x_1 + x_2 &\geq 1 \\ x_1,\quad x_2 &\geq 0 \end{aligned}$$

This problem has no feasible solutions, as shown in Figure 4.4. We shall show that the optimal solution of the big-*M* problem is unbounded. Introduce the slack variables x_3 and x_4 and the artificial variables x_5 and x_6.

z	x_1	x_2	x_3	x_4	x_5	x_6	RHS
1	1	1	0	0	$-M$	$-M$	0
0	1	-1	-1	0	1	0	1
0	-1	1	0	-1	0	1	1

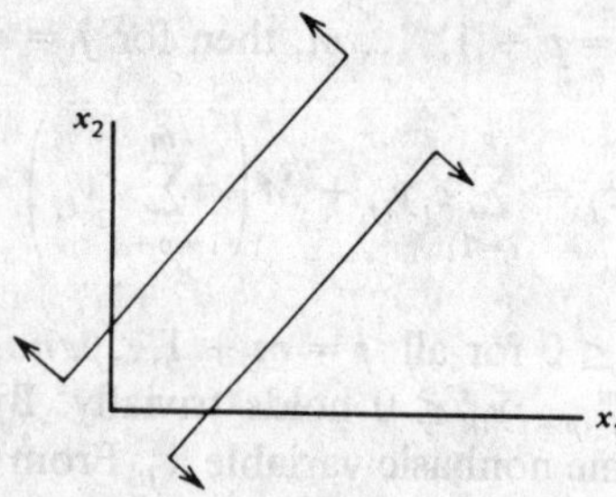

Figure 4.4. Empty feasible region.

Multiply rows 1 and 2 by M and add to row 0.

	z	x_1	x_2	x_3	x_4	x_5	x_6	RHS
z	1	1	1	$-M$	$-M$	0	0	$2M$
x_5	0	1	-1	-1	0	1	0	1
x_6	0	-1	1	0	-1	0	1	1

	z	x_1	x_2	x_3	x_4	x_5	x_6	RHS
z	1	0	2	$1-M$	$-M$	-1	0	$2M-1$
x_1	0	1	-1	-1	0	1	0	1
x_6	0	0	0	-1	-1	1	1	2

The last tableau indicates unboundedness since $z_2 - c_2 > 0$ and $\mathbf{y}_2 \leq \mathbf{0}$. Note, however, that x_6 is positive, so we conclude that the original problem has no feasible solutions. This is also clear by examining the second row, which reads

$$x_5 + x_6 = 2 + x_3 + x_4$$

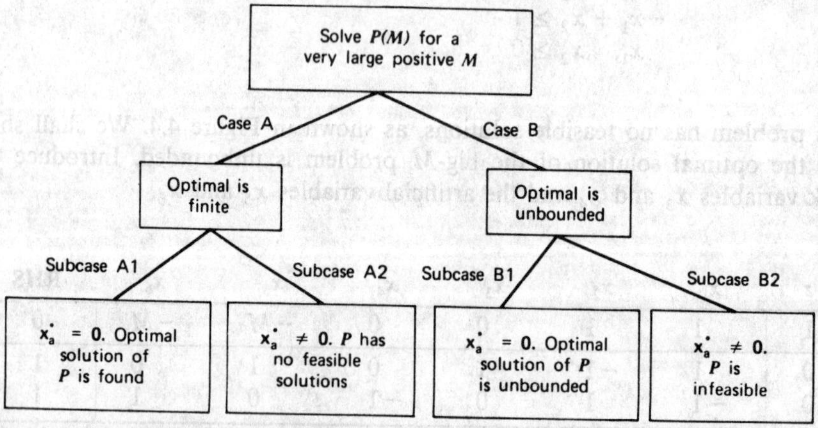

Figure 4.5. Analysis of the big-*M* method.

Since $x_3, x_4 \geq 0$, then $x_5 + x_6$ will always be positive, indicating that the original system is inconsistent.

The possible cases (A1, A2, B1, and B2) that may arise during the course of solving problem P(M) are summarized in Figure 4.5.

4.4 COMPARISON OF THE TWO-PHASE AND THE BIG-*M* METHODS: HOW BIG SHOULD BIG-*M* BE?

The objective function of the big-*M* method is simply the sum of the phase II objective and *M* times the phase I objective of the two-phase method. Hence for any basic feasible solution to the artificial problem, the canonical form of the objective row in the big-*M* method is precisely the sum of the canonical form of the phase II objective and *M* times the canonical form of the phase I objective in the two-phase method (each of these objective functions has simply been rewritten in terms of the nonbasic variables). Consequently, the two-phase method avoids possible round-off computational error problems associated with a large value of *M* by carrying the coefficients of *M* in the big-*M* method separately as phase I objective coefficients and by carrying the other coefficients as phase II objective coefficients. Hence, in the two-phase method, if one were to pick an entering variable according to the one that has the most positive $z_j - c_j$ in the phase I canonical objective row, then one would follow precisely the big-*M* method, provided any persisting ties are broken with the same (arbitrary) choice. In Exercise 4.19, we ask the reader to identify the cases of the big-*M* method given in Figure 4.5 with those for the two-phase method when viewed from the foregoing perspective.

There is another more fundamental issue involved with the big-*M* method, namely how big should *M* be? Clearly, given that the original problem is feasible, *M* should be large enough so that some basic feasible solution to P(M) with all artificial variables equal to zero has objective value strictly better than the best basic feasible solution to P(M) that does not have all the artificial variables equal to zero (why?). Note that a finite value for such an *M* exists (why?). Hence in selecting a value for *M*, it is erroneous to simply look at the magnitude of the objective function coefficients. One also needs to see how small (but positive) the sum of the artificial variables can get at a basic feasible solution to P(M). For example, consider the following problem P and the corresponding artificial problem P(M), where x_3 is the slack variable, x_4 is an artificial variable, and $\epsilon > 0$ is a (small) constant.

$$
\begin{array}{lll}
\text{P:} & \text{Minimize} & x_1 \\
& \text{Subject to} & \epsilon x_1 - x_2 \geq \epsilon \\
& & x_1, \quad x_2 \geq 0
\end{array}
$$

$$
\begin{array}{lll}
\text{P(M):} & \text{Minimize} & x_1 + Mx_4 \\
& \text{Subject to} & \epsilon x_1 - x_2 - x_3 + x_4 = \epsilon \\
& & x_1, \quad x_2, \quad x_3, \quad x_4 \geq 0
\end{array}
$$

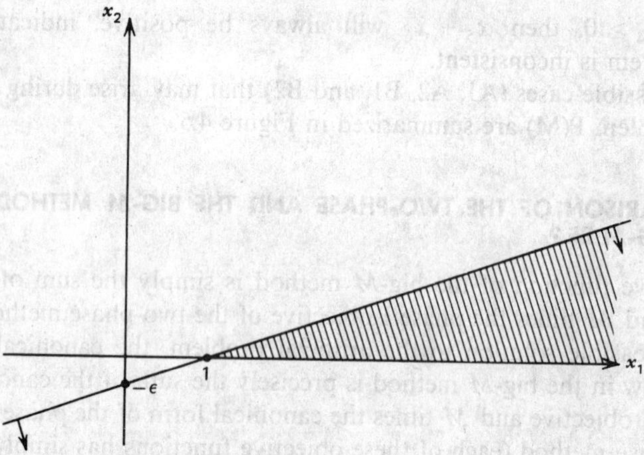

Figure 4.6. Selecting a value for *M*.

Figure 4.6 depicts the feasible region in the (x_1, x_2) space. Problem P(M) has two basic feasible solutions $\bar{\mathbf{x}}$ and $\hat{\mathbf{x}}$, where $\bar{\mathbf{x}} = (1, 0, 0, 0)$ and $\hat{\mathbf{x}} = (0, 0, 0, \epsilon)$. The first of these is (basic) feasible to P, although the second is not. In P(M), the objective value of $\bar{\mathbf{x}}$ is 1 and that of $\hat{\mathbf{x}}$ is ϵM. We would therefore like $\epsilon M > 1$, that is, $M > 1/\epsilon$. Note that if we had picked M as, say, some large constant times the largest objective coefficient in P, this would have been inadequate by making $\epsilon > 0$ small enough. In Figure 4.6, we are comparing (0, 0) with (1, 0), and although the former solution is infeasible, its infeasibility reduces with a reduction in ϵ, thus requiring a higher penalty. In Exercises 4.12 and 4.13, we ask the reader to generalize this concept.

4.5 THE SINGLE ARTIFICIAL VARIABLE TECHNIQUE

Thus far we have described two methods to initiate the simplex algorithm by the use of artificial variables. In this section we discuss a procedure that requires only a single artificial variable to get started. Consider the following problem.

$$
\begin{aligned}
&\text{Minimize} \quad && \mathbf{cx} \\
&\text{Subject to} \quad && \mathbf{Ax} = \mathbf{b} \\
& && \mathbf{x} \geq \mathbf{0}
\end{aligned}
$$

Suppose that we can partition the constraint matrix **A** into **A** = [**B**, **N**], where **B** is a basis matrix, not necessarily feasible. **B** possibly corresponds to a set of variables forced into the basis, regardless of feasibility, in the anticipation that these variables are likely to be positive at optimality. (Such a basis is called a *crash basis*.) Index the basic variables from 1 to m.

Multiplying the constraints by $\mathbf{B}^{-1}$, we get

$$\mathbf{I}\mathbf{x}_B + \mathbf{B}^{-1}\mathbf{N}\mathbf{x}_N = \bar{\mathbf{b}}$$

where $\bar{\mathbf{b}} = \mathbf{B}^{-1}\mathbf{b}$ and suppose that $\bar{\mathbf{b}} \ngeq \mathbf{0}$ (if $\bar{\mathbf{b}} \geq \mathbf{0}$, we have a starting basic feasible solution). To this system let us add a single artificial variable, x_a, with a coefficient of -1 in each constraint with $\bar{b}_i < 0$. Call this column $\mathbf{y}_a$.

$$\mathbf{I}\mathbf{x}_B + \mathbf{B}^{-1}\mathbf{N}\mathbf{x}_N + \mathbf{y}_a\mathbf{x}_a = \bar{\mathbf{b}}$$

Now introduce x_a into the basis by selecting the pivot row r as follows:

$$\bar{b}_r = \underset{i:\bar{b}_i<0}{\text{Minimum}} \left\{\bar{b}_i\right\}$$

Note that $\bar{b}_r < 0$. On pivoting in row r (that is, inserting x_a and removing x_r), we get the new right-hand-side values

$$\bar{b}_r' = -\bar{b}_r(\geq 0)$$

$$\bar{b}_i' = \bar{b}_i - \bar{b}_r(\geq 0), \qquad \text{if } \bar{b}_i < 0,\ i \neq r$$

$$\bar{b}_i' = \bar{b}_i(\geq 0) \qquad \text{otherwise.}$$

Thus by entering x_a and exiting x_r we have constructed a basic feasible solution to the enlarged system (the one including the artificial variable). Starting with this solution, the simplex method can be used with either the two-phase or big-M method.

Note that the variable x_a creates one extra dimension for this problem. The column is the sum of the artificial variable columns we would have added for the two-phase or the big-M method. Empirically, however, it has been found that the latter techniques are more efficient than the single artificial variable method, perhaps because in the present method, a simplex path needs to be traced in which all violated constraints are *simultaneously* satisfied.

Example 4.10

$$\begin{aligned} \text{Minimize} \quad & 2x_1 + 3x_2 \\ \text{Subject to} \quad & x_1 + x_2 \geq 3 \\ & -2x_1 + x_2 \geq 2 \\ & x_1,\ x_2 \geq 0 \end{aligned}$$

Subtracting the slack variables x_3 and x_4 and multiplying by -1, we get the basic system

$$\begin{aligned} -x_1 - x_2 + x_3 \quad &= -3 \\ 2x_1 - x_2 + \quad x_4 &= -2 \end{aligned}$$

Appending a single artificial variable x_5 with activity vector $\begin{pmatrix} -1 \\ -1 \end{pmatrix}$ to the initial tableau of the phase I problem, we get the following tableau.

	x_0	x_1	x_2	x_3	x_4	x_5	RHS
x_0	1	0	0	0	0	−1	0
x_3	0	−1	−1	1	0	(−1)	−3
x_4	0	2	−1	0	1	−1	−2

Pivoting in the x_3 row and x_5 column, we get the following tableau.

	x_0	x_1	x_2	x_3	x_4	x_5	RHS
x_0	1	1	1	−1	0	0	3
x_5	0	1	1	−1	0	1	3
x_4	0	3	0	−1	1	0	1

This tableau is ready for application of the two-phase method. Subsequent tableaux are not shown.

An analysis of the two-phase method and the big-M method for the single artificial variable technique discussed in this section can be made similar to the analysis of Sections 4.2 and 4.3. The details are left to the reader in Exercise 4.29.

4.6 DEGENERACY, CYCLING, AND STALLING

Degeneracy causes several conceptual as well as computational difficulties in linear programming. Conceptually, we have seen, for example, in Chapter 3 that in the presence of degeneracy, we have to be cautious about claiming the existence of alternative optimal solutions simply based on observing a zero reduced cost for a nonbasic variable at optimality. We also should be cautious about interpreting $\partial z / \partial b_i = w_i$ as an *actual* (realizable) rate of change, and must realize that not all bases representing an optimal solution necessarily satisfy $z_j - c_j \leq 0$, $j = 1, \ldots, n$ (for a minimization problem).

Computationally, we have seen in Chapter 3 that in the absence of degeneracy, the simplex method terminates finitely, either producing an optimal basic feasible solution or else verifying unboundedness. However, when a degenerate pivot occurs, we simply switch from one basis to another, both representing the same extreme point solution. This is evident in Table 3.1 by examining the situation before and after pivoting when $\bar{b}_r = 0$.

As the process is repeated, it is conceivable that another degenerate pivot is taken, resulting in the same extreme point with a different basis representation. It is therefore possible, though highly unlikely, that we may stay at a nonoptimal extreme point and pivot through a sequence of associated bases $\mathbf{B}_1, \mathbf{B}_2, \ldots, \mathbf{B}_t$ where $\mathbf{B}_t = \mathbf{B}_1$. If the same sequence of pivots is used over and over again, we shall *cycle* forever among the bases $\mathbf{B}_1, \mathbf{B}_2, \ldots, \mathbf{B}_t = \mathbf{B}_1$ without reaching an optimal solution.

Hence we present in the sequel some cycling prevention rules that guarantee finite convergence of the simplex algorithm by ensuring that no basis can be repeated. By doing so, such rules constructively demonstrate the following result which has so far been evident only in the absence of degeneracy.

***Theorem* 4.1**

Given an optimal extreme point solution (for a minimization linear program), there exists an associated (optimal) basis for which $z_j - c_j \leq 0$ for all $j = 1, \ldots, n$.

We begin our discussion by first presenting an example that demonstrates the phenomenon of cycling.

Example 4.11

(Cycling)

Consider the following example given by Beale.

$$\begin{array}{llllllllll}
\text{Minimize} & & & & -\tfrac{3}{4}x_4 & + 20x_5 & - \tfrac{1}{2}x_6 & + 6x_7 & \\
\text{Subject to} & x_1 & & & + \tfrac{1}{4}x_4 & - 8x_5 & - x_6 & + 9x_7 & = 0 \\
& & x_2 & & + \tfrac{1}{2}x_4 & - 12x_5 & - \tfrac{1}{2}x_6 & + 3x_7 & = 0 \\
& & & x_3 & & & + x_6 & & = 1 \\
& x_1, & x_2, & x_3, & x_4, & x_5, & x_6, & x_7 & \geq 0
\end{array}$$

The optimal solution is given by $x_1 = \frac{3}{4}$, $x_4 = x_6 = 1$, and all other variables equal to zero. The optimal objective value is $-\frac{5}{4}$. The following rules are adopted. The entering variable is that with the most positive $z_j - c_j$, and the leaving variable is determined by the minimum ratio test, where ties are broken arbitrarily.

	z	x_1	x_2	x_3	x_4	x_5	x_6	x_7	RHS
z	1	0	0	0	$\frac{3}{4}$	-20	$\frac{1}{2}$	-6	0
x_1	0	1	0	0	($\frac{1}{4}$)	-8	-1	9	0
x_2	0	0	1	0	$\frac{1}{2}$	-12	$-\frac{1}{2}$	3	0
x_3	0	0	0	1	0	0	1	0	1

	z	x_1	x_2	x_3	x_4	x_5	x_6	x_7	RHS
z	1	-3	0	0	0	4	$\frac{7}{2}$	-33	0
x_4	0	4	0	0	1	-32	-4	36	0
x_2	0	-2	1	0	0	$\textcircled{4}$	$\frac{3}{2}$	-15	0
x_3	0	0	0	1	0	0	1	0	1

	z	x_1	x_2	x_3	x_4	x_5	x_6	x_7	RHS
z	1	-1	-1	0	0	0	2	-18	0
x_4	0	-12	8	0	1	0	$\textcircled{8}$	-84	0
x_5	0	$-\frac{1}{2}$	$\frac{1}{4}$	0	0	1	$\frac{3}{8}$	$-\frac{15}{4}$	0
x_3	0	0	0	1	0	0	1	0	1

	z	x_1	x_2	x_3	x_4	x_5	x_6	x_7	RHS
z	1	2	-3	0	$-\frac{1}{4}$	0	0	3	0
x_6	0	$-\frac{3}{2}$	1	0	$\frac{1}{8}$	0	1	$-\frac{21}{2}$	0
x_5	0	$\frac{1}{16}$	$-\frac{1}{8}$	0	$-\frac{3}{64}$	1	0	$\textcircled{\frac{3}{16}}$	0
x_3	0	$\frac{3}{2}$	-1	1	$-\frac{1}{8}$	0	0	$\frac{21}{2}$	1

	z	x_1	x_2	x_3	x_4	x_5	x_6	x_7	RHS
z	1	1	-1	0	$\frac{1}{2}$	-16	0	0	0
x_6	0	$\textcircled{2}$	-6	0	$-\frac{5}{2}$	56	1	0	0
x_7	0	$\frac{1}{3}$	$-\frac{2}{3}$	0	$-\frac{1}{4}$	$\frac{16}{3}$	0	1	0
x_3	0	-2	6	1	$\frac{5}{2}$	-56	0	0	1

	z	x_1	x_2	x_3	x_4	x_5	x_6	x_7	RHS
z	1	0	2	0	$\frac{7}{4}$	-44	$-\frac{1}{2}$	0	0
x_1	0	1	-3	0	$-\frac{5}{4}$	28	$\frac{1}{2}$	0	0
x_7	0	0	$\textcircled{\frac{1}{3}}$	0	$\frac{1}{6}$	-4	$-\frac{1}{6}$	1	0
x_3	0	0	0	1	0	0	1	0	1

	z	x_1	x_2	x_3	x_4	x_5	x_6	x_7	RHS
z	1	0	0	0	$\frac{3}{4}$	-20	$\frac{1}{2}$	-6	0
x_1	0	1	0	0	$\frac{1}{4}$	-8	-1	9	0
x_2	0	0	1	0	$\frac{1}{2}$	-12	$-\frac{1}{2}$	3	0
x_3	0	0	0	1	0	0	1	0	1

We see that the last tableau is identical to the first tableau. All the tableaux correspond to the extreme point $(0, 0, 1, 0, 0, 0, 0)$ with different bases. The foregoing sequence of pivots generated the bases $\mathbf{B}_1$, $\mathbf{B}_2$, $\mathbf{B}_3$, $\mathbf{B}_4$, $\mathbf{B}_5$, $\mathbf{B}_6$, and $\mathbf{B}_7$, where $\mathbf{B}_7 = \mathbf{B}_1 = [\mathbf{a}_1, \mathbf{a}_2, \mathbf{a}_3]$. If the same sequence of pivots are used over and over again, the simplex algorithm will cycle forever among these bases without reaching the optimal point.

Two Rules that Prevent Cycling

Even though cycling is very unlikely and actually it is not an easy task to formulate a problem that cycles, it is of theoretical interest to develop rules that prevent cycling. We give two such rules here, and illustrate them in Example 4.11. Validation of these rules is postponed until the next section. Consider the following linear programming problem.

$$\begin{aligned} &\text{Minimize } \mathbf{cx} \\ &\text{Subject } \quad \mathbf{Ax} = \mathbf{b} \\ &\qquad\qquad\quad \mathbf{x} \geq \mathbf{0} \end{aligned}$$

where $\mathbf{A}$ is an $m \times n$ matrix of rank m. Since the simplex method is usually started with the initial basis as the identity matrix (corresponding to slack and/or artificial variables), we shall assume that the first m columns of $\mathbf{A}$ form the identity. The following rule, which specifies the variable leaving the basis if the simplex minimum ratio test produces several candidates, will guarantee noncycling.

Lexicographic Rule for Selecting an Exiting Variable

Given a basic feasible solution with basis $\mathbf{B}$, suppose that the nonbasic variable x_k is chosen to enter the basis (say, $0 < z_k - c_k = \text{Maximum } z_j - c_j$). The index r of the variable x_{B_r} leaving the basis is determined as follows. Let

$$I_0 = \left\{ r : \frac{\bar{b}_r}{y_{rk}} = \underset{1 \leq i \leq m}{\text{Minimum}} \left\{ \frac{\bar{b}_i}{y_{ik}} : y_{ik} > 0 \right\} \right\}$$

If I_0 is a singleton, namely $I_0 = \{r\}$, then x_{B_r} leaves the basis. Otherwise form I_1 as follows:

$$I_1 = \left\{ r : \frac{y_{r1}}{y_{rk}} = \underset{i \in I_0}{\text{Minimum}} \left\{ \frac{y_{i1}}{y_{ik}} \right\} \right\}$$

If I_1 is singleton, namely $I_1 = \{r\}$, then x_{B_r} leaves the basis. Otherwise form I_2. In general I_j is formed from I_{j-1} as follows:

$$I_j = \left\{ r : \frac{y_{rj}}{y_{rk}} = \underset{i \in I_{j-1}}{\text{Minimum}} \left\{ \frac{y_{ij}}{y_{ik}} \right\} \right\}$$

Eventually, for some $j \leq m$, I_j will be a singleton (why?). If $I_j = \{r\}$, then x_{B_r} leaves the basis.

Before we illustrate the preceding rule, let us briefly discuss its implications. (The choice of the name for this method will become evident when we validate this rule in the next section.) In this rule, we first use the usual minimum ratio test as an exiting criterion. If this test gives a unique index, then the corresponding variable leaves the basis. In case of a tie we try to break it by replacing the right-hand-side in the minimum ratio calculation by the first column $\mathbf{y}_1$, and by only using the rows corresponding to the tie. If the tie is still not broken, the second column is used, and so forth. When or before column m is reached, the tie must be broken, for if this were not the case, we have two rows of the matrix $\mathbf{B}^{-1} = (\mathbf{y}_1, \mathbf{y}_2, \ldots, \mathbf{y}_m)$, which are proportional (why?). This is impossible, however, in view of linear independence of the rows of $\mathbf{B}^{-1}$.

Example 4.12

We now solve the problem of Example 4.11 using the additional rule for exiting from the basis.

	z	x_1	x_2	x_3	x_4	x_5	x_6	x_7	RHS
z	1	0	0	0	$\frac{3}{4}$	-20	$\frac{1}{2}$	-6	0
x_1	0	1	0	0	$\frac{1}{4}$	-8	-1	9	0
x_2	0	0	1	0	($\frac{1}{2}$)	-12	$-\frac{1}{2}$	3	0
x_3	0	0	0	1	0	0	1	0	1

Here $I_0 = \{1, 2\}$, $I_1 = \{2\}$, and therefore $x_{B_2} = x_2$ leaves the basis. Note that in Example 4.11, x_1 left the basis during the first iteration.

	z	x_1	x_2	x_3	x_4	x_5	x_6	x_7	RHS
z	1	0	$-\frac{3}{2}$	0	0	-2	$\frac{5}{4}$	$-\frac{21}{2}$	0
x_1	0	1	$-\frac{1}{2}$	0	0	-2	$-\frac{3}{4}$	$\frac{15}{2}$	0
x_4	0	0	2	0	1	-24	-1	6	0
x_3	0	0	0	1	0	0	(1)	0	1

Here $I_0 = \{3\}$. Therefore $x_{B_3} = x_3$ leaves.

	z	x_1	x_2	x_3	x_4	x_5	x_6	x_7	RHS
z	1	0	$-\frac{3}{2}$	$-\frac{5}{4}$	0	-2	0	$-\frac{21}{2}$	$-\frac{5}{4}$
x_1	0	1	$-\frac{1}{2}$	$\frac{3}{4}$	0	-2	0	$\frac{15}{2}$	$\frac{3}{4}$
x_4	0	0	2	1	1	-24	0	6	1
x_6	0	0	0	1	0	0	1	0	1

The foregoing tableau gives the optimal solution, since $z_j - c_j \leq 0$ for all nonbasic variables.

Bland's Rule for Selecting Entering and Leaving Variables

Another rule that prevents cycling has been suggested by Robert Bland. This is a very simple rule, but one that restricts the choice of both the entering and leaving variables. In this rule, the variables are first ordered in some sequence, say, $x_1, x_2, \ldots, x_n$ without loss of generality. Then of all nonbasic variables with $z_j - c_j > 0$, the one with the smallest index is selected to enter the basis. Similarly, of all the candidates to leave the basis (which tie in the usual minimum ratio test), the one with the smallest index is chosen as the exiting variable.

Using this rule in Example 4.11, we see that the first four tableaux are produced as given. For example, in the first tableau, out of x_4 and x_6 we select x_4 to enter the basis, and out of the candidates x_1 and x_2 for the exiting variable, we choose x_1 to leave the basis. However, in the fourth tableau, x_1 and x_7 are eligible to enter the basis, and the present rule selects x_1 as the entering variable. The leaving variable is uniquely determined as x_5. In Exercise 4.43 we ask the reader to verify that the resulting tableau leads to a nondegenerate pivot and that the pivots continue until optimality is attained.

Some Practical Implementation Remarks and Stalling

Although most real-world problems are known to be degenerate and the foregoing rules guarantee that cycling will not occur, these rules are largely ignored in most commercial codes for solving linear programming problems via the simplex method. There are two main reasons for this stance. First, the rules are either computationally expensive to implement, as in the case of the lexicographic method, or are computationally inefficient with respect to the length of the simplex path generated, as in the case of Bland's rule. Second, because of computer round-off errors, it is usually argued that the updated right-hand-sides are in any case perturbed from their actual values and one rarely encounters exact zero right-hand-side values. In fact, as we ask the reader to explore in Exercise 4.52, the lexicographic rule has an equivalent interpretation as a perturbation technique in which the original right-hand-side values are perturbed slightly to make the polyhedron nondegenerate. For example, a software package called MINOS, which is developed at the Stanford Systems Optimization Laboratory, adopts a "practical" anti-cycling rule based on appropriate perturbations of right-hand-sides of variable bonding constraints. We remark, however, that in the case of network structured linear programs, as we will be seeing later, neither of the foregoing reasons apply, and cycling prevention rules are both easy to implement and are often computationally advantageous.

There is another issue that needs to be mentioned at this point. Although the foregoing rules prevent cycling, it is entirely possible that the simplex algorithm may go through an exceedingly long (though finite) sequence of degenerate pivots. In particular, for some classes of problems having a certain structure, it may be possible that the algorithm performs a number of consecutive degenerate pivots that are exponential in the size (m and n) of the problem. This phenomenon is known as *stalling*. The term arises because with increasing problem size, the algorithm can spend an enormous (though finite) amount of time at a degenerate vertex before finally verifying optimality or moving off from that vertex. Besides preventing cycling, one would also like to prevent stalling by ensuring that the length of a sequence of degenerate pivots is bounded from above by some polynomial in m and n.

The key to preventing stalling appears to lie in what are known as stages. A *stage* is a sequence of degenerate pivots in which no nonbasic variable remains enterable throughout, with no strict subset of this sequence having the same property. One would like both the number of pivots in a stage, which is the *length* of a stage, and the number of stages in a sequence of degenerate pivots to be "small", that is, polynomially bounded. The first of these two conditions is easy to achieve. For example, consider the following rule for choosing an entering variable, which is sometimes referred to as the *Least Recently Considered (LRC)* variable choice rule. Suppose one maintains a circular list of the variables $x_1, x_2, \ldots, x_n$, and at any pivot, if a variable x_k enters, then suppose that the next entering variable is selected as the first eligible candidate (with $z_j - c_j > 0$) from the list $\{x_{k+1}, \ldots, x_n, x_1, \ldots, x_k\}$, where $x_{n+1} \equiv x_1$. Then it is clear that the length of a stage is no more than n, since no variable could have remained enterable and have been ignored throughout a sequence of n (or more) degenerate pivots. Such a rule, by its nature, is aptly called an *affirmative action* policy. Note that Bland's rule is not an affirmative action strategy, since it "discriminates" in favor of variables with lower indices. In fact, there exist examples that demonstrate that Bland's rule admits stalling (see the Notes and References section of this chapter). On the other hand, it is an open question whether the lexicographic rule plus Dantzig's rule of entering the variable with the most positive $z_j - c_j$ admits or prevents stalling for general linear programs. (For network structured problems, this is known to prevent stalling.)

The issue of ensuring a polynomial bound on the number of stages in a sequence of degenerate pivots is harder to achieve. For general linear programming problems, a rule that provides such a guarantee has not as yet been discovered. However, as we will see in Chapter 9 for network structured problems, a lexicographic cycling prevention rule does exist that is easy to implement, is computationally advantageous, and prevents stalling when used in conjunction with an appropriate affirmative action entering rule.

4.7 VALIDATION OF THE TWO CYCLING PREVENTION RULES

In this section we show that the rules adopted in the previous section indeed prevent cycling. We do this by showing that none of the previous bases visited by the simplex method are repeated. In view of the finite number of bases, this automatically guarantees stopping in a finite number of iterations.

Lexicographic Rule

In order to facilitate the proof of finite convergence under the lexicographic rule, consider the following notion of a lexicographically positive vector. A vector $\mathbf{x}$ is called *lexicographically positive* (denoted by $\mathbf{x} \succ \mathbf{0}$) if the following two requirements hold:

1. $\mathbf{x}$ is not identically zero.
2. The first nonzero component of $\mathbf{x}$ is positive.

For example, $(0, 2, -1, 3)$, $(2, 1, -3, 1)$, and $(0, 0, 1, -1)$ are lexicographically positive vectors whereas $(-1, 1, 2, 3)$, $(0, 0, 0, 0)$, and $(0, 0, -2, 1)$ are not. A *lexicographically nonnegative* vector, denoted by $\succcurlyeq \mathbf{0}$, is either the zero vector or else a lexicographically positive vector. In order to prove that none of the bases generated by the simplex method is repeated, we first show that each row of the $m \times (m+1)$ matrix $(\bar{\mathbf{b}}, \mathbf{B}^{-1})$ is lexicographically positive at each iteration, where $\bar{\mathbf{b}} = \mathbf{B}^{-1}\mathbf{b}$. Basic feasible solutions that satisfy this property are sometimes called *strongly feasible* solutions. Indeed, in the absence of degeneracy $\bar{\mathbf{b}} > \mathbf{0}$, and therefore each row $(\bar{\mathbf{b}}, \mathbf{B}^{-1})$ is clearly lexicographically positive.

First recall that the original basis is $\mathbf{I}$, and since $\mathbf{b} \geq \mathbf{0}$, then each row of the matrix $(\bar{\mathbf{b}}, \mathbf{B}^{-1}) = (\mathbf{b}, \mathbf{I})$ is lexicographically positive. (If a feasible basis that is different from the identity is available, we still have a starting solution that satisfies the lexicographic positive condition. See Exercise 4.50.) In view of this, the preceding result will be proved if we can show the following: if each row of $(\bar{\mathbf{b}}, \mathbf{B}^{-1})$ is $\succ \mathbf{0}$, then each row of $(\hat{\mathbf{b}}, \hat{\mathbf{B}}^{-1})$ is $\succ \mathbf{0}$ where $\hat{\mathbf{B}}$ is the new basis obtained after pivoting and $\hat{\mathbf{b}} = \hat{\mathbf{B}}^{-1}\mathbf{b}$. Consider the following two tableaux before and after introducing x_k, and recall that the first m columns (ignoring the z column) in these tableaux represent $\mathbf{B}^{-1}$ and $\hat{\mathbf{B}}^{-1}$ respectively (since the first m columns of the original problem form the identity). Here $\bar{z}$ denotes the objective value $\mathbf{c}_B\bar{\mathbf{b}}$ before pivoting.

Before Pivoting

	z	x_1	$\cdots$	x_j	$\cdots$	x_m	x_{m+1} $\cdots$	x_k	$\cdots$ x_n	RHS
z	1	$z_1 - c_1$	$\cdots$	$z_j - c_j$	$\cdots$	$z_m - c_m$	$\cdots$	$z_k - c_k$	$\cdots$	$\bar{z}$
x_{B_1}	0	y_{11}	$\cdots$	y_{1j}	$\cdots$	y_{1m}	$\cdots$	y_{1k}	$\cdots$	$\bar{b}_1$
$\vdots$	$\vdots$	$\vdots$		$\vdots$				$\vdots$		$\vdots$
x_{B_i}	0	y_{i1}	$\cdots$	y_{ij}	$\cdots$	y_{im}	$\cdots$	y_{ik}	$\cdots$	$\bar{b}_i$
$\vdots$	$\vdots$	$\vdots$		$\vdots$		$\vdots$		$\vdots$		$\vdots$
x_{B_r}	0	y_{r1}	$\cdots$	y_{rj}	$\cdots$	y_{rm}	$\cdots$	$\boxed{y_{rk}}$	$\cdots$	$\bar{b}_r$
$\vdots$	$\vdots$	$\vdots$		$\vdots$		$\vdots$		$\vdots$		$\vdots$
x_{B_m}	0	y_{m1}	$\cdots$	y_{mj}	$\cdots$	y_{mm}	$\cdots$	y_{mk}	$\cdots$	$\bar{b}_m$

After Pivoting

	z	x_1	$\cdots$	x_j	$\cdots$	x_m	$x_{m+1} \cdots x_k \cdots x_n$	RHS
z	1	$(z_1 - c_1) - \dfrac{y_{r1}}{y_{rk}}(z_k - c_k)$	$\cdots$	$(z_j - c_j) - \dfrac{y_{rj}}{y_{rk}}(z_k - c_k)$	$\cdots$	$(z_m - c_m) - \dfrac{y_{rm}}{y_{rk}}(z_k - c_k)$	$\cdots 0 \cdots$	$\bar{z} - \dfrac{\bar{b}_r}{y_{rk}}(z_k - c_k)$
x_{B_1}	0	$y_{11} - \dfrac{y_{r1}}{y_{rk}} y_{1k}$	$\cdots$	$y_{1j} - \dfrac{y_{rj}}{y_{rk}} y_{1k}$	$\cdots$	$y_{1m} - \dfrac{y_{rm}}{y_{rk}} y_{1k}$	$\cdots 0 \cdots$	$\bar{b}_1 - \dfrac{\bar{b}_r}{y_{rk}} y_{1k}$
$\vdots$	$\vdots$	$\vdots$		$\vdots$		$\vdots$	$\vdots$	$\vdots$
x_{B_i}	0	$y_{i1} - \dfrac{y_{r1}}{y_{rk}} y_{ik}$	$\cdots$	$y_{ij} - \dfrac{y_{rj}}{y_{rk}} y_{ik}$	$\cdots$	$y_{im} - \dfrac{y_{rm}}{y_{rk}} y_{ik}$	$\cdots 0 \cdots$	$\bar{b}_i - \dfrac{\bar{b}_r}{y_{rk}} y_{ik}$
$\vdots$	$\vdots$	$\vdots$		$\vdots$		$\vdots$	$\vdots$	$\vdots$
x_k	0	$\dfrac{y_{r1}}{y_{rk}}$	$\cdots$	$\dfrac{y_{rj}}{y_{rk}}$	$\cdots$	$\dfrac{y_{rm}}{y_{rk}}$	$\cdots 1 \cdots$	$\dfrac{\bar{b}_r}{y_{rk}}$
$\vdots$	$\vdots$	$\vdots$		$\vdots$		$\vdots$	$\vdots$	$\vdots$
x_{B_m}	0	$y_{m1} - \dfrac{y_{r1}}{y_{rk}} y_{mk}$	$\cdots$	$y_{mj} - \dfrac{y_{rj}}{y_{rk}} y_{mk}$	$\cdots$	$y_{mm} - \dfrac{y_{rm}}{y_{rk}} y_{mk}$	$\cdots 0 \cdots$	$\bar{b}_m - \dfrac{\bar{b}_r}{y_{rk}} y_{mk}$

Consider a typical row i of $(\hat{\mathbf{b}}, \hat{\mathbf{B}}^{-1})$. From the foregoing tableau this row is given by

$$\left(\bar{b}_i - \frac{\bar{b}_r}{y_{rk}} y_{ik}, y_{i1} - \frac{y_{r1}}{y_{rk}} y_{ik}, \ldots, y_{im} - \frac{y_{rm}}{y_{rk}} y_{ik}\right) \qquad i \neq r \tag{4.3}$$

$$\left(\frac{\bar{b}_r}{y_{rk}}, \frac{y_{r1}}{y_{rk}}, \ldots, \frac{y_{rm}}{y_{rk}}\right) \qquad i = r \tag{4.4}$$

Since $y_{rk} > 0$ and the rth row is $\succ \mathbf{0}$ before pivoting, then from Equation (4.4) the rth row after pivoting is also $\succ \mathbf{0}$. Now consider $i \neq r$. There are two mutually exclusive cases: either $i \notin I_0$ or else $i \in I_0$. First suppose that $i \notin I_0$. If $y_{ik} \leq 0$, then from Equation (4.3) we see that the ith row after pivoting is given by

$$(\bar{b}_i, y_{i1}, \ldots, y_{im}) - \frac{y_{ik}}{y_{rk}}(\bar{b}_r, y_{r1}, \ldots, y_{rm})$$

which is the sum of two vectors that are lexicographically positive and lexicographically nonnegative (why?), and hence $\succ \mathbf{0}$. Now suppose that $y_{ik} > 0$. By definition of I_0 and since $i \notin I_0$, then $\bar{b}_r/y_{rk} < \bar{b}_i/y_{ik}$ and hence $\bar{b}_i - (\bar{b}_r/y_{rk})y_{ik} > 0$. From Equation (4.3) the ith row is $\succ \mathbf{0}$. Now consider the case $i \in I_0$. Then $y_{ik} > 0$ and $\bar{b}_i - (\bar{b}_r/y_{rk})y_{ik} = 0$. There are two mutually exhaustive cases: either $i \notin I_1$, or else $i \in I_1$. In the former case, by definition of I_1, $y_{i1} - (y_{r1}/y_{rk})y_{ik} > 0$ and from Equation (4.3) the ith row is $\succ \mathbf{0}$. If, on the other hand, $i \in I_1$, then $y_{i1} - (y_{r1}/y_{rk})y_{ik} = 0$ and we examine whether $i \in I_2$ or not. This process is continued until termination in at most $m + 1$ steps, with the conclusion that each row of $(\hat{\mathbf{b}}, \hat{\mathbf{B}}^{-1})$ is $\succ \mathbf{0}$.

The foregoing analysis shows that each row of $(\mathbf{B}^{-1}\mathbf{b}, \mathbf{B}^{-1})$ is lexicographically positive at any given iteration. This fact will be used shortly to prove finite convergence. First note, by examining row 0 before and after pivoting, that

$$\left(\mathbf{c}_B\mathbf{B}^{-1}\mathbf{b}, \mathbf{c}_B\mathbf{B}^{-1}\right) - \left(\mathbf{c}_{\hat{B}}\hat{\mathbf{B}}^{-1}\mathbf{b}, \mathbf{c}_{\hat{B}}\hat{\mathbf{B}}^{-1}\right) = \frac{z_k - c_k}{y_{rk}}(\bar{b}_r, y_{r1}, y_{r2}, \ldots, y_{rm})$$

Note that $(\bar{b}_r, y_{r1}, \ldots, y_{rm})$ is the rth row of $(\bar{\mathbf{b}}, \mathbf{B}^{-1})$ and is therefore $\succ \mathbf{0}$. Since $z_k - c_k > 0$ and $y_{rk} > 0$, it is therefore evident that

$$\left(\mathbf{c}_B\mathbf{B}^{-1}\mathbf{b}, \mathbf{c}_B\mathbf{B}^{-1}\right) - \left(\mathbf{c}_{\hat{B}}\hat{\mathbf{B}}^{-1}\mathbf{b}, \mathbf{c}_{\hat{B}}\hat{\mathbf{B}}^{-1}\right) \succ \mathbf{0}$$

Now we are ready to show that the lexicographic rule of Section 4.6 will indeed prevent cycling. We do this by showing that the bases developed by the simplex method are distinct. Suppose by contradiction that a sequence of bases $\mathbf{B}_1, \mathbf{B}_2, \ldots, \mathbf{B}_t$ is generated where $\mathbf{B}_1 = \mathbf{B}_t$. From the preceding analysis we have

$$\left(\mathbf{c}_{B_j}\mathbf{B}_j^{-1}\mathbf{b}, \mathbf{c}_{B_j}\mathbf{B}_j^{-1}\right) - \left(\mathbf{c}_{B_{j+1}}\mathbf{B}_{j+1}^{-1}\mathbf{b}, \mathbf{c}_{B_{j+1}}\mathbf{B}_{j+1}^{-1}\right) \succ \mathbf{0} \qquad \text{for } j = 1, 2, \ldots, t-1$$

Adding over $j = 1, 2, \ldots, t - 1$ and noting that $\mathbf{B}_1$ is assumed equal to $\mathbf{B}_t$, we get $\mathbf{0} \succ \mathbf{0}$, which is impossible. This contradiction asserts that the bases visited by the simplex algorithm are distinct. Since there are but a finite number of bases, convergence in a finite number of steps is established.

Bland's Rule

In the absence of degeneracy, the objective function itself was a strictly decreasing monotone function that guaranteed that no basis would be repeated by the simplex algorithm. On the other hand, the lexicographic rule ensured finite convergence by showing that while $\mathbf{c}_B\mathbf{B}^{-1}\mathbf{b}$ may remain constant in the presence of degeneracy, the vector $(\mathbf{c}_B\mathbf{B}^{-1}\mathbf{b}, \mathbf{c}_B\mathbf{B}^{-1})$ is lexicographically monotone decreasing. Bland's rule has the following monotone feature: In a sequence of degenerate pivots, if some variable x_q enters the basis, then x_q cannot leave the basis until some other variable with a *higher* index than q, which was nonbasic when x_q entered, also enters the basis. If this holds, then cycling cannot occur because in a cycle any variable that enters must also leave the basis, which means that there exists some highest indexed variable that enters and leaves the basis. This contradicts the foregoing monotone feature (why?).

To show that this monotone feature holds true, consider the basic feasible solution at which x_q enters. Let J_1 and J_2 be the sets of the current nonbasic variables with indices respectively less than and greater than q. As in Equation (3.5), we can write the representation of the linear program in the current canonical form as follows.

$$\begin{aligned} \text{Minimize} \quad & \sum_{j\in J_1} \bar{c}_j x_j + \bar{c}_q x_q + \sum_{j\in J_2} \bar{c}_j x_j \\ \text{Subject to} \quad & \bar{\mathbf{A}}\mathbf{x}_N + \mathbf{x}_B = \bar{\mathbf{b}} \\ & \mathbf{x}_N, \quad \mathbf{x}_B \geq \mathbf{0} \end{aligned} \tag{4.5}$$

where $\mathbf{x}_B$ and $\mathbf{x}_N$ are the current basic and nonbasic variables and $\bar{c}_j = c_j - z_j$ for all j. Note that since x_q is an entering variable by Bland's rule, we have that $\bar{c}_q < 0$ and $\bar{c}_j \geq 0$ for $j \in J_1$.

Now suppose on the contrary that we go through a sequence of degenerate pivots in which x_j, $j \in J_2$ remain nonbasic, and we reach a pivot in which some x_p enters and x_q leaves the basis. Now treat Equation (4.5) as an "original" linear program, that is, define variable columns $\mathbf{a}_j$, bases, and cost coefficients with respect to this representation. In particular, let the basis in Equation (4.5) at the pivot in which x_p enters and x_q leaves be denoted by $\mathbf{B}_1$ with the basic cost coefficient vector being $\bar{\mathbf{c}}_{B_1}$. Hence $\bar{\mathbf{c}}_{B_1}\mathbf{B}_1^{-1}\mathbf{a}_p - \bar{c}_p > 0$ since x_p enters. Because $\bar{c}_p \geq 0$ (why?), this means that $\bar{\mathbf{c}}_{B_1}\mathbf{B}_1^{-1}\mathbf{a}_p > 0$. Let $\mathbf{y}_p = \mathbf{B}_1^{-1}\mathbf{a}_p$ and denote the pivot element in the row for the currently basic variable x_q by y_{qp}. Note that $y_{qp} > 0$ and $\bar{c}_q < 0$, so that in the positive inner product $\bar{\mathbf{c}}_{B_1}\mathbf{y}_p$, the component $y_{qp}\bar{c}_q$ gives a negative contribution. Hence there must be another basic variable x_r that has $\bar{c}_r > 0$ and $y_{rp} > 0$, since other than $\bar{c}_q$ all other components in $\bar{\mathbf{c}}_{B_1}$ correspond either to x_j, $j \in J_1$ or to the $\mathbf{x}_B$ variables, all of which have nonnegative $\bar{c}$ values. In particular, we must have $r \in J_1$ (why?), which also means that x_r is currently zero since we have been performing degenerate pivots. But this means that x_r is also an eligible candidate to leave the basis in the minimum ratio test.

Since $r < q$, the variable x_q cannot be the leaving variable by Bland's rule. This contradiction proves the desired result.

EXERCISES

4.1 Solve the following problem by the two-phase simplex method.

$$\begin{array}{ll} \text{Maximize} & 2x_1 - x_2 + x_3 \\ \\ \text{Subject to} & 2x_1 + x_2 - 2x_3 \leq 8 \\ & 4x_1 - x_2 + 2x_3 \geq 2 \\ & 2x_1 + 3x_2 - x_3 \geq 4 \\ & x_1, \quad x_2, \quad x_3 \geq 0 \end{array}$$

4.2 Consider the following linear programming problem.

$$\begin{array}{ll} \text{Maximize} & -x_1 + 8x_2 \\ \\ \text{Subject to} & x_1 + x_2 \geq 1 \\ & -x_1 + 6x_2 \leq 3 \\ & x_2 \leq 2 \\ & x_1, \quad x_2 \geq 0 \end{array}$$

a. Solve the problem geometrically.
b. Solve the problem by the two-phase simplex method. Show that the points generated by phase I correspond to basic solutions of the original system.

4.3 Phase I of the two-phase method can be made use of to check redundancy. Suppose that we have the following three constraints:

$$x_1 - x_2 \geq 2$$
$$2x_1 + 3x_2 \geq 4$$
$$3x_1 + 2x_2 \geq 6$$

Note that the third constraint can be obtained by adding the first two constraints. Would phase I detect this kind of redundancy? If not, what kind of redundancy will it detect? Does the type of redundancy in which some inequality holds whenever the other inequalities hold imply degeneracy? Discuss.

4.4 Solve the following problem by the two-phase simplex method.

$$\begin{array}{ll} \text{Minimize} & x_1 + 3x_2 - x_3 \\ \\ \text{Subject to} & 2x_1 + x_2 + x_3 \geq 3 \\ & -x_1 + x_2 \geq 2 \\ & -x_1 + 5x_2 + x_3 \leq 4 \\ & x_1, \quad x_2, \quad x_3 \geq 0 \end{array}$$

4.5 Show how phase I of the simplex method can be used to solve n simultaneous linear equations in n unknowns. Show how the following cases can be detected:

a. Inconsistency of the system.
b. Redundancy of the equations.
c. Unique solution.

Also show how the inverse matrix corresponding to the system of equations can be found in part (c). Illustrate using the following system.

$$\begin{aligned} x_1 + 2x_2 + x_3 &= 4 \\ -x_1 - x_2 + 2x_3 &= 3 \\ 3x_1 + 5x_2 \quad &= 5 \end{aligned}$$

4.6 Solve the following problem by the two-phase method.

Maximize $-x_1 - 2x_2$

Subject to
$$\begin{aligned} 3x_1 + 4x_2 &\leq 12 \\ 2x_1 - x_2 &\geq 2 \\ x_1, \quad x_2 &\geq 0 \end{aligned}$$

4.7 Solve the following problem by the two-phase method.

Maximize $5x_1 - 2x_2 + x_3$

Subject to
$$\begin{aligned} 2x_1 + 4x_2 + x_3 &\leq 6 \\ 2x_1 + x_2 + 3x_3 &\geq 2 \\ x_1, \quad x_2 \quad &\geq 0 \\ x_3 \text{ unrestricted} \end{aligned}$$

4.8 Solve the following problem by the two-phase method.

Maximize $4x_1 + 5x_2 - 3x_3$

Subject to
$$\begin{aligned} x_1 + x_2 + x_3 &= 10 \\ x_1 - x_2 \quad &\geq 1 \\ x_1 + 3x_2 + x_3 &\leq 14 \\ x_1, \quad x_2, \quad x_3 &\geq 0 \end{aligned}$$

4.9 Use the big-M method to solve the following problem.

Minimize $-2x_1 + 2x_2 + x_3 + x_4$

Subject to
$$\begin{aligned} x_1 + 2x_2 + x_3 + x_4 &\leq 2 \\ x_1 - x_2 + x_3 + 2x_4 &\geq 3 \\ 2x_1 - x_2 + x_3 \quad &\geq 2 \\ x_1, \quad x_2, \quad x_3, \quad x_4 &\geq 0 \end{aligned}$$

4.10 Use the big-M method to solve the following problem.

$$\begin{array}{ll} \text{Maximize} & x_1 - 2x_2 + x_3 \\ \text{Subject to} & x_1 + x_2 - x_3 \geq 4 \\ & x_1 - 4x_2 + x_3 \leq 2 \\ & x_1, \quad x_2, \quad x_3 \geq 0 \end{array}$$

4.11 Is it possible that the optimal solution value of the big-M problem is unbounded and in the meantime the optimal solution of the original problem is bounded? Discuss in detail.

4.12 Consider the problem: Maximize $\mathbf{cx}$ subject to $\mathbf{Ax} = \mathbf{b}$, $\mathbf{x} \geq \mathbf{0}$, where $\mathbf{A}$ is an $m \times n$ matrix and $\mathbf{b} \geq \mathbf{0}$. Add the artificial vector $\mathbf{x}_a$ and consider the following big-M problem: Maximize $\mathbf{cx} - M\mathbf{1x}_a$ subject to $\mathbf{Ax} + \mathbf{x}_a = \mathbf{b}$, $\mathbf{x} \geq \mathbf{0}$, $\mathbf{x}_a \geq \mathbf{0}$. Show that there exists an $\overline{M} > 0$ such that for all $M \geq \overline{M}$ either the basic feasible solution $(\mathbf{x}^*, \mathbf{x}_a^*)$ solves the big-M problem or else the solution of the big-M problem is unbounded along the ray $\{(\mathbf{x}^*, \mathbf{x}_a^*) + \lambda(\mathbf{d}^*, \mathbf{d}_a^*) : \lambda \geq 0\}$ where $(\mathbf{d}^*, \mathbf{d}_a^*)$ is an extreme direction. Interpret both cases. (This establishes the fact that for M larger than some number, regardless of whether the original problem has a feasible region or not, the solution of the big-M problem is achieved at some fixed extreme point or along some fixed ray.)

4.13 Let $X = \{\mathbf{x} : \mathbf{Ax} = \mathbf{b}, \mathbf{x} \geq \mathbf{0}\}$ be nonempty, where $\mathbf{A}$ is $m \times n$. Consider the *preemptive priority* multiple objective problem of maximizing $\mathbf{c}_1\mathbf{x}$ over $\mathbf{x} \in X$, and among all alternative optimal solutions, maximizing $\mathbf{c}_2\mathbf{x}$, and among all continuing alternative optimal solutions, maximizing $\mathbf{c}_3\mathbf{x}$, and so on, until the final problem of maximizing $\mathbf{c}_r\mathbf{x}$ over continuing alternative optimal solutions. Call any such resulting optimal solution $\mathbf{x}^*$ a *preemptive optimal solution*, and assume that one exists. Show that there exists a scalar $M_0 > 0$ such that for any $M \geq M_0$, a solution $\mathbf{x}^*$ is a preemptive optimal solution if and only if it solves the problem

$$\text{Maximize}\left\{\sum_{i=1}^{r} M^{r-i}\mathbf{c}_i\mathbf{x} : \mathbf{x} \in X\right\}.$$

Relate this result to the problem in Exercise 4.12.

4.14 Suppose that a linear programming problem admits feasible points. Utilize the result of Exercise 4.12 to show that if M is large enough, then a finite optimal solution of the big-M problem must have all artificials equal to zero. Give all details.

4.15 Consider the linear programming problem to minimize $\mathbf{cx}$ subject to $\mathbf{x} \in X = \{\mathbf{x} : \mathbf{Ax} = \mathbf{b}, \mathbf{x} \geq \mathbf{0}\}$, where X is a nonempty, bounded polyhedral set in E^n. Define $\theta_1 < \theta_2 < \cdots < \theta_r$ as the distinct values taken on by $\mathbf{cx}$ over the extreme points of X and let S_i be the set of extreme points of X for which the objective value is θ_i, for $i = 1, \ldots, r$. An arrangement of the extreme points in S_1 followed by those in S_2, then $S_3, \ldots, S_r$ is said to be a *ranking* of the extreme points of X with respect to the objective function $\mathbf{cx}$. Suppose that $\mathbf{x}_1, \ldots, \mathbf{x}_l$ is a partial ranking of the extreme points. Show that a candidate for the next ranked extreme point $\mathbf{x}_{l+1}$ is adjacent to some vertex in $\mathbf{x}_1, \ldots, \mathbf{x}_l$. How would you use this result to produce all the sets $S_1, \ldots, S_r$?

4.16 Is it possible that the region in E^n given by

$$\begin{aligned} \mathbf{Ax} &= \mathbf{b} \\ \mathbf{x} &\geq \mathbf{0} \end{aligned}$$

is bounded, whereas the region in E^{n+m} given by

$$\begin{aligned} \mathbf{Ax} + \mathbf{x}_a &= \mathbf{b} \\ \mathbf{x}, \quad \mathbf{x}_a &\geq \mathbf{0} \end{aligned}$$

is unbounded? What are the implications of your answer on using the big-M method as a solution procedure?

4.17 Solve the following problem by the big-M method.

$$\begin{array}{ll} \text{Minimize} & x_1 + 4x_2 - x_4 \\ \text{Subject to} & -x_1 + 2x_2 - x_3 + x_4 \leq 2 \\ & 2x_1 + x_2 + 2x_3 - 2x_4 = 4 \\ & x_1 - 3x_3 + x_4 \geq 2 \\ & x_1, \quad x_2, \quad x_4 \geq 0 \\ & x_3 \quad \text{unrestricted} \end{array}$$

4.18 Solve the following problem by the big-M method.

$$\begin{array}{ll} \text{Maximize} & 2x_1 + 4x_2 + 4x_3 - 3x_4 \\ \text{Subject to} & 2x_1 + x_2 + x_3 = 4 \\ & x_1 + 4x_2 + 4x_4 = 8 \\ & x_1, \quad x_2, \quad x_3, \quad x_4 \geq 0 \end{array}$$

4.19 Suppose that in the two-phase method, one picks an entering variable according to the one that has the most positive $z_j - c_j$ in the phase I canonical objective row, while breaking ties using the phase II canonical objective row (with any persisting ties broken arbitrarily). Identify the cases shown in Figure 4.5 for the big-M method with this implementation of the two-phase method.

4.20 Suppose that either phase I is completed or the bounded optimal solution of the big-M problem is found. Further suppose that there exists at least one artificial at a positive level indicating that the original system $\mathbf{Ax} = \mathbf{b}$ and $\mathbf{x} \geq \mathbf{0}$ has no solution. Can you differentiate between the following two cases?

a. The system $\mathbf{Ax} = \mathbf{b}$ is inconsistent.
b. The system $\mathbf{Ax} = \mathbf{b}$ is consistent but $\mathbf{Ax} = \mathbf{b}$ implies that $\mathbf{x} \ngeq \mathbf{0}$.

4.21 Suppose that the big-M method is used to solve a linear programming problem. Further suppose that $z_k - c_k = \text{Maximum}\,(z_j - c_j) > 0$. Show that the original problem is infeasible if not all artificials are equal to zero and $y_{ik} \leq 0$ for each i such that x_{B_i} is an artificial variable. (Note that this gives a more general result than Subcase B2 of Section 4.3.)

4.22 Show that problem P could have an unbounded optimal solution even if during the solution of the big-M problem a nonbasic variable x_j is found such that $z_j - c_j > 0$, $\mathbf{y}_j \leq \mathbf{0}$ and not all artificials are zero. (This shows that it is important to examine the most positive $z_j - c_j$.)

4.23 *Geometric redundancy* occurs when deletion of a constraint does not alter the feasible set. How can geometric redundancy be detected? (*Hint.* Consider the objective of minimizing x_s where x_s is a particular slack variable.)

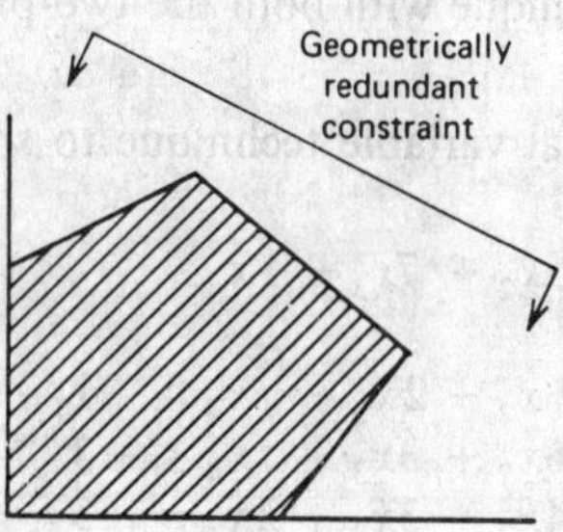

4.24 Suppose that at some iteration of the simplex method the slack variable x_s is basic in the ith row. Show that if $y_{ij} \leq 0$, $j = 1, \ldots, n$, $j \neq s$, then the constraint associated with x_s is geometrically redundant. Also show that the ith constraint and variable x_s may be deleted from the current tableau without changing the feasible set.

4.25 Solve the following linear program by both the two-phase method and the big-M method.

$$\begin{aligned} \text{Minimize} \quad & 3x_1 - 3x_2 + x_3 \\ \text{Subject to} \quad & x_1 + 2x_2 - x_3 \geq 5 \\ & -3x_1 - x_2 + x_3 \leq 4 \\ & x_1, \quad x_2, \quad x_3 \geq 0 \end{aligned}$$

4.26 Solve the following problem by the big-M method.

$$\begin{aligned} \text{Minimize} \quad & 3x_1 + 2x_2 + 4x_3 + 8x_4 \\ \text{Subject to} \quad & x_1 + 2x_2 + 5x_3 + 6x_4 \geq 8 \\ & -2x_1 + 5x_2 + 3x_3 - 5x_4 \leq 3 \\ & x_1, \quad x_2, \quad x_3, \quad x_4 \geq 0 \end{aligned}$$

4.27 Solve the following problem by the big-M method.

$$\begin{aligned} \text{Maximize} \quad & 2x_1 - x_2 \\ \text{Subject to} \quad & x_1 + x_2 \leq 3 \\ & -x_1 + x_2 \geq 1 \\ & x_1, \quad x_2 \geq 0 \end{aligned}$$

4.28 Solve the following problem by the big-M method.

$$\begin{array}{ll} \text{Maximize} & 5x_1 - 2x_2 + x_3 \\ \text{Subject to} & x_1 + 4x_2 + x_3 \leq 6 \\ & 2x_1 + x_2 + 3x_3 \geq 2 \\ & x_1, \quad x_3 \geq 0 \\ & x_2 \quad \text{unrestricted} \end{array}$$

4.29 Discuss in detail all the possible cases that may arise when using the single artificial variable technique with both the two-phase method and the big-M method.

4.30 Use the single artificial variable technique to solve the following problem.

$$\begin{array}{ll} \text{Maximize} & 4x_1 + 5x_2 + 7x_3 - x_4 \\ \text{Subject to} & x_1 + x_2 + 2x_3 - x_4 \geq 1 \\ & 2x_1 - 6x_2 + 3x_3 + x_4 \leq -3 \\ & -2x_1 + 4x_2 + 3x_3 + 2x_4 = -5 \\ & x_1, \quad x_2, \quad x_4 \geq 0 \\ & x_3 \quad \text{unrestricted} \end{array}$$

4.31 Use the single artificial variable technique to solve the following linear programming problem.

$$\begin{array}{ll} \text{Minimize} & -x_1 - 2x_2 + x_3 \\ \text{Subject to} & x_1 + x_2 + x_3 \geq 4 \\ & 2x_1 - x_3 \geq 3 \\ & x_2 + x_3 \leq 2 \\ & x_1, \quad x_2, \quad x_3 \geq 0 \end{array}$$

4.32 Discuss the advantages and disadvantages of using a single artificial variable compared with a method using several artificial variables.

4.33 Suppose it is possible to get the constraints of a linear program to the form $\mathbf{I}\mathbf{x}_B + \mathbf{B}^{-1}\mathbf{N}\mathbf{x}_N = \bar{\mathbf{b}}$, where $\bar{\mathbf{b}} = \mathbf{B}^{-1}\mathbf{b} \not\geq \mathbf{0}$. Show that a single artificial variable x_a with activity vector $\hat{\mathbf{b}}$ (where $\hat{\mathbf{b}} \leq \bar{\mathbf{b}}$) can be added and a basic feasible solution would be easily obtained.

4.34 Construct detailed flow diagrams of the two-phase method and the big-M method. Using FORTRAN (or another language), code either of the two methods.

4.35 A manufacturer wishes to find the optimal weekly production of items A, B, and C that maximizes the profit. The unit profit and the minimal weekly production of these items are respectively \$2.00, \$2.00, and \$4.00 and 100 units, 60 units, and 60 units. Products A, B, and C are processed on three machines. The hours required per item per machine are summarized below.

MACHINE	ITEM A	B	C
1	0	1	2
2	1	1	1
3	2	1	1

The numbers of hours of machines 1, 2, and 3 available per week are 240, 400, and 360 respectively. Find an optimal production schedule.

4.36 A farmer has 200 acres and 18,000 man-hours available. He wishes to determine the acreage allocated to the following products: corn, wheat, okra, tomatoes, and green beans. The farmer must produce at least 250 tons of corn to feed his hogs and cattle, and he must produce at least 80 tons of wheat, which he precontracted. The tonnage and labor in man-hours per acre of the different products are summarized.

	CORN	WHEAT	OKRA	TOMATOES	BEANS
Tons/acre	10	4	4	8	6
Man-hours/acre	120	150	100	80	120

The corn, wheat, okra, tomatoes, and beans can be sold for \$120.00, \$150.00, \$60.00, \$80.00, and \$55.00 per ton. Find the optimal solution.

4.37 A company manufactures stoves and ovens. The company has three warehouses and two retail stores. Sixty, 80, and 50 stoves and 80, 50, and 50 ovens are available at the three warehouses respectively. One hundred and 90 stoves and 60 and 120 ovens are required at the retail stores respectively. The unit shopping costs, which apply to both the stoves and ovens, from the warehouses to the retail stores are given.

WAREHOUSE	STORE 1	2
1	3	5
2	2	3
3	6	3

Find the shipping pattern that minimizes the total transportation cost by the simplex method.

4.38 A manufacturer wishes to plan the production of two items A and B for the months of March, April, May, and June. The demands that must be met are given.

	MARCH	APRIL	MAY	JUNE
Item A	400	500	600	400
Item B	600	600	700	600

Suppose that the inventory of A and B at the end of February is 100 and 150 respectively. Further suppose that at least 150 units of item B must be available at the end of June. The inventory holding costs of items A and B during any month are given by $1.00 and $0.80 times the inventory of the item at the end of the month. Furthermore, because of space limitation, the sum of items A and B in stock cannot exceed 250 during any month. Finally, the maximum number of items A and B that can be produced during any given month is 500 and 650 respectively.

a. Formulate the production problem as a linear program. The objective is to minimize the total inventory cost (the production cost is assumed constant).
b. Find an optimal production/inventory pattern.
c. Management is considering installing a new manufacturing system for item B at the end of April. This would raise the maximum items that can be produced per month from 650 to 700 and meanwhile would reduce the unit manufacturing cost from $8.00 to $6.50. Assess the benefits of this system in reducing the total manufacturing costs plus inventory costs. If you were a member of the management team, discuss how you would assess whether the new system is cost-effective.
d. Suppose that management decided to introduce the new system. Market research indicated that item B can be backlogged without serious dissatisfaction of customers. It was the management's assessment that each unit of unsatisfied demand during any month must be charged an additional $1.00. Formulate the production/inventory problem and find an optimal solution by the simplex method.

4.39 A trucking company owns three types of trucks, type I, type II, and type III. These trucks are equipped to haul three different types of machines per load according to the following chart.

	TRUCK TYPE		
	I	II	III
Machine A	1	1	1
Machine B	0	1	2
Machine C	2	1	1

Trucks of type I, II, and III cost \$400, \$600, and \$900 per trip respectively. We are interested to find how many trucks of each type should be sent to haul 12 machines of type A, 10 machines of type B, and 16 machines of type C. Formulate the problem and then solve it by the simplex method. (This is an integer programming problem; you may ignore the integer requirements.)

4.40 A company produces refrigerators, stoves, and dishwashers. During the coming year, sales are expected to be the following:

PRODUCT	QUARTER 1	2	3	4
Refrigerators	1500	1000	2000	1200
Stoves	1500	1500	1200	1500
Dishwashers	1000	2000	1500	2500

The company wants a production schedule that meets the demand requirements. Management also has decided that the inventory level for each product must be at least 150 units at the end of each quarter. There is no inventory of any product at the start of the first quarter.

During a quarter only 18,000 hours of production time are available. A refrigerator requires 2 hours, a stove 4 hours, and a dishwasher 3 hours of production time. Refrigerators cannot be manufactured in the fourth quarter because the company plans to modify tooling for a new product line.

Assume that each item left in inventory at the end of a quarter incurs a holding cost of \$5. The company wants to plan its production schedule over the year, in such a way that meets the quarterly demands and minimizes the total inventory cost. Formulate the problem and then solve it by the simplex method.

4.41 A manufacturer of metal sheets received an order for producing 2000 sheets of size 2′ × 4′ and 1000 sheets of size 4′ × 7′. Two standard sheets are available of sizes 11′ × 3000′ and 10′ × 2000′. The engineering staff decided

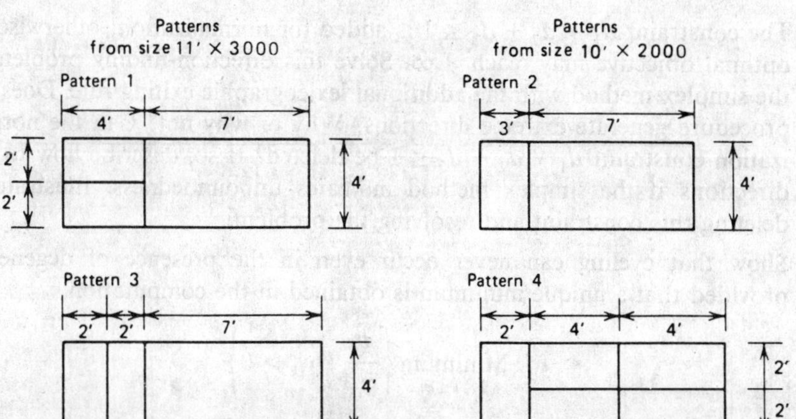

that the above four cutting patterns are suitable for this order. Formulate the problem of meeting the order and minimizing the waste as a linear program and solve it by the simplex method.

4.42 Solve the following problem, using the lexicographic rule for noncycling. Repeat using Bland's rule.

$$\begin{array}{ll} \text{Maximize} & x_1 + 2x_2 + x_3 \\ \text{Subject to} & x_1 + 4x_2 + 6x_3 \le 4 \\ & -x_1 + x_2 + 4x_3 \le 1 \\ & x_1 + 3x_2 + x_3 \le 6 \\ & x_1, \quad x_2, \quad x_3 \ge 0 \end{array}$$

4.43 Solve the problem in Example 4.11 using Bland's cycling prevention rule.

4.44 Consider the following region.

$$\begin{aligned} 2x_1 - 2x_2 + x_3 &\le 4 \\ -x_1 + x_2 - x_3 &\le 3 \\ 4x_1 + x_2 - 3x_3 &\le 2 \\ x_1, \quad x_2, \quad x_3 &\ge 0 \end{aligned}$$

Recall that $\mathbf{d}$ is a direction of the region if $\mathbf{Ad} \le \mathbf{0}$, $\mathbf{d} \ge \mathbf{0}$, and $\mathbf{d}$ is nonzero. Thus in order to find directions of the region, we may solve the following problem.

$$\begin{array}{ll} \text{Maximize} & d_1 + d_2 + d_3 \\ \text{Subject to} & 2d_1 - 2d_2 + d_3 \le 0 \\ & -d_1 + d_2 - d_3 \le 0 \\ & 4d_1 + d_2 - 3d_3 \le 0 \\ & d_1 + d_2 + d_3 \le 1 \\ & d_1, \quad d_2, \quad d_3 \ge 0 \end{array}$$

The constraint $d_1 + d_2 + d_3 \le 1$ is added for normalization; otherwise the optimal objective may reach $+\infty$. Solve this direction-finding problem by the simplex method with the additional lexicographic exiting rule. Does this procedure generate extreme directions? Why or why not? Can the normalization constraint $d_1 + d_2 + d_3 \le 1$ be deleted? If so, describe how to find directions if the simplex method indicates unboundedness. Illustrate by deleting this constraint and resolving the problem.

4.45 Show that cycling can never occur even in the presence of degeneracy provided that a unique minimum is obtained in the computation

$$\underset{1 \le i \le m}{\text{Minimum}} \left\{ \frac{\bar{b}_i}{y_{ik}} : y_{ik} > 0 \right\}$$

where $\bar{\mathbf{b}} = \mathbf{B}^{-1}\mathbf{b}$, $\mathbf{y}_k = \mathbf{B}^{-1}\mathbf{a}_k$, and x_k is the entering variable.

4.46 Consider the following problem.

$$\begin{aligned}\text{Maximize} \quad & 3x_1 + 4x_2 \\ \text{Subject to} \quad & 2x_1 + x_2 \leq 8 \\ & -x_1 + 2x_2 \leq 6 \\ & x_1 + x_2 \leq 6 \\ & x_1, \quad x_2 \geq 0\end{aligned}$$

a. Solve the problem geometrically and verify that the optimal point is a degenerate basic feasible solution.
b. Solve the problem by the simplex method.
c. From part (a), identify the constraint that causes degeneracy and resolve the problem after throwing this constraint away. Note that degeneracy disappears and the same optimal solution is obtained.
d. Is it true in general that degenerate basic feasible solutions can be made nondegenerate by throwing some constraints away without affecting the feasible region?

4.47 The lexicographic rule for exiting from the basis is designed to specify which variable leaves the basis if the minimum ratio test results in a tie or several ties. Show in detail that this exiting rule would specify a unique variable to leave the basis after at most $m + 1$ columns are examined, namely $\bar{\mathbf{b}}, \mathbf{y}_1, \mathbf{y}_2, \ldots,$ and $\mathbf{y}_m$.

4.48 We showed that the row vector $(\mathbf{c}_B\mathbf{B}^{-1}\mathbf{b}, \mathbf{c}_B\mathbf{B}^{-1})$ is lexicographically decreasing from one iteration to another using the lexicographic rule. Give an economic interpretation of this fact. (*Hint*. Note that $z = \mathbf{c}_B\mathbf{B}^{-1}\mathbf{b}$ and that $\partial z/\partial \mathbf{b} = \mathbf{c}_B\mathbf{B}^{-1}$.)

4.49 Suppose that we have an optimal extreme point of a minimization linear programming problem. In the presence of degeneracy, is it possible that this extreme point corresponds to a basic feasible solution such that $z_j - c_j > 0$ for at least one nonbasic variable? If this were the case, are we guaranteed of another basic feasible solution corresponding to the same extreme point where $z_j - c_j \leq 0$ for all nonbasic variables? Why or why not? Illustrate by a numerical example.

4.50 In order to prove finite convergence of the simplex method using the lexicographic rule, we assumed that the first m columns of the constraint matrix form the identity. Show that this assumption can be relaxed provided that we have any basic feasible solution. (*Hint*. Let $\mathbf{B}$ be the starting basis and consider the following equivalent problem.)

$$\begin{aligned}\text{Minimize} \quad & \mathbf{0}\mathbf{x}_B + (\mathbf{c}_N - \mathbf{c}_B\mathbf{B}^{-1}\mathbf{N})\mathbf{x}_N \\ \text{Subject to} \quad & \mathbf{I}\mathbf{x}_B + \mathbf{B}^{-1}\mathbf{N}\,\mathbf{x}_N = \mathbf{B}^{-1}\mathbf{b} \\ & \mathbf{x}_B, \quad \mathbf{x}_N \geq \mathbf{0}\end{aligned}$$

4.51 Consider the following problem.

Minimize $(\mathbf{cx}, \mathbf{cY})$

Subject to $\mathbf{A}(\mathbf{x}, \mathbf{Y}) = (\mathbf{b}, \mathbf{I})$
$(\mathbf{x}, \mathbf{Y}) \succcurlyeq \mathbf{0}$

where $\mathbf{A}$ is an $m \times n$ matrix, $\mathbf{c}$ is an n vector, and the variables are the n vector $\mathbf{x}$ and the $n \times m$ matrix $\mathbf{Y}$. The objective function is a row vector, and the minimization is taken in the lexicographic sense, that is, $(\mathbf{cx}_2, \mathbf{cY}_2) \prec (\mathbf{cx}_1, \mathbf{cY}_1)$ if and only if $(\mathbf{cx}_1, \mathbf{cY}_1) - (\mathbf{cx}_2, \mathbf{cY}_2) \succ \mathbf{0}$. Each row of the matrix $(\mathbf{x}, \mathbf{Y})$ is restricted to be lexicographically nonnegative, which means that each row is zero or $\succ \mathbf{0}$.

a. Let $\mathbf{x}$ be a basic feasible solution of the system $\mathbf{Ax} = \mathbf{b}$, $\mathbf{x} \geq \mathbf{0}$ with basis $\mathbf{B}$. Show that $\mathbf{x} = \begin{pmatrix} \mathbf{B}^{-1}\mathbf{b} \\ \mathbf{0} \end{pmatrix}$ and $\mathbf{Y} = \begin{pmatrix} \mathbf{B}^{-1} \\ \mathbf{0} \end{pmatrix}$ is a feasible solution of the foregoing problem provided that $(\mathbf{B}^{-1}\mathbf{b}, \mathbf{B}^{-1}) \succ \mathbf{0}$.
b. Show that the simplex method with the lexicographic exiting rule of Section 4.6 generates the sequence $(\mathbf{x}_1, \mathbf{Y}_1)$, $(\mathbf{x}_2, \mathbf{Y}_2) \ldots$, where $(\mathbf{cx}_{j-1}, \mathbf{cY}_{j-1}) - (\mathbf{cx}_j, \mathbf{cY}_j) \succ \mathbf{0}$ for all j. Interpret this fact emphasizing the relationship between the bases generated by the simplex method and the foregoing problem.

4.52 Consider the following problem.

Minimize $\mathbf{cx}$

Subject to $\mathbf{Ax} = \mathbf{b}$
$\mathbf{x} \geq \mathbf{0}$

Assume that the first m columns of $\mathbf{A}$ form the identity and assume that $\mathbf{b} \geq \mathbf{0}$. Given a basis $\mathbf{B}$ the corresponding feasible solution is nondegenerate if $\mathbf{B}^{-1}\mathbf{b} > \mathbf{0}$. Consider the following perturbation procedure of Charnes. Replace $\mathbf{b}$ by $\mathbf{b} + \sum_{j=1}^{m} \mathbf{a}_j \epsilon^j$ where ϵ is a very small positive number. Now suppose that we have a basis $\mathbf{B}$, where $\mathbf{B}^{-1}(\mathbf{b} + \sum_{j=1}^{m} \mathbf{a}_j \epsilon^j) = \bar{\mathbf{b}} + \sum_{j=1}^{m} \mathbf{y}_j \epsilon^j > \mathbf{0}$. Suppose that x_k is chosen to enter the basis and the following minimum ratio test is made:

$$\underset{1 \leq i \leq m}{\text{Minimum}} \left\{ \frac{\bar{b}_i + \sum_{j=1}^{m} y_{ij} \epsilon^j}{y_{ik}} : y_{ik} > 0 \right\}$$

a. Show that the minimum ratio occurs at a unique index r. Show that the method of finding this index is precisely the rule of Section 4.6.
b. Show that the new right-hand-side after pivoting is positive and the objective function strictly improves even in the presence of degeneracy in the original problem.

c. Show that cycling will not occur if the rule in part (a) is adopted. Interpret this in terms of the perturbed problem.
d. Show that all the computations can be carried without explicitly replacing the right-hand-side with $\mathbf{b} + \sum_{j=1}^{m} \mathbf{a}_j \epsilon^j$ and without explicitly assigning a value to ϵ.

NOTES AND REFERENCES

1. The use of artificial variables to obtain a starting basic feasible solution was first published by Dantzig [1951a].
2. The single artificial variable technique of Section 4.5 can be viewed as the dual of a similar technique that adds a new row to obtain a starting basic dual feasible solution. The latter is discussed in Section 6.6. Computational results appear in Wolfe and Cutler [1963].
3. A general phase I method in which the sum of infeasibilities is reduced without regard to the feasibility of individual variables has been suggested by Maros [1986].
4. The cycling example of Section 4.6 is due to Beale [1955]. The proof of the cycling prevention rule via lexicographic ordering was published by Dantzig, Orden, and Wolfe [1955]. The cycling prevention rule can also be interpreted as a perturbation technique, as briefly described in Exercise 4.52. This technique was independently devised by Charnes [1952].
5. Bland's cycling prevention rules [1977] are simple and elegant, although computationally unattractive. The notion of stages and an example of a network structured linear program for which Bland's rule admits stalling has been provided by Cunningham [1979]. The term "affirmative action" for entering variable selection strategies is from Fathi and Tovey [1986]. For stalling prevention in network flow problems using Dantzig's entering variable choice and the lexicographic cycling prevention rule, see Orlin [1985]. Marshall and Suurballe [1969] show that the smallest examples which can exhibit cycling have $m = 2$, $n = 6$, $(n - m) \geq 3$, and cycle length equal to 6. Related results for network structured problems are studied by Cunningham [1979]. For a practical anti-cycling rule, see Gill et al. [1988].
6. Exercise 4.13 addresses the notion of *equivalent weights* for preemptive priority multiple objective programs, as discussed in Sherali and Soyster [1983] and in Sherali [1982].
7. Exercise 4.15 states the basic result of Murty [1968] used for ranking extreme points of polytopes. An alternative ranking technique appears in Sherali and Dickey [1986]. For the special case of ranking vertices of an assignment polytope, see Bazaraa and Sherali [1981]. Also, for related work on enumerating all vertices of polytopes, see Mattheiss [1973] and Mattheiss and Rubin [1980].

FIVE: SPECIAL SIMPLEX IMPLEMENTATIONS AND OPTIMALITY CONDITIONS

In this chapter we describe some special implementations of the simplex procedure or slight modifications of it. The formats considered here will prove advantageous in later chapters. The revised simplex method, which proceeds through the same steps as the simplex method but keeps all pertinent information in a smaller array, is described in Section 5.1. Numerically stable forms of this method are also discussed. In Section 5.2 we describe a slight modification of the simplex method for dealing implicitly with lower and upper bounds on the variables without increasing the size of the basis. The remainder of the chapter is devoted to some geometric aspects of the simplex method. In particular, Farkas' lemma and the Karush–Kuhn–Tucker optimality conditions are discussed.

5.1 THE REVISED SIMPLEX METHOD

The *revised simplex* method is a systematic procedure for implementing the steps of the simplex method in a smaller array, thus saving storage space. Let us begin by reviewing the steps of the simplex method.

Steps of the Simplex Method (Minimization Problem)

Suppose that we are given a basic feasible solution with basis $\mathbf{B}$ (and basis inverse $\mathbf{B}^{-1}$). Then:

1. The basic feasible solution is given by $\mathbf{x}_B = \mathbf{B}^{-1}\mathbf{b} = \bar{\mathbf{b}}$ and $\mathbf{x}_N = \mathbf{0}$. The objective $z = \mathbf{c}_B\mathbf{B}^{-1}\mathbf{b} = \mathbf{c}_B\bar{\mathbf{b}}$.
2. Calculate the simplex multipliers $\mathbf{w} = \mathbf{c}_B\mathbf{B}^{-1}$. For each nonbasic variable, calculate $z_j - c_j = \mathbf{c}_B\mathbf{B}^{-1}\mathbf{a}_j - c_j = \mathbf{w}\mathbf{a}_j - c_j$. Let $z_k - c_k = \text{Maximum } z_j - c_j$. If $z_k - c_k \le 0$, then stop; the current solution is optimal. Otherwise go to step 3.
3. Calculate $\mathbf{y}_k = \mathbf{B}^{-1}\mathbf{a}_k$. If $\mathbf{y}_k \le \mathbf{0}$, then stop; the optimal solution is unbounded. Otherwise determine the index of the variable x_{B_r} leaving the

basis as follows:

$$\frac{\bar{b}_r}{y_{rk}} = \underset{1 \le i \le m}{\text{Minimum}} \left\{ \frac{\bar{b}_i}{y_{ik}} : y_{ik} > 0 \right\}$$

Update $\mathbf{B}$ by replacing $\mathbf{a}_{B_r}$ with $\mathbf{a}_k$ and go to step 1.

Examining the preceding steps, it becomes clear that the simplex method can be executed using a smaller array. Suppose that we have a basic feasible solution with a known $\mathbf{B}^{-1}$. The following array is constructed where $\mathbf{w} = \mathbf{c}_B\mathbf{B}^{-1}$ and $\bar{\mathbf{b}} = \mathbf{B}^{-1}\mathbf{b}$.

BASIS INVERSE	RHS
$\mathbf{w}$	$\mathbf{c}_B\bar{\mathbf{b}}$
$\mathbf{B}^{-1}$	$\bar{\mathbf{b}}$

Note that the foregoing tableau, called the *revised simplex tableau*, is also present in the usual simplex tableau, provided that we start with the BASIS INVERSE section having a zero vector in the objective row and an identity matrix in the other rows. (This may correspond to the slack variables if available as a starting basis or may simply be a section of this type maintained for bookkeeping purposes.) The RHS is the usual tableau right-hand-side. Now imagine that we are performing the simplex algorithm on the entire tableau, but we are keeping hidden all but this array and any other required information. Hence only the "exposed" information is explicitly calculated. The first piece of information that we need to see is the $z_j - c_j$ values. Since $\mathbf{w}$ is known, these values can be explicitly calculated as in step 2 to check for optimality. Suppose that $z_k - c_k > 0$. Then we wish to examine the updated column of x_k. Using $\mathbf{B}^{-1}$, we may compute $\mathbf{y}_k = \mathbf{B}^{-1}\mathbf{a}_k$. If $\mathbf{y}_k \le \mathbf{0}$, we stop with the indication that the optimal solution value is unbounded. Otherwise the updated column of x_k can be appended to the revised simplex tableau as shown, while the rest of the tableau is still kept "hidden."

BASIS INVERSE	RHS		x_k
$\mathbf{w}$	$\mathbf{c}_B\bar{\mathbf{b}}$		$z_k - c_k$
	$\bar{b}_1$		y_{1k}
	$\bar{b}_2$		y_{2k}
	$\vdots$		$\vdots$
$\mathbf{B}^{-1}$	$\bar{b}_r$		$\boxed{y_{rk}}$
	$\vdots$		$\vdots$
	$\bar{b}_m$		y_{mk}

The index r of step 3 can now be calculated by the usual minimum ratio test. More importantly, pivoting at y_{rk} gives as usual the new values of $\mathbf{w}$, $\mathbf{B}^{-1}$, $\bar{\mathbf{b}}$, and

$\mathbf{c}_B\bar{\mathbf{b}}$, and the process is repeated. We leave it as an exercise to the reader to rigorously verify that pivoting indeed updates the $(m + 1) \times (m + 1)$ revised simplex tableau.

The revised simplex method converges in a finite number of steps provided that a noncycling rule is adopted. This is obvious since the revised simplex method conducts exactly the same steps of the simplex method, with the exception that only a part of the tableau is presented and other information is generated only as required. The following is a summary of the revised simplex method.

Summary of the Revised Simplex Method in Tableau Format (Minimization Problem)

INITIALIZATION STEP

Find an initial basic feasible solution with basis inverse $\mathbf{B}^{-1}$. Calculate $\mathbf{w} = \mathbf{c}_B\mathbf{B}^{-1}$, $\bar{\mathbf{b}} = \mathbf{B}^{-1}\mathbf{b}$, and form the following array (revised simplex tableau).

BASIS INVERSE	RHS
$\mathbf{w}$	$\mathbf{c}_B\bar{\mathbf{b}}$
$\mathbf{B}^{-1}$	$\bar{\mathbf{b}}$

MAIN STEP

For each nonbasic variable, calculate $z_j - c_j = \mathbf{w}\mathbf{a}_j - c_j$. Let $z_k - c_k = \text{Maximum } z_j - c_j$. If $z_k - c_k \leq 0$, stop; the current basic feasible solution is optimal. Otherwise calculate $\mathbf{y}_k = \mathbf{B}^{-1}\mathbf{a}_k$. If $\mathbf{y}_k \leq \mathbf{0}$, stop; the optimal solution value is unbounded. If $\mathbf{y}_k \nleq \mathbf{0}$, insert the column $\left[\frac{z_k - c_k}{\mathbf{y}_k}\right]$ to the right of the tableau leading to the following tableau.

BASIS INVERSE	RHS		x_k
$\mathbf{w}$	$\mathbf{c}_B\bar{\mathbf{b}}$		$z_k - c_k$
$\mathbf{B}^{-1}$	$\bar{\mathbf{b}}$		$\mathbf{y}_k$

Determine the index r as follows:

$$\frac{\bar{b}_r}{y_{rk}} = \underset{1 \leq i \leq m}{\text{Minimum}} \left\{ \frac{\bar{b}_i}{y_{ik}} : y_{ik} > 0 \right\}$$

Pivot at y_{rk}. This updates the tableau. Now the column corresponding to x_k is completely eliminated from the tableau and the main step is repeated.

Example 5.1

$$\begin{array}{lrl} \text{Minimize} & -x_1 - 2x_2 + x_3 - x_4 - 4x_5 + 2x_6 & \\ \text{Subject to} & x_1 + x_2 + x_3 + x_4 + x_5 + x_6 & \leq 6 \\ & 2x_1 - x_2 - 2x_3 + x_4 & \leq 4 \\ & x_3 + x_4 + 2x_5 + x_6 & \leq 4 \\ & x_1, \quad x_2, \quad x_3, \quad x_4, \quad x_5, \quad x_6 & \geq 0 \end{array}$$

Introduce the slack variables x_7, x_8, and x_9. The initial basis is $\mathbf{B} = [\mathbf{a}_7, \mathbf{a}_8, \mathbf{a}_9] = \mathbf{I}_3$. Also, $\mathbf{w} = \mathbf{c}_B\mathbf{B}^{-1} = (0, 0, 0)$ and $\bar{\mathbf{b}} = \mathbf{b}$.

Iteration 1

	BASIS INVERSE			RHS
z	0	0	0	0
x_7	1	0	0	6
x_8	0	1	0	4
x_9	0	0	1	4

Here $\mathbf{w} = (0, 0, 0)$. Noting that $z_j - c_j = \mathbf{w}\mathbf{a}_j - c_j$, we get

$$z_1 - c_1 = 1,\ z_2 - c_2 = 2,\ z_3 - c_3 = -1,$$
$$z_4 - c_4 = 1,\ z_5 - c_5 = 4,\ z_6 - c_6 = -2$$

Therefore $k = 5$ and x_5 enters the basis:

$$\mathbf{y}_5 = \mathbf{B}^{-1}\mathbf{a}_5 = \begin{bmatrix} 1 & 0 & 0 \\ 0 & 1 & 0 \\ 0 & 0 & 1 \end{bmatrix}\begin{bmatrix} 1 \\ 0 \\ 2 \end{bmatrix} = \begin{bmatrix} 1 \\ 0 \\ 2 \end{bmatrix}$$

Insert the vector

$$\left[\frac{z_5 - c_5}{\mathbf{y}_5}\right] = \begin{bmatrix} 4 \\ 1 \\ 0 \\ 2 \end{bmatrix}$$

to the right of the tableau and pivot at $y_{35} = 2$.

	BASIS INVERSE			RHS	x_5
z	0	0	0	0	4
x_7	1	0	0	6	1
x_8	0	1	0	4	0
x_9	0	0	1	4	(2)

	BASIS INVERSE			RHS
z	0	0	-2	-8
x_7	1	0	$-\frac{1}{2}$	4
x_8	0	1	0	4
x_5	0	0	$\frac{1}{2}$	2

Iteration 2

Here $\mathbf{w} = (0, 0, -2)$. Noting that $z_j - c_j = \mathbf{w}\mathbf{a}_j - c_j$, we get

$$z_1 - c_1 = 1,\ z_2 - c_2 = 2,\ z_3 - c_3 = -3,$$
$$z_4 - c_4 = -1,\ z_6 - c_6 = -4,$$
$$z_9 - c_9 = -2$$

Therefore $k = 2$ and x_2 enters the basis:

$$\mathbf{y}_2 = \mathbf{B}^{-1}\mathbf{a}_2 = \begin{bmatrix} 1 & 0 & -\frac{1}{2} \\ 0 & 1 & 0 \\ 0 & 0 & \frac{1}{2} \end{bmatrix} \begin{bmatrix} 1 \\ -1 \\ 0 \end{bmatrix} = \begin{bmatrix} 1 \\ -1 \\ 0 \end{bmatrix}$$

Insert the vector

$$\begin{bmatrix} z_2 - c_2 \\ \hline \mathbf{y}_2 \end{bmatrix} = \begin{bmatrix} 2 \\ 1 \\ -1 \\ 0 \end{bmatrix}$$

to the right of the tableau and pivot at y_{12}.

	BASIS INVERSE			RHS	x_2
z	0	0	-2	-8	2
x_7	1	0	$-\frac{1}{2}$	4	①
x_8	0	1	0	4	-1
x_5	0	0	$\frac{1}{2}$	2	0

	BASIS INVERSE			RHS
z	-2	0	-1	-16
x_2	1	0	$-\frac{1}{2}$	4
x_8	1	1	$-\frac{1}{2}$	8
x_5	0	0	$\frac{1}{2}$	2

Iteration 3

Here $\mathbf{w} = (-2, 0, -1)$. Noting that $z_j - c_j = \mathbf{w}\mathbf{a}_j - c_j$, we get

$$z_1 - c_1 = -1,\ z_3 - c_3 = -4,\ z_4 - c_4 = -2,$$
$$z_6 - c_6 = -5,\ z_9 - c_9 = -1$$

Since $z_j - c_j \leq 0$ for all nonbasic variables (x_7 just left the basis and so $z_7 - c_7 < 0$), we stop; the basic feasible solution of the foregoing tableau is optimal.

Comparison Between the Simplex and the Revised Simplex Methods

It may be helpful to give a brief comparison between the simplex and the revised simplex methods. For the revised method we need an $(m + 1) \times (m + 1)$ array as opposed to an $(m + 1) \times (n + 1)$ array for the simplex method. If n is significantly larger than m, this would result in a substantial saving in computer core storage. The number of multiplications (division is considered a multiplication) and additions (subtraction is considered an addition) per iteration of both procedures are given in Table 5.1. In Exercise 5.3 we ask the reader to verify the validity of the entries of the table.

From Table 5.1 we see that the number of operations required during an iteration of the simplex method is slightly less than those required for the revised simplex method. Note, however, that most practical problems are *sparse*, that is, the *density* d (number of nonzero elements divided by total number of elements) of nonzero elements in the constraint matrix is usually small (in many cases $d \leq 0.05$). The revised simplex method can take advantage of this situation while calculating $z_j - c_j$. Note that $z_j = \mathbf{w}\mathbf{a}_j$ and we can skip zero elements of $\mathbf{a}_j$ while performing the calculation $\mathbf{w}\mathbf{a}_j = \sum_{j=1}^{m} w_i a_{ij}$. Therefore the number of operations in the revised simplex method for calculating the $z_j - c_j$'s is given by d times the entries of Table 5.1, substantially reducing the total number of operations. While pivoting, for both the simplex and the revised simplex methods, no operations are skipped because the current tableaux usually fill quickly with nonzero entries, even if the original constraint matrix was sparse. (However, see the section on Implementation Remarks and the Notes and References section on preserving sparsity in a factored-form implementation of the revised simplex method.)

Empirically, it is often suggested that on the average in most instances, the simplex method requires roughly on the order of m to $3m$ iterations. (However, it has also been suggested that the number of iterations is often proportional to n or even $\log n$.) Examining Table 5.1, since the number of operations per iteration is of order $O(mn)$ (that is, it is bounded from above by some constant times mn), the average empirical complexity of the simplex method is $O(m^2n)$. However, sparsity is usually present and is exploited in simplex implementations. This is done by storing data in *packed form* (in which only nonzeros are stored with appropriate pointers and are alone used in arithmetic calculations), as well

Table 5.1 Comparison of the Simplex and the Revised Simplex Methods

		OPERATION		
METHOD		PIVOTING	$z_j - c_j$'s	TOTAL
Simplex	Multiplications	$(m + 1)(n - m + 1)$		$m(n - m) + n + 1$
	Additions	$m(n - m + 1)$		$m(n - m + 1)$
Revised Simplex	Multiplications	$(m + 1)^2$	$m(n - m)$	$m(n - m) + (m + 1)^2$
	Additions	$m(m + 1)$	$m(n - m)$	$m(n + 1)$

as by maintaining the basis in factored form as we shall see shortly. Because of this, a regression equation of the form $Km^a nd^b$, where $\alpha \cong 1.25 - 2.5$ and $b \cong 0.33$, usually provides a better fit for overall computational effort.

To summarize, if n is significantly larger than m and if the density d is small, the computational effort of the revised simplex method is significantly smaller than that of the simplex method. Also, in the revised simplex method, the use of the original data for calculating the $z_j - c_j$'s and the updated column $\mathbf{y}_k$ tends to reduce the cumulative round-off error.

Product Form of the Inverse

We now discuss another implementation of the revised simplex method where the inverse of the basis is stored as the product of elementary matrices. (An *elementary matrix* is a square matrix that differs from the identity in only one row or one column.) For sparse problems, this leads to fewer storage and computational burdens, and it provides greater numerical stability by reducing accumulated round-off errors.

Consider a basis $\mathbf{B}$ composed of the columns $\mathbf{a}_{B_1}, \mathbf{a}_{B_2}, \ldots, \mathbf{a}_{B_m}$ and suppose that $\mathbf{B}^{-1}$ is known. Now suppose that the nonbasic column $\mathbf{a}_k$ replaces $\mathbf{a}_{B_r}$, resulting in the new basis $\hat{\mathbf{B}}$. We wish to find $\hat{\mathbf{B}}^{-1}$ in terms of $\mathbf{B}^{-1}$. Noting that $\mathbf{a}_k = \mathbf{B}\mathbf{y}_k$ and $\mathbf{a}_{B_i} = \mathbf{B}\mathbf{e}_i$ where $\mathbf{e}_i$ is a vector of zeros except for 1 at the ith position, we have

$$
\begin{aligned}
\hat{\mathbf{B}} &= (\mathbf{a}_{B_1}, \mathbf{a}_{B_2}, \ldots, \mathbf{a}_{B_{r-1}}, \mathbf{a}_k, \mathbf{a}_{B_{r+1}}, \ldots, \mathbf{a}_{B_m}) \\
&= (\mathbf{B}\mathbf{e}_1, \mathbf{B}\mathbf{e}_2, \ldots, \mathbf{B}\mathbf{e}_{r-1}, \mathbf{B}\mathbf{y}_k, \mathbf{B}\mathbf{e}_{r+1}, \ldots, \mathbf{B}\mathbf{e}_m) \\
&= \mathbf{B}\mathbf{T}
\end{aligned}
$$

where $\mathbf{T}$ is the identity with the rth column replaced by $\mathbf{y}_k$. The inverse of $\mathbf{T}$, which we shall denote by $\mathbf{E}$, is given as follows.

$$
\mathbf{E} = \begin{bmatrix}
1 & 0 & \cdots & 0 & -y_{1k}/y_{rk} & 0 & \cdots & 0 \\
0 & 1 & \cdots & 0 & -y_{2k}/y_{rk} & 0 & \cdots & 0 \\
\vdots & \vdots & & \vdots & \vdots & \vdots & & \vdots \\
0 & 0 & \cdots & 0 & 1/y_{rk} & 0 & \cdots & 0 \\
\vdots & \vdots & & \vdots & \vdots & \vdots & & \vdots \\
0 & 0 & \cdots & 0 & -y_{mk}/y_{rk} & 0 & \cdots & 1
\end{bmatrix}
\begin{matrix} \\ \\ \\ \leftarrow r\text{th row} \\ \\ \\ \end{matrix}
$$

(rth column marked by ↓ above the column $-y_{1k}/y_{rk}, \ldots$)

Therefore $\hat{\mathbf{B}}^{-1} = \mathbf{T}^{-1}\mathbf{B}^{-1} = \mathbf{E}\mathbf{B}^{-1}$ where the elementary matrix $\mathbf{E}$ is specified. To summarize, the basis inverse at a new iteration can be obtained by premultiplying the basis inverse at the previous iteration by an elementary matrix $\mathbf{E}$. Needless to say, only the nonidentity column $\mathbf{g}$, known as the *eta vector*, and its position r need be stored to specify $\mathbf{E}$.

Let the basis $\mathbf{B}_1$ at the first iteration be the identity $\mathbf{I}$. Then the basis inverse $\mathbf{B}_2^{-1}$ at iteration 2 is $\mathbf{B}_2^{-1} = \mathbf{E}_1\mathbf{B}_1^{-1} = \mathbf{E}_1\mathbf{I} = \mathbf{E}_1$ where $\mathbf{E}_1$ is the elementary matrix corresponding to the first iteration. Similarly $\mathbf{B}_3^{-1} = \mathbf{E}_2\mathbf{B}_2^{-1} = \mathbf{E}_2\mathbf{E}_1$, and in general

$$\mathbf{B}_t^{-1} = \mathbf{E}_{t-1}\mathbf{E}_{t-2}\cdots\mathbf{E}_2\mathbf{E}_1 \tag{5.1}$$

Equation (5.1), which specifies the basis inverse as the product of elementary matrices through their associated eta vectors, is called the *product form of the inverse*. Using this form, all the steps of the simplex method can be performed without pivoting. First, it will be helpful to elaborate on multiplying a vector by an elementary matrix.

POST MULTIPLYING

Let $\mathbf{E}$ be an elementary matrix with nonidentity column $\mathbf{g}$ appearing at the rth position. Let $\mathbf{c}$ be a row vector. Then

$$\mathbf{cE} = (c_1, c_2, \ldots, c_m)\begin{bmatrix} 1 & 0 & \cdots & \overset{\substack{\text{position r}\\ \downarrow}}{g_1} & \cdots & 0 \\ 0 & 1 & \cdots & g_2 & \cdots & 0 \\ \vdots & \vdots & & \vdots & & \vdots \\ 0 & 0 & \cdots & g_m & \cdots & 1 \end{bmatrix}$$

$$= \left(c_1, c_2, \ldots, c_{r-1}, \sum_{i=1}^{m} c_i g_i, c_{r+1}, \ldots, c_m\right)$$

$$= (c_1, c_2, \ldots, c_{r-1}, \mathbf{cg}, c_{r+1}, \ldots, c_m) \tag{5.2}$$

In other words, $\mathbf{cE}$ is equal to $\mathbf{c}$ except that the rth component is replaced by $\mathbf{cg}$.

PREMULTIPLYING

Let $\mathbf{a}$ be an m vector. Then

$$\mathbf{Ea} = \begin{bmatrix} 1 & \cdots & g_1 & \cdots & 0 \\ \vdots & & \vdots & & \vdots \\ 0 & \cdots & g_r & \cdots & 0 \\ \vdots & & \vdots & & \vdots \\ 0 & \cdots & g_m & \cdots & 1 \end{bmatrix}\begin{bmatrix} a_1 \\ \vdots \\ a_r \\ \vdots \\ a_m \end{bmatrix}$$

$$= \begin{bmatrix} a_1 + g_1 a_r \\ \vdots \\ g_r a_r \\ \vdots \\ a_m + g_m a_r \end{bmatrix} = \begin{bmatrix} a_1 \\ \vdots \\ 0 \\ \vdots \\ a_m \end{bmatrix} + a_r \begin{bmatrix} g_1 \\ \vdots \\ g_r \\ \vdots \\ g_m \end{bmatrix}$$

In other words,

$$\mathbf{Ea} = \hat{\mathbf{a}} + a_r\mathbf{g} \tag{5.3}$$

where $\hat{\mathbf{a}}$ is equal to $\mathbf{a}$ except that the rth component a_r is replaced by zero.

With the foregoing formulas for post- and premultiplying a vector by an elementary matrix, the revised simplex method can be executed without pivoting. The following discussion elaborates on the simplex calculations.

COMPUTING THE SIMPLEX MULTIPLIER VECTOR $\mathbf{w} = \mathbf{c}_B\mathbf{B}^{-1}$

At iteration t we wish to calculate the vector $\mathbf{w}$. Note that

$$\mathbf{w} = \mathbf{c}_B\mathbf{B}_t^{-1} = \mathbf{c}_B\mathbf{E}_{t-1}\mathbf{E}_{t-2} \cdots \mathbf{E}_2\mathbf{E}_1$$

Computing $\mathbf{w}$ can be iteratively performed as follows. First compute $\mathbf{c}_B\mathbf{E}_{t-1}$ according to Equation (5.2). Then apply Equation (5.2) to calculate $(\mathbf{c}_B\mathbf{E}_{t-1})\mathbf{E}_{t-2}$, and so forth. This backward transformation process is sometimes referred to as a BTRAN process. After $\mathbf{w}$ is computed, we can calculate $z_j - c_j = \mathbf{wa}_j - c_j$ for nonbasic variables. From this we either stop or decide to introduce a nonbasic variable x_k.

COMPUTING THE UPDATED COLUMN $\mathbf{y}_k$ AND THE RIGHT-HAND-SIDE $\bar{\mathbf{b}}$

If x_k is to enter the basis at iteration t, then $\mathbf{y}_k$ is calculated as follows:

$$\mathbf{y}_k = \mathbf{B}_t^{-1}\mathbf{a}_k = \mathbf{E}_{t-1}\mathbf{E}_{t-2} \cdots \mathbf{E}_2\mathbf{E}_1\mathbf{a}_k$$

This computation can be executed by successively applying Equation (5.3) in the order $\mathbf{E}_1, \mathbf{E}_2, \ldots, \mathbf{E}_{t-1}$. This forward transformation process is sometimes referred to as a FTRAN process. If $\mathbf{y}_k \leq \mathbf{0}$, we stop with the conclusion that the optimal solution is unbounded. Otherwise the usual minimum ratio test determines the index r of the variable x_{B_r} leaving the basis. Thus, x_k enters and x_{B_r} leaves the basis. A new elementary matrix $\mathbf{E}_t$ is generated where the nonidentity column $\mathbf{g}$ is given by

$$\begin{bmatrix} \dfrac{-y_{1k}}{y_{rk}} \\ \vdots \\ \dfrac{1}{y_{rk}} \\ \vdots \\ \dfrac{-y_{mk}}{y_{rk}} \end{bmatrix}$$

and appears at position r. The new right-hand-side is given by

$$\mathbf{B}_{t+1}^{-1}\mathbf{b} = \mathbf{E}_t\mathbf{B}_t^{-1}\mathbf{b}$$

Since $\mathbf{B}_t^{-1}\mathbf{b}$ is known from the last iteration, then a single application of Equation (5.3) updates the right-hand-side vector $\bar{\mathbf{b}}$.

UPDATING THE BASIS INVERSE

The basis inverse is updated by generating $\mathbf{E}_t$ as discussed previously. It is worthwhile noting that the number of elementary matrices required to represent the basis inverse increases by 1 at each iteration. If this number becomes large, it would be necessary to reinvert the basis and represent it as the product of m elementary matrices (see Exercise 5.8). It is emphasized that each elementary matrix is completely described by its nonidentity column and its position. Therefore an elementary matrix $\mathbf{E}$ could be stored as $\begin{bmatrix}\mathbf{g}\\ r\end{bmatrix}$ where $\mathbf{g}$ is the nonidentity column and r is its position.

Example 5.2

$$\begin{aligned}\text{Minimize}\quad & -x_1 - 2x_2 + x_3\\ \text{Subject to}\quad & x_1 + x_2 + x_3 \le 4\\ & -x_1 + 2x_2 - 2x_3 \le 6\\ & 2x_1 + x_2 \le 5\\ & x_1, \quad x_2, \quad x_3 \ge 0\end{aligned}$$

Introduce the slack variables x_4, x_5, and x_6. The original basis consists of x_4, x_5, and x_6.

Iteration 1

$$\bar{\mathbf{b}} = \begin{bmatrix}4\\6\\5\end{bmatrix}$$

$$\mathbf{x}_B = \begin{bmatrix}x_{B_1}\\x_{B_2}\\x_{B_3}\end{bmatrix} = \begin{bmatrix}x_4\\x_5\\x_6\end{bmatrix} = \begin{bmatrix}4\\6\\5\end{bmatrix} \qquad \mathbf{x}_N = \begin{bmatrix}x_1\\x_2\\x_3\end{bmatrix} = \begin{bmatrix}0\\0\\0\end{bmatrix}$$

$$z = 0$$

$$\mathbf{w} = \mathbf{c}_B = (0, 0, 0)$$

Note that $z_j - c_j = \mathbf{w}\mathbf{a}_j - c_j$. Therefore

$$z_1 - c_1 = 1,\ z_2 - c_2 = 2,\ z_3 - c_3 = -1$$

Thus $k = 2$ and x_2 enters the basis.

$$\mathbf{y}_2 = \mathbf{a}_2 = \begin{bmatrix}1\\2\\1\end{bmatrix}$$

Here x_{B_r} leaves the basis where r is determined by

$$\text{Minimum}\left\{\frac{\bar{b}_1}{y_{12}}, \frac{\bar{b}_2}{y_{22}}, \frac{\bar{b}_3}{y_{32}}\right\} = \text{Minimum}\left\{\frac{4}{1}, \frac{6}{2}, \frac{5}{1}\right\} = 3$$

Therefore $r = 2$; that is, $x_{B_2} = x_5$ leaves the basis and x_2 enters the basis. The nonidentity column of $\mathbf{E}_1$ is given by

$$\mathbf{g} = \begin{bmatrix} -\dfrac{y_{12}}{y_{22}} \\ \dfrac{1}{y_{22}} \\ -\dfrac{y_{32}}{y_{22}} \end{bmatrix} = \begin{bmatrix} -\dfrac{1}{2} \\ \dfrac{1}{2} \\ -\dfrac{1}{2} \end{bmatrix}$$

and $\mathbf{E}_1$ is represented by $\begin{bmatrix} \mathbf{g} \\ 2 \end{bmatrix}$.

Iteration 2

Update $\bar{\mathbf{b}}$. Noting Equation (5.3), we have

$$\bar{\mathbf{b}} = \mathbf{E}_1 \begin{bmatrix} 4 \\ 6 \\ 5 \end{bmatrix} = \begin{bmatrix} 4 \\ 0 \\ 5 \end{bmatrix} + 6 \begin{bmatrix} -\frac{1}{2} \\ \frac{1}{2} \\ -\frac{1}{2} \end{bmatrix} = \begin{bmatrix} 1 \\ 3 \\ 2 \end{bmatrix}$$

$$\mathbf{x}_B = \begin{bmatrix} x_{B_1} \\ x_{B_2} \\ x_{B_3} \end{bmatrix} = \begin{bmatrix} x_4 \\ x_2 \\ x_6 \end{bmatrix} = \begin{bmatrix} 1 \\ 3 \\ 2 \end{bmatrix} \qquad \mathbf{x}_N = \begin{bmatrix} x_1 \\ x_5 \\ x_3 \end{bmatrix} = \begin{bmatrix} 0 \\ 0 \\ 0 \end{bmatrix}$$

$$z = 0 - \bar{b}_2(z_2 - c_2) = -6$$

$$\mathbf{w} = \mathbf{c}_B\mathbf{E}_1 = (0, -2, 0)\mathbf{E}_1$$

Noting Equation (5.2), then $\mathbf{w} = (0, -1, 0)$. Note that $z_j - c_j = \mathbf{w}\mathbf{a}_j - c_j$. Therefore

$$z_1 - c_1 = 2, \; z_3 - c_3 = 1$$

Thus $k = 1$ and x_1 enters the basis. Noting Equation (5.3),

$$\mathbf{y}_1 = \mathbf{E}_1\mathbf{a}_1 = \mathbf{E}_1 \begin{bmatrix} 1 \\ -1 \\ 2 \end{bmatrix} = \begin{bmatrix} 1 \\ 0 \\ 2 \end{bmatrix} - \begin{bmatrix} -\frac{1}{2} \\ \frac{1}{2} \\ -\frac{1}{2} \end{bmatrix} = \begin{bmatrix} \frac{3}{2} \\ -\frac{1}{2} \\ \frac{5}{2} \end{bmatrix}$$

Then x_{B_r} leaves the basis where r is determined by

$$\text{Minimum} \left\{ \frac{\bar{b}_1}{y_{11}}, \frac{\bar{b}_3}{y_{31}} \right\} = \text{Minimum} \left\{ \frac{1}{\frac{3}{2}}, \frac{2}{\frac{5}{2}} \right\} = \frac{2}{3}$$

Therefore, $r = 1$; that is, $x_{B_1} = x_4$ leaves and x_1 enters the basis. The nonidentity column of $\mathbf{E}_2$ is given by

$$\mathbf{g} = \begin{bmatrix} \frac{1}{y_{11}} \\ -\frac{y_{21}}{y_{11}} \\ -\frac{y_{31}}{y_{11}} \end{bmatrix} = \begin{bmatrix} \frac{2}{3} \\ \frac{1}{3} \\ -\frac{5}{3} \end{bmatrix}$$

Also, $\mathbf{E}_2$ is represented by $\begin{bmatrix} \mathbf{g} \\ 1 \end{bmatrix}$.

Iteration 3

Update $\bar{\mathbf{b}}$. Noting Equation (5.3), we have

$$\bar{\mathbf{b}} = \mathbf{E}_2 \begin{bmatrix} 1 \\ 3 \\ 2 \end{bmatrix} = \begin{bmatrix} 0 \\ 3 \\ 2 \end{bmatrix} + 1 \begin{bmatrix} \frac{2}{3} \\ \frac{1}{3} \\ -\frac{5}{3} \end{bmatrix} = \begin{bmatrix} \frac{2}{3} \\ \frac{10}{3} \\ \frac{1}{3} \end{bmatrix}$$

$$\mathbf{x}_B = \begin{bmatrix} x_{B_1} \\ x_{B_2} \\ x_{B_3} \end{bmatrix} = \begin{bmatrix} x_1 \\ x_2 \\ x_6 \end{bmatrix} = \begin{bmatrix} \frac{2}{3} \\ \frac{10}{3} \\ \frac{1}{3} \end{bmatrix} \qquad \mathbf{x}_N = \begin{bmatrix} x_4 \\ x_5 \\ x_3 \end{bmatrix} = \begin{bmatrix} 0 \\ 0 \\ 0 \end{bmatrix}$$

$$z = -6 - \bar{b}_1(z_1 - c_1) = -\tfrac{22}{3}$$

$$\mathbf{w} = \mathbf{c}_B\mathbf{E}_2\mathbf{E}_1 = (-1, -2, 0)\mathbf{E}_2\mathbf{E}_1$$

Applying Equation (5.2) twice, we get

$$\mathbf{c}_B\mathbf{E}_2 = \left(-\tfrac{4}{3}, -2, 0\right)$$

$$\mathbf{w} = (\mathbf{c}_B\mathbf{E}_2)\mathbf{E}_1 = \left(-\tfrac{4}{3}, -\tfrac{1}{3}, 0\right)$$

Note that $z_j - c_j = \mathbf{w}\mathbf{a}_j - c_j$. Therefore

$$z_3 - c_3 = -\tfrac{5}{3}, \; z_5 - c_5 = -\tfrac{1}{3}$$

Since $z_j - c_j \le 0$ for all nonbasic variables, then the optimal solution is at hand. The objective value is $-\frac{22}{3}$ and

$$(x_1, x_2, x_3, x_4, x_5, x_6) = \left(\tfrac{2}{3}, \tfrac{10}{3}, 0, 0, 0, \tfrac{1}{3}\right)$$

LU Decomposition or Factorization of the Basis

The product form of the inverse essentially records the history of pivots performed through a collection of the eta vectors. Although this method is conceptually important and useful it is computationally somewhat obsolete because it is based on a complete Gauss–Jordan elimination technique for solving systems of equations. A more popular technique used by most modern computer packages is the *LU factorization method*, which is based on the more efficient Gaussian triangularization strategy. It derives its name from its use of lower and upper triangular factors of the basis **B**. This method is most useful when the problem is large-scale and sparse, and it is accurate and numerically stable (round-off errors are controlled and do not tend to accumulate).

In implementing the simplex algorithm, the systems of equations that need to be solved are $\mathbf{Bx}_B = \mathbf{b}$, $\mathbf{wB} = \mathbf{c}_B$, and $\mathbf{By}_k = \mathbf{a}_k$, once an entering variable x_k is determined. The first of these equations gives the basic variable values. Since the variable values are simply updated at every iteration, it only needs to be solved initially or whenever **B** is periodically refactored from scratch in order to maintain numerical accuracy. The second system computes the simplex multiplier vector **w** which is used to price the nonbasic variables via the relationship $z_j - c_j = \mathbf{wa}_j - \mathbf{c}_j$. Once a variable x_k with $z_k - c_k > 0$ has been chosen to enter the basis, its updated column $\mathbf{y}_k$ needs to be computed. This is done via the third system of equations. The vector $\mathbf{y}_k$ provides the information on how the basic variables change when x_k is increased and determines the exiting variable via the minimum ratio test. Hence a new basis is obtained along with the (possibly) revised basic variable values. This process is then repeated.

Now, if **B** happens to be an upper triangular matrix, say, then the foregoing systems of equations can be solved easily and accurately. The systems $\mathbf{Bx}_B = \mathbf{b}$ and $\mathbf{By}_k = \mathbf{a}_k$ can be solved by a backward substitution process in which the last component of the solution vector is given directly by the last equation. This is substituted in the next to last equation to obtain the next to last component of the solution vector, and so on. The system $\mathbf{wB} = \mathbf{c}_B$ can be solved similarly by a forward substitution process in which the components of **w** are obtained sequentially in the order $w_1, w_2, \ldots, w_m$.

On the other hand, if **B** is not (upper) triangular, then it can be made so through a series of row operations. Algebraically, this process can be represented as a premultiplication of **B** by a nonsingular, row-operation transformation matrix **R** such that $\mathbf{RB} = \mathbf{U}$, an upper triangular matrix. If this is done, then to solve the system $\mathbf{By}_k = \mathbf{a}_k$, we can premultiply both sides by **R** to get $\mathbf{RBy}_k = \mathbf{Ra}_k$, that is, $\mathbf{Uy}_k = \mathbf{a}'_k$, where $\mathbf{a}'_k \equiv \mathbf{Ra}_k$. Hence the upper triangular system $\mathbf{Uy}_k = \mathbf{a}'_k$ can now be solved by a backward substitution process. An identical strategy can be used to solve the system $\mathbf{Bx}_B = \mathbf{b}$. In order to solve the system $\mathbf{wB} = \mathbf{c}_B$, consider the affine transformation $\mathbf{w} = \mathbf{w}'\mathbf{R}$. Then the foregoing system becomes $\mathbf{w}'\mathbf{RB} = \mathbf{c}_B$ or $\mathbf{w}'\mathbf{U} = \mathbf{c}_B$, from which $\mathbf{w}'$ can be obtained via a forward substitution process. Knowing $\mathbf{w}'$, we can now compute **w** as $\mathbf{w}'\mathbf{R}$. Hence if the row operations involved are not too computationally intensive and are well conditioned (do not involve divisions by small magnitude numbers, for instance), then the systems of equations can again be solved conveniently and accurately.

Observe that if the row operations required to upper triangularize **B** involve a simple row by row Gaussian reduction process, then **R** would be a lower triangular matrix. In other words, the first row of **B** would be left as is, the second row would have a multiple of the first row added to it in order to zero out the (2, 1) element of **B**, the third row of **B** would have a multiple of the first row and a multiple of the second row added to it in order to zero out the (3, 1) and the (3, 2) elements, and so on. In this case, we would have $\mathbf{B} = \mathbf{R}^{-1}\mathbf{U} \equiv \mathbf{LU}$, a product of a lower triangular matrix $\mathbf{L} \equiv \mathbf{R}^{-1}$ and an upper triangular matrix **U**. Hence such a factorization is called a **LU** factorization, even in instances when **R** is not lower triangular.

Obtaining the LU Factorization of B

Typically, in addition to the Gaussian pivots or reductions, one also needs to perform permutations of rows in the factorization in order to obtain desirable pivot elements. For example, to begin, one may wish to make that row the first row that has the largest absolute valued coefficient in the first column, so that the Gaussian pivots associated with zeroing out the first column under the (1, 1) element are numerically well conditioned. (Of course, if the starting (1, 1) element in $\mathbf{B}$ is zero, then some such permutation is imperative.) Hence the row operation matrix $\mathbf{R}$ may involve a (row interchange) permutation matrix $\mathbf{P}_1$ followed by a Gaussian reduction matrix $\mathbf{G}_1$ used to triangularize the first column (or the first nontriangular column), then possibly another permutation matrix $\mathbf{P}_2$ followed by a Gaussian triangularization matrix $\mathbf{G}_2$, and so on. Hence if we use r such operations, we will have

$$\mathbf{R} = \mathbf{G}_r\mathbf{P}_r \cdots \mathbf{G}_2\mathbf{P}_2\mathbf{G}_1\mathbf{P}_1 = \prod_{i=1}^{r} (\mathbf{G}_i\mathbf{P}_i)$$

Accordingly, a forward transformation process called FTRAN is used to compute $\mathbf{a}'_k = \mathbf{R}\mathbf{a}_k$ while solving the system $\mathbf{B}\mathbf{y}_k = \mathbf{a}_k$ by employing the operators $(\mathbf{G}_i\mathbf{P}_i)$ in the order $1, \ldots, r$ on $\mathbf{a}_k$. Similarly, a backward transformation process called BTRAN is used to compute $\mathbf{w} = \mathbf{w}'\mathbf{R}$ while solving the system $\mathbf{w}\mathbf{B} = \mathbf{c}_B$ by postoperatively using $\mathbf{G}_i\mathbf{P}_i$ on $\mathbf{w}'$ in the order $r, r-1, \ldots, 1$.

Now let us discuss the construction of the operators defining $\mathbf{R}$. Suppose that so far we have upper triangularized columns $1, 2, \ldots, k-1$ of $\mathbf{B}$, and we are now considering the triangularization of a nontriangular column k using, for example, the operator $\mathbf{G}_i\mathbf{P}_i$. Let $\mathbf{B}_k$ be the current partially triangular matrix. Then we first determine the largest absolute valued element in rows k, $k+1, \ldots, m$ for column k in $\mathbf{B}_k$. If this happens to be the (k, k) element itself, then no row interchange (permutation) is necessary, and in this case $\mathbf{P}_i = \mathbf{I}$. Otherwise if this happens to be the element (t, k), then we need to interchange rows k and t. Hence the permutation matrix that accomplishes this is given as follows, where the elements not shown are zeros.

$$\mathbf{P}_i = \begin{array}{cc}
 & \begin{array}{ccccccccccc} \; & \; & \; & k & \; & \; & \; & t & \; & \; & \; \end{array} \\
\begin{array}{c} \\ \\ \\ k \\ \\ \\ \\ t \\ \\ \\ \\ \end{array} & \left[\begin{array}{ccc|c|ccc|c|ccc}
1 & & & & & & & & & & \\
 & \ddots & & & & & & & & & \\
 & & 1 & & & & & & & & \\
\hline
 & & & 0 & & & & 1 & & & \\
\hline
 & & & & 1 & & & & & & \\
 & & & & & \ddots & & & & & \\
 & & & & & & 1 & & & & \\
\hline
 & & & 1 & & & & 0 & & & \\
\hline
 & & & & & & & & 1 & & \\
 & & & & & & & & & \ddots & \\
 & & & & & & & & & & 1
\end{array}\right]
\end{array} \tag{5.4}$$

Having done this, suppose that the matrix $\mathbf{B}_k$ now appears as follows, where the blank section has all zeros.

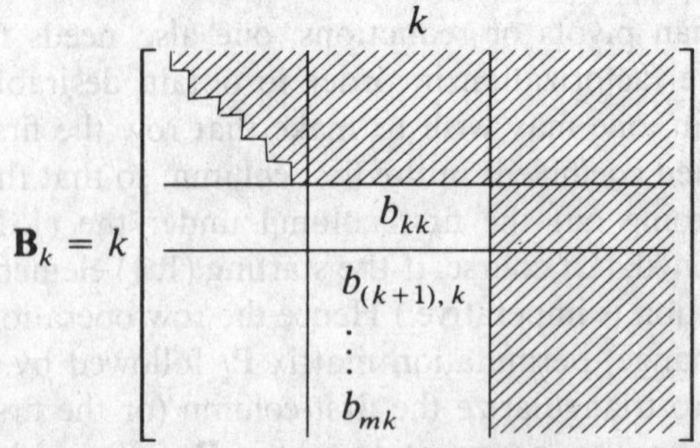

$$= \mathbf{P}_i(\mathbf{G}_{i-1}\mathbf{P}_{i-1}) \cdots (\mathbf{G}_1\mathbf{P}_1)\mathbf{B} \quad (5.5)$$

Now in order to zero out the elements in column k and in rows $(k+1), \ldots, m$, we need to premultiply $\mathbf{B}_k$ by the following Gaussian pivot matrix $\mathbf{G}_i$, where again, all elements not shown are zeros.

$$\mathbf{G}_i = \begin{array}{c} \\ \\ \\ k \\ \\ \\ \\ \end{array} \left[\begin{array}{ccc|c|ccc} 1 & & & & & & \\ & \ddots & & & & & \\ & & 1 & & & & \\ \hline & & & 1 & & & \\ \hline & & & -b_{(k+1),k}/b_{kk} & 1 & & \\ & & & \vdots & & \ddots & \\ & & & -b_{mk}/b_{kk} & & & 1 \end{array}\right] \quad (5.6)$$

Observing $\mathbf{G}_i$, note the need for not having b_{kk} too small. Also compare $\mathbf{G}_i$ with the Gauss–Jordan pivot or elementary matrix $\mathbf{E}$ used in the product form of the inverse. Needless to say, the matrices $\mathbf{P}_i$ and $\mathbf{G}_i$ are stored in compact or packed form. Namely, only the relevant permutation performed by $\mathbf{P}_i$ is stored, and similarly from $\mathbf{G}_i$, only the nonzeros in rows $k+1, \ldots, m$ in column k, along with their positions are stored. This operation now produces $\mathbf{G}_i\mathbf{B}_k = \prod_{l=1}^{i}(\mathbf{G}_l\mathbf{P}_l)\mathbf{B}$ from Equation 5.5. This process may now be continued until one obtains $\mathbf{RB} = \prod_{l=1}^{r}(\mathbf{G}_l\mathbf{P}_l)\mathbf{B} \equiv \mathbf{U}$.

Example 5.3

Suppose that the following matrix $\mathbf{B}$ needs to be factored and we need to solve the system $\mathbf{B}\mathbf{x}_B = \mathbf{b}$ and $\mathbf{wB} = \mathbf{c}_B$, where

$$\mathbf{B} = \begin{bmatrix} 1 & 3 & 2 \\ 2 & 1 & -1 \\ 1 & 2 & 2 \end{bmatrix}, \qquad \mathbf{b} = \begin{bmatrix} 9 \\ 3 \\ 7 \end{bmatrix}, \qquad \text{and } \mathbf{c}_B = (1, 1, 2)$$

Beginning with column 1, we first perform an interchange of rows 1 and 2 in order to make the $(1,1)$ element the largest (in absolute value) in column 1. This corresponds to the permutation matrix $\mathbf{P}_1$ which yields the matrix $\mathbf{B}_1$.

$$\mathbf{P}_1 = \begin{bmatrix} 0 & 1 & 0 \\ 1 & 0 & 0 \\ 0 & 0 & 1 \end{bmatrix}, \qquad \mathbf{B}_1 = \mathbf{P}_1\mathbf{B} = \begin{bmatrix} 2 & 1 & -1 \\ 1 & 3 & 2 \\ 1 & 2 & 2 \end{bmatrix}$$

Next we triangularize column 1 of $\mathbf{B}_1$ via the Gaussian pivot matrix $\mathbf{G}_1$.

$$\mathbf{G}_1 = \begin{bmatrix} 1 & 0 & 0 \\ -\frac{1}{2} & 1 & 0 \\ -\frac{1}{2} & 0 & 1 \end{bmatrix}$$

which gives

$$(\mathbf{G}_1\mathbf{P}_1)\mathbf{B} = \begin{bmatrix} 2 & 1 & -1 \\ 0 & \frac{5}{2} & \frac{5}{2} \\ 0 & \frac{3}{2} & \frac{5}{2} \end{bmatrix}$$

Continuing, we can now take $\mathbf{P}_2$ as the identity matrix that gives $\mathbf{B}_2 \equiv \mathbf{P}_2(\mathbf{G}_1\mathbf{P}_1)\mathbf{B} = (\mathbf{G}_1\mathbf{P}_1)\mathbf{B}$ as previously. We then get from Equation (5.6),

$$\mathbf{G}_2 = \begin{bmatrix} 1 & 0 & 0 \\ 0 & 1 & 0 \\ 0 & -\frac{3}{5} & 1 \end{bmatrix},$$

which gives

$$(\mathbf{G}_2\mathbf{P}_2)(\mathbf{G}_1\mathbf{P}_1)\mathbf{B} = \begin{bmatrix} 2 & 1 & -1 \\ 0 & \frac{5}{2} & \frac{5}{2} \\ 0 & 0 & 1 \end{bmatrix} \equiv \mathbf{U}.$$

Hence $\mathbf{R} = \mathbf{G}_2\mathbf{P}_2\mathbf{G}_1\mathbf{P}_1$ and $\mathbf{U}$ is as previously given.

Now to solve $\mathbf{B}\mathbf{x}_B = \mathbf{b}$, we first compute $\mathbf{R}\mathbf{b}$ as follows using a FTRAN process.

$$\mathbf{R}\mathbf{b} = \mathbf{G}_2\mathbf{P}_2\mathbf{G}_1\mathbf{P}_1\mathbf{b} = \mathbf{G}_2\mathbf{P}_2\mathbf{G}_1\begin{bmatrix} 3 \\ 9 \\ 7 \end{bmatrix} = \mathbf{G}_2\mathbf{P}_2\begin{bmatrix} 3 \\ \frac{15}{2} \\ \frac{11}{2} \end{bmatrix} = \mathbf{G}_2\begin{bmatrix} 3 \\ \frac{15}{2} \\ \frac{11}{2} \end{bmatrix} = \begin{bmatrix} 3 \\ \frac{15}{2} \\ 1 \end{bmatrix}$$

Therefore, since $\mathbf{R}\mathbf{B}\mathbf{x}_B = \mathbf{R}\mathbf{b}$, that is, $\mathbf{U}\mathbf{x}_B = \mathbf{R}\mathbf{b}$, where $\mathbf{U}$ is as given, we compute $\mathbf{x}_B = (1, 2, 1)^t$ by backward substitution.

Similarly, in order to solve $\mathbf{w}\mathbf{B} = \mathbf{c}_B$, we first solve $\mathbf{w}'\mathbf{U} = \mathbf{c}_B$ by forward substitution and then compute $\mathbf{w} = \mathbf{w}'\mathbf{R}$. This gives $\mathbf{w}' = (\frac{1}{2}, \frac{1}{5}, 2)$. Consequently, applying a BTRAN process, we obtain

$$\begin{aligned}\mathbf{w} = \mathbf{w}'\mathbf{R} &= (\tfrac{1}{2}, \tfrac{1}{5}, 2)[\mathbf{G}_2\mathbf{P}_2\mathbf{G}_1\mathbf{P}_1] = (\tfrac{1}{2}, -1, 2)\mathbf{P}_2\mathbf{G}_1\mathbf{P}_1 = (\tfrac{1}{2}, -1, 2)\mathbf{G}_1\mathbf{P}_1 \\ &= (0, -1, 2)\mathbf{P}_1 = (-1, 0, 2)\end{aligned}$$

The reader may wish to observe how the foregoing calculations are performed with $\mathbf{P}_i$ and $\mathbf{G}_i$ stored in the previously mentioned compact form.

Updating the LU Factors

Suppose that we have selected x_k as the entering variable and have determined x_r as the leaving variable in the usual manner. The task now is to update $\mathbf{R}$ so that when operated on the new basis, it produces an upper triangular matrix. Toward this end, suppose that we delete from $\mathbf{B}$ the leaving column, say, column

r, then move the columns $(r+1), \ldots, m$ leftwards, that is, make the current column i as the new $(i-1)$th column for $i = (r+1), \ldots, m$, and insert the entering column a_k of x_k as the last (mth) column. Let this new basis be denoted by $\mathbf{B}_{\text{new}}$. (One would need to accordingly permute the updated right-hand-side so that the new basic variable values appear in the same order as their columns are arranged in B_{new}). Noting that $\mathbf{RB}$ is an upper triangular matrix, we now have $\mathbf{RB}_{\text{new}}$ appearing as shown, where the blank section has all zeros.

$\mathbf{RB}_{\text{new}} =$ 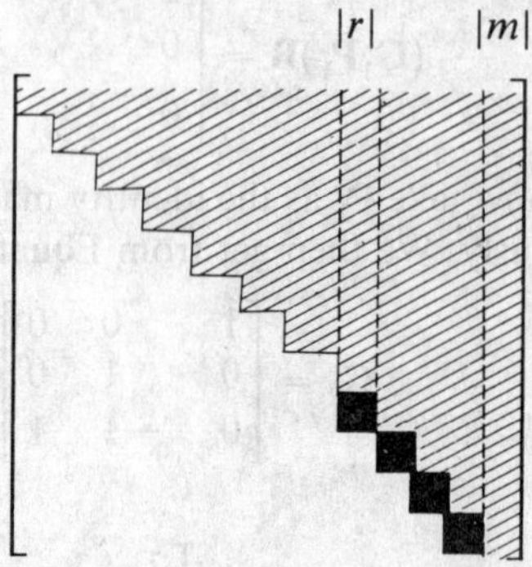 (5.7)

Note that the columns $1, \ldots, r-1$ of $\mathbf{RB}_{\text{new}}$ are as in $\mathbf{U}$, the columns $r, \ldots, m-1$ of $\mathbf{RB}_{\text{new}}$ are precisely the columns $(r+1), \ldots, m$ of $\mathbf{U}$, and the final column in $\mathbf{RB}_{\text{new}}$ is $\mathbf{Ra}_k$. Note that $\mathbf{Ra}_k$ is already available from the solution of the system $\mathbf{By}_k = \mathbf{a}_k$ (why?). Hence we delete column r from $\mathbf{U}$, move the columns that are to the right of it one position leftwards, and insert $\mathbf{Ra}_k$ as the last column in $\mathbf{U}$, in order to obtain $\mathbf{RB}_{\text{new}}$. The type of matrix depicted in Equation (5.7) is called an *upper Hessenberg* matrix. Note that the elements in the shaded region in Equation (5.7) used to be along the diagonal of $\mathbf{U}$, and hence are nonzeros since $\mathbf{U}$ is nonsingular. Now we perform additional permutations and Gaussian pivots embodied in Equations (5.4) and (5.6) as before in order to triangularize $\mathbf{RB}_{\text{new}}$. The additional factors of the type $(\mathbf{G}_i\mathbf{P}_i)$ used are appended to $\mathbf{R}$ in order to derive the new factor $\mathbf{R}$.

Example 5.4

Consider Example 5.3. Suppose that we price out the nonbasic variables and select some variable x_4 as the entering variable. Let $\mathbf{a}_4 = (1, -1, 0)^t$ and $c_4 = -2$. Hence $z_4 - c_4 = \mathbf{wa}_4 - c_4 = (-1, 0, 2)(1, -1, 0)^t - (-2) = 1 > 0$. Furthermore, the system $\mathbf{By}_4 = \mathbf{a}_4$ gives $\mathbf{RBy}_4 = \mathbf{Ra}_4$, where $\mathbf{RB} = \mathbf{U}$, and

$$\mathbf{Ra}_4 = \mathbf{G}_2\mathbf{P}_2\mathbf{G}_1\mathbf{P}_1\mathbf{a}_4 = \mathbf{G}_2\mathbf{P}_2\mathbf{G}_1\begin{bmatrix} -1 \\ 1 \\ 0 \end{bmatrix} = \mathbf{G}_2\mathbf{P}_2\begin{bmatrix} -1 \\ \frac{3}{2} \\ \frac{1}{2} \end{bmatrix} = \mathbf{G}_2\begin{bmatrix} -1 \\ \frac{3}{2} \\ \frac{1}{2} \end{bmatrix} = \begin{bmatrix} -1 \\ \frac{3}{2} \\ -\frac{2}{5} \end{bmatrix} \tag{5.8}$$

Hence $\mathbf{Uy}_4 = \mathbf{Ra}_4$ gives $\mathbf{y}_4 = (-\frac{6}{5}, 1, -\frac{2}{5})^t$. Consequently, performing the minimum ratio test, the basic variable corresponding to the second column of $\mathbf{B}$ leaves the basis. Eliminating this second column of $\mathbf{B}$ and appending $\mathbf{a}_4$ as the final column of $\mathbf{B}$ results in the new basis $\mathbf{B}_{\text{new}}$. From $\mathbf{U}$ and Equation (5.8), we get (as in Equation (5.7))

$$\mathbf{RB}_{\text{new}} = \begin{bmatrix} 2 & -1 & -1 \\ 0 & \frac{5}{2} & \frac{3}{2} \\ 0 & 1 & -\frac{2}{5} \end{bmatrix}$$

Furthermore, on pivoting, the right-hand-side updates to $(\frac{17}{5}, 2, \frac{9}{5})'$. Because of the rearrangement of the columns of $\mathbf{B}_{\text{new}}$, the new right-hand-side is $(\frac{17}{5}, \frac{9}{5}, 2)'$. To triangularize the second column of $\mathbf{RB}_{\text{new}}$, we use $\mathbf{P}_3 = \mathbf{I}$ and, from Equation (5.6), we use

$$\mathbf{G}_3 = \begin{bmatrix} 1 & 0 & 0 \\ 0 & 1 & 0 \\ 0 & -\frac{2}{5} & 1 \end{bmatrix},$$

which gives

$$(\mathbf{G}_3\mathbf{P}_3)(\mathbf{G}_2\mathbf{P}_2)(\mathbf{G}_1\mathbf{P}_1)\mathbf{B}_{\text{new}} \equiv (\mathbf{G}_3\mathbf{P}_3)\mathbf{RB}_{\text{new}} = \begin{bmatrix} 2 & -1 & -1 \\ 0 & \frac{5}{2} & \frac{3}{2} \\ 0 & 0 & -1 \end{bmatrix} \equiv \mathbf{U}_{\text{new}}$$

Hence the new factors $\mathbf{R}$ and $\mathbf{U}$ are, respectively, $\mathbf{R}_{\text{new}} \equiv \mathbf{G}_3\mathbf{P}_3\mathbf{G}_2\mathbf{P}_2\mathbf{G}_1\mathbf{P}_1$ and $\mathbf{U}_{\text{new}}$ as given previously. We now compute the new simplex multiplier vector $\mathbf{w}$, price out the nonbasic variables, and then continue.

Some Implementation Remarks

It should be evident from the foregoing discussion that the $\mathbf{LU}$ factorization technique is particularly well suited for sparse problems, and it can benefit greatly if the factors $\mathbf{R}$ and $\mathbf{U}$ are themselves sparse. Toward this end, one can initially try to permute the rows and columns of the (sparse) matrix $\mathbf{B}$ to make it as "upper triangular" as possible, while trying to reduce any "bumps" and "spikes" that protrude below the diagonal. An attempt is also usually made to preserve sparsity in the factors in this fashion, so that although $\mathbf{B}^{-1}$ may tend to fill up (become dense relative to $\mathbf{B}$), the number of nonzeros in the Gaussian pivot matrices and in $\mathbf{U}$ tend to be of the same order as the number of nonzeros in $\mathbf{B}$ itself. Of course, in order to maintain accuracy, sparse storage requirements, and to reduce computational effort, the current basis should be periodically refactored from scratch.

Most problems also benefit by *scaling*, in which both rows and columns are sequentially scaled by dividing throughout with, for example, the average magnitude of the nonzeros in order to make the coefficient magnitudes in the rows and columns compatible with each other. This enhances the numerical accuracy of the algorithm and can also dramatically reduce the solution effort by virtue of the simplex path generated.

Motivated by the same concepts is the use of a *steepest edge* entering variable selection strategy. Here, instead of selecting the variable with simply the most positive $z_j - c_j$ value to enter the basis (Dantzig's Rule), one may base the selection on the most positive value of $(z_j - c_j)/\gamma_j$, where $\gamma_j = \sqrt{1 + \Sigma_r y_{rj}^2}$.

Note that γ_j is the Euclidean norm of the direction vector along the edge associated with increasing the nonbasic variable x_j by a unit. Hence $(z_j - c_j)/\gamma_j$ is the negative rate of change in the objective function value per unit distance along this direction. Of course, in order to conserve solution effort per iteration, the γ_j-quantities need to be updated rather than recomputed from one iteration to the next (see the Notes and References section). This rule significantly enhances the performance of the simplex algorithm over the use of Dantzig's Rule, particularly on unscaled problems.

Another related strategy is *partial pricing* or *suboptimization*. Thus far, we have suggested that one conduct a total pricing of all the nonbasic variables at each iteration. This is convenient and advantageous for problems with no more than 1000 or so variables. However, for larger-sized problems, such a total pricing operation can be computationally expensive. One usually performs a partial pricing or a so-called suboptimization process. We describe here one particular implementation of this process using Dantzig's Rule for selecting an entering variable. To begin, all the nonbasic variables are priced and the largest $z_j - c_j$, say, $z_k - c_k = \Delta_{\max} > 0$ is found. A list L is constructed of the current basic variables and the nonbasic variables having $z_j - c_j$ within some p_1 percent of $\Delta_{\max}$. At this point, the variables not in L are ignored, and the iterations continue by selecting the nonbasic variable with the most positive $z_j - c_j$ in L for entering the basis. However, when the most positive $z_j - c_j$ falls below some p_2 percent of $\Delta_{\max}$ or when the method goes through some N consecutive degenerate iterations, the method pauses and reconstructs the list L. If this list were not reconstructed, then the algorithm would eventually solve the problem where the ignored nonbasic variables are permanently fixed at zero, hence the name of this technique. Typical values of p_1 and p_2 are, respectively, 0.8 and 0.67 for sparse problems ($d \le 0.05 - 0.1$) and 0.7 and 0.4 for dense problems. (Here d is the density of the matrix $\mathbf{A}$.) A typical value of N is about 10–30.

Finally, we mention that for large-scale problems, one may need to write a *matrix generator* program in order to automatically generate the constraints and the objective function based on the structure of the application instead of manually inputting the data. The software would then store the nonzero coefficients in packed form as noted earlier, and then use them appropriately to perform the foregoing computations. The interested reader is referred to the Notes and References section for further reading on these subjects.

5.2 THE SIMPLEX METHOD FOR BOUNDED VARIABLES

In most practical problems the variables are usually bounded. A typical variable x_j is bounded from below by l_j and from above by u_j, where $l_j < u_j$. If we denote the lower and upper bound vectors by $\mathbf{l}$ and $\mathbf{u}$ respectively, we get the following linear program with bounded variables.

$$\begin{array}{ll} \text{Minimize} & \mathbf{cx} \\ \text{Subject to} & \mathbf{Ax} = \mathbf{b} \\ & \mathbf{l} \le \mathbf{x} \le \mathbf{u} \end{array}$$

By standard transformations, we can assume without loss of generality that l_j is finite for each $j = 1, \ldots, n$ (why?). If $\mathbf{l} = \mathbf{0}$, the usual nonnegativity restrictions are obtained. In fact, any lower bound vector can be transformed into the zero vector by using the change of variables $\mathbf{x}' = \mathbf{x} - \mathbf{l}$. The most straightforward (and the least efficient) method of handling the constraints $\mathbf{l} \leq \mathbf{x} \leq \mathbf{u}$ is to introduce the slack vectors $\mathbf{x}_1$ and $\mathbf{x}_2$, leading to the constraints $\mathbf{x} + \mathbf{x}_1 = \mathbf{u}$ and $\mathbf{x} - \mathbf{x}_2 = \mathbf{l}$. This increases the number of equality constraints from m to $m + 2n$ and the number of variables from n to $3n$. Even if $\mathbf{l} = \mathbf{0}$ or is transformed into $\mathbf{0}$ as discussed, the slack vector $\mathbf{x}_1$ is needed, which increases both the constraints and variables by n.

From the foregoing discussion it is clear that the problem size (and hence the computational effort) would increase significantly if the constraints $\mathbf{l} \leq \mathbf{x} \leq \mathbf{u}$ are treated in the usual manner by introducing slack vectors. The simplex method with bounded variables handles these constraints implicitly in a fashion similar to that used by the simplex method to handle the constraints $\mathbf{x} \geq \mathbf{0}$. Therefore, the so-called "*working basis*" remains of size $m \times m$, and hence such a method is known as a *compact basis* method. As in the simplex method, the algorithm of this section moves from a basic feasible solution to an improved basic feasible solution of the system $\mathbf{Ax} = \mathbf{b}$, $\mathbf{l} \leq \mathbf{x} \leq \mathbf{u}$ (at a nondegenerate pivot step), until optimality is reached or unboundedness is verified.

Now observe that the feasible region of the foregoing problem is a polyhedron in E^n, and as before, its extreme points are feasible points at which some n linearly independent hyperplanes are binding. Since $\mathbf{Ax} = \mathbf{b}$ gives m linearly independent hyperplanes that are binding at each feasible solution we seek $p = n - m$ appropriate additional binding hyperplanes from the inequalities $\mathbf{l} \leq \mathbf{x} \leq \mathbf{u}$ in order to make up the required set of n linearly independent binding equations for defining each extreme point solution. (Throughout we only consider constraints with finite upper or lower bounds to be binding.) Consequently, there are $p = n - m$ degrees of freedom remaining after imposing the constraints $\mathbf{Ax} = \mathbf{b}$, and so the given polyhedron is embedded in E^p. We therefore need to associate with an extreme point some p (independent) (*nonbasic*) *variables* fixed either at their lower or their upper bounds such that the remaining m (dependent) (*basic*) *variables* are uniquely determined via the system $\mathbf{Ax} = \mathbf{b}$. If more than p, say, q, of the inequalities $\mathbf{l} \leq \mathbf{x} \leq \mathbf{u}$ are binding at an extreme point, then such an extreme point is *degenerate* and has $q - p \geq 1$ of the associated basic variables at one of their bounds. Hence we define a basic feasible solution as follows, so that a solution $\mathbf{x}$ is an extreme point if and only if it is a basic feasible solution. Details of the proof based on this exposition are very similar to the case of the set $X = \{\mathbf{x} : \mathbf{Ax} = \mathbf{b}, \mathbf{x} \geq \mathbf{0}\}$ and are left to the reader in Exercise 5.13.

Definition (Basic Feasible Solutions)

Consider the system $\mathbf{Ax} = \mathbf{b}$ and $\mathbf{l} \leq \mathbf{x} \leq \mathbf{u}$, where $\mathbf{A}$ is an $m \times n$ matrix of rank m. The solution $\bar{\mathbf{x}}$ to the equations $\mathbf{Ax} = \mathbf{b}$ is a *basic solution* of this system if $\mathbf{A}$ can be partitioned into $[\mathbf{B}, \mathbf{N}_1, \mathbf{N}_2]$ where the (square) matrix $\mathbf{B}$ has rank m such that with $\mathbf{x}$ partitioned accordingly as $(\mathbf{x}_B, \mathbf{x}_{N_1}, \mathbf{x}_{N_2})$ we have $\bar{\mathbf{x}}_{N_1} = \mathbf{l}_{N_1}$, $\bar{\mathbf{x}}_{N_2} = \mathbf{u}_{N_2}$, and accordingly, $\bar{\mathbf{x}}_B = \mathbf{B}^{-1}\mathbf{b} - \mathbf{B}^{-1}\mathbf{N}_1\mathbf{l}_{N_1} - \mathbf{B}^{-1}\mathbf{N}_2\mathbf{u}_{N_2}$.

Furthermore, if $\mathbf{l}_B \leq \bar{\mathbf{x}}_B \leq \mathbf{u}_B$, then $\bar{\mathbf{x}}$ is said to be a *basic feasible solution*. The matrix $\mathbf{B}$ is called the (*working*) *basis*, $\mathbf{x}_B$ are the *basic variables*, and $\mathbf{x}_{N_1}$ and $\mathbf{x}_{N_2}$ are the *nonbasic variables* at their lower and upper limits respectively. If, in addition, $\mathbf{l}_B < \bar{\mathbf{x}}_B < \mathbf{u}_B$, then $\bar{\mathbf{x}}$ is called a *nondegenerate basic feasible solution*; otherwise it is called a *degenerate basic feasible solution*.

The partition $[\mathbf{B}, \mathbf{N}_1, \mathbf{N}_2]$ corresponding to a basic (feasible) solution is said to be a *basic* (*feasible*) *partition*. Two basic (feasible) partitions are said to be *adjacent* if all but one nonbasic variable coincides in the two partitions, being fixed at the same bounds. Note that for a degenerate extreme point, $\bar{\mathbf{x}}$, each choice of $p = (n - m)$ defining hyperplanes from the set $\mathbf{l} \leq \mathbf{x} \leq \mathbf{u}$ that, along with the m constraints $\mathbf{Ax} = \mathbf{b}$, yield n linearly independent hyperplanes binding at $\bar{\mathbf{x}}$, corresponds to a basic feasible partition associated with $\bar{\mathbf{x}}$.

Example 5.5

Consider the region given by

$$\begin{aligned} x_1 + x_2 &\leq 5 \\ -x_1 + 2x_2 &\leq 4 \\ 0 \leq x_1 &\leq 4 \\ -1 \leq x_2 &\leq 4 \end{aligned}$$

First introduce the slack variables x_3 and x_4. This gives the following system (note that close examination of the system shows that u_3 and u_4 can be replaced by 6 and 10 respectively, if so desired):

$$\begin{aligned} x_1 + x_2 + x_3 \qquad &= 5 \\ -x_1 + 2x_2 \qquad + x_4 &= 4 \\ 0 \leq x_1 &\leq 4 \\ -1 \leq x_2 &\leq 4 \\ 0 \leq x_3 &\leq \infty \\ 0 \leq x_4 &\leq \infty \end{aligned}$$

We would like to find all the basic feasible solutions of this system. This can be accomplished by extracting a basis of the first two constraints, solving the basic variables in terms of the nonbasic variables, and then assigning the nonbasic variables at their lower or upper bounds. To illustrate the method select, say, $\mathbf{a}_2$ and $\mathbf{a}_4$ as the basic vectors. Then

$$\mathbf{B} = [\mathbf{a}_2, \mathbf{a}_4] = \begin{bmatrix} 1 & 0 \\ 2 & 1 \end{bmatrix} \qquad \mathbf{B}^{-1} = \begin{bmatrix} 1 & 0 \\ -2 & 1 \end{bmatrix}$$

Multiplying the first two constraints by $\mathbf{B}^{-1}$ and transferring x_1 and x_3 to the right-hand-side, we get

$$\begin{aligned} x_2 &= 5 - x_1 - x_3 \\ x_4 &= -6 + 3x_1 + 2x_3 \end{aligned}$$

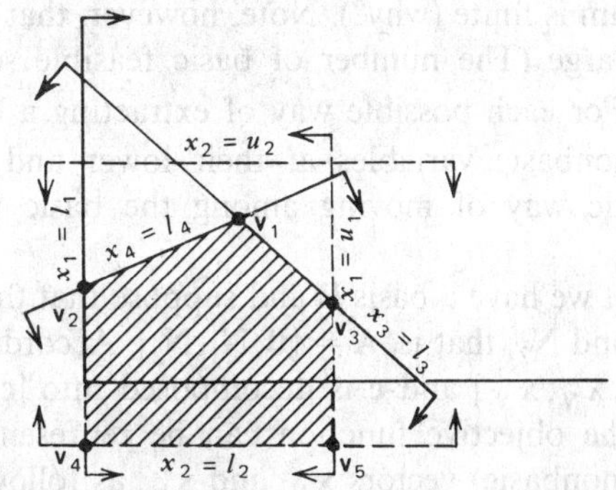

Figure 5.1. Basic feasible solutions.

Now assign x_1 at its lower or upper bound, x_3 at its lower bound, and solve for x_2 and x_4.

1. $x_1 = 0, x_3 = 0 \Rightarrow x_2 = 5$, and $x_4 = -6$. Since $x_4 < 0$, this basic solution is not a basic feasible solution.
2. $x_1 = 4, \quad x_3 = 0 \Rightarrow x_2 = 1$, and $x_4 = 6$. Therefore $(x_1, x_2, x_3, x_4) = (4, 1, 0, 6)$ is a basic feasible solution.

The other basic solutions can be obtained in a similar manner. If all of the possible bases were enumerated we would see that the basic feasible solutions are $(2, 3, 0, 0)$, $(0, 2, 3, 0)$, $(4, 1, 0, 6)$, $(0, -1, 6, 6)$, and $(4, -1, 2, 10)$. Projecting these points in the (x_1, x_2) space, we get the extreme points $\mathbf{v}_1, \ldots, \mathbf{v}_5$ shown in Figure 5.1. In this example, $n = 4$, $m = 2$, and so $p = n - m = 2$. In fact, the representation shown in Figure 5.1 is in the $p = 2$-dimensional space corresponding to x_1 and x_2 as the independent variables. Note how each extreme point is formed by some $p = 2$ linearly independent defining hyperplanes binding in this space. In particular, $(x_3 = l_3, x_4 = l_4)$ gives $\mathbf{v}_1$, $(x_1 = l_1, x_4 = l_4)$ gives $\mathbf{v}_2$, $(x_1 = u_1, x_3 = l_3)$ gives $\mathbf{v}_3$, $(x_1 = l_1, x_2 = l_2)$ gives $\mathbf{v}_4$, and $(x_1 = u_1, x_2 = l_2)$ gives $\mathbf{v}_5$. Hence, as noted previously the basic feasible solutions and the extreme points coincide. Observe also that the extreme points $\mathbf{v}_4$ and $\mathbf{v}_5$ have the same working basis associated with them, namely the columns of the variables x_3 and x_4. However, they differ in the bound values assigned to the (independent) nonbasic variables x_1 and x_2, and hence produce distinct basic feasible partitions and distinct basic feasible solutions. Furthermore, since all extreme points are nondegenerate here, each has a unique basic feasible partition associated with it and the adjacent basic feasible partitions correspond to adjacent extreme points. We ask the reader to explore a generalization of this in Exercise 5.14 by imitating the analysis for the nonnegatively constrained case.

Improving a Basic Feasible Solution

We now know how to characterize a basic feasible solution. We also know that an optimal basic feasible solution exists provided that the feasible region is not empty and the optimum is finite (why?). Note, however, that the number of basic feasible solutions is large (The number of basic feasible solutions is bounded above by $\binom{n}{m}2^{n-m}$. For each possible way of extracting a basis there are 2^{n-m} ways of fixing the nonbasic variables at their lower and/or upper bounds). Therefore a systematic way of moving among the basic feasible solutions is needed.

Now suppose that we have a basis $\mathbf{B}$ and suppose that the nonbasic matrix is decomposed into $\mathbf{N}_1$ and $\mathbf{N}_2$, that is, $\mathbf{A} = [\mathbf{B}, \mathbf{N}_1, \mathbf{N}_2]$. Accordingly, the vector $\mathbf{x}$ is decomposed into $[\mathbf{x}_B, \mathbf{x}_{N_1}, \mathbf{x}_{N_2}]$ and $\mathbf{c}$ is decomposed into $[\mathbf{c}_B, \mathbf{c}_{N_1}, \mathbf{c}_{N_2}]$. Both the basic variables and the objective function can be represented in terms of the independent (that is, nonbasic) vectors $\mathbf{x}_{N_1}$ and $\mathbf{x}_{N_2}$ as follows:

$$\mathbf{x}_B = \mathbf{B}^{-1}\mathbf{b} - \mathbf{B}^{-1}\mathbf{N}_1\mathbf{x}_{N_1} - \mathbf{B}^{-1}\mathbf{N}_2\mathbf{x}_{N_2} \tag{5.9}$$

$$\begin{aligned} z &= \mathbf{c}_B\mathbf{x}_B + \mathbf{c}_{N_1}\mathbf{x}_{N_1} + \mathbf{c}_{N_2}\mathbf{x}_{N_2} \\ &= \mathbf{c}_B\left(\mathbf{B}^{-1}\mathbf{b} - \mathbf{B}^{-1}\mathbf{N}_1\mathbf{x}_{N_1} - \mathbf{B}^{-1}\mathbf{N}_2\mathbf{x}_{N_2}\right) + \mathbf{c}_{N_1}\mathbf{x}_{N_1} + \mathbf{c}_{N_2}\mathbf{x}_{N_2} \\ &= \mathbf{c}_B\mathbf{B}^{-1}\mathbf{b} + \left(\mathbf{c}_{N_1} - \mathbf{c}_B\mathbf{B}^{-1}\mathbf{N}_1\right)\mathbf{x}_{N_1} + \left(\mathbf{c}_{N_2} - \mathbf{c}_B\mathbf{B}^{-1}\mathbf{N}_2\right)\mathbf{x}_{N_2} \end{aligned} \tag{5.10}$$

Suppose that we have a current basic feasible solution where $\mathbf{x}_{N_1} = \mathbf{l}_{N_1}$, $\mathbf{x}_{N_2} = \mathbf{u}_{N_2}$, and $\mathbf{l}_B \leq \mathbf{x}_B \leq \mathbf{u}_B$. This solution is represented by the following tableau. The right-hand-side column gives the *true values* of z and $\mathbf{x}_B$ (denoted by $\hat{z}$ and $\hat{\mathbf{b}}$) when $\mathbf{x}_{N_1} = \mathbf{l}_{N_1}$ and $\mathbf{x}_{N_2} = \mathbf{u}_{N_2}$ are substituted in Equations (5.9) and (5.10). We emphasize that this column does *not* give $\mathbf{c}_B\mathbf{B}^{-1}\mathbf{b}$ and $\mathbf{B}^{-1}\mathbf{b}$, which are the right-hand-sides of Equations (5.10) and (5.9) respectively.

	z	$\mathbf{x}_B$	$\mathbf{x}_{N_1}$	$\mathbf{x}_{N_2}$	RHS
z	1	$\mathbf{0}$	$\mathbf{c}_B\mathbf{B}^{-1}\mathbf{N}_1 - \mathbf{c}_{N_1}$	$\mathbf{c}_B\mathbf{B}^{-1}\mathbf{N}_2 - \mathbf{c}_{N_2}$	$\hat{z}$
$\mathbf{x}_B$	$\mathbf{0}$	$\mathbf{I}$	$\mathbf{B}^{-1}\mathbf{N}_1$	$\mathbf{B}^{-1}\mathbf{N}_2$	$\hat{\mathbf{b}}$

Now we try to improve the objective by investigating the possibility of modifying the nonbasic variables. From Equation (5.10) and noting that $\mathbf{c}_{N_1} - \mathbf{c}_B\mathbf{B}^{-1}\mathbf{N}_1$ and $\mathbf{c}_{N_2} - \mathbf{c}_B\mathbf{B}^{-1}\mathbf{N}_2$ give the reduced costs $c_j - z_j$ of the lower and upper bounded nonbasic variables respectively, we get

$$z = \mathbf{c}_B\mathbf{B}^{-1}\mathbf{b} - \sum_{j \in R_1} (z_j - c_j)x_j - \sum_{j \in R_2} (z_j - c_j)x_j \tag{5.11}$$

where R_1 is the set of indices of nonbasic variables at their lower bounds and R_2 is the set of indices of nonbasic variables at their upper bounds. For $j \in R_1$, if $z_j - c_j > 0$, it would be to our benefit to increase x_j from its current value of l_j. Similarly, for $j \in R_2$, if $z_j - c_j < 0$, it would be to our benefit to decrease x_j from its current value of u_j. As in the simplex method, we shall modify the value of only one nonbasic variable while all other nonbasic variables are fixed, and hence obtain a movement along an edge of the underlying polyhedral set. The index k of this nonbasic variable is determined as follows. First examine

$$\text{Maximum}\left(\underset{j\in R_1}{\text{Maximum}}\, z_j - c_j,\ \underset{j\in R_2}{\text{Maximum}}\, c_j - z_j\right)$$

If this maximum is positive, then let k be the index for which the maximum is achieved. If it corresponds to R_1, then x_k is increased from its current level of l_k, and if it corresponds to R_2, then x_k is decreased from its current level of u_k. If the maximum is ≤ 0, then $z_j - c_j \leq 0$ for all $j \in R_1$ and $z_j - c_j \geq 0$ for all $j \in R_2$. Examining Equation (5.11), this indicates that the current solution is optimal.

To summarize, given a basic feasible solution, if $z_j - c_j \leq 0$ for all nonbasic variables at their lower bounds, and if $z_j - c_j \geq 0$ for all nonbasic variables at their upper bounds, then we stop with the conclusion that the current solution is optimal. Otherwise we choose a nonbasic variable x_k according to the foregoing rule. If x_k is at its lower bound, then it is increased; otherwise it is decreased. These two cases are discussed in detail next.

Increasing x_k from its Current Level l_k

Let $x_k = l_k + \Delta_k$ where Δ_k is the increase in x_k (currently $\Delta_k = 0$). Noting that all other nonbasic variables are fixed and that the current value of $\mathbf{x}_B$ and z are respectively $\hat{\mathbf{b}}$ and $\hat{z}$, substituting $x_k = l_k + \Delta_k$ in Equations (5.9) and (5.11), we get

$$\begin{aligned}\mathbf{x}_B &= \mathbf{B}^{-1}\mathbf{b} - \mathbf{B}^{-1}\mathbf{N}_1\mathbf{l}_{N_1} - \mathbf{B}^{-1}\mathbf{N}_2\mathbf{u}_{N_2} - \mathbf{B}^{-1}\mathbf{a}_k\Delta_k \\ &= \hat{\mathbf{b}} - \mathbf{y}_k\Delta_k \qquad (5.12)\end{aligned}$$

$$\begin{aligned}z &= \mathbf{c}_B\mathbf{B}^{-1}\mathbf{b} - \sum_{j\in R_1}(z_j - c_j)l_j - \sum_{j\in R_2}(z_j - c_j)u_j - (z_k - c_k)\Delta_k \\ &= \hat{z} - (z_k - c_k)\Delta_k \qquad (5.13)\end{aligned}$$

Since $z_k - c_k > 0$ (why?), then from Equation (5.13) it is to our benefit to increase Δ_k as much as possible. However, as Δ_k increases, the basic variables are modified according to Equation (5.12). The increase in Δ_k may be blocked as follows.

1. A BASIC VARIABLE DROPS TO ITS LOWER BOUND

Denote the value of Δ_k at which a basic variable drops to its lower bound by γ_1. From Equation (5.12) we have $\mathbf{l}_B \leq \mathbf{x}_B = \hat{\mathbf{b}} - \mathbf{y}_k\Delta_k$. Therefore $\mathbf{y}_k\Delta_k \leq \hat{\mathbf{b}} - \mathbf{l}_B$. If $\mathbf{y}_k \leq \mathbf{0}$, then Δ_k can be made arbitrarily large without violating this inequality and so $\gamma_1 = \infty$ (that is, no basic variable drops to its lower bound). Otherwise γ_1 is given by the minimum ratio shown.

$$\gamma_1 = \begin{cases} \underset{1\leq i\leq m}{\text{Minimum}}\left\{\dfrac{\hat{b}_i - l_{B_i}}{y_{ik}} : y_{ik} > 0\right\} = \dfrac{\hat{b}_r - l_{B_r}}{y_{rk}} & \text{if } \mathbf{y}_k \not\leq \mathbf{0} \\ \infty \cdots\ \cdots\ \cdots\ \cdots\ \cdots\ \cdots & \text{if } \mathbf{y}_k \leq \mathbf{0}\end{cases} \qquad (5.14)$$

The basic variable that reaches its lower bound is a candidate for x_{B_r}.

2. A BASIC VARIABLE REACHES ITS UPPER BOUND

Denote the value of Δ_k at which a basic variable reaches its upper bound by γ_2. From Equation (5.12) $\mathbf{b} - \mathbf{y}_k\Delta_k = \mathbf{x}_B \le \mathbf{u}_B$ and hence $-\mathbf{y}_k\Delta_k \le \mathbf{u}_B - \hat{\mathbf{b}}$. If $\mathbf{y}_k \ge \mathbf{0}$, then Δ_k can be made arbitrarily large without violating this inequality and so $\gamma_2 = \infty$ (that is, no basic variable reaches its upper bound). Otherwise γ_2 is given by the minimum ratio shown.

$$\gamma_2 = \begin{cases} \underset{1 \le i \le m}{\text{Minimum}} \left\{ \dfrac{u_{B_i} - \hat{b}_i}{-y_{ik}} : y_{ik} < 0 \right\} = \dfrac{u_{B_r} - \hat{b}_r}{-y_{rk}} & \text{if } \mathbf{y}_k \ngeq \mathbf{0} \\ \infty & \text{if } \mathbf{y}_k \ge \mathbf{0} \end{cases} \tag{5.15}$$

The basic variable that reaches its upper bound is a candidate for x_{B_r}.

3. x_k ITSELF REACHES ITS UPPER BOUND

The value of Δ_k at which x_k reaches its upper bound u_k is obviously $u_k - l_k$.

These three cases give the maximum increase in Δ_k before being blocked by a variable or by x_k itself. Obviously Δ_k is given by

$$\Delta_k = \text{Minimum}\,(\gamma_1, \gamma_2, u_k - l_k) \tag{5.16}$$

If $\Delta_k = \infty$, then the increase in x_k is *unblocked* and by Equation (5.13) the optimal solution value is unbounded. If, on the other hand, $\Delta_k < \infty$, a new basic feasible solution is obtained where $x_k = l_k + \Delta_k$ and the basic variables are modified according to Equation (5.12).

Updating the Tableau When the Nonbasic Variable Increases

The current tableau must be updated to reflect the new basic feasible solution. If $\Delta_k = u_k - l_k$, then no change of the working basis is made and x_k is still nonbasic, except this time it is nonbasic at its upper bound. Only the RHS column is changed to reflect the new value of the objective function and the new values of the basic variables. According to Equations (5.13) and (5.12), $\hat{z}$ is replaced by $\hat{z} - (z_k - c_k)\Delta_k$ and $\hat{\mathbf{b}}$ is replaced by $\hat{\mathbf{b}} - \mathbf{y}_k\Delta_k$. On the other hand, if Δ_k is given by γ_1 or γ_2, then x_k enters the basis and x_{B_r} leaves the basis, where the index r is determined according to Equation (5.14) if $\Delta_k = \gamma_1$ or according to (5.15) if $\Delta_k = \gamma_2$. The tableau except the RHS column is updated by pivoting at y_{rk}. Note that y_{rk} may be either positive or negative. Since the right-hand-side is computed separately, this should cause no alarm. The right-hand-side column is updated according to Equations (5.13) and (5.12) except that the rth component of the new vector $\hat{\mathbf{b}}$ is replaced by $l_k + \Delta_k$ to reflect the value of x_k, which has just entered the basis.

Alternately, the right-hand-side vector can be updated directly with the rest of the tableau. This, however, requires three distinct operations (which may be performed in any order). First, we multiply the nonbasic entering column by its current value (either l_k or u_k) and add the result to the RHS vector. Next, we multiply the basic leaving column by the value it will assume (either l_{B_r} or u_{B_r}) and subtract the result from the RHS. Finally, we perform a normal pivot operation on the adjusted RHS vector.

Decreasing x_k from its Current Level u_k

This case is very similar to that of increasing x_k and is discussed only briefly. In this case $z_k - c_k < 0$ and $x_k = u_k - \Delta_k$, where $\Delta_k \geq 0$ denotes the decrease in x_k. Noting Equations (5.9) and (5.11), we get

$$\mathbf{x}_B = \hat{\mathbf{b}} + \mathbf{y}_k \Delta_k \tag{5.17}$$

$$z = \hat{z} + (z_k - c_k)\Delta_k \tag{5.18}$$

The maximum increase in Δ_k is given by Equation (5.16) where γ_1 and γ_2 are specified:

$$\gamma_1 = \begin{cases} \underset{1 \leq i \leq m}{\text{Minimum}} \left\{ \dfrac{\hat{b}_i - l_{B_i}}{-y_{ik}} : y_{ik} < 0 \right\} = \dfrac{\hat{b}_r - l_{B_r}}{-y_{rk}} & \text{if } \mathbf{y}_k \not\geq \mathbf{0} \\ \infty \cdots \cdots \cdots \cdots & \text{if } \mathbf{y}_k \geq \mathbf{0} \end{cases} \tag{5.19}$$

$$\gamma_2 = \begin{cases} \underset{1 \leq i \leq m}{\text{Minimum}} \left\{ \dfrac{u_{B_i} - \hat{b}_i}{y_{ik}} : y_{ik} > 0 \right\} = \dfrac{u_{B_r} - \hat{b}_r}{y_{rk}} & \text{if } \mathbf{y}_k \not\leq \mathbf{0} \\ \infty \cdots \cdots \cdots \cdots \cdots \cdots & \text{if } \mathbf{y}_k \leq \mathbf{0} \end{cases} \tag{5.20}$$

If $\Delta_k = \infty$, then the decrease of x_k is unblocked and by Equation (5.18) the optimal solution value is unbounded. If $\Delta_k < \infty$, then a new basic feasible solution is obtained where $x_k = u_k - \Delta_k$ and the basic variables are modified according to Equation (5.17).

Updating the Tableau When the Nonbasic Variable Decreases

If $\Delta_k = u_k - l_k$, then x_k is still nonbasic but at its lower bound. The tableau is unchanged except for the RHS column, which is updated according to Equations (5.18) and (5.17). If Δ_k is given by γ_1 or γ_2, then x_k enters the basis and x_{B_r} leaves the basis where r is determined by Equation (5.19) if $\Delta_k = \gamma_1$ and by (5.20) if $\Delta_k = \gamma_2$. The tableau except for the RHS is updated by pivoting at y_{rk}. Again y_{rk} could be either positive or negative. The RHS column is updated according to Equations (5.18) and (5.17) except that the rth component of the new vector $\hat{\mathbf{b}}$ is replaced by $u_k - \Delta_k$ to reflect the value of x_k that has just entered the basis. We may also again utilize the alternative method, described before, to update the RHS vector.

Getting Started

If no basic feasible solution is conveniently available, we may start the lower-upper bound simplex method with artificial variables. This is accomplished by (1) setting all of the original variables to one of their bounds, (2) adjusting the

RHS values accordingly, (3) multiplying rows, as necessary, by -1 to get $\hat{b}_i \geq 0$, and (4) adding artificial columns. The two-phase or the big-M method may be employed to drive the artificial variables out of the basis.

Finite Convergence: Degeneracy and Cycling

In the absence of degeneracy, at any iteration, the procedure described previously either moves from one basic feasible solution to an improved basic feasible solution, or declares unboundedness. Hence it must terminate finitely, either with an optimal solution or with an indication of unboundedness. However, in the presence of degeneracy, it is possible to perform a *degenerate pivot* in which γ_1 or γ_2 is zero. Hence the working basis changes, but we still remain at the same extreme point solution. Consequently, it is possible to *cycle*, that is, go through a sequence of consecutive degenerate pivots, and loop through the same set of working bases. Therefore one needs some cycling prevention rule as for the nonnegatively constrained case. For example, the following lexicographic rule may be used.

Suppose that for all basic feasible solutions encountered by the algorithm, the rows of $\mathbf{B}^{-1}$ that correspond to degenerate basic variables at their lower bounds are lexicographic positive, and the rows of $\mathbf{B}^{-1}$ that correspond to degenerate basic variables at their upper bounds are lexicographic negative. A basic feasible partition $[\mathbf{B}, \mathbf{N}_1, \mathbf{N}_2]$ that satisfies this property is called a *strongly feasible basic partition*. This is readily achieved to start with using artificial variables if necessary. In order to maintain strongly feasible basic partitions, the following rule can be adopted whenever $\Delta_k < \infty$. (The reader is asked to justify this rule in Exercise 5.18.) Suppose that the entering variable x_k is increased from its lower bound. If x_k blocks itself ($\Delta_k = u_k - l_k$) *and* if the resulting partition with the same basis remains strongly feasible, then no pivot is required. Otherwise (whether x_k blocks itself or not), let $\mathbf{S}_1$ be the set of basic variable indices i for which $y_{ik} < 0$ and x_i is at its upper bound when $x_k = l_k + \Delta_k$. Let $\mathbf{S}_2$ be the set of basic variable indices i for which $y_{ik} > 0$ and x_i is at its lower bound when $x_k = l_k + \Delta_k$. Letting $\mathbf{B}_i^{-1}$ denote the row of $\mathbf{B}^{-1}$ corresponding to the basic variable x_i, determine the unique index r for which $\mathbf{B}_i^{-1}/y_{ik}$, $i \in \mathbf{S}_1 \cup \mathbf{S}_2$ is lexicographically minimum, and pivot x_r out of the (working) basis. Similarly, suppose that the entering variable x_k is being decreased from its upper bound. If x_k blocks itself *and* if the resulting partition with the same basis remains strongly feasible, then no pivot is required. Otherwise, let $S_1 = \{i : x_i \text{ is basic}, y_{ik} < 0, \text{ and } x_i = l_i \text{ when } x_k = u_k - \Delta_k\}$. Let $\mathbf{S}_2 = \{i : x_i \text{ is basic}, y_{ik} > 0, \text{ and } x_i = u_i \text{ when } x_k = u_k - \Delta_k\}$. Then select x_r to exit from the basis where r is the unique index for which $\mathbf{B}_i^{-1}/y_{ik}$, $i \in \mathbf{S}_1 \cup \mathbf{S}_2$ is lexicographically maximum. Then whenever a sequence of *degenerate pivots* is performed the vector $\mathbf{c}_B\mathbf{B}^{-1}$ can be readily verified to be lexicographically decreasing, and hence cycling cannot occur. (We ask the reader to provide a proof for this in Exercise 5.18.)

In a similar manner, one may generalize Bland's cycling prevention rule to the present case. Suppose that by performing suitable transformations if necessary, we have the lower bound vector $\mathbf{l} = \mathbf{0}$. Furthermore, for convenience, assume that $0 < u_i < \infty$ for each $i = 1, \ldots, n$, and suppose that the upper bounding constraints are written as $\mathbf{x} + \mathbf{s} = \mathbf{u}$ in the problem, where $\mathbf{s} \geq \mathbf{0}$ are the slack variables. Note that given a basic feasible partition $[\mathbf{B}, \mathbf{N}_1, \mathbf{N}_2]$ for the system $\mathbf{Ax} = \mathbf{b}$, $\mathbf{0} \leq \mathbf{x} \leq \mathbf{u}$, there is an associated basis for the nonnegatively contained system $\mathbf{Ax} = \mathbf{b}$, $\mathbf{x} + \mathbf{s} = \mathbf{u}$, $\mathbf{x}$ and $\mathbf{s} \geq \mathbf{0}$, in which the $(m + n)$ basic variables are $\mathbf{x}_B$, $\mathbf{x}_{N_2}$, $\mathbf{s}_B$, and $\mathbf{s}_{N_1}$ and the $(n - m)$ nonbasic variables (at value zero) are $\mathbf{x}_{N_1}$ and $\mathbf{s}_{N_2}$, and vice versa. But Bland's rule applied to the latter system would require, as in Section 4.5, that the variables be arranged in some order, say, $x_1, \ldots, x_n, s_1, \ldots, s_n$, and ties for the entering and leaving candidates be broken by picking the first candidate from this list. The translation of this rule in the present context is now readily evident using the foregoing equivalence of partitions into sets of basic and nonbasic variables. (We leave the details to the reader in Exercise 5.19.)

Summary of the Simplex Method for Bounded Variables (Minimization Problem)

INITIALIZATION STEP

Find a starting basic feasible solution (use artificials if necessary). Let $\mathbf{x}_B$ be the basic variables and let $\mathbf{x}_{N_1}$ and $\mathbf{x}_{N_2}$ be the nonbasic variables at their lower and upper bounds respectively. Form the following tableau where $\hat{z} = \mathbf{c}_B\mathbf{B}^{-1}\mathbf{b} + (\mathbf{c}_{N_1} - \mathbf{c}_B\mathbf{B}^{-1}\mathbf{N}_1)\mathbf{l}_{N_1} + (\mathbf{c}_{N_2} - \mathbf{c}_B\mathbf{B}^{-1}\mathbf{N}_2)\mathbf{u}_{N_2}$ and $\hat{\mathbf{b}} = \mathbf{B}^{-1}\mathbf{b} - \mathbf{B}^{-1}\mathbf{N}_1\mathbf{l}_{N_1} - \mathbf{B}^{-1}\mathbf{N}_2\mathbf{u}_{N_2}$.

	z	$\mathbf{x}_B$	$\mathbf{x}_{N_1}$	$\mathbf{x}_{N_2}$	RHS
z	1	$\mathbf{0}$	$\mathbf{c}_B\mathbf{B}^{-1}\mathbf{N}_1 - \mathbf{c}_{N_1}$	$\mathbf{c}_B\mathbf{B}^{-1}\mathbf{N}_2 - \mathbf{c}_{N_2}$	$\hat{z}$
$\mathbf{x}_B$	$\mathbf{0}$	$\mathbf{I}$	$\mathbf{B}^{-1}\mathbf{N}_1$	$\mathbf{B}^{-1}\mathbf{N}_2$	$\hat{\mathbf{b}}$

MAIN STEP

1. If $z_j - c_j \leq 0$ for nonbasic variables at their lower bound and $z_j - c_j \geq 0$ for nonbasic variables at their upper bound, then the current solution is optimal. Otherwise if one of these conditions is violated for the index k, then go to step 2 if x_k is at its lower bound and step 3 if x_k is at its upper bound.
2. The variable x_k is increased from its current value of l_k to $l_k + \Delta_k$. The value of Δ_k is given by Equation by (5.16) where γ_1 and γ_2 are given by Equations (5.14) and (5.15). If $\Delta_k = \infty$, stop; the optimal solution value is unbounded. Otherwise the tableau is updated, as described previously. Repeat step 1.
3. The variable x_k is decreased from its current value of u_k to $u_k - \Delta_k$. The value of Δ_k is given by Equation (5.16) where γ_1 and γ_2 are given by

Equations (5.19) and (5.20). If $\Delta_k = \infty$, stop; the optimal solution value is unbounded. Otherwise the tableau is updated as described previously. Repeat step 1.

It will be helpful to distinguish between nonbasic variables at their lower and upper bounds during the simplex iterations. This is done by flagging the corresponding columns by l and u respectively.

Example 5.6

$$\begin{array}{lrl} \text{Minimize} & -2x_1 - 4x_2 - x_3 & \\ \text{Subject to} & 2x_1 + x_2 + x_3 & \leq 10 \\ & x_1 + x_2 - x_3 & \leq 4 \\ & 0 \leq x_1 & \leq 4 \\ & 0 \leq x_2 & \leq 6 \\ & 1 \leq x_3 & \leq 4 \end{array}$$

Introduce the slack variables x_4 and x_5. These are bounded below by 0 and bounded above by ∞. Initially, the basic variables are x_4 and x_5, and the nonbasic variables at their lower bound are $x_1 = x_2 = 0$ and $x_3 = 1$. Note that the objective is -1 and the values of the basic variables x_4 and x_5 are given by $10 - 1 = 9$ and $4 + 1 = 5$.

Iteration 1

		l	l	l			
	z	x_1	x_2	x_3	x_4	x_5	RHS
z	1	2	4	1	0	0	-1
x_4	0	2	1	1	1	0	9
x_5	0	1	(1)	-1	0	1	5

The maximum value of $z_j - c_j$ for lower bounded nonbasic variables is 4 corresponding to x_2. Therefore $x_k = x_2$ is increased. Then $\mathbf{y}_2 = \begin{bmatrix} 1 \\ 1 \end{bmatrix}$ and Δ_2 is given by Minimum$(\gamma_1, \gamma_2, u_2 - l_2) =$ Minimum$(\gamma_1, \gamma_2, 6)$. Also γ_1 and γ_2 are given according to Equations (5.14) and (5.15) as follows. First,

$$\gamma_1 = \text{Minimum}\left\{\frac{9-0}{1}, \frac{5-0}{1}\right\} = 5$$

corresponding to $x_{B_2} = x_5$, that is, $r = 2$. This means that Δ_2 can be increased to value 5 before a basic variable drops to its lower bound. Second, $\gamma_2 = \infty$, which means that Δ_2 can be increased indefinitely without any basic variable reaching its upper bound.

Therefore, $\Delta_2 = \text{Minimum}(5, \infty, 6) = 5$. The objective is replaced by $-1 - (z_2 - c_2)\Delta_2 = -1 - 4 \times 5 = -21$ and

$$\begin{bmatrix} x_4 \\ x_5 \end{bmatrix} = \begin{bmatrix} 9 \\ 5 \end{bmatrix} - \mathbf{y}_2\Delta_2 = \begin{bmatrix} 9 \\ 5 \end{bmatrix} - \begin{bmatrix} 1 \\ 1 \end{bmatrix} 5 = \begin{bmatrix} 4 \\ 0 \end{bmatrix}.$$

The value of x_2 is given by $\Delta_2 = 5$. Then x_2 enters and x_5 leaves. The tableau is updated by pivoting at y_{22}.

Iteration 2

	z	x_1 (l)	x_2	x_3 (l)	x_4	x_5 (l)	RHS
z	1	−2	0	5	0	−4	−21
x_4	0	1	0	2	1	−1	4
x_2	0	1	1	(−1)	0	1	5

All nonbasic variables are at their lower bounds and the maximum value of $z_j - c_j$ is 5, corresponding to x_3. Therefore x_3 enters, $\mathbf{y}_3 = \begin{bmatrix} 2 \\ -1 \end{bmatrix}$, and Δ_3 is given by

$$\Delta_3 = \text{Minimum}(\gamma_1, \gamma_2, u_3 - l_3) = \text{Minimum}(\gamma_1, \gamma_2, 3)$$

The values of γ_1 and γ_2 are obtained from Equations (5.14) and (5.15) as follows.

$$\gamma_1 = \frac{4 - 0}{2} = 2$$

corresponding to $x_{B_r} = x_{B_1} = x_4$, that is, $r = 1$. This means that x_4 drops to its lower limit as x_3 is increased. Second,

$$\gamma_2 = \frac{6 - 5}{1} = 1$$

corresponding to $x_{B_r} = x_{B_2} = x_2$, that is, $r = 2$. This means that x_2 reaches its upper bound as x_3 increased. Therefore $\Delta_3 = \text{Minimum}(2, 1, 3) = 1 = \gamma_2$. Now $x_3 = 1 + \Delta_3 = 2$.

The objective is replaced by $-21 - (z_3 - c_3)\Delta_3 = -21 - 5 \times 1 = -26$.

$$\begin{bmatrix} x_4 \\ x_2 \end{bmatrix} = \begin{bmatrix} 4 \\ 5 \end{bmatrix} - \mathbf{y}_3\Delta_3 = \begin{bmatrix} 4 \\ 5 \end{bmatrix} - \begin{bmatrix} 2 \\ -1 \end{bmatrix} 1 = \begin{bmatrix} 2 \\ 6 \end{bmatrix}$$

Here x_3 enters the basis and x_2 reaches its upper bound and leaves the basis. The tableau (except the RHS, which was updated separately) is updated by pivoting at $y_{r3} = y_{23} = -1$. If we had chosen to use the alternative method for updating the RHS column, we would first replace the RHS column of iteration 2 by

$$\begin{bmatrix} -21 \\ 4 \\ 5 \end{bmatrix} + 1 \begin{bmatrix} 5 \\ 2 \\ -1 \end{bmatrix} - 6 \begin{bmatrix} 0 \\ 0 \\ 1 \end{bmatrix} = \begin{bmatrix} -16 \\ 6 \\ -2 \end{bmatrix}$$

Upon pivoting at $y_{23}(=-1)$, we would obtain the right-hand-side column shown.

Iteration 3

	z	x_1 (l)	x_2 (u)	x_3	x_4	x_5 (l)	RHS
z	1	3	5	0	0	1	-26
x_4	0	(3)	2	0	1	1	2
x_3	0	-1	-1	1	0	-1	2

The maximum value of $z_j - c_j$ for nonbasic variables at their lower bound is 3, corresponding to x_1. Therefore x_1 is increased. Here $\mathbf{y}_1 = \begin{bmatrix} 3 \\ -1 \end{bmatrix}$ and Δ_1 is given by

$$\Delta_1 = \text{Minimum}\,(\gamma_1, \gamma_2, u_1 - l_1) = \text{Minimum}\,(\gamma_1, \gamma_2, 4)$$

The values of γ_1 and γ_2 are given by Equations (5.14) and (5.15) as follows. First,

$$\gamma_1 = \frac{2-0}{3} = \frac{2}{3}$$

corresponding to $x_{B_r} = x_{B_1} = x_4$, that is, $r = 1$. This means that as Δ_1 is increased to $\frac{2}{3}$, x_4 drops to its lower limit and drops from the basis. Second,

$$\gamma_2 = \frac{4-2}{1} = 2$$

corresponding to $x_{B_r} = x_{B_2} = x_3$, that is, $r = 3$. This means that as Δ_1 is increased to 2, x_3 reaches its upper limit and drops from the basis.

Therefore $\Delta_1 = \text{Minimum}(\frac{2}{3}, 2, 4) = \frac{2}{3}$. So $x_1 = \frac{2}{3}$. The objective is replaced by $-26 - (z_1 - c_1)\Delta_1 = -26 - 3 \times \frac{2}{3} = -28$.

$$\begin{bmatrix} x_4 \\ x_3 \end{bmatrix} = \begin{bmatrix} 2 \\ 2 \end{bmatrix} - \mathbf{y}_1\Delta_1 = \begin{bmatrix} 2 \\ 2 \end{bmatrix} - \begin{bmatrix} 3 \\ -1 \end{bmatrix}\frac{2}{3} = \begin{bmatrix} 0 \\ \frac{8}{3} \end{bmatrix}$$

Here x_1 enters the basis and x_4 leaves the basis.

The tableau is updated by pivoting at $y_{11} = 3$, and the RHS is updated separately where $z = -28$, $x_1 = \frac{2}{3}$, and $x_3 = \frac{8}{3}$.

Iteration 4

	z	x_1	x_2 (u)	x_3	x_4 (l)	x_5 (l)	RHS
z	1	0	3	0	-1	0	-28
x_1	0	1	$\frac{2}{3}$	0	$\frac{1}{3}$	$\frac{1}{3}$	$\frac{2}{3}$
x_3	0	0	$-\frac{1}{3}$	1	$\frac{1}{3}$	$-\frac{2}{3}$	$\frac{8}{3}$

Since $z_j - c_j \geq 0$ for nonbasic variables at their upper bound and $z_j - c_j \leq 0$ for nonbasic variables at their lower bound, then the foregoing tableau gives an optimal solution (is it unique?). The variables are given by $(x_1, x_2, x_3, x_4, x_5) = (\frac{2}{3}, 6, \frac{8}{3}, 0, 0)$ and the objective is -28.

5.3 FARKAS' LEMMA VIA THE SIMPLEX METHOD

Farkas' lemma is a key ingredient to establishing the Karush–Kuhn–Tucker optimality conditions that are both necessary and sufficient for linear programming problems. The result deals with two linear systems of equations and inequalities, exactly one of which has a solution in any instance. It turns out that the simplex method can be used to provide a constructive proof for this result. We state the lemma, give an insightful proof, and then discuss its geometric interpretation.

***Lemma* 5.1** (Farkas' Lemma)

One and only one of the following two systems has a solution.

$$\textit{System } 1: \quad \mathbf{Ax} \geq \mathbf{0} \quad \text{and} \quad \mathbf{cx} < 0$$

$$\textit{System } 2: \quad \mathbf{wA} = \mathbf{c} \quad \text{and} \quad \mathbf{w} \geq \mathbf{0}$$

where $\mathbf{A}$ is a given $m \times n$ matrix and $\mathbf{c}$ is a given n vector.

Proof. The variables in the two systems are $\mathbf{x}$ and $\mathbf{w}$ respectively. The theorem can be restated as follows. If there exists an $\mathbf{x}$ with $\mathbf{Ax} \geq \mathbf{0}$ and $\mathbf{cx} < 0$, then there is no $\mathbf{w} \geq \mathbf{0}$ with $\mathbf{wA} = \mathbf{c}$. Conversely, if there exists no $\mathbf{x}$ with $\mathbf{Ax} \geq \mathbf{0}$ and $\mathbf{cx} < 0$, then there exists a $\mathbf{w} \geq \mathbf{0}$ such that $\mathbf{wA} = \mathbf{c}$.

Suppose that System 1 has a solution $\mathbf{x}$. If System 2 also has a solution $\mathbf{w}$, then $\mathbf{cx} = \mathbf{wAx} \geq 0$, since $\mathbf{w} \geq \mathbf{0}$ and $\mathbf{Ax} \geq \mathbf{0}$. This contradicts $\mathbf{cx} < 0$; System 2 cannot have a solution.

Next suppose that System 1 does not have a solution. Consider the problem P: Minimize $\{\mathbf{cx} : \mathbf{Ax} \geq \mathbf{0}\}$. Note that zero is the optimal value for Problem P (why?). Putting P in standard form by writing $\mathbf{x} = \mathbf{x}' - \mathbf{x}''$, where $\mathbf{x}', \mathbf{x}'' \geq \mathbf{0}$, and denoting the surplus variable vector in the inequality constraints as $\mathbf{s}$, we obtain the equivalent problem:

$$\text{P}': \quad \text{Minimize}\{\mathbf{cx}' - \mathbf{cx}'' : \mathbf{Ax}' - \mathbf{Ax}'' - \mathbf{s} = \mathbf{0}, \mathbf{x}', \mathbf{x}'', \mathbf{s} \geq \mathbf{0}\}$$

Note that $\mathbf{x}' = \mathbf{x}'' = \mathbf{0}$, $\mathbf{s} = \mathbf{0}$ is an optimal extreme point solution to P′. Starting with $\mathbf{s}$ as the initial set of basic variables, we can use a cycling prevention rule to obtain an optimal basis for P′ such that "$z_j - c_j$" ≤ 0 for all the variables. (Such a basis exists by Theorem 4.1.) Let $\mathbf{w} =$ "$\mathbf{c}_B\mathbf{B}^{-1}$" be the set of simplex multipliers associated with this basis $\mathbf{B}$, say. Since "$z_j - c_j = \mathbf{wa}_j - c_j$" ≤ 0 for all the variables, we get $\mathbf{wA} - \mathbf{c} \leq \mathbf{0}$, $-\mathbf{wA} + \mathbf{c} \leq \mathbf{0}$ and $-\mathbf{w} \leq \mathbf{0}$ from the columns of the variables $\mathbf{x}'$, $\mathbf{x}''$, and $\mathbf{s}$ respectively. Hence we have constructively obtained a $\mathbf{w} \geq \mathbf{0}$ such that $\mathbf{wA} = \mathbf{c}$. The proof is complete.

Alternate Forms of Farkas' Lemma

Several variants of Lemma 5.1 exist that deal with other pairs of systems of equations and inequalities, exactly one of which has a solution in any instance. These results are collectively known as *Theorems of the Alternative*. We state one such variant as Corollary 5.1 for the sake of illustration and present another form of this result in Exercise 5.37.

***Corollary* 5.1** (Alternate Form of Farkas' Lemma)
One and only one of the following two systems has a solution.

$$\textit{System 1:}\quad \mathbf{Ay} \leq \mathbf{0}, \mathbf{y} \leq \mathbf{0} \qquad \text{and} \qquad \mathbf{cy} > 0$$

$$\textit{System 2:}\quad \mathbf{wA} \leq \mathbf{c} \qquad \text{and} \qquad \mathbf{w} \geq \mathbf{0}$$

This form can be easily deduced from the format in Lemma 5.1 by changing System 2 into equality form. Since $\mathbf{wA} \leq \mathbf{c}$ and $\mathbf{w} \geq \mathbf{0}$ is equivalent to $(\mathbf{w}, \mathbf{v})\begin{pmatrix}\mathbf{A}\\ \mathbf{I}\end{pmatrix} = \mathbf{c}$ and $(\mathbf{w}, \mathbf{v}) \geq (\mathbf{0}, \mathbf{0})$, then from Lemma 5.1, System 1 must read $\begin{pmatrix}\mathbf{A}\\ \mathbf{I}\end{pmatrix}\mathbf{x} \geq \begin{pmatrix}\mathbf{0}\\ \mathbf{0}\end{pmatrix}$ and $\mathbf{cx} < 0$, that is, $\mathbf{Ax} \geq \mathbf{0}$, $\mathbf{x} \geq \mathbf{0}$, and $\mathbf{cx} < 0$. Putting $\mathbf{y} = -\mathbf{x}$ gives the desired result.

Geometric Interpretation of Farkas' Lemma

Denote the ith row of $\mathbf{A}$ by $\mathbf{a}^i$, $i = 1, \ldots, m$, and consider System 2. Here $\mathbf{wA} = \mathbf{c}$ and $\mathbf{w} \geq \mathbf{0}$ simply means that $\mathbf{c} = \sum_{i=1}^{m} w_i \mathbf{a}^i$, $w_i \geq \mathbf{0}$ for $i = 1, \ldots, m$. In other words, System 2 has a solution if and only if $\mathbf{c}$ belongs to the cone generated by the rows of $\mathbf{A}$, namely the vectors $\mathbf{a}^1, \ldots, \mathbf{a}^m$. Hence Farkas' Lemma asserts that either $\mathbf{c}$ belongs to the cone generated by the rows $\mathbf{A}$ or it does not. In the former case, System 1 has no solution, and in the latter case, it does. For convenience in illustration, let us equivalently restate System 1 as requiring to find a solution $\mathbf{y}$ satisfying $\mathbf{Ay} \leq \mathbf{0}$ and $\mathbf{cy} > 0$. (Simply put $\mathbf{y} = -\mathbf{x}$.) Geometrically, such a vector $\mathbf{y}$ should make an angle $\geq 90°$ with each row of $\mathbf{A}$ since $\mathbf{a}^i\mathbf{y} \leq 0$ for $i = 1, \ldots, m$, and it should make an angle $< 90°$ with the vector $\mathbf{c}$ since $\mathbf{cy} > 0$. Therefore System 1 has a solution if and only if the intersection of the *cone* $\{\mathbf{y} : \mathbf{Ay} \leq \mathbf{0}\}$ and the *open halfspace* $\{\mathbf{y} : \mathbf{cy} > 0\}$ is not empty. Figure 5.2 shows a case where System 1 has a solution (with any $\mathbf{y}$ in the shaded region yielding a solution $\mathbf{x} = -\mathbf{y}$ to System 1). Figure 5.3 shows a case where System 2 has a solution. Accordingly, System 2 has no solution in Figure 5.2, and System 1 has no solution in Figure 5.3.

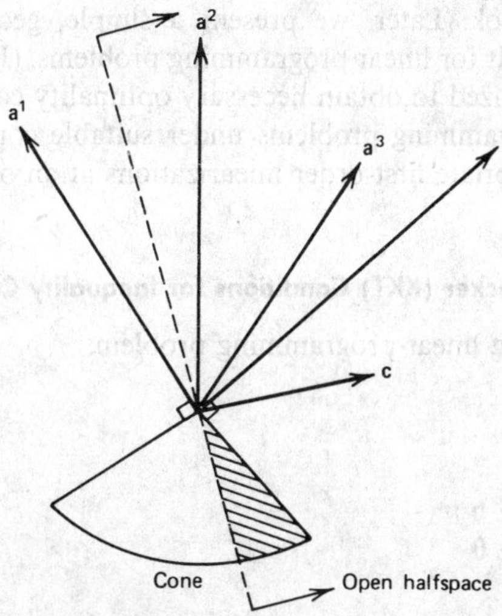

Figure 5.2. **System 1 has a solution.**

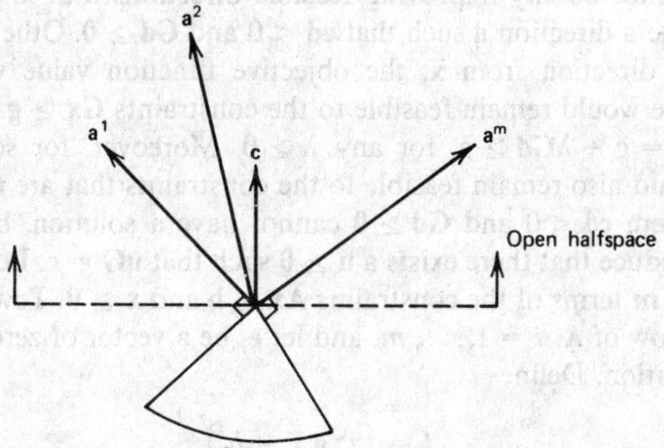

Figure 5.3. **System 2 has a solution.**

5.4 THE KARUSH – KUHN – TUCKER OPTIMALITY CONDITIONS

The Karush–Kuhn–Tucker (KKT) conditions form the backbone of nonlinear programming. These conditions are necessary for optimality for differentiable nonlinear programming problems that satisfy certain regularity conditions (known as constraint qualifications) and suffice to indicate optimality for certain differentiable nonlinear programming problems that satisfy various generalized convexity properties. For linear programming problems, these conditions are both necessary and sufficient, as we shall be seeing, and hence form an important characterization of optimality. These conditions will be widely used in the remainder of this book. Later, we present a simple, geometrically motivated derivation of this result for linear programming problems. (In fact, this derivation can be readily generalized to obtain necessary optimality conditions for differentiable nonlinear programming problems under suitable constraint qualifications by considering appropriate first-order linearizations at an optimal solution.)

The Karush – Kuhn – Tucker (KKT) Conditions for Inequality Constrained Problems

Consider the following linear programming problem.

$$\begin{array}{ll} \text{Minimize} & \mathbf{cx} \\ \text{Subject to} & \mathbf{Ax} \geq \mathbf{b} \\ & \mathbf{x} \geq \mathbf{0} \end{array}$$

where $\mathbf{c}$ is an n vector, $\mathbf{b}$ is an m vector, and $\mathbf{A}$ is an $m \times n$ matrix. Let $\bar{\mathbf{x}}$ be any feasible solution to this problem. Denote by $\mathbf{Gx} \geq \mathbf{g}$ the set of inequalities from $\mathbf{Ax} \geq \mathbf{b}$ and $\mathbf{x} \geq \mathbf{0}$ that are binding at $\bar{\mathbf{x}}$. Observe that if $\bar{\mathbf{x}}$ is an optimal solution, then there cannot be any improving feasible directions $\mathbf{d}$ at $\bar{\mathbf{x}}$. Mathematically, there cannot be a direction $\mathbf{d}$ such that $\mathbf{cd} < 0$ and $\mathbf{Gd} \geq \mathbf{0}$. Otherwise by moving along such a direction from $\bar{\mathbf{x}}$, the objective function value would fall since $\mathbf{cd} < 0$, and we would remain feasible to the constraints $\mathbf{Gx} \geq \mathbf{g}$ since $\mathbf{G}(\bar{\mathbf{x}} + \lambda\mathbf{d}) = \mathbf{G}\bar{\mathbf{x}} + \lambda\mathbf{Gd} = \mathbf{g} + \lambda\mathbf{Gd} \geq \mathbf{g}$, for any $\lambda \geq 0$. Moreover, for some step length $\lambda > 0$, we would also remain feasible to the constraints that are nonbinding at $\bar{\mathbf{x}}$. Since the system $\mathbf{cd} < 0$ and $\mathbf{Gd} \geq \mathbf{0}$ cannot have a solution, by using Farkas' Lemma, we deduce that there exists a $\mathbf{u} \geq \mathbf{0}$ such that $\mathbf{uG} = \mathbf{c}$. Let us now rewrite this statement in terms of the constraints $\mathbf{Ax} \geq \mathbf{b}$ and $\mathbf{x} \geq \mathbf{0}$. Toward this end, let $\mathbf{a}^i$ be the ith row of $\mathbf{A}$, $i = 1, \ldots, m$, and let $\mathbf{e}_j$ be a vector of zeros except for a 1 in the jth position. Define

$$I = \left\{ i : \mathbf{a}^i\bar{\mathbf{x}} = b_i \right\}$$

as the set of binding constraints at $\bar{\mathbf{x}}$ from the set $\mathbf{Ax} \geq \mathbf{b}$ and define

$$J = \left\{ j : \bar{x}_j = 0 \right\}$$

as the set of binding constraints at $\bar{\mathbf{x}}$ from the set $\mathbf{x} \geq \mathbf{0}$. Hence the rows of $\mathbf{G}$ are comprised of $\mathbf{a}^i$, $i \in I$, and $\mathbf{e}_j$, $j \in J$. Accordingly, letting $\mathbf{u} = (w_i \text{ for } i \in I,\ v_j \text{ for } j \in J) \geq 0$, we can rewrite the conditions $\mathbf{uG} = \mathbf{c}$, $\mathbf{u} \geq \mathbf{0}$ as

$$\sum_{i \in I} w_i \mathbf{a}^i + \sum_{j \in J} v_j \mathbf{e}_j = \mathbf{c} \tag{5.21}$$

$$w_i \geq 0 \text{ for } i \in I, \quad \text{and} \quad v_j \geq 0 \text{ for } j \in J \tag{5.22}$$

Equations (5.21) and (5.22) along with the feasibility of $\bar{\mathbf{x}}$ are the KKT conditions for the foregoing linear programming problem. So far, we have shown that these conditions must necessarily hold at an optimal solution $\bar{\mathbf{x}}$ to this problem. *Geometrically*, Equations (5.21) and (5.22) assert that the objective gradient $\mathbf{c}$ can be represented as a nonnegative linear combination of the gradients of the binding constraints at an optimal solution, where $\mathbf{a}^i$ is the gradient of the constraint $\mathbf{a}^i\mathbf{x} \geq b_i$ and $\mathbf{e}_j$ is the gradient of the constraint $x_j \geq 0$. In other words, Equations (5.21) and (5.22) say that the objective gradient $\mathbf{c}$ must lie in the cone generated by the gradients of the binding constraints.

Conversely, suppose that Equations (5.21) and (5.22) hold at some feasible point $\bar{\mathbf{x}}$ with I and J as defined. Consider any other feasible solution $\hat{\mathbf{x}}$ to the linear program. Then multiplying both sides of Equation (5.21) by $(\hat{\mathbf{x}} - \bar{\mathbf{x}})$ and using the fact that $\mathbf{a}^i\bar{\mathbf{x}} = b_i$ for $i \in I$ and $\mathbf{e}_j\bar{\mathbf{x}} = 0$ for $j \in J$, we obtain

$$\mathbf{c}\hat{\mathbf{x}} - \mathbf{c}\bar{\mathbf{x}} = \sum_{i \in I} w_i\left(\mathbf{a}^i\hat{\mathbf{x}} - b_i\right) + \sum_{j \in J} v_j\mathbf{e}_j\hat{\mathbf{x}} \geq 0$$

since $\mathbf{a}^i\hat{\mathbf{x}} \geq b_i$ for $i \in I$, $\mathbf{e}_j\hat{\mathbf{x}} \geq 0$ for $j \in J$, and Equation (5.22) holds true. Hence $\mathbf{c}\hat{\mathbf{x}} \geq \mathbf{c}\bar{\mathbf{x}}$ for any feasible solution $\hat{\mathbf{x}}$ (in fact, feasible simply to the inequalities binding at $\bar{\mathbf{x}}$), and so, $\bar{\mathbf{x}}$ is an optimal solution.

Consequently, we have shown that a feasible solution $\bar{\mathbf{x}}$ is optimal to the foregoing linear program if and only if Equations (5.21) and (5.22) hold, that is, *if and only if the objective gradient* $\mathbf{c}$ *lies in the cone generated by the gradients of the binding constraints at* $\bar{\mathbf{x}}$. Hence the KKT conditions given by Equations (5.21) and (5.22) along with the feasibility of $\bar{\mathbf{x}}$ are both necessary and sufficient for the optimality of $\bar{\mathbf{x}}$. These conditions are usually written in the following convenient form. Define the vectors $\mathbf{w} = (w_1, \ldots, w_m) \geq \mathbf{0}$ and $\mathbf{v} = (v_1, \ldots, v_n) \geq \mathbf{0}$. Note that w_i and v_j must be zero corresponding to the nonbinding constraints. Under this restriction, Equation (5.21) can be written as $\mathbf{wA} + \mathbf{vI} = \mathbf{c}$, or simply, $\mathbf{wA} + \mathbf{v} = \mathbf{c}$. Hence the KKT conditions can be equivalently stated as requiring a solution $(\mathbf{x}, \mathbf{w}, \mathbf{v})$ to the following system.

$$\mathbf{Ax} \geq \mathbf{b}, \qquad \mathbf{x} \geq \mathbf{0} \tag{5.23}$$

$$\mathbf{wA} + \mathbf{v} = \mathbf{c}, \qquad \mathbf{w} \geq \mathbf{0}, \mathbf{v} \geq \mathbf{0} \tag{5.24}$$

$$\mathbf{w}(\mathbf{Ax} - \mathbf{b}) = 0, \qquad \mathbf{vx} = 0 \tag{5.25}$$

The first condition (5.23) merely states that the candidate point must be feasible; that is, it must satisfy the constraints of the problem. This is usually referred to as *primal feasibility*. The second condition (5.24) is usually referred to as *dual feasibility*, since it corresponds to feasibility of a problem closely related to the original problem. This problem is called the *dual* problem and will be discussed in detail in Chapter 6. Here $\mathbf{w}$ and $\mathbf{v}$ are called the *Lagrangian multipliers* (or *dual variables*) corresponding to the constraints $\mathbf{Ax} \geq \mathbf{b}$ and $\mathbf{x} \geq \mathbf{0}$ respectively. Finally, the third condition (5.25) is usually referred to as *complementary slackness*. Since $\mathbf{w} \geq \mathbf{0}$ and $\mathbf{Ax} \geq \mathbf{b}$, then $\mathbf{w}(\mathbf{Ax} - \mathbf{b}) = 0$ if and only if either w_i is 0 or else the ith slack variable is 0, that is, the ith constraint in $\mathbf{Ax} \geq \mathbf{b}$ is binding.

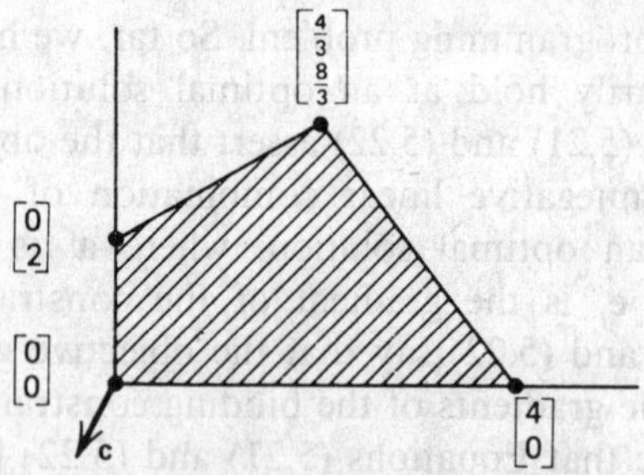

Figure 5.4. Verification of the KKT conditions.

Similarly $\mathbf{vx} = 0$ if and only if either x_j is 0 or else v_j is 0. Hence Equations (5.24) and (5.25) are equivalent to Equations (5.21) and (5.22).

The following example illustrates the Karush–Kuhn–Tucker conditions.

Example 5.7

$$\begin{array}{lrcrcr} \text{Minimize} & -x_1 & - & 3x_2 & & \\ \text{Subject to} & x_1 & - & 2x_2 & \geq & -4 \\ & -x_1 & - & x_2 & \geq & -4 \\ & x_1, & & x_2 & \geq & 0 \end{array}$$

Equations (5.23), (5.24) and (5.25) represent a useful tool in verifying whether a certain point is optimal. To illustrate, suppose we were told that the optimal solution of the foregoing problem is the point $(0, 0)$. We see geometrically in Figure 5.4 or by using the simplex method that $(0, 0)$ is not the optimal point. First, inequality (5.23) holds since $(0, 0)$ is indeed a feasible point. Since none of the first two constraints is binding (that is, $+x_1 - 2x_2 > -4$ and $-x_1 - x_2 > -4$), then $w_1 = w_2 = 0$, in order to satisfy Equation (5.25). Since $\mathbf{w} = \mathbf{0}$, then from Equation (5.24) we have to have $\mathbf{c} = \mathbf{v}$, that is, $\mathbf{v} = (-1, -3)$. This violates nonnegativity of $\mathbf{v}$, however. Therefore $(0, 0)$ could not be an optimal solution of this problem.

Now suppose that we were told the the optimal point is $(\frac{4}{3}, \frac{8}{3})$. In order to check whether this is a true statement, we can use the KKT conditions. Since $x_1, x_2 > 0$, then $v_1 = v_2 = 0$ in order to satisfy complementary slackness. From Equation (5.24), $\mathbf{w}$ must satisfy the equations $\mathbf{c} - \mathbf{wA} = \mathbf{0}$, that is,

$$w_1 - w_2 = -1$$

$$-2w_1 - w_2 = -3$$

and hence $w_1 = \frac{2}{3}$ and $w_2 = \frac{5}{3}$. Note that $\mathbf{w} \geq \mathbf{0}$ and $\mathbf{Ax} = \mathbf{b}$ and hence $\mathbf{w(Ax - b)} = 0$. Therefore conditions (5.23), (5.24), and (5.25) hold and $(\frac{4}{3}, \frac{8}{3})$ is indeed an optimal point.

The following example illustrates the geometric interpretation of the KKT conditions.

Example 5.8

$$\begin{aligned} \text{Minimize} \quad & -x_1 - 3x_2 \\ \text{Subject to} \quad & x_1 - 2x_2 \geq -4 \\ & -x_1 - x_2 \geq -4 \\ & x_1, \quad x_2 \geq 0 \end{aligned}$$

The gradients of the objective function and the constraints are given:

$$\mathbf{c} = (-1, -3)$$

$$\mathbf{a}^1 = (1, -2)$$

$$\mathbf{a}^2 = (-1, -1)$$

$$\mathbf{e}_1 = (1, 0)$$

$$\mathbf{e}_2 = (0, 1)$$

Let us consider the four extreme points of Figure 5.5.

1. The extreme point $\mathbf{x} = (0, 0)$. The binding constraints are the nonnegativity constraints. We see from Figure 5.5 that $\mathbf{c}$ does not belong to the cone of the gradients $\mathbf{e}_1$ and $\mathbf{e}_2$ of the binding constraints. Therefore $(0, 0)$ is not optimal.
2. The extreme point $\mathbf{x} = (0, 2)$. The binding constraints are $x_1 - 2x_2 \geq -4$ and $x_1 \geq 0$. Here $\mathbf{c}$ does not belong to the cone generated by the gradients $\mathbf{a}^1$ and $\mathbf{e}_1$ of the binding constraints. So $(0, 2)$ is not optimal.
3. The extreme point $(\frac{4}{3}, \frac{8}{3})$. The binding constraints are $x_1 - 2x_2 \geq -4$ and $-x_1 - x_2 \geq -4$. Here $\mathbf{c}$ belongs to the cone of the gradients $\mathbf{a}^1$ and $\mathbf{a}^2$ of the binding constraints. Therefore $(\frac{4}{3}, \frac{8}{3})$ is an optimal point.
4. The extreme point $(4, 0)$. The binding constraints are $-x_1 - x_2 \geq -4$ and $x_2 \geq 0$. Here $\mathbf{c}$ does not belong to the cone generated by the gradients $\mathbf{a}^2$ and $\mathbf{e}_2$ of the binding constraints and hence is not optimal.

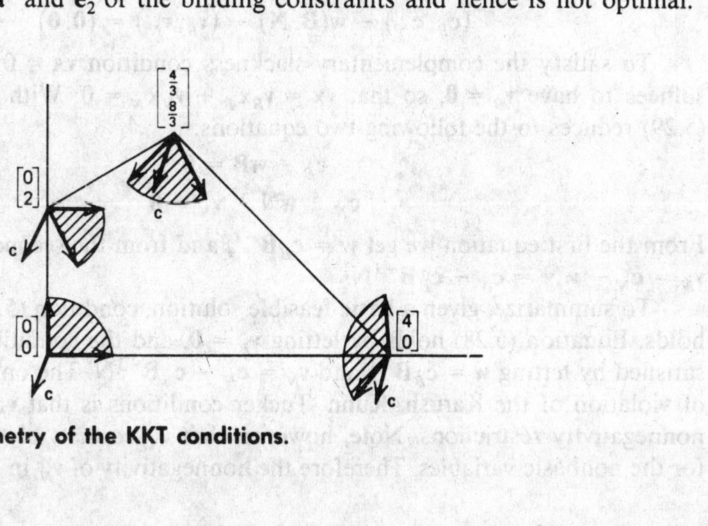

Figure 5.5. Geometry of the KKT conditions.

The Karush – Kuhn – Tucker Conditions for Equality Constraints

Consider the following linear programming problem in equality form.

$$\begin{array}{ll} \text{Minimize} & \mathbf{cx} \\ \text{Subject to} & \mathbf{Ax} = \mathbf{b} \\ & \mathbf{x} \geq \mathbf{0} \end{array}$$

By changing the equality into two inequalities of the form $\mathbf{Ax} \geq \mathbf{b}$ and $-\mathbf{Ax} \geq -\mathbf{b}$, the KKT conditions developed earlier would simplify to

$$\mathbf{Ax} = \mathbf{b}, \qquad \mathbf{x} \geq \mathbf{0} \tag{5.26}$$

$$\mathbf{wA} + \mathbf{v} = \mathbf{c}, \qquad \mathbf{w} \text{ unrestricted}, \qquad \mathbf{v} \geq \mathbf{0} \tag{5.27}$$

$$\mathbf{vx} = 0 \tag{5.28}$$

The main difference between these conditions and the conditions for the inequality problem is that the Lagrangian multiplier vector (or dual vector) $\mathbf{w}$ corresponding to the constraint $\mathbf{Ax} = \mathbf{b}$ is unrestricted in sign.

Optimality at a Basic Feasible Solution

Consider the following problem.

$$\begin{array}{ll} \text{Minimize} & \mathbf{cx} \\ \text{Subject to} & \mathbf{Ax} = \mathbf{b} \\ & \mathbf{x} \geq \mathbf{0} \end{array}$$

Assume that rank $(\mathbf{A}) = m$. Let us reinvestigate how the simplex method recognizes an optimal basic feasible solution. Suppose that we have a basic feasible solution $\mathbf{x}$ with basis $\mathbf{B}$ and let us examine conditions (5.26) (5.27), and (5.28). Obviously (5.26) holds. The condition $\mathbf{c} - \mathbf{wA} - \mathbf{v} = \mathbf{0}$ can be rewritten as follows, where $\mathbf{v}$ is decomposed into $\mathbf{v}_B$ and $\mathbf{v}_N$:

$$(\mathbf{c}_B, \mathbf{c}_N) - \mathbf{w}(\mathbf{B}, \mathbf{N}) - (\mathbf{v}_B, \mathbf{v}_N) = (\mathbf{0}, \mathbf{0}) \tag{5.29}$$

To satisfy the complementary slackness condition $\mathbf{vx} = 0$, since $\mathbf{x}_N = \mathbf{0}$, it suffices to have $\mathbf{v}_B = \mathbf{0}$, so that $\mathbf{vx} = \mathbf{v}_B\mathbf{x}_B + \mathbf{v}_N\mathbf{x}_N = 0$. With $\mathbf{v}_B = \mathbf{0}$, Equation (5.29) reduces to the following two equations:

$$\mathbf{c}_B - \mathbf{wB} = \mathbf{0}$$

$$\mathbf{c}_N - \mathbf{wN} - \mathbf{v}_N = \mathbf{0}$$

From the first equation we get $\mathbf{w} = \mathbf{c}_B\mathbf{B}^{-1}$, and from the second equation we get $\mathbf{v}_N = \mathbf{c}_N - \mathbf{wN} = \mathbf{c}_N - \mathbf{c}_B\mathbf{B}^{-1}\mathbf{N}$.

To summarize, given a basic feasible solution, condition (5.26) automatically holds. Equation (5.28) holds by letting $\mathbf{v}_B = \mathbf{0}$, and the condition $\mathbf{wA} + \mathbf{v} = \mathbf{c}$ is satisfied by letting $\mathbf{w} = \mathbf{c}_B\mathbf{B}^{-1}$ and $\mathbf{v}_N = \mathbf{c}_N - \mathbf{c}_B\mathbf{B}^{-1}\mathbf{N}$. The only possible source of violation of the Karush–Kuhn–Tucker conditions is that $\mathbf{v}_N$ may violate the nonnegativity restrictions. Note, however, that $\mathbf{v}_N$ consists of the $c_j - z_j$ values for the nonbasic variables. Therefore the nonnegativity of $\mathbf{v}_N$ in Equation (5.27) is violated if $c_j - z_j < 0$ for some nonbasic variable. Of course, if $c_j - z_j \geq 0$ for each nonbasic variable, then $\mathbf{v}_N \geq \mathbf{0}$ and all the Karush–Kuhn–Tucker conditions are met. These are precisely the simplex termination criteria.

Reinterpretation of the Simplex Method

From the previous discussion, the simplex method can be interpreted as a systematic procedure for approaching an optimal extreme point satisfying the Karush–Kuhn–Tucker conditions. At each iteration, feasibility (called primal feasibility) is satisfied, and hence condition (5.26) always holds. Also complementary slackness is always satisfied since either a variable x_j is nonbasic and has value zero, or else $v_j = c_j - z_j = 0$, and hence $v_j x_j = 0$ for all j and $\mathbf{vx} = 0$. So condition (5.28) is always satisfied during the simplex method. Condition (5.27) (called dual feasibility; more on this in Chapter 6) is partially violated during the iterations of the simplex method. Condition (5.27) has two portions, namely $\mathbf{wA} + \mathbf{v} = \mathbf{c}$ and $\mathbf{v} \geq \mathbf{0}$. The first portion always holds by letting $\mathbf{w} = \mathbf{c}_B\mathbf{B}^{-1}$ and $\mathbf{v} = (\mathbf{v}_B, \mathbf{v}_N) = (\mathbf{0}, \mathbf{c}_N - \mathbf{c}_B\mathbf{B}^{-1}\mathbf{N})$. However, the second portion, namely nonnegativity of $\mathbf{c}_N - \mathbf{c}_B\mathbf{B}^{-1}\mathbf{N}$, which is part of dual feasibility, is violated, until of course, the optimal solution is reached. To summarize, the simplex method always satisfies primal feasibility and complementary slackness. Dual feasibility is violated, and the violation is used to improve the objective function, by increasing the nonbasic variable with the most negative $c_j - z_j$, until dual feasibility is also satisfied.

Finding the Lagrangian Multipliers From the Simplex Tableau

We already know that $\mathbf{v}_B = \mathbf{0}$ and $\mathbf{v}_N = \mathbf{c}_N - \mathbf{c}_B\mathbf{B}^{-1}\mathbf{N}$. Therefore the Lagrangian multiplier v_j corresponding to the constraint $x_j \geq 0$ can be easily obtained from row 0 of the simplex tableau. More precisely, v_j is the negative of the $z_j - c_j$ entry in row 0 under the x_j column.

Now we turn to the Lagrangian vector $\mathbf{w} = \mathbf{c}_B\mathbf{B}^{-1}$ corresponding to the constraints $\mathbf{Ax} = \mathbf{b}$. The method for obtaining $\mathbf{w}$ from the simplex tableau was discussed in Chapter 3. We shall elaborate on this further. Recall that row 0 of the simplex method consists of $z_j - c_j$ for $j = 1, \ldots, n$, which is given by the vector $\mathbf{c}_B\mathbf{B}^{-1}\mathbf{A} - \mathbf{c}$. If the matrix $\mathbf{A}$ has an identity matrix as a portion of its columns, then in row 0 under these columns, we have $\mathbf{c}_B\mathbf{B}^{-1}\mathbf{I} - \hat{\mathbf{c}} = \mathbf{w} - \hat{\mathbf{c}}$, where $\hat{\mathbf{c}}$ is the part of the cost vector $\mathbf{c}$ corresponding to the identity columns in the original problem. By simply adding the vector $\hat{\mathbf{c}}$ to $\mathbf{w} - \hat{\mathbf{c}}$ in row 0, we get $\mathbf{w}$.

EXERCISES

5.1 Solve the following linear program by the revised simplex method.

$$
\begin{array}{lrcrcrcr}
\text{Maximize} & & - & 2x_2 & + & x_3 & & \\
\text{Subject to} & x_1 & - & 2x_2 & + & x_3 & \geq & -4 \\
 & x_1 & + & x_2 & + & x_3 & \leq & 9 \\
 & 2x_1 & - & x_2 & - & x_3 & \leq & 5 \\
 & x_1, & & x_2, & & x_3 & \geq & 0
\end{array}
$$

5.2 Solve the following problem using the revised simplex method.

$$\begin{array}{ll} \text{Minimize} & x_1 + 6x_2 - 7x_3 + x_4 + 5x_5 \\ \text{Subject to} & x_1 - \frac{3}{4}x_2 + 2x_3 - \frac{1}{4}x_4 = 5 \\ & - \frac{1}{4}x_2 + 3x_3 - \frac{3}{4}x_4 + x_5 = 5 \\ & x_1, \quad x_2, \quad x_3, \quad x_4, \quad x_5 \geq 0 \end{array}$$

5.3 Verify the entries in Table 5.1.

5.4 Solve Exercise 5.2 using the product form of the inverse.

5.5 Solve the following problem by the revised simplex method using the product form of the inverse.

$$\begin{array}{ll} \text{Maximize} & 3x_1 + 4x_2 + x_3 + x_4 \\ \text{Subject to} & 8x_1 + 3x_2 + 4x_3 + x_4 \leq 7 \\ & 2x_1 + 6x_2 + x_3 + 5x_4 \leq 3 \\ & x_1 + 4x_2 + 5x_3 + 2x_4 \leq 8 \\ & x_1, \quad x_2, \quad x_3, \quad x_4 \geq 0 \end{array}$$

5.6 Repeat Exercise 5.5 using an **LU**-factorization of the basis.

5.7 Apply the revised simplex method, with and without the lexicographic cycling prevention rule, to the following problem (this is the example problem of cycling in Chapter 4).

$$\begin{array}{ll} \text{Minimize} & -\frac{3}{4}x_4 + 20x_5 - \frac{1}{2}x_6 + 6x_7 \\ \text{Subject to} & x_1 \quad + \frac{1}{4}x_4 - 8x_5 - x_6 + 9x_7 = 0 \\ & x_2 \quad + \frac{1}{2}x_4 - 12x_5 - \frac{1}{2}x_6 + 3x_7 = 0 \\ & x_3 \quad + x_6 \quad = 1 \\ & x_1, x_2, x_3, \quad x_4, \quad x_5, \quad x_6, \quad x_7 \geq 0 \end{array}$$

5.8 In the revised simplex method with product form of the inverse, the number of elementary matrices increases by 1 at each iteration. If the number of elementary matrices becomes excessive, it would be necessary to reinvert **B**. Let **B** be a basis. Show how can **B** be reinverted such that $\mathbf{B}^{-1}$ is represented as the product of m elementary matrices. Illustrate by an example.

5.9 Determine the number of multiplications and additions needed per iteration of the revised simplex method using the product form of the inverse. How can we take advantage of sparsity of nonzero elements in the matrix **A**? Give a detailed comparison between the simplex method and the revised simplex method using the product form of the inverse. Repeat using an **LU**-factorization of the basis.

5.10 Use the simplex method for bounded variables to solve the following problem.

$$\begin{aligned} \text{Maximize} \quad & 2x_1 + x_2 + 3x_3 \\ \text{Subject to} \quad & 3x_1 + x_2 + x_3 \le 12 \\ & -x_1 + x_2 \le 5 \\ & x_2 + 2x_3 \le 8 \\ & 0 \le x_1 \le 3 \\ & 0 \le x_2 \le 6 \\ & 0 \le x_3 \le 4 \end{aligned}$$

5.11 Use the simplex method for bounded variables to solve the following problem.

$$\begin{aligned} \text{Minimize} \quad & x_1 + 2x_2 + 3x_3 - x_4 \\ \text{Subject to} \quad & 2x_1 - x_2 + x_3 - 2x_4 \le 6 \\ & -x_1 + 2x_2 - x_3 + x_4 \le 8 \\ & 2x_1 + x_2 - x_3 \ge 2 \\ & 0 \le x_1 \le 3 \\ & 1 \le x_2 \le 4 \\ & 0 \le x_3 \le 10 \\ & 2 \le x_4 \le 5 \end{aligned}$$

5.12 How would you use the revised simplex method to solve a bounded variables linear programming problem? Consider the bounded variables linear program

$$\begin{aligned} \text{Maximize} \quad & z = x_1 - x_2 \\ \text{Subject to} \quad & -x_1 - x_2 \le 1 \\ & 2x_1 - x_2 \le 1 \\ & -2 \le x_1 \le 1 \\ & x_2 \le 2 \end{aligned}$$

a. Draw the feasible region in the (x_1, x_2) space and identify the optimal solution. Is this degenerate? Why or why not?

b. For the extreme point $(x_1, x_2) = (1, 1)$, identify the working basis $\mathbf{B}$ and set up the revised simplex tableau by determining $\mathbf{B}^{-1}$, $\mathbf{c}_B\mathbf{B}^{-1}$, the objective function, and the basic variable values. Continue solving the problem from this basic feasible solution using the bounded variables *revised* simplex algorithm.

c. Denote the right-hand-sides of the first two constraints by b_1 and b_2. What conditions must b_1 and b_2 satisfy for the optimal partition to remain optimal?

5.13 Consider the problem: Minimize $\mathbf{cx}$ subject to $\mathbf{Ax} = \mathbf{b}$, $\mathbf{l} \leq \mathbf{x} \leq \mathbf{u}$. Show in detail that the collection of extreme points and the collection of basic feasible solutions as defined in Section 5.2 are nonempty and equal, given that the problem is feasible and that $\mathbf{l}$ is finite.

5.14 For the bounded variables linear program, what is the relationship between basic feasible partitions $[\mathbf{B}, \mathbf{N}_1, \mathbf{N}_2]$ and extreme points? What is the relationship between adjacent basic feasible partitions and adjacent extreme points, both for the nondegenerate and the degenerate cases?

5.15 Consider the problem: Minimize $\mathbf{cx}$ subject to $\mathbf{Ax} = \mathbf{b}$, $\mathbf{0} \leq \mathbf{x} \leq \mathbf{u}$. Show that the basic feasible solutions defined in Section 5.2, and the basic feasible solutions that would be obtained if the constraint $\mathbf{x} \leq \mathbf{u}$ is transformed into $\mathbf{x} + \mathbf{x}_s = \mathbf{u}$ and $\mathbf{x}_s \geq \mathbf{0}$, are equivalent.

5.16 Solve the following problem by the simplex method for bounded variables.

$$\begin{array}{ll} \text{Maximize} & 2x_1 + 3x_2 - 2x_3 \\ \text{Subject to} & x_1 + 2x_2 + x_3 \leq 8 \\ & 2x_1 + x_2 - x_3 \geq 3 \\ & x_1 \leq 4 \\ & -2 \leq x_2 \leq 3 \\ & x_3 \geq 2 \end{array}$$

5.17 Compare the simplex method of Chapter 3 with the lower-upper bound simplex method. Indicate the number of operations (additions, subtractions, and so on) when each of the two methods is applied to the same lower-upper bounded linear program.

5.18 Show in detail that the simplex method for bounded variables discussed in Section 5.2 converges in a finite number of steps in the absence of degeneracy. Discuss in detail the problem of degeneracy and prove that the lexicographic rule given in Section 5.2 maintains strongly feasible basic partitions and prevents cycling.

5.19 Using the characterization in Exercise 5.15 for the problem: Minimize $\mathbf{cx}$ subject to $\mathbf{Ax} = \mathbf{b}$, $\mathbf{0} \leq \mathbf{x} \leq \mathbf{u}$, and the discussion on Bland's cycling prevention rule in Section 4.5, devise a version of Bland's rule for this bounded variables linear program.

5.20 Illustrate the lexicographic as well as Bland's cycling prevention rules on the following bounded variables linear program.

$$\begin{array}{ll} \text{Maximize} & z = 2x_1 + x_2 \\ \text{Subject to} & -2x_1 + x_2 \leq 0 \\ & 2x_1 + 3x_2 \leq 12 \\ & x_1 - x_2 \leq 1 \\ & 0 \leq x_1 \leq 1, 0 \leq x_2 \leq 3 \end{array}$$

5.21 a. Solve the following (knapsack) problem.

$$\begin{array}{ll} \text{Maximize} & 2x_1 + 3x_2 + 8x_3 + x_4 + 5x_5 \\ \text{Subject to} & 3x_1 + 7x_2 + 12x_3 + 2x_4 + 7x_5 \leq 10 \\ & x_1, \quad x_2, \quad x_3, \quad x_4, \quad x_5 \geq 0 \end{array}$$

b. Give a generalized closed form solution for the following problem.

$$\begin{array}{ll} \text{Maximize} & c_1x_1 + \cdots + c_nx_n \\ \text{Subject to} & a_1x_1 + \cdots + a_nx_n \leq b \\ & x_1, \quad \cdots, \quad x_n \geq 0 \end{array}$$

where c_j and a_j are positive scalars for each j.

c. What is the form of the optimal solution if c_j and a_j are allowed to be any scalars in part (b)?

d. Extend parts (b) and (c) to include upper bounds on the variables.

5.22 Let $Q = \{1, 2, \ldots, n\}$, $P_i \subset Q$ with $P_i \cap P_j = \Phi$ for $i, j = 1, 2, \ldots, r$ and $i \neq j$, and

$$\bigcup_{i=1}^{r} P_i = Q$$

Develop an efficient method to solve the following problem where $c_j \geq 0$ for each j.

$$\begin{array}{lll} \text{Maximize} & \sum_{j \in Q} c_j x_j & \\ \text{Subject to} & b_0' \leq \sum_{j \in Q} x_j \leq b_0'' & \\ & b_i' \leq \sum_{j \in P_i} x_j \leq b_i'' & i = 1, \ldots, r \\ & 0 \leq x_j \leq u_j & j \in Q \end{array}$$

Apply the method to the following problem.

$$\begin{array}{ll} \text{Maximize} & 10x_1 + 6x_2 + 3x_3 + 5x_4 + 8x_5 \\ \text{Subject to} & 30 \leq x_1 + x_2 + x_3 + x_4 + x_5 \leq 100 \\ & 2 \leq x_1 + x_2 \leq 50 \\ & 70 \leq x_3 + x_4 + x_5 \leq 80 \\ & 0 \leq x_1, x_4, x_5 \leq 30 \\ & 0 \leq x_2, x_3 \leq 25 \end{array}$$

5.23 Solve the following problem by the simplex method for bounded variables.

$$\begin{array}{ll} \text{Minimize} & 2x_1 + 6x_2 - x_3 - 4x_4 + x_5 \\ \text{Subject to} & 2x_1 + x_2 + 4x_3 + x_4 + x_5 = 10 \\ & 3x_1 + 8x_2 - 3x_3 + x_4 \quad = 7 \\ & 0 \le x_1 \le 3 \\ & 1 \le x_2 \le 4 \\ & 0 \le x_3 \le 8 \\ & 1 \le x_4 \le 2 \\ & 0 \le x_5 \le 20 \end{array}$$

5.24 Consider the following problem.

$$\begin{array}{ll} \text{Minimize} & x_1 + 3x_2 + 4x_3 \\ \text{Subject to} & -x_1 - 2x_2 - x_3 \le -12 \\ & -x_1 - x_2 + 2x_3 \le -6 \\ & +2x_1 - x_2 - 4x_3 \le -24 \\ & x_1, \quad x_2, \quad x_3 \ge 0 \end{array}$$

Let x_a be an artificial variable with an activity vector $\hat{\mathbf{b}} \le \mathbf{b}$. Introducing the restrictions $0 \le x_a \le 1$ and letting $x_a = 1$ would lead to a starting basic feasible solution of the new system. Use the bounded simplex method to find a basic feasible solution of the original system.

5.25 Solve the following problem by the simplex method for bounded variables.

$$\begin{array}{ll} \text{Maximize} & 6x_1 + 4x_2 + 2x_3 \\ \text{Subject to} & 4x_1 - 3x_2 + x_3 \le 8 \\ & 3x_1 + 2x_2 + 4x_3 \le 10 \\ & 0 \le x_1 \le 3 \\ & 0 \le x_2 \le 2 \\ & 0 \le x_3 \end{array}$$

5.26 Solve the following problem by the simplex method for bounded variables.

$$\begin{array}{ll} \text{Minimize} & 6x_1 + 2x_2 \\ \text{Subject to} & x_1 + x_2 \ge 4 \\ & x_1 - 2x_2 \le 0 \\ & x_1 \ge 2 \\ & x_2 \ge 1 \end{array}$$

5.27 Solve the following problem by the simplex method for bounded variables. Begin with x_1 at its lower bound and x_2 at its upper bound.

$$\begin{array}{ll} \text{Maximize} & 6x_1 + 4x_2 \\ \text{Subject to} & x_1 + 2x_2 \leq 8 \\ & 2x_1 + 2x_2 \leq 10 \\ & 0 \leq x_1 \leq 3 \\ & 0 \leq x_2 \leq 4 \end{array}$$

5.28 Show that the following two problems are equivalent.

$$\begin{array}{ll} \mathrm{P}_1\text{: Minimize} & \mathbf{cx} \\ \text{Subject to} & \mathbf{b}_1 \leq \mathbf{Ax} \leq \mathbf{b}_2 \\ & \mathbf{x} \geq \mathbf{0} \end{array} \qquad \begin{array}{ll} \mathrm{P}_2\text{: Minimize} & \mathbf{cx} \\ \text{Subject to} & \mathbf{Ax} + \mathbf{s} = \mathbf{b}_2 \\ & \mathbf{x} \geq \mathbf{0} \\ & \mathbf{0} \leq \mathbf{s} \leq \mathbf{b}_2 - \mathbf{b}_1 \end{array}$$

Use the lower-upper bound simplex method to solve the following.

$$\begin{array}{ll} \text{Minimize} & 3x_1 - 4x_2 \\ \text{Subject to} & 3 \leq x_1 + x_2 \leq 5 \\ & -15 \leq 2x_1 - 5x_2 \leq 2 \\ & x_1, \quad x_2 \geq 0 \end{array}$$

5.29 A government has allocated \$1.5 billion of its budget for military purposes. Sixty percent of the military budget will be used to purchase tanks, planes, and missile batteries. These can be acquired at a unit cost of \$600,000, \$2 million, and \$800,000 respectively. It is decided that at least 200 tanks and 200 planes must be acquired. Because of the shortage of experienced pilots, it is also decided not to purchase more than 300 planes. For strategic purposes the ratio of the missile batteries to the planes purchased must fall in the range from $\frac{1}{4}$ to $\frac{1}{2}$. The objective is to maximize the overall utility of these weapons where the individual utilities are given as 1, 3, and 2 respectively. Find the optimal solution.

5.30 Solve Exercise 1.14.

5.31 A farmer who raises chickens would like to determine the amounts of the available ingredients that will meet certain nutritional requirements. The available ingredients and their cost, the nutrients in the ingredients, and the daily requirements are summarized.

NUTRIENT	INGREDIENT CORN	LIME	ALFALFA	MINIMUM DAILY REQUIREMENT
Protein	8	4	4	10
Carbohydrates	4	2	4	6
Vitamins	2	3	4	5
\$cost	0.10	0.06	0.04	

Find the optimal mix using the revised simplex method with the product form of the inverse. Use only one artificial variable.

5.32 An automobile manufacturer has contracted to export 400 cars of model A and 500 cars of model B overseas. The model A car occupies a volume of 12 cubic meters, and the model B car occupies a volume of 15 cubic meters. Three ships for transporting the automobiles are available. These arrive at the port of destination at the beginning of January, the middle of February, and the end of March respectively. The first ship only transports model A cars at \$450 per automobile. The second and third ships transport both types at a cost of \$35 and \$40 per cubic meter respectively. The first ship can only accommodate 200 cars, and the second and third ships have available volumes of 4500 and 6000 cubic meters. If the manufacturer has contracted to deliver at least 250 and 200 cars of model A and B by the middle of February and the remainder by the end of March, what is the shipping pattern that minimizes the total cost? Use the revised simplex method.

5.33 A manufacturing firm would like to plan its production/inventory policy for the months of August, September, October, and November. The product under consideration is seasonal, and its demand over the particular months is estimated to be 500, 600, 800, and 1200 units respectively. Presently the monthly production capacity is 600 units with a unit cost of \$25. Management has decided to install a new production system with monthly capacity of 1100 units at a unit cost of \$30. However, the new system cannot be installed until the middle of November. Assume that the starting inventory is 250 units and that at most 400 units can be stored during any given month. If the holding inventory cost per month per item is \$3, find the production schedule that minimizes the total production and inventory cost using the bounded simplex method. Assume that demand must be satisfied and that 100 units are required in inventory at the end of November.

5.34 Consider the problem: Minimize $\mathbf{cx}$ subject to $\mathbf{Ax} = \mathbf{b}$, $\mathbf{x} \geq \mathbf{0}$. Let $\mathbf{B}$ be a basis. After adding the redundant constraints $\mathbf{x}_N - \mathbf{x}_N = \mathbf{0}$, the following equations represent all the variables in terms of the independent variables $\mathbf{x}_N$:

	$\mathbf{x}_N$	RHS
z	$\mathbf{c}_B\mathbf{B}^{-1}\mathbf{N} - \mathbf{c}_N$	$\mathbf{c}_B\bar{\mathbf{b}}$
$\mathbf{x}_B$	$\mathbf{B}^{-1}\mathbf{N}$	$\bar{\mathbf{b}}$
$\mathbf{x}_N$	$-\mathbf{I}$	$\mathbf{0}$

The simplex method proceeds by choosing the most positive $z_j - c_j$, say, $z_k - c_k$. Then x_k enters the basis and the usual minimum ratio test indicates that x_{B_r} leaves the basis. The foregoing array can be updated by *column pivoting* at y_{rk} as follows.

1. Divide the kth column (pivot column) by $-y_{rk}$.
2. Multiply the kth column (pivot column) by y_{rj} and add to the jth column.
3. Multiply the kth column by $\bar{b}_r$ and add to the right-hand-side.
4. Remove the variable x_k from the list of nonbasic variables and add x_{B_r} instead in its place. Note that no row designations are changed.

This method of displaying the tableau and updating it is usually called the *column simplex method*. Show that pivoting gives the representation of all the variables in terms of the new nonbasic variables. In particular, show that pivoting updates the tableau such that the new $\mathbf{c}_B\mathbf{B}^{-1}\mathbf{N} - \mathbf{c}_N$, $\mathbf{B}^{-1}\mathbf{N}$, $\mathbf{B}^{-1}\mathbf{b}$, and $\mathbf{c}_B\mathbf{B}^{-1}\mathbf{b}$ are immediately available.

5.35 Referring to Problem 5.34, solve the following problem using the column simplex method.

$$\begin{array}{ll} \text{Maximize} & x_1 + 2x_2 + 3x_3 \\ \text{Subject to} & 3x_1 + 2x_2 + 2x_3 \le 6 \\ & -x_1 + 2x_2 + 4x_3 \le 8 \\ & 2x_1 + x_2 - x_3 \le 2 \\ & x_1, \quad x_2, \quad x_3 \ge 0 \end{array}$$

5.36 Referring to Problem 5.34, is it possible to extract the inverse basis from a typical column simplex tableau? If so, how can this be done?

5.37 Suppose that the following system has no solution.

$$\mathbf{Ax} = \mathbf{0}, \qquad \mathbf{x} \ge \mathbf{0}, \qquad \text{and} \quad \mathbf{cx} > 0$$

Devise another system that must have a solution. (*Hint*. Use Farkas' lemma.)

5.38 Let $\mathbf{A} = \begin{bmatrix} 1 & 1 \\ 0 & 2 \\ -1 & 4 \end{bmatrix}$ and $\mathbf{c} = (1, 4)$. Which of the following two systems has a solution?

$$\text{System 1:} \quad \mathbf{Ax} \le \mathbf{0}, \quad \mathbf{cx} > 0$$

$$\text{System 2:} \quad \mathbf{wA} = \mathbf{c}, \quad \mathbf{w} \ge \mathbf{0}$$

Illustrate geometrically.

5.39 Write the KKT optimality conditions for each of the following problems.

a. Maximize $\mathbf{cx}$
   Subject to $\mathbf{Ax} \le \mathbf{b}$
   $\mathbf{x} \ge \mathbf{0}$

b. Maximize $\mathbf{cx}$
   Subject to $\mathbf{Ax} \ge \mathbf{b}$
   $\mathbf{x} \ge \mathbf{0}$

c. Minimize $\mathbf{cx}$
   Subject to $\mathbf{Ax} \le \mathbf{b}$
   $\mathbf{x} \ge \mathbf{0}$

d. Minimize $\mathbf{cx}$
Subject to $\mathbf{A}_1\mathbf{x} = \mathbf{b}_1$
$\mathbf{A}_2\mathbf{x} \geq \mathbf{b}_2$
$\mathbf{x} \geq \mathbf{0}$

e. Minimize $\mathbf{cx}$
Subject to $\mathbf{Ax} = \mathbf{b}$
$\mathbf{l} \leq \mathbf{x} \leq \mathbf{u}$

5.40 Consider the problem: Minimize $\mathbf{cx}$ subject to $\mathbf{Ax} = \mathbf{b}$, $\mathbf{x} \geq \mathbf{0}$. Suppose that $\mathbf{x}$ is an optimal solution. Further suppose that $x_j > 0$ for $j = 1, 2, \ldots, p$, and $x_j = 0$ for $j = p + 1, \ldots, n$. Show that the system $\mathbf{Ad} = \mathbf{0}$, $d_{p+1}, \ldots, d_n \geq 0$, and $\mathbf{cd} < 0$ has no solution $\mathbf{d}$ in E^n. Hence use Farkas' lemma to derive the KKT conditions for this problem. Also, *directly* show that any solution to these KKT conditions is optimal for this problem.

5.41 Consider the problem: Maximize $\mathbf{cx}$ subject to $\mathbf{Ax} \leq \mathbf{b}$, $\mathbf{x} \geq \mathbf{0}$. Introducing the slack vector $\mathbf{x}_s$, we get the equivalent problem: Maximize $\mathbf{cx}$ subject to $\mathbf{Ax} + \mathbf{x}_s = \mathbf{b}$, $\mathbf{x} \geq \mathbf{0}$, $\mathbf{x}_s \geq \mathbf{0}$. Write the Karush–Kuhn–Tucker optimality conditions for both problems. Show equivalence of the optimality conditions, and show that the Lagrangian multipliers corresponding to the nonnegativity $\mathbf{x}_s \geq \mathbf{0}$ are equal to the Lagrangian multipliers of the constraints $\mathbf{Ax} + \mathbf{x}_s = \mathbf{b}$ and $\mathbf{Ax} \leq \mathbf{b}$.

5.42 Consider the following problem.

$$
\begin{array}{lrl}
\text{Maximize} & x_1 + 2x_2 & \\
\text{Subject to} & -x_1 + x_2 & \leq 1 \\
& x_2 & \leq 2 \\
& x_1 + x_2 & \leq 3 \\
& x_1, \quad x_2 & \geq 0
\end{array}
$$

Starting at the origin, solve the problem by the simplex method, and show that the solution satisfies the KKT conditions. At each iteration, point out the source of violation of the optimality conditions. Interpret each iteration geometrically.

5.43 Consider the linear programming problem: Minimize $\mathbf{cx}$ subject to $\mathbf{Ax} = \mathbf{b}$, $\mathbf{x} \geq \mathbf{0}$. It is well known that a feasible point $\mathbf{x}$ is an optimal solution if and only if there exist vectors $\mathbf{w}$ and $\mathbf{v}$ such that

$$
\begin{aligned}
\mathbf{c} - \mathbf{wA} - \mathbf{v} &= \mathbf{0} \\
\mathbf{v} &\geq \mathbf{0} \\
\mathbf{vx} &= 0
\end{aligned}
$$

Is it possible that $\mathbf{x}$ be optimal if $v_i \geq 0$ for all $i \neq j$ whereas $v_j < 0$ and the corresponding $x_j = 0$? In other words, is it possible to have an optimal solution with one of the Lagrangian multipliers (shadow prices) of the nonnegativity constraints being negative? Explain why or why not. Illustrate by a numerical example. (*Hint*. Construct a linear program with an optimal degenerate basic feasible solution. Try different basis representations. See Exercise 3.18b).

5.44 Consider the problem: Minimize $\mathbf{cx}$ subject to $\mathbf{Ax} \geq \mathbf{b}$, $\mathbf{x} \geq \mathbf{0}$. Let $\mathbf{x}^*$ be an optimal solution. Suppose that $\mathbf{A}$ is decomposed into $\begin{bmatrix} \mathbf{A}_1 \\ \mathbf{A}_2 \end{bmatrix}$ and $\mathbf{b}$ is decomposed into $\begin{bmatrix} \mathbf{b}_1 \\ \mathbf{b}_2 \end{bmatrix}$ such that $\mathbf{A}_1\mathbf{x}^* = \mathbf{b}_1$ and $\mathbf{A}_2\mathbf{x}^* > \mathbf{b}_2$. Show that $\mathbf{x}^*$ is also an optimal solution of the problem: Minimize $\mathbf{cx}$ subject to $\mathbf{A}_1\mathbf{x} \geq \mathbf{b}_1$, $\mathbf{x} \geq \mathbf{0}$, and to the problem: Minimize $\mathbf{cx}$ subject to $\mathbf{A}_1\mathbf{x} = \mathbf{b}_1$, $\mathbf{x} \geq \mathbf{0}$.

5.45 A manufacturer produces two items with unit profits \$10.00 and \$15.00. Each unit of item 1 uses 4 man-hours and 3 machine-hours. Each unit of item 2 uses 7 man-hours and 6 machine-hours. If the total man-hours and machine-hours available are 300 and 500 respectively, find the optimal solution and verify optimality by the KKT conditions. Interpret the optimality conditions geometrically. Give an economic interpretation of the KKT conditions at the optimal point. (*Hint*. Recall the economic interpretation of w_1 and w_2.)

5.46 Convert the KKT conditions for the linear programming problem: Minimize $\{\mathbf{cx} : \mathbf{Ax} \geq \mathbf{b}, \mathbf{x} \geq \mathbf{0}\}$ to equivalently requiring a solution to a system of linear inequalities. Is there a computational advantage in such a transformation?

5.47 Consider the problem: Minimize $\mathbf{cx}$ subject to $\mathbf{Ax} = \mathbf{b}$, $\mathbf{x} \geq \mathbf{0}$. Given a feasible point $\mathbf{x}$, formulate a problem that will find a feasible direction which best improves the objective function for a unit step length, if one exists. How does this problem relate to the original problem?

5.48 Consider the following problem.

$$\begin{aligned} \text{Maximize} \quad & x_1 - 2x_2 + x_3 \\ \text{Subject to} \quad & x_1 + x_2 + x_3 \leq 6 \\ & 2x_1 + x_2 \leq 4 \\ & -x_1 + 2x_2 - x_3 \leq 4 \\ & x_1, \quad x_2, \quad x_3 \geq 0 \end{aligned}$$

Solve the problem by the simplex method, illustrating at each iteration any source of violation of the KKT conditions.

5.49 Consider the following problem.

$$\begin{aligned} \text{Maximize} \quad & 2x_1 + 3x_2 \\ \text{Subject to} \quad & x_1 + x_2 \leq 8 \\ & -2x_1 + 3x_2 \leq 12 \\ & x_1, \quad x_2 \geq 0 \end{aligned}$$

a. Solve the problem geometrically. Starting at the origin, at each iteration identify the variable that enters and the variable that leaves the basis.
b. Write the KKT conditions and show that they hold at optimality.

5.50 The following is an idea of a graphical example of the simplex method at work.

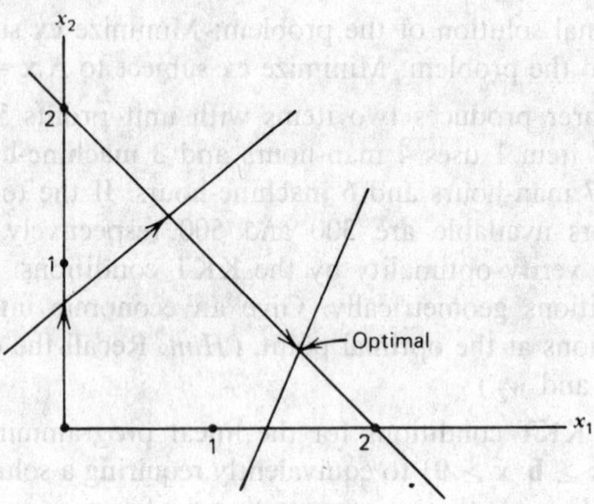

a. Give the starting basis and each succeeding basis until the optimal point is reached. Specify the entering and leaving vectors.
b. If the optimal point is unique, could the simplex method have gone in the direction it did assuming that the entering variable is that with the most positive $z_j - c_j$?

5.51 Consider the linear programming problem: Minimize $\mathbf{cx}$ subject to $\mathbf{Ax} = \mathbf{b}$, $\mathbf{x} \geq \mathbf{0}$. Let $\mathbf{x}$ be a basic feasible solution with basis $\mathbf{B}$. The simplex method proceeds by increasing the nonbasic variable with the most positive $z_j - c_j$.
a. Devise a procedure in which all nonbasic variables with positive $z_j - c_j$'s are increased. How are the basic variables modified? By how much would you increase the nonbasic variables? Interpret increasing several variables simultaneously.
b. At later iterations we may have more than m positive variables. How would you represent the corresponding point in a tableau format? (*Hint.* Let the basis consist of the largest m variables. Other nonbasic variables may be either positive or zero.)
c. At later iterations you may have a positive nonbasic variable with negative $z_j - c_j$. What happens if you decrease x_j?
d. Use the ideas of parts (a), (b), and (c) to construct a complete algorithm for solving linear programs where several nonbasic variables are simultaneously modified. How do you recognize reaching the optimal solution? What are the advantages and disadvantages of your procedure?
e. Illustrate your procedure by solving the following problem.

$$\begin{array}{lrl} \text{Minimize} & -x_1 - 2x_2 - x_3 & \\ \text{Subject to} & x_1 + x_2 + 3x_3 & \leq 12 \\ & x_1 + 2x_2 & \leq 6 \\ & x_1 + x_3 & \leq 8 \\ & x_1, \quad x_2, \quad x_3 & \geq 0 \end{array}$$

f. Consider the following alternative procedure for modifying the nonbasic variables. For each nonbasic variable x_j let

$$d_j = \begin{cases} z_j - c_j & \text{if } x_j > 0 \\ \text{Maximum}\,(0, z_j - c_j) & \text{if } x_j = 0 \end{cases}$$

Modify the nonbasic variables according to the d_j's and the basic variables according to the relationship

$$\mathbf{x}_B = \hat{\mathbf{b}} - \lambda \sum_{j \in R} \mathbf{y}_j d_j$$

where the vector $\hat{\mathbf{b}}$ represents the current values of the basic variables and $\lambda \geq 0$ is to be determined. Interpret this method and compare it with the method in part (d). Solve the problem in part (e) by this procedure.

5.52 The accompanying diagram depicts the region given by $a_1x_2 + a_2x_2 \leq b$ and $x_1, x_2 \geq 0$. Let (x_1, x_2) be the shown point. The distance from the x_1 and x_2 axes give the values x_1 and x_2, respectively. Indicate on the diagram the value of the slack variable. How can you generalize the result to n variables?

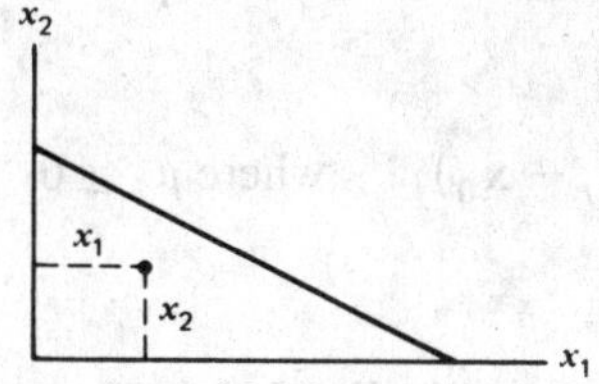

5.53 Suppose that we are given an extreme point $\mathbf{x}$ of a polyhedral set X. Consider the following alternative definition to that given in Section 2.6. An extreme point $\mathbf{y} \neq \mathbf{x}$ is called *adjacent* to $\mathbf{x}$ if there exists a hyperplane that supports X and its intersection with X is the line segment joining $\mathbf{x}$ and $\mathbf{y}$. In the accompanying diagram, obviously $\mathbf{x}_2$ and $\mathbf{x}_5$ are extreme points of X that are adjacent to $\mathbf{x}_1$, whereas $\mathbf{x}_3$ and $\mathbf{x}_4$ are not adjacent to $\mathbf{x}_1$. Now let X consist of all points satisfying $\mathbf{Ax} = \mathbf{b}$ and $\mathbf{x} \geq \mathbf{0}$, where $\mathbf{A}$ is a $m \times n$ matrix with rank m. Further suppose that X is bounded. Let $\mathbf{x}$ be a nondegenerate basic feasible solution (extreme point). Characterize the collection of adjacent extreme points. What is their number? What happens if the nondegeneracy assumption is relaxed? In each case justify your answer. Also, formally establish the equivalence of this definition with that in Section 2.6.

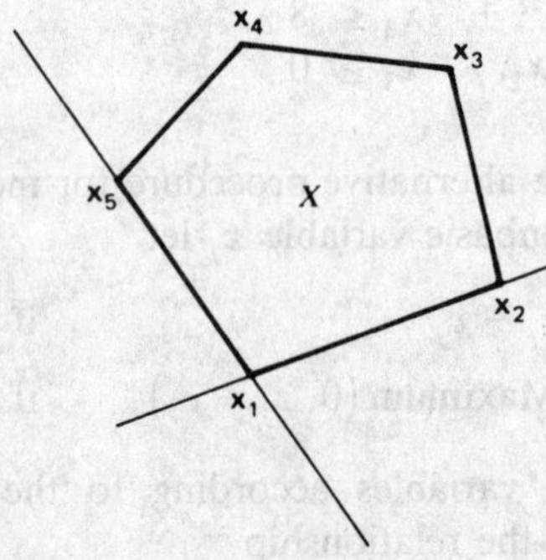

5.54 Referring to Exercise 5.53, show that the simplex method moves from an extreme point to an adjacent extreme point in a nondegenerate pivot. (*Hint*. Suppose you have a basic feasible solution $\mathbf{x}$ with basis $\mathbf{B}$ consisting of the first m columns of $\mathbf{A}$. Further suppose that x_k entered the basis. Consider the hyperplane passing through $\mathbf{x}$ and whose normal vector is $(\mathbf{p}, \mathbf{p}\mathbf{B}^{-1}\mathbf{N} + (1, 1, \ldots, 1, 0, 1, \ldots, 1))$, where $\mathbf{p}$ is an arbitrary m vector and the zero component of $(1, 1, \ldots, 1, 0, 1, 1, \ldots, 1)$ appears at position $k - m$.)

5.55 Consider the collection of points satisfying $\mathbf{x} \geq \mathbf{0}$ and $\mathbf{Ax} = \mathbf{b}$, where $\mathbf{A}$ is an $m \times n$ matrix with rank m. Further suppose that the region is bounded. Let $\mathbf{x}_0$ be an extreme point of the region, and let $\mathbf{x}_1, \mathbf{x}_2, \ldots, \mathbf{x}_k$ be its adjacent extreme points in the region (refer to Exercise 5.53). Let $\mathbf{x}$ be any point in the region. Show that $\mathbf{x}$ can be represented as

$$\mathbf{x} = \mathbf{x}_0 + \sum_{j=1}^{k} \mu_j(\mathbf{x}_j - \mathbf{x}_0) \qquad \text{where } \mu_j \geq 0 \qquad \text{for } j = 1, 2, \ldots, k$$

Interpret this result geometrically. [*Hint*. Let $\mathbf{x}_0 = \begin{bmatrix} \mathbf{0} \\ \mathbf{B}^{-1}\mathbf{b} \end{bmatrix} = \begin{bmatrix} \mathbf{0} \\ \bar{\mathbf{b}} \end{bmatrix}$ (the nonbasic variables are placed first for convenience). Show that

$$\mathbf{x}_j = \begin{bmatrix} 0 \\ \vdots \\ \lambda_j \\ \vdots \\ 0 \\ \hline \bar{\mathbf{b}} - \lambda_j \mathbf{y}_j \end{bmatrix}$$

for $j = 1, 2, \ldots, k = n - m$ where $\mathbf{y}_j = \mathbf{B}^{-1}\mathbf{a}_j$ and

$$\lambda_j = \underset{1 \le i \le m}{\text{Minimum}} \left\{ \frac{\bar{b}_i}{y_{ij}} : y_{ij} > 0 \right\} \Bigg].$$

5.56 Suppose that the boundedness restriction in Exercise 5.55 is dropped. Can you generalize the foregoing result? Interpret your result geometrically. (*Hint*. Introduce the notion of an adjacent direction. Then $\mathbf{x}$ in the region can be represented as

$$\mathbf{x} = \mathbf{x}_0 + \sum_{j \in I} \mu_j (\mathbf{x}_j - \mathbf{x}_0) + \sum_{j \in J} \mu_j \mathbf{d}_j$$

where $\mu_j \ge 0$ for $j \in I \cup J$, $\mathbf{x}_j$ for $j \in I$ are adjacent extreme points and $\mathbf{d}_j$ for $j \in J$ are adjacent extreme directions.)

5.57 Show that an extreme point of a bounded polyhedral set has a minimal objective if and only if it has an objective that is smaller than or equal to that of any adjacent extreme point. Can you generalize the result to the unbounded case? (*Hint*. Use Exercises 5.55 and 5.56.)

5.58 Consider the problem: Minimize $\mathbf{cx}$ subject to $\mathbf{Ax} = \mathbf{b}$, $\mathbf{x} \ge \mathbf{0}$. Let $\mathbf{x}^*$ be an unique optimal extreme point. Show that the second best extreme point must be adjacent to $\mathbf{x}^*$. What happens if the uniqueness assumption is relaxed?

NOTES AND REFERENCES

1. The revised simplex method was devised by Dantzig and Orchard–Hays [1953] (also see Dantzig and Orchard–Hays [1954] for the product form of the inverse). For further reading on this topic, refer to Dantzig [1963a].
2. The experimentation related to fitting the regression equation Km^and^b for computational effort in solving linear programs appears in Knolmayer [1982]. For information on *theoretical* bounds on computational effort using the simplex method, see Chapter 8.
3. The simplex method using the **LU**-factorization of the basis was introduced by Bartels and Golub [1969] and is said to employ the "elimination form of the inverse." For more information on this subject, see Gill *et al.* [1974] and Saunders [1976]. For factorizations designed for use in a parallel computing environment, refer to Helgason *et al.* [1987].
4. The steepest edge variations of the rules for selecting entering variables are discussed in Harris [1973] and in Goldfarb and Reid [1977]. For problem scaling, see Kuhn and Quandt [1962], Tomlin [1975], and Sherali *et al.* [1983]. The concept of partial pricing or suboptimization discussed in Section 5.1 is standard and in widespread use. The particular description given here has been adapted from Frendewey and Glover's [1981] experiments on large-scale linear programs with embedded networks. Also, see Gibbs *et al.* [1983]. Finally, for model and input/output management of large-scale linear programs, see Greenberg [1983] and Murphy and Stohr [1986].

5. In Chapter 7 we describe the use of the revised simplex method in solving large-scale problems in the context of decomposition by generating columns at each iteration.
6. The simplex method for bounded variables was developed by Dantzig [1955] at the RAND corporation to provide a shortcut routine for solving a problem of assigning personnel. The method was independently developed by Charnes and Lemke [1954].
7. The strongly feasible partition lexicographic cycling prevention rule for bounded variables linear programming problems was proposed by Cunningham [1976] and Murty [1978]. For further developments involving generalized upper bounds (GUB's) and variable upper bounds (VUB's), the interested reader is referred to Dantzig and Van Slyke [1965] and Schrage [1975, 1978].
8. The Karush–Kuhn–Tucker optimality conditions for nonlinear programs were developed independently by Karush [1939] in his master's thesis, and by Kuhn and Tucker [1950]. Specialization of these conditions for linear programs is given in Section 5.4. For further reader reading on the Theorems of the Alternative and the KKT conditions the reader may refer to Bazaraa and Shetty [1979], Mangasarian [1969], and Zangwill [1969]. The KKT conditions in linear programming and the subject of duality are very closely associated. This fact will become apparent after the reader studies Chapter 6.

SIX: DUALITY AND SENSITIVITY ANALYSIS

For every linear program we solve, there is another associated linear program which we happen to be simultaneously solving. This new linear program satisfies some very important properties. It may be used to obtain the solution to the original program. Its variables provide extremely useful information about the optimal solution to the original linear program.

This leads to rich economic interpretations related to the original linear programming problem. In fact, the roots of this problem lie in the characterization of the optimality conditions for the original linear program. For the sake of expository reference, we shall call the original linear programming problem the *primal* (linear programming) problem, and we shall call this related linear program the *dual* (linear programming) problem. Although the term "dual" comes from linear algebra, the term "primal" was suggested by Dantzig's father, Tobias Dantzig (who was a mathematician), as an appropriate Latin antonym to substitute for the cumbersome phrase, "the original problem of which this is the dual." Actually, the terms primal and dual for this related pair of linear programming problems are only relative, because the dual of the "dual" is the "primal" itself.

We shall begin by formulating this new *dual* (linear programming) problem and proceeding to develop some of its important properties. These properties will lead to two new algorithms, the dual simplex method and the primal-dual algorithm, for solving linear programs. Finally, we shall discuss the effect of variation in the data, that is, the cost coefficients, the right-hand-side coefficients, and the constraint coefficients on the optimal solution to a linear program.

6.1 FORMULATION OF THE DUAL PROBLEM

Associated with each linear programming problem there is another linear programming problem called the dual. The dual linear program possesses many important properties relative to the original primal linear program. There are two important forms (definitions) of duality: the *canonical* form of duality and the *standard* form of duality. These two forms are completely equivalent. They arise respectively from the canonical and the standard representation of linear programming problems.

Canonical Form of Duality

Suppose that the primal linear program is given in the (canonical) form:

$$\begin{aligned} \text{P:}\quad & \text{Minimize} && \mathbf{cx} \\ & \text{Subject to} && \mathbf{Ax} \geq \mathbf{b} \\ & && \mathbf{x} \geq \mathbf{0} \end{aligned}$$

Then the *dual* linear program is defined by:

$$\begin{aligned} \text{D:}\quad & \text{Maximize} && \mathbf{wb} \\ & \text{Subject to} && \mathbf{wA} \leq \mathbf{c} \\ & && \mathbf{w} \geq \mathbf{0} \end{aligned}$$

Note that there is exactly one dual variable for each primal constraint and exactly one dual constraint for each primary variable. We shall say more about this later.

Example 6.1

Consider the following linear program and its dual.

$$\begin{aligned} \text{P:}\quad & \text{Minimize} && 6x_1 + 8x_2 \\ & \text{Subject to} && 3x_1 + x_2 \geq 4 \\ & && 5x_1 + 2x_2 \geq 7 \\ & && x_1,\ x_2 \geq 0 \end{aligned}$$

$$\begin{aligned} \text{D:}\quad & \text{Maximize} && 4w_1 + 7w_2 \\ & \text{Subject to} && 3w_1 + 5w_2 \leq 6 \\ & && w_1 + 2w_2 \leq 8 \\ & && w_1,\ w_2 \geq 0 \end{aligned}$$

Before proceeding further, try solving both problems and comparing their optimal objective values. This will provide a hint of things to come.

In the canonical definition of duality it is important for problem P to have a "Minimization" objective with all "greater than or equal to" constraints and all "nonnegative" variables. In theory, to apply the canonical definition of duality we must first convert the primal linear program to the foregoing format. However, in practice it is possible to immediately write down the dual of any linear program. We shall discuss this shortly.

Standard Form of Duality

Another equivalent definition of duality may be given with the primal linear program stated in the following standard form:

P: Minimize $\mathbf{cx}$

Subject to $\mathbf{Ax} = \mathbf{b}$
$\mathbf{x} \geq \mathbf{0}$

Then the *dual* linear program is defined by:

D: Maximize $\mathbf{wb}$

Subject to $\mathbf{wA} \leq \mathbf{c}$
$\mathbf{w}$ unrestricted

Example 6.2

Consider the following linear program and its dual.

P: Minimize $6x_1 + 8x_2$

Subject to
$$\begin{aligned} 3x_1 + x_2 - x_3 \quad &= 4 \\ 5x_1 + 2x_2 \quad - x_4 &= 7 \\ x_1, \quad x_2, \quad x_3, \quad x_4 &\geq 0 \end{aligned}$$

D: Maximize $4w_1 + 7w_2$

Subject to
$$\begin{aligned} 3w_1 + 5w_2 &\leq 6 \\ w_1 + 2w_2 &\leq 8 \\ -w_1 \quad &\leq 0 \\ -\quad w_2 &\leq 0 \\ w_1, \quad w_2 \quad &\text{unrestricted} \end{aligned}$$

Given one of the definitions, canonical or standard, it is easy to demonstrate that the other definition is valid. For example, suppose that we accept the standard form as a definition and wish to demonstrate that the canonical form is correct. By adding slack variables to the canonical form of a linear program, we may apply the standard form of duality to obtain the dual problem.

P: Minimize $\mathbf{cx}$

Subject to $\mathbf{Ax} - \mathbf{Ix}_s = \mathbf{b}$
$\mathbf{x}, \quad \mathbf{x}_s \geq \mathbf{0}$

D: Maximize $\mathbf{wb}$

Subject to $\mathbf{wA} \leq \mathbf{c}$
$-\mathbf{wI} \leq \mathbf{0}$
$\mathbf{w}$ unrestricted

But since $-\mathbf{wI} \leq \mathbf{0}$ is the same as $\mathbf{w} \geq \mathbf{0}$, we obtain the canonical form of the dual problem.

Dual of the Dual

Since the dual linear program is itself a linear program, we may wonder what its dual might be. Consider the dual in canonical form:

$$\begin{array}{ll} \text{Maximize} & \mathbf{wb} \\ \\ \text{Subject to} & \mathbf{wA} \leq \mathbf{c} \\ & \mathbf{w} \geq \mathbf{0} \end{array}$$

Applying the transformation techniques of Chapter 1, we may rewrite this problem in the form:

$$\begin{array}{ll} \text{Minimize} & (-\mathbf{b}^t)\mathbf{w}^t \\ \\ \text{Subject to} & (-\mathbf{A}^t)\mathbf{w}^t \geq (-\mathbf{c}^t) \\ & \mathbf{w}^t \geq \mathbf{0} \end{array}$$

The dual linear program for this linear program is given by (letting $\mathbf{x}^t$ play the role of the row vector of dual variables):

$$\begin{array}{ll} \text{Maximize} & \mathbf{x}^t(-\mathbf{c}^t) \\ \\ \text{Subject to} & \mathbf{x}^t(-\mathbf{A}^t) \leq (-\mathbf{b}^t) \\ & \mathbf{x}^t \geq \mathbf{0} \end{array}$$

But this is the same as

$$\begin{array}{ll} \text{Minimize} & \mathbf{cx} \\ \\ \text{Subject to} & \mathbf{Ax} \geq \mathbf{b} \\ & \mathbf{x} \geq \mathbf{0} \end{array}$$

which is precisely the original primal problem. Thus we have the following lemma, which is known as the *involutory* property of duality.

***Lemma* 6.1**

The dual of the dual is the primal.

This lemma indicates that the definitions may be applied in reverse. The terms "primal" and "dual" are relative to the frame of reference we choose.

Mixed Forms of Duality

In practice, many linear programs contain some constraints of the "less than or equal to" type, some of the "greater than or equal to" type, and some of the "equal to" type. Also, variables may be "≥ 0," "≤ 0," or "unrestricted." In theory, this presents no problem since we may apply the transformation techniques of Chapter 1 to convert any "mixed" problem to one of the primal or dual forms discussed, after which the dual can be readily obtained. In practice such conversions can be tedious. Fortunately, it is not necessary actually to make these conversions, and it is possible to give immediately the dual of any linear program.

Consider the following linear program.

$$
\begin{aligned}
\text{P:}\quad & \text{Minimize} && \mathbf{c}_1\mathbf{x}_1 + \mathbf{c}_2\mathbf{x}_2 + \mathbf{c}_3\mathbf{x}_3 \\
& \text{Subject to} && \mathbf{A}_{11}\mathbf{x}_1 + \mathbf{A}_{12}\mathbf{x}_2 + \mathbf{A}_{13}\mathbf{x}_3 \geq \mathbf{b}_1 \\
& && \mathbf{A}_{21}\mathbf{x}_1 + \mathbf{A}_{22}\mathbf{x}_2 + \mathbf{A}_{23}\mathbf{x}_3 \leq \mathbf{b}_2 \\
& && \mathbf{A}_{31}\mathbf{x}_1 + \mathbf{A}_{32}\mathbf{x}_2 + \mathbf{A}_{33}\mathbf{x}_3 = \mathbf{b}_3 \\
& && \mathbf{x}_1 \geq 0,\ \mathbf{x}_2 \leq 0,\ \mathbf{x}_3 \ \text{ unrestricted}
\end{aligned}
$$

Converting this problem to canonical form by multiplying the second set of inequalities by -1, writing the equality constraint set equivalently as two inequalities, and substituting $\mathbf{x}_2 = -\mathbf{x}_2'$, $\mathbf{x}_3 = \mathbf{x}_3' - \mathbf{x}_3''$, we get

$$
\begin{aligned}
& \text{Minimize} && \mathbf{c}_1\mathbf{x}_1 - \mathbf{c}_2\mathbf{x}_2' + \mathbf{c}_3\mathbf{x}_3' - \mathbf{c}_3\mathbf{x}_3'' \\
& \text{Subject to} && \mathbf{A}_{11}\mathbf{x}_1 - \mathbf{A}_{12}\mathbf{x}_2' + \mathbf{A}_{13}\mathbf{x}_3' - \mathbf{A}_{13}\mathbf{x}_3'' \geq \mathbf{b}_1 \\
& && -\mathbf{A}_{21}\mathbf{x}_1 + \mathbf{A}_{22}\mathbf{x}_2' - \mathbf{A}_{23}\mathbf{x}_3' + \mathbf{A}_{23}\mathbf{x}_3'' \geq -\mathbf{b}_2 \\
& && \mathbf{A}_{31}\mathbf{x}_1 - \mathbf{A}_{32}\mathbf{x}_2' + \mathbf{A}_{33}\mathbf{x}_3' - \mathbf{A}_{33}\mathbf{x}_3'' \geq \mathbf{b}_3 \\
& && -\mathbf{A}_{31}\mathbf{x}_1 + \mathbf{A}_{32}\mathbf{x}_2' - \mathbf{A}_{33}\mathbf{x}_3' + \mathbf{A}_{33}\mathbf{x}_3'' \geq -\mathbf{b}_3 \\
& && \mathbf{x}_1 \geq 0,\ \mathbf{x}_2' \geq 0,\ \mathbf{x}_3' \geq 0,\ \mathbf{x}_3'' \geq 0
\end{aligned}
$$

Denoting the dual variables associated with the four constraint sets as $\mathbf{w}_1$, $\mathbf{w}_2'$, $\mathbf{w}_3'$, and $\mathbf{w}_3''$ respectively, we obtain the dual to this problem as follows.

$$
\begin{aligned}
& \text{Maximize} && \mathbf{w}_1\mathbf{b}_1 - \mathbf{w}_2'\mathbf{b}_2 + \mathbf{w}_3'\mathbf{b}_3 - \mathbf{w}_3''\mathbf{b}_3 \\
& \text{Subject to} && \mathbf{w}_1\mathbf{A}_{11} - \mathbf{w}_2'\mathbf{A}_{21} + \mathbf{w}_3'\mathbf{A}_{31} - \mathbf{w}_3''\mathbf{A}_{31} \leq \mathbf{c}_1 \\
& && -\mathbf{w}_1\mathbf{A}_{12} + \mathbf{w}_2'\mathbf{A}_{22} - \mathbf{w}_3'\mathbf{A}_{32} + \mathbf{w}_3'\mathbf{A}_{32} \leq -\mathbf{c}_2 \\
& && \mathbf{w}_1\mathbf{A}_{13} - \mathbf{w}_2'\mathbf{A}_{23} + \mathbf{w}_3'\mathbf{A}_{33} - \mathbf{w}_3''\mathbf{A}_{33} \leq \mathbf{c}_3 \\
& && -\mathbf{w}_1\mathbf{A}_{13} + \mathbf{w}_2'\mathbf{A}_{23} - \mathbf{w}_3'\mathbf{A}_{33} + \mathbf{w}_3''\mathbf{A}_{33} \leq -\mathbf{c}_3 \\
& && \mathbf{w}_1 \geq 0,\ \mathbf{w}_2' \geq 0,\ \mathbf{w}_3' \geq 0,\ \mathbf{w}_3'' \geq 0
\end{aligned}
$$

Table 6.1 Relationships Between Primal and Dual Problems

	MINIMIZATION PROBLEM		MAXIMIZATION PROBLEM	
Variables	≥ 0	$\longleftrightarrow$	$\leq$	Constraints
	≤ 0	$\longleftrightarrow$	$\geq$	
	Unrestricted	$\longleftrightarrow$	$=$	
Constraints	$\geq$	$\longleftrightarrow$	≥ 0	Variables
	$\leq$	$\longleftrightarrow$	≤ 0	
	$=$	$\longleftrightarrow$	Unrestricted	

Finally, using $\mathbf{w}_2 = -\mathbf{w}_2'$ and $\mathbf{w}_3 = \mathbf{w}_3' - \mathbf{w}_3''$, the foregoing problem may be equivalently stated as follows.

$$\begin{array}{lll} \text{D:} & \text{Maximize} & \mathbf{w}_1\mathbf{b}_1 + \mathbf{w}_2\mathbf{b}_2 + \mathbf{w}_3\mathbf{b}_3 \\ & \text{Subject to} & \mathbf{w}_1\mathbf{A}_{11} + \mathbf{w}_2\mathbf{A}_{21} + \mathbf{w}_3\mathbf{A}_{31} \leq \mathbf{c}_1 \\ & & \mathbf{w}_1\mathbf{A}_{12} + \mathbf{w}_2\mathbf{A}_{22} + \mathbf{w}_3\mathbf{A}_{32} \geq \mathbf{c}_2 \\ & & \mathbf{w}_1\mathbf{A}_{13} + \mathbf{w}_2\mathbf{A}_{23} + \mathbf{w}_3\mathbf{A}_{33} = \mathbf{c}_3 \\ & & \mathbf{w}_1 \geq \mathbf{0}, \ \mathbf{w}_2 \leq \mathbf{0}, \ \mathbf{w}_3 \ \text{ unrestricted} \end{array}$$

Note how this form of the dual D relates to the original primal problem P, where $\mathbf{w}_1$, $\mathbf{w}_2$, and $\mathbf{w}_3$ are the dual variables associated with the three sets of constraints in P; similarly, $\mathbf{x}_1$, $\mathbf{x}_2$, and $\mathbf{x}_3$ are the "dual" variables associated with the three sets of constraints in D. Furthermore, note that the first set of constraints in P were already in the required form, and hence $\mathbf{w}_1 \geq \mathbf{0}$ in D. The second constraint set needed to be multiplied by -1 in order to be put in the required form, and hence $\mathbf{w}_2 \leq \mathbf{0}$ in D. Similarly, because of the transformation needed on the third set of constraints in P, we obtain $\mathbf{w}_3$ as unrestricted in D. In a likewise manner, the transformations needed on the variables in P to get P in the required canonical form dictate the form of the constraints in D. With these observations, the dual D to the problem P can now be written *directly* without using the intermediate transformation steps. These results are summarized in Table 6.1. In Exercise 6.5, we ask the reader to repeat this derivation using the standard form of duality.

Example 6.3

Consider the following linear program.

$$\begin{array}{ll} \text{Maximize} & 8x_1 + 3x_2 - 2x_3 \\ \text{Subject to} & x_1 - 6x_2 + x_3 \geq 2 \\ & 5x_1 + 7x_2 - 2x_3 = -4 \\ & x_1 \leq 0, \ x_2 \geq 0, \ x_3 \ \text{ unrestricted} \end{array}$$

Applying the results of the table, we can immediately write down the dual.

$$\begin{array}{ll} \text{Minimize} & 2w_1 - 4w_2 \\ \text{Subject to} & w_1 + 5w_2 \le 8 \\ & -6w_1 + 7w_2 \ge 3 \\ & w_1 - 2w_2 = -2 \\ & w_1 \le 0,\ w_2 \text{ unrestricted} \end{array}$$

6.2 PRIMAL-DUAL RELATIONSHIPS

The definition we have selected for the dual problem leads to many important relationships between the primal and dual linear programs.

The Relationship Between Objective Values

Consider the canonical form of duality and let $\mathbf{x}_0$ and $\mathbf{w}_0$ be any feasible solutions to the primal and dual programs respectively. Then $\mathbf{Ax}_0 \ge \mathbf{b}$, $\mathbf{x}_0 \ge \mathbf{0}$, $\mathbf{w}_0\mathbf{A} \le \mathbf{c}$, and $\mathbf{w}_0 \ge \mathbf{0}$. Multiplying $\mathbf{Ax}_0 \ge \mathbf{b}$ on the left by $\mathbf{w}_0 \ge \mathbf{0}$ and $\mathbf{w}_0\mathbf{A} \le \mathbf{c}$ on the right by $\mathbf{x}_0 \ge \mathbf{0}$, we get

$$\mathbf{cx}_0 \ge \mathbf{w}_0\mathbf{Ax}_0 \ge \mathbf{w}_0\mathbf{b}$$

The result is the following. This is known as the *weak duality* property.

Lemma 6.2

The objective function value for any feasible solution to the minimization problem is always greater than or equal to the objective function value for any feasible solution to the maximization problem. In particular, the objective value of any feasible solution of the minimization problem gives an upper bound on the optimal objective of the maximization problem. Similarly, the objective value of any feasible solution of the maximization problem is a lower bound on the optimal objective of the minimization problem.

As an illustration of the application of this lemma, suppose that in Example 6.1 we select the feasible primal and dual solutions $\mathbf{x}_0 = (\frac{7}{5}, 0)'$ and $\mathbf{w}_0 = (2, 0)$. Then $\mathbf{cx}_0 = \frac{42}{5} = 8.4$ and $\mathbf{w}_0\mathbf{b} = 8$. Thus the optimal solution for either problem has objective value between 8 and 8.4. This allows us to stop a linear programming solution procedure with a near optimal solution.

The following corollaries are immediate consequences of Lemma 6.2.

Corollary 6.1

If $\mathbf{x}_0$ and $\mathbf{w}_0$ are feasible solutions to the primal and dual problems such that $\mathbf{cx}_0 = \mathbf{w}_0\mathbf{b}$, then $\mathbf{x}_0$ and $\mathbf{w}_0$ are optimal solutions to their respective problems.

Corollary 6.2

If either problem has an unbounded objective value, then the other problem possesses no feasible solution.

This corollary indicates that unboundedness in one problem implies infeasibility in the other problem. Is this property symmetric? Does infeasibility in one problem imply unboundedness in the other? The answer is "not necessarily." This is best illustrated by the following example.

Example 6.4

Consider the following primal and dual problems.

P: Minimize $-x_1 - x_2$

Subject to
$$\begin{aligned} x_1 - x_2 &\geq 1 \\ -x_1 + x_2 &\geq 1 \\ x_1, \quad x_2 &\geq 0 \end{aligned}$$

D: Maximize $w_1 + w_2$

Subject to
$$\begin{aligned} w_1 - w_2 &\leq -1 \\ -w_1 + w_2 &\leq -1 \\ w_1, \quad w_2 &\geq \ \ 0 \end{aligned}$$

Upon graphing both problems (in Figure 6.1) we find that neither problem possesses a feasible solution.

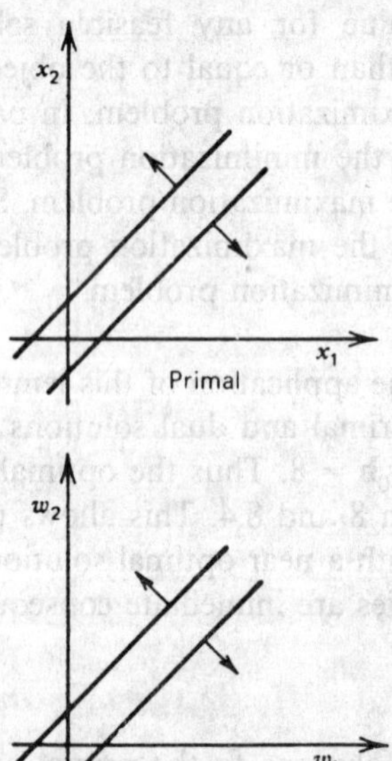

Figure 6.1. **An example of infeasible primal and dual problems.**

Origin of Duality and the Karush–Kuhn–Tucker (KKT) Optimality Conditions

Recall from Chapter 5 that the optimality conditions for a linear program state that a necessary and sufficient condition for $\mathbf{x}^*$ to be an optimal point to the linear program: Minimize $\mathbf{cx}$ subject to $\mathbf{Ax} \geq \mathbf{b}$, $\mathbf{x} \geq \mathbf{0}$ is that there exists a vector $\mathbf{w}^*$ such that

$$1.\ \mathbf{Ax}^* \geq \mathbf{b},\ \mathbf{x}^* \geq \mathbf{0}$$

$$2.\ \mathbf{w}^*\mathbf{A} \leq \mathbf{c},\ \mathbf{w}^* \geq \mathbf{0}$$

$$3.\ \mathbf{w}^*(\mathbf{Ax}^* - \mathbf{b}) = 0$$

$$(\mathbf{c} - \mathbf{w}^*\mathbf{A})\mathbf{x}^* = 0$$

Condition 1 simply requires that the optimal point $\mathbf{x}^*$ must be feasible to the primal. In light of our discussion of duality we can now interpret condition 2. This condition indicates that the vector $\mathbf{w}^*$ must be a feasible point for the dual problem. From the complementary slackness condition 3, we find that $\mathbf{cx}^* = \mathbf{w}^*\mathbf{b}$. Hence $\mathbf{w}^*$ must be an optimal solution to the dual problem. Consequently, the feasible region of the dual problem arises from condition 2. Its objective function arises from the fact that any $\mathbf{w}$ feasible to the dual must satisfy $\mathbf{wb} \leq \mathbf{cx}^*$ by the weak duality property. Conditions 2 and 3 state however, that the KKT solution provides a $\mathbf{w}^*$ which is feasible to the dual and for which $\mathbf{w}^*\mathbf{b} = \mathbf{cx}^*$. Hence $\mathbf{w}^*$ must maximize $\mathbf{wb}$ over the dual feasible region. Symmetrically, the KKT optimality conditions for the dual problem imply the existence of a primal feasible solution whose objective is equal to that of the optimal dual (why?). This leads to the following lemma, which is known as the *strong duality* property.

***Lemma* 6.3**

If one problem possesses an optimal solution, then both problems possess optimal solutions and the two optimal objective values are equal.

It is also possible to see how the dual problem arises naturally in the context of the simplex algorithm. Each tableau in canonical form represents a situation in which some multiples $w_1, \ldots, w_m$ of the m rows in $\mathbf{Ax} - \mathbf{x}_s = \mathbf{b}$ have been added to the objective row, where $\mathbf{x}_s \geq \mathbf{0}$ is the vector of slack variables. This means that the objective row coefficients for the $\mathbf{x}$ variables are $-\mathbf{c} + \mathbf{wA}$, those for the $\mathbf{x}_s$ variables are $-\mathbf{w}$, and the right-hand-side is $\mathbf{wb}$. Note that the dual feasibility condition 2 requires a set of these so-called *simplex multipliers* $\mathbf{w}^*$ such that the canonical objective row coefficients are nonpositive for all the $\mathbf{x}$ and $\mathbf{x}_s$ variables. Indeed, if the primal problem has been solved to optimality, then the available simplex multipliers $\mathbf{w}^* \equiv \mathbf{c}_B\mathbf{B}^{-1}$, where $\mathbf{B}$ is the optimal basis matrix, satisfy these dual feasibility conditions. Moreover, if $\mathbf{x}^*$ is a primal optimal solution, its objective value $\mathbf{cx}^*$ coincides with $\mathbf{w}^*\mathbf{b}$, the right-hand-side value in the objective row of the optimal tableau. In view of the weak duality property of Lemma 6.2, we have that, at optimality, the simplex multipliers $\mathbf{w}^*$ must be a maximizing solution to $\mathbf{wb}$ subject to the constraints $\mathbf{wA} \leq \mathbf{c}$ and $\mathbf{w} \geq \mathbf{0}$, that is, to the so-called dual linear programming problem.

Lemma 6.3 provides an insightful result that is related to Corollary 6.2. For the linear program P: Minimize $\mathbf{cx}$, subject to $\mathbf{Ax} \geq \mathbf{b}$, $\mathbf{x} \geq \mathbf{0}$, consider the *homogeneous form* of its dual problem, namely HD: Maximize $\mathbf{wb}$, subject to $\mathbf{wA} \leq \mathbf{0}$, $\mathbf{w} \geq \mathbf{0}$. Note that the dual to HD has the same feasible region as P. Hence if HD is unbounded, then by Corollary 6.2 P is infeasible. Conversely, if P is infeasible, then HD must be unbounded, or else, since HD is feasible ($\mathbf{w} = \mathbf{0}$ is a feasible solution), it must have an optimal solution. But Lemma 6.3 tells us that the dual to HD then has an optimal solution, that is, P is feasible, a contradiction. Therefore we have the following result. We ask the reader to verify this using Example 6.4. It is insightful to note that the homogeneous form of the dual arises from the dual problem via a characterization of its recession directions as in Chapter 2.

***Corollary* 6.3**

The primal problem is infeasible if and only if the homogeneous form of the dual problem is unbounded (and vice versa).

By utilizing the foregoing results we obtain two important basic theorems of duality. These two theorems will permit us to use the dual problem to solve the primal problem and, also, to develop new algorithms to solve both problems.

The Fundamental Theorem of Duality

Combining the results of the lemmas, corollaries, and examples of the previous section we obtain the following.

***Theorem* 6.1** (Fundamental Theorem of Duality)

With regard to the primal and dual linear programming problems, exactly one of the following statements is true.

1. Both possess optimal solutions $\mathbf{x}^*$ and $\mathbf{w}^*$ with $\mathbf{cx}^* = \mathbf{w}^*\mathbf{b}$.
2. One problem has unbounded objective value, in which case the other problem must be infeasible.
3. Both problems are infeasible.

From this theorem we see that duality is not completely symmetric. The best we can say is that (here optimal means finite optimal, and unbounded means having an unbounded optimal objective value):

P	OPTIMAL	$\Leftrightarrow$	D	OPTIMAL
P(D)	UNBOUNDED	$\Rightarrow$	D(P)	INFEASIBLE
P(D)	INFEASIBLE	$\Rightarrow$	D(P)	UNBOUNDED OR INFEASIBLE
P(D)	INFEASIBLE	$\Leftrightarrow$	D(P)	UNBOUNDED IN HOMOGENEOUS FORM

Complementary Slackness and the Supervisor's Principle

We have seen that if $\mathbf{x}$ and $\mathbf{w}$ are any pair of primal and dual feasible solutions to the linear program: Minimize $\mathbf{cx}$, subject to $\mathbf{Ax} \geq \mathbf{b}$, $\mathbf{x} \geq \mathbf{0}$, then they are both optimal to their respective problems if and only if they have the same objective value (that is, $\mathbf{cx} = \mathbf{wb}$). This is known as the *supervisor's principle* because it provides a simple check by which a "supervisor" can verify the (simultaneous) optimality of a pair of candidate primal and dual feasible solutions. The KKT conditions tell us that another equivalent supervisor's principle is to check for the complementary slackness conditions (written as condition 3). Indeed, if $\mathbf{x}^*$ and $\mathbf{w}^*$ respectively satisfy the primal and dual feasibility conditions 1 and 2, then $\mathbf{cx}^* = \mathbf{w}^*\mathbf{b}$, that is, $\mathbf{w}^*(\mathbf{Ax}^* - \mathbf{b}) + (\mathbf{c} - \mathbf{w}^*\mathbf{A})\mathbf{x}^* = 0$ holds, if and only if condition 3 holds, since $\mathbf{w}^* \geq \mathbf{0}$, $\mathbf{Ax}^* - \mathbf{b} \geq \mathbf{0}$, $\mathbf{c} - \mathbf{w}^*\mathbf{A} \geq \mathbf{0}$, and $\mathbf{x}^* \geq \mathbf{0}$ In expanded form, since $\mathbf{w}^* \geq \mathbf{0}$ and $\mathbf{Ax}^* - \mathbf{b} \geq \mathbf{0}$, then $\mathbf{w}^*(\mathbf{Ax}^* - \mathbf{b}) = 0$ implies $w_i^*(\mathbf{a}^i\mathbf{x}^* - b_i) = 0$ for $i = 1, \ldots, m$. Similarly, $(\mathbf{c} - \mathbf{w}^*\mathbf{A})\mathbf{x}^* = 0$ implies $(c_j - \mathbf{w}^*\mathbf{a}_j)x_j^* = 0$ for $j = 1, \ldots, n$.

Thus we have the following theorem.

***Theorem* 6.2** (Complementary Slackness Theorem)

Let $\mathbf{x}^*$ and $\mathbf{w}^*$ be *any* feasible solutions to the primal and dual problems in the canonical form. Then they are respectively optimal if and only if

$$(c_j - \mathbf{w}^*\mathbf{a}_j)x_j^* = 0 \qquad j = 1, \ldots, n$$

and

$$w_i^*(\mathbf{a}^i\mathbf{x}^* - b_i) = 0 \qquad i = 1, \ldots, m$$

This is a very important theorem relating the primal and dual problems. It obviously indicates that at least one of the two terms in each expression must be zero. In particular,

$$x_j^* > 0 \Rightarrow \mathbf{w}^*\mathbf{a}_j = c_j$$

$$\mathbf{w}^*\mathbf{a}_j < c_j \Rightarrow x_j^* = 0$$

$$w_i^* > 0 \Rightarrow \mathbf{a}^i\mathbf{x}^* = b_i$$

$$\mathbf{a}^i\mathbf{x}^* > b_i \Rightarrow w_i^* = 0$$

Hence at optimality, if a variable in one problem is positive, then the corresponding constraint in the other problem must be *tight*. If a constraint in one problem is not tight, then the corresponding variable in the other problem must be zero.

Suppose that we let $x_{n+i} = \mathbf{a}^i\mathbf{x} - b_i \geq 0$, $i = 1, \ldots, m$ be the m slack variables in the primal problem and let $w_{m+j} = c_j - \mathbf{wa}_j \geq 0$, $j = 1, \ldots, n$ be the n slack variables in the dual problem (in Chapter 5 while stating the KKT conditions, w_{m+j} was denoted by v_j). Then we may rewrite the complementary slackness conditions as follows:

$$x_j^* w_{m+j}^* = 0 \qquad j = 1, \ldots, n$$

$$w_i^* x_{n+i}^* = 0 \qquad i = 1, \ldots, m$$

This relates variables in one problem to slack variables in the other problem. In particular, x_j and w_{m+j} for $j = 1, \ldots, n$, and similarly, x_{n+i} and w_i for $i = 1, \ldots, n$, are known as *complementary* pairs of variables.

Using the Dual to Solve the Primal

We now have at hand powerful analysis tools, in the form of the two theorems of this section, to utilize the dual problem in solving the primal problem. Let us illustrate this potential usefulness by the following example.

Example 6.5

Consider the following primal and dual problems.

$$\begin{array}{lll} \text{P:} & \text{Minimize} & 2x_1 + 3x_2 + 5x_3 + 2x_4 + 3x_5 \\ & \text{Subject to} & x_1 + x_2 + 2x_3 + x_4 + 3x_5 \geq 4 \\ & & 2x_1 - 2x_2 + 3x_3 + x_4 + x_5 \geq 3 \\ & & x_1, \quad x_2, \quad x_3, \quad x_4, \quad x_5 \geq 0 \end{array}$$

$$\begin{array}{lll} \text{D:} & \text{Maximize} & 4w_1 + 3w_2 \\ & \text{Subject to} & w_1 + 2w_2 \leq 2 \\ & & w_1 - 2w_2 \leq 3 \\ & & 2w_1 + 3w_2 \leq 5 \\ & & w_1 + w_2 \leq 2 \\ & & 3w_1 + w_2 \leq 3 \\ & & w_1, \quad w_2 \geq 0 \end{array}$$

Since the dual has only two variables, we may solve it graphically as shown in Figure 6.2. The optimal solution to the dual is $w_1^* = \frac{4}{5}$, $w_2^* = \frac{3}{5}$ with objective 5. Right away we know that $z^* = 5$. Utilizing the theorem of complementary slackness, we further know that $x_2^* = x_3^* = x_4^* = 0$ since none of the corresponding complementary dual constraints are tight. Since $w_1^*, w_2^* > 0$, then $x_1^* + 3x_5^* = 4$ and $2x_1^* + x_5^* = 3$. From these two equations we get $x_1^* = 1$ and $x_5^* = 1$. Thus the primal optimal point is obtained from the duality theorems and the dual optimal point.

6.3 ECONOMIC INTERPRETATION OF THE DUAL

Consider the following linear program and its dual.

$$\begin{array}{lllll} \text{P:} & \text{Minimize} & \mathbf{cx} & \text{D:} & \text{Maximize} \quad \mathbf{wb} \\ & \text{Subject to} & \mathbf{Ax} \geq \mathbf{b} & & \text{Subject to} \quad \mathbf{wA} \leq \mathbf{c} \\ & & \mathbf{x} \geq \mathbf{0} & & \qquad\qquad\quad\;\; \mathbf{w} \geq \mathbf{0} \end{array} \tag{6.1}$$

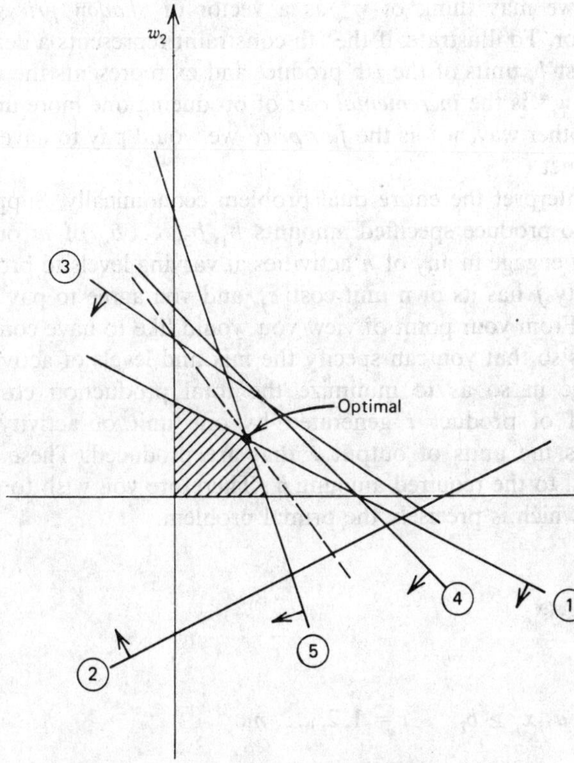

Figure 6.2. Solving the dual problem graphically.

Let **B** be an optimal basis for the primal problem with $\mathbf{c}_B$ as its associated cost vector. Suppose that this optimal basic feasible solution $\mathbf{x}^*$ is *nondegenerate*. Denoting R as the index set for the nonbasic variables x_j, which may include both slack and structural variables (that is, $R \subseteq \{1, \ldots, n + m\}$), we have

$$z = \mathbf{c}_B \mathbf{B}^{-1}\mathbf{b} - \sum_{j \in \mathbf{R}} (z_j - c_j)x_j = \mathbf{w}^*\mathbf{b} - \sum_{j \in \mathbf{R}} (z_j - c_j)x_j. \qquad (6.2)$$

If the ith right-hand-side b_i is perturbed slightly (positively or negatively), since the current basic feasible solution remains feasible, we maintain its optimality. Hence if z^* is the optimal objective function value and $\mathbf{B}_i^{-1}$ is the ith column of $\mathbf{B}^{-1}$, we have

$$\frac{\partial z^*}{\partial b_i} = \mathbf{c}_B \mathbf{B}_i^{-1} = w_i^* \qquad (6.3)$$

Thus w_i^* is the rate of change of the optimal objective value with a unit increase in the ith right-hand-side value, given that the current nonbasic variables are held at zero (regardless of feasibility). Since $w_i^* \geq 0$, z^* will increase (respectively, decrease) or stay constant as b_i increases (respectively, decreases).

Economically, we may think of $\mathbf{w}^*$ as a vector of *shadow prices* for the right-hand-side vector. To illustrate, if the ith constraint represents a demand for production of at least b_i units of the ith product and $\mathbf{cx}$ represents the total cost of production, then w_i^* is the *incremental cost* of producing one more unit of the ith product. Put another way, w_i^* is the *fair price* we would pay to have an extra unit of the ith product.

We may also interpret the entire dual problem economically. Suppose that you engage a firm to produce specified amounts $b_1, b_2, \ldots, b_m$ of m outputs or goods. The firm may engage in any of n activities at varying levels to produce the outputs. Each activity j has its own unit cost c_j, and you agree to pay the total cost of production. From your point of view you would like to have control over the firm's operations so that you can specify the mix and levels of activities that the firm will engage in so as to minimize the total production cost. If a_{ij} denotes the amount of product i generated by one unit of activity j, then $\sum_{j=1}^{n} a_{ij}x_j$ represents the units of output i that are produced. These must be greater than or equal to the required amount b_i. Therefore you wish to solve the following problem, which is precisely the primal problem.

$$\begin{array}{lll}
\text{Minimize} & \displaystyle\sum_{j=1}^{n} c_j x_j & \\
\text{Subject to} & \displaystyle\sum_{j=1}^{n} a_{ij}x_j \geq b_i & i = 1, 2, \ldots, m \\
& x_j \geq 0 & j = 1, 2, \ldots, n
\end{array}$$

Instead of trying to control the operation of the firm to obtain the most desirable mix of activities, suppose that you agree to pay the firm unit prices $w_1, w_2, \ldots, w_m$ for each of the m outputs. However, you stipulate that these prices announced by the firm must be *fair*. Since a_{ij} is the number of units of output i produced by one unit of activity j and w_i is the unit price of output i, then $\sum_{i=1}^{m} a_{ij}w_i$ can be interpreted as the unit price of activity j consistent with the prices $w_1, w_2, \ldots, w_m$. Therefore you ask the firm that the implicit price of activity j, namely $\sum_{i=1}^{m} a_{ij}w_i$, does not exceed the actual cost c_j. Therefore the firm must observe the constraints $\sum_{i=1}^{m} a_{ij}w_i \leq c_j$ for $j = 1, 2, \ldots, n$. Within these constraints the firm would like to choose a set of prices that maximizes its return $\sum_{i=1}^{m} w_i b_i$. This leads to the following dual problem of the firm.

$$\begin{array}{lll}
\text{Maximize} & \displaystyle\sum_{i=1}^{m} w_i b_i & \\
\text{Subject to} & \displaystyle\sum_{i=1}^{m} a_{ij}w_i \leq c_j & j = 1, 2, \ldots, n \\
& w_i \geq 0 & i = 1, \ldots, m
\end{array}$$

The main duality theorem states that there is an equilibrium set of activities and set of prices where the minimal production cost is equal to the maximal return. That the two objectives are equal at optimality becomes intuitively clear by noting that they represent the fair charge to the customer, where the primal objective is derived by cost considerations and the dual objective is arrived at by a pricing mechanism. These shadow prices are often used by regulated industries, such as electric utilities, to set and *justify* prices.

Observe that the complementary slackness theorem says that if, at optimality, $\sum_{j=1}^{n} a_{ij}x_j^* > b_i$, then $w_i^* = 0$, that is, if the optimal level of activities that meets all demand requirements automatically produces an excess of product i, then the incremental cost associated with marginally increasing b_i is naturally zero. Similarly, if $\sum_{i=1}^{m} a_{ij}w_i^* < c_j$, then $x_j^* = 0$, that is, if the total revenue generated via the items produced by a unit level of activity j is less than the associated production cost, then the level of activity j should be zero at optimality.

Interchanging the roles of the previous primal and dual problems, there is another interesting economic interpretation that can be given. Consider the following pair of dual problems, where the problems D and P have been rewritten as P and D, respectively.

$$\begin{aligned} \text{P:}\quad &\text{Maximize} && \sum_{j=1}^{n} c_j x_j \\ &\text{Subject to} && \sum_{j=1}^{n} a_{ij}x_j \le b_i \quad \text{for } i = 1, \ldots, m \\ &&& \mathbf{x} \ge \mathbf{0} \end{aligned}$$

$$\begin{aligned} \text{D:}\quad &\text{Minimize} && \sum_{i=1}^{m} b_i w_i \\ &\text{Subject to} && \sum_{i=1}^{m} a_{ij}w_i \ge c_j \quad \text{for } j = 1, \ldots, n \\ &&& \mathbf{w} \ge \mathbf{0} \end{aligned}$$

Assume that problem P is a product-mix problem in which n products are being produced using some m types of resources. Here x_j represents the number of units (per year) produced of product j and b_i represents the number of units available for resource i for $j = 1, \ldots, n$ and $i = 1, \ldots, m$. A unit of product j fetches a profit of c_j units and consumes a_{ij} units of resource i for $i = 1, \ldots, m$.

As before, let w_i denote the *shadow price* or the *inputed value* or the *fair market price/rent* for a unit of resource i. Then the dual constraints have the following interpretation. Observe that if the manufacturer considers renting out the m resources at unit prices $w_1, \ldots, w_m$ instead of manufacturing the mix of products, then every unit of product j not manufactured would result in a loss of profit of c_j units, but would fetch a rent of $\Sigma_{i=1}^m a_{ij}w_i$ units by virtue of the resources this releases. The dual constraints ensure that this venture is at least as profitable as manufacturing the products. However, in order to maintain fair or competitive prices while guaranteeing this profitability, the dual objective seeks to minimize the total rent. The main duality result tells us that the resulting total rent will match the total optimal profit obtainable by manufacturing the products. Hence w_i imputes a value that the manufacturer ascribes to a unit of the ith resource from the viewpoint of optimally operating the production process. In particular, at optimality, the complementary slackness theorem says that if not all available units of resource i are utilized, then the marginal worth w_i^* of an extra unit of resource i is naturally zero. Similarly, if the net value $\Sigma_{i=1}^m a_{ij}w_i^*$ of the resources consumed by a unit level of production of product j exceeds the realizable profit c_j, then the optimal production level of product j should be zero.

Shadow Prices Under Degeneracy

Now consider the pair of primal and dual problems given in Equation (6.1), and assume that we have a degenerate optimal basic feasible solution. Then while the derivative of z in Equation (6.2) with respect to b_i remains w_i^* in the usual sense of the derivative where all other parameters b_k, $k \neq i$ and x_j, $j \in R$ are held constant at their present values, this may not reflect the true "shadow price." That is, Equation (6.3) may not be valid. In fact, as will be seen later, z^* may not be a differentiable function of b_i. In other words, a perturbation in b_i may make the current optimal basis infeasible and a change in basis may be required.

To be more precise, let us assume that D in Equation (6.1) is feasible so that P is not unbounded for *any* value of $\mathbf{b}$. Let us denote by $z^*(b_i)$ the optimal objective function value as a function of some right-hand-side (parameter) b_i. Then, as long as b_i varies in a region that leaves P feasible, by the extreme point optimality property and the duality theorem, we have $z^*(b_i) =$ maximum$\{\mathbf{w}^j\mathbf{b} : j = 1,\ldots, E\}$, where $\mathbf{w}^j$, $j = 1,\ldots, E$ are the extreme points of the dual feasible region. Note that this admissible region for b_i, which leaves P feasible, is a polyhedral set $F \subseteq E^1$, given by $\mathbf{bd} \leq 0$ for all extreme directions $\mathbf{d}$ of the dual feasible region (why?). Denoting $\Sigma_{k \neq i} w_k^j b_k$ as w_0^j for $j = 1,\ldots, E$, we can rewrite $z^*(b_i)$ as

$$z^*(b_i) = \text{Maximum}\left\{ w_0^j + w_i^j b_i : j = 1,\ldots, E\right\} \tag{6.4}$$

Observe that $z^*(b_i)$ is the pointwise maximum of a finite number of affine functions in the variable b_i, as long as $b_i \in F$. Hence, as is shown in Figure 6.3, $z^*(b_i)$ is piecewise linear and convex over F. Also, since an increase in b_i makes the feasible region of P more restrictive, $z^*(b_i)$ is also nondecreasing as b_i increases.

Referring to Figure 6.3 and assuming that $z^*(\cdot)$ is composed of four affine functions with the associated dual extreme points as shown in the figure, consider

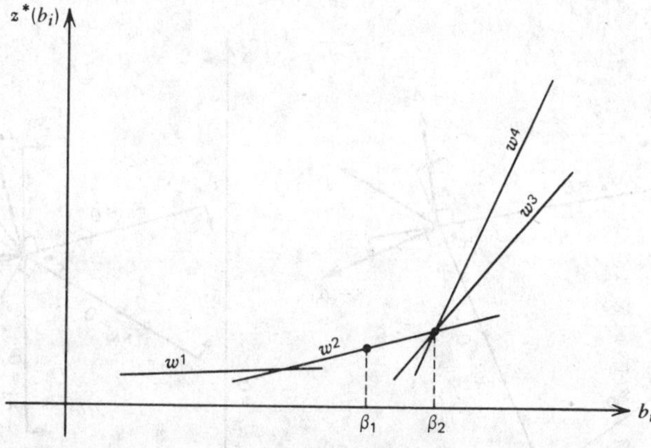

Figure 6.3. **Variation in z^* as a function of b_i.**

$b_i = \beta_1$. At this point $z^*(b_i)$ is differentiable, and in fact, $\partial z^*/\partial b_i = w_i^2$, which agrees with Equation (6.3). However, when $b_i = \beta_2$, for example, $z^*(b_i)$ has a kink at this point and is nondifferentiable. Note that when $b_i = \beta_2$, the dual has alternative optimal extreme point solutions $\mathbf{w}^2$, $\mathbf{w}^3$, and $\mathbf{w}^4$. This implies degeneracy at optimality in the primal problem, for if an optimal primal basic feasible solution $\mathbf{x}^*$ with basis $\mathbf{B}$ is nondegenerate, then by the complementary slackness theorem, *any* dual optimal solution $\mathbf{w}$ must satisfy the dual constraints with respect to the basic variable columns as equalities, that is, $\mathbf{wB} = \mathbf{c}_B$ must hold. This gives $\mathbf{w} = \mathbf{c}_B\mathbf{B}^{-1}$ as the unique dual optimum, a contradiction. In fact, since any optimal dual extreme point solution must be complementary to $\mathbf{x}^*$, the dual constraints corresponding to the $q < m$ positive basic variables in $\mathbf{x}^*$ must be binding. This gives q linearly independent binding hyperplanes at optimality in the dual from the $(m + n)$ inequalities defining the dual feasible region. Each choice of $(m - q)$ additional (linearly independent) hyperplanes that gives some optimal dual extreme point corresponds to a choice of the $(m - q)$ columns of the variables (structural or slacks) in P associated with these constraints. Therefore, because of the linear independence, this corresponds to a choice of a basis in P, and since the nonbasic variables for this basis are zero in $\mathbf{x}^*$, this is a basis associated with $\mathbf{x}^*$. Hence *the alternative optimal dual extreme point solutions all correspond to alternative bases associated with the primal degenerate vertex* $\mathbf{x}^*$.

Now from Equation (6.4), we can define the one-sided (right-hand and left-hand) directional derivatives of $z^*(b_i)$ with respect to b_i. The right-hand derivative, denoted $\partial^+ z^*/\partial b_i$, gives the rate of change in z^* with an *increase* in b_i. As evident from Equation (6.4) (see Figure 6.3)

$$\frac{\partial^+ z^*}{\partial b_i} = \text{Maximum}\left\{w_i^j : \mathbf{w}^j \text{ is an optimal dual vertex at } b_i\right\} \tag{6.5}$$

Similarly, the left-hand derivative, denoted $\partial^- z^*/\partial b_i$, gives the negative of the rate of change in z^* as b_i *decreases*. Again, from Equation (6.4), referring to

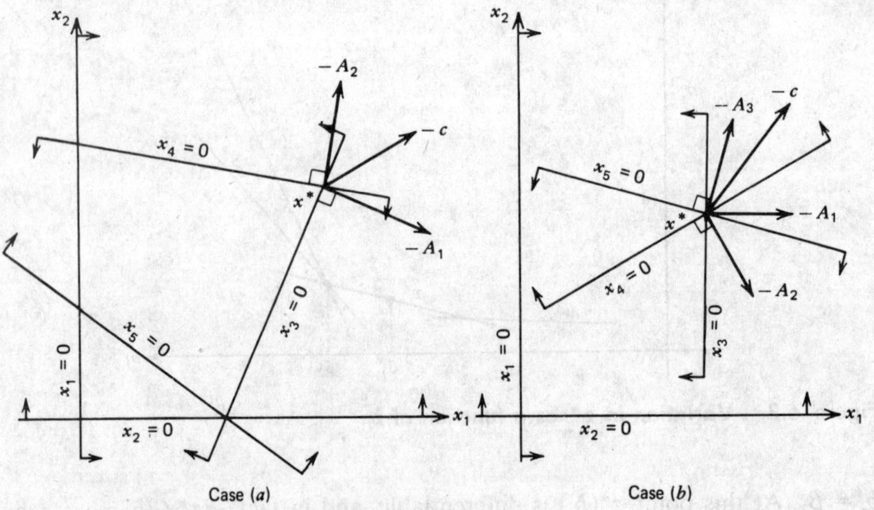

Figure 6.4. Illustration of shadow prices.

Figure 6.3, this is given by

$$\frac{\partial^- z^*}{\partial b_i} = \text{Minimum}\left\{ w_i^j : \mathbf{w}^j \text{ is an optimal dual vertex at } b_i \right\} \tag{6.6}$$

Equations (6.5) and (6.6) respectively give the *right-hand* and *left-hand shadow prices* associated with the ith right-hand-side. Observe that if $\mathbf{x}^*$ is nondegenerate, so that there is a unique dual optimal extreme point $\mathbf{w}^*$, we get $\partial^+ z^*/\partial b_i = \partial^- z^*/\partial b_i = w_i^*$ as in Equation (6.3).

Example 6.6

Consider the (generic) instance of Problem P of Equation (6.1) as shown in Figure 6.4. Here $m = 3$, $n = 2$, x_3, x_4, and x_5 are the slack variables associated with the three primal constraints, and $\mathbf{A}_1$, $\mathbf{A}_2$, and $\mathbf{A}_3$ denote the gradients of the three rows of $\mathbf{A}$. Observe that since the feasible halfspaces are $\mathbf{Ax} \geq \mathbf{b}$, the normal vectors $-\mathbf{A}_1$, $-\mathbf{A}_2$, and $-\mathbf{A}_3$ are as shown in the figure. Note that each dual basis has two basic variables.

First, consider Figure 6.4a. This problem has a unique primal optimal solution $\mathbf{x}^*$. The KKT conditions at $\mathbf{x}^*$ require $-\mathbf{c}$ to be written as $-w_1\mathbf{A}_2 - w_2\mathbf{A}_2$, and hence w_1 and w_2 are both positive and are uniquely determined. That is, there is a unique dual extreme point solution given by $\mathbf{w}^* =$ "$\mathbf{c}_B\mathbf{B}^{-1}$". ($w_1$ and w_2 are the basic variables, and w_3, w_4, and w_5 are nonbasic, where w_4 and w_5 are the two slack variables in the dual constraints.) Because the maximum in Equation (6.4) is unique, Equation (6.3) holds and we have $\partial z^*/\partial b_1 = w_1^* > 0$, $\partial z^*/\partial b_2 = w_2^* > 0$, and $\partial z^*/\partial b_3 = w_3^* = 0$. It is instructive to note that in the optimal tableau representing $\mathbf{x}^*$, $\partial z/\partial x_3 = -(z_3 - c_3) = w_1^*$. Note that the edge along which we move when we hold $x_4 = 0$ and increase x_3 is precisely the direction in which the optimal solution moves when b_1 is (marginally) increased.

Second, in Figure 6.4*a*, rotate $-\mathbf{c}$ so that it aligns with the vector $-\mathbf{A}_1$. We now have alternative optimal primal solutions, one of which is degenerate. However, since $\mathbf{x}^*$ is still optimal, and since by the KKT conditions, all optimal dual solutions must be complementary to $\mathbf{x}^*$, we only need to examine any one primal optimal solution, say, $\mathbf{x}^*$, in order to capture *all* dual optimal solutions. Applying the KKT conditions at $\mathbf{x}^*$, we find that again there is only one dual optimal solution $\mathbf{w}^*$ in which $w_1^* > 0$ is a solution to $-w_1\mathbf{A}_1 = -\mathbf{c}$, and $w_2^* = w_3^* = 0$. Hence Equation (6.3) again holds, although the primal has a degenerate optimum.

Last, consider Figure 6.4*b*. $\mathbf{x}^*$ is a unique optimum. However, it is degenerate and there are two optimal bases, having either x_3 and x_5 as nonbasic variables, or x_4 and x_5 as nonbasic variables. (Note that not all $z_j - c_j \le 0$ for the choice of nonbasic variables x_3 and x_4; in fact, for this choice, $z_4 - c_4 > 0$ (why?).) Hence each of these two bases gives a dual optimal extreme point. The first basis gives a dual optimal vertex $\mathbf{w}^1$ in which w_1 and w_3 are basic, with $w_1^1 > 0$, $w_3^1 > 0$, and $w_2^1 = 0$. The second basis gives a dual optimal vertex $\mathbf{w}^2$ in which w_2 and w_3 are basic, with $w_1^2 = 0$, $w_2^2 > 0$, and $w_3^2 > 0$. In the first case, $-\mathbf{c}$ is written (uniquely) in terms of $-\mathbf{A}_1$ and $-\mathbf{A}_3$ as $-w_1^1\mathbf{A}_1 - w_3^1\mathbf{A}_3$; in the second case, $-\mathbf{c}$ is written (uniquely) in terms of $-\mathbf{A}_2$ and $-\mathbf{A}_3$ as $-w_2^2\mathbf{A}_2 - w_3^2\mathbf{A}_3$. By completing the parallelogram in Figure 6.4*b* to compute the foregoing vector representations, note that $w_3^1 \le w_3^2$. Hence, from Equations (6.5) and (6.6), we get

$$\frac{\partial^+ z}{\partial b_1} = w_1^1 > 0, \quad \frac{\partial^- z}{\partial b_1} = w_1^2 = 0, \quad \frac{\partial^+ z}{\partial b_2} = w_2^2 > 0, \quad \frac{\partial^- z}{\partial b_2} = w_2^1 = 0$$

and

$$\frac{\partial^+ z}{\partial b_3} = w_3^2 > 0 \quad \text{and is greater than} \quad \frac{\partial^- z}{\partial b_3} = w_3^1 > 0.$$

We ask the reader to verify these right-hand and left-hand shadow prices by actually perturbing the respective constraints in Figure 6.4*b* in the appropriate directions. (For example, to realize $\partial^+ z/\partial b_1$ and $\partial^- z/\partial b_1$, perturb the first constraint parallel to itself in the direction of $\mathbf{A}_1$ and $-\mathbf{A}_1$, respectively.) In Exercise 6.24, we ask the reader to repeat the analysis with $-\mathbf{c}$ aligned along $-\mathbf{A}_1$.

6.4 THE DUAL SIMPLEX METHOD

In this section we describe the *dual simplex* method, which solves the dual problem directly on the (primal) simplex tableau. At each iteration we move from a basic feasible solution of the dual problem to an improved basic feasible solution until optimality of the dual (and also the primal) is reached, or else until we conclude that the dual is unbounded and that the primal is infeasible.

Interpretation of Dual Feasibility on the Primal Simplex Tableau

Consider the following linear programming problem.

$$\begin{array}{ll} \text{Minimize} & \mathbf{cx} \\ \text{Subject to} & \mathbf{Ax} \geq \mathbf{b} \\ & \mathbf{x} \geq \mathbf{0} \end{array}$$

Let $\mathbf{B}$ be a basis that is not necessarily feasible and consider the following tableau.

	z	x_1	x_2	$\cdots$	x_n	x_{n+1} (SLACK VARIABLES)	$\cdots$	x_{n+m}	RHS
z	1	$z_1 - c_1$	$z_2 - c_2$	$\cdots$	$z_n - c_n$	$z_{n+1} - c_{n+1}$	$\cdots$	$z_{n+m} - c_{n+m}$	$\mathbf{c}_B\bar{\mathbf{b}}$
x_{B_1}	0	y_{11}	y_{12}	$\cdots$	y_{1n}	$y_{1,n+1}$	$\cdots$	$y_{1,n+m}$	$\bar{b}_1$
x_{B_2}	0	y_{21}	y_{22}	$\cdots$	y_{2n}	$y_{2,n+1}$	$\cdots$	$y_{2,n+m}$	$\bar{b}_2$
$\vdots$	$\vdots$	$\vdots$	$\vdots$		$\vdots$	$\vdots$		$\vdots$	$\vdots$
x_{B_m}	0	y_{m1}	y_{m2}	$\cdots$	y_{mn}	$y_{m,n+1}$	$\cdots$	$y_{m,n+m}$	$\bar{b}_m$

The tableau presents a primal feasible solution if $\bar{b}_i \geq 0$ for $i = 1, 2, \ldots, m$; that is, if $\bar{\mathbf{b}} = \mathbf{B}^{-1}\mathbf{b} \geq \mathbf{0}$. Furthermore, the tableau is optimal if $z_j - c_j \leq 0$ for $j = 1, 2, \ldots, n + m$. Define $\mathbf{w} = \mathbf{c}_B\mathbf{B}^{-1}$. For $j = 1, 2, \ldots, n$, we have

$$z_j - c_j = \mathbf{c}_B\mathbf{B}^{-1}\mathbf{a}_j - c_j = \mathbf{wa}_j - c_j$$

Hence $z_j - c_j \leq 0$ for $j = 1, 2, \ldots, n$ implies that $\mathbf{wa}_j - c_j \leq 0$ for $j = 1, 2, \ldots, n$, which in turn implies that $\mathbf{wA} \leq \mathbf{c}$. Furthermore, note that $\mathbf{a}_{n+i} = -\mathbf{e}_i$ and $c_{n+i} = 0$ for $i = 1, 2, \ldots, m$ and so we have

$$\begin{aligned} z_{n+i} - c_{n+i} &= \mathbf{wa}_{n+i} - c_{n+i} \\ &= \mathbf{w}(-\mathbf{e}_i) - 0 \\ &= -w_i \qquad i = 1, 2, \ldots, m \end{aligned}$$

In addition, if $z_{n+i} - c_{n+i} \leq 0$ for $i = 1, 2, \ldots, m$, then $w_i \geq 0$ for $i = 1, 2, \ldots, m$ and so $\mathbf{w} \geq \mathbf{0}$. We have just shown that $z_j - c_j \leq 0$ for $j = 1, 2, \ldots, n + m$ implies that $\mathbf{wA} \leq \mathbf{c}$ and $\mathbf{w} \geq \mathbf{0}$, where $\mathbf{w} = \mathbf{c}_B\mathbf{B}^{-1}$. In other words, dual feasibility is precisely the simplex optimality criteria $z_j - c_j \leq 0$ for all j. At optimality $\mathbf{w}^* = \mathbf{c}_B\mathbf{B}^{-1}$ and the dual objective $\mathbf{w}^*\mathbf{b} = (\mathbf{c}_B\mathbf{B}^{-1})\mathbf{b} = \mathbf{c}_B(\mathbf{B}^{-1}\mathbf{b}) = \mathbf{c}_B\bar{\mathbf{b}} = z^*$; that is, the primal and dual objectives are equal. Thus we have the following result.

Lemma 6.4

At optimality of the primal minimization problem in the canonical form (that is, $z_j - c_j \leq 0$ for all j), $\mathbf{w}^* = \mathbf{c}_B\mathbf{B}^{-1}$ is an optimal solution to the dual problem. Furthermore $w_i^* = -(z_{n+i} - c_{n+i}) = -z_{n+i}$ for $i = 1, 2, \ldots, m$. Note that a symmetric analysis holds for the equality constrained case. For an optimal tableau with basis $\mathbf{B}$, taking $\mathbf{w}^* = \mathbf{c}_B\mathbf{B}^{-1}$, we again have that the primal optimality conditions $z_j - c_j \le 0$ imply $\mathbf{w}^*\mathbf{a}_j - c_j \le 0$ for $j = 1, \ldots, n$, that is, $\mathbf{w}^*$ is dual feasible. Moreover, the optimal primal objective value z^* equals $\mathbf{c}_B\mathbf{B}^{-1}\mathbf{b} = \mathbf{w}^*\mathbf{b}$, and so the solution $\mathbf{w}^* = \mathbf{c}_B\mathbf{B}^{-1}$ is optimal in the dual problem.

The Dual Simplex Method

Consider the following linear programming problem.

$$\begin{aligned} \text{Minimize} \quad & \mathbf{cx} \\ \text{Subject to} \quad & \mathbf{Ax} = \mathbf{b} \\ & \mathbf{x} \ge \mathbf{0} \end{aligned}$$

In certain instances it is difficult to find a starting basic solution that is feasible (that is, all $\bar{b}_i \ge 0$) to a linear program without adding artificial variables. In these same instances it is often possible to find a starting basis, which is not necessarily feasible, but which is dual feasible (that is, all $z_j - c_j \le 0$ for a minimization problem). In such cases it is useful to develop a variant of the simplex method that would produce a series of simplex tableaux that maintain dual feasibility and complementary slackness and strive toward primal feasibility.

	z	x_1	$\cdots$	x_j	$\cdots$	x_k	$\cdots$	x_n	RHS
z	1	$z_1 - c_1$	$\cdots$	$z_j - c_j$	$\cdots$	$z_k - c_k$	$\cdots$	$z_n - c_n$	$\mathbf{c}_B\bar{\mathbf{b}}$
x_{B_1}	0	y_{11}	$\cdots$	y_{1j}	$\cdots$	y_{1k}	$\cdots$	y_{1n}	$\bar{b}_1$
x_{B_2}	0	y_{21}	$\cdots$	y_{2j}	$\cdots$	y_{2k}	$\cdots$	y_{2n}	$\bar{b}_2$
$\vdots$	$\vdots$	$\vdots$		$\vdots$	$\vdots$				$\vdots$
x_{B_r}	0	y_{r1}	$\cdots$	y_{rj}	$\cdots$	$\textcircled{y_{rk}}$	$\cdots$	y_{rn}	$\bar{b}_r$
$\vdots$	$\vdots$	$\vdots$		$\vdots$		$\vdots$			$\vdots$
x_{B_m}	0	y_{m1}	$\cdots$	y_{mj}	$\cdots$	y_{mk}	$\cdots$	y_{mn}	$\bar{b}_m$

Consider the tableau representing a basic solution at some iteration. Suppose that the tableau is dual feasible (that is, $z_j - c_j \le 0$ for a minimization problem). If the tableau is also primal feasible (that is, all $\bar{b}_i \ge 0$), then we have the optimal solution. Otherwise, consider some $\bar{b}_r < 0$. By selecting row r as a pivot row and some column k such that $y_{rk} < 0$ as a pivot column we can make the new right-hand-side $\bar{b}'_r > 0$. Through a series of such pivots we hope to make all $\bar{b}_i \ge 0$ while maintaining all $z_j - c_j \le 0$ and thus achieve optimality. The question that remains is how do we select the pivot column so as to maintain dual feasibility after pivoting. The pivot column k is determined by the following minimum ratio test.

$$\frac{z_k - c_k}{y_{rk}} = \text{Minimum}\left\{\frac{z_j - c_j}{y_{rj}} : y_{rj} < 0\right\} \tag{6.7}$$

Note that the new entries in row 0 after pivoting are given by

$$(z_j - c_j)' = (z_j - c_j) - \frac{y_{rj}}{y_{rk}}(z_k - c_k)$$

If $y_{rj} \geq 0$, and since $z_k - c_k \leq 0$ and $y_{rk} < 0$, then $(y_{rj}/y_{rk})(z_k - c_k) \geq 0$ and hence $(z_j - c_j)' \leq z_j - c_j$. Since the previous solution was dual feasible, then $z_j - c_j \leq 0$ and hence $(z_j - c_j)' \leq 0$. Now consider the case where $y_{rj} < 0$. By Equation (6.1) we have

$$\frac{z_k - c_k}{y_{rk}} \leq \frac{z_j - c_j}{y_{rj}}$$

Multiplying both sides by $y_{rj} < 0$, we get $(z_j - c_j) - (y_{rj}/y_{rk})(z_k - c_k) \leq 0$, that is, $(z_j - c_j)' \leq 0$. To summarize, if the pivot column is chosen according to Equation (6.7), then the new basis obtained by pivoting at y_{rk} is still dual feasible. Moreover, the dual objective after pivoting is given by $\mathbf{c}_B\mathbf{B}^{-1}\mathbf{b} - (z_k - c_k)\bar{b}_r/y_{rk}$. Since $z_k - c_k \leq 0$, $\bar{b}_r < 0$, and $y_{rk} < 0$, then $-(z_k - c_k)\bar{b}_r/y_{rk} \geq 0$ and the dual objective improves over the current value of $\mathbf{c}_B\mathbf{B}^{-1}\mathbf{b} = \mathbf{wb}$.

We have just described a procedure that moves from a dual basic feasible solution to an improved (at least not worse) basic dual feasible solution. To complete the analysis we must consider the case when $y_{rj} \geq 0$ for all j and hence no column is eligible to be the pivot column. In this case the ith row reads: $\sum_j y_{rj}x_j = \bar{b}_r$. Since $y_{rj} \geq 0$ for all j and x_j is required to be nonnegative, then $\sum_j y_{rj}x_j \geq 0$ for any feasible solution. However, $\bar{b}_r < 0$. This contradiction shows that the primal is infeasible and the dual is unbounded (why?). In Exercise 6.34 we ask the reader to show directly that the dual is unbounded by constructing a direction satisfying the unboundedness criterion.

Summary of the Dual Simplex Method (Minimization Problem)

INITIALIZATION STEP

Find a basis $\mathbf{B}$ of the primal such that $z_j - c_j = \mathbf{c}_B\mathbf{B}^{-1}\mathbf{a}_j - c_j \leq 0$ for all j (in Section 6.6 we describe a procedure for finding such a basis if it is not immediately available).

MAIN STEP

1. If $\bar{\mathbf{b}} = \mathbf{B}^{-1}\mathbf{b} \geq \mathbf{0}$, stop; the current solution is optimal. Otherwise select the pivot row r with $\bar{b}_r < 0$; say, $\bar{b}_r = \text{Minimum}\{\bar{b}_i\}$.
2. If $y_{rj} \geq 0$ for all j, stop; the dual is unbounded and the primal is infeasible. Otherwise select the pivot column k by the following minimum ratio test:

$$\frac{z_k - c_k}{y_{rk}} = \underset{j}{\text{Minimum}}\left\{\frac{z_j - c_j}{y_{rj}} : y_{rj} < 0\right\}$$

3. Pivot at y_{rk} and return to step 1.

In Exercise 6.58, we ask the reader to develop a similar algorithm for the bounded variables case.

Example 6.7

Consider the following problem.

$$\begin{aligned} \text{Minimize} \quad & 2x_1 + 3x_2 + 4x_3 \\ \text{Subject to} \quad & x_1 + 2x_2 + x_3 \geq 3 \\ & 2x_1 - x_2 + 3x_3 \geq 4 \\ & x_1, \ x_2, \ x_3 \geq 0 \end{aligned}$$

A starting basic solution that is dual feasible can be obtained by utilizing the slack variables x_4 and x_5. This results from the fact that the cost vector is nonnegative. Applying the dual simplex method, we obtain the following series of tableaux.

	z	x_1	x_2	x_3	x_4	x_5	RHS
z	1	-2	-3	-4	0	0	0
x_4	0	-1	-2	-1	1	0	-3
x_5	0	$\boxed{-2}$	1	-3	0	1	-4

	z	x_1	x_2	x_3	x_4	x_5	RHS
z	1	0	-4	-1	0	-1	4
x_4	0	0	$\boxed{-\frac{5}{2}}$	$\frac{1}{2}$	1	$-\frac{1}{2}$	-1
x_1	0	1	$-\frac{1}{2}$	$\frac{3}{2}$	0	$-\frac{1}{2}$	2

	z	x_1	x_2	x_3	x_4	x_5	RHS
z	1	0	0	$-\frac{9}{5}$	$-\frac{8}{5}$	$-\frac{1}{5}$	$\frac{28}{5}$
x_2	0	0	1	$-\frac{1}{5}$	$-\frac{2}{5}$	$\frac{1}{5}$	$\frac{2}{5}$
x_1	0	1	0	$\frac{7}{5}$	$-\frac{1}{5}$	$-\frac{2}{5}$	$\frac{11}{5}$

Since $\bar{\mathbf{b}} \geq \mathbf{0}$ and $z_j - c_j \leq 0$ for all j, the optimal primal and dual solutions are at hand. In particular,

$$(x_1^*, x_2^*, x_3^*, x_4^*, x_5^*) = \left(\tfrac{11}{5}, \tfrac{2}{5}, 0, 0, 0\right)$$

$$(w_1^*, w_2^*) = \left(\tfrac{8}{5}, \tfrac{1}{5}\right)$$

Note that w_1^* and w_2^* are respectively the negatives of the $z_j - c_j$ entries under the slack variables x_4 and x_5. Also note that in each subsequent tableau the value of the objective function is increasing, as it should, for the dual (maximization) problem.

The Complementary Basic Dual Solution

Consider the pair of primal and dual linear programs in canonical form, where **A** is $m \times n$.

P: Minimize	$\mathbf{cx}$	D: Maximize	$\mathbf{wb}$
Subject to	$\mathbf{Ax} \geq \mathbf{b}$	Subject to	$\mathbf{wA} \leq \mathbf{c}$
	$\mathbf{x} \geq \mathbf{0}$		$\mathbf{w} \geq \mathbf{0}$

Recall that the variables $x_1, \ldots, x_n$ in the primal are complementary to the slack variables $w_{m+1}, \ldots, w_{m+n}$ in the dual, and the slack variables $x_{n+1}, \ldots, x_{n+m}$ in the primal are complementary to the variables $w_1, \ldots, w_m$ in the dual. Now consider any basis B for the primal problem (feasible or not). The corresponding tableau for this basis has $(w_1, \ldots, w_m) = \mathbf{c}_B\mathbf{B}^{-1}$ as the simplex multipliers that have been used with the constraint rows when adding them to the objective row. Accordingly, for this solution $\mathbf{w}$, the $z_j - c_j$ values for the x_j variables are $\mathbf{wa}_j - c_j = -w_{m+j}$ for $j = 1, \ldots, n$, and the $z_{n+i} - c_{n+i}$ values for the x_{n+i} variables are $-w_i$ for $i = 1, \ldots, m$, as seen at the beginning of this section. Now $z_j - c_j = 0$ for x_j basic, $j \in \{1, \ldots, n + m\}$, and in fact, making these $z_j - c_j$ equal to zero uniquely determines the $z_j - c_j$ values for the nonbasic variables. In other words, putting the dual variables complementary to the basic x_j variables equal to zero, uniquely determines from the dual equality system the other dual variables that are complementary to the nonbasic x_j variables. Hence the dual solution, the negative of which appears in the objective row of the current tableau, is a *basic* dual solution. Here the basic dual variables are complementary to the nonbasic primal variables, and correspondingly, the nonbasic dual variables are complementary to the basic primal variables. Hence letting $\mathbf{x}_{OB}$ and $\mathbf{x}_{ON}$ be the basic and nonbasic variables respectively from the variables $x_1, \ldots, x_n$ and letting $\mathbf{x}_{SB}$ and $\mathbf{x}_{SN}$ be the basic and nonbasic variables respectively from the slack variables $(x_{n+1}, \ldots, x_{n+m})$, we can rewrite the primal constraints as

$$\begin{aligned} \mathbf{A}_{11}\mathbf{x}_{OB} + \mathbf{A}_{12}\mathbf{x}_{ON} - \mathbf{x}_{SN} &= \mathbf{b}_1 \\ \mathbf{A}_{21}\mathbf{x}_{OB} + \mathbf{A}_{22}\mathbf{x}_{ON} - \mathbf{x}_{SB} &= \mathbf{b}_2 \end{aligned} \tag{6.8}$$

Then the primal basis **B** is given by the columns of $\mathbf{x}_{OB}$ and $\mathbf{x}_{SB}$ as

$$\mathbf{B} = \begin{array}{c} \begin{matrix} \mathbf{x}_{OB} & \mathbf{x}_{SB} \end{matrix} \\ \left[\begin{array}{c|c} \mathbf{A}_{11} & \mathbf{0} \\ \hline \mathbf{A}_{21} & -\mathbf{I} \end{array}\right] \end{array} \tag{6.9}$$

Now denote by $\mathbf{w}_{OB}$ and $\mathbf{w}_{ON}$ the dual variables associated with the two constraint sets in Equation (6.8). Let $\mathbf{w}_{SN}$ and $\mathbf{w}_{SB}$ be the slack variables in the dual constraints written with respect to the columns of the variables $\mathbf{x}_{OB}$ and $\mathbf{x}_{ON}$ respectively. Then with $\mathbf{c}$ partitioned accordingly as $(\mathbf{c}_{OB}, \mathbf{c}_{ON})$, the dual constraints can be written as follows.

$$\mathbf{A}_{11}^t\mathbf{w}_{OB} + \mathbf{A}_{21}^t\mathbf{w}_{ON} + \mathbf{w}_{SN} = \mathbf{c}_{OB}^t$$

$$\mathbf{A}_{12}^t\mathbf{w}_{OB} + \mathbf{A}_{22}^t\mathbf{w}_{ON} + \mathbf{w}_{SB} = \mathbf{c}_{ON}^t$$

Hence the complementary dual basis associated with the primal basis $\mathbf{B}$ has $\mathbf{w}_{OB}$ and $\mathbf{w}_{SB}$ as the basic variables and is given by

$$\begin{array}{cc} \mathbf{w}_{OB} & \mathbf{w}_{SB} \end{array}$$
$$\left[\begin{array}{c|c} \mathbf{A}_{11}^t & \mathbf{0} \\ \hline \mathbf{A}_{12}^t & \mathbf{I} \end{array}\right] \tag{6.10}$$

We ask the reader to verify in Exercise 6.37 that the matrix (6.10) is invertible given that $\mathbf{B}^{-1}$ in Equation (6.9) exists, and in fact, the dual basic solution corresponding to Equation (6.10) gives $(w_1, \ldots, w_m) = \mathbf{c}_B\mathbf{B}^{-1}$, where $\mathbf{c}_B \equiv (\mathbf{c}_{OB}, \mathbf{0})$. Note that since the basic variables in one problem are complementary to the nonbasic variables in the other problem, the complementary slackness condition holds for this primal-dual pair of basic solutions.

Next, consider the following pair of primal and dual problems in standard form.

P:	Minimize	$\mathbf{cx}$	D:	Maximize	$\mathbf{wb}$
	Subject to	$\mathbf{Ax} = \mathbf{b}$		Subject to	$\mathbf{wA} \le \mathbf{c}$
		$\mathbf{x} \ge \mathbf{0}$			$\mathbf{w}$ unrestricted

Given any primal basis $\mathbf{B}$, there is as previously an associated complementary dual basis. To illustrate, introduce the dual slack vector $\mathbf{w}_s$ so that $\mathbf{wA} + \mathbf{w}_s = \mathbf{c}$. The dual constraints can be rewritten in the following more convenient form:

$$\mathbf{A}^t\mathbf{w}^t + \mathbf{I}\mathbf{w}_s^t = \mathbf{c}^t$$

$$\mathbf{w}^t \text{ unrestricted}$$

$$\mathbf{w}_s^t \ge \mathbf{0}$$

Given the primal basis $\mathbf{B}$, recall that $\mathbf{w} = \mathbf{c}_B\mathbf{B}^{-1}$. Substituting in the dual constraints, we get

$$\begin{aligned} \mathbf{w}_s^t &= \mathbf{c}^t - \mathbf{A}^t\mathbf{w}^t \\ &= \mathbf{c}^t - \mathbf{A}^t(\mathbf{B}^{-1})^t\mathbf{c}_B^t \\ &= \begin{pmatrix} \mathbf{c}_B^t \\ \mathbf{c}_N^t \end{pmatrix} - \begin{pmatrix} \mathbf{B}^t \\ \mathbf{N}^t \end{pmatrix}(\mathbf{B}^{-1})^t\mathbf{c}_B^t \\ &= \begin{pmatrix} \mathbf{0} \\ \mathbf{c}_N^t - \mathbf{N}^t(\mathbf{B}^{-1})^t\mathbf{c}_B^t \end{pmatrix} \end{aligned} \tag{6.11}$$

Note that $\mathbf{w} = \mathbf{c}_B\mathbf{B}^{-1}$ and Equation (6.11) lead naturally to a dual basis. Since both $\mathbf{c}_B\mathbf{B}^{-1}$ and $\mathbf{c}_N^t - \mathbf{N}^t(\mathbf{B}^{-1})^t\mathbf{c}_B^t$ are not necessarily zero, then the vector $\mathbf{w}$ and the last $n - m$ components of $\mathbf{w}_s$, which are complementary to the primal nonbasic variables, form the dual basis. In particular the dual basis corresponding to the primal basis $\mathbf{B}$ is given by

$$\left[\begin{array}{c|c} \mathbf{B}^t & \mathbf{0} \\ \hline \mathbf{N}^t & \mathbf{I}_{n-m} \end{array}\right]$$

The rank of the preceding matrix is n. The primal basis is feasible if $\mathbf{B}^{-1}\mathbf{b} \geq \mathbf{0}$ and the dual basis is feasible if $\mathbf{w}_s \geq \mathbf{0}$; that is, if $\mathbf{c}_N - \mathbf{c}_B\mathbf{B}^{-1}\mathbf{N} \geq \mathbf{0}$ (see Equation 6.11). Even if these conditions do not hold, the primal and dual bases are *complementary* in the sense that the complementary slackness condition $(\mathbf{wA} - \mathbf{c})\mathbf{x} = 0$ holds as shown:

$$(\mathbf{wA} - \mathbf{c})\mathbf{x} = \mathbf{w}_s\mathbf{x} = (\mathbf{0}, \mathbf{c}_N - \mathbf{c}_B\mathbf{B}^{-1}\mathbf{N})\begin{pmatrix} \mathbf{B}^{-1}\mathbf{b} \\ \mathbf{0} \end{pmatrix} = 0$$

To summarize, during any dual simplex iteration we have a primal basis that is not necessarily feasible and a complementary dual feasible basis. At termination primal feasibility is attained, and so all the KKT optimality conditions hold.

Finite Convergence of the Dual Simplex Method

First, let us show finite convergence in the absence of dual degeneracy. Note that the dual simplex method moves among dual feasible bases. Also recall that the difference in the dual objective between two successive iterations is $-(z_k - c_k)\bar{b}_r/y_{rk}$. Note that $\bar{b}_r < 0$, $y_{rk} < 0$,and $z_k - c_k \leq 0$ and hence $-(z_k - c_k)\bar{b}_r/y_{rk} \geq 0$. In particular, if $z_k - c_k < 0$, then the dual objective strictly increases and hence no basis can be repeated and the algorithm must converge in a finite number of steps. By the foregoing characterization of the complementary dual basis, and since x_k is a nonbasic primal variable, then the dual slack of the constraint $\mathbf{wa}_k \leq c_k$ is basic. Assuming dual nondegeneracy, this dual slack variable must be positive so that $\mathbf{wa}_k < c_k$, that is, $z_k - c_k < 0$. As discussed, this would guarantee finite convergence since the dual objective strictly increases at each iteration.

In the presence of dual degeneracy, however, one would need a cycling prevention rule in order to guarantee finite convergence. From the discussion in Section 4.6 and noting that the dual simplex method is essentially applying the ordinary simplex algorithm to the dual problem, we can readily devise cycling prevention rules. In particular, the following lexicographic cycling prevention rule may be used. Suppose that we have a dual feasible tableau for which $\begin{bmatrix} c_j - z_j \\ \mathbf{y}_j \end{bmatrix}$ is lexicographically positive for each nonbasic variable column. Note that this is the updated nonbasic variable column, with the negative of the objective row coefficient. A technique to obtain a starting tableau satisfying this property is presented in Section 6.6. Then having selected a pivot row r in the dual simplex tableau with $\bar{b}_r < 0$, one can scale each of these lexicographically positive nonbasic columns that have a negative y_{rj} coefficient in row r by dividing it by

$-y_{rj}$ and picking the resulting lexicographically minimum column from among these as a pivot column. We ask the reader in Exercise 6.36 to show that this will maintain the lexicographic positivity property and, in fact, prevent cycling. Similarly, noting that every entering and leaving basic variable in the primal problem has a corresponding leaving and entering complementary variable in the complementary dual basis, Bland's rule may be readily adapted to prevent cycling in the dual simplex method. Identical to the discussion in Section 4.6, of all the candidates for pivot rows and similarly for pivot columns, one can break ties (if any) by favoring the x_j variable with the smallest index. We defer the details to Exercise 6.36.

6.5 THE PRIMAL-DUAL METHOD

Recall that in the dual simplex method we begin with a basic (not necessarily feasible) solution to the primal problem and a complementary basic feasible solution to the dual problem. The dual simplex method proceeds, by pivoting, through a series of dual basic feasible solutions until the associated complementary primal basic solution is feasible, thus satisfying all of the KKT conditions for optimality.

In this section we describe a method, called the *primal-dual* algorithm, similar to the dual simplex method, which begins with dual feasibility and proceeds to obtain primal feasibility while maintaining complementary slackness. An important difference between the dual simplex method and the primal-dual method is that the primal-dual algorithm does not require a dual feasible solution to be basic. Given a dual feasible solution, the primal variables that correspond to tight dual constraints (so that complementary slackness is satisfied) are determined. Using phase I of the simplex method, we attempt to attain primal feasibility using only these variables. If we are unable to obtain primal feasibility, we change the dual feasible solution in such a way as to admit at least one new variable to the phase I problem. This is continued until either the primal becomes feasible or the dual becomes unbounded.

Development of the Primal-Dual Method

Consider the following primal and dual problems in standard form where $\mathbf{b} \geq \mathbf{0}$.

P: Minimize $\mathbf{cx}$	D: Maximize $\mathbf{wb}$
Subject to $\mathbf{Ax} = \mathbf{b}$	Subject to $\mathbf{wA} \leq \mathbf{c}$
$\mathbf{x} \geq \mathbf{0}$	$\mathbf{w}$ unrestricted

Let $\mathbf{w}$ be an initial dual feasible solution, that is, $\mathbf{wa}_j \leq c_j$ for all j. By complementary slackness, if $\mathbf{wa}_j = c_j$, then x_j is allowed to be positive and we attempt to attain primal feasibility from among these variables. Let $Q = \{j : \mathbf{wa}_j - c_j = 0\}$, that is, the set of indices of primal variables allowed to be positive. Then the phase I problem that attempts to find a feasible solution to the primal problem among variables in the set Q becomes:

$$\text{Minimize} \quad \sum_{j \in Q} 0x_j + \mathbf{1x}_a$$

$$\text{Subject to} \quad \sum_{j \in Q} \mathbf{a}_j x_j + \mathbf{Ix}_a = \mathbf{b}$$
$$x_j \geq 0 \quad \text{for } j \in Q$$
$$\mathbf{x}_a \geq \mathbf{0}$$

We utilize the artificial vector $\mathbf{x}_a$ to obtain a starting basic feasible solution to the phase I problem. The phase I problem is sometimes called the *restricted primal* problem.

Denote the optimal objective value of the foregoing problem by x_0. At optimality of the phase I problem either $x_0 = 0$ or $x_0 > 0$. When $x_0 = 0$, we have a feasible solution to the primal problem since all artificials are zeros. Furthermore, we have a dual feasible solution, and the complementary slackness condition $(\mathbf{wa}_j - c_j)x_j = 0$ holds because either $j \in Q$ in which case $\mathbf{wa}_j - c_j = 0$, or else $j \notin Q$ in which case $x_j = 0$. Therefore we have an optimal solution of the overall problem whenever $x_0 = 0$. If $x_0 > 0$, primal feasibility is not achieved and we must construct a new dual solution that would admit a new variable to the restricted primal problem in such a way that x_0 might be decreased. We shall modify the dual vector $\mathbf{w}$ such that all the basic primal variables in the restricted problem remain in the new restricted primal problem, and in addition, at least one primal variable that did not belong to the set Q would get passed to the restricted primal problem. Furthermore, this variable would reduce x_0 if introduced in the basis. In order to construct such a dual vector, consider the following dual of the phase I problem.

$$\text{Maximize} \quad \mathbf{vb}$$

$$\begin{aligned} \text{Subject to} \quad & \mathbf{va}_j \leq 0 \qquad j \in Q \\ & \mathbf{v} \quad\ \leq \mathbf{1} \\ & \mathbf{v} \qquad \text{unrestricted} \end{aligned}$$

Let $\mathbf{v}^*$ be an optimal solution to the foregoing problem. Then if a real variable x_j is a member of the optimal basis for the restricted primal, the associated dual constraint must be tight, that is, $\mathbf{v}^*\mathbf{a}_j = 0$. Also the criterion for basis entry in the restricted primal problem is that the associated dual constraint be violated, that is, $\mathbf{v}^*\mathbf{a}_j > 0$. However, no variable presently in the restricted primal has this property since the restricted primal is optimal. For $j \notin Q$, compute $\mathbf{v}^*\mathbf{a}_j$. If $\mathbf{v}^*\mathbf{a}_j > 0$, then if x_j could be passed to the restricted primal problem it would be a candidate to enter the basis with the potential of a further decrease in x_0. Therefore we must find a way to force some variable x_j with $\mathbf{v}^*\mathbf{a}_j > 0$ into the set Q.

Construct the following dual vector $\mathbf{w}'$, where $\theta > 0$:

$$\mathbf{w}' = \mathbf{w} + \theta \mathbf{v}^*$$

Then

$$\begin{aligned}\mathbf{w}'\mathbf{a}_j - c_j &= (\mathbf{w} + \theta\mathbf{v}^*)\mathbf{a}_j - c_j \\ &= (\mathbf{w}\mathbf{a}_j - c_j) + \theta(\mathbf{v}^*\mathbf{a}_j) \qquad (6.12)\end{aligned}$$

Note that $\mathbf{w}\mathbf{a}_j - c_j = 0$ and $\mathbf{v}^*\mathbf{a}_j \le 0$ for $j \in Q$. Thus Equation (6.12) implies that $\mathbf{w}'\mathbf{a}_j - c_j \le 0$ for $j \in Q$. In particular, if x_j with $j \in Q$ is a basic variable in the restricted primal, then $\mathbf{v}^*\mathbf{a}_j = 0$ and $\mathbf{w}'\mathbf{a}_j - c_j = 0$, permitting j in the new restricted primal problem. If $j \notin Q$ and $\mathbf{v}^*\mathbf{a}_j \le 0$, then from Equation (6.12) and noting that $\mathbf{w}\mathbf{a}_j - c_j < 0$, we have $\mathbf{w}'\mathbf{a}_j - c_j < 0$. Finally consider $j \notin Q$ with $\mathbf{v}^*\mathbf{a}_j > 0$. Examining Equation (6.12) and noting that $\mathbf{w}\mathbf{a}_j - c_j < 0$ for $j \notin Q$, it is evident that we can choose a $\theta > 0$ such that $\mathbf{w}'\mathbf{a}_j - c_j \le 0$ for $j \notin Q$ with at least one component equal to zero. In particular, define θ as follows:

$$\theta = \frac{-(\mathbf{w}\mathbf{a}_k - c_k)}{\mathbf{v}^*\mathbf{a}_k} = \underset{j}{\text{Minimum}} \left\{ \frac{-(\mathbf{w}\mathbf{a}_j - c_j)}{\mathbf{v}^*\mathbf{a}_j} : \mathbf{v}^*\mathbf{a}_j > 0 \right\} > 0 \qquad (6.13)$$

By definition of θ and from Equation (6.12), we see that $\mathbf{w}'\mathbf{a}_k - c_k = 0$. Furthermore, for each j with $\mathbf{v}^*\mathbf{a}_j > 0$, and noting Equations (6.12) and (6.13), we have $\mathbf{w}'\mathbf{a}_j - c_j \le 0$.

To summarize, modifying the dual vector as detailed previously leads to a new feasible dual solution where $\mathbf{w}'\mathbf{a}_j - c_j \le 0$ for all j. Furthermore, all the variables that belonged to the restricted primal basis are passed to the new restricted primal. In addition, a new variable x_k that is a candidate to enter the basis, is passed to the restricted primal problem. Hence we continue from the present restricted primal basis by entering x_k, which leads to a potential reduction in x_0.

Case of Unbounded Dual

The foregoing process is continued until either $x_0 = 0$ in which case we have an optimal solution or else $x_0 > 0$ and $\mathbf{v}^*\mathbf{a}_j \le 0$ for all $j \notin Q$. In this case consider $\mathbf{w}' = \mathbf{w} + \theta\mathbf{v}^*$. Since $\mathbf{w}\mathbf{a}_j - c_j \le 0$ for all j and by assumption $\mathbf{v}^*\mathbf{a}_j \le 0$ for all j, then from Equation (6.12) $\mathbf{w}'$ is a dual feasible solution for all $\theta > 0$. Furthermore, the dual objective is

$$\mathbf{w}'\mathbf{b} = (\mathbf{w} + \theta\mathbf{v}^*)\mathbf{b} = \mathbf{w}\mathbf{b} + \theta\mathbf{v}^*\mathbf{b}$$

Since $\mathbf{v}^*\mathbf{b} = x_0$ (why?) and the latter is positive, then $\mathbf{w}'\mathbf{b}$ can be increased indefinitely by choosing θ arbitrarily large. Therefore the dual is unbounded and hence the primal is infeasible. Alternatively, note that the *total* (usual) phase I problem has the current restricted phase I solution as an optimum (why?). Since $x_0 > 0$, the primal is infeasible.

Summary of the Primal-Dual Algorithm (Minimization Problem)

INITIALIZATION STEP

Choose a vector $\mathbf{w}$ such that $\mathbf{w}\mathbf{a}_j - c_j \le 0$ for all j.

MAIN STEP

1. Let $Q = \{j : \mathbf{w}\mathbf{a}_j - c_j = 0\}$ and solve the following restricted primal problem.

$$\begin{aligned} \text{Minimize} \quad & \sum_{j\in Q} 0x_j + \mathbf{1}\mathbf{x}_a \\ \text{Subject to} \quad & \sum_{j\in Q} \mathbf{a}_j x_j + \mathbf{x}_a = \mathbf{b} \\ & x_j \geq 0 \quad \text{for } j \in Q \\ & \mathbf{x}_a \geq \mathbf{0} \end{aligned}$$

   Denote the optimal objective by x_0. If $x_0 = 0$, stop; an optimal solution is obtained. Otherwise let $\mathbf{v}^*$ be the optimal dual solution to the foregoing restricted primal problem.
2. If $\mathbf{v}^*\mathbf{a}_j \leq 0$ for all j, then stop; the dual is unbounded and the primal is infeasible. Otherwise let

$$\theta = \underset{j}{\text{Minimum}} \left\{ \frac{-(\mathbf{w}\mathbf{a}_j - c_j)}{\mathbf{v}^*\mathbf{a}_j} : \mathbf{v}^*\mathbf{a}_j > 0 \right\} > 0$$

   and replace $\mathbf{w}$ by $\mathbf{w} + \theta\mathbf{v}^*$. Repeat step 1.

Example 6.8

Consider the following problem.

$$\begin{aligned} \text{Minimize} \quad & 3x_1 + 4x_2 + 6x_3 + 7x_4 + x_5 \\ \text{Subject to} \quad & 2x_1 - x_2 + x_3 + 6x_4 - 5x_5 - x_6 = 6 \\ & x_1 + x_2 + 2x_3 + x_4 + 2x_5 - x_7 = 3 \\ & x_1, x_2, x_3, x_4, x_5, x_6, x_7 \geq 0 \end{aligned}$$

The dual problem is given by the following.

$$\begin{aligned} \text{Maximize} \quad & 6w_1 + 3w_2 \\ \text{Subject to} \quad & 2w_1 + w_2 \leq 3 \\ & -w_1 + w_2 \leq 4 \\ & w_1 + 2w_2 \leq 6 \\ & 6w_1 + w_2 \leq 7 \\ & -5w_1 + 2w_2 \leq 1 \\ & -w_1 \leq 0 \\ & -w_2 \leq 0 \\ & w_1, w_2 \text{ unrestricted} \end{aligned}$$

An initial dual feasible solution is given by $\mathbf{w} = (w_1, w_2) = (0, 0)$. Substituting $\mathbf{w}$ in each dual constraint, we find that the last two dual constraints are tight so that $Q = \{6, 7\}$. Denoting the artificial variables by x_8 and x_9, the restricted primal problem becomes as follows.

$$\begin{array}{llll} \text{Minimize} & x_8 + x_9 & & \\ \text{Subject to} & -x_6 \quad + x_8 & = 6 \\ & \quad - x_7 \quad + x_9 & = 3 \\ & x_6, \quad x_7, \quad x_8, \quad x_9 & \geq 0 \end{array}$$

The optimal solution to this restricted primal is clearly $(x_6, x_7, x_8, x_9) = (0, 0, 6, 3)$ and the optimal objective $x_0 = 9$. The dual of the foregoing restricted primal is the following.

$$\begin{array}{ll} \text{Maximize} & 6v_1 + 3v_2 \\ \text{Subject to} & -v_1 \leq 0 \\ & \quad - v_2 \leq 0 \\ & v_1 \leq 1 \\ & \quad v_2 \leq 1 \\ & v_1, \quad v_2 \quad \text{unrestricted} \end{array}$$

Utilizing complementary slackness, we see that since x_8 and x_9 are basic, the last two dual constraints must be tight and $\mathbf{v}^* = (v_1^*, v_2^*) = (1, 1)$. Computing $\mathbf{v}^*\mathbf{a}_j$ for each column j, we have $\mathbf{v}^*\mathbf{a}_1 = 3$, $\mathbf{v}^*\mathbf{a}_2 = 0$, $\mathbf{v}^*\mathbf{a}_3 = 3$, $\mathbf{v}^*\mathbf{a}_4 = 7$, and $\mathbf{v}^*\mathbf{a}_5 = -3$. Then θ is determined as follows:

$$\theta = \text{Minimum}\left\{-\left(-\tfrac{3}{3}\right), -\left(-\tfrac{6}{3}\right), -\left(-\tfrac{7}{7}\right)\right\} = 1$$

and $\mathbf{w}' = (0, 0) + 1(1, 1) = (1, 1)$.

With the new dual solution $\mathbf{w}'$ we recompute Q and obtain $Q = \{1, 4\}$, giving the following restricted primal:

$$\begin{array}{ll} \text{Minimize} & x_8 + x_9 \\ \text{Subject to} & 2x_1 + 6x_4 + x_8 \quad = 6 \\ & x_1 + \quad x_4 \quad + x_9 = 3 \\ & x_1, \quad x_4, \quad x_8, \quad x_9 \geq 0 \end{array}$$

This time an optimal solution to the restricted problem is given by

$$(x_1, x_4, x_8, x_9) = (3, 0, 0, 0)$$

with $x_0 = 0$. Thus we have an optimal solution to the original problem with the optimal primal and dual solutions being

$$(x_1^*, x_2^*, x_3^*, x_4^*, x_5^*, x_6^*, x_7^*) = (3, 0, 0, 0, 0, 0, 0)$$

and

$$(w_1^*, w_2^*) = (1,1).$$

Tableau Form of the Primal-Dual Method

Let $z_j - c_j$ be the row zero coefficients for the original primal problem and let $\hat{z}_j - \hat{c}_j$ be the row zero coefficients for the restricted primal problem. Then for each real variable x_j we have

$$z_j - c_j = \mathbf{wa}_j - c_j \qquad \text{and} \qquad \hat{z}_j - \hat{c}_j = \mathbf{va}_j - 0 = \mathbf{va}_j$$

We also have

$$\frac{\mathbf{wa}_j - c_j}{\mathbf{va}_j} = \frac{z_j - c_j}{\hat{z}_j - \hat{c}_j}$$

and

$$(\mathbf{wa}_j - c_j) + \theta \mathbf{va}_j = (z_j - c_j) + \theta(\hat{z}_j - \hat{c}_j)$$

We can carry out all of the necessary operations directly in one tableau. In this tableau we have two objective rows; the first gives the $z_j - c_j$'s and the second gives the $\hat{z}_j - \hat{c}_j$'s. We shall apply this tableau method to the foregoing problem. The initial tableau is displayed. In this example $\mathbf{w}$ is initially $(0,0)$, so that $z_j - c_j = \mathbf{wa}_j - c_j = -c_j$ and the right-hand-side entry in the z-row is zero. When $\mathbf{w} \neq \mathbf{0}$, we still compute $z_j - c_j = \mathbf{wa}_j - c_j$, but also initialize the RHS entry of the z-row to $\mathbf{wb}$ instead to zero. [Try starting the tableau with $\mathbf{w} = (1,0)$.]

x_1	x_2	x_3	x_4	x_5	x_6	x_7	x_8	x_9	RHS
−3	−4	−6	−7	−1	0	0	0	0	0
0	0	0	0	0	0	0	−1	−1	0
2	−1	1	6	−5	−1	0	1	0	6
1	1	2	1	2	0	−1	0	1	3

Since we begin with x_8 and x_9 in the basis for the restricted primal, we must perform some preliminary pivoting to zero their cost coefficients in the phase I objective. We do this by adding the first and second constraint rows to the restricted primal objective row. Then $\hat{z}_j - \hat{c}_j = 0$ for the two basic variables x_8 and x_9. Let □ indicate the variables in the restricted primal problem, that is, those for which $z_j - c_j = 0$. As the restricted primal problem is solved, only the variables signaled with □ are allowed to enter the basis.

						□	□	□	□	
	x_1	x_2	x_3	x_4	x_5	x_6	x_7	x_8	x_9	RHS
z	−3	−4	−6	−7	−1	0	0	0	0	0
x_0	3	0	3	7	−3	−1	−1	0	0	9
x_8	2	−1	1	6	−5	−1	0	1	0	6
x_9	1	1	2	1	2	0	−1	0	1	3

Since $\hat{z}_j - \hat{c}_j \leq 0$ for all variables in the restricted problem, we have an optimal solution for phase I. Then θ is given by

$$\theta = \text{Minimum}\left\{\frac{-(z_j - c_j)}{\hat{z}_j - \hat{c}_j} : \hat{z}_j - \hat{c}_j > 0\right\}$$

$$\text{Minimum}\left\{-\left(-\tfrac{3}{3}\right), -\left(-\tfrac{6}{3}\right), -\left(-\tfrac{7}{7}\right)\right\} = 1$$

Thus we add one times the phase I objective row to the original objective row. This leads to the following tableau. The phase I problem is solved by only utilizing the variables in the set Q, that is, those with $z_j - c_j = 0$.

	□ x_1	x_2	x_3	□ x_4	x_5	x_6	x_7	□ x_8	□ x_9	RHS
z	0	−4	−3	0	−4	−1	−1	0	0	9
x_0	3	0	3	7	−3	−1	−1	0	0	9
x_8	2	−1	1	(6)	−5	−1	0	1	0	6
x_9	1	1	2	1	2	0	−1	0	1	3

	□ x_1	x_2	x_3	□ x_4	x_5	x_6	x_7	□ x_8	□ x_9	RHS
z	0	−4	−3	0	−4	−1	−1	0	0	9
x_0	$\frac{4}{6}$	$\frac{7}{6}$	$\frac{11}{6}$	0	$\frac{17}{6}$	$\frac{1}{6}$	−1	$-\frac{7}{6}$	0	2
x_4	$\frac{2}{6}$	$-\frac{1}{6}$	$\frac{1}{6}$	1	$-\frac{5}{6}$	$-\frac{1}{6}$	0	$\frac{1}{6}$	0	1
x_9	($\frac{4}{6}$)	$\frac{7}{6}$	$\frac{11}{6}$	0	$\frac{17}{6}$	$\frac{1}{6}$	−1	$-\frac{1}{6}$	1	2

	□ x_1	x_2	x_3	□ x_4	x_5	x_6	x_7	□ x_8	□ x_9	RHS
z	0	−4	−3	0	−4	−1	−1	0	0	9
x_0	0	0	0	0	0	0	0	−1	−1	0
x_4	0	$-\frac{3}{4}$	$-\frac{3}{4}$	1	$-\frac{9}{4}$	$-\frac{1}{4}$	$\frac{2}{4}$	$\frac{1}{4}$	$-\frac{2}{4}$	0
x_1	1	$\frac{7}{4}$	$\frac{11}{4}$	0	$\frac{17}{4}$	$\frac{1}{4}$	$-\frac{6}{4}$	$-\frac{1}{4}$	$\frac{6}{4}$	3

Since $x_0 = 0$, then the optimal solution is found, namely

$$(x_1^*, x_2^*, x_3^*, x_4^*, x_5^*, x_6^*, x_7^*) = (3, 0, 0, 0, 0, 0, 0)$$

whose objective is 9.

Finite Convergence of the Primal-Dual Method and an Insight

Recall that at each iteration an improving variable is added to the restricted primal problem. Therefore, in the absence of degeneracy in the restricted primal problem, the optimal objective x_0 strictly decreases at each iteration. This means that the set Q generated at any iteration is distinct from all those generated at previous iterations. Since there is only a finite number of sets of the form Q (recall $Q \subset \{1, 2, \ldots, n\}$) and none of them can be repeated, then the algorithm terminates in a finite number of steps.

Let us provide an insight into the primal-dual algorithm that readily resolves the convergence issue under degeneracy, following Section 4.6. Note that the primal-dual method is essentially working on the *total* phase I problem of minimizing the sum of the artificial variables $\mathbf{1x}_a$ subject to $\mathbf{Ax} + \mathbf{Ix}_a = \mathbf{b}$, $\mathbf{x} \geq \mathbf{0}$ and $\mathbf{x}_a \geq \mathbf{0}$. However, it is restricting the nonbasic variables that can be chosen to enter the basis via the dual vector $\mathbf{w}$. At any tableau, either we have $\hat{z}_j - \hat{c}_j = \mathbf{va}_j \leq 0$ for all j, in which case the total phase I problem is solved, or we have $\hat{z}_j - \hat{c}_j = \mathbf{va}_j > 0$ for some nonbasic variables. In the former case, if $x_0 > 0$, then since the (total) phase I problem has a positive optimal solution, the primal is infeasible. On the other hand, if $x_0 = 0$, then the current primal-dual pair $(\mathbf{x}, \mathbf{w})$ satisfies the KKT conditions and is hence primal-dual optimal. Otherwise if the total phase I problem has not been solved as yet, the manipulation of $\mathbf{w}$ guarantees that at each iteration, some nonbasic variable x_j with $\hat{z}_j - \hat{c}_j = \mathbf{va}_j > 0$ is in the set Q and is hence enterable. Therefore we are essentially using the primal simplex algorithm on the (total) phase I problem along with a restricted entering variable criterion. Consequently, by employing the usual lexicographic rule of Section 4.6, which is independent of how one selects an entering variable, we can obtain finite convergence even in the presence of degeneracy. The minor details of this implementation are left to the reader in Exercise 6.52.

6.6 FINDING AN INITIAL DUAL FEASIBLE SOLUTION: THE ARTIFICIAL CONSTRAINT TECHNIQUE

Both the dual simplex method and the primal-dual method require an initial dual feasible solution. In the primal tableau this requirement of dual feasibility translates to $z_j - c_j \leq 0$ for all j for a minimization problem. We shall now see that this can be accommodated by adding a single new primal constraint.

Suppose that the first m columns constitute the initial basis and consider adding the constraint $\Sigma_{j=m+1}^{n} x_j \leq M$, where $M > 0$ is large. The initial tableau is displayed, where x_{n+1} is the slack variable of the additional constraint.

	z	x_1	x_2	$\cdots$	x_m	x_{m+1}	$\cdots$	x_n	x_{n+1}	RHS
z	1	0	0		0	$z_{m+1} - c_{m+1}$	$\cdots$	$z_n - c_n$	0	$\mathbf{c}_B\bar{\mathbf{b}}$
x_{n+1}	0	0	0	$\cdots$	0	1	$\cdots$	1	1	M
x_1	0	1	0	$\cdots$	0	$y_{1,m+1}$	$\cdots$	y_{1n}	0	$\bar{b}_1$
x_2	0	0	1	$\cdots$	0	$y_{2,m+1}$	$\cdots$	y_{2n}	0	$\bar{b}_2$
$\vdots$	$\vdots$	$\vdots$	$\vdots$		$\vdots$	$\vdots$		$\vdots$	$\vdots$	$\vdots$
x_m	0	0	0	$\cdots$	1	$y_{m,m+1}$	$\cdots$	y_{mn}	0	$\bar{b}_m$

This additional constraint bounds the nonbasic variables and thus indirectly bounds the basic variables and thereby the overall primal problem. To obtain a dual feasible solution in the new tableau we let

$$z_k - c_k = \underset{j}{\text{Maximum}}\ \{z_j - c_j\}$$

Once column k has been selected, perform a single pivot with column k as an entry column and column $n + 1$ as an exit column. In particular, to zero $z_k - c_k$ we shall subtract $z_k - c_k$ times the new row from the objective function row. Note that the choice of k and the single pivot described ensure that all new entries in row 0 are nonpositive, and thus we have a (basic) feasible dual solution. With this solution available either the dual simplex method or the primal-dual simplex method can be applied, eventually leading to one of the following three cases.

1. The dual solution is unbounded.
2. The optimal primal and dual solutions are obtained with $x^*_{n+1} > 0$.
3. The optimal primal and dual solutions are obtained with $x^*_{n+1} = 0$.

In case 1 the primal problem is infeasible. In case 2 we have the optimal solution to the primal problem. In case 3, the new bounding constraint is tight at optimality. If $z_{n+1} - c_{n+1} < 0$, then x_{n+1} is nonbasic and this new bounding constraint limits the primal solution. As M increases, the objective will continue to reduce indefinitely; hence the primal problem is unbounded. However, if $z_{n+1} - c_{n+1} = 0$, then the current primal solution is optimal (though not basic to the original problem). In Exercise 6.45 we ask the reader to explore this case in detail.

Furthermore, in Exercise 6.44 we ask the reader to show that applying the artificial constraint technique to the primal problem is equivalent to applying the single artificial variable technique (described in Chapter 4) with the big-M method to the dual problem and vice versa.

Example 6.9

Suppose that we wish to apply the dual simplex method to the following tableau.

	z	x_1	x_2	x_3	x_4	x_5	RHS
z	1	0	1	5	-1	0	0
x_1	0	1	2	-1	1	0	4
x_5	0	0	3	4	-1	1	3

Adding the artificial constraint $x_2 + x_3 + x_4 \leq M$ whose slack is x_6, we get the following tableau.

	z	x	x_2	x_3	x_4	x_5	x_6	RHS
z	1	0	1	5	-1	0	0	0
x_6	0	0	1	①	1	0	1	M
x_1	0	1	2	-1	1	0	0	4
x_5	0	0	3	4	-1	1	0	3

From the tableau we find that Maximum $z_j - c_j = z_3 - c_3 = 5$. Pivoting in the x_3 column and the x_6 row, we get a new tableau that is dual feasible. The dual simplex method can now be carried out in standard form.

	z	x_1	x_2	x_3	x_4	x_5	x_6	RHS
z	1	0	-4	0	-6	0	-5	$-5M$
x_3	0	0	1	1	1	0	1	M
x_1	0	1	3	0	2	0	1	$M+4$
x_5	0	0	-1	0	-5	1	-4	$-4M+3$

6.7 SENSITIVITY ANALYSIS

In most practical applications, some of the problem data are not known exactly and hence are estimated as well as possible. It is important to be able to find the new optimal solution of the problem as other estimates of some of the data become available, without the expensive task of resolving the problem from scratch. Also at early stages of problem formulation some factors may be overlooked. It is important to update the current solution in a way that takes care of these factors. Furthermore, in many situations the constraints are not very rigid. For example, a constraint may reflect the availability of some resource. This availability can be increased by extra purchase, overtime, buying new equipment, and the like. It is desirable to examine the effect of relaxing some of the constraints on the value of the optimal objective without having to resolve the problem. These and other related topics constitute *sensitivity analysis*. We shall discuss some methods for updating the optimal solution under different problem variations.

Consider the following problem.

$$\begin{aligned} \text{Minimize} \quad & \mathbf{cx} \\ \text{Subject to} \quad & \mathbf{Ax} = \mathbf{b} \\ & \mathbf{x} \geq \mathbf{0} \end{aligned}$$

Suppose that the simplex method produced an optimal basis **B**. We shall describe how to make use of the optimality conditions (primal-dual relationships) in order to find the new optimal solution, if some of the problem data change, without resolving the problem from scratch. In particular, the following variations in the problem will be considered. In doing so, we will also provide some further *insights* into duality.

Change in the cost vector **c**.
Change in the right-hand-side vector **b**.
Change in the constraint matrix **A**.
Addition of a new activity.
Addition of a new constraint.

Change in the Cost Vector

Given an optimal basic feasible solution, suppose that the cost coefficient of one (or more) of the variables is changed from c_k to c'_k. The effect of this change on the final tableau will occur in the cost row; that is, dual feasibility may be lost. Consider the following two cases.

Case I: x_k Is Nonbasic

In this case $\mathbf{c}_B$ is not affected, and hence $z_j = \mathbf{c}_B\mathbf{B}^{-1}\mathbf{a}_j$ is not changed for any j. Thus $z_k - c_k$ is replaced by $z_k - c'_k$. Note that $z_k - c_k \leq 0$ since the current point was an optimal solution of the original problem. If $z_k - c'_k = (z_k - c_k) + (c_k - c'_k)$ is positive, then x_k must be introduced into the basis and the (primal) simplex method is continued as usual. Otherwise the old solution is still optimal with respect to the new problem.

Case II: x_k Is Basic, Say, $x_k \equiv x_{B_t}$

Here c_{B_t} is replaced by c'_{B_t}. Let the new value of z_j be z'_j. Then $z'_j - c_j$ is calculated as follows:

$$\begin{aligned} z'_j - c_j &= \mathbf{c}'_B\mathbf{B}^{-1}\mathbf{a}_j - c_j = \left(\mathbf{c}_B\mathbf{B}^{-1}\mathbf{a}_j - c_j\right) + \left(0, 0, \dots, c'_{B_t} - c_{B_t}, 0, \dots, 0\right)\mathbf{y}_j \\ &= \left(z_j - c_j\right) + \left(c'_{B_t} - c_{B_t}\right) y_{tj} \qquad \text{for all } j \end{aligned}$$

In particular for $j = k$, $z_k - c_k = 0$, and $y_{tk} = 1$, and hence $z'_k - c_k = c'_k - c_k$. As we should expect, $z'_k - c'_k$ is still equal to zero. Therefore the cost row can be updated by adding the net change in the cost of $x_{B_t} \equiv x_k$ times the current t row of the final tableau, to the original cost row. Then $z'_k - c_k$ is updated to $z'_k - c'_k = 0$. Of course the new objective value $\mathbf{c}'_B\mathbf{B}^{-1}\mathbf{b} = \mathbf{c}_B\mathbf{B}^{-1}\mathbf{b} + (c'_{B_t} - c_{B_t})\bar{b}_t$ will be obtained in the process.

Example 6.10

Consider the following problem.

$$\begin{array}{lrl} \text{Minimize} & -2x_1 + x_2 - x_3 & \\ \text{Subject to} & x_1 + x_2 + x_3 & \leq 6 \\ & -x_1 + 2x_2 \quad\; & \leq 4 \\ & x_1, \quad x_2, \quad x_3 & \geq 0 \end{array}$$

The optimal tableau is given by the following.

	z	x_1	x_2	x_3	x_4	x_5	RHS
z	1	0	−3	−1	−2	0	−12
x_1	0	1	1	1	1	0	6
x_5	0	0	3	1	1	1	10

Suppose that $c_2 = 1$ is replaced by -3. Since x_2 is nonbasic, then $z_2 - c'_2 =$

$(z_2 - c_2) + (c_2 - c_2') = -3 + 4 = 1$, and all other $z_j - c_j$ are unaffected. Hence x_2 enters the basis.

	z	x_1	x_2	x_3	x_4	x_5	RHS
z	1	0	1	-1	-2	0	-12
x_1	0	1	1	1	1	0	6
x_5	0	0	③	1	1	1	10

The subsequent tableaux are not shown. Next suppose that $c_1 = -2$ is replaced by zero. Since x_1 is basic, then the new cost row, except $z_1 - c_1$, is obtained by multiplying the row of x_1 by the net change in c_1 [that is, $0 - (-2) = 2$] and adding to the old cost row. The new $z_1 - c_1$ remains zero. Note that the new $z_3 - c_3$ is now positive and so x_3 enters the basis.

	z	x_1	x_2	x_3	x_4	x_5	RHS
z	1	0	-1	1	0	0	0
x_1	0	1	1	①	1	0	6
x_5	0	0	3	1	1	1	10

The subsequent tableaux are not shown.

Change in the Right-Hand-Side

If the right-hand-side vector $\mathbf{b}$ is replaced by $\mathbf{b}'$, then $\mathbf{B}^{-1}\mathbf{b}$ will be replaced by $\mathbf{B}^{-1}\mathbf{b}'$. The new right-hand-side can be calculated without explicitly evaluating $\mathbf{B}^{-1}\mathbf{b}'$. This is evident by noting that $\mathbf{B}^{-1}\mathbf{b}' = \mathbf{B}^{-1}\mathbf{b} + \mathbf{B}^{-1}(\mathbf{b}' - \mathbf{b})$. If the first m columns originally form the identity, then $\mathbf{B}^{-1}(\mathbf{b}' - \mathbf{b}) = \sum_{j=1}^{m} \mathbf{y}_j(b_j' - b_j)$ and hence $\mathbf{B}^{-1}\mathbf{b}' = \bar{\mathbf{b}} + \sum_{j=1}^{m} \mathbf{y}_j(b_j' - b_j)$. Since $z_j - c_j \le 0$ for all nonbasic variables (for a minimum problem), the only possible violation of optimality is that the new vector $\mathbf{B}^{-1}\mathbf{b}'$ may have some negative entries. If $\mathbf{B}^{-1}\mathbf{b}' \ge \mathbf{0}$, then the same basis remains optimal, and the values of the basic variables are $\mathbf{B}^{-1}\mathbf{b}'$ and the objective has value $\mathbf{c}_B\mathbf{B}^{-1}\mathbf{b}'$. Otherwise the dual simplex method is used to find the new optimal solution by restoring feasibility.

Note that this case is similar to the previous case in that one is changing the cost vector in the dual problem. Hence if $\mathbf{B}^{-1}\mathbf{b}' \ge \mathbf{0}$, then dual optimality is maintained. Otherwise in the available dual basic feasible solution, the nonbasic dual variable complementary to a negative basic primal variable is enterable, and one proceeds by applying the simplex algorithm to the dual as previously.

Example 6.11

Suppose that the right-hand-side of Example 6.10 is replaced by $\begin{pmatrix} 3 \\ 4 \end{pmatrix}$. Note that $\mathbf{B}^{-1} = \begin{bmatrix} 1 & 0 \\ 1 & 1 \end{bmatrix}$ and hence $\mathbf{B}^{-1}\mathbf{b}' = \begin{bmatrix} 1 & 0 \\ 1 & 1 \end{bmatrix}\begin{bmatrix} 3 \\ 4 \end{bmatrix} = \begin{pmatrix} 3 \\ 7 \end{pmatrix}$. Then $\mathbf{B}^{-1}\mathbf{b}' \ge \mathbf{0}$ and hence the new optimal solution is $x_1 = 3$, $x_5 = 7$, $x_2 = x_3 = x_4 = 0$.

Change in the Constraint Matrix

We now discuss the effect of changing some of the entries of the constraint matrix **A**. Two cases, namely, changes involving nonbasic columns and changes involving basic columns are discussed. The dual operations of changing rows in **A** are also discussed.

Case I: Changes in Activity Vectors for Nonbasic Columns (Changes in rows with basic slack variables)

Suppose that the nonbasic column $\mathbf{a}_j$ is modified to $\mathbf{a}'_j$. Then the new updated column is $\mathbf{B}^{-1}\mathbf{a}'_j$ and $z'_j - c_j = \mathbf{c}_B\mathbf{B}^{-1}\mathbf{a}'_j - c_j$. If $z'_j - c_j \leq 0$, then the old solution is optimal; otherwise the simplex method is continued, after column j of the tableau is updated, by introducing the nonbasic variable x_j.

Changing a nonbasic variable column in the primal corresponds to changing a dual constraint for which the slack variable is basic. Hence the dual operation for this case identifies with changing the row (in a "primal" problem) where the associated slack variable is basic. In this case, the new row simply replaces the old row and is then updated to bring the tableau into canonical form. If the slack variable is negative when this is accomplished, then one may continue by using the dual simplex algorithm, starting by pivoting on this row, and hence entering the complementary dual variable into the basis.

Case II: Changes in Activity Vectors for Basic Columns (Changes in rows with nonbasic slack variables)

Suppose that a basic column $\mathbf{a}_j$ is modified to $\mathbf{a}'_j$. Now it is possible that the current set of basic vectors no longer form a basis after the change. Even if this does not occur, a change in the activity vector for a single basic column will change $\mathbf{B}^{-1}$ and thus the entries in every column.

This change is handled very easily in two steps. First, assume that a new activity x'_j is being added to the problem with column $\mathbf{a}'_j$ (and objective coefficient c_j). Second, eliminate the old variable x_j from the problem. The first step is accomplished by computing $\mathbf{y}'_j = \mathbf{B}^{-1}\mathbf{a}'_j$ and $z'_j - c_j = \mathbf{c}_B\mathbf{B}^{-1}\mathbf{a}'_j - c_j$, where **B** is the current basis matrix. This gives the updated column for x'_j. If the element y'_{jj} in the basic row for x_j and the column of x'_j is not zero, then x'_j can be exchanged for x_j in the basis and the column of the old variable x_j can be eliminated from the problem (why?). This pivot may destroy either or both primal and dual feasibility, but we can restore primal or dual feasibility using artificial variables if necessary and reoptimize. On the other hand, if $y'_{jj} = 0$, then the current set of basic vectors with the new column for x_j no longer forms a basis (why?). In this case, one way to eliminate x_j from the problem is to treat it as an artificial variable that is basic in that row and to resort to the two-phase or the big-M method.

The dual operation for this case is to change a row in the problem where the corresponding slack variable is no nbasic. Since several coefficients in the updated slack variable column can be nonzero, this row may have multiples of it added to several other rows. For convenience, let us refer to the slacks in the old and the new form of this constraint as x_{si} and x'_{si} respectively. Then the dual operations to the ones given previously are as follows: First, add this new row with x'_{si} basic and update the tableau into canonical form. (Some related mathematical details are given later in the discussion on adding a new constraint.) If the element in this row and the column of x_{si} is nonzero, then exchange x'_{si} for x_{si} in the basis. (This corresponds to the case $y'_{jj} \neq 0$.) Delete the representation of x_{si} in terms of the nonbasic variables from the problem, since the old form of the constraint is no longer needed. Otherwise, if the exchange of x'_{si} for x_{si} in the basis is not possible, one can change the right-hand-side of the old constraint to a big M value in order to make this old form of the constraint essentially redundant to the problem and continue the optimization by using the dual simplex method. This corresponds to the handling of the case $y'_{jj} = 0$.

Example 6.12(a).

Suppose that in Example 6.10, $\mathbf{a}_2$ is changed from $\begin{pmatrix}1\\2\end{pmatrix}$ to $\begin{pmatrix}2\\5\end{pmatrix}$. Then

$$\mathbf{y}'_2 = \mathbf{B}^{-1}\mathbf{a}'_2 = \begin{pmatrix}1 & 0\\1 & 1\end{pmatrix}\begin{pmatrix}2\\5\end{pmatrix} = \begin{pmatrix}2\\7\end{pmatrix}$$

$$\mathbf{c}_B\mathbf{B}^{-1}\mathbf{a}'_2 - c_2 = (-2, 0)\begin{pmatrix}2\\7\end{pmatrix} - 1 = -5$$

Thus the current optimal tableau remains optimal with column x_2 replaced by $(-5, 2, 7)'$.

Next suppose that column $\mathbf{a}_1$ is changed from $\begin{pmatrix}1\\-1\end{pmatrix}$ to $\begin{pmatrix}0\\-1\end{pmatrix}$. Then

$$\mathbf{y}'_1 = \mathbf{B}^{-1}\mathbf{a}'_1 = \begin{pmatrix}1 & 0\\1 & 1\end{pmatrix}\begin{pmatrix}0\\-1\end{pmatrix} = \begin{pmatrix}0\\-1\end{pmatrix}$$

$$\mathbf{c}_B\mathbf{B}^{-1}\mathbf{a}'_1 - c_1 = (-2, 0)\begin{pmatrix}0\\-1\end{pmatrix} - (-2) = 2$$

Here the entry in the x_1 row of $\mathbf{y}'_1$ is zero, and so the current basic columns no longer span the space. Adding the column $(2, 0, -1)'$ of x'_1 and making x_1 an artificial variable in the basis with a big-M penalty in the objective function, we get the following tableau.

	z	x_1	x'_1	x_2	x_3	x_4	x_5	RHS
z	1	$-M$	2	-3	-1	-2	0	-12
x_1	0	①	0	1	1	1	0	6
x_5	0	0	-1	3	1	1	1	10

After preliminary pivoting at row x_1 and column x_1 to gt $z_1 - c_1 = 0$, that is, to get the tableau in canonical form, we may proceed with the big-M method.

Finally, suppose that column $\mathbf{a}_1$ is changed from $\begin{pmatrix} 1 \\ -1 \end{pmatrix}$ to $\begin{pmatrix} 3 \\ 6 \end{pmatrix}$. Then

$$\mathbf{y}_1' = \mathbf{B}^{-1}\mathbf{a}_1' = \begin{pmatrix} 1 & 0 \\ 1 & 1 \end{pmatrix}\begin{pmatrix} 3 \\ 6 \end{pmatrix} = \begin{pmatrix} 3 \\ 9 \end{pmatrix}$$

$$\mathbf{c}_B\mathbf{B}^{-1}\mathbf{a}_1' - c_1 = (-2, 0)\begin{pmatrix} 3 \\ 9 \end{pmatrix} - (-2) = -4$$

In this case the entry in the x_1 row of $\mathbf{y}_1'$ is nonzero and so we add the column $(-4, 3, 9)'$ of x_1', pivot in the x_1' column and the row for x_1, and proceed with x_1 eliminated.

	z	x_1'	x_2	x_3	x_4	x_5	RHS
z	1	-4	-3	-1	-2	0	-12
x_1	0	③	1	1	1	0	6
x_5	0	9	3	1	1	1	10

The subsequent tableaux are not shown.

Example 6.12(b)

Suppose that in Example 6.10 the first constraint is changed to $x_2 - x_3 \leq 6$. Call the new slack variable x_4' and add this constraint to the optimal tableau with x_4' basic in the new row. This gives the following tableau (in canonical form):

	z	x_1	x_2	x_3	x_4	x_5	x_4'	RHS
z	1	0	-3	-1	-2	0	0	-12
x_1	0	1	1	1	1	0	0	6
x_5	0	0	3	1	1	1	0	10
x_4'	0	0	1	-1	0	0	1	6

Note that the element in row x_4' and column x_4 is zero. (Otherwise, we could have pivoted on this element and deleted the resulting representation of x_4 in terms of the nonbasic variables in the last row.) Hence we change the right-hand-side of the old first constraint to M. This makes the updated right-hand-side vector equal to $(-2M, M, M + 4, 6)$. Note that we still have optimality in this case, with the objective value of order $-M$. (The old constraint is restricting the optimum with $z_4 - c_4 < 0$.) Hence the new problem is unbounded.

Adding a New Activity

Suppose that a new activity x_{n+1} with unit cost c_{n+1} and *consumption* column $\mathbf{a}_{n+1}$ is considered for possible production. Without resolving the problem, we can easily determine whether producing x_{n+1} is worthwhile. First calculate $z_{n+1} - c_{n+1}$. If $z_{n+1} - c_{n+1} \leq 0$ (for a minimization problem), then $x_{n+1}^* = 0$ and the current solution is optimal. On the other hand, if $z_{n+1} - c_{n+1} > 0$, then x_{n+1} is introduced into the basis and the simplex method continues to find the new optimal solution.

Example 6.13

Consider Example 6.10. We wish to find the new optimal solution if a new activity $x_6 \geq 0$ with $c_6 = 1$ and $\mathbf{a}_6 = \begin{pmatrix} -1 \\ 2 \end{pmatrix}$ is introduced. First we calculate $z_6 - c_6$:

$$z_6 - c_6 = \mathbf{w}\mathbf{a}_6 - c_6$$

$$= (-2, 0)\begin{pmatrix} -1 \\ 2 \end{pmatrix} - 1 = 1$$

$$\mathbf{y}_6 = \mathbf{B}^{-1}\mathbf{a}_6 = \begin{bmatrix} 1 & 0 \\ 1 & 1 \end{bmatrix}\begin{bmatrix} -1 \\ 2 \end{bmatrix} = \begin{bmatrix} -1 \\ 1 \end{bmatrix}$$

Therefore x_6 is introduced in the basis by pivoting at the x_5 row and the x_6 column.

	z	x_1	x_2	x_3	x_4	x_5	x_6	RHS
z	1	0	-3	-1	-2	0	1	-12
x_1	0	1	1	1	1	0	-1	6
x_5	0	0	3	1	1	1	(1)	10

The subsequent tableaux are not shown.

Adding a New Constraint

Suppose that a new constraint is added to the problem. As discussed in the section on modifying the **A** matrix, this operation is dual to that of adding a new activity. If the optimal solution to the original problem satisfies the added constraint, it is then obvious that the point is also an optimal solution of the new problem (why?). If, on the other hand, the point does not satisfy the new constraint, that is, if the constraint "cuts away" the optimal point, we can use the dual simplex method to find the new optimal solution. These two cases are illustrated in Figure 6.5, and they correspond to the two cases related to adding a new activity.

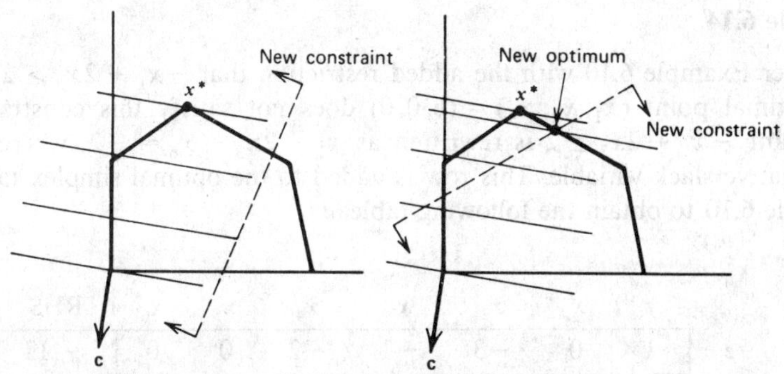

Figure 6.5. **Addition of a new constraint.**

Suppose that $\mathbf{B}$ is the optimal basis before the constraint $\mathbf{a}^{m+1}\mathbf{x} \leq b_{m+1}$ is added. The corresponding canonical system is as follows.

$$z + (\mathbf{c}_B\mathbf{B}^{-1}\mathbf{N} - \mathbf{c}_N)\mathbf{x}_N = \mathbf{c}_B\mathbf{B}^{-1}\mathbf{b}$$

$$\mathbf{x}_B + \mathbf{B}^{-1}\mathbf{N}\mathbf{x}_N = \mathbf{B}^{-1}\mathbf{b} \tag{6.14}$$

The constraint $\mathbf{a}^{m+1}\mathbf{x} \leq b_{m+1}$ is rewritten as $\mathbf{a}_B^{m+1}\mathbf{x}_B + \mathbf{a}_N^{m+1}\mathbf{x}_N + x_{n+1} = b_{m+1}$, where $\mathbf{a}^{m+1}$ is decomposed into $(\mathbf{a}_B^{m+1}, \mathbf{a}_N^{m+1})$ and x_{n+1} is a nonnegative slack variable. Multiplying Equation (6.14) by $\mathbf{a}_B^{m+1}$ and subtracting from the new constraint gives the following system:

$$z + (\mathbf{c}_B\mathbf{B}^{-1}\mathbf{N} - \mathbf{c}_N)\mathbf{x}_N = \mathbf{c}_B\mathbf{B}^{-1}\mathbf{b}$$

$$\mathbf{x}_B + \mathbf{B}^{-1}\mathbf{N}\mathbf{x}_N = \mathbf{B}^{-1}\mathbf{b}$$

$$(\mathbf{a}_N^{m+1} - \mathbf{a}_B^{m+1}\mathbf{B}^{-1}\mathbf{N})\mathbf{x}_N + x_{n+1} = b_{m+1} - \mathbf{a}_B^{m+1}\mathbf{B}^{-1}\mathbf{b}$$

These equations give us a basic solution of the new system (why?). The only possible violation of optimality of the new problem is the sign of $b_{m+1} - \mathbf{a}_B^{m+1}\mathbf{B}^{-1}\mathbf{b}$. So if $b_{m+1} - \mathbf{a}_B^{m+1}\mathbf{B}^{-1}\mathbf{b} \geq 0$, then the current solution is optimal. Otherwise, if $b_{m+1} - \mathbf{a}_B^{m+1}\mathbf{B}^{-1}\mathbf{b} < 0$, then the dual simplex method is used to restore feasibility.

Note that if an equality constraint is added, after bringing the tableau to canonical form one may then use an artificial variable as a basic variable in this row and continue with the primal two-phase or the big-M method. The dual operation adds an unrestricted variable x_j. The equivalent dual technique is to add the updated version of this column to the tableau. Since the variable is unrestricted, we can use its positive or negative column and assume that the variable x_j has $z_j - c_j \leq 0$. Now we impose the bound restriction $x_j \geq -M$. Using the transformation $x_j' = x_j + M \geq 0$ to obtain nonnegatively constrained variables, one may continue optimizing with the dual simplex method. Alternatively, one could have used the transformation $x_j = x_j' - x_j''$, with $x_j', x_j'' \geq 0$, and added the two columns of x_j' and x_j'' to the problem. (What is the equivalent dual technique?)

Example 6.14

Consider Example 6.10 with the added restriction that $-x_1 + 2x_3 \geq 2$. Clearly the optimal point $(x_1, x_2, x_3) = (6, 0, 0)$ does not satisfy this constraint. The constraint $-x_1 + 2x_3 \geq 2$ is rewritten as $x_1 - 2x_3 + x_6 = -2$, where x_6 is a nonnegative slack variable. This row is added to the optimal simplex tableau of Example 6.10 to obtain the following tableau.

	z	x_1	x_2	x_3	x_4	x_5	x_6	RHS
z	1	0	−3	−1	−2	0	0	−12
x_1	0	1	1	1	1	0	0	6
x_5	0	0	3	1	1	1	0	10
x_6	0	1	0	−2	0	0	1	−2

Multiply row 1 by −1 and add to row 3 in order to restore column x_1 to a unit vector. The dual simplex method can then be applied to the resulting tableau below.

	z	x_1	x_2	x_3	x_4	x_5	x_6	RHS
z	1	0	−3	−1	−2	0	0	−12
x_1	0	1	1	1	1	0	0	6
x_5	0	0	3	1	1	1	0	10
x_6	0	0	−1	(−3)	−1	0	1	−8

Subsequent tableaux are not shown.

An Application of Adding Constraints in Integer Programming

The linear integer programming problem may be stated as follows.

$$\begin{aligned} &\text{Minimize} && \mathbf{cx} \\ &\text{Subject to} && \mathbf{Ax} = \mathbf{b} \\ &&& \mathbf{x} \geq \mathbf{0} \\ &&& \mathbf{x} \quad \text{integer} \end{aligned}$$

A natural method to solve this problem is to ignore the last condition, **x** integer, and solve the problem as a linear program. At optimality, if all of the variables have integer values, then we have the optimal solution to the original integer program (why?). Otherwise consider adding a new constraint to the linear program. This additional constraint should "cut off" the current optimal noninteger linear programming solution without cutting off any feasible integer solution. Adding the new constraint to the optimal tableau, we apply the dual simplex to reoptimize the new linear program. The new solution is either integer or not. The procedure of adding constraints is repeated until either an all integer solution is found or infeasibility results (indicating no integer solution). How, then, can such a cutting constraint be generated?

Consider the optimal simplex tableau when a noninteger solution results. Let $\bar{b}_r$ be noninteger. Assume that the basic variables are indexed from 1 to m. The equation associated with $\bar{b}_r$ is

$$x_r + \sum_{j=m+1}^{n} y_{rj}x_j = \bar{b}_r$$

Let I_{rj} be the greatest integer that is less than or equal to y_{rj} (I_{rj} is called the integer part of y_{rj}). Similarly, let I_r be the integer part of $\bar{b}_r$. Let F_{rj} and F_r be the respective fractional parts, that is,

$$F_{rj} = y_{rj} - I_{rj} \qquad \text{and} \qquad F_r = \bar{b}_r - I_r$$

Then $0 \le F_{rj} < 1$ and $0 < F_r < 1$ (why?). Using this, we may rewrite the basic equation for x_r as

$$x_r + \sum_{j=m+1}^{n} (I_{rj} + F_{rj})x_j = I_r + F_r$$

Rearranging terms, we get

$$x_r + \sum_{j=m+1}^{n} I_{rj}x_j - I_r = F_r - \sum_{j=m+1}^{n} F_{rj}x_j$$

Now the left-hand-side of this equation will be integer for any feasible integer solution (why?). The right-hand-side is strictly less than 1, since $F_r < 1$, $F_{rj} \ge 0$, and $x_j \ge 0$. But since the right-hand-side must also be integer because it equals the left-hand-side, we may conclude that it must be less than or equal to zero (there are no integers greater than zero and less than one). Thus we may write

$$F_r - \sum_{j=m+1}^{n} F_{rj}x_j \le 0$$

However, since x_j is currently nonbasic (and hence $x_j = 0$) for $j = m + 1, \ldots, n$ and $F_r > 0$, the current optimal (noninteger) linear programming solution does not satisfy this additional constraint. In other words, this new constraint will cut off the current optimal solution if added to the current optimal tableau. The dual simplex method is then applied to obtain a new optimal linear programming solution, which is again tested for integrality. Such a procedure is called a *cutting plane* algorithm. This cut is called *Gomory's dual fractional cut*.

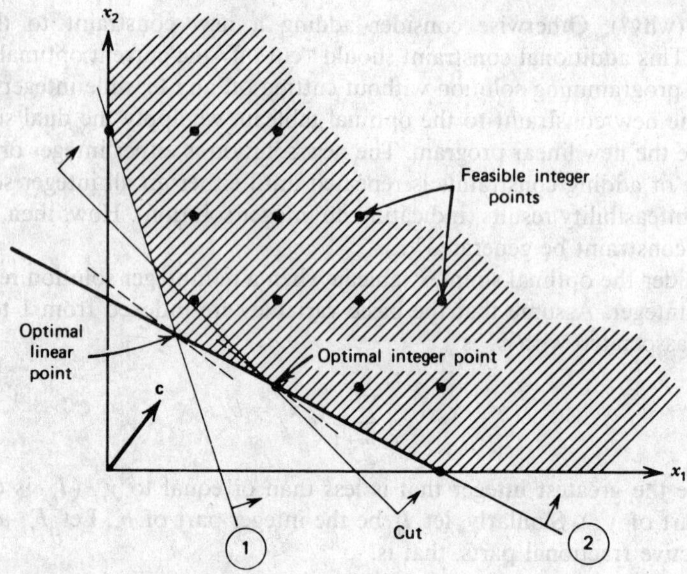

Figure 6.6. The graphical solution.

Example 6.15

Consider the following integer program.

$$\begin{aligned} \text{Minimize} \quad & 3x_1 + 4x_2 \\ \text{Subject to} \quad & 3x_1 + x_2 \geq 4 \\ & x_1 + 2x_2 \geq 4 \\ & x_1, \quad x_2 \geq 0 \\ & x_1, \quad x_2 \quad \text{integer} \end{aligned}$$

In Figure 6.6 we show the optimal linear programming and integer programming solutions respectively.

Ignoring the integer conditions, the following tableau gives the optimal linear programming solution.

	z	x_1	x_2	x_3	x_4	RHS
z	1	0	0	$-\frac{2}{5}$	$-\frac{9}{5}$	$\frac{44}{5}$
x_1	0	1	0	$-\frac{2}{5}$	$\frac{1}{5}$	$\frac{4}{5}$
x_2	0	0	1	$\frac{1}{5}$	$-\frac{3}{5}$	$\frac{8}{5}$

Since this solution is noninteger, we may select a noninteger variable for generating a cut (including z). Select x_2. (*Note*. Selecting different variables may generate different cuts.) The equation for the basic variable x_2 is

$$x_2 + \tfrac{1}{5}x_3 - \tfrac{3}{5}x_4 = \tfrac{8}{5}$$

From this we get

$$I_{23} = 0, \quad I_{24} = -1, \quad I_2 = 1$$

$$F_{23} = \tfrac{1}{5}, \quad F_{24} = \tfrac{2}{5}, \quad F_2 = \tfrac{3}{5}$$

and the additional constraint becomes

$$\tfrac{1}{5}x_3 + \tfrac{2}{5}x_4 \geq \tfrac{3}{5} \qquad \text{(cut)}$$

Adding this constraint with slack variable x_5 to the tableau and applying the dual simplex method, we get the following tableaux.

	z	x_1	x_2	x_3	x_4	x_5	RHS
z	1	0	0	$-\frac{2}{5}$	$-\frac{9}{5}$	0	$\frac{44}{5}$
x_1	0	1	0	$-\frac{2}{5}$	$\frac{1}{5}$	0	$\frac{4}{5}$
x_2	0	0	1	$\frac{1}{5}$	$-\frac{3}{5}$	0	$\frac{8}{5}$
x_5	0	0	0	$-\frac{1}{5}$ (circled)	$-\frac{2}{5}$	1	$-\frac{3}{5}$

	z	x_1	x_2	x_3	x_4	x_5	RHS
z	1	0	0	0	-1	-2	10
x_1	0	1	0	0	1	-2	2
x_2	0	0	1	0	-1	1	1
x_3	0	0	0	1	2	-5	3

Hence we have obtained the optimal integer solution $\mathbf{x}^* = (2, 1)^t$ with only one cut. In other integer programs we might have to repeat the cutting plane process many times. If in the foregoing tableau some variable had turned out noninteger, we would have generated a new cut and continued.

It is interesting to examine the cut in terms of the original variables. Substituting $x_3 = 3x_1 + x_2 - 4$ and $x_4 = x_1 + 2x_2 - 4$ into $\frac{1}{5}x_3 + \frac{2}{5}x_4 \geq \frac{3}{5}$ and simplifying, we get

$$x_1 + x_2 \geq 3 \qquad \text{(cut in terms of } x_1 \text{ and } x_2\text{)}$$

It can easily be seen that the addition of this constraint to Figure 6.6 will yield the required integer optimum. Finite convergence of such a procedure may be achieved by using an appropriate dual lexicographic pivot rule (see the Notes and References Section).

Some Additional Issues in Sensitivity Analysis

There are three pertinent issues related to sensitivity analysis in linear programming that need to be discussed. The first is performing sensitivity analysis in a revised simplex implementation. Here the cases discussed previously are treated similarly, except that one only needs to update $\mathbf{c}_B\mathbf{B}^{-1}$, $\mathbf{B}^{-1}$, and $\mathbf{B}^{-1}\mathbf{b}$, as the case may be, in order to obtain the new revised simplex tableau. We hence leave the details to the reader in Exercise 6.54, but point out one important construct. Suppose that one has added a new row to the problem that has the coefficient vector $\mathbf{a}_B^{m+1}$ in the columns of the current basic variables. Also, suppose that the slack or artificial variable x_{n+1} whose column is a unit vector has been made basic in this new row. Then the new basis $\mathbf{B}_{\text{new}}$ and its inverse appear as given.

$$\mathbf{B}_{\text{new}} = \begin{array}{c} \quad\;\; \mathbf{x}_B \quad x_{n+1} \\ \left[\begin{array}{c|c} \mathbf{B} & 0 \\ \hline \mathbf{a}_B^{m+1} & 1 \end{array}\right] \end{array}, \qquad \mathbf{B}_{\text{new}}^{-1} = \left[\begin{array}{c|c} \mathbf{B}^{-1} & 0 \\ \hline -\mathbf{a}_B^{m+1}\mathbf{B}^{-1} & 1 \end{array}\right]$$

This construct is useful in updating the revised simplex tableau in such a case.

The second issue relates to performing sensitivity analysis for bounded variables linear programming problems. Again, the various cases discussed in this section (including changes in bounds on the variables) may be handled similarly, except that one needs to make sure in the revision that the designated nonbasic variables are at either their lower or upper bounds and that the basic variables are feasible to their bounds. Although the former condition is readily maintained, the following construct may be used in order to maintain the latter condition. Suppose that an update has been made while maintaining the nonbasic variables at their lower or upper bounds, but ignoring the bound restrictions on the basic variables. Let $\Delta\mathbf{b}$ be the vector with the smallest norm that needs to be added to the basic variable values in order to make them feasible to their bounds. Hence $\Delta b_i = l_{Bi} - x_{Bi}$ if $x_{Bi} < l_{Bi}$, $\Delta b_i = u_{Bi} - x_{Bi}$ if $x_{Bi} > u_{Bi}$, and $\Delta b_i = 0$ otherwise. Then construct an artificial variable x_{n+1} with $-\Delta b$ as its *updated* column and with $0 \le x_{n+1} \le 1$ as its bound restriction. Observe that by making x_{n+1} nonbasic at its upper bound we obtain a basic feasible solution to the artificial problem (why?). Hence we may now proceed with the two-phase or the big-M method. (Alternatively, one may use a bounded variables version of the dual simplex method, if applicable (see Exercise 6.58).) We leave the details of the various sensitivity analysis cases to the reader in Exercise 6.57.

Finally, a third issue related to sensitivity analysis is the so-called *tolerance approach*. One is concerned with a tolerance on the simultaneous and independent variations in the coefficients of the objective function or of the right-hand-side so that the current optimal basis remains optimal. More precisely, consider the original and the perturbed linear programs P and P′, where c'_j, $j = 1, \ldots, n$ are given constants and α_j, $j = 1, \ldots, n$ denote perturbation parameters.

$$\text{P:}\quad \text{Minimize} \quad \sum_{j=1}^{n} c_j x_j$$

$$\text{Subject to} \quad \sum_{j=1}^{n} a_{ij} x_j = b_i \quad \text{for } i = 1, \ldots, m$$

$$\mathbf{x} \geq \mathbf{0}$$

$$\text{P}'\text{:}\quad \text{Minimize} \quad \sum_{j=1}^{n} (c_j + \alpha_j c'_j) x_j \tag{6.15}$$

$$\text{Subject to} \quad \sum_{j=1}^{n} a_{ij} x_j = b_i \quad \text{for } i = 1, \ldots, m$$

$$\mathbf{x} \geq \mathbf{0}$$

Observe that if $c'_j = c_j$ for $j = 1, \ldots, n$, then $100\alpha_j$ gives a percentage variation (positive or negative) in the objective coefficient c_j of the variable x_j. Given an optimal basis **B** for P, the question raised in this context is what is the maximum value $\alpha_0^* \geq 0$ such that whenever $-\alpha_0^* \leq \alpha_j \leq \alpha_0^*$ for each $j = 1, \ldots, n$, we still have **B** as an optimal basis in P′. The value α_0^* is then a *maximum allowable tolerance on variations in the objective coefficients* with respect to $\mathbf{c}'$. Note that when $\mathbf{c}' = \mathbf{c}$, α_0^* gives a (maximum) permissible accuracy tolerance on estimating the (nonzero) objective coefficients so that current basis remains optimal.

A symmetric question can be addressed with respect to right-hand-side perturbations using the original and perturbed linear programs P and P′.

$$\text{P:}\quad \text{Minimize} \quad \sum_{j=1}^{n} c_j x_j$$

$$\text{Subject to} \quad \sum_{j=1}^{n} a_{ij} x_j = b_i \quad \text{for } i = 1, \ldots, m$$

$$\mathbf{x} \geq \mathbf{0}$$

$$\text{P}'\text{:}\quad \text{Minimize} \quad \sum_{j=1}^{n} c_j x_j \tag{6.16}$$

$$\text{Subject to} \quad \sum_{j=1}^{n} a_{ij} x_j = b_i + \beta_i b'_i \quad \text{for } i = 1, \ldots, m$$

$$\mathbf{x} \geq \mathbf{0}$$

Here one is interested in determining a *maximum allowable tolerance* $\beta_0^* \geq 0$ *on variations in the right-hand-side values* with respect to $\mathbf{b}'$, given an optimal basis $\mathbf{B}$ for P such that whenever $-\beta_0^* \leq \beta_i \leq \beta_0^*$ for each $i = 1, \ldots, m$, we still have $\mathbf{B}$ as an optimal basis in P′.

In order to determine α_0^*, let $\mathbf{B}$ denote an optimal basis in P, let c_{Bi}, c'_{Bi}, and α_{Bi} denote the corresponding basic variable coefficients in the vectors $\mathbf{c}$, $\mathbf{c}'$, and $\boldsymbol{\alpha}$ respectively, let $R \neq \phi$ denote the index set for the nonbasic variables, let $\mathbf{y}_j = \mathbf{B}^{-1}\mathbf{a}_j$ denote the updated column for x_j, $j \in R$, with coefficients y_{ij}, $i = 1, \ldots, m$ corresponding to the rows of the basic variables x_{Bi}, $i = 1, \ldots, m$, and as usual, let $z_j - c_j = \mathbf{c}_B\mathbf{B}^{-1}\mathbf{a}_j - c_j$ for $j \in R$. Then $\mathbf{B}$ remains an optimal basis in P′ if and only if the objective row coefficients in P′ are nonpositive for the canonical representation with respect to the basis $\mathbf{B}$. In other words, we must have

$$\sum_{i=1}^{m} (c_{Bi} + \alpha_{Bi}c'_{Bi})y_{ij} - (c_j + \alpha_j c'_j) \leq 0 \qquad \text{for each } j \in R$$

that is,

$$\sum_{i=1}^{m} c'_{Bi}y_{ij}\alpha_{Bi} - \alpha_j c'_j \leq (c_j - z_j) \qquad \text{for each } j \in R \tag{6.17}$$

We now need to obtain α_0^* as the largest nonnegative value such that so long as $-\alpha_0^* \leq \alpha_j \leq \alpha_0^*$, the value of the left-hand-side in Equation (6.17) is no more than $(c_j - z_j) \geq 0$, for each $j \in R$. But the largest that the left-hand-side in Equation (6.17) can ever get subject to $-\alpha_0 \leq \alpha_j \leq \alpha_0$, $j = 1, \ldots, n$ for any $\alpha_0 \geq 0$ is $(\sum_{i=1}^m |c'_{Bi}y_{ij}| + |c'_j|)\alpha_0$. Therefore we have

$$\alpha_0^* = \text{Maximum}\left\{\alpha_0 \geq 0: \left(\sum_{i=1}^{m} |c'_{Bi}y_{ij}| + |c'_j|\right)\alpha_0 \leq (c_j - z_j) \qquad \text{for each } j \in R\right\} \tag{6.18}$$

Denoting $R^+ = \{j \in R: \sum_{i=1}^m |c'_{Bi}y_{ij}| + |c'_j| > 0\}$, we get in closed form

$$\alpha_0^* = \underset{j \in R^+}{\text{Minimum}} \frac{c_j - z_j}{\sum_{i=1}^{m} |c'_{Bi}y_{ij}| + |c'_j|} \tag{6.19}$$

Similarly, denoting the components of $\mathbf{B}^{-1}$ as B_{ij}^{-1}, $i, j = 1, \ldots, m$, and letting $\bar{\mathbf{b}} = \mathbf{B}^{-1}\mathbf{b}$, we have that $\mathbf{B}$ is an optimal basis in P′ of Equation (6.16) if and only if it is feasible, that is, $\sum_{j=1}^m (B_{ij}^{-1}(b_j + \beta_j b'_j) \geq 0$ for each $i = 1, \ldots, m$. This can be rewritten as requiring that

$$\sum_{j=1}^{m} \left(-B_{ij}^{-1}b'_j\right)\beta_j \leq \bar{b}_i \qquad \text{for each } i = 1, \ldots, m \tag{6.20}$$

Given any $\beta_0 \geq 0$, the largest possible value of the left-hand-side in Equation (6.20) subject to $-\beta_0 \leq \beta_j \leq \beta_0$ for $j = 1, \ldots, m$, is $(\sum_{j=1}^m |B_{ij}^{-1}b'_j|)\beta_0$ (why?).

Hence we obtain

$$\beta_0^* = \text{Maximum}\left\{\beta_0: \left(\sum_{j=1}^{m}\left|B_{ij}^{-1}b_j'\right|\right)\beta_0 \le \bar{b}_i \quad \text{for } i = 1,\ldots,m\right\} \quad (6.21)$$

Denoting $I^+ = \{i \in \{1,\ldots,m\}: \Sigma_{j=1}^m |B_{ij}^{-1}b_j'| > 0\}$, we get

$$\beta_0^* = \underset{i \in I^+}{\text{Minimum}} \frac{\bar{b}_i}{\sum_{j=1}^{m}\left|B_{ij}^{-1}b_j'\right|} \quad (6.22)$$

In closing this section, we remark that the foregoing analysis is only one type of tolerance issue that may be addressed. For example, one may wish to determine the maximum tolerances subject to other imposed restrictions on variations in objective or right-hand-side parameters. In other words, certain bounds on some of these parameters or certain interrelationships in their variations may be known *a priori* in terms of additional restrictions on the α_j or the β_j values. Such cases are readily analyzed by including these constraints along with $-\alpha_0 \le \alpha_j \le \alpha_0$, $j = 1,\ldots,n$ (or $-\beta_0 \le \beta_j \le \beta_0$, $j = 1,\ldots,m$) when determining the maximum possible value of the left-hand-side in Equation (6.17) [or in Equation (6.20)] before formulating and solving Equation (6.18) for α_0^* [or Equation (6.21) for β_0^*]. The following example illustrates this point.

Example 6.16

Consider Example 6.10 with the optimal basis $\mathbf{B} = (\mathbf{a}_1, \mathbf{a}_5)$ as given there. Suppose that $\mathbf{c}' = \mathbf{c}$. Hence $R = \{2,3,4\}$, $\mathbf{c}_B' = (-2,0)$, and we obtain $\{(c_j - z_j), j \in R\} = \{3,1,2\}$, $\{|c_j'|, j \in R\} = \{1,1,0\}$, and $\{\Sigma_{i=1}^m |c_{Bi}' y_{ij}|, j \in R\} = \{2,2,2\}$. Hence from Equation (6.19), we get $\alpha_0^* = \text{Minimum}\{3/3, 1/3, 2/2\} = 1/3$. Consequently, the objective coefficients can vary within a tolerance of $33\frac{1}{3}$ percent and the given basis will remain optimal.

Now suppose that the objective coefficients of x_1, x_4, and x_5 are known with certainty to be -2, 0, and 0 respectively, but variations in c_2 and c_3 are permissible. Then $\mathbf{c}' = (0,1,-1,0,0)$, and so $\{|c_j'|, j \in R\} = \{1,1,0\}$ and $\mathbf{c}_B' = (0,0)$. This gives $R^+ = \{2,3\}$, and we get from Equation (6.19) that $\alpha_0^* = \text{Minimum}\{3/1, 1/1\} = 1$. Therefore, subject to the stated restrictions, a 100 percent tolerance in variation is permissible.

On the other hand, suppose that $\mathbf{c}' = \mathbf{c}$ but we known that the objective coefficient for x_1 will vary from its given value by no more than ± 25 percent. Now the relationships (6.17) require that $-2\alpha_1 - \alpha_2 \le 3$, $-2\alpha_1 + \alpha_3 \le 1$, and $-2\alpha_1 \le 2$. Hence given that $-\alpha_0 \le \alpha_j \le \alpha_0$ for $j = 1,\ldots,5$ *and* that $1/4 \le \alpha_1 \le 1/4$, the maximum values of the left-hand-sides of the foregoing three inequalities are respectively $2[\text{Min}\{\alpha_0, 1/4\}] + \alpha_0$, $2[\text{Min}\{\alpha_0, 1/4\}] + \alpha_0$, and $2[\text{Min}\{\alpha_0, 1/4\}]$. We need to find the largest value α_0^{**} of α_0 such that these quantities are respectively no more than 3, 1, and 2. Since we know that $\alpha_0^{**} \ge 1/3$ (why?), we have $\text{Min}\{\alpha_0^{**}, 1/4\} = 1/4$, and so, $\alpha_0^{**} = \text{Max}\{\alpha_0: 1/2 + \alpha_0 \le 3,\ 1/2 + \alpha_0 \le 1 \text{ and } 1/2 \le 2\}$. This gives $\alpha_0^{**} = \text{Minimum}\{5/2, 1/2\} = 1/2$. Therefore, a 50 percent, instead of only a $33\frac{1}{3}$

percent, variation in the objective coefficients is permissible, given a maximum variation of 25 percent in c_1.

Finally, let us illustrate the case of variations in the right-hand-side. Assuming that $\mathbf{b}' = \mathbf{b}$, we have $\bar{\mathbf{b}} = (6, 10)'$, $\mathbf{B} = \begin{bmatrix} 1 & 0 \\ 1 & 1 \end{bmatrix}$, and $\Sigma_{j=1}^{m} |B_{ij}^{-1} b_j'| = 6$ and 10 for the first and second basic variable rows, respectively. Hence from Equation (6.22), $\beta_0^* = \text{Minimum}\{6/6, 10/10\} = 1$, that is, a 100 percent variation in the right-hand-sides is permissible for the given basis to remain optimal.

6.8 PARAMETRIC ANALYSIS

Parametric analysis is used quite often in large-scale linear programming and in nonlinear optimization, where one often finds a direction along which the objective function or the constraints are perturbed, and then seeks to move along this direction. So we seek the optimal solutions to a class of problems by perturbing either the objective vector or the RHS vector along a fixed direction.

Perturbation of the Cost Vector

Consider the following problem.

$$\begin{aligned} \text{Minimize} \quad & \mathbf{cx} \\ \text{Subject to} \quad & \mathbf{Ax} = \mathbf{b} \\ & \mathbf{x} \geq \mathbf{0} \end{aligned}$$

Assume that $\mathbf{B}$ is an optimal basis. Suppose that the cost vector $\mathbf{c}$ is perturbed along the cost direction $\mathbf{c}'$, that is, $\mathbf{c}$ is replaced by $\mathbf{c} + \lambda \mathbf{c}'$ where $\lambda \geq 0$. We are interested in finding the optimal points and corresponding objective values as a function of $\lambda \geq 0$. Decomposing $\mathbf{A}$ into $[\mathbf{B}, \mathbf{N}]$, $\mathbf{c}$ into $(\mathbf{c}_B, \mathbf{c}_N)$, and $\mathbf{c}'$ into $(\mathbf{c}_B', \mathbf{c}_N')$, we get

$$\begin{aligned} z - (\mathbf{c}_B + \lambda \mathbf{c}_B')\mathbf{x}_B - (\mathbf{c}_N + \lambda \mathbf{c}_N')\mathbf{x}_N &= 0 \\ \mathbf{B}\mathbf{x}_B + \mathbf{N}\mathbf{x}_N &= \mathbf{b} \end{aligned}$$

Updating the tableau and denoting $\mathbf{c}_B'\mathbf{y}_j$ by z_j', we get

$$z + \sum_{j \in R} \left[(z_j - c_j) + \lambda (z_j' - c_j') \right] x_j = \mathbf{c}_B \bar{\mathbf{b}} + \lambda \mathbf{c}_B' \bar{\mathbf{b}}$$

$$\mathbf{x}_B + \sum_{j \in R} \mathbf{y}_j x_j = \bar{\mathbf{b}}$$

where R is the set of current indices associated with the nonbasic variables. The current tableau has $\lambda = 0$ and gives an optimal basic feasible solution of the original problem without perturbation. We would like to find out how far we can move in the direction $\mathbf{c}'$ while still maintaining optimality of the current point. Let $S = \{ j\colon (z_j' - c_j') > 0\}$. If $S = \phi$, then the current solution is optimal for all values of $\lambda \geq 0$ (why?). Otherwise, calculate $\hat{\lambda}$ as follows:

$$\hat{\lambda} = \underset{j \in S}{\text{Minimum}} \left\{ \frac{-(z_j - c_j)}{z_j' - c_j'} \right\} = \frac{-(z_k - c_k)}{z_k' - c_k'} \tag{6.23}$$

Let $\lambda_1 = \hat{\lambda}$. For $\lambda \in [0, \lambda_1]$ the current solution is optimal and the optimal objective value is given by $\mathbf{c}_B\bar{\mathbf{b}} + \lambda\mathbf{c}'_B\bar{\mathbf{b}} = \mathbf{c}_B\mathbf{B}^{-1}\mathbf{b} + \lambda\mathbf{c}'_B\mathbf{B}^{-1}\mathbf{b}$. For $\lambda \in [0, \lambda_1]$, the objective row coefficients in the simplex tableau are replaced by $(z_j - c_j) + \lambda(z'_j - c'_j)$. At $\lambda = \lambda_1$, x_k is introduced into the basis (if a blocking variable exists). After the tableau is updated, the process is repeated by recalculating S and $\hat{\lambda}$ and letting $\lambda_2 = \hat{\lambda}$. For $\lambda \in [\lambda_1, \lambda_2]$ the new current solution is optimal and its objective value is given by $\mathbf{c}_B\bar{\mathbf{b}} + \lambda\mathbf{c}'_B\bar{\mathbf{b}} = \mathbf{c}_B\mathbf{B}^{-1}\mathbf{b} + \lambda\mathbf{c}'_B\mathbf{B}^{-1}\mathbf{b}$ where $\mathbf{B}$ is the current basis. The process is repeated until S becomes empty. If there is no blocking variable when x_k enters the basis, then the problem is unbounded for all values of λ greater than the current value.

Example 6.17

Consider the following problem.

$$\begin{array}{lrl} \text{Minimize} & -x_1 - 3x_2 & \\ \text{Subject to} & x_1 + x_2 & \leq 6 \\ & -x_1 + 2x_2 & \leq 6 \\ & x_1, \quad x_2 & \geq 0 \end{array}$$

It is desired to find the optimal solutions and optimal objective values of the class of problems whose objective function is $(-1 + 2\lambda, -3 + \lambda)$ for $\lambda \geq 0$; that is, we perturb the cost vector along the vector $(2, 1)$. First we solve the problem with $\lambda = 0$ where x_3 and x_4 are the slack variables. The optimal tableau for $\lambda = 0$ is given by the following.

	z	x_1	x_2	x_3	x_4	RHS
z	1	0	0	$-\frac{5}{3}$	$-\frac{2}{3}$	-14
x_1	0	1	0	$\frac{2}{3}$	$-\frac{1}{3}$	2
x_2	0	0	1	$\frac{1}{3}$	$\frac{1}{3}$	4

In order to find the range over which this tableau is optimal, first find $\mathbf{c}'_B\mathbf{B}^{-1}\mathbf{N} - \mathbf{c}'_N$:

$$\begin{aligned} \mathbf{c}'_B\mathbf{B}^{-1}\mathbf{N} - \mathbf{c}'_N &= \mathbf{c}'_B(\mathbf{y}_3, \mathbf{y}_4) - (c'_3, c'_4) \\ &= (2, 1)\begin{bmatrix} \frac{2}{3} & -\frac{1}{3} \\ \frac{1}{3} & \frac{1}{3} \end{bmatrix} - (0, 0) \\ &= \left(\tfrac{5}{3}, -\tfrac{1}{3}\right) \end{aligned} \tag{6.24}$$

Therefore $S = \{3\}$ and from Equation (6.23) $\hat{\lambda}$ is given by

$$\hat{\lambda} = \frac{-(z_3 - c_3)}{z'_3 - c'_3} = \frac{-\left(-\frac{5}{3}\right)}{\frac{5}{3}} = 1$$

Therefore $\lambda_1 = 1$ and for $\lambda \in [0, 1]$ the basis $[\mathbf{a}_1, \mathbf{a}_2]$ remains optimal. The optimal objective value $z(\lambda)$ in this interval is given by

$$z(\lambda) = \mathbf{c}_B\bar{\mathbf{b}} + \lambda\mathbf{c}'_B\bar{\mathbf{b}}$$

$$= -14 + \lambda(2,1)\begin{pmatrix}2\\4\end{pmatrix} = -14 + 8\lambda$$

Noting Equation (6.24), the objective row coefficients of the nonbasic variables x_3 and x_4 are given by

$$(z_3 - c_3) + \lambda(z'_3 - c'_3) = -\tfrac{5}{3} + \tfrac{5}{3}\lambda$$

$$(z_4 - c_4) + \lambda(z'_4 - c'_4) = -\tfrac{2}{3} - \tfrac{1}{3}\lambda$$

Hence the optimal solution for any λ in the interval $[0,1]$ is given by the following tableau.

	z	x_1	x_2	x_3	x_4	RHS
z	1	0	0	$-\frac{5}{3} + \frac{5}{3}\lambda$	$-\frac{2}{3} - \frac{1}{3}\lambda$	$-14 + 8\lambda$
x_1	0	1	0	$\frac{2}{3}$	$-\frac{1}{3}$	2
x_2	0	0	1	$\frac{1}{3}$	$\frac{1}{3}$	4

At $\lambda = 1$ the coefficient of x_3 in row 0 is equal to 0 and x_3 is introduced in the basis leading to the following new tableau.

	z	x_1	x_2	x_3	x_4	RHS
z	1	0	0	0	-1	-6
x_3	0	$\frac{3}{2}$	0	1	$-\frac{1}{2}$	3
x_2	0	$-\frac{1}{2}$	1	0	$\frac{1}{2}$	3

We would now like to find the interval $[1, \lambda_2]$ over which the foregoing tableau is optimal. Note that

$$z_1 - c_1 = \mathbf{c}_B\mathbf{y}_1 - c_1 = (0,-3)\begin{pmatrix}\frac{3}{2}\\-\frac{1}{2}\end{pmatrix} + 1 = \tfrac{5}{2}$$

$$z_4 - c_4 = \mathbf{c}_B\mathbf{y}_4 - c_4 = (0,-3)\begin{pmatrix}-\frac{1}{2}\\\frac{1}{2}\end{pmatrix} - 0 = -\tfrac{3}{2}$$

$$z'_1 - c'_1 = \mathbf{c}'_B\mathbf{y}_1 - c'_1 = (0,1)\begin{pmatrix}\frac{3}{2}\\-\frac{1}{2}\end{pmatrix} - 2 = -\tfrac{5}{2}$$

$$z'_4 - c'_4 = \mathbf{c}'_B\mathbf{y}_4 - c'_4 = (0,1)\begin{pmatrix}-\frac{1}{2}\\\frac{1}{2}\end{pmatrix} - 0 = \tfrac{1}{2}$$

Therefore the objective row coefficients for the nonbasic variables x_1 and x_4 are given by

$$(z_1 - c_1) + \lambda(z_1' - c_1') = \tfrac{5}{2} - \tfrac{5}{2}\lambda$$
$$(z_4 - c_4) + \lambda(z_4' - c_4') = -\tfrac{3}{2} + \tfrac{1}{2}\lambda$$

Therefore for λ in the interval $[1, 3]$ these values are nonpositive and the basis consisting of $\mathbf{a}_3$ and $\mathbf{a}_2$ is optimal. Note that $\lambda = 3$ can also be determined as follows:

$$S = \{4\} \quad \text{and} \quad \lambda = \frac{-(z_4 - c_4)}{z_4' - c_4'} = \frac{\frac{3}{2}}{\frac{1}{2}} = 3$$

Over the interval $[1, 3]$ the objective function $z(\lambda)$ is given by

$$z(\lambda) = \mathbf{c}_B\bar{\mathbf{b}} + \lambda\mathbf{c}_B'\bar{\mathbf{b}}$$
$$= (0, -3)\begin{pmatrix} 3 \\ 3 \end{pmatrix} + \lambda(0, 1)\begin{pmatrix} 3 \\ 3 \end{pmatrix} = -9 + 3\lambda$$

Hence the optimal tableau for $\lambda \in [1, 3]$ is given as follows.

	z	x_1	x_2	x_3	x_4	RHS
z	1	$\frac{5}{2} - \frac{5}{2}\lambda$	0	0	$-\frac{3}{2} + \frac{1}{2}\lambda$	$-9 + 3\lambda$
x_3	0	$\frac{3}{2}$	0	1	$-\frac{1}{2}$	3
x_2	0	$-\frac{1}{2}$	1	0	$\frac{1}{2}$	3

At $\lambda = 3$ the coefficients of x_4 in row 0 is equal to zero, and x_4 is introduced in the basis leading to the following tableau.

	z	x_1	x_2	x_3	x_4	RHS
z	1	-5	0	0	0	0
x_3	0	1	1	1	0	6
x_4	0	-1	2	0	1	6

We would like to calculate the interval over which the foregoing tableau is optimal. First calculate

$$z_1' - c_1' = \mathbf{c}_B'\mathbf{y}_1 - c_1' = -2$$
$$z_2' - c_2' = \mathbf{c}_B'\mathbf{y}_2 - c_2' = -1$$

Therefore $S = \phi$ and hence the basis $[\mathbf{a}_3, \mathbf{a}_4]$ is optimal for all $\lambda \in [3, \infty)$. Figure 6.7 shows the optimal points and the optimal objective as a function of λ. Note that this function is piecewise-linear and concave. In Exercise 6.69 the reader is asked to show that this is always true. The break points correspond to the values of λ at which alterative optimal solutions exist.

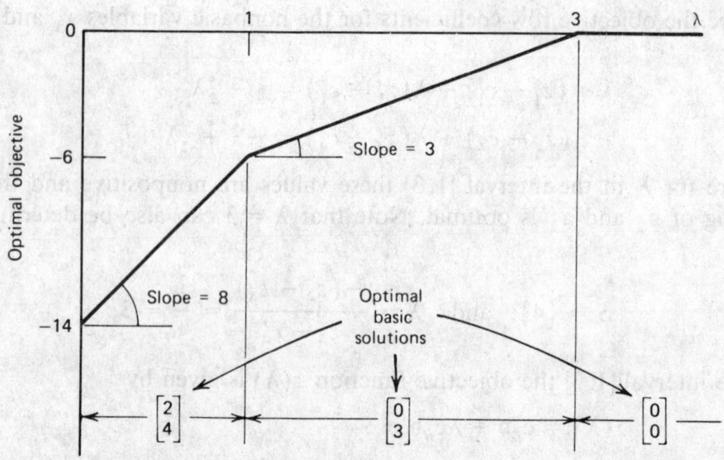

Figure 6.7. **Optimal objectives and optimal points as a function of λ.**

Perturbation of the Right-Hand-Side

Suppose that the right-hand-side vector $\mathbf{b}$ is replaced by $\mathbf{b} + \lambda\mathbf{b}'$ where $\lambda \geq 0$. This means that the right-hand-side is perturbed along the vector $\mathbf{b}'$. Since the right-hand-side of the primal problem is the objective of the dual problem, perturbing the right-hand-side can be analyzed as perturbing the objective function of the dual problem. We shall now handle the perturbation directly by considering the primal problem. Suppose that we have an optimal basis $\mathbf{B}$ of the original problem, that is, with $\lambda = 0$. The corresponding tableau is given by

$$z + (\mathbf{c}_B\mathbf{B}^{-1}\mathbf{N} - \mathbf{c}_N)\mathbf{x}_N = \mathbf{c}_B\mathbf{B}^{-1}\mathbf{b}$$

$$\mathbf{x}_B + \mathbf{B}^{-1}\mathbf{N}\mathbf{x}_N = \mathbf{B}^{-1}\mathbf{b}$$

where $\mathbf{c}_B\mathbf{B}^{-1}\mathbf{N} - \mathbf{c}_N \leq \mathbf{0}$. If $\mathbf{b}$ is replaced by $\mathbf{b} + \lambda\mathbf{b}'$, the vector $\mathbf{c}_B\mathbf{B}^{-1}\mathbf{N} - \mathbf{c}_N$ will not be affected; that is, dual feasibility will not be affected. The only change is that $\mathbf{B}^{-1}\mathbf{b}$ will be replaced by $\mathbf{B}^{-1}(\mathbf{b} + \lambda\mathbf{b}')$, and accordingly the objective becomes $\mathbf{c}_B\mathbf{B}^{-1}(\mathbf{b} + \lambda\mathbf{b}')$. As long as $\mathbf{B}^{-1}(\mathbf{b} + \lambda\mathbf{b}')$ is nonnegative, the current basis remains optimal. The value of λ at which another basis becomes optimal, can therefore, be determined as follows. Let $S = \{i\colon \bar{b}'_i < 0\}$ where $\bar{\mathbf{b}}' = \mathbf{B}^{-1}\mathbf{b}'$. If $S = \phi$, then the current basis is optimal for all values of $\lambda \geq 0$. Otherwise let

$$\hat{\lambda} = \underset{i \in S}{\text{Minimum}} \left\{ \frac{\bar{b}_i}{-\bar{b}'_i} \right\} = \frac{\bar{b}_r}{-\bar{b}'_r} \tag{6.25}$$

Let $\lambda_1 = \hat{\lambda}$. For $\lambda \in [0, \lambda_1]$ the current basis is optimal, where $\mathbf{x}_B = \mathbf{B}^{-1}(\mathbf{b} + \lambda\mathbf{b}')$ and the optimal objective is $\mathbf{c}_B\mathbf{B}^{-1}(\mathbf{b} + \lambda\mathbf{b}')$. At λ_1 the right-hand-side is replaced by $\mathbf{B}^{-1}(\mathbf{b} + \lambda_1\mathbf{b}')$, x_{B_r} is removed from the basis, and an appropriate variable (according to dual simplex method criterion) enters the basis. After the tableau is updated, the process is repeated in order to find the range $[\lambda_1, \lambda_2]$ over which the new basis is optimal, where $\lambda_2 = \hat{\lambda}$ from Equation (6.25). The process is terminated when either S is empty, in which case the current basis is optimal for all values of λ greater than or equal to the last value of λ, or else when all the entries in the row whose right-hand-side dropped to zero, are nonnegative. In this latter case no feasible solutions exist for all values of λ greater than the current value (why?).

Example 6.18

Consider the following problem.

$$\begin{array}{ll} \text{Minimize} & -x_1 - 3x_2 \\ \text{Subject to} & x_1 + x_2 \leq 6 \\ & -x_1 + 2x_2 \leq 6 \\ & x_1, \quad x_2 \geq 0 \end{array}$$

It is desired to find the optimal solution and optimal basis as the right-hand-side is perturbed along the direction $\begin{pmatrix} -1 \\ 1 \end{pmatrix}$, that is, if $\mathbf{b} = \begin{pmatrix} 6 \\ 6 \end{pmatrix}$ is replaced by $\mathbf{b} + \lambda\mathbf{b}' = \begin{pmatrix} 6 \\ 6 \end{pmatrix} + \lambda\begin{pmatrix} -1 \\ 1 \end{pmatrix}$ for $\lambda \geq 0$. The optimal solution with $\lambda = 0$ is shown, where x_3 and x_4 are the slack variables.

	z	x_1	x_2	x_3	x_4	RHS
z	1	0	0	$-\frac{5}{3}$	$-\frac{2}{3}$	-14
x_1	0	1	0	$\frac{2}{3}$	$-\frac{1}{3}$	2
x_2	0	0	1	$\frac{1}{3}$	$\frac{1}{3}$	4

To find the range over which the basis is optimal, first calculate $\bar{\mathbf{b}}'$:

$$\bar{\mathbf{b}}' = \mathbf{B}^{-1}\mathbf{b}'$$

$$= \begin{bmatrix} \frac{2}{3} & -\frac{1}{3} \\ \frac{1}{3} & \frac{1}{3} \end{bmatrix} \begin{bmatrix} -1 \\ 1 \end{bmatrix} = \begin{bmatrix} -1 \\ 0 \end{bmatrix}$$

Therefore $S = \{1\}$ and from Equation (6.25) λ_1 is given by

$$\lambda_1 = \frac{\bar{b}_1}{-\bar{b}'_1} = \frac{2}{-(-1)} = 2$$

Therefore the basis $[\mathbf{a}_1, \mathbf{a}_2]$ remains optimal over the interval $[0, 2]$. In particular, for any $\lambda \in [0, 2]$ the objective value and the right-hand-side are given by

$$\begin{aligned} z(\lambda) &= \mathbf{c}_B\bar{\mathbf{b}} + \lambda\mathbf{c}_B\bar{\mathbf{b}}' \\ &= (-1, -3)\begin{pmatrix} 2 \\ 4 \end{pmatrix} + \lambda(-1, -3)\begin{pmatrix} -1 \\ 0 \end{pmatrix} \\ &= -14 + \lambda \\ \bar{\mathbf{b}} + \lambda\bar{\mathbf{b}}' &= \begin{pmatrix} 2 \\ 4 \end{pmatrix} + \lambda\begin{pmatrix} -1 \\ 0 \end{pmatrix} \\ &= \begin{pmatrix} 2 - \lambda \\ 4 \end{pmatrix} \end{aligned}$$

	z	x_1	x_2	x_3	x_4	RHS
z	1	0	0	$-\frac{5}{3}$	$-\frac{2}{3}$	$-14 + \lambda$
x_1	0	1	0	$\frac{2}{3}$	$-\frac{1}{3}$	$2 - \lambda$
x_2	0	0	1	$\frac{1}{3}$	$\frac{1}{3}$	4

At $\lambda = 2$, $x_{B_r} = x_1$ drops to zero. A dual simplex pivot is performed so that x_1 leaves the basis and x_4 enters the basis leading to the following tableau.

	z	x_1	x_2	x_3	x_4	RHS
z	1	-2	0	-3	0	-12
x_4	0	-3	0	-2	1	0
x_2	0	1	1	1	0	4

To find the range $[2, \lambda_2]$ over which this tableau is optimal, first find $\bar{\mathbf{b}}$ and $\bar{\mathbf{b}}'$:

$$\bar{\mathbf{b}} = \mathbf{B}^{-1}\mathbf{b} = \begin{bmatrix} -2 & 1 \\ 1 & 0 \end{bmatrix}\begin{bmatrix} 6 \\ 6 \end{bmatrix} = \begin{bmatrix} -6 \\ 6 \end{bmatrix}$$

$$\bar{\mathbf{b}}' = \mathbf{B}^{-1}\mathbf{b}' = \begin{bmatrix} -2 & 1 \\ 1 & 0 \end{bmatrix}\begin{bmatrix} -1 \\ 1 \end{bmatrix} = \begin{bmatrix} 3 \\ -1 \end{bmatrix}$$

Therefore $S = \{2\}$ and λ_2 is given by

$$\lambda_2 = \frac{\bar{b}_2}{-\bar{b}'_2} = \frac{6}{-(-1)} = 6$$

For λ in the interval $[2, 6]$ the optimal objective value and the right-hand-side are given by

$$\begin{aligned} z(\lambda) &= \mathbf{c}_B\bar{\mathbf{b}} + \lambda\mathbf{c}_B\bar{\mathbf{b}}' \\ &= (0, -3)\begin{pmatrix} -6 \\ 6 \end{pmatrix} + \lambda(0, -3)\begin{pmatrix} 3 \\ -1 \end{pmatrix} \\ &= -18 + 3\lambda \\ \bar{\mathbf{b}} + \lambda\bar{\mathbf{b}}' &= \begin{bmatrix} -6 \\ 6 \end{bmatrix} + \lambda\begin{bmatrix} 3 \\ -1 \end{bmatrix} \\ &= \begin{bmatrix} -6 + 3\lambda \\ 6 - \lambda \end{bmatrix} \end{aligned}$$

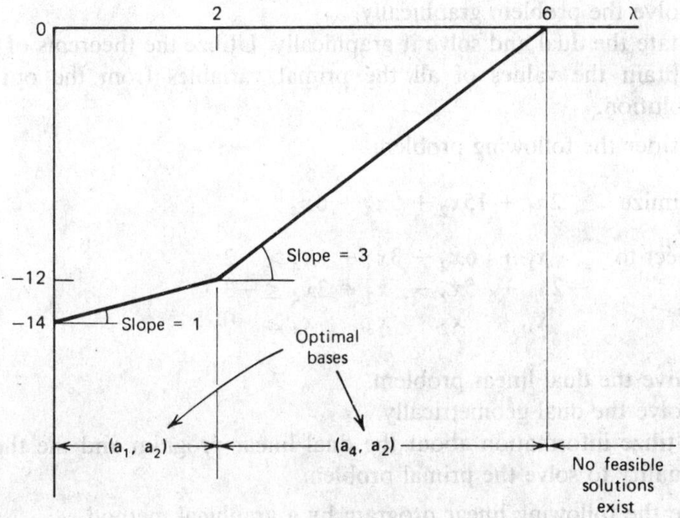

Figure 6.8. Optimal objectives and bases as a function of λ.

The optimal tableau over the interval [2, 6] is depicted here.

	z	x_1	x_2	x_3	x_4	RHS
z	1	-2	0	-3	0	$-18 + 3\lambda$
x_4	0	-3	0	-2	1	$-6 + 3\lambda$
x_2	0	1	1	1	0	$6 - \lambda$

At $\lambda = 6$, x_2 drops to zero. Since all entries in the x_2 row are nonnegative, we stop with the conclusion that for $\lambda > 6$ there exist no feasible solutions. Figure 6.8 summarizes the optimal bases and corresponding objectives for $\lambda \geq 0$. Note that the optimal objective as a function of λ is piecewise-linear and convex. In Exercise 6.70 we ask the reader to show that this is always true. The break points correspond to the values of λ for which alternative optimal dual solutions exist.

EXERCISES

6.1 Consider the following problem.

$$\begin{aligned} \text{Maximize} \quad & -x_1 + 3x_2 \\ \text{Subject to} \quad & 2x_1 + 3x_2 \leq 6 \\ & x_1 - 2x_2 \geq -2 \\ & x_1, \quad x_2 \geq 0 \end{aligned}$$

a. Solve the problem graphically.
b. State the dual and solve it graphically. Utilize the theorems of duality to obtain the values of all the primal variables from the optimal dual solution.

6.2 Consider the following problem.

$$\begin{array}{ll} \text{Minimize} & 2x_1 + 15x_2 + 5x_3 + 6x_4 \\ \text{Subject to} & x_1 + 6x_2 + 3x_3 + x_4 \geq 2 \\ & -2x_1 + 5x_2 - x_3 + 3x_4 \leq -3 \\ & x_1, \quad x_2, \quad x_3, \quad x_4 \geq 0 \end{array}$$

a. Give the dual linear problem.
b. Solve the dual geometrically.
c. Utilize information about the dual linear program and the theorems of duality to solve the primal problem.

6.3 Solve the following linear program by a graphical method.

$$\begin{array}{ll} \text{Maximize} & 3x_1 + 3x_2 + 21x_3 \\ \text{Subject to} & 6x_1 + 9x_2 + 25x_3 \leq 25 \\ & 3x_1 + 2x_2 + 25x_3 \leq 20 \\ & x_1, \quad x_2, \quad x_3 \geq 0 \end{array}$$

(*Hint*. Utilize the dual problem.)

6.4 Give the dual of the following problem.

$$\begin{array}{ll} \text{Minimize} & -2x_1 + 3x_2 + 5x_3 \\ \text{Subject to} & -2x_1 + x_2 + 3x_3 + x_4 \geq 5 \\ & 2x_1 \qquad + x_3 \qquad \leq 4 \\ & \qquad 2x_2 + x_3 + x_4 = 6 \\ & x_1 \leq 0 \\ & x_2, \quad x_3 \geq 0 \\ & x_4 \quad \text{unrestricted} \end{array}$$

6.5 Use the standard form of duality to obtain the dual of the following problem. Also verify the relationships in Table 6.1.

$$\begin{array}{ll} \text{Minimize} & \mathbf{c}_1\mathbf{x}_1 + \mathbf{c}_2\mathbf{x}_2 + \mathbf{c}_3\mathbf{x}_3 \\ \text{Subject to} & \mathbf{A}_{11}\mathbf{x}_1 + \mathbf{A}_{12}\mathbf{x}_2 + \mathbf{A}_{13}\mathbf{x}_3 \geq \mathbf{b}_1 \\ & \mathbf{A}_{21}\mathbf{x}_1 + \mathbf{A}_{22}\mathbf{x}_2 + \mathbf{A}_{23}\mathbf{x}_3 \leq \mathbf{b}_2 \\ & \mathbf{A}_{31}\mathbf{x}_1 + \mathbf{A}_{32}\mathbf{x}_2 + \mathbf{A}_{33}\mathbf{x}_3 = \mathbf{b}_3 \\ & \mathbf{x}_1 \geq \mathbf{0} \\ & \mathbf{x}_2 \leq \mathbf{0} \\ & \mathbf{x}_3 \quad \text{unrestricted} \end{array}$$

6.6 Consider the following problem.

$$\text{Maximize} \quad 10x_1 + 24x_2 + 20x_3 + 20x_4 + 25x_5$$

$$\begin{aligned} \text{Subject to} \quad x_1 + x_2 + 2x_3 + 3x_4 + 5x_5 &\leq 19 \\ 2x_1 + 4x_2 + 3x_3 + 2x_4 + x_5 &\leq 57 \\ x_1, \quad x_2, \quad x_3, \quad x_4, \quad x_5 &\geq 0 \end{aligned}$$

a. Write the dual problem and verify that $(w_1, w_2) = (4, 5)$ is a feasible solution.

b. Use the information in part (a) to derive an optimal solution to both the primal and the dual problems.

6.7 The following refer to a primal-dual (min-max) pair P and D of linear programming problems in canonical form. Provide a brief explanation with your answers.

a. If a basic solution to the primal is infeasible and has an objective value less than the optimal value, then the associated complementary dual basic solution is feasible. True or False?

b. For the linear program Minimize $\{x_1 : 2x_1 - x_2 \geq 0, -2x_1 + 3x_2 \geq -6, \mathbf{x} \geq \mathbf{0}\}$, consider the basic feasible solution with a basis comprised of the columns of x_1 and the slack variable in the second constraint. Give the associated complementary dual basic solution. What can you say about this pair of primal-dual basic solutions?

c. If P has alternative optimal solutions and if $\mathbf{w}^*$ is any optimal basic feasible solution for D, then $\mathbf{w}^*$ must be degenerate. True or False?

d. Let z^* be the common (finite) optimal value of P and D. Suppose that $\bar{\mathbf{x}}$ is a basic infeasible solution to P whose complementary dual basic solution is feasible. Is it possible that the common objective value of this pair of primal-dual basic solutions is z^*?

e. If P is unbounded, it is possible to change its right-hand-side and make it have a finite optimum. True or False?

f. Referring to Figure 6.4*a*, suppose that $-\mathbf{c}$ is aligned along $-\mathbf{A}_1$. Consider the basic solution with basis $\mathbf{B} = [\mathbf{a}_2, \mathbf{a}_4, \mathbf{a}_5]$. Comment on the corresponding pair of complementary primal and dual basic solutions with respect to feasibility, optimality, and degeneracy.

6.8 The following simplex tableau shows the optimal solution of a linear programming problem. It is known that x_4 and x_5 are the slack variables in the first and second constraints of the original problem. The constraints are of the $\leq$ type.

	z	x_1	x_2	x_3	x_4	x_5	RHS
z	1	0	-2	0	-3	-2	-35
x_3	0	0	$\frac{1}{4}$	1	$\frac{1}{2}$	0	$\frac{5}{2}$
x_1	0	1	$-\frac{1}{2}$	0	$-\frac{1}{6}$	$\frac{1}{3}$	$\frac{5}{2}$

a. Write the original problem.
b. What is the dual of the original problem?
c. Obtain the optimal solution of the dual problem from the tableau.

6.9 Consider the following linear programming problem.

$$\begin{array}{ll} \text{Maximize} & 2x_1 + 3x_2 + 6x_3 \\ \text{Subject to} & x_1 + 2x_2 + 3x_3 \le 10 \\ & x_1 - 2x_2 + 2x_3 \le 6 \\ & x_1, \quad x_2, \quad x_3 \ge 0 \end{array}$$

a. Write the dual problem.
b. Solve the foregoing problem by the simplex method. At each iteration, identify the dual variables and show which dual constraints are violated.
c. At each iteration, identify the 3×3 dual basis that goes with the simplex iteration. Identify the dual basic and nonbasic variables.
d. Show that at each iteration of the simplex method, the dual objective is "worsened."
e. Verify that at termination, feasible solutions of both problems are at hand, with equal objectives and with complementary slackness.

6.10 Consider the following bounded variables linear program

$$\begin{array}{ll} \text{Maximize} & x_1 + x_2 \\ \text{Subject to} & -2x_1 + x_2 \le 2 \\ & x_1 - x_2 \le 0 \\ & -2 \le x_1 \le 2 \\ & -1 \le x_2 \le 2 \end{array}$$

a. Solve the problem graphically in the (x_1, x_2) space.
b. Give all optimal basic feasible partitions. (Specify sets of basic and nonbasic variables at optimality.)
c. For the extreme point $(x_1, x_2) = (0, 2)$, construct the bounded variables simplex tableau and perform one iteration. Is the resulting tableau optimal?
d. Graphically verify whether the following is true or false. Starting at the point where the slack from the second constraint and x_2 are nonbasic at their lower bounds, if one introduces x_2 into the basis, then the resulting basic feasible solution is optimal.
e. Write the dual to the foregoing problem by associating one dual variable with each of the six inequality constraints.
f. Using the graph of part (a), compute the set of dual optimal solutions and determine why or why not the dual has alternative optimal solutions.
g. Graphically add the constraint $x_1 + x_2 \le 4$ to the problem. Is there a degenerate optimal dual basic solution? Is there a nondegenerate optimal dual basic solution?

6.11 Consider the problem: Minimize $\mathbf{cx}$ subject to $\mathbf{Ax} = \mathbf{b}$, $\mathbf{x} \geq \mathbf{0}$ where $m = n$, $\mathbf{c} = \mathbf{b}'$ and $\mathbf{A} = \mathbf{A}'$. Show that if there exists an $\mathbf{x}_0$ such that $\mathbf{Ax}_0 = \mathbf{b}$, $\mathbf{x}_0 \geq \mathbf{0}$, then $\mathbf{x}_0$ is an optimal point. (*Hint.* Use duality.)

6.12 Consider the problem: Minimize z subject to $z - \mathbf{cx} = 0$, $\mathbf{Ax} = \mathbf{b}$, $\mathbf{x} \geq \mathbf{0}$.
a. State the dual.
b. At optimality what will be the value of the first dual variable? Explain!

6.13 Consider the problem: Minimize $\mathbf{cx}$ subject to $\mathbf{Ax} = \mathbf{b}$, $\mathbf{x} \geq \mathbf{0}$. Let $\mathbf{B}$ be a basis that is neither primal nor dual feasible. Indicate how one can solve this problem starting with the basis $\mathbf{B}$.

6.14 Prove that if a given basic feasible solution to some linear programming problem is optimal, the same basic vectors will yield an optimal solution for any requirements vector that lies in the cone spanned by these basic vectors.

6.15 The Sewel Manufacturing Company produces two types of reclining chairs for sale in the Northeast. Two basic types of skilled labor are involved —assembly and finishing. One unit of the top-of-the-line recliner requires $\frac{3}{2}$ hours of assembly, 1 hour of finishing, and sells for a profit of \$20. A unit of the second-line recliner requires $\frac{1}{2}$ hour of assembly, $\frac{1}{2}$ hour of finishing, and sells for a profit of \$12. Presently there are 100 assembly hours and 80 finishing hours available to the company. The company is involved in labor negotiations concerning salary modifications for the coming year. You are asked to provide the company with indications of the worth of an hour of assembly worker's time and an hour of finishing worker's time.

6.16 Show by duality that if the problem Minimize $\mathbf{cx}$ subject to $\mathbf{Ax} = \mathbf{b}$ and $\mathbf{x} \geq \mathbf{0}$ has a finite optimal solution, then the new problem Minimize $\mathbf{cx}$ subject to $\mathbf{Ax} = \mathbf{b}'$ and $\mathbf{x} \geq \mathbf{0}$ cannot be unbounded, no matter what value the vector $\mathbf{b}'$ might take.

6.17 Show directly that if the primal problem Minimize $\mathbf{cx}$ subject to $\mathbf{Ax} \geq \mathbf{b}$ and $\mathbf{x} \geq \mathbf{0}$ has no feasible solutions, and if the dual problem has a feasible solution, then the dual problem is unbounded. (*Hint.* Use Farkas' lemma. If the system $\mathbf{Ax} \geq \mathbf{b}$ and $\mathbf{x} \geq \mathbf{0}$ has no solutions, then the system $\mathbf{wA} \leq \mathbf{0}$, $\mathbf{w} \geq \mathbf{0}$, $\mathbf{wb} > 0$ has a solution.)

6.18 Consider the problem: Minimize $\mathbf{cx}$ subject to $\mathbf{Ax} = \mathbf{b}$, $\mathbf{l} \leq \mathbf{x} \leq \mathbf{u}$, where $\mathbf{l}$ and $\mathbf{u}$ are finite.
a. Give the dual.
b. Show that the dual always possesses a feasible solution.
c. If the primal problem possesses a feasible solution, what conclusions would you reach?

6.19 Consider the problem: Maximize $\mathbf{cx}$ subject to $\mathbf{Ax} = \mathbf{b}$, $\mathbf{x} \geq \mathbf{0}$. Let $z_j - c_j$, y_{ij}, and $\bar{b}_i$ be the updated entries at some iteration of the simplex algorithm. Indicate whether each of the following statements is true or false. Discuss.

a. $y_{ij} = -\dfrac{\partial x_{B_i}}{\partial x_j}$

b. $z_j - c_j = -\dfrac{\partial z}{\partial x_j}$

c. Dual feasibility is the same as primal optimality.
d. Performing row operations on inequality systems yields equivalent systems.
e. Adding artificial variables to the primal serves to restrict variables that are really unrestricted in the dual.
f. Linear programming by the simplex method is essentially a gradient search.
g. A linear problem can be solved by the two-phase method if it can be solved by the big-M method.
h. There is a *duality gap* (difference in optimal objective values) when both the primal and the dual programs have no feasible solutions.
i. Converting a maximization problem to a minimization problem changes the sign of the dual variables.
j. A linear program with some variables required to be greater than or equal to zero can always be converted into one where all variables are unrestricted, without adding any new constraints.

6.20 Use the main duality theorem to prove Farkas' lemma. (*Hint.* Consider the following pair of primal and dual problems.)

Minimize	$\mathbf{0x}$	Maximize	$\mathbf{wb}$
Subject to	$\mathbf{Ax} = \mathbf{b}$	Subject to	$\mathbf{wA} \leq \mathbf{0}$
	$\mathbf{x} \geq \mathbf{0}$		$\mathbf{w}$ unrestricted

6.21 Consider a pair of primal and dual linear programs in standard form.
a. What happens to the dual solution if the kth primal constraint is multiplied by a nonzero scalar λ?
b. What happens to the dual solution if a scalar multiple of one primal constraint is added to another primal constraint?
c. What happens to the primal and dual solutions if a scalar multiple of one primal column is added to another primal column?

6.22 Show that if a set of constraints is redundant, then the corresponding dual variables can only be specified within a constant of addition (that is, if one dual variable in the set is changed by an amount θ, then all dual variables in the set would change by appropriate multiples of θ).

6.23 For the example in Exercise 1.24, what dual variable in the optimal linear programming solution will yield the equilibrium price p^*? Interpret this using the figure of Exercise 1.24 that shows the supply–demand interaction.

6.24 In Figure 6.4b suppose that $-\mathbf{c}$ is aligned along $-\mathbf{A}_1$. In terms of dual optimal solution values, derive the right-hand and the left-hand shadow prices with respect to perturbations in b_1, b_2, and b_3.

6.25 Two players are involved in a competitive game. One player, called the row player, has two strategies available; the other player, called the column player, has three strategies available. If the row player selects strategy i and the column player selects strategy j, the payoff to the row player is c_{ij} and the payoff to the column player is $-c_{ij}$. Thus the column player loses what the row player wins and vice versa—a *two-person zero-sum game*. The following matrix gives the payoffs to the row player.

	1	2	3
1	2	−1	0
2	−3	1	1

Let x_1, x_2, and x_3 be probabilities with which the column player will select the various strategies over many plays of the game. Thus $x_1 + x_2 + x_3 = 1$, $x_1, x_2, x_3 \geq 0$. If the column player applies these probabilities to the selection of his strategy for any play of the game, consider the row player's options. If the row player selects row 1, then his expected payoff is $2x_1 - x_2$. If the row player selects row 2, his payoff is $-3x_1 + x_2 + x_3$. Wishing to minimize the maximum expected payoff to the row player, the column player should solve the following linear program.

$$\begin{aligned} &\text{Minimize} \quad z \\ &\text{Subject to} \quad \begin{aligned} x_1 + x_2 + x_3 &= 1 \\ 2x_1 - x_2 \qquad &\leq z \\ -3x_1 + x_2 + x_3 &\leq z \\ x_1, \quad x_2, \quad x_3 &\geq 0 \\ z \quad &\text{unrestricted} \end{aligned} \end{aligned}$$

Transposing the variable z to the left-hand-side, we get the column player's problem:

$$\begin{aligned} &\text{Minimize} \quad z \\ &\text{Subject to} \quad \begin{aligned} x_1 + x_2 + x_3 &= 1 \\ z - 2x_1 + x_2 \qquad &\geq 0 \\ z + 3x_1 - x_2 - x_3 &\geq 0 \\ x_1, \quad x_2, \quad x_3 &\geq 0 \\ z \quad &\text{unrestricted} \end{aligned} \end{aligned}$$

a. Give the dual of this linear program.
b. Interpret the dual problem in part (a). (*Hint.* Consider the row player's problem.)
c. Solve the dual problem of part (a). (*Hint.* This problem may be solved graphically.)
d. Use the optimal dual solution of part (c) to compute the column player's probabilities.
e. Interpret the complementary slackness conditions for this two-person zero-sum game.

6.26 Show that discarding a redundant constraint is equivalent to setting the corresponding dual variable to zero.

6.27 Let $\mathbf{x}^*$ be an optimal solution to the problem: Minimize $\mathbf{cx}$ subject to $\mathbf{a}^i\mathbf{x} = b_i$, $i = 1, \ldots, m$, $\mathbf{x} \geq \mathbf{0}$. Let $\mathbf{w}^*$ be an optimal dual solution. Show that $\mathbf{x}^*$ is also an optimal to the problem: Minimize $(\mathbf{c} - w_k^* \mathbf{a}^k)\mathbf{x}$ subject to $\mathbf{a}^i\mathbf{x} = b_i$, $i = 1, \ldots, m$, $i \neq k$, $\mathbf{x} \geq \mathbf{0}$, where w_k^* is the kth component of $\mathbf{w}^*$. Discuss!

6.28 The following are the initial and current tableaux of a linear programming problem.

	z	x_1	x_2	x_3	x_4	x_5	x_6	x_7	RHS
z	1	1	6	-7	a	5	0	0	0
x_6	0	5	-4	13	b	1	1	0	20
x_7	0	1	-1	5	c	1	0	1	8

	z	x_1	x_2	x_3	x_4	x_5	x_6	x_7	RHS
z	1	$\frac{72}{7}$	0	0	$\frac{11}{7}$	$\frac{8}{7}$	$\frac{23}{7}$	$-\frac{50}{7}$	$\frac{60}{7}$
x_3	0	$-\frac{1}{7}$	0	1	$-\frac{2}{7}$	$\frac{3}{7}$	$-\frac{1}{7}$	$\frac{4}{7}$	$\frac{12}{7}$
x_2	0	$-\frac{12}{7}$	1	0	$-\frac{3}{7}$	$\frac{8}{7}$	$-\frac{5}{7}$	$\frac{13}{7}$	$\frac{4}{7}$

a. Find a, b, and c.
b. Find $\mathbf{B}^{-1}$.
c. Find $\partial x_2/\partial x_5$.
d. Find $\partial x_3/\partial b_2$.
e. Find $\partial z/\partial x_6$.
f. Find the complementary dual solution.

6.29 The following is an optimal simplex tableau (maximization and all $\leq$ constraints).

	z	x_1	x_2	x_3	x_4 (SLACKS)	x_5 (SLACKS)	x_6 (SLACKS)	RHS
z	1	0	0	0	2	0	5	θ
x_1	0	1	1	0	2	0	1	2
x_3	0	0	0	1	1	0	4	$\frac{3}{2}$
x_5	0	0	-2	0	-1	1	6	1

a. Give the optimal solution.
b. Give the optimal dual solution.
c. Find $\partial z/\partial b_1$. Interpret this number.
d. Find $\partial x_1/\partial x_6$. Interpret this number.
e. If you could buy an additional unit of the first resource for a cost of $\frac{5}{2}$ would you do this? Why?

f. Another firm wishes to purchase one unit of the third resource from you. How much is such a unit worth to you? Why?

g. Are there any alternative optimal solutions? If not, why not? If so, give one.

h. Find the objective value θ.

6.30 Solve the following problem by the dual simplex method.

$$\begin{array}{ll} \text{Maximize} & -4x_1 - 6x_2 - 18x_3 \\ \text{Subject to} & 2x_1 + 3x_3 \geq 3 \\ & 3x_2 + 2x_3 \geq 5 \\ & x_1, \quad x_2, \quad x_3 \geq 0 \end{array}$$

6.31 Consider the following linear programming problem.

$$\begin{array}{ll} \text{Maximize} & 2x_1 - 3x_2 \\ \text{Subject to} & x_1 + x_2 \geq 3 \\ & 3x_1 + x_2 \leq 6 \\ & x_1, \quad x_2 \geq 0 \end{array}$$

You are told that the optimal solution is $x_1 = \frac{3}{2}$ and $x_2 = \frac{3}{2}$. Verify this statement by duality. Describe two procedures for modifying the problem in such a way that the dual simplex method can be used. Use one of these procedures for solving the problem by the dual simplex method.

6.32 Solve the following linear program by the dual simplex method.

$$\begin{array}{ll} \text{Minimize} & 2x_1 + 3x_2 + 5x_3 + 6x_4 \\ \text{Subject to} & x_1 + 2x_2 + 3x_3 + x_4 \geq 2 \\ & -x_1 + x_2 - x_3 + 3x_4 \leq -3 \\ & x_1, \quad x_2, \quad x_3, \quad x_4 \geq 0 \end{array}$$

6.33 Consider the following problem.

$$\begin{array}{ll} \text{Minimize} & 3x_1 - 5x_2 - x_3 + 2x_4 - 4x_5 \\ \text{Subject to} & x_1 + x_2 + x_3 + 3x_4 + x_5 \leq 6 \\ & -x_1 - x_2 + 2x_3 + x_4 - x_5 \geq 3 \\ & x_1, \quad x_2, \quad x_3, \quad x_4, \quad x_5 \geq 0 \end{array}$$

a. Give the dual problem.

b. Solve the dual problem using the artificial constraint technique.

c. Find the primal solution from the dual solution.

6.34 Show that, in the dual simplex method, if $\bar{b}_r < 0$ and $y_{rj} \geq 0$ for $j = 1, \ldots, n$, then the dual is unbounded (and the primal is infeasible) by constructing a suitable direction. (*Hint*. Consider $\mathbf{w} = \mathbf{c}_B\mathbf{B}^{-1} + \lambda\mathbf{B}^r$, where $\mathbf{B}^r$ is the rth row of $\mathbf{B}^{-1}$.)

6.35 Show that the dual simplex algorithm is precisely the primal simplex algorithm applied to the dual problem. Be explicit.

6.36 Provide the details and the justification for the lexicographic method and Bland's method discussed in Section 6.4 for preventing cycling in the dual simplex method.

6.37 Given the primal basis in Equation (6.9), show that the complementary dual basis in Equation (6.10) is invertible and corresponds to the dual solution $(w_1, \ldots, w_m) = \mathbf{c}_B\mathbf{B}^{-1}$.

6.38 In Section 6.5 we showed that the complementary dual basis matrix for the linear program in standard form is given by

$$\left[\begin{array}{c|c} \mathbf{B}^t & \mathbf{0} \\ \hline \mathbf{N}^t & I_{n-m} \end{array}\right]$$

a. Give the complete starting dual tableau.
b. Give the inverse of this basis matrix.
c. Use the result of part (b) to develop the dual tableau associated with this basis matrix.

6.39 Suppose that an optimal solution to the problem Minimize **cx** subject to $\mathbf{Ax} \geq \mathbf{b}$, $\mathbf{x} \geq \mathbf{0}$, exists. Prove the following complementarity theorem.
a. A variable is zero for all primal optimal solutions if and only if its complementary dual variable is positive for some dual optimal solution.
b. A variable is unbounded in the primal feasible set if and only if its complementary dual variable is bounded in the dual feasible set.
c. A variable is unbounded in the primal *optimal* set if and only if its complementary dual variable is zero for all dual feasible solutions.
d. A variable is positive for some primal feasible solution if and only if its complementary dual variable is bounded in the dual *optimal* set.

6.40 Consider the following problem.

$$\begin{array}{lrl} \text{P: Minimize} & 2x_1 - 3x_2 & \\ \text{Subject to} & x_1 + x_2 & \geq 2 \\ & - x_2 & \geq -4 \\ & x_1, \quad x_2 & \geq 0 \end{array}$$

a. Solve P graphically.
b. Give the dual of P. Solve the dual graphically.
c. Illustrate the theorem of the previous exercise for each primal and dual variable, including slacks.

6.41 Consider the following problem.

$$\begin{array}{lrl} \text{P: Minimize} & - x_2 & \\ \text{Subject to} & x_1 + x_2 & \geq 2 \\ & - x_2 & \geq -4 \\ & x_1, \quad x_2 & \geq 0 \end{array}$$

a. Solve P graphically.
b. Give the dual of P. Solve the dual graphically.
c. Illustrate the complementarity theorem of Exercise 6.39 for each primal and dual variable, including slacks.

6.42 Apply the dual simplex method to the following problem.

$$\begin{array}{ll} \text{Minimize} & w_3 \\ \text{Subject to} & -\frac{1}{4}w_1 - \frac{1}{2}w_2 \leq -\frac{3}{4} \\ & 8w_1 + 12w_2 \leq 20 \\ & w_1 + \frac{1}{2}w_2 - w_3 \leq -\frac{1}{2} \\ & -9w_1 - 3w_2 \leq 6 \\ & w_1, \quad w_2, \quad w_3 \geq 0 \end{array}$$

6.43 Apply the perturbation technique to the dual simplex method to ensure finiteness. Specifically consider the perturbed problem

$$\begin{array}{ll} \text{Minimize} & (\mathbf{c} + \boldsymbol{\epsilon})\mathbf{x} \\ \text{Subject to} & \mathbf{Ax} = \mathbf{b} \\ & \mathbf{x} \geq \mathbf{0} \end{array}$$

where $\boldsymbol{\epsilon} = (\epsilon^1, \epsilon^2, \ldots, \epsilon^n)$ is a vector formed from the powers of a positive scalar ϵ. Compare the derived procedure with the one produced by the lexicographic method developed in Exercise 6.36.

6.44 Show that the artificial constraint technique applied to the primal problem is precisely the single artificial variable technique (of Chapter 4) with the big-M method applied to the dual problem and vice versa. (*Hint*: Consider the dual of minimize $\mathbf{0x}_B + (\mathbf{c}_N - \mathbf{c}_B\mathbf{B}^{-1}\mathbf{N})\mathbf{x}_N$ subject to $\mathbf{x}_B + \mathbf{B}^{-1}\mathbf{N}\mathbf{x}_N = \mathbf{B}^{-1}\mathbf{b}$, $\mathbf{x}_B, \mathbf{x}_N \geq \mathbf{0}$.)

6.45 Suppose that the artificial constraint technique (with slack x_{n+1}) is utilized to find a starting dual solution. Show that if this constraint is tight at optimality and $z_{n+1} - c_{n+1} < 0$, then the primal problem is unbounded. Explain in detail the case when this constraint is tight at optimality, but $z_{n+1} - c_{n+1} = 0$. How large should M be in order to reach these conclusions? Illustrate your answer using the problem to minimize $x_1 - x_2$ subject to $-x_1 + x_2 \leq 1$ and $\mathbf{x} \geq \mathbf{0}$.

6.46 Solve the following problem by the primal-dual algorithm.

$$\begin{array}{ll} \text{Maximize} & x_1 + 6x_2 \\ \text{Subject to} & x_1 + x_2 \geq 2 \\ & x_1 + 2x_2 \leq 3 \\ & x_1, \quad x_2 \geq 0 \end{array}$$

6.47 Apply the primal-dual method to the following problem.

$$\begin{array}{ll} \text{Minimize} & 9x_1 + 7x_2 + 4x_3 + 2x_4 + 6x_5 + 10x_6 \\ \text{Subject to} & x_1 + x_2 + x_3 = 8 \\ & x_4 + x_5 + x_6 = 5 \\ & x_1 + x_4 = 6 \\ & x_2 + x_5 = 4 \\ & x_3 + x_6 = 3 \\ & x_1, \ x_2, \ x_3, \ x_4, \ x_5, \ x_6 \geq 0 \end{array}$$

6.48 Solve the following problem by the primal-dual algorithm.

$$\begin{array}{ll} \text{Minimize} & x_1 + 2x_3 - x_4 \\ \text{Subject to} & x_1 + x_2 + x_3 + x_4 \leq 6 \\ & 2x_1 - x_2 + 3x_3 - 2x_4 \geq 5 \\ & x_1, \ x_2, \ x_3, \ x_4 \geq 0 \end{array}$$

6.49 Suppose that at the end of the restricted primal problem we have $x_0 > 0$ and $\mathbf{v}^*\mathbf{a}_j \leq 0$ for all j. Using Farkas' lemma, show that the primal problem has no feasible solution.

6.50 When is the primal-dual algorithm preferred to the dual simplex algorithm and vice versa?

6.51 Apply the primal-dual algorithm to the following problem.

$$\begin{array}{ll} \text{Maximize} & 7x_1 + 2x_2 + x_3 + 4x_4 + 6x_5 \\ \text{Subject to} & 3x_1 + 5x_2 - 6x_3 + 2x_4 + 4x_5 = 27 \\ & x_1 + 2x_2 + 3x_3 - 7x_4 + 6x_5 \geq 2 \\ & 9x_1 - 4x_2 + 2x_3 + 5x_4 - 2x_5 = 16 \\ & x_1, \ x_2, \ x_3, \ x_4, \ x_5 \geq 0 \end{array}$$

6.52 Using the restricted (total) phase I interpretation of the primal-dual algorithm, give the details of a lexicographic cycling prevention rule. (*Hint*. See Section 6.5.)

6.53 Consider the following linear programming problem and its optimal final tableau shown below.

$$\begin{array}{ll} \text{Maximize} & 2x_1 + x_2 - x_3 \\ \text{Subject to} & x_1 + 2x_2 + x_3 \leq 8 \\ & -x_1 + x_2 - 2x_3 \leq 4 \\ & x_1, \ x_2, \ x_3 \geq 0 \end{array}$$

Final Tableau

	z	x_1	x_2	x_3	x_4	x_5	RHS
z	1	0	3	3	2	0	16
x_1	0	1	2	1	1	0	8
x_5	0	0	3	-1	1	1	12

a. Write the dual problem and find the optimal dual variables from the foregoing tableau.
b. Using sensitivity analysis, find the new optimal solution if the coefficient of x_2 in the objective function is changed from 1 to 6.
c. Suppose that the coefficient of x_2 in the first constraint is changed from $+2$ to $\frac{1}{4}$. Using sensitivity, find the new optimal solution.
d. Suppose that the following constraint is added to the problem: $x_2 + x_3 = 3$. Using sensitivity, find the new optimal solution.
e. If you were to choose between increasing the right-hand-side of the first and second constraints, which one would you choose? Why? What is the effect of this increase on the optimal value of the objective function?
f. Suppose that a new activity x_6 is proposed with unit return 4 and *consumption* vector $\mathbf{a}_6 = (1, 2)'$. Find the new optimal solution.

6.54 Develop in detail the rules to conduct sensitivity analyses for the different cases in Section 6.7 using a revised simplex implementation. Solve part (d) of Exercise 6.53 using this implementation. (*Hint*. See the end of Section 6.7.)

6.55 Consider the following optimal tableau of a maximization problem where the constraints are of the $\leq$ type.

	z	x_1	x_2	x_3	x_4	x_5	x_6 (SLACKS)	x_7 (SLACKS)	x_8 (SLACKS)	RHS
z	1	0	0	0	2	0	2	$\frac{1}{10}$	2	θ
x_1	0	1	0	0	-1	0	$\frac{1}{2}$	$\frac{1}{5}$	-1	2
x_2	0	0	1	0	2	1	-1	0	$\frac{1}{2}$	3
x_3	0	0	0	1	-1	-2	5	$-\frac{3}{10}$	2	1

a. Find the optimal objective value θ.
b. Would the solution be altered if a new activity x_9 with coefficients $(2, 0, 3)'$ in the constraints, and price of 5, were added to the problem?
c. How large can b_1 (the first constraint resource) be made without violating feasibility?

6.56 Consider the tableau of Exercise 6.55. Suppose that we add the constraint $x_1 - x_2 + 2x_3 \leq 10$ to the problem. Is the solution still optimal? If not, find the new optimal solution.

6.57 Develop in detail the rules to conduct sensitivity analyses for the different cases in Section 6.7 for the bounded variables simplex algorithm. Include the case of changing bounds on variables. To illustrate, consider Example 5.6 in Section 5.2 and suppose that the column of x_2 is changed from $(1, 1)'$ to $(-1, -1)'$. Update the optimal solution using sensitivity analysis on the given final tableau. Repeat assuming that the upper bound on x_2 is changed from 6 to $\frac{1}{2}$. (*Hint*. See the end of Section 6.7.)

6.58 Develop the details of a dual simplex algorithm for the bounded variables simplex algorithm. Provide convergence arguments. Illustrate by using this algorithm to update the optimal solution for the sensitivity analysis instances of Exercise 6.57. (*Hint*. Choose a row of a basic variable that violates its upper or lower bound as the pivot row, and pick an appropriate pivot column where the change in the nonbasic variable upon pivoting *improves* the bound infeasibility in this pivot row, while maintaining dual feasibility. Change this nonbasic variable until the basic variable in the current row reaches its nearer bound. Proceed with this same row if infeasibility persists, until it is resolved.)

6.59 Assuming $\mathbf{c}' = \mathbf{c}$ and $\mathbf{b}' = \mathbf{b}$, find the maximum allowable tolerances α_0^* and β_0^* on variations in the objective coefficients and on the right-hand-sides respectively, to maintain the (optimal) basis of Exercise 6.53 as optimal. Suppose it is known that the objective coefficient for x_1 can vary no more than ± 30 percent. How does this affect α_0^*? On the other hand, suppose it is known that the variation β_1 in the first constraint is no more than 50 percent of the variation β_2 in the second constraint. How does this affect β_0^*?

6.60 Consider the following algorithm known as the *Criss–Cross* or the *Self-Dual Method*. Suppose that a basic solution to the problem: Minimize $\{\mathbf{cx} : \mathbf{Ax} = \mathbf{b}, \mathbf{x} \geq \mathbf{0}\}$ is given. If this is primal feasible, proceed using the primal simplex algorithm. Similarly, if this is dual feasible, proceed using the dual simplex algorithm. Otherwise perform the following type of iteration. If the elements in a primal infeasible row are all nonnegative, then *stop*; the primal is infeasible. If the elements in a dual infeasible column (with $z_j - c_j > 0$) are all nonpositive, *stop*; the dual is infeasible. Otherwise perform a primal or dual pivot as follows. (One can alternate between primal and dual pivots.) For a primal pivot, pick a nonbasic (dual infeasible) column with $z_j - c_j > 0$ and find a leaving variable by performing the usual minimum ratio test, but using only the primal feasible rows. However, pivot as usual and update the entire tableau. If no such pivot is possible using *any* dual infeasible column in this manner, try a dual pivot. For a dual pivot, pick a primal infeasible row and determine a pivot column by performing the usual (dual simplex) minimum ratio test, but using only the dual feasible columns. Again, if a pivot column is available, then pivot by updating the entire tableau. If no such pivot is possible using *any* primal infeasible row in this manner, then try a primal pivot. However, if the primal pivot was already found to be unperformable, then *stop*; no (finite) optimal solution exists.

a. Justify the three termination statements of the algorithm.
b. Illustrate the algorithm using the slack variables as a starting basis in Example 4.3 of Section 4.2.
c. Show that the algorithm will converge finitely (barring cycling) if a constraint $\Sigma_{j=1}^{n} x_j \leq M$ is added to the problem after a finite number of iterations (typically after $2(n + m)$ iterations), where m is arbitrarily large, and then only the primal pivots are performed until dual feasibility is achieved, when the usual dual simplex algorithm is used.

6.61 A farmer has 500 acres of land and wishes to determine the acreage allocated to the following three crops: wheat, corn, and soybeans. The man-days, preparation cost, and profit per acre of the three crops are summarized.

CROP	MAN-DAYS	PREPARATION COST $	PROFIT $
Wheat	6	100	60
Corn	8	150	100
Soybeans	10	120	80

Suppose that the maximum number of man-days available are 5000 and that the farmer has $60,000 for preparation.

a. Find the optimal solution.
b. Assuming an 8-hour work day, would it be profitable to the farmer to acquire additional help at $3 per hour? Why or why not?
c. Suppose that the farmer has contracted to deliver at least the equivalent of 100 acres of wheat. Use sensitivity analysis to find the new optimal solution.

6.62 A product is assembled from three parts that can be manufactured on two machines A and B. Neither machine can process different parts at the same time. The number of parts processed by each machine per hour are summarized.

	MACHINE A	MACHINE B
Part 1	12	6
Part 2	15	12
Part 3	—	25

Management seeks a daily schedule of the machines so that the number of assemblies is maximized. Currently the company has three machines of type A and five machines of type B.

a. Solve the problem.
b. If only one machine can be acquired, which type would you recommend and why?
c. Management is contemplating the purchase of a type A machine at a cost of $100,000. Suppose that the life of the machine is 10 years and that each year is equivalent to 2000 working hours. Would you recommend the purchase if the unit profit from each assembly is $1? Why or why not?

6.63 The tourism department of a certain country would like to decide which projects to fund during the coming year. The projects were divided into three main categories: religious, historical, and construction (hotels, roads, nightclubs, and so on). Three proposals A, B, and C for restoring religious sites are submitted, with estimated costs of $5, $7, and $3 million respectively. Four proposals D, E, F, and G for the restoration of historical sites are submitted with estimated costs of $15, $12, $5, and $7 million. Finally, five proposals H, I, J, K, and L for constructing new facilities are submitted. These cost $2, $15, $22, $8, and $10 million respectively. In order to determine relative priority of these projects, experts from the tourism department developed a scoring model with the following scores for proposals A, B, C, D, E, F, G, H, I, J, K, L: 5, 6, 2, 8, 11, 1, 7, 2, 10, 9, 5, and 4 respectively. The department decides that at least one project of each category must be funded. Projects E and F represent a continuation of a plan that started during the previous year, and at least one of them must be funded. Furthermore, at most two historical and three construction projects can be chosen. Which projects should the tourism department fund in order to maximize the total score and not to exceed $80 million? (*Hint*. Project j is chosen if $x_j = 1$ and is not chosen if $x_j = 0$. First solve the continuous linear program by the bounded simplex method and then add appropriate cuts.)

6.64 An airline company wishes to assign two types of its aircraft to three routes. Each aircraft can make at most two daily trips. Furthermore, 3 and 4 aircraft of types A and B are available respectively. The capacity of type A aircraft is 140 passengers and that of type B aircraft is 100 passengers. The expected number of daily passengers on the three routes is 300, 700, and 220 respectively.

The operating costs per trip on the different routes are summarized.

AIRCRAFT TYPE	OPERATING COST FOR A GIVEN ROUTE 1	2	3
A	3000	2500	2000
B	2400	2000	1800

a. Find the optimal solution of the continuous linear programming problem. Does this solution make any sense?

b. Using the cutting plane algorithm of Section 6.7, find the optimal integer solution.

6.65 Consider the following optimal tableau of a maximization problem where the constraints are of the $\leq$ type.

	z	x_1	x_2	x_3	x_4	x_5	x_6 (SLACKS)	x_7 (SLACKS)	x_8 (SLACKS)	RHS
z	1	0	0	0	2	0	2	$\frac{1}{10}$	2	17
x_1	0	1	0	0	-1	0	$\frac{1}{2}$	$\frac{1}{5}$	-1	3
x_2	0	0	1	0	2	1	-1	0	$\frac{1}{2}$	1
x_3	0	0	0	1	-1	-2	5	$-\frac{3}{10}$	2	7

Construct the sequence of optimal solutions for $b_1' = b_1 - \theta$ where θ varies between 0 and ∞.

6.66 Consider Exercise 6.1. Suppose that the cost vector is modified in the direction $(-1, +1)$. Using parametric analysis on the cost vector, find the sequence of optimal solutions.

6.67 Consider the following problem.

$$\begin{array}{ll} \text{Minimize} & -x_1 + x_2 - 2x_3 \\ \text{Subject to} & x_1 + x_2 + x_3 \leq 6 \\ & -x_1 + 2x_2 + 3x_3 \leq 9 \\ & x_1, \quad x_2, \quad x_3 \geq 0 \end{array}$$

a. Solve the problem by the simplex method.

b. Suppose that the vector $\mathbf{c} = (-1, 1, -2)$ is replaced by $(-1, 1, -2) + \lambda(2, -1, 1)$ where λ is the real number. Find optimal solutions for all values of λ.

6.68 Consider the following problem.

$$\begin{array}{ll} \text{Maximize} & 2x_1 + 3x_2 + 5x_3 \\ \text{Subject to} & x_1 + x_2 + 2x_3 \leq 8 \\ & x_1 - x_2 + x_3 \leq 4 \\ & x_1, \quad x_2, \quad x_3 \geq 0 \end{array}$$

a. Find the optimal solution.

b. Find the new optimal solution if the cost coefficient c_2 changes from 3 to -4.

c. Find the optimal solution if the cost coefficient c_2 varies over the entire real line $(-\infty, \infty)$.

6.69 Prove that parametric analysis on the cost vector in a minimization problem always produces a piecewise-linear and concave function $z(\lambda)$.

6.70 Prove that parametric analysis on the RHS vector in a minimization problem always produces a piecewise-linear and convex function $z(\lambda)$.

6.71 Prove that if $\mathbf{K}$ is a skew symmetric matrix (that is, $\mathbf{K} = -\mathbf{K}^t$), then the system

$$\mathbf{Kx} \geq \mathbf{0}, \qquad \mathbf{x} \geq \mathbf{0}$$

possesses at least one solution $\bar{\mathbf{x}}$ such that $\mathbf{K}\bar{\mathbf{x}} + \bar{\mathbf{x}} > \mathbf{0}$. (*Hint*. Apply Farkas' lemma to the system $\mathbf{Kx} \geq \mathbf{0}$, $\mathbf{x} \geq \mathbf{0}$, $\mathbf{e}_j\mathbf{x} > 0$. Repeat for each j and combine solutions by summing.)

6.72 Apply the results of the previous problem to the system

$$\begin{aligned} \mathbf{Ax} - r\mathbf{b} &\geq \mathbf{0}, \qquad \mathbf{x} \geq \mathbf{0} \\ -\mathbf{wA} + r\mathbf{c} &\geq \mathbf{0}, \qquad \mathbf{w} \geq \mathbf{0} \\ \mathbf{wb} - \mathbf{cx} &\geq 0, \qquad r \geq 0 \end{aligned}$$

a. Use the results to derive the fundamental theorem of duality.
b. Use the results to prove that at optimality there exists at least one pair of primal and dual optimal points with the property that if a variable in one problem is zero, then the complementary variable in the other problem is positive. That is called the *strong theorem of complementary slackness*, or the *strict complementary slackness* property.
c. Illustrate the strong theorem of complementary slackness geometrically. (*Hint*. Consider a linear program where the objective function contours are parallel to one of the constraints and alternate optimal solutions result.)
d. When must the solution to the strong complementary slackness theorem not occur at an extreme point? (*Hint*. Consider part (c).)

6.73 Consider the following primal problem.

$$\begin{aligned} &\text{Minimize} && \mathbf{cx} \\ &\text{Subject to} && \mathbf{Ax} \geq \mathbf{b} \\ &&& \mathbf{x} \in X \end{aligned}$$

where X is a polyhedral set. (Often the set X consists of constraints that are easy to handle.) Associated with the foregoing primal problem is the following *Lagrangian dual* problem.

$$\begin{aligned} &\text{Maximize} && f(\mathbf{w}) \\ &\text{Subject to} && \mathbf{w} \geq \mathbf{0} \end{aligned}$$

where $f(\mathbf{w}) = \mathbf{wb} + \text{Minimum}_{x \in X}(\mathbf{c} - \mathbf{wA})\mathbf{x}$.

a. Show that if $\mathbf{x}_0$ is feasible to the primal problem, that is, $\mathbf{Ax}_0 \geq \mathbf{b}$ and $\mathbf{x}_0 \in X$, and $\mathbf{w}_0$ is feasible to the Lagrangian dual problem, that is, $\mathbf{w}_0 \geq \mathbf{0}$, then $\mathbf{cx}_0 \geq f(\mathbf{w}_0)$.

b. Suppose that X is nonempty and bounded and that the primal problem possesses a finite optimal solution. Show that

$$\begin{array}{ll} \text{Minimum } \mathbf{cx} = \text{Maximum } f(\mathbf{w}) \\ \mathbf{Ax} \geq \mathbf{b} \qquad\qquad \mathbf{w} \geq \mathbf{0} \\ \mathbf{x} \in X \end{array}$$

6.74 Consider the problem: Minimize $x_1 + 2x_2$ subject to $3x_1 + x_2 \geq 6$, $-x_1 + x_2 \leq 2$, $x_1 + x_2 \leq 8$, and $x_1, x_2 \geq 0$. Let $X = \{\mathbf{x}: -x_1 + x_2 \leq 2,\ x_1 + x_2 \leq 8,\ x_1, x_2 \geq 0\}$.

a. Formulate the Lagrangian dual problem.

b. Show that $f(w) = 6w + \text{Minimum}\{0, 4 - 2w, 13 - 14w, 8 - 24w\}$. (*Hint*. Examine the second term in $f(w)$ in Exercise 6.73 and enumerate the extreme points of X graphically.)

c. Plot $f(w)$ for each value of w.

d. From part (c) locate the optimal solution to the Lagrangian dual problem.

e. From part (d) find the optimal solution to the primal problem.

NOTES AND REFERENCES

1. John von Neumann is credited with having first postulated the existence of a dual linear program. His insights came through his work in game theory and economics together with a strong mathematical capability. Many individuals have continued to develop and extend the basic duality theorems, notably Tucker [1960a] and A. C. Williams [1970]. Also, see Dantzig [1982] for his pioneering contributions to duality theory.
2. For more details on shadow prices and relationships with directional derivatives for nondifferentiable convex functions, see Akgul [1982] and Bazaraa and Shetty [1979]. For a related application to marginal cost pricing for an electric utility, where a closed form derivation for the dual variables is presented, see Sherali *et al.* [1984].
3. The dual simplex method was first developed by Lemke [1954].
4. The primal-dual algorithm was developed by Dantzig, Ford, and Fulkerson [1956]. This development was fostered out of the work of Kuhn [1955] on the assignment problem.
5. The cutting plane application in integer programming presented in Section 6.7 is mainly for illustration. For more up-to-date information on integer programming methods, see Schrijver [1986], Johnson *et al.* [1985], and Crowder *et al.* [1983].
6. The tolerance approach to sensitivity analysis presented in Section 6.7 is from Wendell [1984, 1985]. Also, see Freund [1985] for sensitivity analysis involving simultaneous changes in constraint coefficients.
7. The Criss–Cross algorithm presented in Exercise 6.60 is from Zionts [1969, 1974].

SEVEN: THE DECOMPOSITION PRINCIPLE

This chapter discusses the Dantzig–Wolfe decomposition technique and its relationships with Benders' partitioning and Lagrangian relaxation techniques for dealing with large-scale and/or specially structured linear programming problems. It is not unusual in a corporate management model or in a logistics model to produce a linear program with many thousands of rows and a seemingly unlimited number of columns. In such problems some method must be applied to convert the large problems into one or more appropriately coordinated smaller problems of manageable size. The Dantzig–Wolfe or the Benders' or the Lagrangian relaxation techniques, which are all equivalent for linear programming problems, and which we simply refer to as applying the *decomposition principle*, do exactly this.

Even if a linear program is of manageable size, certain of its constraints may possess special structure that would permit efficient handling. In such cases we would like to separate the linear program into one with general structure and one with special structure where a more efficient method may be applied. Again, the decomposition principle can be applied to such a linear program to achieve the desired effect.

The decomposition principle is a systematic procedure for solving large-scale linear programs or linear programs that contain constraints of special structure. The constraints are divided into two sets: general constraints (or *complicating* constraints) and constraints with special structure. It will become apparent that it is not necessary for either set to have special structure; however, special structure, when available, enhances the efficiency of the procedure.

The strategy of the decomposition procedure is to operate on two separate linear programs: one over the set of general constraints and one over the set of special constraints. Information is passed back and forth between the two linear programs until a point is reached where the solution to the original problem is achieved. The linear program over the general constraints is called the *master problem*, and the linear program over the special constraints is called the *subproblem*. The master problem passes down a new set of cost coefficients to the subproblem and receives a new column based on these cost coefficients. For this reason, such a procedure is also known as a *column generation* technique.

We shall begin by assuming that the special constraint set is bounded. Once the decomposition principle is developed for this case and we have discussed how to get started, we shall relax the boundedness assumption and also extend the procedure to multiple subproblems, and exhibit its relationship with other procedures.

7.1 THE DECOMPOSITION ALGORITHM

Consider the following linear program, where X is a polyhedral set representing constraints of special structure, $\mathbf{A}$ is an $m \times n$ matrix, and $\mathbf{b}$ is an m vector.

$$\begin{aligned} \text{Minimize} \quad & \mathbf{cx} \\ \text{Subject to} \quad & \mathbf{Ax} = \mathbf{b} \\ & \mathbf{x} \in X \end{aligned}$$

To simplify the presentation, assume that X is bounded (this assumption will be relaxed in Section 7.4). Since X is a bounded polyhedral set, then any point $\mathbf{x} \in X$ can be represented as a convex combination of the finite number of extreme points of X. Denoting these points by $\mathbf{x}_1, \mathbf{x}_2, \ldots, \mathbf{x}_t$, any $\mathbf{x} \in X$ can be representing as

$$\mathbf{x} = \sum_{j=1}^{t} \lambda_j \mathbf{x}_j$$

$$\sum_{j=1}^{t} \lambda_j = 1$$

$$\lambda_j \geq 0 \qquad j = 1, 2, \ldots, t$$

Substituting for $\mathbf{x}$, the foregoing optimization problem can be transformed into the following so-called *master problem* in the variables $\lambda_1, \lambda_2, \ldots, \lambda_t$.

$$\text{Minimize} \quad \sum_{j=1}^{t} (\mathbf{cx}_j)\lambda_j$$

$$\text{Subject to} \quad \sum_{j=1}^{t} (\mathbf{Ax}_j)\lambda_j = \mathbf{b} \tag{7.1}$$

$$\sum_{j=1}^{t} \lambda_j = 1 \tag{7.2}$$

$$\lambda_j \geq 0 \qquad j = 1, 2, \ldots, t \tag{7.3}$$

Since t, the number of extreme points of the set X, is usually very large, attempting to explicitly enumerate all the extreme points $\mathbf{x}_1, \mathbf{x}_2, \ldots, \mathbf{x}_t$, and explicitly solving this problem is a very difficult task. Rather, we shall attempt to find an optimal solution of the problem (and hence the original problem) without explicitly enumerating all the extreme points.

Application of the Revised Simplex Method

Consider solving the foregoing problem by the revised simplex method. Suppose that we have a basic feasible solution $\boldsymbol{\lambda} = (\boldsymbol{\lambda}_B, \boldsymbol{\lambda}_N)$. Further suppose that the associated $(m+1) \times (m+1)$ basis inverse $\mathbf{B}^{-1}$ is known (the process of initialization is discussed in detail in Section 7.3). Denoting the dual variables corresponding to Equations (7.1) and (7.2) by $\mathbf{w}$ and α, we get $(\mathbf{w}, \alpha) = \hat{\mathbf{c}}_B\mathbf{B}^{-1}$, where $\hat{\mathbf{c}}_B$ is the cost of the basic variables with $\hat{c}_j = \mathbf{c}\mathbf{x}_j$ for each basic variable λ_j. The basis inverse, the dual variables, the values of the basic variables, and the objective function are displayed, where $\bar{\mathbf{b}} = \mathbf{B}^{-1}\begin{pmatrix}\mathbf{b}\\1\end{pmatrix}$.

BASIS INVERSE	RHS
$(\mathbf{w}, \alpha)$	$\hat{\mathbf{c}}_B\bar{\mathbf{b}}$
$\mathbf{B}^{-1}$	$\bar{\mathbf{b}}$

The revised simplex method proceeds by concluding that the current solution is optimal or else by deciding to increase a nonbasic variable. This is done by first calculating

$$\begin{aligned} z_k - \hat{c}_k &= \underset{1 \le j \le t}{\text{Maximum}}\, z_j - \hat{c}_j = \underset{1 \le j \le t}{\text{Maximum}}\, (\mathbf{w}, \alpha)\begin{bmatrix}\mathbf{A}\mathbf{x}_j\\1\end{bmatrix} - \mathbf{c}\mathbf{x}_j \\ &= \underset{1 \le j \le t}{\text{Maximum}}\, \mathbf{w}\mathbf{A}\mathbf{x}_j + \alpha - \mathbf{c}\mathbf{x}_j \end{aligned} \tag{7.4}$$

Since $z_j - \hat{c}_j = 0$ for basic variables, then the foregoing maximum is ≥ 0. Thus if $z_k - \hat{c}_k = 0$, then $z_j - \hat{c}_j \le 0$ for all nonbasic variables and the optimal solution is at hand. On the other hand, if $z_k - \hat{c}_k > 0$, then the nonbasic variable λ_k may be increased.

Determining the index k using Equation (7.4) directly is computationally infeasible because t is very large and the extreme points $\mathbf{x}_j$'s corresponding to the nonbasic λ_j's are not explicitly known. Therefore an alternative scheme must be devised. Since X is a bounded polyhedral set, the maximum of any linear objective can be achieved at one of the extreme points. Therefore

$$\underset{1 \le j \le t}{\text{Maximum}}\, (\mathbf{w}\mathbf{A} - \mathbf{c})\mathbf{x}_j + \alpha = \underset{\mathbf{x} \in X}{\text{Maximum}}\, (\mathbf{w}\mathbf{A} - \mathbf{c})\mathbf{x} + \alpha$$

To summarize, given a basic feasible solution $(\boldsymbol{\lambda}_B, \boldsymbol{\lambda}_N)$ with dual variables $(\mathbf{w}, \alpha)$, solve the following linear *subproblem*, which is "easy" because of the special structure of X.

$$\begin{aligned} &\text{Maximize} \quad (\mathbf{w}\mathbf{A} - \mathbf{c})\mathbf{x} + \alpha \\ &\text{Subject to} \quad \mathbf{x} \in X \end{aligned}$$

Note that the objective function contains a constant. This is easily handled by initializing the RHS value for z to α instead of the normal value of 0. Let $\mathbf{x}_k$ be an optimal solution to the foregoing subproblem with objective value $z_k - \hat{c}_k$. If $z_k - \hat{c}_k = 0$, then the basic feasible solution (λ_B, λ_N) is optimal. Otherwise if $z_k - \hat{c}_k > 0$, then the variable λ_k enters the basis. As in the revised simplex method the corresponding column $\begin{pmatrix} \mathbf{Ax}_k \\ 1 \end{pmatrix}$ is updated by premultiplying it by $\mathbf{B}^{-1}$ giving $\mathbf{y}_k = \mathbf{B}^{-1}\begin{pmatrix} \mathbf{Ax}_k \\ 1 \end{pmatrix}$. Note that $\mathbf{y}_k \leq \mathbf{0}$ cannot occur since X was assumed bounded, producing a bounded master problem. The updated column $\begin{pmatrix} z_k - \hat{c}_k \\ \mathbf{y}_k \end{pmatrix}$ is adjoined to the revised simplex tableau. The variable λ_{B_r} leaving the basis is determined by the usual minimum ratio test. The basis inverse, dual variables, and RHS are updated by pivoting at y_{rk}. After updating, the process is repeated.

Now we have all the ingredients of the decomposition algorithm, a summary of which is given. Note that the master step gives an improved feasible solution of the overall problem, and the subproblem checks whether $z_j - \hat{c}_j \leq 0$ for all λ_j, or else determines the most positive $z_k - \hat{c}_k$.

Summary of the Decomposition Algorithm

INITIALIZATION STEP

Find an initial basic feasible solution of the system defined by Equations (7.1), (7.2), and (7.3) (getting an initial basic feasible solution is discussed in detail in Section 7.3). Let the basis be $\mathbf{B}$ and form the following *master array* where $(\mathbf{w}, \alpha) = \hat{\mathbf{c}}_B\mathbf{B}^{-1}$ (recall that $\hat{c}_j = \mathbf{cx}_j$), and $\bar{\mathbf{b}} = \mathbf{B}^{-1}\begin{bmatrix} \mathbf{b} \\ 1 \end{bmatrix}$.

BASIS INVERSE	RHS
$(\mathbf{w}, \alpha)$	$\hat{\mathbf{c}}_B\bar{\mathbf{b}}$
$\mathbf{B}^{-1}$	$\bar{\mathbf{b}}$

MAIN STEP

1. Solve the following *subproblem*.

   Maximize $(\mathbf{wA} - \mathbf{c})\mathbf{x} + \alpha$

   Subject to $\mathbf{x} \in X$

   Let $\mathbf{x}_k$ be an optimal basic feasible solution with objective value of $z_k - \hat{c}_k$. If $z_k - \hat{c}_k = 0$ stop; the basic feasible solution of the last master step is an optimal solution of the overall problem. Otherwise go to step 2.

2. Let $\mathbf{y}_k = \mathbf{B}^{-1}\begin{bmatrix}\mathbf{A}\mathbf{x}_k \\ 1\end{bmatrix}$ and adjoin the updated column $\begin{pmatrix} z_k - \hat{c}_k \\ \mathbf{y}_k \end{pmatrix}$ to the master array. Pivot at y_{rk} where the index r is determined as follows:

$$\frac{\bar{b}_r}{y_{rk}} = \underset{1 \le i \le m+1}{\text{Minimum}} \left\{ \frac{\bar{b}_i}{y_{ik}} : y_{ik} > 0 \right\}$$

This updates the dual variables, the basis inverse, and the right-hand-side. After pivoting, the column of λ_k is deleted and step 1 is repeated.

Some Remarks on the Decomposition Algorithm

1. Note that the foregoing algorithm is a direct implementation of the revised simplex method except that the calculation $z_k - \hat{c}_k$ is performed by solving a subproblem. Therefore the algorithm converges in a finite number of iterations provided that a cycling prevention rule is used in both the master step and the subproblem in the presence of degeneracy.
2. At each iteration the master step provides (at a nondegenerate pivot step) a new improved basic feasible solution of the system given by Equations (7.1), (7.2), and (7.3) by introducing the nonbasic variable λ_k, which is generated by the subproblem. At each iteration the subproblem provides an extreme point $\mathbf{x}_k$, which corresponds to an updated column $\begin{bmatrix} z_k - \hat{c}_k \\ \mathbf{y}_k \end{bmatrix}$, and hence, as pointed out previously, this procedure is sometimes referred to as a *column generation scheme*.
3. At each iteration a different dual vector is passed from the master step to the subproblem. Rather than solving the subproblem anew at each iteration, the optimal basis of the last iteration could be utilized by modifying the cost row.
4. At each iteration, the subproblem need not be completely optimized. It is only necessary that the current extreme point $\mathbf{x}_k$ satisfies $z_k - \hat{c}_k = (\mathbf{wA} - \mathbf{c})\mathbf{x}_k + \alpha > 0$. In this case λ_k is a candidate to enter the basis of the master problem.
5. If the master constraints are of the inequality type, then we must check the $z_j - \hat{c}_j$ for nonbasic slack variables in addition to solving the subproblem. For a master constraint i of the $\le$ type with associated slack variables s_i, we get

$$z_{s_i} - c_{s_i} = (\mathbf{w}, \alpha)\begin{pmatrix} \mathbf{e}_i \\ 0 \end{pmatrix} - 0 = w_i$$

Thus, for a minimization problem a slack variable associated with a $\le$ constraint is eligible to enter the basis if $w_i > 0$. (Note that the entry criterion is $w_i < 0$ for constraints of the $\ge$ type.)

It should be clear that if there are other variable columns that are explicitly maintained in the master problem in addition to the λ-variable columns or the slack columns, then like the latter, these columns also may be priced using the simplex multiplier vector $(\mathbf{w}, \alpha)$ in the usual manner in order to test for their eligibility to enter the basis. (See Exercise 7.4).

6. Assume that the linear program is feasible and that **A** is $m \times n$ of rank m, so that the master problem has basic feasible solutions with bases of size $(m + 1) \times (m + 1)$. Since there are m linearly independent constraints $\mathbf{Ax} = \mathbf{b}$ binding at every feasible solution, we are searching for an optimum over some $p = (n - m)$ dimensional subset of X. Hence at an optimal extreme point solution $\mathbf{x}^*$ of the original linear program, (at least) some p linearly independent constraints defining X must be binding. Consequently, $\mathbf{x}^*$ lies on (at most) some $(n - p) = m$ dimensional face of X. From the discussion of Chapter 2 it follows that since this face is itself (at most) an m dimensional polytope with its vertices a subset of the vertices of X, we can manage to represent $\mathbf{x}^*$ using at most $(m + 1)$ vertices of X. Hence we will need at most $(m + 1)$ positive λ-variables at optimality. This of course concurs with the size of the master problem basis.

   On this (at most) m dimensional face of X, there may be several ways of representing $\mathbf{x}^*$ in terms of at most $(m + 1)$ extreme points of X. Computationally, it so turns out that one can usually achieve a convex combination of a set of extreme points of X that brings the resulting solution to within 1–5 percent of optimality fairly quickly. However, it has been observed on many classes of problems that the tail-end convergence rate in obtaining a more exact representation for an optimal solution can be very slow. (The algorithm need not produce an *extreme point* optimal solution for the original problem when alternative optima exist.) Hence one is advised to terminate this procedure once it comes to within 1–5 percent of optimality. The following discussion provides a useful construct for implementing such a termination criterion.

Calculation and Use of Lower Bounds

Recall that the decomposition algorithm stops when Maximum $z_j - \hat{c}_j = 0$. Because of the large number of variables $\lambda_1, \lambda_2, \ldots, \lambda_t$, continuing the computations until this condition is satisfied may be time-consuming for large problems.

We shall develop a lower bound on the objective of any feasible solution of the overall problem, and hence a lower bound on the optimal objective. Since the decomposition algorithm generates feasible points with nonworsening objective values via the master problem, we have a sequence of nonincreasing upper bounds. Hence we may stop when the difference between the objective of the current feasible point and the lower bounds is within an acceptable tolerance. This may not give the true optimal point, but will guarantee good feasible solutions, within any desirable accuracy from the optimal. Consider the following subproblem.

$$\begin{aligned} &\text{Maximize} \quad (\mathbf{wA} - \mathbf{c})\mathbf{x} + \alpha \\ &\text{Subject to} \quad \mathbf{x} \in X \end{aligned}$$

where $\mathbf{w}$ is the dual vector passed from the master step. Let the optimal objective of the foregoing subproblem be $z_k - \hat{c}_k$. Now let $\mathbf{x}$ be any feasible solution of the overall problem, that is, $\mathbf{Ax} = \mathbf{b}$ and $\mathbf{x} \in X$. By definition of $z_k - \hat{c}_k$ and since $\mathbf{x} \in X$, we have

$$(\mathbf{wA} - \mathbf{c})\mathbf{x} + \alpha \leq (z_k - \hat{c}_k)$$

Since $\mathbf{Ax} = \mathbf{b}$, then this inequality implies that

$$\mathbf{cx} \geq \mathbf{wAx} - (z_k - \hat{c}_k) + \alpha = \mathbf{wb} + \alpha - (z_k - \hat{c}_k) = \hat{\mathbf{c}}_B\bar{\mathbf{b}} - (z_k - \hat{c}_k)$$

Since this is true for each $\mathbf{x} \in X$ with $\mathbf{Ax} = \mathbf{b}$, then

$$\underset{\substack{\mathbf{Ax}=\mathbf{b} \\ \mathbf{x}\in X}}{\text{Minimum}} \quad \mathbf{cx} \geq \hat{\mathbf{c}}_B\bar{\mathbf{b}} - (z_k - \hat{c}_k)$$

In other words, $\hat{\mathbf{c}}_B\bar{\mathbf{b}} - (z_k - \hat{c}_k)$ is a lower bound on the optimal objective value of the overall problem. Note that $\hat{\mathbf{c}}_B\bar{\mathbf{b}}$ is the current best upper bound. However, the lower bounds generated need not be monotone and one would need to maintain the best (greatest) lower bound.

7.2 NUMERICAL EXAMPLE

Consider the following problem.

$$\begin{array}{llllll}
\text{Minimize} & -2x_1 - & x_2 - & x_3 + & x_4 & \\
\text{Subject to} & x_1 & & + \; x_3 & & \leq 2 \\
& x_1 + & x_2 & & + \; 2x_4 & \leq 3 \\
& x_1 & & & & \leq 2 \\
& x_1 + 2x_2 & & & & \leq 5 \\
& & - & x_3 + & x_4 & \leq 2 \\
& & & 2x_3 + & x_4 & \leq 6 \\
& x_1, & x_2, & x_3, & x_4 & \geq 0
\end{array}$$

Note that the third and fourth constraints involve only x_1 and x_2, whereas the fifth and sixth constraints involve only x_3 and x_4 (we shall have more to say

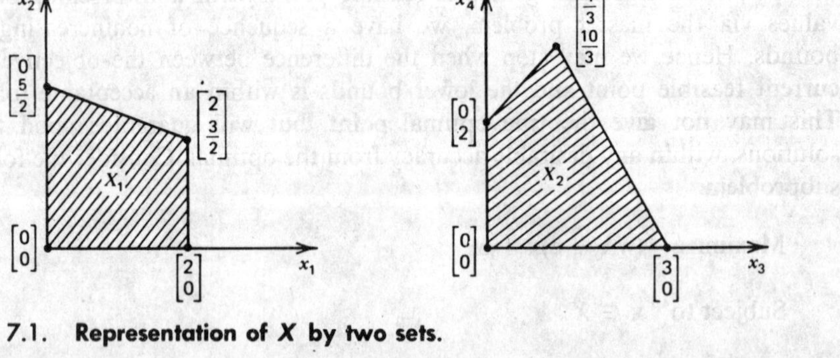

Figure 7.1. **Representation of *X* by two sets.**

about this special structure later). If we let X consist of the last four constraints in addition to the nonnegativity restrictions, then minimizing a linear function over X becomes a simple process, since the subproblem can be decomposed into two subproblems. Therefore we shall handle the first two constraints as $\mathbf{Ax} \leq \mathbf{b}$, where $\mathbf{A} = \begin{bmatrix} 1 & 0 & 1 & 0 \\ 1 & 1 & 0 & 2 \end{bmatrix}$, $\mathbf{b} = \begin{bmatrix} 2 \\ 3 \end{bmatrix}$, and the remaining constraints as X. Note that any point (x_1, x_2, x_3, x_4) in X must have its first two components and its last two components in the respective sets X_1 and X_2 shown in Figure 7.1.

Initialization Step

The problem is reformulated as follows, where $\mathbf{x}_1, \mathbf{x}_2, \ldots, \mathbf{x}_t$ are the extreme points of X, $\hat{c}_j = \mathbf{cx}_j$ for $j = 1, 2, \ldots, t$, and $\mathbf{s} \geq \mathbf{0}$ is the slack vector.

$$\begin{aligned}
\text{Minimize} \quad & \sum_{j=1}^{t} \hat{c}_j \lambda_j \\
\text{Subject to} \quad & \sum_{j=1}^{t} (\mathbf{Ax}_j)\lambda_j + \mathbf{s} = \mathbf{b} \\
& \sum_{j=1}^{t} \lambda_j = 1 \\
& \lambda_j \geq 0 \qquad j = 1, 2, \ldots, t \\
& \mathbf{s} \geq \mathbf{0}
\end{aligned}$$

We need a starting basis with known $\mathbf{B}^{-1}$. Let the starting basis consist of $\mathbf{s}$ and λ_1 where $\mathbf{x}_1 = (0, 0, 0, 0)$ is an extreme point of X with $\mathbf{cx}_1 = 0$. Therefore

$$\mathbf{B} = \begin{bmatrix} 1 & 0 & 0 \\ 0 & 1 & 0 \\ 0 & 0 & 1 \end{bmatrix}$$

The vector $(\mathbf{w}, \alpha) = \hat{\mathbf{c}}_B \mathbf{B}^{-1} = \mathbf{0B}^{-1} = \mathbf{0}$, and $\bar{\mathbf{b}} = \mathbf{B}^{-1}\begin{bmatrix} \mathbf{b} \\ 1 \end{bmatrix} = \begin{bmatrix} \mathbf{b} \\ 1 \end{bmatrix}$. This gives the following tableau. Note that the first three columns give (w_1, w_2, α) in row 0 and $\mathbf{B}^{-1}$ in the remaining rows.

	BASIS INVERSE			RHS
z	0	0	0	0
s_1	1	0	0	2
s_2	0	1	0	3
λ_1	0	0	1	1

Iteration 1

SUBPROBLEM

Solve the following subproblem.

$$\begin{aligned}\text{Maximize} \quad & (\mathbf{wA} - \mathbf{c})\mathbf{x} + \alpha \\ \text{Subject to} \quad & \mathbf{x} \in X\end{aligned}$$

Here $(w_1, w_2) = (0, 0)$ from the foregoing array. Therefore the subproblem is as follows:

$$\begin{aligned}\text{Maximize} \quad & 2x_1 + x_2 + x_3 - x_4 + 0 \\ \text{Subject to} \quad & \mathbf{x} \in X\end{aligned}$$

This problem is separable in the vectors (x_1, x_2) and (x_3, x_4) and can be solved geometrically. Using Figure 7.1, it is easily verified that the optimal solution is $\mathbf{x}_2 = (2, \frac{3}{2}, 3, 0)$ with objective $z_2 - \hat{c}_2 = \frac{17}{2}$. Since $z_2 - \hat{c}_2 = \frac{17}{2} > 0$, then λ_2 corresponding to $\mathbf{x}_2$ is introduced. The lower bound $= \hat{\mathbf{c}}_B\bar{\mathbf{b}} - (z_2 - \hat{c}_2) = 0 - \frac{17}{2}$. Recall that the best objective so far is 0.

MASTER STEP

$z_2 - \hat{c}_2 = \frac{17}{2}$

$$\mathbf{Ax}_2 = \begin{bmatrix} 1 & 0 & 1 & 0 \\ 1 & 1 & 0 & 2 \end{bmatrix}\begin{bmatrix} 2 \\ \frac{3}{2} \\ 3 \\ 0 \end{bmatrix} = \begin{bmatrix} 5 \\ \frac{7}{2} \end{bmatrix}$$

Then

$$\begin{bmatrix} \mathbf{Ax}_2 \\ 1 \end{bmatrix} = \begin{bmatrix} 5 \\ \frac{7}{2} \\ 1 \end{bmatrix}$$

is updated by premultiplying by $\mathbf{B}^{-1}$. So

$$\mathbf{y}_2 = \mathbf{B}^{-1}\begin{bmatrix} 5 \\ \frac{7}{2} \\ 1 \end{bmatrix} = \mathbf{I}\begin{bmatrix} 5 \\ \frac{7}{2} \\ 1 \end{bmatrix} = \begin{bmatrix} 5 \\ \frac{7}{2} \\ 1 \end{bmatrix}$$

Insert the column

$$\begin{bmatrix} z_2 - \hat{c}_2 \\ \mathbf{y}_2 \end{bmatrix} = \begin{bmatrix} \frac{17}{2} \\ 5 \\ \frac{7}{2} \\ 1 \end{bmatrix}$$

into the foregoing array and pivot. This leads to the following two tableaux (the λ_2 column is deleted after pivoting.)

	BASIS INVERSE			RHS		λ_2
z	0	0	0	0		$\frac{17}{2}$
s_1	1	0	0	2		(5)
s_2	0	1	0	3		$\frac{7}{2}$
λ_1	0	0	1	1		1

	BASIS INVERSE			RHS
z	$-\frac{17}{10}$	0	0	$-\frac{17}{5}$
λ_2	$\frac{1}{5}$	0	0	$\frac{2}{5}$
s_2	$-\frac{7}{10}$	1	0	$\frac{8}{5}$
λ_1	$-\frac{1}{5}$	0	1	$\frac{3}{5}$

The best-known feasible solution of the overall problem is given by

$$\begin{aligned}\mathbf{x} &= \lambda_1\mathbf{x}_1 + \lambda_2\mathbf{x}_2\\ &= \tfrac{3}{5}(0,0,0,0) + \tfrac{2}{5}\left(2,\tfrac{3}{2},3,0\right) = \left(\tfrac{4}{5},\tfrac{3}{5},\tfrac{6}{5},0\right)\end{aligned}$$

The objective is $-\frac{17}{5}$. Also $(w_1, w_2, \alpha) = (-\frac{17}{10}, 0, 0)$.

Iteration 2

Since $w_1 < 0$, s_1 is not eligible to enter the basis at this time.

SUBPROBLEM

Solve the following problem.

$$\begin{aligned}&\text{Maximize} \quad (\mathbf{wA}-\mathbf{c})\mathbf{x}+\alpha\\ &\text{Subject to} \quad \mathbf{x}\in X\end{aligned}$$

$$\mathbf{wA}-\mathbf{c} = \left(-\tfrac{17}{10},0\right)\begin{bmatrix}1 & 0 & 1 & 0\\ 1 & 1 & 0 & 2\end{bmatrix} - (-2,-1,-1,1) = \left(\tfrac{3}{10},1,-\tfrac{7}{10},-1\right)$$

Therefore the subproblem is

$$\begin{aligned}&\text{Maximize} \quad \tfrac{3}{10}x_1 + x_2 - \tfrac{7}{10}x_3 - x_4 + 0\\ &\text{Subject to} \quad \mathbf{x}\in X\end{aligned}$$

The problem decomposes into two problems involving (x_1, x_2) and (x_3, x_4). Using Figure 7.1, the optimal solution if $\mathbf{x}_3 = (0, \frac{5}{2}, 0, 0)$ with objective $z_3 - \hat{c}_3 = \frac{5}{2}$. Since $z_3 - \hat{c}_3 > 0$, then λ_3 is introduced.

The lower bound is $\hat{\mathbf{c}}_B\bar{\mathbf{b}} - (z_3 - \hat{c}_3) = -\frac{17}{5} - \frac{5}{2} = -5.9$. (Recall that the best-known objective so far is -3.4.)

MASTER STEP

$z_3 - \hat{c}_3 = \frac{5}{2}$

$$\mathbf{Ax}_3 = \begin{bmatrix} 1 & 0 & 1 & 0 \\ 1 & 1 & 0 & 2 \end{bmatrix} \begin{bmatrix} 0 \\ \frac{5}{2} \\ 0 \\ 0 \end{bmatrix} = \begin{bmatrix} 0 \\ \frac{5}{2} \end{bmatrix}$$

$$\mathbf{y}_3 = \mathbf{B}^{-1} \begin{bmatrix} \mathbf{Ax}_3 \\ 1 \end{bmatrix} = \begin{bmatrix} \frac{1}{5} & 0 & 0 \\ -\frac{7}{10} & 1 & 0 \\ -\frac{1}{5} & 0 & 1 \end{bmatrix} \begin{bmatrix} 0 \\ \frac{5}{2} \\ 1 \end{bmatrix} = \begin{bmatrix} 0 \\ \frac{5}{2} \\ 1 \end{bmatrix}$$

Insert the column $\begin{bmatrix} z_3 - \hat{c}_3 \\ \mathbf{y}_3 \end{bmatrix}$ into the foregoing array and pivot. This leads to the following two tableaux (the λ_3 column is deleted after pivoting).

	BASIS INVERSE			RHS	λ_3
z	$-\frac{17}{10}$	0	0	$-\frac{17}{5}$	$\frac{5}{2}$
λ_2	$\frac{1}{5}$	0	0	$\frac{2}{5}$	0
s_2	$-\frac{7}{10}$	1	0	$\frac{8}{5}$	$\frac{5}{2}$
λ_1	$-\frac{1}{5}$	0	1	$\frac{3}{5}$	①

	BASIS INVERSE			RHS
z	$-\frac{6}{5}$	0	$-\frac{5}{2}$	$-\frac{49}{10}$
λ_2	$\frac{1}{5}$	0	0	$\frac{2}{5}$
s_2	$-\frac{1}{5}$	1	$-\frac{5}{2}$	$\frac{1}{10}$
λ_3	$-\frac{1}{5}$	0	1	$\frac{3}{5}$

The best-known feasible solution of the overall problem is given by

$$\mathbf{x} = \lambda_2\mathbf{x}_2 + \lambda_3\mathbf{x}_3$$
$$= \tfrac{2}{5}\left(2, \tfrac{3}{2}, 3, 0\right) + \tfrac{3}{5}\left(0, \tfrac{5}{2}, 0, 0\right) = \left(\tfrac{4}{5}, \tfrac{21}{10}, \tfrac{6}{5}, 0\right)$$

The objective is -4.9. Also $(w_1, w_2, \alpha) = (-\frac{6}{4}, 0, -\frac{5}{2})$.

Iteration 3

Since $w_1 < 0$, s_1 is not eligible to enter the basis at this time.

SUBPROBLEM

Solve the following subproblem.

$$\text{Maximize} \quad (\mathbf{wA} - \mathbf{c})\mathbf{x} + \alpha$$

$$\text{Subject to} \quad \mathbf{x} \in X$$

$$\mathbf{wA} - \mathbf{c} = \left(-\tfrac{6}{5}, 0\right)\begin{bmatrix} 1 & 0 & 1 & 0 \\ 1 & 1 & 0 & 2 \end{bmatrix} - (-2, -1, -1, 1) = \left(\tfrac{4}{5}, 1, -\tfrac{1}{5}, -1\right)$$

Therefore the subproblem is as follows.

$$\text{Maximize} \quad \tfrac{4}{5}x_1 + x_2 - \tfrac{1}{5}x_3 - x_4 - \tfrac{5}{2}$$

$$\text{Subject to} \quad \mathbf{x} \in X$$

Using Figure 7.1, the optimal solution is $\mathbf{x}_4 = (2, \frac{3}{2}, 0, 0)$ with objective $z_4 - \hat{c}_4 = \frac{3}{5}$, and so λ_4 is introduced.

The lower bound is given by $\hat{\mathbf{c}}_B\bar{\mathbf{b}} - (z_4 - \hat{c}_4) = -\frac{49}{10} - \frac{3}{5} = -5.5$. Recall that the best-known objective so far is -4.9. If we were interested only in an approximate solution, we could have stopped here with the feasible solution $\mathbf{x} = (\frac{4}{5}, \frac{21}{10}, \frac{6}{5}, 0)$ whose objective is -4.9.

MASTER STEP

$z_4 - \hat{c}_4 = \frac{3}{5}$

$$\mathbf{Ax}_4 = \begin{bmatrix} 1 & 0 & 1 & 0 \\ 1 & 1 & 0 & 2 \end{bmatrix}\begin{bmatrix} 2 \\ \frac{3}{2} \\ 0 \\ 0 \end{bmatrix} = \begin{bmatrix} 2 \\ \frac{7}{2} \end{bmatrix}$$

The updated column $\mathbf{y}_4$ is given by

$$\mathbf{y}_4 = \mathbf{B}^{-1}\begin{bmatrix} \mathbf{Ax}_4 \\ 1 \end{bmatrix} = \begin{bmatrix} \frac{1}{5} & 0 & 0 \\ -\frac{1}{5} & 1 & -\frac{5}{2} \\ -\frac{1}{5} & 0 & 1 \end{bmatrix}\begin{bmatrix} 2 \\ \frac{7}{2} \\ 1 \end{bmatrix} = \begin{bmatrix} \frac{2}{5} \\ \frac{3}{5} \\ \frac{3}{5} \end{bmatrix}$$

Insert the column $\begin{pmatrix} z_4 - \hat{c}_4 \\ \mathbf{y}_4 \end{pmatrix}$ in the foregoing array and pivot. This leads to the following two tableaux (the λ_4 column is deleted after pivoting).

	BASIS INVERSE			RHS		λ_4
z	$-\frac{6}{5}$	0	$-\frac{5}{2}$	$\frac{49}{10}$		$\frac{3}{5}$
λ_2	$\frac{1}{5}$	0	0	$\frac{2}{5}$		$\frac{2}{5}$
s_2	$-\frac{1}{5}$	1	$-\frac{5}{2}$	$\frac{1}{10}$		$\boxed{\frac{3}{5}}$
λ_3	$-\frac{1}{5}$	0	1	$\frac{3}{5}$		$\frac{3}{5}$

	BASIS INVERSE			RHS
z	-1	-1	0	-5
λ_2	$\frac{1}{3}$	$-\frac{2}{3}$	$\frac{5}{3}$	$\frac{1}{3}$
λ_4	$-\frac{1}{3}$	$\frac{5}{3}$	$-\frac{25}{6}$	$\frac{1}{6}$
λ_3	0	-1	$\frac{7}{2}$	$\frac{1}{2}$

The best-known feasible solution of the overall problem is given by

$$\begin{aligned}\mathbf{x} &= \lambda_2\mathbf{x}_2 + \lambda_3\mathbf{x}_3 + \lambda_4\mathbf{x}_4 \\ &= \tfrac{1}{3}\left(2, \tfrac{3}{2}, 3, 0\right) + \tfrac{1}{2}\left(0, \tfrac{5}{2}, 0, 0\right) + \tfrac{1}{6}\left(2, \tfrac{3}{2}, 0, 0\right) = (1, 2, 1, 0)\end{aligned}$$

The objective is -5. Also $(w_1, w_2, \alpha) = (-1, -1, 0)$.

Iteration 4

Since $w_1 < 0$ and $w_2 < 0$, s_1 and s_2 are not eligible to enter the basis at this time.

SUBPROBLEM

Maximize $(\mathbf{wA} - \mathbf{c})\mathbf{x} + \alpha$

Subject to $\mathbf{x} \in X$

$$\mathbf{wA} - \mathbf{c} = (-1, -1)\begin{bmatrix} 1 & 0 & 1 & 0 \\ 1 & 1 & 0 & 2 \end{bmatrix} - (-2, -1, -1, 1) = (0, 0, 0, -3)$$

Therefore the subproblem is as follows.

Maximize $0x_1 + 0x_2 + 0x_3 - 3x_4 + 0$

Subject to $\mathbf{x} \in X$

Using Figure 7.1, an optimal solution is $\mathbf{x}_5 = (0, 0, 0, 0)$ with objective $z_5 - \hat{c}_5 = 0$, which is the termination criterion. Also note that the lower bound is $\hat{\mathbf{c}}_B\bar{\mathbf{b}} - (z_5 - \hat{c}_5) = -5 - 0 = -5$, which is equal to the best (and therefore optimal) solution known so far.

To summarize, the optimal solution $(x_1, x_2, x_3, x_4) = (1, 2, 1, 0)$ with objective -5 is at hand. The progress of the lower bounds and the objective values of the primal feasible solutions generated by the decomposition algorithm is shown in Figure 7.2. Optimality is reached at iteration 4. If we were interested in an

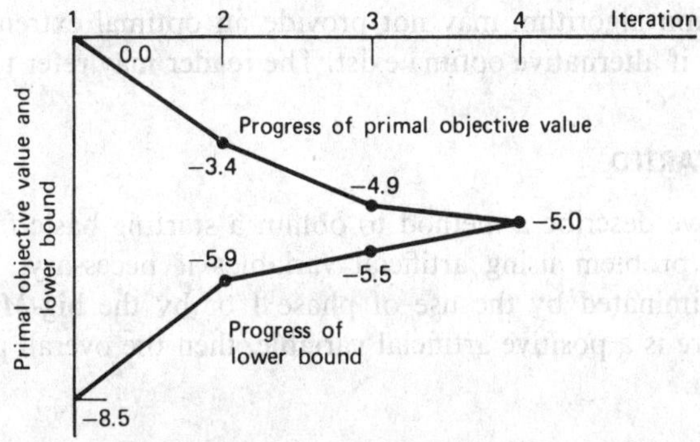

Figure 7.2. **Progress of the primal objective value and the lower bound.**

approximate solution, we could have stopped at iteration 3, since we have a feasible solution with an objective value equal to -4.9, and meanwhile are assured (by the lower bound) that there exist no feasible solutions with an objective less than -5.5.

The optimal point $(x_1, x_2, x_3, x_4) = (1, 2, 1, 0)$ is shown in Figure 7.3 in the two sets X_1 and X_2. Note that $(1, 2)$ is not an extreme point of X_1 and $(1, 0)$ is not an extreme point of X_2. Note, however, that we can map the master constraints

$$\begin{aligned} x_1 \qquad + x_3 \qquad &\leq 2 \\ x_1 + x_2 \qquad + 2x_4 &\leq 3 \end{aligned}$$

into the (x_1, x_2) space by substituting $x_3 = 1$ and $x_4 = 0$. This leads to the two restrictions $x_1 \leq 1$ and $x_1 + x_2 \leq 3$, which are shown in Figure 7.3. We see that $(1, 2)$ is an extreme point of X_1 intersected with these two additional constraints. Similarly, in the (x_3, x_4) space, by substituting the values $x_1 = 1$ and $x_2 = 2$, the master constraints reduce to $x_3 \leq 1$ and $2x_4 \leq 0$. Again $(1, 0)$ is an extreme point of X_2 intersected with these additional constraints. It is worthwhile noting that

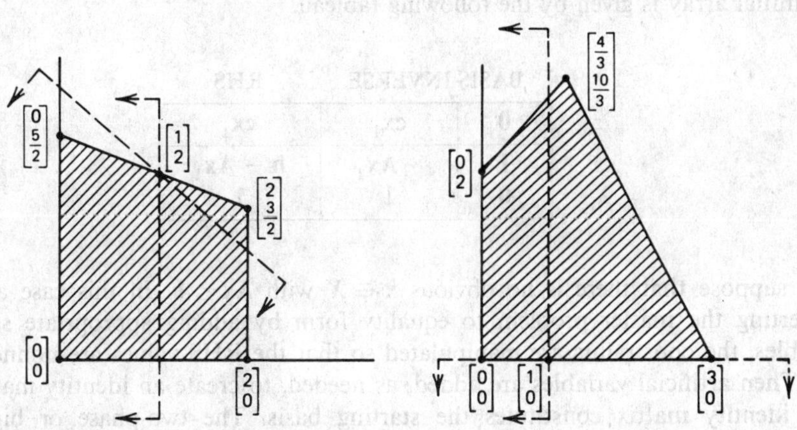

Figure 7.3. **Illustration of the optimal point.**

the decomposition algorithm may not provide an optimal extreme point of the overall problem if alternative optima exist. The reader may refer to Exercise 7.21.

7.3 GETTING STARTED

In this section we describe a method to obtain a starting basic feasible solution for the master problem using artificial variables if necessary. These artificial variables are eliminated by the use of phase I or by the big-M method. If at termination there is a positive artificial variable, then the overall problem has no feasible solution.

Inequality Constraints

Consider the following problem.

$$\begin{aligned}
\text{Minimize} \quad & \sum_{j=1}^{t} (\mathbf{c}\mathbf{x}_j)\lambda_j \\
\text{Subject to} \quad & \sum_{j=1}^{t} (\mathbf{A}\mathbf{x}_j)\lambda_j \leq \mathbf{b} \\
& \sum_{j=1}^{t} \lambda_j = 1 \\
& \lambda_j \geq 0 \qquad j = 1, 2, \ldots, t
\end{aligned}$$

If there is a convenient $\mathbf{x}_1 \in X$ with $\mathbf{A}\mathbf{x}_1 \leq \mathbf{b}$, then the following basis is at hand, where the identity corresponds to the slack vector $\mathbf{s} \geq \mathbf{0}$.

$$\mathbf{B} = \left[\begin{array}{c|c} \mathbf{I} & \mathbf{A}\mathbf{x}_1 \\ \hline \mathbf{0} & 1 \end{array}\right], \qquad \mathbf{B}^{-1} = \left[\begin{array}{c|c} \mathbf{I} & -\mathbf{A}\mathbf{x}_1 \\ \hline \mathbf{0} & 1 \end{array}\right]$$

The initial array is given by the following tableau.

	BASIS INVERSE		RHS
z	$\mathbf{0}$	$\mathbf{c}\mathbf{x}_1$	$\mathbf{c}\mathbf{x}_1$
s	$\mathbf{I}$	$-\mathbf{A}\mathbf{x}_1$	$\mathbf{b} - \mathbf{A}\mathbf{x}_1$
λ_1	$\mathbf{0}$	1	1

Now suppose that there is no obvious $\mathbf{x} \in X$ with $\mathbf{A}\mathbf{x} \leq \mathbf{b}$. In this case after converting the master problem to equality form by adding appropriate slack variables, the constraints are manipulated so that the RHS values are nonnegative. Then artificial variables are added, as needed, to create an identity matrix. This identity matrix constitutes the starting basis. The two-phase or big-M methods can be used to drive the artificial variables out of the basis.

Equality Constraints

In this case $m + 1$ artificial variables can be introduced to form the initial basis. The artificial variables are eliminated by the two-phase or by the big-M method.

7.4 THE CASE OF UNBOUNDED REGION X

For an unbounded set X, the decomposition algorithm must be slightly modified. In this case points in X can no longer be represented as a convex combination of the extreme points, but rather as a convex combination of the extreme points plus a nonnegative combination of the extreme directions. In other words, $\mathbf{x} \in X$ if and only if

$$\mathbf{x} = \sum_{j=1}^{t} \lambda_j \mathbf{x}_j + \sum_{j=1}^{l} \mu_j \mathbf{d}_j$$

$$\sum_{j=1}^{t} \lambda_j = 1$$

$$\lambda_j \geq 0 \qquad j = 1, 2, \ldots, t$$

$$\mu_j \geq 0 \qquad j = 1, 2, \ldots, l$$

where $\mathbf{x}_1, \mathbf{x}_2, \ldots, \mathbf{x}_t$ are the extreme points of X and $\mathbf{d}_1, \mathbf{d}_2, \ldots, \mathbf{d}_l$ are the extreme directions of X. The primal problem can be transformed into a so-called master problem in the variables $\lambda_1, \lambda_2, \ldots, \lambda_t$ and $\mu_1, \mu_2, \ldots, \mu_l$ as follows:

$$\text{Minimize} \quad \sum_{j=1}^{t} (\mathbf{c}\mathbf{x}_j)\lambda_j + \sum_{j=1}^{l} (\mathbf{c}\mathbf{d}_j)\mu_j$$

$$\text{Subject to} \quad \sum_{j=1}^{t} (\mathbf{A}\mathbf{x}_j)\lambda_j + \sum_{j=1}^{l} (\mathbf{A}\mathbf{d}_j)\mu_j = \mathbf{b} \tag{7.5}$$

$$\sum_{j=1}^{t} \lambda_j = 1 \tag{7.6}$$

$$\lambda_j \geq 0 \qquad j = 1, 2, \ldots, t$$
$$\mu_j \geq 0 \qquad j = 1, 2, \ldots, l$$

Since t and l are usually very large, we shall attempt to solve the foregoing problem by the revised simplex method. Suppose that we have a basic feasible solution of the foregoing system with basis $\mathbf{B}$, and let $\mathbf{w}$ and α be the dual variables corresponding to constraints (7.5) and (7.6). Further suppose that $\mathbf{B}^{-1}$, $(\mathbf{w}, \alpha) = \hat{\mathbf{c}}_B\mathbf{B}^{-1}$($\hat{\mathbf{c}}_B$ is the cost of the basic variables), and $\bar{\mathbf{b}} = \mathbf{B}^{-1}\begin{pmatrix}\mathbf{b}\\1\end{pmatrix}$ are known and displayed.

BASIS INVERSE	RHS
$(\mathbf{w}, \alpha)$	$\hat{\mathbf{c}}_B\bar{\mathbf{b}}$
$\mathbf{B}^{-1}$	$\bar{\mathbf{b}}$

Recall that the current solution is optimal to the overall problem if $z_j - \hat{c}_j \le 0$ for each variable. In particular, the following conditions must hold at optimality:

$$\lambda_j \text{ nonbasic} \Rightarrow 0 \ge z_j - \hat{c}_j = (\mathbf{w}, \alpha)\begin{pmatrix} \mathbf{Ax}_j \\ 1 \end{pmatrix} - \mathbf{cx}_j = \mathbf{wAx}_j + \alpha - \mathbf{cx}_j \tag{7.7}$$

$$\mu_j \text{ nonbasic} \Rightarrow 0 \ge z_j - \hat{c}_j = (\mathbf{w}, \alpha)\begin{pmatrix} \mathbf{Ad}_j \\ 0 \end{pmatrix} - \mathbf{cd}_j = \mathbf{wAd}_j - \mathbf{cd}_j \tag{7.8}$$

Since the number of nonbasic variables is very large, checking conditions (7.7) and (7.8) by generating the corresponding extreme points and directions is computationally infeasible. However, we may determine whether or not these conditions hold by solving the following *subproblem*. More importantly, as the subproblem is solved, if conditions (7.7) or (7.8) do not hold, a nonbasic variable with a positive $z_k - \hat{c}_k$, and hence eligible to enter the basis, is found.

$$\begin{aligned} &\text{Maximize} \quad (\mathbf{wA} - \mathbf{c})\mathbf{x} + \alpha \\ &\text{Subject to} \quad \mathbf{x} \in X \end{aligned}$$

First, suppose that the optimal solution value of the subproblem is unbounded. Recall that this is only possible if an extreme direction $\mathbf{d}_k$ is found such that $(\mathbf{wA} - \mathbf{c})\mathbf{d}_k > 0$. This means that condition (7.8) is violated. Moreover, $z_k - \hat{c}_k = (\mathbf{wA} - \mathbf{c})\mathbf{d}_k > 0$ and μ_k is eligible to enter the basis. In this case $\begin{pmatrix} \mathbf{Ad}_k \\ 0 \end{pmatrix}$ is updated by premultiplying by $\mathbf{B}^{-1}$ and the resulting column $\begin{pmatrix} z_k - \hat{c}_k \\ \mathbf{y}_k \end{pmatrix}$ is inserted in the foregoing array and the revised simplex method is continued. Now consider the case where the optimal solution value is bounded. A necessary and sufficient condition for boundedness is that $(\mathbf{wA} - \mathbf{c})\mathbf{d}_j \le 0$ for all extreme directions, and so Equation (7.8) holds. Now we check whether (7.7) holds. Let $\mathbf{x}_k$ be an optimal extreme point and consider the optimal objective, $z_k - \hat{c}_k$, to the subproblem. If $z_k - \hat{c}_k \le 0$, then by optimality of $\mathbf{x}_k$, for each extreme point $\mathbf{x}_j$, we have

$$(\mathbf{wA} - \mathbf{c})\mathbf{x}_j + \alpha \le (\mathbf{wA} - \mathbf{c})\mathbf{x}_k + \alpha = z_k - \hat{c}_k \le 0$$

and hence condition (7.7) holds and we stop with an optimal solution of the overall problem. If, on the other hand, $z_k - \hat{c}_k > 0$, then λ_k is introduced in the basis. This is done by inserting the column $\begin{pmatrix} z_k - \hat{c}_k \\ \mathbf{y}_k \end{pmatrix}$ into the foregoing array and pivoting, where $\mathbf{y}_k = \mathbf{B}^{-1}\begin{pmatrix} \mathbf{Ax}_k \\ 1 \end{pmatrix}$. Note that, as in the bounded case, if the master problem includes slack or other explicitly present variables, then the $z_j - \hat{c}_j$ values for these variables must be checked before deducing optimality. Also, the bounded subproblems yield lower bounds as for the previous case.

To summarize, solving the foregoing subproblem leads us either to terminate the algorithm with an optimal solution, or else to identify an improving variable to enter the basis. We now have all the ingredients for a decomposition algorithm in the case of an unbounded set X. Example 7.1 illustrates such an algorithm. We ask the reader in Exercise 7.17 to write a step-by-step procedure for this case.

Example 7.1

$$\begin{array}{ll} \text{Minimize} & -x_1 - 2x_2 - x_3 \\ \text{Subject to} & x_1 + x_2 + x_3 \le 12 \\ & -x_1 + x_2 \le 2 \\ & -x_1 + 2x_2 \le 8 \\ & x_3 \le 3 \\ & x_1, \quad x_2, \quad x_3 \ge 0 \end{array}$$

The first constraint is handled as $\mathbf{Ax} \le \mathbf{b}$ and the rest of the constraints are treated by X. Note that X decomposes into the two sets of Figure 7.4. The problem is transformed into $\lambda_1, \ldots, \lambda_t$ and $\mu_1, \ldots, \mu_l$ as follows.

$$\begin{array}{ll} \text{Minimize} & \sum_{j=1}^{t} (\mathbf{cx}_j)\lambda_j + \sum_{j=1}^{l} (\mathbf{cd}_j)\mu_j \\ & \sum_{j=1}^{t} (\mathbf{Ax}_j)\lambda_j + \sum_{j=1}^{l} (\mathbf{Ad}_j)\mu_j \le \mathbf{b} \\ & \sum_{j=1}^{t} \lambda_j = 1 \\ & \lambda_j \ge 0 \qquad j = 1, 2, \ldots, t \\ & \mu_j \ge 0 \qquad j = 1, 2, \ldots, l \end{array}$$

Note that $\mathbf{x}_1 = (0, 0, 0)$ belongs to X and $\mathbf{Ax}_1 = 0 + 0 + 0 \le 12$. Therefore the initial basis consists of λ_1 (corresponding to $\mathbf{x}_1$) plus the slack variable s. This leads to the following array, where $w = \alpha = 0$.

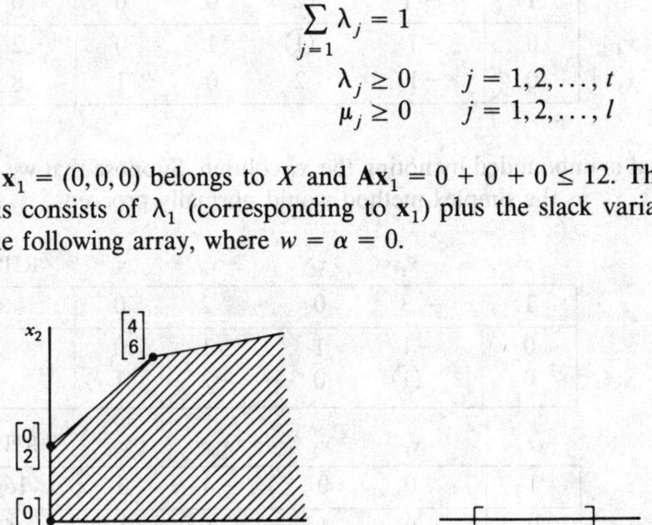

Figure 7.4. Illustration of an unbounded X.

	BASIS INVERSE		RHS
z	0	0	0
s	1	0	12
λ_1	0	1	1

Iteration 1

SUBPROBLEM

Maximize $(\mathbf{wA} - \mathbf{c})\mathbf{x} + \alpha$

Subject to $\mathbf{x} \in X$

Since $w = \alpha = 0$ and $\mathbf{A} = (1, 1, 1)$, this problem reduces to the following.

$$\text{Maximize} \quad x_1 + 2x_2 + x_3 + 0$$

$$\begin{aligned} \text{Subject to} \quad -x_1 + x_2 &\leq 2 \\ -x_1 + 2x_2 &\leq 8 \\ x_3 &\leq 3 \\ x_1, \quad x_2, \quad x_3 &\geq 0 \end{aligned}$$

Note that the problem decomposes into two problems in (x_1, x_2) and x_3. The optimal value of x_3 is 3. The other part of the problem can be solved geometrically or by the simplex method, where x_4 and x_5 are the slack variables.

	z	x_1	x_2	x_4	x_5	RHS
z	1	−1	−2	0	0	0
x_4	0	−1	①	1	0	2
x_5	0	−1	2	0	1	8

The optimal is unbounded by noting the x_1 column. Suppose that we continue by introducing x_2 as the simplex method would normally proceed.

	z	x_1	x_2	x_4	x_5	RHS
z	1	−3	0	2	0	4
x_2	0	−1	1	1	0	2
x_5	0	①	0	−2	1	4

	z	x_1	x_2	x_4	x_5	RHS
z	1	0	0	−4	3	16
x_2	0	0	1	−1	1	6
x_1	0	1	0	−2	1	4

Now as x_4 increases by one unit, z increases by four units, x_1 increases by two units and x_2 increases by one unit; that is, in the (x_1, x_2) space we have found a direction $\mathbf{d}_1 = \begin{pmatrix} 2 \\ 1 \end{pmatrix}$ leading to an unbounded objective value. In the (x_1, x_2, x_3) space, $\mathbf{d}_1$ is given by $(2, 1, 0)'$ (why?). Also $(\mathbf{wA} - \mathbf{c})\mathbf{d}_1 = 4$ (the negative of -4 in row 0 under x_4) and so μ_1 is introduced in the basis.

MASTER STEP

$z_1 - \hat{c}_1 = 4$

$$\mathbf{Ad}_1 = (1, 1, 1)\begin{pmatrix} 2 \\ 1 \\ 0 \end{pmatrix} = 3$$

$$\mathbf{y}_1 = \mathbf{B}^{-1}\begin{pmatrix} \mathbf{Ad}_1 \\ 0 \end{pmatrix}$$

where $\mathbf{B}^{-1}$ is obtained from the initial array of the master problem. In particular,

$$\mathbf{y}_1 = \begin{bmatrix} 1 & 0 \\ 0 & 1 \end{bmatrix}\begin{bmatrix} 3 \\ 0 \end{bmatrix} = \begin{bmatrix} 3 \\ 0 \end{bmatrix}$$

Introduce the column $\begin{pmatrix} z_1 - \hat{c}_1 \\ \mathbf{y}_1 \end{pmatrix}$ in the master array and pivot. The μ_1 column is eliminated after pivoting.

	BASIS INVERSE		RHS	μ_1
z	0	0	0	4
s	1	0	12	(3)
λ_1	0	1	1	0

	BASIS INVERSE		RHS
z	$-\frac{4}{3}$	0	-16
μ_1	$\frac{1}{3}$	0	4
λ_1	0	1	1

Iteration 2

$w = -\frac{4}{3}$ and $\alpha = 0$

Since $w < 0$, s is not a candidate to enter the basis.

SUBPROBLEM

Maximize $(\mathbf{wA} - \mathbf{c})\mathbf{x} + \alpha$

Subject to $\mathbf{x} \in X$

This reduces to the following.

$$\begin{array}{llll} \text{Maximize} & -\frac{1}{3}x_1 + \frac{2}{3}x_2 + 0 & \text{Maximize} & -\frac{1}{3}x_3 \\ \text{Subject to} & -x_1 + x_2 \le 2 & \text{Subject to} & 0 \le x_3 \le 3 \\ & -x_1 + 2x_2 \le 8 & & \\ & x_1, \quad\ x_2 \ge 0 & & \end{array}$$

Here the value $\alpha = 0$ is added to only one of the subproblems. Obviously $x_3 = 0$. The new problem in (x_1, x_2) is solved by utilizing the corresponding tableau of the last iteration, deleting row 0, and introducing the new costs.

	z	x_1	x_2	x_4	x_5	RHS
z	1	$\frac{1}{3}$	$-\frac{2}{3}$	0	0	0
x_2	0	0	1	-1	1	6
x_1	0	1	0	-2	1	4

Multiply row 1 by $\frac{2}{3}$ and row 2 by $-\frac{1}{3}$ and add to row 0.

	z	x_1	x_2	x_4	x_5	RHS
z	1	0	0	0	$\frac{1}{3}$	$\frac{8}{3}$
x_2	0	0	1	-1	1	6
x_1	0	1	0	-2	1	4

The foregoing tableau is optimal (not unique). The optimal objective of the subproblem is $z_2 - \hat{c}_2 = \frac{8}{3} > 0$ and so λ_2 corresponding to $\mathbf{x}_2 = (x_1, x_2, x_3) = (4, 6, 0)$ is introduced. A lower bound on the problem is now available as $\hat{\mathbf{c}}_B\bar{\mathbf{b}} - (z_2 - \hat{c}_2) = -16 - \frac{8}{3} = -\frac{56}{3}$.

MASTER STEP

$z_2 - \hat{c}_2 = \frac{8}{3}$

$$\mathbf{Ax}_2 = 10$$

$$\mathbf{y}_2 = \mathbf{B}^{-1}\begin{pmatrix} \mathbf{Ax}_2 \\ 1 \end{pmatrix} = \begin{bmatrix} \frac{1}{3} & 0 \\ 0 & 1 \end{bmatrix}\begin{bmatrix} 10 \\ 1 \end{bmatrix} = \begin{bmatrix} \frac{10}{3} \\ 1 \end{bmatrix}$$

Introduce $\begin{pmatrix} z_2 - \hat{c}_2 \\ \mathbf{y}_2 \end{pmatrix}$ into the master array and pivot. The λ_2 column is eliminated after pivoting.

	BASIS INVERSE		RHS		λ_2
z	$-\frac{4}{3}$	0	-16		$\frac{8}{3}$
μ_1	$\frac{1}{3}$	0	4		$\frac{10}{3}$
λ_1	0	1	1		①

	BASIS INVERSE		RHS
z	$-\frac{4}{3}$	$-\frac{8}{3}$	$-\frac{56}{3}$
μ_1	$\frac{1}{3}$	$-\frac{10}{3}$	$\frac{2}{3}$
λ_2	0	1	1

Observe that the known lower bound of $-\frac{56}{3}$ matches with the upper bound, and hence the optimum is at hand. Let us verify this nonetheless.

Iteration 3

Note that $w = -\frac{4}{3}$ did not alter from the last iteration. So s is still not a candidate to enter the basis. Also the optimal solution of the last subproblem remains the same (see iteration 2). The objective value of $\frac{8}{3}$ was for the previous dual solution with $\alpha = 0$. For $\alpha = -\frac{8}{3}$ we have $z_3 - \hat{c}_3 = \frac{8}{3} - \frac{8}{3} = 0$, which is the termination criterion, and the optimal solution is at hand. More specifically, the optimal $\mathbf{x}^*$ is given by

$$\mathbf{x}^* = \lambda_2 \mathbf{x}_2 + \mu_1 \mathbf{d}_1$$

$$= 1\begin{bmatrix} 4 \\ 6 \\ 0 \end{bmatrix} + \frac{2}{3}\begin{bmatrix} 2 \\ 1 \\ 0 \end{bmatrix} = \begin{bmatrix} \frac{16}{3} \\ \frac{20}{3} \\ 0 \end{bmatrix}$$

The objective is $-\frac{56}{3}$.

7.5 BLOCK DIAGONAL OR ANGULAR STRUCTURE

In this section we discuss the important special case when the set X has a block diagonal structure. In this case, X can itself be decomposed into several sets $X_1, X_2, \ldots, X_T$, each involving a subset of the variables, which do not appear in any other set. If we decompose the vector $\mathbf{x}$ accordingly into the vectors $\mathbf{x}_1, \mathbf{x}_2, \ldots, \mathbf{x}_T$, the vector $\mathbf{c}$ into $\mathbf{c}_1, \mathbf{c}_2, \ldots, \mathbf{c}_T$, and the matrix $\mathbf{A}$ of the master constraints $\mathbf{Ax} = \mathbf{b}$ into the matrices $\mathbf{A}_1, \mathbf{A}_2, \ldots, \mathbf{A}_T$, we get the following problem.

$$\begin{array}{lllll} \text{Minimize} & \mathbf{c}_1\mathbf{x}_1 + & \mathbf{c}_2\mathbf{x}_2 + \ldots + & \mathbf{c}_T\mathbf{x}_T & \\ \text{Subject to} & \mathbf{A}_1\mathbf{x}_1 + & \mathbf{A}_2\mathbf{x}_2 + \ldots + & \mathbf{A}_T\mathbf{x}_T & = \mathbf{b} \\ & \mathbf{B}_1\mathbf{x}_1 & & & \leq \mathbf{b}_1 \\ & & \mathbf{B}_2\mathbf{x}_2 & & \leq \mathbf{b}_2 \\ & & \quad \ddots & & \vdots \\ & & & \mathbf{B}_T\mathbf{x}_T & \leq \mathbf{b}_T \\ & \mathbf{x}_1, & \mathbf{x}_2, \quad \ldots, & \mathbf{x}_T & \geq \mathbf{0} \end{array}$$

where $X_i = \{\mathbf{x}_i : \mathbf{B}_i\mathbf{x}_i \leq \mathbf{b}_i, \mathbf{x}_i \geq \mathbf{0}\}$ for $i = 1, 2, \ldots, T$.

Problems with the foregoing structure arise frequently in network flows with several commodities and in the allocation of scarce resources among competing activities. Problems of this structure can be solved by the decomposition algorithm of this chapter (see Section 7.2 and Example 7.1). However, this *block diagonal* or *angular* structure of X can be utilized further, as will be discussed in this section.

For subproblem i, $\mathbf{x}_i \in X_i$ if and only if

$$\mathbf{x}_i = \sum_{j=1}^{t_i} \lambda_{ij}\mathbf{x}_{ij} + \sum_{j=1}^{l_i} \mu_{ij}\mathbf{d}_{ij}$$

$$\sum_{j=1}^{t_i} \lambda_{ij} = 1$$

$$\lambda_{ij} \geq 0 \qquad j = 1, 2, \ldots, t_i$$

$$\mu_{ij} \geq 0 \qquad j = 1, 2, \ldots, l_i$$

where the $\mathbf{x}_{ij}$'s and the $\mathbf{d}_{ij}$'s are the extreme points and the extreme directions (if any) of X_i. Replacing each $\mathbf{x}_i$ by the foregoing representation, the original problem can be reformulated as the following *master problem*.

$$\text{Minimize} \quad \sum_{i=1}^{T}\sum_{j=1}^{t_i} (\mathbf{c}_i\mathbf{x}_{ij})\lambda_{ij} + \sum_{i=1}^{T}\sum_{j=1}^{l_i} (\mathbf{c}_i\mathbf{d}_{ij})\mu_{ij}$$

$$\text{Subject to} \quad \sum_{i=1}^{T}\sum_{j=1}^{t_i} (\mathbf{A}_i\mathbf{x}_{ij})\lambda_{ij} + \sum_{i=1}^{T}\sum_{j=1}^{l_i} (\mathbf{A}_i\mathbf{d}_{ij})\mu_{ij} = \mathbf{b} \tag{7.9}$$

$$\sum_{j=1}^{t_i} \lambda_{ij} = 1 \qquad i = 1, 2, \ldots, T \tag{7.10}$$

$$\lambda_{ij} \geq 0 \qquad j = 1, 2, \ldots, t_i, \ i = 1, 2, \ldots, T$$

$$\mu_{ij} \geq 0 \qquad j = 1, 2, \ldots, l_i, \ i = 1, 2, \ldots, T$$

Note the difference in the foregoing formulation and that of Section 7.2. Here we allow different convex combinations and linear combinations for each subproblem i since we have T *convexity constraints* [the constraints of Equation (7.10)]. This adds more flexibility but at the same time increases the number of constraints from $m + 1$ to $m + T$.

Suppose that we have a basic feasible solution of the foregoing system with an $(m + T) \times (m + T)$ basis $\mathbf{B}$. Note that each basis must contain at least one variable λ_{ij} from each block i (see Exercise 7.19). Further suppose that $\mathbf{B}^{-1}, \bar{\mathbf{b}} = \mathbf{B}^{-1}\begin{pmatrix}\mathbf{b}\\ \mathbf{1}\end{pmatrix}$, $(\mathbf{w}, \alpha) = (w_1, \ldots, w_m, \alpha_1, \ldots, \alpha_T) = \hat{\mathbf{c}}_B\mathbf{B}^{-1}$ are known, where $\hat{\mathbf{c}}_B$ is the cost of the basic variables ($\hat{c}_{ij} = \mathbf{c}_i\mathbf{x}_{ij}$ for λ_{ij} and $\hat{c}_{ij} = \mathbf{c}_i\mathbf{d}_{ij}$ for μ_{ij}). These are displayed below.

BASIS INVERSE	RHS
$(\mathbf{w}, \alpha)$	$\hat{\mathbf{c}}_B\bar{\mathbf{b}}$
$\mathbf{B}^{-1}$	$\bar{\mathbf{b}}$

This solution is optimal if $z_{ij} - \hat{c}_{ij} \leq 0$ for each variable (naturally $z_{ij} - \hat{c}_{ij} = 0$ for each basic variable). In particular the following conditions must hold at optimality

$$\lambda_{ij} \text{ nonbasic} \Rightarrow 0 \geq z_{ij} - \hat{c}_{ij} = \mathbf{wA}_i\mathbf{x}_{ij} + \alpha_i - \mathbf{c}_i\mathbf{x}_{ij} \tag{7.11}$$

$$\mu_{ij} \text{ nonbasic} \Rightarrow 0 \geq z_{ij} - \hat{c}_{ij} = \mathbf{wA}_i\mathbf{d}_{ij} - \mathbf{c}_i\mathbf{d}_{ij} \tag{7.12}$$

Whether conditions (7.11) and (7.12) hold or not, can be easily verified by solving the following subproblems.

$$\begin{aligned} &\text{Maximize} && (\mathbf{wA}_i - \mathbf{c}_i)\mathbf{x}_i + \alpha_i \\ &\text{Subject to} && \mathbf{x}_i \in X_i \end{aligned}$$

If the optimal solution value is unbounded, then an extreme direction $\mathbf{d}_{ik}$ is found such that $(\mathbf{wA}_i - \mathbf{c}_i)\mathbf{d}_{ik} > 0$; that is, condition (7.12) is violated and introducing μ_{ik} will improve the objective function since $z_{ik} - \hat{c}_{ik} = (\mathbf{wA}_i - \mathbf{c}_i)\mathbf{d}_{ik} > 0$. If the optimal is bounded, then automatically condition (7.12) holds for subproblem i. Let $\mathbf{x}_{ik}$ be an optimal extreme point. If the optimal objective value $z_{ik} - \hat{c}_{ik} = \mathbf{wA}_i\mathbf{x}_{ik} + \alpha_i - \mathbf{c}_i\mathbf{x}_{ik} \leq 0$, then condition (7.11) holds for subproblem i. Otherwise λ_{ik} can be introduced in the basis. When all subproblems have $z_{ik} - \hat{c}_{ik} \leq 0$, then the optimal solution to the original problem is obtained. If the master problem contains other explicit variables including slack variables, then we must also check the $z_j - \hat{c}_j$ values for these variables (as we did in Section 7.1) before terminating.

To summarize, each subproblem i is solved in turn. If subproblem i yields an unbounded solution value, then an extreme direction $\mathbf{d}_{ik}$ is found that is a candidate to enter the master basis. If subproblem i yields a bounded optimal point and $\mathbf{w}\mathbf{A}_i\mathbf{x}_{ik} + \alpha_i - \mathbf{c}_i\mathbf{x}_{ik} > 0$, then the extreme point is a candidate to enter the master basis. If neither of these conditions occurs, then there is currently no candidate column to enter the master basis from subproblem i. If no subproblem yields a candidate to enter the master basis, then we are optimal. Otherwise, we must select one from among the various candidates to enter the master basis. We may use the rule of the most positive $z_{ik} - \hat{c}_{ik}$, the first positive $z_{ik} - \hat{c}_{ik}$, and so on. If we use the rule of the first positive $z_{ik} - \hat{c}_{ik}$, then we may stop solving the subproblems after the first candidate comes available. On selecting the candidate, we update the entering column, pivot on the master array, and repeat the process.

Calculation of lower bounds for the case of bounded subproblems

Let $\mathbf{x}_1, \mathbf{x}_2, \ldots, \mathbf{x}_T$ represent a feasible solution of the overall problem so that $\mathbf{x}_i \in X_i$ for each i and $\Sigma_i \mathbf{A}_i\mathbf{x}_i = \mathbf{b}$. By definition of $z_{ik} - \hat{c}_{ik}$ we have

$$(\mathbf{w}\mathbf{A}_i - \mathbf{c}_i)\mathbf{x}_i + \alpha_i \leq (z_{ik} - \hat{c}_{ik})$$

or

$$\mathbf{c}_i\mathbf{x}_i \geq \mathbf{w}\mathbf{A}_i\mathbf{x}_i + \alpha_i - (z_{ik} - \hat{c}_{ik}).$$

Summing on i we get

$$\sum_i \mathbf{c}_i\mathbf{x}_i \geq \mathbf{w}\sum_i \mathbf{A}_i\mathbf{x}_i + \sum_i \alpha_i - \sum_i (z_{ik} - \hat{c}_{ik}).$$

But $\Sigma_i \mathbf{c}_i\mathbf{x}_i = \mathbf{c}\mathbf{x}$ and $\Sigma_i \mathbf{A}_i\mathbf{x}_i = \mathbf{b}$. Thus we get

$$\mathbf{c}\mathbf{x} \geq \mathbf{w}\mathbf{b} + \alpha\mathbf{1} - \sum_i (z_{ik} - \hat{c}_{ik})$$

or

$$\mathbf{c}\mathbf{x} \geq \hat{\mathbf{c}}_B\bar{\mathbf{b}} - \sum_i (z_{ik} - \hat{c}_{ik}).$$

The right-hand-side in this inequality provides a lower bound on the problem. This is a natural extension of the case for one bounded subproblem presented earlier.

Example 7.2

$$\begin{array}{lrl}
\text{Minimize} & -2x_1 - x_2 - 3x_3 - x_4 & \\
\text{Subject to} & x_1 + x_2 + x_3 + x_4 & \leq 6 \\
& x_2 + 2x_3 + x_4 & \leq 4 \\
& x_1 + x_2 & \leq 6 \\
& x_2 & \leq 2 \\
& -x_3 + x_4 & \leq 3 \\
& x_3 + x_4 & \leq 5 \\
& x_1, \quad x_2, \quad x_3, \quad x_4 & \geq 0
\end{array}$$

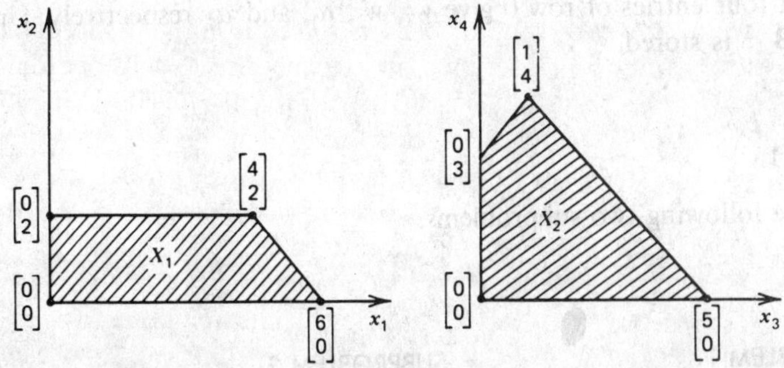

Figure 7.5. The region X.

The first two constraints are handled by $\mathbf{Ax} \le \mathbf{b}$, and the rest of the constraints are treated by X. Note that X decomposes into two sets as shown in Figure 7.5. The problem is transformed into the following.

$$\begin{aligned}
\text{Minimize} \quad & \sum_{j=1}^{t_1} (\mathbf{c}_1\mathbf{x}_{1j})\lambda_{1j} + \sum_{j=1}^{t_2} (\mathbf{c}_2\mathbf{x}_{2j})\lambda_{2j} \\
\text{Subject to} \quad & \sum_{j=1}^{t_1} (\mathbf{A}_1\mathbf{x}_{1j})\lambda_{1j} + \sum_{j=1}^{t_2} (\mathbf{A}_2\mathbf{x}_{2j})\lambda_{2j} \le \mathbf{b} \\
& \sum_{j=1}^{t_1} \lambda_{1j} = 1 \\
& \sum_{j=1}^{t_2} \lambda_{2j} = 1 \\
& \lambda_{1j} \ge 0 \qquad j = 1, 2, \ldots, t_1 \\
& \lambda_{2j} \ge 0 \qquad j = 1, 2, \ldots, t_2
\end{aligned}$$

where $\mathbf{c}_1 = (-2, -1)$, $\mathbf{c}_2 = (-3, -1)$, $\mathbf{A}_1 = \begin{bmatrix} 1 & 1 \\ 0 & 1 \end{bmatrix}$, and $\mathbf{A}_2 = \begin{bmatrix} 1 & 1 \\ 2 & 1 \end{bmatrix}$. Note that $\mathbf{x}_{11} = (x_1, x_2) = (0, 0)$ and $\mathbf{x}_{21} = (x_3, x_4) = (0, 0)$ belong to X_1 and X_2 and satisfy the master constraints. Therefore we have a basic feasible solution of the overall system where the basis consists of s_1, s_2, λ_{11}, and λ_{21} (s_1 and s_2 are the slacks). This leads to the following master array.

	BASIS INVERSE				RHS
z	0	0	0	0	0
s_1	1	0	0	0	6
s_2	0	1	0	0	4
λ_{11}	0	0	1	0	1
λ_{21}	0	0	0	1	1

The first four entries of row 0 give w_1, w_2, α_1, and α_2 respectively. Under these entries $\mathbf{B}^{-1}$ is stored.

Iteration 1

Solve the following two subproblems.

SUBPROBLEM 1

Maximize $(\mathbf{wA}_1 - \mathbf{c}_1)\mathbf{x}_1 + \alpha_1$

Subject to $\mathbf{x}_1 \in X_1$

SUBPROBLEM 2

Maximize $(\mathbf{wA}_2 - \mathbf{c}_2)\mathbf{x}_2 + \alpha_2$

Subject to $\mathbf{x}_2 \in X_2$

Since $\mathbf{w} = (0, 0)$ and $\boldsymbol{\alpha} = (0, 0)$, these reduce to maximizing $2x_1 + x_2 + 0$ and maximizing $3x_3 + x_4 + 0$ over the two regions of Figure 7.5. The optimal solutions are respectively $\mathbf{x}_{12} = (x_1, x_2) = (6, 0)$ with objective 12 and $\mathbf{x}_{22} = (x_3, x_4) = (5, 0)$ with objective 15. Then $(\mathbf{wA}_1 - \mathbf{c}_1)\mathbf{x}_{12} + \alpha_1 = 12$ and $(\mathbf{wA}_2 - \mathbf{c}_2)\mathbf{x}_{22} + \alpha_2 = 15$. Therefore λ_{12} and λ_{22} are both candidates to enter. Select λ_{22} since $z_{22} - \hat{c}_{22} = 15$ is the most positive. Note that a lower bound on the problem is $0 - 12 - 15 = -27$.

MASTER PROBLEM

$$z_{22} - \hat{c}_{22} = 15$$

Form the column

$$\begin{pmatrix} \mathbf{A}_2\mathbf{x}_{22} \\ 0 \\ 1 \end{pmatrix}$$

Note that:

$$\mathbf{A}_2\mathbf{x}_{22} = \begin{bmatrix} 1 & 1 \\ 2 & 1 \end{bmatrix}\begin{bmatrix} 5 \\ 0 \end{bmatrix} = \begin{bmatrix} 5 \\ 10 \end{bmatrix}, \qquad \begin{pmatrix} \mathbf{A}_2\mathbf{x}_{22} \\ 0 \\ 1 \end{pmatrix} = \begin{pmatrix} 5 \\ 10 \\ 0 \\ 1 \end{pmatrix}$$

This column is updated by premultiplying by $\mathbf{B}^{-1} = \mathbf{I}$. Insert this column along with $z_{22} - \hat{c}_{22}$ in the master array. Update the current basic feasible solution by pivoting. After completion, throw column λ_{22} away.

	BASIS INVERSE				RHS		λ_{22}
z	0	0	0	0	0		15
s_1	1	0	0	0	6		5
s_2	0	1	0	0	4		(10)
λ_{11}	0	0	1	0	1		0
λ_{21}	0	0	0	1	1		1

	BASIS INVERSE				RHS
z	0	$-\frac{3}{2}$	0	0	-6
s_1	1	$-\frac{1}{2}$	0	0	4
λ_{22}	0	$\frac{1}{10}$	0	0	$\frac{2}{5}$
λ_{11}	0	0	1	0	1
λ_{21}	0	$-\frac{1}{10}$	0	1	$\frac{3}{5}$

Note that $w_1 = 0$, $w_2 = -\frac{3}{2}$, $\alpha_1 = \alpha_2 = 0$.

Iteration 2

Since s_2 just left the basis, it will not be a candidate to immediately reenter. Solve the following two subproblems.

SUBPROBLEM 1

Maximize $(\mathbf{w}\mathbf{A}_1 - \mathbf{c}_1)\mathbf{x}_1 + \alpha_1$

Subject to $\mathbf{x}_1 \in X_1$

SUBPROBLEM 2

Maximize $(\mathbf{w}\mathbf{A}_2 - \mathbf{c}_2)\mathbf{x}_2 + \alpha_2$

Subject to $\mathbf{x}_2 \in X_2$

These problems reduce to the following.

Maximize $2x_1 - \frac{1}{2}x_2 + 0$

Subject to $(x_1, x_2) \in X_1$

Maximize $0x_3 - \frac{1}{2}x_4 + 0$

Subject to $(x_3, x_4) \in X_2$

The optimal solutions are respectively $\mathbf{x}_{13} = (x_1, x_2) = (6, 0)$ with objective $z_{13} - \hat{c}_{13} = (\mathbf{w}\mathbf{A}_1 - \mathbf{c}_1)\mathbf{x}_{13} + \alpha_1 = 12$ and $\mathbf{x}_{23} = (x_3, x_4) = (5, 0)$ with objective $z_{23} - \hat{c}_{23} = (\mathbf{w}\mathbf{A}_2 - \mathbf{c}_2)\mathbf{x}_{23} + \alpha_2 = 0$. Thus there is no candidate from subproblem 2 at this time and λ_{13} is a candidate to enter the master basis. The lower bound available at this stage is $-6 - 12 - 0 = -18$.

MASTER PROBLEM

$$z_{13} - \hat{c}_{13} = 12$$

Form the column

$$\begin{pmatrix} \mathbf{A}_1\mathbf{x}_{13} \\ 1 \\ 0 \end{pmatrix}$$

Note that

$$\mathbf{A}_1\mathbf{x}_{13} = \begin{bmatrix} 1 & 1 \\ 0 & 1 \end{bmatrix}\begin{bmatrix} 6 \\ 0 \end{bmatrix} = \begin{bmatrix} 6 \\ 0 \end{bmatrix}, \qquad \begin{pmatrix} \mathbf{A}_1\mathbf{x}_{13} \\ 1 \\ 0 \end{pmatrix} = \begin{pmatrix} 6 \\ 0 \\ 1 \\ 0 \end{pmatrix}$$

Updating this column, we get

$$\mathbf{y}_{13} = \mathbf{B}^{-1}\begin{pmatrix} \mathbf{A}_1\mathbf{x}_{13} \\ 1 \\ 0 \end{pmatrix} = \begin{bmatrix} 6 \\ 0 \\ 1 \\ 0 \end{bmatrix}$$

Insert this column along with $z_{13} - \hat{c}_{13} = 12$ into the master array. Update the current basic feasible solution by pivoting and then discard column λ_{13}.

	BASIS INVERSE				RHS	λ_{13}
z	0	$-\frac{3}{2}$	0	0	-6	12
s_1	1	$-\frac{1}{2}$	0	0	4	(6)
λ_{22}	0	$\frac{1}{10}$	0	0	$\frac{2}{5}$	0
λ_{11}	0	0	1	0	1	1
λ_{21}	0	$-\frac{1}{10}$	0	1	$\frac{3}{5}$	0

	BASIS INVERSE				RHS
z	-2	$-\frac{1}{2}$	0	0	-14
λ_{13}	$\frac{1}{6}$	$-\frac{1}{12}$	0	0	$\frac{2}{3}$
λ_{22}	0	$\frac{1}{10}$	0	0	$\frac{2}{5}$
λ_{11}	$-\frac{1}{6}$	$\frac{1}{12}$	1	0	$\frac{1}{3}$
λ_{21}	0	$-\frac{1}{10}$	0	1	$\frac{3}{5}$

Note that $w_1 = -2$, $w_2 = -\frac{1}{2}$, $\alpha_1 = \alpha_2 = 0$.

Iteration 3

Since $w_1 < 0$ and $w_2 < 0$, neither s_1 nor s_2 are candidates to enter the basis. Solve the following two subproblems.

SUBPROBLEM 1	SUBPROBLEM 2
Maximize $(\mathbf{wA}_1 - \mathbf{c}_1)\mathbf{x}_1 + \alpha_1$	Maximize $(\mathbf{wA}_2 - \mathbf{c}_2)\mathbf{x}_2 + \alpha_2$
Subject to $\mathbf{x}_1 \in X_1$	Subject to $\mathbf{x}_2 \in X_2$

These problems reduce to the following.

Maximize $0x_1 - \frac{3}{2}x_2 + 0$	Maximize $0x_3 - \frac{3}{2}x_4 + 0$
Subject to $(x_1, x_2) \in X_1$	Subject to $(x_3, x_4) \in X_2$

From Figure 7.5, $\mathbf{x}_{14} = (x_1, x_2) = (0, 0)$ with objective 0 and $\mathbf{x}_{24} = (x_3, x_4) = (0, 0)$ with objective 0 are optimal solutions. Then $(\mathbf{wA}_1 - \mathbf{c}_1)\mathbf{x}_{14} + \alpha_1 = (\mathbf{wA}_2 - \mathbf{c}_2)\mathbf{x}_{24} + \alpha_2 = 0$ and the optimal solution is reached. The lower and upper bounds now match. From the master problem the optimal point $\mathbf{x}^*$ is given by

$$\begin{pmatrix} x_1 \\ x_2 \end{pmatrix} = \lambda_{11}\mathbf{x}_{11} + \lambda_{13}\mathbf{x}_{13}$$

$$= \tfrac{1}{3}\begin{pmatrix} 0 \\ 0 \end{pmatrix} + \tfrac{2}{3}\begin{pmatrix} 6 \\ 0 \end{pmatrix} = \begin{pmatrix} 4 \\ 0 \end{pmatrix}$$

$$\begin{pmatrix} x_3 \\ x_4 \end{pmatrix} = \lambda_{21}\mathbf{x}_{21} + \lambda_{22}\mathbf{x}_{22}$$

$$= \tfrac{3}{5}\begin{pmatrix} 0 \\ 0 \end{pmatrix} + \tfrac{2}{5}\begin{pmatrix} 5 \\ 0 \end{pmatrix} = \begin{pmatrix} 2 \\ 0 \end{pmatrix}$$

Therefore $\mathbf{x}^* = (x_1, x_2, x_3, x_4) = (4, 0, 2, 0)$ with objective -14.

Economic Interpretation

The decomposition algorithm has an interesting economic interpretation. Consider the case of a large system that is composed of smaller subsystems $1, 2, \ldots, T$. Each subsystem i has its own objective, and the objective function of the overall system is the sum of the objective functions of the subsystems. Each subsystem has its constraints designated by the set X_i, which is assumed to be bounded for the purpose of simplification. In addition, all the subsystems share a few common resources, and hence the consumption of these resources by all the subsystems must not exceed the availability given by the vector $\mathbf{b}$.

Recall the following economic interpretation of the dual variables (Lagrangian multipliers). Here w_i is the rate of change of the objective as a function of b_i; that is, if b_i is replaced by $b_i + \Delta$, holding the current nonbasic variables in the master problem at zero, then the objective is modified by adding $w_i\Delta$. Hence $-w_i$ can be thought of as the price of consuming one unit of the ith common resource. Similarly, $-\alpha_i$ can be thought of as the price of consuming a portion of the ith convexity constraint.

With this in mind, the decomposition algorithm can be interpreted as follows. With the current proposals of the subsystems, the superordinate (total system) obtains the optimal weights of these proposals and announces a set of prices for using the common resources. These prices are passed down to the subsystems, which modify their proposals according to these new prices. A typical subsystem i solves the following subproblem.

$$\begin{aligned} &\text{Maximize} \quad (\mathbf{wA}_i - \mathbf{c}_i)\mathbf{x}_i + \alpha_i \\ &\text{Subject to} \quad \mathbf{x}_i \in X_i \end{aligned}$$

or equivalently

$$\begin{aligned} &\text{Minimize} \quad (\mathbf{c}_i - \mathbf{wA}_i)\mathbf{x}_i - \alpha_i \\ &\text{Subject to} \quad \mathbf{x}_i \in X_i \end{aligned}$$

The original objective of subsystem i is $\mathbf{c}_i\mathbf{x}_i$. The term $-\mathbf{wA}_i\mathbf{x}_i$ reflects the indirect price of using the common resources. Note that $\mathbf{A}_i\mathbf{x}_i$ is the amount of the common resources consumed by the $\mathbf{x}_i$ proposal. Since the price of using these resources is $-\mathbf{w}$, then the indirect cost of using them is $-\mathbf{wA}_i\mathbf{x}_i$, and the total cost is $(\mathbf{c}_i - \mathbf{wA}_i)\mathbf{x}_i$. Note that the term $-\mathbf{wA}_i\mathbf{x}_i$ makes proposals that use much of the common resources unattractive from a cost point of view. Subsystem i announces an optimal proposal $\mathbf{x}_{ik}$. If this proposal is to be considered, then the weight of the older proposals $\mathbf{x}_{ij}$'s must decrease in order to "make room" for this proposal; that is, $\Sigma_j \lambda_{ij}$ must decrease from its present level of 1. The resulting saving is precisely α_i. If the cost of introducing the proposal $\mathbf{x}_{ik}$ is less than the saving realized; that is, if $(\mathbf{c}_i - \mathbf{wA}_i)\mathbf{x}_{ik} - \alpha_i < 0$, or $(\mathbf{wA}_i - \mathbf{c}_i)\mathbf{x}_{ik} + \alpha_i > 0$, then the superordinate would consider this new proposal. After all the subsystems introduce their new proposals, the superordinate calculates the optimum mix of these proposals and passes down new prices. The process is repeated until none of the subsystems has a new attractive proposal; that is, when $(\mathbf{c}_i - \mathbf{wA}_i)\mathbf{x}_{ik} - \alpha_i \geq 0$ for each i.

7.6 DUALITY AND RELATIONSHIPS WITH OTHER DECOMPOSITION PROCEDURES

The Dantzig–Wolfe decomposition procedure discussed in this chapter is equivalent for linear programming problems to two other well known partitioning/decomposition/relaxation techniques, namely the Benders' decomposition and the Lagrangian relaxation methods. To expose this relationship, consider a linear programming problem P of the following form. (The choice of equalities and inequalities are only for the sake of illustration; other forms may be readily handled.)

$$\begin{aligned} \text{P:} \quad &\text{Minimize} \quad \mathbf{cx} \\ &\text{Subject to} \quad \mathbf{Ax} = \mathbf{b} \\ &\qquad\qquad\quad \mathbf{x} \in X = \{\mathbf{x} : \mathbf{Dx} \geq \mathbf{d}, \mathbf{x} \geq \mathbf{0}\} \end{aligned}$$

For convenience, let us assume that X is nonempty and bounded. (We ask the reader in Exercise 7.33 to explore the following connections for the case of unbounded X.)

Now let us write the dual D to problem P. We designate w and v as the dual variables associated with the constraints $\mathbf{Ax} = \mathbf{b}$ and $\mathbf{Dx} \geq \mathbf{d}$ respectively.

$$\begin{array}{lll} \text{D:} & \text{Maximize} & \mathbf{wb} + \mathbf{vd} \\ & \text{Subject to} & \mathbf{wA} + \mathbf{vD} \leq \mathbf{c} \\ & & \mathbf{w} \quad \text{unrestricted}, \ \mathbf{v} \geq \mathbf{0} \end{array}$$

Observe that when $\mathbf{w}$ is fixed at some arbitrary value, we obtain a linear programming problem in the variables $\mathbf{v}$. In particular this linear program may be specially structured or easy to solve. Assuming this to be our motivation, let us proceed by *partitioning* problem D, while treating the variables $\mathbf{w}$ as *complicating* variables as follows:

$$D: \quad \underset{\mathbf{w}\text{ unres}}{\text{Maximize}} \left\{ \begin{array}{l} \mathbf{wb} + \text{Maximum } \mathbf{vd} \\ \qquad \text{Subject to } \mathbf{vD} \leq \mathbf{c} - \mathbf{wA} \\ \qquad\qquad \mathbf{v} \geq \mathbf{0} \end{array} \right\}$$

$$= \underset{\mathbf{w}\text{ unres}}{\text{Maximum}} \left\{ \mathbf{wb} + \underset{\mathbf{x}\in X}{\text{Minimum}} (\mathbf{c} - \mathbf{wA})\mathbf{x} \right\} \tag{7.13}$$

Here we have written the dual to the *inner optimization problem* at the last step, noting the definition of X. Since X is assumed to be nonempty and bounded, the inner minimization problem in Equation (7.13) attains an extreme point optimal solution. Denoting $\mathbf{x}_1, \ldots, \mathbf{x}_t$ as the vertices of X as before, we have that D is equivalent to the problem of maximizing $\{\mathbf{wb} + \text{Minimum}_{j=1,\ldots,t}(\mathbf{c} - \mathbf{wA})\mathbf{x}_j\}$ over unrestricted values of $\mathbf{w}$. Denoting z as the objective function in $\{\cdot\}$, this may be rewritten as the following *Master Problem MP*:

$$\begin{array}{lll} \text{MP:} & \text{Maximize} & z \\ & \text{Subject to} & z \leq \mathbf{wb} + (\mathbf{c} - \mathbf{wA})\mathbf{x}_j \qquad \text{for } j = 1, \ldots, t \\ & & z, \mathbf{w} \quad \text{unrestricted} \end{array} \tag{7.14}$$

Note that MP is inconvenient to solve directly because it typically has far too many constraints. Hence we can adopt a *relaxation strategy*, in which only a few of the constraints in Equation (7.14) are explicitly maintained. Suppose that for such a *relaxed master program* we obtain an optimal solution $(\bar{z}, \bar{\mathbf{w}})$. Then $\bar{z}$ is an upper bound on the optimal value to the original problem (why?). Furthermore $(\bar{z}, \bar{\mathbf{w}})$ is optimal for MP if and only if $(\bar{z}, \bar{\mathbf{w}})$ is feasible to all constraints in Equation (7.14) (why?). Hence in order to check if any constraints in Equation (7.14) are violated, we wish to check if $\bar{z} \leq \bar{\mathbf{w}}\mathbf{b} + (\mathbf{c} - \bar{\mathbf{w}}\mathbf{A})\mathbf{x}_j$ for all $j = 1, \ldots, t$, that is, if $\bar{z} \leq \bar{\mathbf{w}}\mathbf{b} + \text{Minimum}_{j=1,\ldots,t}\{(\mathbf{c} - \bar{\mathbf{w}}\mathbf{A})\mathbf{x}_j\}$. However the latter problem is equivalent to the linear programming *subproblem*

$$\overline{\mathbf{w}}\mathbf{b} + \underset{\mathbf{x} \in \mathbf{X}}{\text{Minimum}} \{(\mathbf{c} - \overline{\mathbf{w}}\mathbf{A})\mathbf{x}\} \tag{7.15}$$

If $\bar{z}$ is less than or equal to the optimal objective value in Equation (7.15), then we are done. (In fact, in this case, $\bar{z}$ will be *equal* to the optimal value in Equation (7.15) since $\bar{z}$ is equal to $\overline{\mathbf{w}}\mathbf{b} + (\mathbf{c} - \overline{\mathbf{w}}\mathbf{A})\mathbf{x}_j$ for some $j \in \{1, \ldots, t\}$ in Equation (7.14).) Otherwise if $\mathbf{x}_k$ solves Problem (7.15), we have $\bar{z} > \overline{\mathbf{w}}\mathbf{b} + (\mathbf{c} - \overline{\mathbf{w}}\mathbf{A})\mathbf{x}_k$, and we can generate the constraint $z \leq \mathbf{wb} + (\mathbf{c} - \mathbf{wA})\mathbf{x}_k$ and add it to the current (relaxed) master program and reoptimize. This process may be repeated until the solution $(\bar{\mathbf{z}}, \overline{\mathbf{w}})$ to some relaxed master problem gives $\bar{z}$ equal to the optimal value in Equation (7.15). This must occur finitely since X has only a finite number of vertices.

The foregoing procedure is known as *Benders' partitioning or decomposition technique*. (Its generalization to nonlinear or discrete problem structures is addressed in Exercise 7.34 and in the Notes and References section.) Problem (7.14) (or its relaxation) is referred to as *Benders' (relaxed) master problem*, and Problem (7.15) is known as *Benders' subproblem*. However note that Problem (7.15) is also the subproblem solved by the Dantzig–Wolfe decomposition method, and that the Benders' master program (7.14) is simply the dual to the Dantzig–Wolfe master problem with constraints (7.1)–(7.3). (Here the nonnegative dual multipliers associated with the constraints (7.14) are λ_j, $j = 1, \ldots, t$; these variables sum to unity by virtue of the column of the variable z in Equation (7.14).) Therefore Benders' algorithm is called a *row generation technique* in contrast with the Dantzig–Wolfe *column generation* procedure. It follows that if at the end of each relaxed master program solution, one maintains only those generated rows in Equation (7.14) that have currently nonbasic slack variables, and so whose complementary dual variables λ_j are basic, then the resulting algorithm is precisely the same as the decomposition algorithm of Section 7.1. In particular, the starting solution technique of Section 7.3 provides an initial bounded Benders' master program. Also when Benders' algorithm terminates, the available optimal primal solution may be computed as the sum of the products of $\mathbf{x}_j$ and the corresponding dual variable λ_j, which are associated with the explicitly present constraints (7.14) in the final Benders' relaxed master problem. (Note that this sum is a convex combination.) Furthermore note that if we did not delete any of the constraints generated by Benders' algorithm and solved each relaxed Benders' master program to optimality, the equivalent implementation for the Dantzig–Wolfe method would be to optimize each master program over all the λ_j-variable columns generated thus far.

Observe that the foregoing discussion also indicates how to recover an optimal set of dual multipliers $(\mathbf{w}^*, \mathbf{v}^*)$ for the problem P after solving it using the Dantzig–Wolfe decomposition algorithm. The vector $\mathbf{w}^*$ is directly available as the optimal set of multipliers associated with the constraints (7.1). The corresponding solution $\mathbf{v}^*$ is obtainable from the first problem in Equation (7.13) after fixing $\mathbf{w}$ at $\mathbf{w}^*$. Hence if the second inner minimization subproblem in Equation (7.13) is solved with $\mathbf{w} = \mathbf{w}^*$, then $\mathbf{v}^*$ is obtained as the set of optimal multipliers associated with the constraints in X. For instance, in the example of Section 7.2, we obtain $(w_1^*, w_2^*) = (-1, -1)$. Furthermore, with $\mathbf{w} = \mathbf{w}^*$ fixed, the (final) subproblem solved is to maximize $\{-3x_4 : \mathbf{x} \in X\}$, which yields an optimal set of dual multipliers $\mathbf{v}^* = (0, 0, 0, 0)$ for constraints 3, 4, 5, and 6 in the example.

It may also be noted that the Dantzig–Wolfe and the Benders' decomposition methods are respectively classified as *price directive* and *resource directive* schemes. The former type of procedure conducts the coordination in the decomposition method by adjusting *prices* or objective function coefficients via Lagrangian multipliers. On the other hand, the latter type of procedure is dual to this scheme, and conducts this coordination by adjusting the common *resource* availabilities (right-hand-sides) by fixing certain variable values, for example.

There is another general price-directive optimization strategy for P that finds an equivalence with the foregoing methods when solving linear programming problems, namely the *Lagrangian relaxation technique*. (See Exercise 6.73.) Observe from Equation (7.13) that if we denote the function in $\{\cdot\}$, which is a function of $\mathbf{w}$, as $\theta(\mathbf{w})$, we can equivalently state D as the problem:

$$\text{Maximize}\{\theta(\mathbf{w}) : \mathbf{w} \text{ unrestricted}\} \tag{7.16}$$

where

$$\theta(\mathbf{w}) = \mathbf{wb} + \text{Minimum}\{(\mathbf{c} - \mathbf{wA})\mathbf{x} : \mathbf{x} \in X\}. \tag{7.17}$$

Problem (7.16) is known as a *Lagrangian dual* to Problem P. Problem (7.17) is the associated *Lagrangian subproblem*. Note that in Problem (7.17), the constraints $\mathbf{Ax} = \mathbf{b}$ have been accommodated into the objective function of P via the Lagrangian multipliers $\mathbf{w}$. By the foregoing discussion, it is evident that the maximum value in Problem (7.16) matches with the optimal objective value for P, and hence $\theta(\mathbf{w})$ provides a lower bound for this optimal value for any arbitrary $\mathbf{w}$. This is why Problem (7.16) is called a "Lagrangian dual" problem. Furthermore since $\theta(\mathbf{w}) = \mathbf{wb} + \text{Minimum}_{j=1,\ldots,t}\{(\mathbf{c} - \mathbf{wA})\mathbf{x}_j\}$, $\theta(\mathbf{w})$ is a piecewise linear and concave function of $\mathbf{w}$ (why?). Hence Problem (7.16) seeks an unconstrained maximum for this function. Although several nonlinear search methods for nonsmooth optimization problems may be used to solve Problem (7.16), the Benders' procedure is evidently one viable solution technique. This cutting plane or row generation scheme can be viewed as a tangential approximation method for solving Problem (7.16), in which only a few of the tangential supports or segments $z \le \mathbf{wb} + (\mathbf{c} - \mathbf{wA})\mathbf{x}_j$, $j = 1, \ldots, t$ that describe $\theta(\cdot)$ are generated. z is the value measured along the $\theta(\cdot)$ axis. Using this viewpoint, the relaxed master problem MP from Equation (7.14) employs an *outer-linearization* of Problem (7.16), using only a subset of the tangential supports that actually describe the piecewise linear function $\theta(\cdot)$. If $\bar{\mathbf{w}}$ solves this problem with objective value $\bar{z}$, then $\bar{z}$ is an upper bound on the optimal value of Problem (7.16). However, if $\bar{z} = \theta(\bar{\mathbf{w}})$ itself, then $\bar{\mathbf{w}}$ is optimal in Problem (7.16) (why?). Otherwise $\bar{z} > \theta(\bar{\mathbf{w}}) \equiv \bar{\mathbf{w}}\mathbf{b} + \text{Minimum}\{(\mathbf{c} - \bar{\mathbf{w}}\mathbf{A})\mathbf{x} : \mathbf{x} \in X\} = \bar{\mathbf{w}}\mathbf{b} + (\mathbf{c} - \bar{\mathbf{w}}\mathbf{A})\mathbf{x}_k$, say, as produced by the Subproblem (7.17), which coincides with the Benders' Subproblem (7.15). Hence the tangential support $z \le \mathbf{wb} + (\mathbf{c} - \mathbf{wA})\mathbf{x}_k$ is generated and included in the outer-linearization of $\theta(\cdot)$, and the procedure repeats until termination is obtained. At termination, the optimal primal solution may be recovered by weighting each $\mathbf{x}_j$ in Equation (7.14) by the corresponding dual multiplier λ_j, if positive at optimality, and summing.

Example 7.3

Consider the linear programming problem given in the example of Section 7.2. Denote the dual variables associated with the constraints as $(w_1, w_2, v_1, v_2, v_3, v_4)$ and write the dual as follows.

$$\begin{array}{llllllll}
\text{Maximize} & 2w_1 + 3w_2 + 2v_1 + 5v_2 + 2v_3 + 6v_4 \\
\text{Subject to} & w_1 + w_2 + v_1 + v_2 & \le -2 \\
& w_2 + 2v_2 & \le -1 \\
& w_1 - v_3 + 2v_4 & \le -1 \\
& 2w_2 + v_3 + v_4 & \le 1 \\
& \mathbf{w} \le \mathbf{0}, \mathbf{v} \le \mathbf{0}
\end{array}$$

Treating $\mathbf{w}$ as the complicating variables, we obtain the Benders' master problem in this case as follows:

$$\begin{array}{ll}
\text{Maximize} & z \\
\text{Subject to} & z \le 2w_1 + 3w_2 + x_{j1}(-2 - w_1 - w_2) \\
& \quad + x_{j2}(-1 - w_2) + x_{j3}(-1 - w_1) + x_{j4}(1 - 2w_2) \\
& \qquad \text{for } j = 1, \ldots, t \\
& z \ \text{unrestricted}, \mathbf{w} \le \mathbf{0}
\end{array}$$

where $\mathbf{x}_j = (x_{j1}, x_{j2}, x_{j3}, x_{j4})$, $j = 1, \ldots, t$ are the vertices of X as defined in Equation (7.2). The corresponding Lagrangian dual problem is to

$$\text{Maximize}\{\theta(\mathbf{w}) : \mathbf{w} \le \mathbf{0}\}$$

where, for a given $\mathbf{w} = (w_1, w_2)$, we have

$$\theta(\mathbf{w}) = 2w_1 + 3w_2 + \underset{\mathbf{x} \in X}{\text{Minimum}}\{(-2 - w_1 - w_2)x_1 + (-1 - w_2)x_2 + (-1 - w_1)x_3 + (1 - 2w_2)x_4\}$$

Note that the evaluation of $\theta(\mathbf{w})$, given $\mathbf{w}$, is precisely the Benders' subproblem.

Suppose that as in Section 7.2 we begin with the vertex $\mathbf{x}_1 = (0, 0, 0, 0)$ of X. Hence the relaxed Benders' master problem is of the form

$$\begin{array}{ll}
\text{Maximize} & z \\
\text{Subject to} & z \le 2w_1 + 3w_2 \\
& \mathbf{w} \le \mathbf{0}
\end{array}$$

The optimal solution is $\bar{z} = 0$, $\bar{\mathbf{w}} = (0, 0)$. Hence using only the tangential support $z \le 2w_1 + 3w_2$ for $\theta(\cdot)$, we obtain $\bar{\mathbf{w}} = (0, 0)$ as the maximizing solution. Solving the Benders' subproblem with $\bar{\mathbf{w}} = (0, 0)$, that is, computing $\theta(\bar{\mathbf{w}})$, we get

$\theta(\overline{\mathbf{w}}) = -\frac{17}{2}$, realized at $\mathbf{x}_2 = (2, \frac{3}{2}, 3, 0)$. Since $\theta(\overline{\mathbf{w}}) < \bar{z}$, we generate a second Benders' cut or tangential support using $\mathbf{x}_2$:

$$z \leq 2w_1 + 3w_2 + 2(-2 - w_1 - w_2) + \tfrac{3}{2}(-1 - w_2) + 3(-1 - w_1)$$
$$= -\tfrac{17}{2} - 3w_1 - \tfrac{1}{2}w_2$$

This leads to the Benders' master program

$$\begin{aligned} &\text{Maximize} && z \\ &\text{Subject to} && z \leq 2w_1 + 3w_2 \\ &&& z \leq -\tfrac{17}{2} - 3w_1 - \tfrac{1}{2}w_2 \\ &&& \mathbf{w} \leq \mathbf{0} \end{aligned}$$

The optimal solution is $\bar{z} = -\frac{17}{5}$, and $\overline{\mathbf{w}} = (-\frac{17}{10}, 0)$. The slack variables in both constraints are nonbasic, and hence both constraints are maintained. (The respective dual variables are $\frac{3}{5}$ and $\frac{2}{5}$.) Solving the Benders' subproblem, we compute $\theta(\overline{\mathbf{w}}) = -\frac{59}{10}$ realized at $\mathbf{x}_3 = (0, \frac{5}{2}, 0, 0)$. Since $\bar{z} > \theta(\overline{\mathbf{w}})$, we generate a third Benders' constraint or tangential support as $z \leq 2w_1 + 3w_2 + \frac{5}{2}(-1 - w_2) = -\frac{5}{2} + 2w_1 + \frac{1}{2}w_2$. Appending this to the master problem, we obtain

$$\begin{aligned} &\text{Maximize} && z \\ &\text{Subject to} && z \leq 2w_1 + 3w_2 \\ &&& z \leq -\tfrac{17}{2} - 3w_1 - \tfrac{1}{2}w_2 \\ &&& z \leq -\tfrac{5}{2} + 2w_1 + \tfrac{1}{2}w_2 \\ &&& \mathbf{w} \leq \mathbf{0} \end{aligned}$$

The optimal solution to this problem is $\bar{z} = -\frac{49}{10}$, and $\overline{\mathbf{w}} = (-\frac{6}{5}, 0)$ with the slack in the first constraint being basic. (The optimal dual multipliers are $\lambda_1 = 0$, $\lambda_2 = \frac{2}{5}$, and $\lambda_3 = \frac{3}{5}$.) Hence the first constraint may be deleted from the current outer-linearization. (A deleted constraint can possibly be regenerated later.) Observe how these computations relate to the calculations in the numerical example of Section 7.2. We ask the reader in Exercise 7.32 to continue the solution process by next computing $\theta(\overline{\mathbf{w}})$ and to recover primal and dual optimal solutions on completion.

EXERCISES

7.1 Consider the following linear programming problem

$$\begin{aligned} &\text{Maximize} && x_1 \\ &\text{Subject to} && x_1 + 3x_2 \leq \tfrac{9}{4} \\ &&& 2x_1 - 3x_2 \leq 0 \\ &&& \mathbf{x} \in X = \{(x_1, x_2) : 0 \leq x_1 \leq 1, 0 \leq x_2 \leq 1\} \end{aligned}$$

a. Solve this problem graphically.
b. Solve using the Dantzig–Wolfe decomposition method. Comment on the number of extreme points of X used at optimality in terms of Remark 6 of Section 7.1
c. Does the Dantzig–Wolfe master program (7.1)–(7.3) have alternative optimal solutions? Does the original problem have alternative optimal solutions? Explain.

7.2 Solve the following problem using the decomposition technique with one convexity constraint. Show the progress of the lower bound and primal objective. Obtain primal and dual solutions.

$$\begin{array}{ll}\text{Minimize} & -x_1 - x_2 - 2x_3 - x_4 \\ \text{Subject to} & x_1 + 2x_2 + 2x_3 + x_4 \leq 40 \\ & -x_1 + x_2 + x_3 + x_4 \leq 10 \\ & x_1 + 3x_2 \leq 30 \\ & 2x_1 + x_2 \leq 20 \\ & x_3 \leq 10 \\ & x_4 \leq 10 \\ & x_3 + x_4 \leq 15 \\ & x_1, \quad x_2, \quad x_3, \quad x_4 \geq 0\end{array}$$

7.3 Solve the following linear programming problem by the decomposition method using one convexity constraint. Show the progress of the lower bound and primal objective. Obtain primal and dual solutions.

$$\begin{array}{ll}\text{Minimize} & -x_1 - 3x_2 + x_3 - x_4 \\ \text{Subject to} & x_1 + x_2 + x_3 + x_4 \leq 8 \\ & x_1 + x_2 \leq 6 \\ & x_3 + 2x_4 \leq 10 \\ & -x_3 + x_4 \leq 4 \\ & x_1, \quad x_2, \quad x_3, \quad x_4 \geq 0\end{array}$$

7.4 Develop the master and the subproblem using the decomposition technique for the following linear programming problem. Assume that X is polyhedral and has a special structure. Formally state the decomposition algorithm for this case.

$$\begin{array}{ll}\text{Maximize} & \mathbf{cx} + \mathbf{dy} \\ \text{Subject to} & \mathbf{Ax} + \mathbf{Dy} \leq \mathbf{b} \\ & \mathbf{x} \in X\end{array}$$

7.5 Consider the feasible and bounded problem: Minimize $\mathbf{cx}$ subject to $\mathbf{Ax} = \mathbf{b}$, $\mathbf{x} \in X$. Suppose that $\mathbf{w}^*\mathbf{A} - \mathbf{c} = \mathbf{0}$ where $\mathbf{w}^*$ is the optimal Lagrange multiplier vector associated with the constraints $\mathbf{Ax} = \mathbf{b}$. Does this imply that there exists an optimal solution $\mathbf{x}^*$ that belongs to the interior of X? Is the converse of this implication true? Interpret your answers geometrically.

7.6 Consider the problem: Minimize $\mathbf{cx}$ subject to $\mathbf{Ax} = \mathbf{b}$, $\mathbf{Dx} = \mathbf{d}$, $\mathbf{x} \geq \mathbf{0}$. Suppose that $\mathbf{w}_1^*$ and $\mathbf{w}_2^*$ are the optimal dual vectors of the constraints $\mathbf{Ax} = \mathbf{b}$ and $\mathbf{Dx} = \mathbf{d}$ respectively. Consider the problem: Maximize $(\mathbf{w}_1^*\mathbf{A} - \mathbf{c})\mathbf{x}$ subject to $\mathbf{x} \in X = \{\mathbf{x} : \mathbf{Dx} = \mathbf{d}, \mathbf{x} \geq \mathbf{0}\}$. Assume that X is bounded. Let $\mathbf{x}_1^*, \mathbf{x}_2^*, \ldots, \mathbf{x}_k^*$ be optimal solutions of the foregoing problem that are extreme points of the set X. Show that an optimal solution of the original problem can be represented as a convex combination of $\mathbf{x}_1^*, \mathbf{x}_2^*, \ldots, \mathbf{x}_k^*$, that is,

$$\mathbf{x}^* = \sum_{j=1}^{k} \lambda_j \mathbf{x}_j^*$$

$$\sum_{j=1}^{k} \lambda_j = 1$$

$$\lambda_j \geq 0 \qquad j = 1, 2, \ldots, k$$

Is this true if X is unbounded? Prove this assertion or give a counterexample and modify the representation statement.

7.7 Consider the following problem.

$$\begin{array}{ll} \text{Minimize} & 2x_1 - 3x_2 - 4x_3 \\ \text{Subject to} & x_1 + 2x_2 + x_3 \leq 6 \\ & x_1 + x_2 - 2x_3 \geq 2 \\ & 0 \leq x_1, \quad x_2, \quad x_3 \leq 3 \end{array}$$

a. Set up the problem so that it can be solved by the decomposition algorithm.
b. Find a starting basis in the λ-space.
c. Find the optimal solution by the decomposition algorithm and compare with the bounded simplex method.

7.8 Consider the following (transportation) problem.

$$\begin{array}{llllllllll} \text{Minimize} & x_{11} + 6x_{12} + 4x_{13} + 2x_{14} + 2x_{21} + 3x_{22} + 3x_{23} + 5x_{24} \\ \text{Subject to} & x_{11} + x_{12} + x_{13} + x_{14} & & & & & & & & = 500 \\ & & & & & x_{21} + & x_{22} + & x_{23} + & x_{24} & = 700 \\ & x_{11} & & & & + x_{21} & & & & = 200 \\ & & x_{12} & & & & + x_{22} & & & = 300 \\ & & & x_{13} & & & & + x_{23} & & = 400 \\ & & & & x_{14} & & & & + x_{24} & = 300 \\ & x_{11}, & x_{12}, & x_{13}, & x_{14}, & x_{21}, & x_{22}, & x_{23}, & x_{24} & \geq 0 \end{array}$$

a. Set up the problem so that it can be solved by the decomposition algorithm using four convexity constraints.
b. Find an optimal solution using the decomposition algorithm.

7.9 Use the decomposition principle to solve the following problem.

$$\begin{array}{ll} \text{Maximize} & x_1 + 5x_2 + 4x_3 + 3x_4 \\ \text{Subject to} & 2x_1 + 4x_2 + 5x_3 + 2x_4 \leq 7 \\ & 2x_1 + 3x_2 \leq 6 \\ & 5x_1 + x_2 \leq 5 \\ & 3x_3 + 4x_4 \geq 12 \\ & x_3 \leq 4 \\ & x_4 \leq 3 \\ & x_1, \quad x_2, \quad x_3, \quad x_4 \geq 0 \end{array}$$

7.10 In the text we develop a lower bound on the optimal objective value when the subproblem is bounded. Examine in detail the case when this assumption is relaxed.

7.11 Solve the following problem by the decomposition technique using two convexity constraints.

$$\begin{array}{ll} \text{Maximize} & x_1 + x_2 + 3x_3 - x_4 \\ \text{Subject to} & 2x_1 + x_2 + x_3 + x_4 \leq 12 \\ & -x_1 + x_2 \leq 2 \\ & 3x_1 - 4x_3 \leq 5 \\ & x_3 + x_4 \leq 4 \\ & -x_3 + x_4 \leq 5 \\ & x_1, \quad x_2, \quad x_3, \quad x_4 \geq 0 \end{array}$$

7.12 Solve the following problem by the decomposition technique.

$$\begin{array}{ll} \text{Minimize} & -x_1 - 2x_2 + 3x_3 + x_4 \\ \text{Subject to} & x_1 + 2x_2 + 2x_3 + x_4 \geq 40 \\ & x_1 + x_2 \leq 2 \\ & -x_1 + 2x_2 \leq 2 \\ & x_3 + x_4 \geq 6 \\ & x_1, \quad x_2, \quad x_3, \quad x_4 \geq 0 \end{array}$$

7.13 Consider the problem: Minimize $\mathbf{cx}$ subject to $\mathbf{Ax} = \mathbf{b}$, $\mathbf{x} \in X$. Suppose that X has a block diagonal structure. The decomposition algorithm can be applied by using either one convexity constraint or several convexity constraints, one for each subproblem. Discuss the advantages and disadvantages of both strategies. Which one would you prefer and why?

7.14 Apply the decomposition algorithm to the following problem.

$$\begin{aligned} \text{Minimize} \quad & -2x_1 + x_2 - 5x_3 \\ \text{Subject to} \quad & 6x_1 - x_2 + 3x_3 \leq 4 \\ & 0 \leq x_1, \quad x_2, \quad x_3 \leq 1 \end{aligned}$$

7.15 Suppose that the columns added during each master step of the decomposition algorithm are not deleted. In particular suppose that the master problem at iteration p is as follows.

$$\begin{aligned} \text{Minimize} \quad & \sum_{j=1}^{p} (\mathbf{cx}_j)\lambda_j \\ \text{Subject to} \quad & \sum_{j=1}^{p} (\mathbf{Ax}_j)\lambda_j = \mathbf{b} \\ & \sum_{j=1}^{p} \lambda_j = 1 \\ & \lambda_j \geq 0 \qquad j = 1, 2, \ldots, p \end{aligned}$$

where $\mathbf{x}_1, \ldots, \mathbf{x}_p$ are the extreme points generated so far. Discuss the details of such a decomposition procedure and compare with that of Section 7.1. Illustrate by solving the problem of Section 7.2.

7.16 Referring to Exercise 7.15, consider the following master problem.

$$\begin{aligned} \text{Minimize} \quad & \sum_{j=1}^{p} (\mathbf{cx}_j)\lambda_j \\ \text{Subject to} \quad & \sum_{j=1}^{p} (\mathbf{Ax}_j)\lambda_j = \mathbf{b} \\ & \sum_{j=1}^{p} \lambda_j = 1 \\ & \lambda_j \geq 0 \qquad j = 1, 2, \ldots, p \end{aligned}$$

Write the dual of this master problem. Suppose that the dual problem has been solved instead of this problem; how does the decomposition algorithm proceed? Give a detailed algorithm statement and relate to the discussion in Section 7.6. Show convergence of the procedure and interpret it geometrically.

7.17 Construct both a flow chart and detailed steps of the decomposition algorithm for solving the problem: Minimize $\mathbf{cx}$ subject to $\mathbf{Ax} = \mathbf{b}$, $\mathbf{x} \in X$, where X is not necessarily bounded. Then code the decomposition algorithm in FORTRAN or another computer language. Use your code to solve some of the numerical problems given here.

7.18 Solve the following problem by the decomposition algorithm.

$$\begin{array}{ll} \text{Minimize} & -x_1 - 2x_2 - x_3 - 2x_4 \\ \text{Subject to} & 3x_1 + x_2 + x_3 + x_4 \leq 12 \\ & -x_1 + x_2 \leq 4 \\ & -x_1 + 2x_2 \leq 12 \\ & x_3 + 2x_4 \leq 8 \\ & x_1, \quad x_2, \quad x_3, \quad x_4 \geq 0 \end{array}$$

7.19 Consider the following problem.

$$\begin{array}{ll} \text{Minimize} & \sum_{i=1}^{T} \mathbf{c}_i \mathbf{x}_i \\ \text{Subject to} & \sum_{i=1}^{T} \mathbf{A}_i \mathbf{x}_i = \mathbf{b} \\ & \mathbf{x}_i \in X_i \qquad i = 1, 2, \ldots, T \end{array}$$

Show that any basis in the λ_{ij} space must contain at least one λ_{ij} for each $i = 1, 2, \ldots, T$.

7.20 Solve the following problem by the decomposition algorithm. Use phase I to get started in the λ-space.

$$\begin{array}{ll} \text{Maximize} & 4x_1 + 3x_2 + x_3 \\ \text{Subject to} & x_1 + x_2 + x_3 \leq 6 \\ & x_1 + x_2 \geq 4 \\ & -x_1 + x_2 \geq 2 \\ & x_3 \geq 3 \\ & x_1, \quad x_2, \quad x_3 \geq 0 \end{array}$$

7.21 Is it possible that the decomposition algorithm would generate an optimal nonextreme point of the overall problem in case of alternative optimal solutions? Discuss.
(*Hint*. Consider the following problem and start with the extreme point $(0, 0)$.)

$$\begin{array}{ll} \text{Maximize} & x_1 + x_2 \\ \text{Subject to} & x_1 + x_2 \leq \frac{3}{2} \\ & 0 \leq x_1, \quad x_2 \leq 1 \end{array}$$

7.22 Many options are available while solving the subproblem(s) and the master problem. These include the following.

a. The subproblem is terminated if an extreme point $\mathbf{x}_k$ is found with $z_k - \hat{c}_k > 0$. Then λ_k is introduced in the master problem.

b. Several columns can be generated from the subproblem(s) at each iteration.

c. At least one additional column is added to the master problem without discarding any of the previously generated columns. In this case the master problem reduces to finding the optimal mix of all columns generated so far.

Discuss in detail the foregoing options and compare and contrast them. Elaborate on the advantages and disadvantages of each option.

7.23 Give a detailed analysis of the cases that may be encountered as artificial variables are used to find a starting basis of the master problem. Discuss both the two-phase method and the big-M method.

7.24 Consider the following problem.

$$\text{Minimize} \quad \mathbf{c}_0\mathbf{x}_0 + \mathbf{c}_1\mathbf{x}_1 + \mathbf{c}_2\mathbf{x}_2 + \cdots + \mathbf{c}_T\mathbf{x}_T$$

$$\begin{array}{llllll} \text{Subject to} & \mathbf{D}_0\mathbf{x}_0 + \mathbf{A}_1\mathbf{x}_1 + \mathbf{A}_2\mathbf{x}_2 + \cdots + \mathbf{A}_T\mathbf{x}_T & = \mathbf{b}_0 \\ & \mathbf{D}_1\mathbf{x}_0 + \mathbf{B}_1\mathbf{x}_1 & = \mathbf{b}_1 \\ & \mathbf{D}_2\mathbf{x}_0 \qquad + \mathbf{B}_2\mathbf{x}_2 & = \mathbf{b}_2 \\ & \vdots \qquad\qquad \ddots & \vdots \\ & \mathbf{D}_T\mathbf{x}_0 \qquad\qquad + \mathbf{B}_T\mathbf{x}_T & = \mathbf{b}_T \\ & \mathbf{x}_0, \quad \mathbf{x}_1, \quad \mathbf{x}_2, \ldots, \quad \mathbf{x}_T & \geq \mathbf{0} \end{array}$$

Describe in detail how the decomposition technique can be used to solve problems of the foregoing structure. (*Hint*. Let the first set of constraints be the constraints of the master problem. The subproblem consists of the remaining constraints. Take the dual of the subproblem and solve it by decomposition. This becomes a "three-level" algorithm. The subproblem is said to have a dual-angular structure.)

7.25 Consider the following staircase structured linear program:

$$\text{Minimize} \quad \mathbf{c}_1\mathbf{x}_1 + \mathbf{c}_2\mathbf{x}_2 + \mathbf{c}_3\mathbf{x}_3 + \mathbf{c}_4\mathbf{x}_4 + \cdots + \mathbf{c}_T\mathbf{x}_T$$

Subject to

$$\begin{array}{ll} \mathbf{A}_1\mathbf{x}_1 & = \mathbf{b}_1 \\ \mathbf{B}_1\mathbf{x}_1 + \mathbf{A}_2\mathbf{x}_2 & = \mathbf{b}_2 \\ \qquad \mathbf{B}_2\mathbf{x}_2 + \mathbf{A}_3\mathbf{x}_3 & = \mathbf{b}_3 \\ \qquad\qquad \mathbf{B}_3\mathbf{x}_3 + \mathbf{A}_4\mathbf{x}_4 & = \mathbf{b}_4 \\ \qquad\qquad\qquad \ddots & \vdots \\ \qquad\qquad\qquad \mathbf{B}_{T-1}\mathbf{x}_{T-1} + \mathbf{A}_T\mathbf{x}_T & = \mathbf{b}_T \\ \mathbf{x}_1, \quad \mathbf{x}_2, \quad \mathbf{x}_3, \quad \mathbf{x}_4, \ldots, \quad \mathbf{x}_{T-1}, \quad \mathbf{x}_T & \geq 0 \end{array}$$

a. Writing $\mathbf{x}_1$ as a convex combination (with weights λ_{1j}, $j = 1, \ldots, E_1$) of vertices in $X_1 = \{\mathbf{x} : \mathbf{A}_1\mathbf{x}_1 = \mathbf{b}_1, \mathbf{x}_1 \geq \mathbf{0}\}$, assumed to be nonempty and bounded, use the decomposition principle on the problem.

b. For the master program obtained in part (a), define X_2 as the set of λ_{1j}, $j = 1, \ldots, E_1$ and $\mathbf{x}_2$ that satisfy the constraints involving only these variables (that is, not involving the variables $\mathbf{x}_3, \ldots, \mathbf{x}_T$). Use the decomposition principle on this master problem with subproblems solved over X_2 to obtain a *second-level nested decomposition* of the problem.

c. Continuing in this fashion, propose a multilevel nested decomposition algorithm for this problem.

7.26 Consider the following problem.

$$\begin{array}{ll} \text{Minimize} & \mathbf{c}_1\mathbf{x}_1 + \mathbf{c}_2\mathbf{x}_2 + \cdots + \mathbf{c}_T\mathbf{x}_T \\ \text{Subject to} & \mathbf{A}_1\mathbf{x}_1 + \mathbf{A}_2\mathbf{x}_2 + \cdots + \mathbf{A}_T\mathbf{x}_T \leq \mathbf{b} \\ & \mathbf{B}_1\mathbf{x}_1 \qquad\qquad\qquad\qquad\quad \leq \mathbf{b}_1 \\ & \qquad\quad \mathbf{B}_2\mathbf{x}_2 \qquad\qquad\quad \leq \mathbf{b}_2 \\ & \qquad\qquad\qquad \ddots \qquad\qquad \vdots \\ & \qquad\qquad\qquad\qquad \mathbf{B}_T\mathbf{x}_T \leq \mathbf{b}_T \\ & \mathbf{x}_1, \quad \mathbf{x}_2, \ldots, \quad \mathbf{x}_T \geq \mathbf{0} \end{array}$$

The following implementation of the decomposition principle is a possibility. The subproblem constraints are

$$\begin{array}{c} \mathbf{A}_1\mathbf{x}_1 + \mathbf{A}_2\mathbf{x}_2 + \cdots + \mathbf{A}_T\mathbf{x}_T \leq \mathbf{b} \\ \mathbf{x}_1, \quad \mathbf{x}_2, \ldots, \quad \mathbf{x}_T \geq \mathbf{0} \end{array}$$

and the master constraints are $\mathbf{B}_1\mathbf{x}_1 \leq \mathbf{b}_1$, $\mathbf{B}_2\mathbf{x}_2 \leq \mathbf{b}_2, \ldots, \mathbf{B}_T\mathbf{x}_T \leq \mathbf{b}_T$. Describe the details of such an algorithm. What are the advantages and the disadvantages of this procedure? Does this scheme have an economic interpretation? Use the scheme to solve Exercise 7.12.

7.27 Consider the following problem.

$$\begin{array}{lll} \text{Minimize} & \displaystyle\sum_{j=1}^{T} \mathbf{c}_j\mathbf{x}_j + \sum_{j=1}^{T} \mathbf{d}_j\mathbf{y}_j & \\ \text{Subject to} & \mathbf{x}_{j-1} - \mathbf{x}_j + \mathbf{A}_j\mathbf{y}_j = \mathbf{b}_j & j = 1, 2, \ldots, T \\ & \mathbf{x}_T = \mathbf{b} & \\ & \mathbf{0} \leq \mathbf{x}_j \leq \mathbf{u}_j & j = 1, 2, \ldots, T \\ & \mathbf{0} \leq \mathbf{y}_j \leq \mathbf{u}'_j & j = 1, 2, \ldots, T \end{array}$$

where $\mathbf{x}_0$ is a known vector.

a. What class of problems lend themselves to this general structure? What is the interpretation of the vectors $\mathbf{x}_j$ and $\mathbf{y}_j$? (*Hint*. Examine a discrete control system.)

b. How can the decomposition algorithm be applied to this system? (*Hint*. Choose every other constraint to form the master constraints.)

c. Apply the procedure in part (b) to solve the following problem, using $x_0 = 80$.

$$
\begin{array}{lrl}
\text{Minimize} & x_1 + x_2 + x_3 + x_4 + 3y_1 + 5y_2 + 4y_3 + 6y_4 & \\
& x_0 - x_1 \qquad\qquad\qquad + y_1 & = 40 \\
& x_1 - x_2 \qquad\qquad + y_2 & = 50 \\
& x_2 - x_3 \qquad + y_3 & = 60 \\
& x_3 - x_4 + y_4 & = 40 \\
& x_4 & = 30 \\
& 0 \le x_1, x_2, x_3, x_4 & \le 40 \\
& 0 \le y_1, y_2 & \le 45 \\
& 0 \le y_3, y_4 & \le 50
\end{array}
$$

7.28 Consider the following *cutting stock problem*. We have standard rolls of length l and an order is placed requiring b_i units of length l_i where $i = 1, 2, \ldots, m$. It is desired to find the minimum number of rolls which satisfy the order.

a. Formulate the problem.
b. Apply the decomposition algorithm to solve the problem, neglecting integrality requirements. Discuss in detail. (*Hint*. Consider the column $\mathbf{a}_j$ representing the jth cutting pattern. Here $\mathbf{a}_j$ is a vector of nonnegative integers; a_{ij}, the ith component, is the number of rolls of length l_i and the jth cutting pattern. Develop a scheme for generating these $\mathbf{a}_j$ columns. What is the master problem and the subproblem?)

7.29 An agricultural mill produces cattle feed and chicken feed. These products are composed of three main ingredients, namely corn, lime, and fish meal. The ingredients contain two main types of nutrients, namely protein and calcium. The following table gives the nutrients' contents per pound of each ingredient.

	INGREDIENT		
NUTRIENT	CORN	LIME	FISH MEAL
Protein	25	15	25
Calcium	15	30	20

The protein content must lie in the interval [18, 22] per pound of cattle feed. Also the calcium content must be greater than or equal to 20 per pound of the cattle feed. Similarly, the protein content and the calcium content must be in the intervals [20, 23] and [20, 25], respectively, per pound of the chicken feed. Suppose that 3000, 2500, and 1000 pounds of corn, lime, and fish meal are available. Also suppose that it is required to product 4000 and 2000 pounds of the cattle and chicken feed respectively. Let the price per pound of the corn, lime, fish meal be respectively \$0.10, \$0.10, and \$0.08. Formulate the blending problem with an objective of minimizing the cost.

Solve the problem by the decomposition algorithm using two convexity constraints. Extra corn and fish meal can be obtained but, because of shortages, at the higher prices of \$0.12 and \$0.10 per pound. Would you advise the mill to consider extra corn and fish meal and modify their blending at these prices? Why or why not?

7.30 A company owns two refineries in Dallas and New York. The company can purchase two types of crude oil, light crude oil and heavy crude oil at the prices of \$11 and \$9 per barrel respectively. Because of shortages, the maximum amounts of these crudes that can be purchased are 2 million and 1 million barrels respectively. The following quantities of gasoline, kerosene, and jet fuel are produced per barrel of each type of oil.

	GASOLINE	KEROSENE	JET FUEL
Light crude oil	0.40	0.20	0.35
Heavy crude oil	0.32	0.40	0.20

Note that 5 percent and 8 percent of the crude are lost during the refining process, respectively. The company has contracted to deliver these products to three consumers in Kansas City, Los Angeles, and Detroit. The demands of these products are given

	GASOLINE	KEROSENE	JET FUEL
Kansas City	200,000	400,000	—
Los Angeles	300,000	200,000	600,000
Detroit	500,000	100,000	300,000

It is desired to find the amounts that must be purchased by the company of each crude at each of its refining facilities, and the shipping pattern of the products to Kansas City, Los Angeles, and Detroit, that satisfy the demands and minimize the total cost (purchase + shipping). The shipping and handling of a barrel of any finished product from the refineries to the consumers is given.

	KANSAS CITY	LOS ANGELES	DETROIT
Dallas Refinery	\$0.60	\$0.40	\$0.50
New York Refinery	\$0.35	\$0.80	\$0.30

a. Formulate the problem.
b. Suggest a decomposition scheme for solving the problem.
c. Solve the problem using your scheme in part (b).

7.31 A company has two manufacturing facilities in Atlanta and Los Angeles. The two facilities produce refrigerators and washer/dryers. The production capacities of these items in Atlanta are 5000 and 7000 respectively. Similarly the capacity of the Los Angeles facility is 8000 refrigerators and 4000 washer/dryers. The company delivers these products to three major customers in New York City, Seattle, and Miami. The customer's demand is given.

DEMAND / CUSTOMER	NEW YORK	SEATTLE	MIAMI
Refrigerators	4000	5000	4000
Washer/dryers	3000	3000	4000

The items are transported from the manufacturing facilities to the customers via a railroad network. The unit transportation costs (no distinction is made between the two items) are summarized. Also, because of limited space, the maximum number of refrigerators and/or washer/dryers that can be transported from a facility to a customer is given.

FACILITY		CUSTOMER NEW YORK	SEATTLE	MIAMI
Atlanta	Unit shipping cost $	6	14	7
	Maximum number of units	6000	3000	8000
Los Angeles	Unit Shipping cost $	10	8	15
	Maximum number of units	3000	9000	3000

It is desired to find the shipping pattern that minimizes the total transportation cost.

a. Formulate the problem.
b. Use the decomposition technique with two convexity constraints to solve the problem.

(*Note*. This problem is called a *multicommodity transportation problem*. The subproblem decomposes into two transportation problems. If you are familiar with the transportation algorithm, you can use it to solve the subproblems. Otherwise use the simplex method to solve the subproblems.)

7.32 Complete the solution of Example 7.3 in Section 7.6, and interpret the solution process as an application of the Benders' and the Lagrangian relaxation techniques. Provide both primal and dual solutions.

7.33 Provide the development of Benders' partitioning algorithm and the Lagrangian relaxation method given in Section 7.6 for the case of unbounded X (*without* artificially bounding this set). Give a formal and complete statement of the algorithm you develop and establish its convergence. Illustrate by applying this algorithm to Example 7.1, and identify the step-by-step relationship of this procedure with the Dantzig–Wolfe decomposition method of Section 7.4.

7.34 Consider the generalization of Benders' partitioning algorithm for the following nonlinear and/or discrete problem.

$$\begin{array}{ll} \text{Minimize} & \mathbf{cx} + f(\mathbf{y}) \\ \text{Subject to} & \mathbf{Ax} + \mathbf{By} = \mathbf{b} \\ & \mathbf{x} \geq \mathbf{0} \\ & \mathbf{y} \in Y \end{array}$$

where $\mathbf{c}, \mathbf{b}, \mathbf{x}, \mathbf{y}$ are vectors, $\mathbf{A}, \mathbf{B}$ are matrices, f is an arbitrary function, and Y is an arbitrary set.

a. Show that the problem can be reformulated as follows.

$$\begin{array}{lll} \text{Minimize} & z & \\ \text{Subject to} & z \geq f(\mathbf{y}) + \mathbf{w}(\mathbf{b} - \mathbf{By}) & \text{for each } \mathbf{w} \in W \\ & \mathbf{y} \in Y & \end{array}$$

where $W = \{\mathbf{w} \quad \text{unrestricted: } \mathbf{wA} \leq \mathbf{c}\}$. (*Hint*. The original problem can be reformulated as follows.

$$\underset{\mathbf{y} \in Y}{\text{Minimize}} \left(f(\mathbf{y}) + \underset{\substack{\mathbf{Ax} = \mathbf{b} - \mathbf{By} \\ \mathbf{x} \geq \mathbf{0}}}{\text{Minimum}} \, \mathbf{cx} \right)$$

Take the dual of

$$\begin{array}{ll} \text{Minimize} & \mathbf{cx} \\ \text{Subject to} & \mathbf{Ax} = \mathbf{b} - \mathbf{By} \\ & \mathbf{x} \geq \mathbf{0}) \end{array}$$

b. Show that $z \geq f(\mathbf{y}) + \mathbf{w}(\mathbf{b} - \mathbf{B}\mathbf{y})$ for each $\mathbf{wA} \leq \mathbf{c}$ if and only if $z \geq f(\mathbf{y}) + \mathbf{w}_j(\mathbf{b} - \mathbf{B}\mathbf{y})$ and $\mathbf{d}_j(\mathbf{b} - \mathbf{B}\mathbf{y}) \leq 0$ for each extreme point $\mathbf{w}_j$ and every extreme direction $\mathbf{d}_j$ of the region W.

c. Make use of part (b) to reformulate the problem in part (a) as follows:

$$\begin{array}{lll} \text{Minimize} & z & \\ \text{Subject to} & z \geq f(\mathbf{y}) + \mathbf{w}_j(\mathbf{b} - \mathbf{B}\mathbf{y}) & j = 1, 2, \ldots, t \\ & \mathbf{d}_j(\mathbf{b} - \mathbf{B}\mathbf{y}) \leq 0 & j = 1, 2, \ldots, l \\ & \mathbf{y} \in Y & \end{array}$$

where $\mathbf{w}_1, \ldots, \mathbf{w}_t$ and $\mathbf{d}_1, \ldots, \mathbf{d}_l$ are the extreme points and the extreme directions of W.

d. Without explicitly enumerating $\mathbf{w}_1, \ldots, \mathbf{w}_t$ and $\mathbf{d}_1, \ldots, \mathbf{d}_l$ beforehand, devise a decomposition algorithm for solving the problem in part (c). (*Hint*.

Master Problem:

$$\begin{array}{lll} \text{Minimize} & z & \\ \text{Subject to} & z \geq f(\mathbf{y}) + \mathbf{w}_j(\mathbf{b} - \mathbf{B}\mathbf{y}) & j = 1, 2, \ldots, t' \\ & \mathbf{d}_j(\mathbf{b} - \mathbf{B}\mathbf{y}) \leq 0 & j = 1, 2, \ldots, l' \\ & \mathbf{y} \in Y & \end{array}$$

where $\mathbf{w}_1, \ldots, \mathbf{w}_{t'}$ and $\mathbf{d}_1, \ldots, \mathbf{d}_{l'}$ are the extreme points and extreme directions generated so far.

Subproblem:

$$\begin{array}{ll} \text{Maximize} & \mathbf{w}(\mathbf{b} - \mathbf{B}\mathbf{y}) \\ \text{Subject to} & \mathbf{wA} \leq \mathbf{c} \end{array}$$

where $\mathbf{y}$ is obtained from the optimal of the master problem.)

e. How would you obtain the optimal $(\mathbf{x}, \mathbf{y})$ at termination of the decomposition algorithm in part (d)?

(Note. This algorithm is *Benders' partitioning procedure*. Note that the set Y can be discrete and so the procedure can be used for solving mixed integer problems. In this case the master problem in part (d) is a pure integer programming problem and the subproblem is a linear program.)

7.35 Apply Benders' partitioning procedure of Exercise 7.34 to solve the following problem [let (x_1, x_2) be $\mathbf{x}$ and (x_3, x_4) be $\mathbf{y}$].

$$\begin{array}{ll} \text{Minimize} & -x_1 - 2x_2 - 3x_3 - x_4 \\ \text{Subject to} & x_1 + x_2 + 2x_3 + x_4 \leq 12 \\ & -x_1 + x_2 \leq 4 \\ & 2x_1 + x_2 \leq 6 \\ & x_3 + x_4 \leq 8 \\ & x_1, \quad x_2, \quad x_3, \quad x_4 \geq 0 \end{array}$$

7.36 A company is planning to build several warehouses for storing a certain product. These warehouses would serve two major customers with monthly demands of 3000 and 5000 units. Three candidate warehouses with capacities 4000, 5000, and 6000 can be constructed. Using the estimated construction cost of the warehouses, their useful life, and time value of money, the construction cost per month for the three warehouses is estimated as \$8000, \$12,000, and \$7000. The unit transportation cost from the three candidate warehouses to the customers is given.

	CUSTOMER	
WAREHOUSE	1	2
1	1.50	2.00
2	2.00	1.50
3	2.50	2.25

Use Benders' partitioning procedure of Exercise 7.34 to determine which warehouses to construct and the corresponding shipping pattern.

7.37 Solve the following linear program entirely graphically using decomposition.

$$\begin{array}{ll} \text{Minimize} & 2x_1 + 5x_2 + x_3 - 2x_4 + 3x_5 \\ \text{Subject to} & x_1 + x_2 + x_3 + x_4 \geq 2 \\ & 3x_1 + x_2 + 5x_3 + x_4 - 2x_5 \geq 5 \\ & -x_1 + 2x_3 + x_4 \geq 2 \\ & x_1, \quad x_2, \quad x_3, \quad x_4, \quad x_5 \geq 0 \end{array}$$

(*Hint*. Let the first constraint denote $\mathbf{Ax} \geq \mathbf{b}$ and the next two constraints represent X. Then take the dual of each set.)

7.38 Indicate how the results of the previous problem may be generalized to any number m of constraints. (*Hint*. Let the first constraint denote $\mathbf{Ax} = \mathbf{b}$ and the remaining $m - 1$ constraints be part of the subproblem. Then reapply decomposition to the subproblem in a similar way.)

7.39 Assume that a linear program requires $3m/2$ iterations for a solution. Also, assume that standard techniques of pivoting are used to update the basis inverse and RHS vector [in total an $(m + 1) \times (m + 1)$ matrix if we ignore the z column]. If there is no special structure to the constraint matrix, then is there an optimal split for decomposition? That is, find $m_1 + m_2 = m$ such that the first m_1 constraints form the master problem and the next m_2 constraints are subproblem constraints, and the total "effort" is minimized. Let the "effort" be defined by the number of elementary operations (additions, subtractions, multiplications, and divisions).

7.40 In the previous problem suppose that m_1 and m_2 are given and that the second m_2 constraints are of a special structure. Specifically, suppose that the subproblem requires only 5 percent of the normal effort to yield a solution when treated by itself.

a. Should the problem be decomposed for efficiency?
b. Is there a critical value of the percentage effort required?

7.41 Solve the following problem by decomposition.

$$\begin{array}{ll} \text{Minimize} & -2x_1 + 5x_2 - 4x_3 \\ \text{Subject to} & x_1 + 2x_2 + a_1x_3 \le 6 \\ & 3x_1 - 6x_2 + a_2x_3 \le 5 \\ & 2a_1 + 3a_2 = 4 \\ & x_1, x_2, x_3, a_1, a_2 \ge 0 \end{array}$$

[*Hint*. Let $X = \{(a_1, a_2) : 2a_1 + 3a_2 = 4,\ a_1,\ a_2 \ge 0\}$. This is called a *generalized linear programming problem*.]

NOTES AND REFERENCES

1. The decomposition algorithm of this chapter is an adaptation of the Dantzig–Wolfe decomposition principle [1960, 1961]. The latter was inspired by the suggestions of Ford and Fulkerson [1958b] for solving the special case of multicommodity network flow problems.
2. The decomposition method presented in this chapter is closely associated with the concepts of generalized Lagrangian multipliers, tangential approximation of the Lagrangian dual function, and the dual cutting plane algorithm, as discussed in Section 7.6. For further reading on these topics the reader may refer to Bazaraa and Shetty [1979], Everett [1963], Geoffrion [1971], Lasdon [1970], Kelley [1960], and Zangwill [1969].
3. In addition to the Dantzig–Wolfe and similar decomposition algorithms, the literature has a great deal of other decomposition methods. These can be classified as *price-directive* and *resource-directive* algorithms. In the former, a direction for modifying the Lagrangian multipliers of the coupling constraints is found and then a suitable step size is taken along this direction. See, for example, Geoffrion [1970], Lasdon [1970], Grinold [1972], Balas [1966b], Held, Wolfe, and Crowder [1974], and Bazaraa and Goode [1979]. The resource-directive algorithms proceed by finding a direction for modifying the shares of

the common resources among the subproblems and then determining the step size. The reader may refer to Geoffrion [1970], Lasdon [1970], and Abadie [1963].

4. In Exercise 7.34 we describe the partitioning scheme of Benders [1962] that was specialized in Section 7.6. This scheme is particularly suited for solving mixed integer programming problems.
5. For an in-depth geometric analysis of the Dantzig–Wolfe decomposition algorithm, see Todd [1983]. For implementation issues, see Ho and Loute [1981, 1983]. Birge [1985] provides some further insights into this procedure, and Wittrock [1985] applies it to the dual of the staircase structured problem of Exercise 7.25. (The approach sketched in Exercise 7.25 is from Glassey [1973].) For generalizations of Dantzig–Wolfe/Benders' algorithms, see Geoffrion [1972] and Burkard *et al*. [1985].

EIGHT: COMPLEXITY OF THE SIMPLEX ALGORITHM AND POLYNOMIAL ALGORITHMS

In this chapter we discuss some fundamental computational complexity issues and theoretically efficient, that is to say, polynomial-time algorithms for solving linear programming problems. We begin by discussing the issue of exponential versus polynomial-time algorithms based on a worst-case computational performance analysis. According to this classification scheme, the simplex algorithm is shown to be "bad," because it exhibits an exponential growth in effort on certain classes of problems as problem size increases. Fortunately, there are some average-case theoretical results that help resolve the discrepancy between the worst-case theoretical results and the observed practical efficiency of the simplex algorithm. However, there do exist theoretically efficient algorithms for linear programming problems that exhibit a growth in solution effort that is polynomial in the size of the problem. The two well-publicized algorithms of Khachian and Karmarkar fall in this category. Although Khachian's algorithm has failed to be of practical computational value, Karmarkar's algorithm is far more promising. We therefore relegate the development of Khachian's algorithm to the exercises, but treat Karmarkar's algorithm in detail in this chapter. Curiously, both of these algorithms are nonlinear approaches to linear programming problems. Although a preparation in nonlinear programming theory and algorithms will provide additional insights to the reader, we maintain a level of presentation, aided by geometric motivations, that will require a knowledge of only some basic linear algebra and calculus on the part of the reader. The reader is also encouraged to review the material in Chapter 2.

8.1 POLYNOMIAL COMPLEXITY ISSUES

As the field of Operations Research continued to develop following the birth of linear programming, a rich collection of various types of problems were introduced along with a host of competing solution algorithms. Consequently, there arose a need to be able to formally classify and compare problems and algorithms for computational tractability. To address this need, in the 1970s computer scientists and operations research analysts introduced the issue of computational complexity of problems and algorithms. The adopted viewpoint is to conduct an evaluation of the problem or algorithm based on the performance in a worst-case type of situation. The idea is to ascertain how difficult a problem is to solve in the worst case, or what is the growth in computational effort of an algorithm as a function of the size of the problem in the worst case. Such an analysis provides a *performance guarantee* for an algorithm that solves a given class of problems.

Since our primary aim here is to relate this development with the study of linear programming problems, let us present some relevant issues with respect to this particular class of problems. Consider the linear programming optimization problem

$$\text{Minimize } \{\mathbf{cx} : \mathbf{Ax} = \mathbf{b}, \mathbf{x} \geq \mathbf{0}\}$$

where $\mathbf{A}$ is $m \times n$ and $m, n \geq 2$. We assume here and throughout this chapter that the *data is all integer*, perhaps converted to this form from rational data. By an *instance* of this problem, we mean a particular member of this class of problems with specific parameter values or data. Hence, the specification of values for m, n, $\mathbf{c}$, $\mathbf{A}$, and $\mathbf{b}$ defines an instance of a linear programming problem. The *size* of an instance of this problem is represented by the entities (m, n, L), where L is the number of binary bits required to record all the data of the problem and is known as the *input length* of an instance of the problem. Note that the number of binary bits required to represent a positive integer Δ is $\lceil \log(1 + \Delta) \rceil$, where $\lceil \cdot \rceil$ denotes the rounded-up value and $\log(\cdot)$ denotes logarithm to the base 2. For example, in order to represent any integer $\Delta \in [2^r, 2^{r+1})$ for an integer $r \geq 1$, we require $(r + 1)$ binary bits. If Δ is of either sign, we can use an extra bit to record its sign, and therefore we can represent Δ using $1 + \lceil \log(1 + |\Delta|) \rceil$ bits. Consequently, the data of the foregoing linear program can be recorded by a *binary encoding scheme* using the following number of bits:

$$L = \lceil 1 + \log(1 + m) \rceil + \lceil 1 + \log(1 + n) \rceil + \sum_j \left\{1 + \lceil \log(1 + |c_j|) \rceil \right\}$$

$$+ \sum_i \sum_j \left\{1 + \lceil \log(1 + |a_{ij}|) \rceil \right\} + \sum_i \left\{1 + \lceil \log(1 + |b_i|) \rceil \right\}$$

Consider an algorithm that finitely solves a linear programming problem. In order to analyze its *computational complexity*, we are required to determine an upper bound on the effort required to solve any instance of this problem. This effort may be measured in terms of the number of elementary operations such as additions, multiplications, and comparisons that are required to solve the problem as a function of the size of the problem. Fortunately, an exact count of the total number of such operations is unnecessary. We are only required to determine a function $g(m, n, L)$ in terms of the size of the problem, such that for some sufficiently large constant $\tau > 0$ the total number of elementary operations required by the algorithm to solve the problem is no more than $\tau g(m, n, L)$. In such a case, we say that the algorithm is of *order of complexity* $O(g(m, n, L))$.

When the function $g(m, n, L)$ is a polynomial in m, n, and L, the algorithm is said to be of *polynomial complexity* or to be *polynomially bounded*. For example, if an algorithm actually involves a maximum of, say, $6m^2n + 15mn + 12m$ elementary operations to solve any instance of some given class of problems, where the problem size is determined by m and n, we can simply say that it is of complexity $O(m^2n)$ (why?). Hence we are only required to ascertain that the growth in effort for this algorithm is dictated dominantly by the function $g(m, n) = m^2n$.

Since we are only required to determine an upper bound on the solution effort, we need not compute L exactly in defining the size of an instance of a problem. In other words, if we take L as some *lower bound* on the actual number of bits required to record the data and exhibit a polynomial growth in effort as a monotone increasing function of this quantity, then this would also imply a polynomial bound on the effort in terms of the actual size of the problem. Hence, for instance, we may take L to be the following quantity in our analysis of linear programming problems.

$$L = \left[1 + \log m + \log n + \sum_j [1 + \log(1 + |c_j|)] + \sum_i \sum_j [1 + \log(1 + |a_{ij}|)] + \sum_i [1 + \log(1 + |b_i|)]\right] \quad (8.1)$$

Our obsession with a polynomial-time algorithm can be well appreciated by comparing the growth of n^2 versus 2^n. Let "n" represent the size of some problem, and let A_1 and A_2 be two algorithms for this problem. Further suppose that algorithm A_1 is of complexity $O(n^2)$ and algorithm A_2 is of complexity $O(2^n)$. This means that there exist constants τ_1 and τ_2 such that the effort with algorithms A_1 and A_2 are respectively bounded by the expressions $\tau_1 n^2$ and $\tau_2 2^n$. Consequently, for a problem of size $n = 50$, assuming $\tau_1 = \tau_2 = 1$, although algorithm A_1 will perform no more than 2500 operations, the number of operations required by algorithm A_2 can get as high as 1.1259×10^{15}—an astronomical figure! There are two obvious anomalies with this system of evaluation. First, we would prefer *low-order* polynomials. A polynomial-time algorithm with complexity $O(n^{50})$ would clearly be of no greater practical value than algorithm A_2. Second, we would like the polynomial *coefficient magnitudes* to be relatively small. If $\tau_1 = 2^{50}$ and $\tau_2 = 1$, then although the effort with algorithm A_2 blows up considerably faster with an increase in n than with algorithm A_1, from a practical viewpoint algorithm A_1 is unmanageable for any value of $n \geq 1$. However, algorithm A_2 is at least manageable for some sufficiently small values of n.

It is worthwhile to note the role played by the binary encoding scheme in defining a polynomial-time algorithm. Suppose that an algorithm was proposed for linear programming problems that was of order $O(mn\Delta)$ in complexity, where Δ is the largest absolute value coefficient from **c**, **b**, and **A**. Then this algorithm is *not* of polynomial complexity. In fact, defining $L_1 = \log(\Delta)$ as representing the size of the problem (for example, for fixed m and n), we see that the algorithm is of an exponential complexity of order $O(mn2^{L_1})$. Suppose that we had adopted to measure the size of the problem in terms of a *stroke encoding* scheme, that is, a scheme that uses one stroke for every unit of data. To store the integer $\Delta > 0$ we would require Δ strokes or spaces, so that the problem size is determined by the triplet (m, n, Δ) and not by $(m, n, \log \Delta)$. In this case, the algorithmic effort is indeed bounded above by a polynomial in (m, n, Δ). Such an algorithm is referred to as being *pseudopolynomial*. We will see later in Chapter 10 that the Hungarian algorithm for solving linear assignment problems is pseudopolynomial in complexity.

On the other hand, an algorithm whose complexity depends only on the number of problem parameters, as determined for linear programs by m and n for example and is independent of the magnitudes of these parameters, is said to be *genuinely* or *strongly polynomial*. Whereas no strongly polynomial algorithm is known to exist for general linear programming problems, such algorithms indeed exist for the special class of network structured linear programs.

In concluding this section, let us place the foregoing discussion in the context of the approach taken by computer scientists. In the domain of computer science, problems are usually posed as *decision problems* rather than *optimization problems*. For example, the linear programming *optimization* problem has as its counterpart the following linear programming *decision* problem.

> *Given* **c**, **b**, *and* **A** (*of the appropriate dimensions*) *and given a rational number* K, *does there exist a rational vector* **x** *such that* $\mathbf{Ax} = \mathbf{b}$, $\mathbf{x} \geq \mathbf{0}$, *and* $\mathbf{cx} \leq K$?

A *polynomial-time algorithm* for a decision problem may be (informally) defined as one that can "solve" any instance of the problem using a number of elementary operations bounded above by some polynomial in the size of the problem. By "solve," one means halt with an answer yes, given a yes-instance of the decision problem. Hence if one has a polynomial-time algorithm and is given a no-instance of a problem, it is sufficient to exceed the established polynomial bound without halting to conclude that the answer is no. Therefore the polynomial bound holds for all instances. Problems for which some polynomial-time algorithm is known are said to belong to the distinguished *Class P* of problems.

It is fruitful to see the relationship between polynomial-time algorithms for optimization problems and for decision problems. A polynomial-time algorithm for an optimization problem can clearly solve a decision problem in polynomial time (how?). Conversely, suppose that one has a polynomial-time algorithm for the linear programming *decision* problem. Consider an instance of the optimization problem that is feasible and bounded. Then there exists a basic feasible solution of objective value $\leq 2^L$ in magnitude (see Exercise 8.9). Hence, the optimal value lies in the interval $[-2^L, 2^L]$. Assuming that the algorithm for solving the decision problem also actually produces a solution in a yes-instance, one can use this algorithm to solve the optimization problem by repeatedly performing a bisection search on this interval. This can be done as follows.

First, using $K = 0$, which is the midpoint of the current *interval of uncertainty* $[-2^L, 2^L]$, it can be found in polynomial-time whether or not the optimal value lies in the interval $[-2^L, 0]$ or $[0, 2^L]$. Using the appropriate interval which contains the optimal value, this interval of uncertainty can again be bisected; this process can be continued until, say, the optimal value is known to lie in the interval $[v_1, v_2]$, where $v_2 - v_1 < 2^{-2L}$. Since the initial interval of uncertainty is of length 2^{L+1} and this is halved at each bisection, we can obtain this reduced interval of uncertainty after using $3L + 2$ applications of the decision problem algorithm (why?). This means that we will have used a polynomial-time algorithm a polynomial number of times, and hence we will have accomplished this in polynomial-time. If a corresponding solution in this interval is not yet available, one can be obtained by applying the decision problem with the task of finding a solution $\mathbf{x}$ with $\mathbf{cx} \le v_2$. As will be shown later in Section 8.5, given a solution with this accuracy in objective value, one can use a polynomial-time "rounding scheme" in order to determine an optimal solution to the optimization problem. Hence, the optimization problem can be solved in polynomial time by using a polynomial-time algorithm for the decision problem.

We now proceed to demonstrate that the simplex algorithm is of exponential complexity, and then present in detail a particular polynomial-time algorithm for solving linear programming problems.

8.2 COMPUTATIONAL COMPLEXITY OF THE SIMPLEX ALGORITHM

When Dantzig first introduced the simplex algorithm, the intuition-based reaction of the research community was that this algorithm would not prove to be very efficient. By its nature, the simplex algorithm crawls along the edges of a polyhedron and makes no attempt to skip through its interior or over higher dimensional faces. In fact, it occasionally stumbles through a sequence of degenerate pivots at some extreme points. However, researchers were pleasantly surprised when in *practice*, this method performed exceedingly well. For most practical problems, this method has been empirically observed to take roughly $3m/2$ iterations, and seldom more than $3m$ iterations, where the coefficient matrix $\mathbf{A}$ is $m \times n$. As pointed out in Chapter 5, a regression equation of the type $Km^{2.5}nd^{0.33}$, for example, where d is the density of the matrix $\mathbf{A}$, may typically predict the performance of the simplex algorithm on a given problem quite well. However, the fact remains that the algorithm is indeed entrapped in the potentially combinatorial aspect of up to $\binom{n}{m}$ vertices that the algorithm could possibly visit. Since

$$\binom{n}{m} = \frac{n!}{m!(n-m)!} = \left(\frac{n}{m}\right)\left(\frac{n-1}{m-1}\right)\left(\frac{n-2}{m-2}\right)\cdots\left[\frac{n-(m-1)}{m-(m-1)}\right] > \left(\frac{n}{m}\right)^m$$

which is at least 2^m whenever $n \ge 2m$, the potential of an exponential order of effort for some problems is quite plausible.

The fear of exponential effort with the simplex method was confirmed in 1971 when Victor Klee and George Minty produced a class of problems defined by some $m = n$ equality constraints in $2n$ nonnegative variables, for which the simplex algorithm requires $2^n - 1$ iterations, traversing *all* the vertices of the problem. This class of problems results by defining the feasible region as a suitable distortion of the n-dimensional hypercube in E^n, which has 2^n vertices. Mathematically, the problem in E^n may be stated as follows, where ϵ is some rational number in the interval $(0, \frac{1}{2})$.

$$\begin{aligned}
&\text{Maximize} \quad x_n \\
&\text{Subject to} \quad 0 \le x_1 \le 1 \\
&\qquad \epsilon x_{j-1} \le x_j \le 1 - \epsilon x_{j-1} \qquad \text{for } j = 2, \ldots, n \\
&\qquad x_j \ge 0, \; j = 1, \ldots, n
\end{aligned}$$

After suitably transforming this problem, we exhibit that, using Dantzig's rule of entering the nonbasic variable with the largest reduced cost and starting at the origin, the simplex algorithm traverses $2^n - 1$ edges of this polytope and visits all the 2^n vertices.

In order to put the problem in the desired form, consider the linear transformation

$$y_1 = x_1, \; y_j = (x_j - \epsilon x_{j-1})/\epsilon^{j-1} \qquad \text{for } j = 2, \ldots, n \tag{8.2}$$

Using this transformation and putting $\theta = 1/\epsilon$, we obtain the following equivalent problem. (The algebraic details are left to the reader in Exercise 8.2.)

$$\begin{aligned}
&\text{Maximize} \quad \sum_{j=1}^{n} y_j \\
&\text{Subject to} \quad y_1 \le 1 \\
&\qquad y_j + 2\sum_{k=1}^{j-1} y_k \le \theta^{j-1} \qquad \text{for } j = 2, \ldots, n \\
&\qquad y_1, \ldots, y_n \ge 0
\end{aligned} \tag{8.3}$$

For convenience, define $s_1, \ldots, s_n$ as the *slack variables* associated with the above n constraints. Observe that the basic solution with basic variables $(s_1, \ldots, s_{n-1}, y_n)$ is feasible, yielding $\mathbf{y} = (0, \ldots, 0, \theta^{n-1})$ with objective function value θ^{n-1}. This corresponds to the solution $\mathbf{x} = (0, \ldots, 0, 1)$ for the original problem. Noting the final constraint and the objective function in the original problem, this clearly corresponds to the optimal solution to this problem (why?). In Exercise 8.3 we ask the reader to verify this by computing the $z_j - c_j$ values for the nonbasic variables.

Let us now consider the case $n = 2$. The problem in this case is of the form to maximize $y_1 + y_2$, subject to $y_1 \le 1$, $y_2 + 2y_1 \le \theta$, and $y_1, y_2 \ge 0$. Figure 8.1a shows the feasible region using $\theta = 3$. Note that for the origin the basic

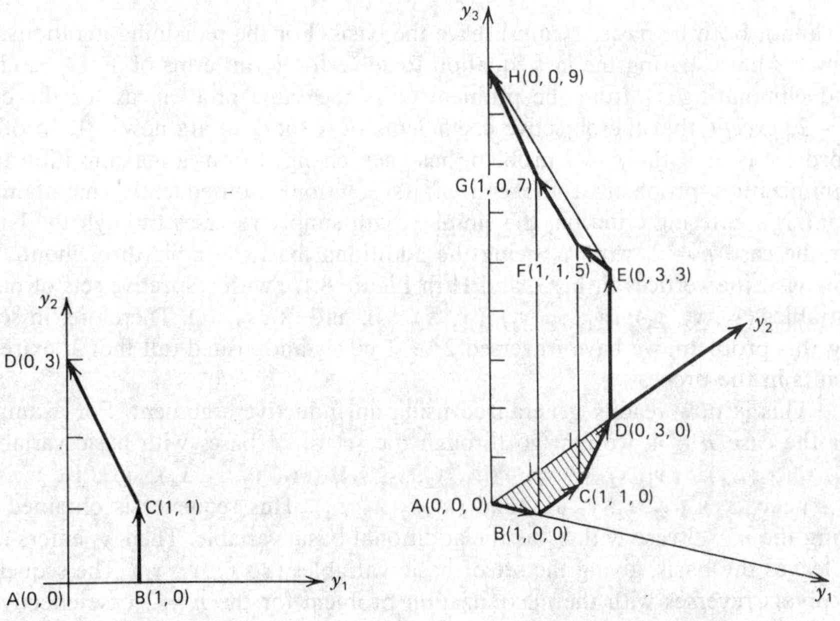

(*a*) Case for $n = 2$ (*b*) Case for $n = 3$

Figure 8.1. Illustration of the Klee – Minty type polytopes for $n = 2$ and 3.

variables are (s_1, s_2). Hence, the $z_j - c_j$ values for y_1 and y_2 are both -1. Selecting y_1 as the entering variable takes us to the adjacent extreme point solution with basis (y_1, s_2). Continuing, we now follow the simplex path A, B, C, D in Figure 8.1*a*. In doing so, we traverse the 2^2 basic feasible solutions with respective basic variables (s_1, s_2), (y_1, s_2), (y_1, y_2), and finally (s_1, y_2). The reader may verify that $z_j - c_j = -1$ for the entering variable at each iteration. Note that if we had broken the *tie* at the first iteration by selecting y_2 as the entering variable, we would have solved this problem in one iteration.

Next consider the case $n = 3$. The problem in this case is of the form to maximize $y_1 + y_2 + y_3$, subject to $y_1 \leq 1$, $y_2 + 2y_1 \leq \theta$, $y_3 + 2y_1 + 2y_2 \leq \theta^2$, and $y_1, y_2, y_3 \geq 0$. Figure 8.1*b* shows the feasible region using $\theta = 3$ as previously. Now let us make the following observation. First, notice that with $y_3 = 0$, the extreme points in the (y_1, y_2) space obtained for the case $n = 2$ are still extreme points in the present case; s_3 is the third basic variable. The corresponding sets of basic variables are (s_1, s_2, s_3), (y_1, s_2, s_3), (y_1, y_2, s_3), and (s_1, y_2, s_3). Second, observe that in moving from each one of these bases to the next, the $z_j - c_j$ value for the entering variable is -1 as for the case $n = 2$. The $z_j - c_j$ value for the new nonbasic variable y_3 also remains at -1 throughout, since the last component of the simplex multiplier vector $\mathbf{c_B B}^{-1}$ is zero for all these bases $\mathbf{B}$. Hence, the simplex path A, B, C, D for the case $n = 2$ remains a simplex path for the case $n = 3$ under the Dantzig entering rule. Moreover, y_3 remains enterable throughout. Third, at the vertex D, we must enter y_3 and since s_3 and

y_3 cannot both be basic, s_3 must leave the basis. For the remaining iterations, y_3 remains basic. Using the last equation to solve for y_3 in terms of y_1, y_2, and s_3 and eliminating y_3 from the problem gives the same problem as for the case $n = 2$, *except* that the objective coefficients of y_1 and y_2 are now -1. In other words, it is as if the $n = 2$ problem has been changed from a maximization to a minimization problem at the end of its solution! Consequently, maintaining Dantzig's entering criterion, the simplex path simply reverses through the bases for the case $n = 2$, with y_3 being the additional basic variable throughout. We now visit the vertices E, F, G, and H in Figure 8.1*b*, with respective sets of basic variables (s_1, y_2, y_3), (y_1, y_2, y_3), (y_1, s_2, y_3), and (s_1, s_2, y_3). Therefore, in solving this problem, we have traversed $2^3 - 1$ edges and visited (all the) 2^3 extreme points in the process.

This is now readily generalized using an inductive argument. For example, for the case $n = 4$, we first go through the set of 2^3 bases with basic variables (s_1, s_2, s_3, s_4), (y_1, s_2, s_3, s_4), (y_1, y_2, s_3, s_4), (s_1, y_2, s_3, s_4), (s_1, y_2, y_3, s_4), (y_1, y_2, y_3, s_4), (y_1, s_2, y_3, s_4), and (s_1, s_2, y_3, s_4). This sequence is obtained by using the $n = 3$ case with s_4 as the additional basic variable. Then y_4 enters and s_4 leaves the basis, giving the set of basic variables (s_1, s_2, y_3, y_4). The sequence of pivots reverses with the maximization problem for the $n = 3$ case effectively turning into a minimization problem when y_4 is held basic. We next generate the 2^3 bases with basic variables (s_1, s_2, y_3, y_4), (y_1, s_2, y_3, y_4), (y_1, y_2, y_3, y_4), (s_1, y_2, y_3, y_4), (s_1, y_2, s_3, y_4), (y_1, y_2, s_3, y_4), (y_1, s_2, s_3, y_4), and (s_1, s_2, s_3, y_4). This results in $2 \times 2^3 = 2^4$ bases over $2^4 - 1$ pivots. The construction of the formal inductive proof is now straightforward and is left to the reader in Exercise 8.4. It is interesting to observe that for the general case n, all throughout the first 2^{n-1} basic feasible solutions the variable y_n remains enterable. Recall the definition of a stage from Chapter 4! Also notice that at the end of 2^{n-1} *pivots*, the basic variables are $(s_1, s_2, \ldots, s_{n-2}, y_{n-1}, y_n)$. If s_{n-1} is entered at this pivot, then the optimum results. (This corresponds to going from E to H in Figure 8.1*b*.) However, both y_1 and s_{n-1} have $z_j - c_j = -1$ here, and the choice of entering y_1 takes us through the reverse sequence of pivots.

Notice the exorbitant amount of effort as problem size increases. Assume that the problem with n constraints in $2n$ nonnegative variables requires $0.25n^2$ elementary operations per pivot, and that each operation takes one nanosecond (10^{-9} secs) to perform. Then the computational effort for $n = 50$ is 22.3 years; for $n = 60$ it is over 328 centuries! One may argue that the simplex algorithm did have an opportunity to solve the foregoing problem in one iteration. In fact, by scaling the columns, for example, using the transformation $\gamma_i = 2^{-i}y_i$, we can eliminate the ties in the entering criterion at the first iteration and thereby solve the problem in one iteration unarguably. Observe also that the rule of entering that variable that gives the maximum improvement in the objective function value also solves this problem unambiguously in one iteration. Moreover, this criterion is invariant with respect to scaling and problem representation, since it is intimately connected with the structure of the polytope itself. In 1973, however, Jeroslow showed the existence of problem classes that yield an exponential effort with the maximum-improvement entering criterion as well. In fact, for some fixed integer $r \geq 1$, if one were to perform all r next pivots and pick that r-pivot simplex path that gives the best improvement in objective value, then the algorithm would still exhibit an exponential trend on some classes of problems. The root of the problem is therefore in the myopic or local viewpoint of the simplex algorithm in which the decisions are based on the local combinatorial structure of the polytope and the motion is restricted to an edge path.

8.3 KHACHIAN'S ELLIPSOID ALGORITHM

Confronted with the exponential worst-case behavior of the simplex algorithm, the question of whether a polynomial-time algorithm for solving linear programming problems did exist remained open until 1979, when it was answered affirmatively by L. G. Khachian. Khachian proposed a polynomial-time algorithm for determining a solution (if one exists) to the open set of linear inequalities $S = \{\mathbf{x} : \mathbf{G}\mathbf{x} < \mathbf{g}\}$, where $\mathbf{G}$ is $r \times q$, $r, q \geq 2$, and where $\mathbf{G}$ and $\mathbf{g}$ have all integer components. This algorithm either finds a solution in S, if one exists, or else concludes that S is empty. Its basic mode of operation is as follows.

First, the method defines a ball or sphere in E^p, centered at the origin and with a radius large enough to encompass a sufficiently large volume of S, assuming that S is nonempty. If the center of this ball is in S, then the algorithm terminates. Otherwise it proceeds to construct a suitable sequence of ellipsoids which are monotonically shrinking in volume, but all of which contain that region of S which was captured by the initial ball. Whenever the center of any ellipsoid is found to lie in S, the algorithm terminates. The principal result states that if S is nonempty, then the center of some ellipsoid will lie in S within a number of iterations bounded above by a polynomial in the size of the problem. Moreover, if this bound on the number of iterations is exceeded, then the algorithm may terminate with the assertion that S is indeed empty. Exercises 8.10–8.15 deal with the details of this algorithm. How such an algorithm helps solve the linear programming (optimization) problem in polynomial-time is addressed in Exercises 8.18–8.21. In particular, it is shown that the linear programming problem to minimize $\{\mathbf{cx} : \mathbf{Ax} \geq \mathbf{b}, \mathbf{x} \geq \mathbf{0}\}$, where $\mathbf{A}$ is $m \times n$, can be solved by this algorithm within an effort of $O[(m + n)^6 L]$, where L is given by Equation (8.1).

Unfortunately, computational experience with Khachian's algorithm and its variants has been very disappointing. This has contributed to the realization that one ought to consider as "efficient" only low-order polynomial algorithms, perhaps those which are also strongly or genuinely polynomial. The *practical* performance of Khachian's algorithm is strongly connected with its theoretical worst-case bound. The example in Exercise 8.22 exhibits the strong dependency of algorithmic effort on the magnitudes of the input data. Hence the initial excitement that this algorithm generated, both from the theoretical viewpoint and from the simplicity of its steps (which are programmable even on a hand calculator), fizzled out as it was realized that even problems with about 100 variables can require an enormous amount of effort. Nonetheless we encourage the reader who is interested in the theoretical aspects of this algorithm, either from a linear or a nonlinear programming viewpoint, to study Exercises 8.10–8.15 and 8.18–8.21, as well as the relevant papers cited in the Notes and References section.

8.4 KARMARKAR'S PROJECTIVE ALGORITHM

The simplex method, having undergone considerable refinement and sophistication in implementation, had no serious competition until 1984 when N. Karmarkar proposed a new polynomial-time algorithm for linear programming problems. This algorithm addresses linear programming problems of the form

$$\begin{aligned} \text{Minimize} \quad & \mathbf{cx} \\ \text{Subject to} \quad & \mathbf{Ax} = \mathbf{0} \\ & \mathbf{1x} = 1 \\ & \mathbf{x} \geq \mathbf{0} \end{aligned} \tag{8.4}$$

where $\mathbf{A}$ is $m \times n$ of rank m, $n \geq 2$, $\mathbf{c}$ and $\mathbf{A}$ are all integers, $\mathbf{1}$ is a row vector of n ones, and the following assumptions (A1) and (A2) hold:

(A1) The point $\mathbf{x}_0 = \left(\frac{1}{n}, \ldots, \frac{1}{n}\right)$ is feasible in Problem (8.4).

(A2) The optimal objective value of Problem (8.4) is zero.

At first glance the form of the linear program (8.4) and the accompanying assumptions (A1) and (A2) may appear to be unduly restrictive. However, as shown later, any general linear programming problem can be cast in this form through the use of artificial variables, an artificial bounding constraint, and through variable redefinitions, if necessary. Moreover, we will show that the analysis under these conditions can be readily extended to polynomially solve the general linear programming problem. Let us proceed with the derivation of a polynomial-time algorithm to solve Problem (8.4) under the assumptions (A1) and (A2). Note that under these assumptions, the problem (8.4) is feasible and bounded, and hence has an optimum.

Starting with $x_0 = (1/n, \ldots, 1/n)$ and $k = 0$, the algorithm performs the following iterative step, given any feasible solution $\mathbf{x}_k > \mathbf{0}$. Define the diagonal matrix $\mathbf{D}_k = \operatorname{diag}\{x_{k1}, \ldots, x_{kn}\}$, where $\mathbf{x}_k = (x_{k1}, \ldots, x_{kn})$, and consider the following variable transformation.

$$\mathbf{y} = \frac{\mathbf{D}_k^{-1}\mathbf{x}}{\mathbf{1}\mathbf{D}_k^{-1}\mathbf{x}}$$

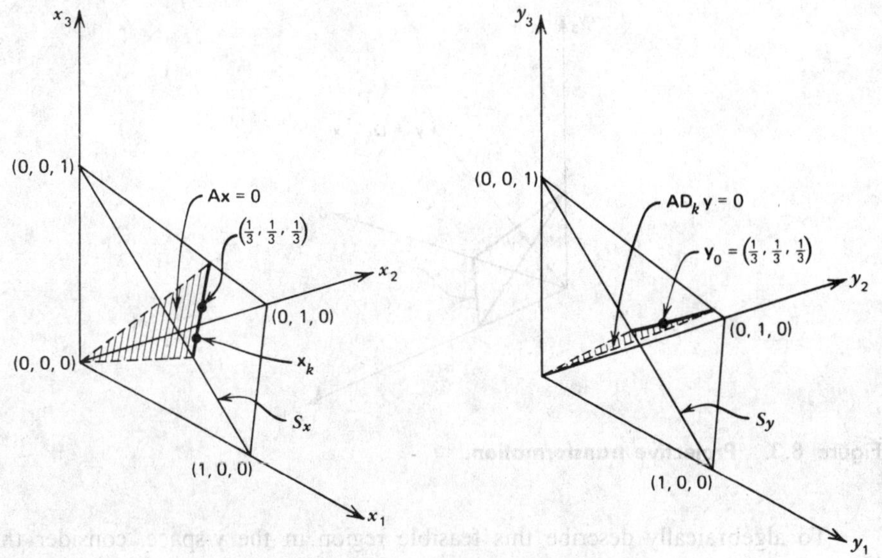

Figure 8.2. Projective transformation of the feasible region.

that is,

$$y_i = \frac{x_i/x_{ki}}{\sum_{j=1}^{n} x_j/x_{kj}} \quad \text{for } i = 1, \ldots, n. \tag{8.5a}$$

Observe that the feasible region in Problem (8.4) is described by the intersection of the $(n - m)$ dimensional linear subspace defined by the homogeneous set of equalities $\mathbf{Ax} = \mathbf{0}$ with the $(n - 1)$ dimensional simplex $S_x = \{\mathbf{x} : \mathbf{1x} = 1, \mathbf{x} \geq \mathbf{0}\}$. The intersection is some $(n - m - 1)$ dimensional region. Figure 8.2 illustrates a situation in $n = 3$ dimensions with a single $(m = 1)$ homogeneous equality constraint. Under the transformation (8.5a), any point in S_x is transformed into a point in the $(n - 1)$ dimensional simplex $S_y = \{\mathbf{y} : \mathbf{1y} = 1, \mathbf{y} \geq \mathbf{0}\}$. In particular, the current point $\mathbf{x}_k$ is transformed into the point $\mathbf{y}_0 = (1/n, \ldots, 1/n)$, the center of the simplex S_y. (See Figure 8.2.) Note that if one were to have used the *affine transformation* $\mathbf{y}' = \mathbf{D}_k^{-1}\mathbf{x}$, which is simply a scaling operation, then a point $\mathbf{x} \in S_x$ would, for example, transform into a point $\mathbf{y}'$ as shown in Figure 8.3. If this point $\mathbf{y}'$ is *projected* onto S_y along the ray defined by the origin and $\mathbf{y}'$, then one would obtain the point $\mathbf{y}$ given by transformation (8.5a), as shown in Figure 8.3. This amounts to dividing $\mathbf{y}'$ by the sum of its components as in transformation (8.5a), so that the components of $\mathbf{y}$ sum to one. For this reason the transformation (8.5a) is known as a *projective transformation*. Performing this operation on the feasible region in the $\mathbf{x}$-space results in the feasible region in the $\mathbf{y}$-space as illustrated in Figure 8.2.

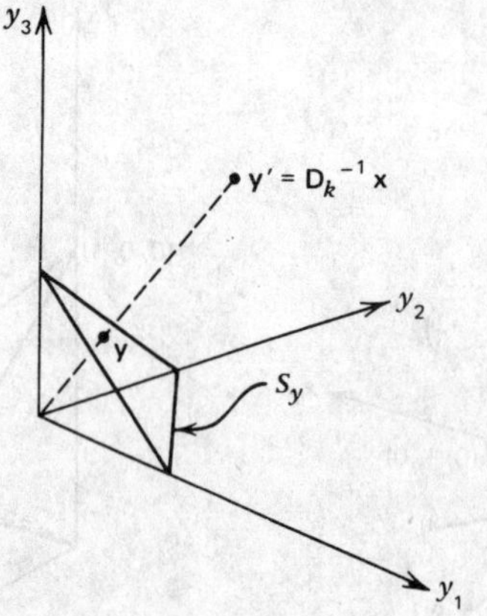

Figure 8.3. **Projective transformation.**

To algebraically describe this feasible region in the **y**-space, consider the inverse projective transformation obtained by solving for **x** in S_x from transformation (8.5*a*). This yields $\mathbf{x} = (\mathbf{D}_k\mathbf{y})(\mathbf{1D}_k^{-1}\mathbf{x})$. Since $\mathbf{1x} = 1$, we have $(\mathbf{1D}_k\mathbf{y})(\mathbf{1D}_k^{-1}\mathbf{x}) = 1$. Hence $(\mathbf{1D}_k^{-1}\mathbf{x}) = 1/(\mathbf{1D}_k\mathbf{y})$. Substituting this into the previous expression for **x** yields

$$\mathbf{x} = \frac{\mathbf{D}_k\mathbf{y}}{\mathbf{1D}_k\mathbf{y}} \tag{8.5b}$$

Therefore the transformation (8.5) maps points from the simplex S_x onto the simplex S_y and vice versa.

Under transformation (8.5), the problem (8.4) gets transformed into the following problem in the **y**-space over a subset of the simplex S_y.

$$\text{Minimize}\left\{\frac{\mathbf{cD}_k\mathbf{y}}{\mathbf{1D}_k\mathbf{y}} : \mathbf{AD}_k\mathbf{y} = \mathbf{0},\ \mathbf{1y} = 1,\ \mathbf{y} \geq \mathbf{0}\right\} \tag{8.6}$$

Note that although the constraints remain linear because $\mathbf{Ax} = \mathbf{0}$ is homogeneous, the objective function has become a quotient of linear functions. This type of problem is known as a *linear fractional programming problem*. However by assumption (A2), the optimal objective value in Equation (8.6) is zero. Hence we can equivalently minimize the numerator in the problem, since the denominator is positive and bounded away from zero for all $\mathbf{y} \in S_y$. This leads to the following problem.

$$\text{Minimize}\ \{\bar{\mathbf{c}}\mathbf{y} : \mathbf{Py} = \mathbf{P}_0,\ \mathbf{y} \geq \mathbf{0}\} \tag{8.7}$$

where

$$\bar{\mathbf{c}} = \mathbf{cD}_k, \qquad \mathbf{P} = \begin{bmatrix} \mathbf{AD}_k \\ \mathbf{1} \end{bmatrix} \quad \text{and} \quad \mathbf{P}_0 = \begin{bmatrix} \mathbf{0} \\ 1 \end{bmatrix}$$

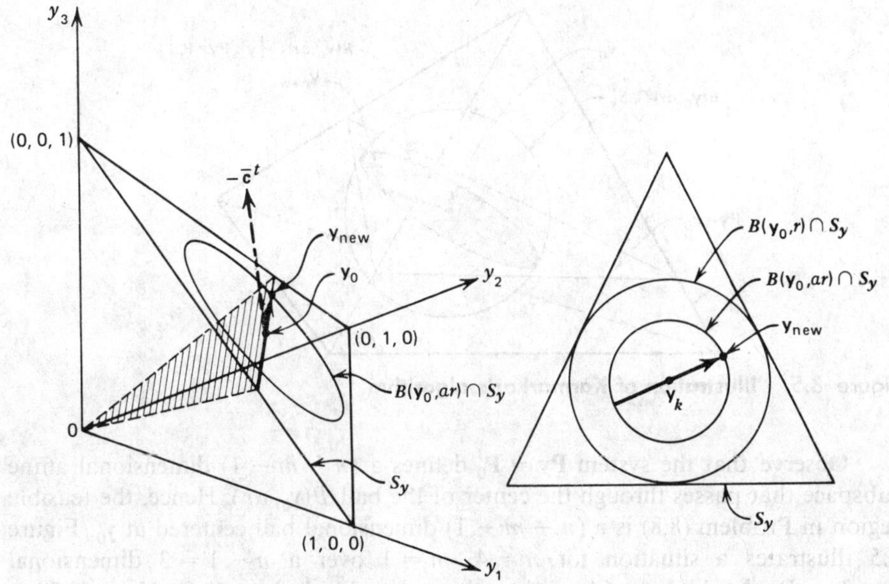

Figure 8.4. Main step of Karmarkar's algorithm.

Rather than solve Problem (8.7), which is equivalent to solving the original problem, we will optimize a particular simpler *restriction* of this problem. This will yield a new feasible solution. By iteratively repeating the step, we will be able to polynomially obtain an optimal solution to the original problem.

To define this restriction, consider the n-dimensional sphere or ball $B(\mathbf{y}_0, r)$ with center at $\mathbf{y}_0 = (1/n, \ldots, 1/n)$ and with the appropriate radius r such that the intersection of this ball with the simplex constraint $\{\mathbf{y} : \mathbf{1y} = 1\}$ is an $(n-1)$ dimensional ball with the same center and radius, which is inscribed within S_y (see Figure 8.4). Note that this radius r is the distance from the center $(1/n, \ldots, 1/n)$ of the simplex to the center of one of its facets. But the facets of S_y are simply one lower dimensional simplices. Hence considering the facet at which the component y_1 is zero, for example, we obtain r as the distance from $(1/n, \ldots, 1/n)$ to $(0, 1/(n-1), \ldots, 1/(n-1))$. This gives $r = 1/\sqrt{n(n-1)}$.

Consider a restriction of Equation (8.7) in which one seeks to minimize $\bar{\mathbf{c}}\mathbf{y}$ subject to $\mathbf{Py} = \mathbf{P}_0$, $\mathbf{y} \geq \mathbf{0}$, and $\mathbf{y} \in B(\mathbf{y}_0, \alpha r)$, where $0 < \alpha < 1$ is some constant. The specific choice of α is addressed later. Note that we have shrunk the inscribed ball by the factor α, and so its radius is αr. Moreover by remaining feasible to this restriction, the iterates remain strictly positive. Since $\mathbf{y} \geq \mathbf{0}$ is implied by the intersection of the ball and the simplex constraint $\{\mathbf{y} : \mathbf{1y} = 1\}$, this restriction is equivalent to the problem

$$\text{Minimize}\left\{\bar{\mathbf{c}}\mathbf{y} : \mathbf{Py} = \mathbf{P}_0,\ (\mathbf{y} - \mathbf{y}_0)^t(\mathbf{y} - \mathbf{y}_0) \leq \alpha^2 r^2\right\} \tag{8.8}$$

where $\{\mathbf{y} : (\mathbf{y} - \mathbf{y}_0)^t(\mathbf{y} - \mathbf{y}_0) \leq \alpha^2 r^2\}$ describes the ball $B(\mathbf{y}_0, \alpha r)$. Figure 8.4 shows the feasible region over the two-dimensional simplex.

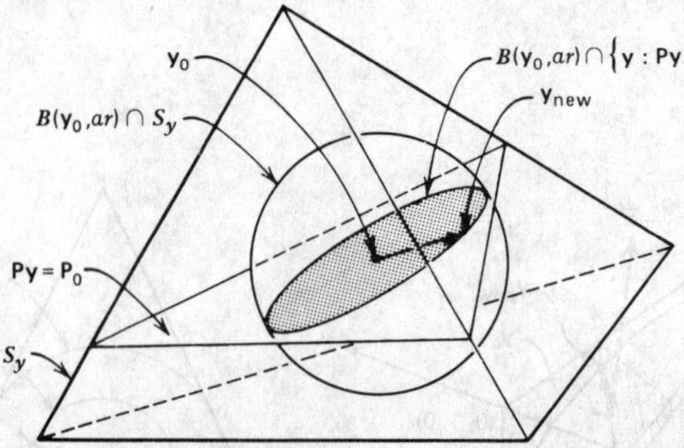

Figure 8.5. Illustration of Karmarkar's algorithm.

Observe that the system $\mathbf{Py} = \mathbf{P}_0$ defines a $(n-m-1)$ dimensional affine subspace that passes through the center of the ball $B(\mathbf{y}_0, \alpha r)$. Hence, the feasible region in Problem (8.8) is a $(n-m-1)$ dimensional ball centered at $\mathbf{y}_0$. Figure 8.5 illustrates a situation for $n = 4$, $m = 1$ over a $n - 1 = 3$ dimensional simplex. It should be evident that the optimal solution to Problem (8.8) is obtained by simply projecting the negative gradient of the objective function $-\bar{\mathbf{c}}^t$, centered at $\mathbf{y}_0$, onto the null space or the constraint surface of $\mathbf{Py} = \mathbf{P}_0$ and moving from $\mathbf{y}_0$ along this projected direction to the boundary of the ball $B(\mathbf{y}_0, \alpha r)$. (See Figures 8.4 and 8.5.) Observe that the simplicity of this solution process is a consequence of the fact that we are optimizing a linear function over a ball, and that we are currently located at the center of this ball. Denoting the projection of the gradient vector $\bar{\mathbf{c}}^t$ as $\mathbf{c}_p$ and the optimum to Problem (8.8) as $\mathbf{y}_{\text{new}}$, we get

$$\mathbf{y}_{\text{new}} = \mathbf{y}_0 - \alpha r \frac{\mathbf{c}_p}{\|\mathbf{c}_p\|} \tag{8.9a}$$

[Note that if $\mathbf{c}_p = \mathbf{0}$, then any feasible solution is optimal (why?), and one may terminate with $\mathbf{x}_k$ as optimal in Problem (8.4).]

In order to calculate $\mathbf{c}_p$, observe that since $\mathbf{c}_p$ lies on the constraint surface of $\mathbf{Py} = \mathbf{P}_0$, the vector $(\bar{\mathbf{c}}^t - \mathbf{c}_p)$ is contained in the space spanned by the gradients to $\mathbf{Py} = \mathbf{P}_0$. Hence there exists a vector $\bar{\mathbf{w}}$ such that $\mathbf{P}^t\bar{\mathbf{w}} = \bar{\mathbf{c}}^t - \mathbf{c}_p$. Multiplying both sides by $\mathbf{P}$, we get $\mathbf{PP}^t\bar{\mathbf{w}} = \mathbf{P}\bar{\mathbf{c}}^t$, since $\mathbf{Pc}_p = \mathbf{0}$ by definition. Since $\mathbf{A}$ is of full row rank and $\mathbf{x}_k > \mathbf{0}$, the square matrix $\mathbf{PP}^t$ is invertible (Exercise 8.23), which gives $\bar{\mathbf{w}} = (\mathbf{PP}^t)^{-1}\mathbf{P}\bar{\mathbf{c}}^t$. Hence we have $\mathbf{c}_p = \bar{\mathbf{c}}^t - \mathbf{P}^t\bar{\mathbf{w}}$ given as

$$\mathbf{c}_p = \left[\mathbf{I} - \mathbf{P}^t(\mathbf{PP}^t)^{-1}\mathbf{P}\right]\bar{\mathbf{c}}^t \tag{8.9b}$$

In Exercise (8.24), we ask the reader to use an alternative approach to show that the optimal solution to Problem (8.8) is given by Equations (8.9). We remark here that the solution $\bar{\mathbf{w}}$ is the solution to the *least-squares problem* of minimizing $\|\bar{\mathbf{c}}^t - \mathbf{P}^t\mathbf{w}\|^2$ over $\mathbf{w} \in E^{(m+1)}$ (see Exercise 8.25). In lieu of explicitly computing $\mathbf{c}_p$ using Equation (8.9b), $\mathbf{c}_p$ may be obtained by deriving a solution $\bar{\mathbf{w}}$ to this problem by any method and then setting $\mathbf{c}_p = \bar{\mathbf{c}}^t - \mathbf{P}^t\bar{\mathbf{w}}$. In either case, the formula (8.9b) gives $\mathbf{c}_p$.

Given $\mathbf{y}_{\text{new}}$ in Equation (8.9a), the corresponding revised vector $\mathbf{x}_{k+1}$ in the x-space is obtained via the inverse transformation (8.5b) as

$$\mathbf{x}_{k+1} = \frac{\mathbf{D}_k \mathbf{y}_{\text{new}}}{\mathbf{1D}_k \mathbf{y}_{\text{new}}} \tag{8.10}$$

Observe that $\mathbf{y}_{\text{new}} > \mathbf{0}$ since it lies in the interior of the $(n-1)$ dimensional sphere inscribed in S_y. Hence $\mathbf{x}_{k+1} > 0$ in Equation (8.10). This completes one iteration. The step may be repeated after incrementing k by one. As shown in Section 8.5, the objective value in Problem (8.4) of the sequence of iterates thus generated approaches the optimal objective value of zero in the limit as $k \to \infty$, although not necessarily monotonically. Hence for practical purposes, this process may be terminated when the objective value gets sufficiently close to zero.

For theoretical purposes, consider the following lower bound L on the input length for Problem (8.4):

$$L = \left\lceil 1 + \log\left(1 + |c_{j\,\max}|\right) + \log\left(|\det{}_{\max}|\right)\right\rceil$$

where $|c_{j\,\max}|$ is the largest numerical value of any cost coefficient c_j and $|\det_{\max}|$ is the largest numerical value of the determinant of any basis for Problem (8.4). It follows from a property of the determinants stated in Exercise 8.7 that L is indeed a lower bound on the input length for Problem (8.4), since $\log(|\det_{\max}|) \le \log(1 + m) + \Sigma_i \Sigma_j \log(1 + |a_{ij}|)$. (Note that an alternative, although larger, value of L is given by replacing $\log(|\det_{\max}|)$ with the latter term, as illustrated in Example 8.2.) In the next section we will show that by using a value of $\alpha = (n-1)/3n$, the algorithm produces a solution of objective value smaller than 2^{-L} within $10nL$ iterations, with an overall effort of polynomial complexity $O(n^{3.5}L)$. Once this occurs, an exact *extreme point* optimal solution to Problem (8.4) may be obtained via the following polynomial procedure.

Starting with the final iterate $\mathbf{x}_k$ obtained with objective value $\mathbf{cx}_k < 2^{-L}$, this procedure finds an extreme point solution with at least as good an objective value using the following method known as a *purification scheme*. If n linearly independent constraints are binding at $\mathbf{x}_k$, then it is already a basic feasible solution. Otherwise there exists a direction $\mathbf{d} \neq \mathbf{0}$ lying in the null space of the binding constraints, that is, satisfying the homogeneous equality system corresponding to the binding constraints. The method now moves the current iterate along the direction $\mathbf{d}$ if $\mathbf{cd} < 0$ and along the direction $-\mathbf{d}$ otherwise until some constraint blocks any further motion by feasibility considerations. This must happen since the feasible region is bounded. Note that at the new solution the objective value is no more than $\mathbf{cx}_k < 2^{-L}$ and at least one additional linearly independent constraint is binding. Proceeding in this fashion, a basic feasible solution $\bar{\mathbf{x}}$ to Problem (8.4) can be obtained with objective value strictly less than 2^{-L}.

Note that this process involves at most $n - (m + 1)$ such steps, since it begins with the $(m + 1)$ linearly independent equality constraints in Problem

(8.4) binding and then adds at least one additional linearly independent binding constraint to this set at each step. Moreover, note that each step is of polynomial complexity (why?). Hence $\bar{\mathbf{x}}$ is determined from the final iterate $\mathbf{x}_k$ in polynomial time. (See Exercise 8.26 for a suggestion for efficiently implementing this process with a polynomial bound of $O(m^2n)$.)

We now show that this resulting extreme point solution $\bar{\mathbf{x}}$ is indeed optimal for Problem (8.4). For this reason, the process is known as an *optimal rounding routine*. Since the data is all integer, any basic feasible solution for Program (8.4) with basis $\mathbf{B}$ has objective value "$\mathbf{c}_B\mathbf{B}^{-1}\mathbf{b}$" given by N/D, where N and D are integers, and where $D \equiv |\det \mathbf{B}|$ (why?). But $D < 2^L$, where L is defined as previously. Hence by assumption (A2), any *nonoptimal* extreme point solution for Problem (8.4) must have a positive objective value $N/D \geq 1/D > 2^{-L}$. Since $\mathbf{c}\bar{\mathbf{x}} < 2^{-L}$, we must therefore have $\bar{\mathbf{x}}$ optimal for Problem (8.4) with $\mathbf{c}\bar{\mathbf{x}} = 0$.

Summary of Karmarkar's Algorithm

INITIALIZATION

Compute $r = 1/\sqrt{n(n-1)}$, $L = \lceil 1 + \log(1 + |c_{j\,\max}|) + \log(|\det_{\max}|) \rceil$, and select $\alpha = (n-1)/3n$. Let $\mathbf{x}_0 = (1/n, \ldots, 1/n)$ and put $k = 0$.

MAIN STEP

If $\mathbf{cx}_k < 2^{-L}$, use the optimal rounding routine to determine an optimal solution, and stop. (Practically, since 2^{-L} may be very small, one may terminate when $\mathbf{cx}_k$ is less than some other desired tolerance.) Otherwise define

$$\mathbf{D}_k = \operatorname{diag}\{x_{k1}, \ldots, x_{kn}\}, \qquad \mathbf{y}_0 = \left(\frac{1}{n}, \ldots, \frac{1}{n}\right)^t,$$

$$\mathbf{P} = \begin{bmatrix} \mathbf{AD}_k \\ \mathbf{1} \end{bmatrix} \quad \text{and} \quad \bar{\mathbf{c}} = \mathbf{cD}_k$$

and compute

$$\mathbf{y}_{\text{new}} = \mathbf{y}_0 - \alpha r \frac{\mathbf{c}_p}{\|\mathbf{c}_p\|}, \qquad \text{where } \mathbf{c}_p = \left[\mathbf{I} - \mathbf{P}^t(\mathbf{PP}^t)^{-1}\mathbf{P}\right]\bar{\mathbf{c}}^t$$

Hence obtain $\mathbf{x}_{k+1} = (\mathbf{D}_k\mathbf{y}_{\text{new}})/(\mathbf{1D}_k\mathbf{y}_{\text{new}})$. Increment k by one and repeat the Main Step.

OPTIMAL ROUNDING ROUTINE

Starting with $\mathbf{x}_k$, determine an extreme point solution $\bar{\mathbf{x}}$ for Problem (8.4) with $\mathbf{c}\bar{\mathbf{x}} \leq \mathbf{cx}_k < 2^{-L}$ using the earlier *purification scheme*. Terminate with $\bar{\mathbf{x}}$ as an optimal solution to Problem (8.4).

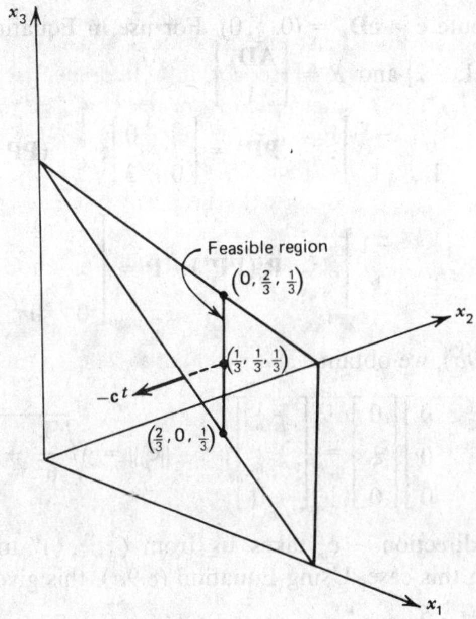

Figure 8.6. **Illustration for Example 8.1.**

Example 8.1

Consider the linear programming problem

$$\begin{aligned} &\text{Minimize} \quad x_2 \\ &\text{Subject to} \quad x_1 + x_2 - 2x_3 = 0 \\ &\qquad\qquad\quad\; x_1 + x_2 + \;\; x_3 = 1 \\ &\qquad\qquad\quad\; x_1, \;\; x_2, \;\;\; x_3 \geq 0 \end{aligned}$$

Here $n = 3$ and $m = 1$. The $n - m - 1 = 1$ dimensional feasible region is illustrated in Figure 8.6. Clearly, the point $\mathbf{x}_0 = (\frac{1}{3}, \frac{1}{3}, \frac{1}{3})^t$ is feasible and the solution $(\frac{2}{3}, 0, \frac{1}{3})^t$ is optimal with objective value zero. Hence assumptions (A1) and (A2) are satisfied. We also have $r = 1/\sqrt{6}$ and $\alpha = \frac{2}{9}$.

Iteration 1

Starting with $\mathbf{x}_0$ and $k = 0$, we define $\mathbf{D}_0 = \text{diag}\{\frac{1}{3}, \frac{1}{3}, \frac{1}{3}\}$. Since $\mathbf{1D}_k^{-1}\mathbf{x} = \Sigma 3x_i = 3$, the projective transformation (8.5a) gives $\mathbf{y} \equiv \mathbf{x}$. Hence the problem in the $\mathbf{y}$-space coincides with that of Figure 8.6. The current point in the $\mathbf{y}$-space, as always, is $\mathbf{y}_0 = (\frac{1}{3}, \frac{1}{3}, \frac{1}{3})^t$.

We now compute $\bar{\mathbf{c}} = \mathbf{c}\mathbf{D}_k = (0, \frac{1}{3}, 0)$. For use in Equation (8.9b), we compute using $\mathbf{A} = [1, 1, -2]$ and $P = \begin{bmatrix} \mathbf{A}\mathbf{D}_k \\ \mathbf{1} \end{bmatrix}$.

$$\mathbf{P} = \begin{bmatrix} \frac{1}{3} & \frac{1}{3} & -\frac{2}{3} \\ 1 & 1 & 1 \end{bmatrix}, \qquad \mathbf{P}\mathbf{P}^t = \begin{bmatrix} \frac{2}{3} & 0 \\ 0 & 3 \end{bmatrix}, \qquad (\mathbf{P}\mathbf{P}^t)^{-1} = \begin{bmatrix} \frac{3}{2} & 0 \\ 0 & \frac{1}{3} \end{bmatrix}$$

$$(\mathbf{P}\mathbf{P}^t)^{-1}\mathbf{P} = \begin{bmatrix} \frac{1}{2} & \frac{1}{2} & -1 \\ \frac{1}{3} & \frac{1}{3} & \frac{1}{3} \end{bmatrix}, \qquad \mathbf{P}^t(\mathbf{P}\mathbf{P}^t)^{-1}\mathbf{P} = \begin{bmatrix} \frac{1}{2} & \frac{1}{2} & 0 \\ \frac{1}{2} & \frac{1}{2} & 0 \\ 0 & 0 & 1 \end{bmatrix}$$

From Equation (8.9b), we obtain

$$\mathbf{c}_p = \begin{bmatrix} \frac{1}{2} & -\frac{1}{2} & 0 \\ -\frac{1}{2} & \frac{1}{2} & 0 \\ 0 & 0 & 0 \end{bmatrix} \begin{bmatrix} 0 \\ \frac{1}{3} \\ 0 \end{bmatrix} = \begin{bmatrix} -\frac{1}{6} \\ \frac{1}{6} \\ 0 \end{bmatrix}, \qquad \|\mathbf{c}_p\| = \sqrt{\frac{1}{6^2} + \frac{1}{6^2} + 0} = \frac{\sqrt{2}}{6}$$

Observe that the direction $-\mathbf{c}_p$ takes us from $(\frac{1}{3}, \frac{1}{3}, \frac{1}{3})^t$ toward the optimal solution $(\frac{2}{3}, 0, \frac{1}{3})^t$ in this case. Using Equation (8.9a), this gives us

$$\mathbf{y}_{\text{new}} = \begin{bmatrix} \frac{1}{3} \\ \frac{1}{3} \\ \frac{1}{3} \end{bmatrix} - \left(\frac{2}{9}\right)\left(\frac{1}{\sqrt{6}}\right)\left(\frac{6}{\sqrt{2}}\right) \begin{bmatrix} -\frac{1}{6} \\ \frac{1}{6} \\ 0 \end{bmatrix} = \begin{bmatrix} 0.397484 \\ 0.269183 \\ 0.333333 \end{bmatrix}$$

Transforming into the $\mathbf{x}$-space via Equation (8.10), we obtain $\mathbf{x}_1 = \mathbf{y}_{\text{new}}$ for this first iteration. The present objective value is $\mathbf{c}\mathbf{x}_1 = 0.269183$.

Iteration 2

We now have $k = 1$. We define

$$\mathbf{D}_1 = \text{diag}\{0.397484, 0.269183, 0.333333\}, \; \bar{\mathbf{c}} = \mathbf{c}\mathbf{D}_k = (0, 0.269183, 0)$$

and

$$\mathbf{P} = \begin{bmatrix} \mathbf{A}\mathbf{D}_k \\ \mathbf{1} \end{bmatrix} = \begin{bmatrix} 0.397484 & 0.269183 & -0.666666 \\ 1 & 1 & 1 \end{bmatrix}$$

From Equation (8.9b), we compute

$$\mathbf{c}_p = \left[\mathbf{I} - \mathbf{P}^t(\mathbf{P}\mathbf{P}^t)^{-1}\mathbf{P}\right]\bar{\mathbf{c}}^t = \begin{bmatrix} -0.1324029 \\ 0.1505548 \\ -0.0181517 \end{bmatrix} \quad \text{and} \quad \|\mathbf{c}_p\| = 0.2013121$$

Substituting into Equation (8.9a) gives, using $\alpha r/\|\mathbf{c}_p\| = 0.4506524$, that

$$\mathbf{y}_{\text{new}} = \begin{bmatrix} \frac{1}{3} \\ \frac{1}{3} \\ \frac{1}{3} \end{bmatrix} - 0.4506524 \begin{bmatrix} -0.1324029 \\ 0.1505548 \\ -0.0181517 \end{bmatrix} = \begin{bmatrix} 0.3930009 \\ 0.2654855 \\ 0.3415134 \end{bmatrix}.$$

From Equation (8.10), we obtain

$$\mathbf{D}_1\mathbf{y}_{\text{new}} = \begin{bmatrix} 0.1562112 \\ 0.0714641 \\ 0.1138378 \end{bmatrix}, \qquad \mathbf{1D}_1\mathbf{y}_{\text{new}} = 0.3415131,$$

$$\mathbf{x}_2 = \frac{\mathbf{D}_1\mathbf{y}_{\text{new}}}{\mathbf{1D}_1\mathbf{y}_{\text{new}}} = \begin{bmatrix} 0.457409 \\ 0.209258 \\ 0.333333 \end{bmatrix}.$$

This completes the second iteration. The current objective function value is $\mathbf{cx}_2 = 0.209258$.

It should be evident that the iterates $\mathbf{x}_k$ being generated are of the type $(\frac{1}{3} + \theta, \frac{1}{3} - \theta, \frac{1}{3})'$, where θ is gradually increasing. In fact, by carrying out these computations, it may be verified that θ approaches $\frac{1}{3}$ in the limit as $k \to \infty$. Note that $L = \lceil 1 + \log(2) + \log(3) \rceil = 4$, and so, $2^{-L} = 0.0625$ for our problem. Hence theoretically, we may halt the computations when $\mathbf{cx}_k < 0.0625$. This happens for $k = 6$, from where we obtain $\mathbf{x}_6 = (0.606509, 0.060158, 0.333333)'$, with $\mathbf{cx}_6 = 0.060158$.

Finally let us optimally round this solution. Note that only the two (linearly independent) equality constraints in the problem are binding at $\mathbf{x}_6$. Hence we need to first determine a solution $\mathbf{d} = (d_1, d_2, d_3) \neq \mathbf{0}$ to the homogeneous system $d_1 + d_2 - 2d_3 = 0$ and $d_1 + d_2 + d_3 = 0$. Solving for d_1 and d_3 in terms of d_2, we get $d_1 = -d_2$ and $d_3 = 0$. Arbitrarily taking $d_2 = 1$ gives the direction $\mathbf{d} = (-1, 1, 0)'$. Since $\mathbf{cd} = 1 > 0$, we move along the direction $-\mathbf{d} = (1, -1, 0)'$. The constraint $x_2 \geq 0$ blocks the motion along this direction after a step length of 0.060158 and yields the required third linearly independent constraint. The new solution $\bar{\mathbf{x}} = (0.666667, 0, 0.333333)'$ is therefore an extreme point solution, which is moreover optimal.

Example 8.2

As another illustration of the computations and the progress of the iterates in Karmarkar's algorithm, consider the linear program of the form (8.4):

$$\begin{array}{lrl} \text{Minimize} & -x_1 - 2x_2 + 4x_5 & \\ \text{Subject to} & x_1 - x_2 + 2x_3 \qquad - 2x_5 &= 0 \\ & x_1 + 2x_2 \qquad + x_4 - 4x_5 &= 0 \\ & x_1 + x_2 + x_3 + x_4 + x_5 &= 1 \\ & \mathbf{x} & \geq \mathbf{0} \end{array}$$

Note that the solution $\mathbf{x}_0 = \left(\frac{1}{5}, \frac{1}{5}, \frac{1}{5}, \frac{1}{5}, \frac{1}{5}\right)'$ is feasible here. Observe from the second constraint that the objective value is nonnegative for all feasible solutions, while it is zero for the feasible solution $\mathbf{x}^* = (0, \frac{2}{5}, \frac{2}{5}, 0, \frac{1}{5})'$. Hence $\mathbf{x}^*$ is optimal with objective value zero, and so, the required assumptions (A1) and (A2) hold.

We now have $n = 5$, so $r = 1/\sqrt{20}$ and $\alpha = \frac{4}{15}$. Following the earlier suggestion of using $\log(1 + m) + \Sigma_i\Sigma_j \log(1 + |a_{ij}|)$ in lieu of having to compute $\log(|\det_{\max}|)$, we have $L = \lceil 1 + \log(5) + \log(3) + \{4\log(2) + 3\log(3) + \log(5)\}\rceil = 16$, so that $2^{-L} = 0.00001526$. We start with the known feasible, positive solution $\mathbf{x}_0$ with objective value 0.2.

Iteration 1

The following quantities are computed:

$$\mathbf{D}_0 = \text{diag}\{0.2, 0.2, 0.2, 0.2, 0.2\}, \qquad \mathbf{y}_0 = (0.2, 0.2, 0.2, 0.2, 0.2)',$$

$$\bar{\mathbf{c}} = (-0.2, -0.4, 0, 0, 0.8),$$

$$\mathbf{P} = \begin{bmatrix} 0.2 & -0.2 & 0.4 & 0.0 & -0.4 \\ 0.2 & 0.4 & 0.0 & 0.2 & -0.8 \\ 1.0 & 1.0 & 1.0 & 1.0 & 1.0 \end{bmatrix},$$

$$\left[\mathbf{I} - \mathbf{P}'(\mathbf{P}\mathbf{P}')^{-1}\mathbf{P}\right] = \begin{bmatrix} 0.6947 & -0.1474 & -0.3754 & -0.2175 & 0.0456 \\ -0.1474 & 0.2737 & 0.2211 & -0.3579 & 0.0105 \\ -0.3754 & 0.2211 & 0.2854 & -0.1181 & -0.0129 \\ -0.2175 & -0.3579 & -0.1181 & 0.7415 & -0.0480 \\ 0.0456 & 0.0105 & -0.0129 & -0.0480 & 0.0047 \end{bmatrix}$$

$$\mathbf{c}_p = (-0.0435, -0.0716, -0.0236, 0.1483, -0.0096)', \qquad \|\mathbf{c}_p\| = 0.1722$$

$$\mathbf{y}_{\text{new}} = \mathbf{y}_0 - \frac{\alpha r \mathbf{c}_p}{\|\mathbf{c}_p\|} = (0.2151, 0.2248, 0.2082, 0.1487, 0.2033)',$$

$$\mathbf{x}_1 = \frac{\mathbf{D}_0 \mathbf{y}_{\text{new}}}{\mathbf{1}\mathbf{D}_0 \mathbf{y}_{\text{new}}} \equiv \mathbf{y}_{\text{new}}$$

for this first iteration, and $\mathbf{c}\mathbf{x}_1 = 0.1487$. This is now repeated with $\mathbf{x}_1$ leading to the sequence

$$\mathbf{x}_2 = (0.2268, 0.2446, 0.2149, 0.1078, 0.2059)' \qquad \text{with } \mathbf{c}\mathbf{x}_2 = 0.1078,$$

$$\mathbf{x}_3 = (0.2355, 0.2596, 0.2200, 0.0769, 0.2079)' \qquad \text{with } \mathbf{c}\mathbf{x}_3 = 0.0769,$$

until we obtain

$$\mathbf{x}_{26} = (0.2569681, 0.2972037, 0.2329655, 0.0000150, 0.2128477)' \qquad \text{with } \mathbf{c}\mathbf{x}_{26} = 0.000015$$

Note that $\mathbf{c}\mathbf{x}_{26} < 2^{-L} = 0.00001526$ at termination.

To perform the optimal rounding, note that corresponding to $\bar{\mathbf{x}} = \mathbf{x}_{26}$, the three (linearly independent) equality constraints in the problem are binding. Hence we need to determine a nonzero solution to the homogeneous system $d_1 - d_2 + 2d_3 - 2d_5 = 0$, $d_1 + 2d_2 + d_4 - 4d_5 = 0$, and $d_1 + d_2 + d_3 + d_4 + d_5 = 0$. Solving for d_1, d_2, and d_3 in terms of d_4 and d_5, we get $d_1 = d_4 + 20d_5$, $d_2 = -d_4 - 8d_5$, and $d_3 = -d_4 - 13d_5$. Arbitrarily setting $d_4 = 1$ and $d_5 = 0$ gives the direction $\mathbf{d} = (1, -1, -1, 1, 0)'$. Since $\mathbf{c}\mathbf{d} = 1 > 0$, we move along the direction $-\mathbf{d}$. This motion is blocked by the constraint $x_4 \geq 0$ at a step length of

0.000015, which leads to the new solution $\bar{\mathbf{x}} = (0.2569531, 0.2972187, 0.2329805, 0, 0.2128477)'$. Observe that $\mathbf{c}\bar{\mathbf{x}} = 0$, and so we are incidentally already at an optimal, but nonextreme point solution. Continuing, we now have four linearly independent constraints binding. Adding the constraint $d_4 = 0$ to this homogeneous system, we obtain $d_1 = 20d_5$, $d_2 = -8d_5$, and $d_3 = -13d_5$, in terms of d_5. Arbitrarily setting $d_5 = 1$ gives the direction $\mathbf{d} = (20, -8, -13, 0, 1)'$. Since $\mathbf{cd} = 0$, we can move along the direction $\mathbf{d}$. The constraint $x_3 \geq 0$ blocks the motion along this direction at a step length of 0.017921576, which yields the revised solution $\bar{\mathbf{x}} = (0.6153846, 0.1538461, 0, 0, 0.2307692)'$. Since there are five linearly independent constraints binding at this new solution $\bar{\mathbf{x}}$, it is an optimal extreme point solution. Note that if we moved along the direction $-\mathbf{d}$ at the previous step, we would have reached the solution $(0, 0.4, 0.4, 0, 0.2)'$ which happens to be the other of the two alternative optimal extreme point solutions for this problem.

Converting a General Linear Program into Karmarkar's Form

Consider a general linear programming problem to

$$\text{Minimize } \{\mathbf{cx} : \mathbf{Ax} = \mathbf{b}, \mathbf{x} \geq \mathbf{0}\} \tag{8.11}$$

where $\mathbf{A}$ is $m \times n$ of rank m and the data is all integer. We will now show how to convert this problem into the form of Linear Program (8.4) required by Karmarkar, while satisfying assumptions (A1) and (A2). (See Exercises 8.41 and 8.43 for an alternative technique.)

Toward this end, let us begin by first *regularizing* problem (8.11), that is, adding a bounding constraint $\sum_{j=1}^{n} x_j \leq Q$. Here Q may be taken as some known (hopefully small) integer bound on the sum of the variables, derived from feasibility and/or optimality considerations. Noting Exercise 8.8, one can take $Q = 2^L$ in the worst case, where L is given by Equation (8.1). If this constraint is binding at optimality with the objective value of magnitude $-2^{O(L)}$, then the given problem (8.11) can be deduced to be unbounded.

Adding this constraint along with a slack variable x_{n+1}, we may rewrite Equation (8.11) as follows:

$$\text{Minimize } \{\mathbf{cx} : \mathbf{Ax} = \mathbf{b}, \mathbf{1x} + x_{n+1} = Q, \mathbf{x} \geq \mathbf{0}, x_{n+1} \geq 0\}$$

At this stage, we can homogenize the constraints $\mathbf{Ax} = \mathbf{b}$ by equivalently replacing the right-hand-side $\mathbf{b}$ with $\mathbf{b}(\mathbf{1x} + x_{n+1})/Q$, noting the foregoing final equality constraint. However, this will ruin any sparsity structure in $\mathbf{A}$ since $\mathbf{b}$ is usually dense. Alternatively, let us introduce a dummy variable x_{n+2} along with the constraint $x_{n+2} = 1$ to equivalently rewrite the constraints as $\mathbf{Ax} - \mathbf{b}x_{n+2} = \mathbf{0}$, $x_{n+2} = 1$, $\mathbf{1x} + x_{n+1} + x_{n+2} = Q + 1$, plus the nonnegativity restrictions. Using the final equality constraint to homogenize the constraint $x_{n+2} = 1$, we can equivalently write the problem as follows.

$$\begin{aligned}\text{Minimize } \{\mathbf{cx} : \mathbf{Ax} - \mathbf{b}x_{n+2} = \mathbf{0}, \mathbf{1x} + x_{n+1} - Qx_{n+2} = 0,\\ \mathbf{1x} + x_{n+1} + x_{n+2} = (Q + 1), \mathbf{x} \geq \mathbf{0}, x_{n+1} \geq 0, x_{n+2} \geq 0\}\end{aligned}$$

Next let us use the transformation $x_j = (Q+1)y_j$, $j = 1, \ldots, n+2$ in order to obtain unity as the right-hand-side of the final equality constraint and hence derive the following equivalent problem.

$$\text{Minimize } \{\mathbf{cy} : \mathbf{Ay} - \mathbf{b}y_{n+2} = 0,\ \mathbf{1y} + y_{n+1} - Qy_{n+2} = 0,\ \mathbf{1y} + y_{n+1} + y_{n+2} = 1, \mathbf{y} \geq \mathbf{0},\ y_{n+1} \geq 0,\ y_{n+2} \geq 0\} \quad (8.12)$$

The constraints in Equation (8.12) are now of the form in Problem (8.4). In order to satisfy assumption (A1), let us introduce an artificial variable y_{n+3} into Equation (8.12) with objective coefficient M and constraint coefficients such that the solution $(y_1, \ldots, y_{n+3}) = [1/(n+3), \ldots, 1/(n+3)]$ is feasible. For this to occur, the coefficients of the resulting homogeneous constraints must add to zero, and the coefficient of y_{n+3} in the final equality constraint must be unity. Note that a value of M of magnitude $2^{O(L)}$ exists that ensures that the artificial variable y_{n+3} will be zero at optimality, provided that Equation (8.12) is feasible. (See Exercise 8.27.) Hence the input length of Equation (8.12) with the artificial variable is still of $O(L)$. This artificial problem is then given as follows.

$$\begin{aligned} \text{Minimize} \quad & \mathbf{cy} + My_{n+3} \\ \text{Subject to} \quad & \mathbf{Ay} - \mathbf{b}y_{n+2} - [\mathbf{A1}' - \mathbf{b}]y_{n+3} = \mathbf{0} \\ & \mathbf{1y} + y_{n+1} - Qy_{n+2} - (n+1-Q)y_{n+3} = 0 \\ & \mathbf{1y} + y_{n+1} + y_{n+2} + y_{n+3} = 1 \\ & y_j \geq 0,\ j = 1, \ldots, n+3 \end{aligned} \quad (8.13)$$

Problem (8.13) is of the form (8.4) in $(n+3)$ variables, with $[1/(n+3), \ldots, 1/(n+3)]$ a feasible solution. Moreover, its size is polynomially related to that of Problem (8.11), that is, its input length is of $O(L)$.

Finally consider assumption (A2). In the following section, we show how one can solve a linear program of the form (8.4) that satisfies assumption (A1), but dispenses with assumption (A2). Using this technique, Problem (8.11) may be solved via the Problem (8.13). However a simple though computationally inadvisable way of satisfying assumption (A2) in the earlier derivation is as follows.

By linear programming duality, note that the optimality conditions for Problem (8.11) require a solution, if one exists, to the system

$$\mathbf{Ax} = \mathbf{b},\ \mathbf{x} \geq \mathbf{0},\ \mathbf{wA} \leq \mathbf{c}, \quad \text{and} \quad \mathbf{cx} = \mathbf{wb} \quad (8.14)$$

Putting these constraints in a standard form of the type $\mathbf{A'x'} = \mathbf{b'}$ and $\mathbf{x'} \geq \mathbf{0}$, the foregoing transformations maybe used to obtain a problem of the type (8.13), where the objective function is simply the minimization of the artificial variable y_{n+3}. If the original problem has an optimum, then this problem (8.13) has an optimal objective value of zero. Moreover, its solution yields both primal and dual optimal solutions to the original problem (8.11). Otherwise the objective value of Problem (8.13) is positive at optimality. This will be revealed by the inability of the algorithm to drive the objective value below $2^{-L'}$ within the polynomial bound $10nL'$ on the number of iterations as established in the following section, where $L' \equiv O(L)$ represents the input length of the transformed problem (8.13). In this case, the original problem is either infeasible or unbounded, and this may be resolved by similarly testing for primal and dual feasibility.

Example 8.3

To illustrate the use of duality in concert with the foregoing transformations, consider the following linear programming problem.

$$\text{Maximize} \quad 2x_1 + x_2$$

$$\begin{aligned} \text{Subject to} \quad x_1 - x_2 &\le 2 \\ x_1 + 2x_2 &\le 4 \\ \mathbf{x} &\ge \mathbf{0} \end{aligned}$$

The dual to this problem is given as follows.

$$\text{Minimize } \{2w_1 + 4w_2 : w_1 + w_2 \ge 2, -w_1 + 2w_2 \ge 1, \mathbf{w} \ge \mathbf{0}\}$$

Adding slack variables (x_3, x_4) and (w_3, w_4) in the primal and dual problems, respectively, we need to determine a solution to the following system in order to find a pair of primal and dual optimal solutions.

$$\begin{aligned} &x_1 - x_2 + x_3 = 2, &&x_1 + 2x_2 + x_4 = 4, \\ &w_1 + w_2 - w_3 = 2, &&-w_1 + 2w_2 - w_4 = 1, \\ &2x_1 + x_2 = 2w_1 + 4w_2, &&\text{and } \mathbf{x} \ge \mathbf{0}, \mathbf{w} \ge \mathbf{0} \end{aligned}$$

Let us now introduce the bounding constraint $\Sigma x_i + \Sigma w_i \le Q$. (An examination of the foregoing system suggests that $Q = 15$ is sufficiently large.) Adding a slack variable s_1, this constraint becomes $\Sigma x_i + \Sigma w_i + s_1 = Q$. Next we introduce a new dummy variable s_2 along with the restriction that $s_2 = 1$ and use this variable to homogenize the problem constraints. Further we replace $s_2 = 1$ and $\Sigma x_i + \Sigma w_i + s_1 = Q$ with the equivalent constraints $\Sigma x_i + \Sigma w_i + s_1 - Qs_2 = 0$ and $\Sigma x_i + \Sigma w_i + s_1 + s_2 = (Q + 1)$. This yields the system:

$$\begin{aligned} &x_1 - x_2 + x_3 - 2s_2 = 0, &&x_1 + 2x_2 + x_4 - 4s_2 = 0, \\ &w_1 + w_2 - w_3 - 2s_2 = 0, &&-w_1 + 2w_2 - w_4 - s_2 = 0, \\ &&2x_1 + x_2 - 2w_1 - 4w_2 = 0,& \\ &\sum x_i + \sum w_i + s_1 - Qs_2 = 0, &&\sum x_i + \sum w_i + s_1 + s_2 = (Q + 1), \\ &\text{and } \mathbf{x} \ge \mathbf{0}, \quad \mathbf{w} \ge \mathbf{0}, &&s_1 \ge 0, \quad s_2 \ge 0 \end{aligned}$$

Next we use a change of variables defined by $x_j = (Q + 1)y_j$ for $j = 1, \ldots, 4$, $w_j = (Q + 1)y_{4+j}$ for $j = 1, \ldots, 4$, $s_1 = (Q + 1)y_9$, and $s_2 = (Q + 1)y_{10}$. This gives the system

$$\begin{aligned} &y_1 - y_2 + y_3 - 2y_{10} = 0, &&y_1 + 2y_2 + y_4 - 4y_{10} = 0, \\ &y_5 + y_6 - y_7 - 2y_{10} = 0, &&-y_5 + 2y_6 - y_8 - y_{10} = 0, \\ &&2y_1 + y_2 - 2y_5 - 4y_6 = 0,& \\ &\sum_{i=1}^{9} y_i - Qy_{10} = 0, &&\sum_{i=1}^{10} y_i = 1, \quad \text{and} \quad \mathbf{y} \ge \mathbf{0} \end{aligned}$$

Finally introducing the artificial variable y_{11} with constraint coefficients such that the sum of the coefficients in each homogeneous constraint is zero and accommodating y_{11} in the final equality constraint as well (so that assumption A1 holds), we get the following problem.

$$\begin{aligned}
\text{Minimize} \quad & y_{11} \\
\text{Subject to} \quad & y_1 - y_2 + y_3 - 2y_{10} + y_{11} = 0 \\
& y_1 + 2y_2 + y_4 - 4y_{10} = 0 \\
& y_5 + y_6 - y_7 - 2y_{10} + y_{11} = 0 \\
& -y_5 + 2y_6 - y_8 - y_{10} + y_{11} = 0 \\
& 2y_1 + y_2 - 2y_5 - 4y_6 + 3y_{11} = 0 \\
& \sum_{i=1}^{9} y_i - Qy_{10} + (Q-9)y_{11} = 0 \\
& \sum_{i=1}^{11} y_i = 1, \quad \text{and} \quad \mathbf{y} \geq \mathbf{0}
\end{aligned}$$

The problem is now in the required form (8.4), and it satisfies assumptions (A1) and (A2). Using Karmarkar's algorithm to solve this problem will yield a pair of primal and dual optimal solutions to the original linear program.

8.5 ANALYSIS OF KARMARKAR'S ALGORITHM: CONVERGENCE, COMPLEXITY, SLIDING OBJECTIVE METHOD, AND BASIC OPTIMAL SOLUTIONS

In this section we begin with a convergence and complexity analysis of Karmarkar's algorithm applied to Problem (8.4) under assumptions (A1) and (A2). Next we present a sliding objective function technique that dispenses with assumption (A2) and polynomially solves Problem (8.4) under assumption (A1) alone. As before, the polynomial rounding routine we adopt is a purification scheme that yields an optimal *basic* solution. Using transformation (8.13) of the previous section provides the capability of obtaining an optimal basic solution to a general linear program (8.11), given that one exists. For such a solution, a primal basis **B** whose complementary dual basis is also feasible may be obtained in polynomial time. As expounded in Chapter 6, the latter capability is important from the viewpoint of deriving insightful duality interpretations in terms of shadow prices and for performing useful sensitivity analyses.

Constructs for the Convergence and Complexity Analysis

In order to prove the convergence of the algorithm and establish its polynomial complexity, there are two constructs that we employ. The first is the consideration of a *relaxation* of Problem (8.7), which complements its restriction (8.8). The second is the use of a novel function known as the *potential function*, that measures the progress of the algorithm in a very clever fashion and helps establish its polynomial complexity.

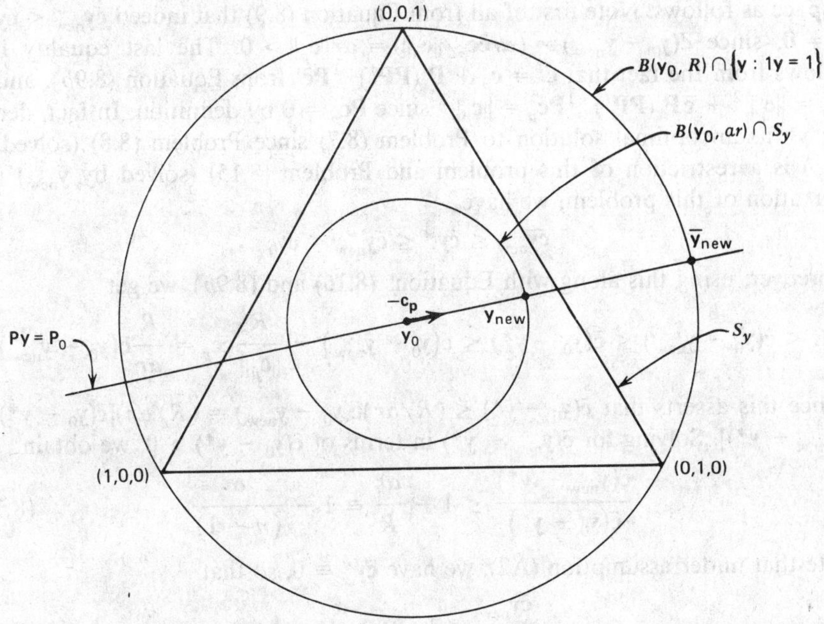

Figure 8.7. **Relaxation over the circumscribed ball.**

To construct the required relaxation of Problem (8.7), consider the n-dimensional ball $B(\mathbf{y}_0, R)$ in the $\mathbf{y}$-space that has its center at $\mathbf{y}_0$, coinciding with that of the simplex S_y, and that has radius R such that $B(\mathbf{y}_0, R)$ intersected with $\{\mathbf{y} : \mathbf{1y} = 1\}$ is an $(n-1)$ dimensional ball that circumscribes S_y (see Figure 8.7). Hence R is the distance from the point $(1/n, \ldots, 1/n)$ to any vertex of S_y, say, $(1, 0, \ldots, 0)$. This gives $R = \sqrt{(n-1)/n}$. Now consider the problem obtained by adding the redundant constraint $\mathbf{y} \in B(\mathbf{y}_0, R)$ in Problem (8.7), but relaxing (deleting) the nonnegativity restrictions:

$$\text{Minimize } \left\{ \bar{\mathbf{c}}\mathbf{y} : \mathbf{Py} = \mathbf{P}_0, (\mathbf{y} - \mathbf{y}_0)^t(\mathbf{y} - \mathbf{y}_0) \le R^2 \right\} \tag{8.15}$$

Again, the feasible region in Equation (8.15) is an $(n-m-1)$ dimensional ball centered at $\mathbf{y}_0$, defined by the intersection of the $(n-m-1)$ dimensional affine subspace $\{\mathbf{y} : \mathbf{Py} = \mathbf{P}_0\}$ with $B(\mathbf{y}_0, R)$. Figure 8.7 shows this intersection for a case with $n = 3$ and $m = 1$. (For an instance with $n = 4$ and $m = 1$, a similar intersection with respect to the inscribed (shrunk) sphere is shown in Figure 8.5.) Hence the optimal solution $\bar{\mathbf{y}}_{\text{new}}$ to Problem (8.15) is obtained in a similar fashion to Equation (8.9), and is given by

$$\bar{\mathbf{y}}_{\text{new}} = \mathbf{y}_0 - \frac{R\mathbf{c}_p}{\|\mathbf{c}_p\|} \tag{8.16}$$

where $\mathbf{c}_p$ is defined in Equation (8.9b). (See Figure 8.7.) One can now obtain an estimate for the progress made with respect to the objective function $\bar{\mathbf{c}}\mathbf{y}$ in the y-space as follows. Note first of all from Equation (8.9) that indeed $\bar{\mathbf{c}}\mathbf{y}_{\text{new}} < \bar{\mathbf{c}}\mathbf{y}_0$ if $\mathbf{c}_p \neq \mathbf{0}$, since $\bar{\mathbf{c}}(\mathbf{y}_0 - \mathbf{y}_{\text{new}}) = (\alpha r \bar{\mathbf{c}}\mathbf{c}_p/\|\mathbf{c}_p\|) = \alpha r \|\mathbf{c}_p\| > 0$. The last equality here follows from the fact that $\bar{\mathbf{c}}' = \mathbf{c}_p + \mathbf{P}'(\mathbf{P}\mathbf{P}')^{-1}\mathbf{P}\bar{\mathbf{c}}'$ from Equation (8.9b), and so $\bar{\mathbf{c}}\mathbf{c}_p = \|\mathbf{c}_p\|^2 + \bar{\mathbf{c}}\mathbf{P}'(\mathbf{P}\mathbf{P}')^{-1}\mathbf{P}\mathbf{c}_p = \|\mathbf{c}_p\|^2$ since $\mathbf{P}\mathbf{c}_p = \mathbf{0}$ by definition. In fact, denoting $\mathbf{y}^*$ as an optimal solution to Problem (8.7) since Problem (8.8) (solved by $\mathbf{y}_{\text{new}}$) is a restriction of this problem and Problem (8.15) (solved by $\bar{\mathbf{y}}_{\text{new}}$) is a relaxation of this problem, we have

$$\bar{\mathbf{c}}\bar{\mathbf{y}}_{\text{new}} \leq \bar{\mathbf{c}}\mathbf{y}^* \leq \bar{\mathbf{c}}\mathbf{y}_{\text{new}} < \bar{\mathbf{c}}\mathbf{y}_0$$

Moreover, using this along with Equations (8.16) and (8.9b), we get

$$0 < \bar{\mathbf{c}}(\mathbf{y}_0 - \mathbf{y}_{\text{new}}) \leq \bar{\mathbf{c}}(\mathbf{y}_0 - \mathbf{y}^*) \leq \bar{\mathbf{c}}(\mathbf{y}_0 - \bar{\mathbf{y}}_{\text{new}}) = \frac{R}{\|\mathbf{c}_p\|}\bar{\mathbf{c}}\mathbf{c}_p = \frac{R}{\alpha r}\bar{\mathbf{c}}(\mathbf{y}_0 - \mathbf{y}_{\text{new}})$$

Hence this asserts that $\bar{\mathbf{c}}(\mathbf{y}_0 - \mathbf{y}^*) \leq (R/\alpha r)\bar{\mathbf{c}}(\mathbf{y}_0 - \mathbf{y}_{\text{new}}) = (R/\alpha r)[\bar{\mathbf{c}}(\mathbf{y}_0 - \mathbf{y}^*) - \bar{\mathbf{c}}(\mathbf{y}_{\text{new}} - \mathbf{y}^*)]$. Solving for $\bar{\mathbf{c}}(\mathbf{y}_{\text{new}} - \mathbf{y}^*)$ in terms of $\bar{\mathbf{c}}(\mathbf{y}_0 - \mathbf{y}^*) > 0$, we obtain

$$\frac{\bar{\mathbf{c}}(\mathbf{y}_{\text{new}} - \mathbf{y}^*)}{\bar{\mathbf{c}}(\mathbf{y}_0 - \mathbf{y}^*)} \leq 1 - \frac{\alpha r}{R} \equiv 1 - \frac{\alpha}{(n-1)} \tag{8.17}$$

Note that under assumption (A2), we have $\bar{\mathbf{c}}\mathbf{y}^* \equiv 0$, so that

$$\frac{\bar{\mathbf{c}}\mathbf{y}_{\text{new}}}{\bar{\mathbf{c}}\mathbf{y}_0} \leq 1 - \frac{\alpha}{(n-1)} \tag{8.18}$$

Equation (8.17) tells us that at any iteration, with respect to the objective function $\bar{\mathbf{c}}\mathbf{y}$, the gap to optimality is reduced by $100\alpha/(n-1)$ percent. However $\bar{\mathbf{c}} = \mathbf{c}\mathbf{D}_k$ is dependent on the particular iteration k, and moreover, Equation (8.17) only guarantees a decrease in the numerator of Problem (8.6). In fact, the fractional objective in Problem (8.6), and hence the original objective function in Problem (8.4), may not fall and may actually increase. Fortunately, there is another function that preserves strict monotonicity and hence assures convergence to optimality. In fact, it assures polynomial-time convergence. This function is known as a *potential function*, and is given by

$$f(x) \equiv \sum_{j=1}^{n} \ln\left[\frac{\mathbf{c}\mathbf{x}}{x_j}\right] = n\ln(\mathbf{c}\mathbf{x}) - \sum_{j=1}^{n}\ln(x_j) \tag{8.19a}$$

where $\ln(\cdot)$ denotes logarithm to the Naperian (natural) base e. {Taking the antilog of Equation (8.19a) note that one can equivalently employ the *multiplicative potential function* $\mathbf{c}\mathbf{x}/[\prod_{j=1}^{n} x_j]^{1/n}$ in lieu of Equation (8.19a). The logarithmic transformation is taken here only for mathematical convenience.}

Observe that while the projective transformation (8.5) does not preserve the linearity of a function as seen by the objectives in Equations (8.4) and (8.6), it does preserve *ratios* of linear functions. Indeed, under Equation (8.5), $\mathbf{c}\mathbf{x}/x_j = \mathbf{c}\mathbf{D}_k\mathbf{y}/x_{kj}y_j \equiv \bar{\mathbf{c}}\mathbf{y}/x_{ky}y_j$, so that the potential function (8.19a) transforms as

$$F(\mathbf{y}) \equiv f\left[\frac{\mathbf{D}_k\mathbf{y}}{\mathbf{1}\mathbf{D}_k\mathbf{y}}\right] = \sum_{j=1}^{n}\ln\left[\frac{\bar{\mathbf{c}}\mathbf{y}}{x_{kj}y_j}\right] = n\ln(\bar{\mathbf{c}}\mathbf{y}) - \sum_{j=1}^{n}\ln(y_j) - \sum_{j=1}^{n}\ln(x_{kj}). \tag{8.19b}$$

Let us now measure the decrease in the value of the potential function (8.19a), or equivalently in the potential function (8.19b) in the **y**-space at iteration k. Using the potential function (8.19b), we obtain

$$F(\mathbf{y}_{\text{new}}) - F(\mathbf{y}_0) = n \ln \left[\frac{\bar{\mathbf{c}}\mathbf{y}_{\text{new}}}{\bar{\mathbf{c}}\mathbf{y}_0} \right] - \sum_{j=1}^{n} \ln [n y_{(\text{new})j}]$$

since $y_{0j} = 1/n$ for $j = 1, \ldots, n$. However, from Equation (8.18), we deduce that

$$\ln \left[\frac{\bar{\mathbf{c}}\mathbf{y}_{\text{new}}}{\bar{\mathbf{c}}\mathbf{y}_0} \right] \le \ln \left[1 - \frac{\alpha}{(n-1)} \right] \le -\frac{\alpha}{(n-1)}$$

by a property of the $\ln(\cdot)$ function. Using this equation, we derive

$$F(\mathbf{y}_{\text{new}}) - F(\mathbf{y}_0) \le -\frac{n\alpha}{(n-1)} - \sum_{j=1}^{n} \ln [n y_{(\text{new})j}] \tag{8.20}$$

Consequently, if $-\Sigma_{j=1}^{n} \ln [n y_{(\text{new})j}]$ does not increase too much in relation to $n\alpha/(n-1)$, we will obtain a sufficient drop in the potential function. This will enable us to guarantee the desired convergence results. In fact, defining $\bar{\alpha} = n\alpha/(n-1)$ and selecting $0 < \alpha < 1$ small enough so that $\sqrt{(n-1)/n}\,\bar{\alpha} = \sqrt{n/(n-1)}\,\alpha < 1$, we will show that

$$\|n\mathbf{y} - \mathbf{1}\| \le \sqrt{\frac{n-1}{n}}\,\bar{\alpha} < 1, \qquad \mathbf{1y} = 1, \qquad \text{and} \qquad \mathbf{y} > \mathbf{0}$$

implies that

$$0 \le -\sum_{j=1}^{n} \ln [n y_j] \le \frac{\bar{\alpha}^2}{2(1-\bar{\alpha})^2} \tag{8.21}$$

Observe that $\mathbf{y}_{\text{new}}$ satisfies the conditions in Equation (8.21) since $\mathbf{1y}_{\text{new}} = 1$, $\mathbf{y}_{\text{new}} > \mathbf{0}$, and from Equation (8.9$a$), $\|n\mathbf{y}_{\text{new}} - \mathbf{1}\| = n\alpha r = \sqrt{(n-1)/n}\,\bar{\alpha}$. Using Equation (8.21) in Equation (8.20), we obtain

$$F(\mathbf{y}_{\text{new}}) - F(\mathbf{y}_0) \le -\bar{\alpha} + \frac{\bar{\alpha}^2}{2(1-\bar{\alpha})^2} \le -\frac{1}{5}, \qquad \text{when } \bar{\alpha} \equiv \frac{n\alpha}{(n-1)} = \frac{1}{3} \tag{8.22}$$

Therefore when $\bar{\alpha} = \frac{1}{3}$, the function $F(\cdot)$ and consequently the potential function $f(\cdot)$ falls by $\frac{1}{5}$ every iteration. This means that over k iterations, $f(\mathbf{x}_k) - f(\mathbf{x}_0) \le -k/5$. But

$$f(\mathbf{x}_k) - f(\mathbf{x}_0) = n \ln \left[\frac{\mathbf{cx}_k}{\mathbf{cx}_0} \right] - \sum_{j=1}^{n} \ln (n x_{kj})$$

since $\mathbf{x}_0 = (\frac{1}{n}, \ldots, \frac{1}{n})$. Hence

$$n \ln \left[\frac{\mathbf{cx}_k}{\mathbf{cx}_0} \right] \le \sum_{j=1}^{n} \ln (n x_{kj}) - \frac{k}{5}$$

But using the fact that the geometric mean $[\prod_{j=1}^{n}(x_{kj})]^{1/n}$ is no more than the arithmetic mean $\mathbf{1x}_k/n$ and that $\mathbf{1x}_k = 1$, we have

$$\sum_{j=1}^{n} \ln(nx_{kj}) = n \ln\left[n\left(\prod_{j=1}^{n} x_{kj}\right)^{1/n}\right] \leq n \ln[\mathbf{1x}_k] = 0$$

This therefore means from the previous inequality that

$$\ln\left[\frac{\mathbf{cx}_k}{\mathbf{cx}_0}\right] \leq -\frac{k}{5n} \qquad \text{for all } k = 0, 1, 2, \ldots \tag{8.23}$$

Hence, although the objective function value may increase from one iteration to the next, a sufficient *overall* decrease from the original objective value at $k = 0$ is maintained as the iterations progress. In fact, from Equation (8.23), we will have $\mathbf{cx}_k < 2^{-L}$ when $\mathbf{cx}_k \leq (\mathbf{cx}_0)e^{-k/5n} < 2^{-L}$. Therefore when $k = 10nL$, we will have

$$\mathbf{cx}_k \leq (\mathbf{cx}_0)e^{-k/5n} = \left(\sum_{j=1}^{n} c_j/n\right)(e^{-2L}) < (2^L)(2^{-2L}) = 2^{-L}$$

Furthermore each iteration requires no more than $O(n^3)$ computations, and so the algorithm is of polynomial complexity $O(n^4L)$.

More strongly, observe that

$$\mathbf{PP}^t = \begin{bmatrix} \mathbf{AD}_k^2\mathbf{A}^t & \mathbf{0} \\ \mathbf{0} & n \end{bmatrix}$$

and that the only change in this matrix from one iteration to the next is in the elements of the diagonal matrix $\mathbf{D}_k$. Accordingly, Karmarkar shows how a slight modification of the algorithm, based on *updating* rather than recomputing an appropriate inverse in the gradient projection operation, can be made to run with $O(n^{2.5})$ effort per iteration. This results in an overall reduced polynomial complexity of $O(n^{3.5}L)$ for the modified algorithm.

To conclude this subsection, let us sketch the derivation of Equation (8.21). Toward this end, consider the following optimization problem, which enables us to bound $-\sum_{j=1}^{n} \ln[ny_j]$ subject to the conditions in Equation (8.21):

$$\text{Maximize}\left\{-\sum_{j=1}^{n} \ln[ny_j] : (n\mathbf{y} - \mathbf{1})^t(n\mathbf{y} - \mathbf{1}) \leq \left(\frac{n-1}{n}\right)\bar{\alpha}^2, \mathbf{1y} = 1, \mathbf{y} \geq \mathbf{0}\right\} \tag{8.24}$$

Observe from the first constraint that since its right-hand-side is less than one, we must have $\mathbf{y} > \mathbf{0}$ for feasibility in Problem (8.24) (why?). Since the objective function is equivalent to minimizing the product $[y_1, y_2, \ldots, y_n]$, it can be shown that at optimality in Problem (8.24) (see Exercise 8.32), the first constraint is binding. Moreover, it can be shown (see Exercise 8.32) that some $1 \leq q \leq (n-1)$ components of an optimal solution are the same and are equal to the value U, say, while the remaining $(n-q)$ components are equal to some common value V, where $0 < U < 1/n < V < 1$. Since $\mathbf{1}y = 1$ and

$(n\mathbf{y} - \mathbf{1})^t(n\mathbf{y} - \mathbf{1}) = [(n-1)/n]\bar{\alpha}^2$ must be satisfied, we get $qU + (n-q)V = 1$ and $q(nU-1)^2 + (n-q)(nV-1)^2 = [(n-1)/n]\bar{\alpha}^2$. This gives

$$nU = (1 - \theta_1),\ nV = (1 + \theta_2)$$

where

$$\theta_1 = \sqrt{\frac{(n-1)(n-q)}{q}}\ \frac{\bar{\alpha}}{n} \quad \text{and} \quad \theta_2 \equiv \frac{q\theta_1}{(n-q)} \tag{8.25}$$

Note that the optimal solution is positive, as required. Hence we get for some $1 \le q \le (n-1)$ that the objective value in Equation (8.24) satisfies

$$-\sum_{j=1}^{n} \ln[ny_j] \le -q\ln[1-\theta_1] - (n-q)\ln[1+\theta_2] \tag{8.26}$$

By the well-known second-order Taylor or Mean Value Theorem applied to the function $\ln(\cdot)$, we get for any $0 < \theta < 1$ that $\ln[1-\theta] = -\theta - \theta^2/2\Delta^2$, where $(1-\theta) < \Delta < 1$. This in turn means that $-\ln[1-\theta] \le \theta + \theta^2/2(1-\theta)^2$. Similarly, $-\ln[1+\theta] \le -\theta + \theta^2/2$ for any $0 < \theta < 1$. Using these inequalities in Equation (8.26) along with the fact that $0 < \theta_i < 1$ for $i = 1, 2$ and using Equation (8.25), we obtain

$$\begin{aligned}
-\sum_{j=1}^{n} \ln[ny_j] &\le q\left[\theta_1 + \frac{\theta_1^2}{2(1-\theta_1)^2}\right] + (n-q)\left[-\theta_2 + \frac{\theta_2^2}{2}\right] \\
&= \frac{q\theta_1^2}{2(1-\theta_1)^2}\left[\frac{n - q\theta_1(2-\theta_1)}{(n-q)}\right] \\
&= \frac{(n-1)}{n}\,\frac{\bar{\alpha}^2}{2(1-\theta_1)^2}\left[1 - \frac{q\theta_1(2-\theta_1)}{n}\right] \\
&\le \frac{\bar{\alpha}^2}{2(1-\theta_1)^2} \le \frac{\bar{\alpha}^2}{2(1-\bar{\alpha})^2}
\end{aligned}$$

since $0 < \theta_1 < \bar{\alpha} < 1$. This establishes Equation (8.21) and completes our derivation.

Dealing With an Unknown Optimal Objective Value and Obtaining a Basic Optimal Solution

We now describe a method for solving Problem (8.4) under assumption (A1), but dispense with the assumption (A2) which requires the optimal objective value of Problem (8.4) to be zero. The method conducts an interval search for the optimal objective function value, and is known as a *sliding objective function* method. As before, we augment this method with a purification scheme that determines an optimal basic feasible solution to Problem (8.4). Hence given any general Linear Program (8.11), we can transform it into the form (8.13) and determine an optimal basic feasible solution, if one exists, to the given problem.

Toward this end, let us first show how one can (polynomially) determine whether the optimal value v^*, say, in Problem (8.4) is positive or nonpositive. Suppose that $v^* \le 0$, so that $\bar{\mathbf{c}}\mathbf{y}^* \le 0$, where $\mathbf{y}^*$ solves Problem (8.7). Let us proceed by *assuming* that $v^* = 0$. Now if $\bar{\mathbf{c}}\mathbf{y}_{\text{new}} \le 0$ at iteration k, then we know that $v^* \le 0$ and we stop. Otherwise if both $\bar{\mathbf{c}}\mathbf{y}_0 > 0$ and $\bar{\mathbf{c}}\mathbf{y}_{\text{new}} > 0$, then noting that Equation (8.17) holds irrespective of the sign on v^*, we have

$$\bar{\mathbf{c}}\mathbf{y}_{\text{new}} \le \bar{\mathbf{c}}\mathbf{y}_0\left[1 - \frac{\alpha}{(n-1)}\right] + \bar{\mathbf{c}}\mathbf{y}^*\left[\frac{\alpha}{(n-1)}\right] \le \bar{\mathbf{c}}\mathbf{y}_0\left[1 - \frac{\alpha}{(n-1)}\right]$$

so that Equation (8.18) continues to hold. Consequently, we obtain the usual decrease of $\frac{1}{5}$ in the potential function value as in Equation (8.22) at this iteration. If we do not stop before $k = 10nL$ iterations with the indication that $v^* \le 0$, then we will have obtained a feasible solution $\mathbf{x}_k$ satisfying $\mathbf{c}\mathbf{x}_k < 2^{-L}$ by this time. This will then indicate to us that there does exist a feasible solution with a nonpositive objective value (why?), and such a solution may be found using the optimal rounding routine of Section 8.4.

On the other hand, if $v^* > 0$, then denoting $\mathbf{B}$ as an optimal basis for Problem (8.4), we have $v^* = N/\det(\mathbf{B})$ for some integer N (why?). Hence we know that $v^* \ge 1/|\det(\mathbf{B})| > 2^{-L}$ by the definition of L. Consequently, within $k = 10nL$ iterations, we must experience a violation in the promised decrease of $\frac{1}{5}$ in the potential function value at some iteration, or else we will produce an iterate $\mathbf{x}_k$ with $\mathbf{c}\mathbf{x}_k < 2^{-L}$, as before. This contradicts $v^* > 2^{-L}$. Therefore, within $10nL$ iterations, we can recognize that $v^* > 0$.

Let us now use this technique for recognizing the sign on the optimal value of Problem (8.4) in order to solve this problem. To begin with, suppose that we have some integer lower and upper bounds l and u respectively on v^*. Note that one can take l as the smallest coefficient c_j (why?) and u as $\lceil \mathbf{c}\mathbf{x}_0 \rceil$. Clearly $[l, u] \subseteq [-2^L, 2^L]$. We will now perform a bisection search starting with this interval of uncertainty.

Suppose that we have some current interval of uncertainty, and that z^* is the midpoint of this interval. Consider the following modification of Problem (8.4):

$$\text{Minimize } \{\mathbf{c}\mathbf{x} - z^*(\mathbf{1}\mathbf{x}) : \mathbf{A}\mathbf{x} = 0, \mathbf{1}\mathbf{x} = 1, \mathbf{x} \ge \mathbf{0}\} \tag{8.27}$$

Note that the objective function value in Problem (8.27) is a constant z^* smaller than that in Problem (8.4) for all feasible solutions (why?). Hence the two problems are equivalent. Moreover, the objective value in Problem (8.27) is nonpositive if and only if $z^* \ge v^*$. Furthermore, at the Kth bisection, $z^* = l + [N(u - l)/2^K]$, where N is some integer satisfying $0 < N < 2^K$. Also, the current interval of uncertainty is of length $(u - l)/2^{K-1}$. As long as the interval of uncertainty is of length $\ge 2^{-2L}$, we have $(u - l)/2^{K-1} \ge 2^{-2L}$, or $2^{K-1} \le (u - l)2^{2L} \le 2^{3L+1}$, that is, we have $K \le 3L + 2$. This means that as long as the interval of uncertainty is of length $\ge 2^{-2L}$, we have N and 2^K bounded above by 2^{3L+2}. Therefore after integerizing the objective in Problem (8.27) by multiplying with 2^K, we still have the size of the integer input length in Problem (8.27) of order $O(L)$. Consequently, in $O(nL)$ iterations, we can recognize the sign on the objective function in Problem (8.27) as before, and therefore ascertain whether

$z^* \geq v^*$ or $z^* < v^*$. This enables us to halve the current interval of uncertainty. (The minor details of this argument follow readily and are left to the reader in Exercise 8.33. Note that in practice one can use the best-known solution generated at any bisection stage to further reduce the interval of uncertainty.) In this fashion, one can continue bisecting the interval of uncertainty until it is reduced to length less than 2^{-2L}. This happens after at most $3L + 2$ bisections.

Denote now the final interval of uncertainty as $[l, u]$. By applying the optimal rounding routine following the solution of Problem (8.27) with $z^* = u$ if necessary, we can assume that we also have a solution $\bar{\mathbf{x}}$ to Problem (8.4) with objective value $\bar{z} \leq u$. Starting with the solution $\bar{\mathbf{x}}$, we now use the purification scheme to obtain in polynomial time a revised solution $\bar{\mathbf{x}}$, which is an extreme point of Problem (8.4) with objective value $\mathbf{c}\bar{\mathbf{x}} \leq \bar{z}$. This resulting basic feasible solution $\bar{\mathbf{x}}$ is indeed optimal to Problem (8.4). If this is not so, then there exists another basic feasible solution $\hat{\mathbf{x}}$ of objective value $\mathbf{c}\hat{\mathbf{x}} < \mathbf{c}\bar{\mathbf{x}}$. Denoting $\mathbf{B}_1$ and $\mathbf{B}_2$ as the bases corresponding to $\bar{\mathbf{x}}$ and $\hat{\mathbf{x}}$, respectively, we have (with obvious notation) $0 < \mathbf{c}\bar{\mathbf{x}} - \mathbf{c}\hat{\mathbf{x}} = N_1/D_1 - N_2/D_2$, where $D_i = |\det \mathbf{B}_i|$, $i = 1, 2$ and where N_1 and N_2 are integers. Hence $\mathbf{c}\bar{\mathbf{x}} - \mathbf{c}\hat{\mathbf{x}} \geq 1/D_1 D_2 > 2^{-2L}$ by the definition of L. This means that $\mathbf{c}\hat{\mathbf{x}} < \mathbf{c}\bar{\mathbf{x}} - 2^{-2L}$, which contradicts that $(u - l) < 2^{-2L}$ and that $\bar{z}$ and $v^* \in [l, u]$.

This method provides a technique for obtaining an optimal extreme point solution for a general linear program of the form (8.11). After converting it to the equivalent form (8.13), we can use this procedure and determine a *basic* optimal solution to Problem (8.13). Of course, if $y_{n+3} > 0$ at optimality, then Problem (8.11) is infeasible. Furthermore if $y_{n+1} = 0$ at optimality, being nonbasic with a negative "$z_j - c_j$" value, then Problem (8.11) is unbounded (why?). Otherwise $y_{n+3} = 0$, $y_{n+1} \geq 0$ is basic, and $y_{n+2} = 1/(Q + 1)$ is basic at an optimal extreme point of Problem (8.13). Therefore the corresponding solution to Problem (8.11) obtained via the transformation $x_j = (Q + 1)y_j$, $j = 1, \ldots, n$ gives an optimal basic solution to Problem (8.11). Furthermore using the corresponding optimal basis $\mathbf{B}$, one can (polynomially) solve the system $\mathbf{wB} = \mathbf{c}_B$ (in the usual notation) in order to obtain the associated complementary dual basic solution. If this is feasible, then it is also optimal for the dual problem. However, in the presence of degeneracy, some degenerate simplex pivots may be required in order to obtain an optimal basis $\mathbf{B}$ whose complementary dual basis is also feasible. Although this latter step is not of polynomial complexity, such a basis $\mathbf{B}$ can indeed be obtained polynomially as follows.

Suppose that in addition to the foregoing primal vertex, an optimal dual vertex with an associated basis is also available (via the use of a similar polynomial scheme, if necessary). Denote by v_j, $j = 1, \ldots, n$, the slack variables in the constraints for the dual to Problem (8.11). Let $\bar{v}_j$, $j = 1, \ldots, n$ be the values of these variables at the given optimal dual vertex. Define $J = \{ j \in \{1, \ldots, n\} : \bar{v}_j = 0\}$. Note that the set J includes the indices of the m nonbasic v_j variables and those of any degenerate basic v_j variables. Hence, the columns of x_j for $j \in J$ have rank m. Moreover, the latter columns include those for the currently positive x_j variables, which are themselves linearly independent. Hence we can polynomially construct a primal basis $\mathbf{B}$ by Gaussian elimination, using the columns of the x_j variables for $j \in J$, including those for the positive x_j variables. This basis $\mathbf{B}$ represents the given primal vertex (why?). Furthermore, the given dual solution is the (unique) complementary dual basic solution corresponding to $\mathbf{B}$, and so the basis $\mathbf{B}$ satisfies the desired property.

Example 8.4

Consider the linear programming problem

$$\begin{array}{llllllll}
\text{Minimize} & -5x_1 & -6x_2 & -4x_3 & -4x_4 & & & \\
\text{Subject to} & x_1 & -x_2 & +2x_3 & & -2x_5 & = & 0 \\
 & x_1 & +2x_2 & & +x_4 & -4x_5 & = & 0 \\
 & x_1 & +x_2 & +x_3 & +x_4 & +x_5 & = & 1 \\
 & & & & & \mathbf{x} & \geq & \mathbf{0}
\end{array}$$

Notice that if we add $4(x_1 + x_2 + x_3 + x_4 + x_5)$ to the objective function, we obtain the problem of Example 8.2. Hence, the optimal objective value for this example is $v^* = -4$, attained at $(0, \frac{2}{5}, \frac{2}{5}, 0, \frac{1}{5})'$, as in Example 8.2.

Suppose that we assume the optimal objective value to be $z^* = -3$. Subtracting $-3(\Sigma_j x_j)$ from the objective function modifies this function to

$$\text{Minimize } -2x_1 - 3x_2 - x_3 - x_4 + 3x_5$$

Notice that since $z^* > v^*$, the optimal value with respect to the foregoing objective is negative (in fact, -1). Applying Karmarkar's algorithm while assuming this value to be zero results in an objective value of -0.8 corresponding to the starting solution itself. Hence we immediately recognize that $v^* < z^* = -3$.

On the other hand, suppose that we assume the optimal objective value to be $z^* = -5$. Subtracting $-5(\Sigma_j x_j)$ from the objective function modifies this function to

$$\text{Minimize } -x_2 + x_3 + x_4 + 5x_5$$

Since $z^* < v^*$, the foregoing objective has a positive optimal value (in fact, $+1$). Applying Karmarkar's algorithm and assuming this value to be zero gives at the very first iteration an improvement in the potential function of the amount

$$F(\mathbf{y}_{\text{new}}) - F(\mathbf{y}_0) = n \ln\left[\frac{\bar{\mathbf{c}}\mathbf{y}_{\text{new}}}{\bar{\mathbf{c}}\mathbf{y}_0}\right] - \sum_{j=1}^{n} \ln\left[ny_{(\text{new})j}\right]$$

$$= (5)\ln\left[\frac{0.221606}{0.22973}\right] - \ln(0.9505476) = -0.12932$$

(The reader may verify these values by performing one iteration of Karmarkar's algorithm.) Since the guaranteed descent of 0.2 in the potential function value is not obtained at this iteration, we recognize that $v^* > z^* = -5$. The bisection search uses such a technique to iteratively reduce the interval of uncertainty to a sufficiently small length.

Example 8.5

Consider the problem of Example 8.4. Let us illustrate the purification procedure by arbitrarily applying it to the starting solution $\mathbf{x}_0 = (0.2, 0.2, 0.2, 0.2, 0.2)'$ itself. Here the three equality constraints alone are binding. A direction **d** lying in the null space of these constraints should satisfy $d_1 - d_2 + 2d_3 - 2d_5 = 0$, $d_1 + 2d_2 + d_4 - 4d_5 = 0$, and $d_1 + d_2 + d_3 + d_4 + d_5 = 0$. Solving this system for d_1, d_3, and d_4, say, we obtain $d_1 = -d_2 - 8d_5$, $d_3 = d_2 + 5d_5$, and $d_4 = -d_2 + 12d_5$. Hence, for example, arbitrarily putting $d_2 = 1$ and $d_5 = 0$ gives a direction $\mathbf{d} = (-1, 1, 1, -1, 0)$ that keeps the binding constraints tight. Note that $\mathbf{cd} = -1$ and so this direction is improving. Moving along this direction, we are blocked by both x_1 and x_4 dropping to zero simultaneously at a step length of 0.2 (why?). This results in the new (improved) feasible solution $(0.2, 0.2, 0.2, 0.2, 0.2)' + 0.2(-1, 1, 1, -1, 0)' = (0, 0.4, 0.4, 0, 0.2)'$. It may be verified that the three equality constraints in the problem along with $x_1 = 0$ and $x_4 = 0$, which are binding at the current solution are linearly independent. Hence we have reached an extreme point solution. For this problem, we have coincidentally attained optimality by purifying the starting solution itself. Since this optimum is nondegenerate, the associated complementary dual basic solution is also optimal.

As far as computational experience with the basic algorithm and the transformation (8.13) is concerned, it has been shown to yield very competitive results with respect to the simplex algorithm. (Many of the variants of Karmarkar's algorithm mentioned in the Notes and References section have been shown to be substantially faster than the simplex algorithm, particularly as problem size increases.) It may be noted that Karmarkar's algorithm typically converges within 50 iterations (almost independently of problem size!), making very rapid initial improvements in objective value in relation to its tail-end convergence rate. This implies the following two noteworthy points.

First, the algorithmic efficiency can be improved by reducing the effort per iteration. Note that the major task in an iteration of the algorithm is to compute the projection $\mathbf{c}_p$ in Equation (8.9*b*). This step also requires a very careful implementation in order to avoid numerical errors from a rapid loss in feasibility of the iterates. (See the Notes and References Section for a variant that permits approximate projections.) Recall that $\mathbf{c}_p$ in Equation (8.9*b*) is given by $\bar{\mathbf{c}}' - \mathbf{P}'\bar{\mathbf{w}}$, where $\bar{\mathbf{w}}$ solves the least-squares problem $\|\bar{\mathbf{c}}' - \mathbf{P}'\mathbf{w}\|^2$ over $\mathbf{w} \in E^{m+1}$. A factored-form implementation is recommended to control the numerical accuracy while solving this problem (see Exercise 8.38). Hence the effort for computing $\mathbf{c}_p$ can be reduced by solving this associated least-squares problem more efficiently. Alternatively, an appropriate technique may be used for directly computing $(\mathbf{PP}')^{-1}$ in Equation (8.9*b*), which exploits the fact that the only change from one iteration to the next in computing $\mathbf{c}_p$ is in the diagonal matrix $\mathbf{D}_k$.

Second, it also has been suggested that one use Karmarkar's algorithm for some initial iterations, "purify" the available solution to an extreme point with at least as good an objective value, and then switch over to the simplex algorithm in order to complete the optimization process. This has yielded some encouraging results. Another suggestion that appears worthwhile is to drop variables that are approaching zero from the problem as the iterations progress, and hence reduce the dimension of the problem as well as avoid the possible ill-conditioning of the matrix $\mathbf{PP}'$, associated with its determinant getting too close to zero. This is done at the expense of somehow later resurrecting some deleted variables if necessary. Strategies of this type may be used to enhance the computational performance of the algorithm.

EXERCISES

8.1 Provide the details of an approach that uses a polynomial-time algorithm for the linear programming decision problem in order to polynomially solve the linear programming optimization problem. What is the complexity of the latter in terms of the complexity of the former?

8.2 Demonstrate that the transformation (8.2) reduces the Klee–Minty problem to the form (8.3).

8.3 With respect to the problem of Equation (8.3), show that the $z_j - c_j$ values corresponding to the basis with basic variables $(s_1, s_2, \ldots, s_{n-1}, y_n)$ are all nonnegative.

8.4 Following the arguments in Section 8.2, give a detailed proof by induction that for Problem (8.3) in n-dimensions there exists a simplex path of length $2^n - 1$ from the origin to the optimum using (Dantzig's) most negative $z_j - c_j$ entering rule.

8.5 Plot the (original) Klee–Minty polytope for $n = 2$ and $n = 3$. What is the relationship between these polytopes and the ones shown in Figure 8.1? Analogous to the statement of Exercise 8.4, what precise complexity assertion can you make with respect to the original Klee–Minty problem?

8.6 Consider the linear programming problem to

$$\text{Maximize} \quad \sum_{j=1}^{n} 10^{n-j} x_j$$

$$\text{Subject to} \quad \left[2 \sum_{j=1}^{i-1} 10^{i-j} x_j \right] + x_i \leq 100^{i-1} \quad \text{for } i = 1, \ldots, n$$

$$x_j \geq 0 \quad \text{for } j = 1, \ldots, n$$

a. Demonstrate graphically for $n = 2$ and $n = 3$ that using (Dantzig's) most negative $z_j - c_j$ entering rule, the simplex method can go through $2^n - 1$ pivots, starting at the origin.

b. Provide a proof by induction for the statement in part (a) for general n.

8.7 Given a square matrix $\mathbf{B}$ with components B_{ij}, $i = 1, \ldots, m$, $j = 1, \ldots, m$, show by induction that $|\det \mathbf{B}| \leq \prod_{i,j=1}^{n} (1 + |B_{ij}|)$. (*Hint:* Use the column (row) wise expansion formula for $\det \mathbf{B}$.)

8.8 In the usual notation, let $\bar{\mathbf{x}} = (\bar{\mathbf{x}}_B, \bar{\mathbf{x}}_N)$ be a basic feasible solution of the system $\mathbf{Ax} = \mathbf{b}$, $\mathbf{x} \geq \mathbf{0}$, where $\mathbf{A}$ is $m \times n$ of rank m and where the data is all integer. Show that the kth component $\bar{x}_{Bk}$ of $\bar{\mathbf{x}}_B$ satisfies $\bar{x}_{Bk} < 2^L/(mn)$, where

$$L = \left\lceil 1 + \log m + \log n + \sum_i \sum_j [1 + \log(1 + |a_{ij}|)] + \sum_i [1 + \log(1 + |b_i|)] \right\rceil$$

Here a_{ij} and b_i are the components of $\mathbf{A}$ and $\mathbf{b}$, respectively. Hence show that $\|\bar{\mathbf{x}}\| < 2^L/n$.

8.9 For the linear programming problem to minimize $\mathbf{cx}$ subject to $\mathbf{Ax} = \mathbf{b}$, $\mathbf{x} \geq \mathbf{0}$, where $\mathbf{A}$ is $m \times n$ and where the data is all integer, define L as in Equation (8.1). Using Exercise 8.8, and assuming the linear program to be feasible and bounded, show that the optimal objective value is no more than 2^L.

The following exercises 8.10–8.15 *relate to Khachian's algorithm for polynomially determining if there exists an* $\mathbf{x}$ *belonging to the set*

$$S = \{\mathbf{x} : \mathbf{Gx} < \mathbf{g}\}$$

where $\mathbf{G}$ *is* $r \times q$, $\mathbf{g}$ *is* $r \times 1$, r *and* $q \geq 2$, *and where the data is all integer, or conclude that* $S = \phi$.

8.10 Define

$$L = \left\lceil 1 + \log r + \log q + \sum_i \sum_j [1 + \log(1 + |G_{ij}|)] + \sum_i [1 + \log(1 + |g_i|)] \right\rceil$$

and let $E_0 = \{\mathbf{x} : \|\mathbf{x}\| \leq 2^L\}$. Show that if $S \neq \phi$, then

$$\text{vol}[E_0 \cap S] \geq 2^{-(q+1)L}$$

where vol$(\cdot)$ denotes the q-dimensional volume. (*Hint:* The volume of a q-dimensional simplex with vertices $\mathbf{v}_1, \ldots, \mathbf{v}_{q+1}$ is

$$1/q! \left| \det \begin{bmatrix} 1 & 1 & \cdots & 1 \\ \mathbf{v}_1 & \mathbf{v}_2 & \cdots & \mathbf{v}_{q+1} \end{bmatrix} \right|).$$

8.11 (*Khachian's Main Step* 1) Starting with $k = 0$, suppose that at iteration k, the ellipsoid E_k defined here is given.

$$E_k = \left\{\mathbf{x} : (\mathbf{x} - \mathbf{x}_k)^t \mathbf{Q}_k^{-1} (\mathbf{x} - \mathbf{x}_k) \leq 1\right\}$$

where $\mathbf{x}_k$ is the center and $\mathbf{Q}_k$ is a $q \times q$ symmetric positive definite matrix (that is, $\mathbf{y}^t\mathbf{Q}_k\mathbf{y} > \mathbf{0}$ for all $\mathbf{y} \neq \mathbf{0}$. In particular, $\det(\mathbf{Q}_k) \neq 0$). Given E_k, if $\mathbf{x}_k \in S$, then halt. Otherwise let the most violated constraint be $\mathbf{G}_v\mathbf{x} < g_v$.

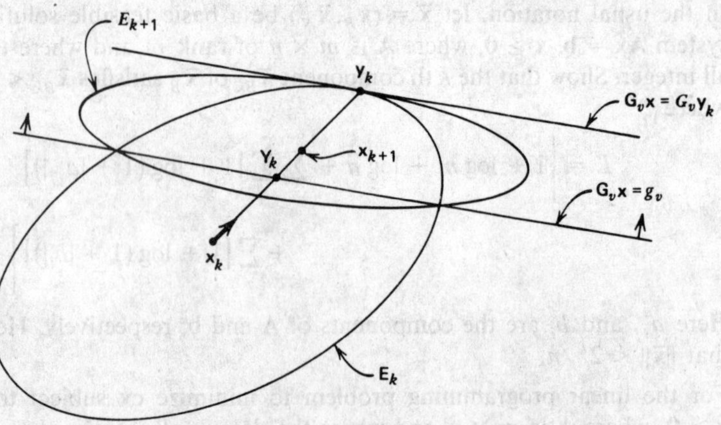

Figure 8.8. **Geometry of the main step of the ellipsoid algorithm.**

Hence $\mathbf{G}_v\mathbf{x}_k \geq g_v$. Geometrically, translate the hyperplane $\mathbf{G}_v\mathbf{x} = g_v$ parallel to itself in the (feasible) direction of $-\mathbf{G}_v^t$ until it becomes tangential to the ellipsoid E_k at the point $\mathbf{y}_k$ (see Figure 8.8). Show that

$$\mathbf{y}_k = \mathbf{x}_k + \mathbf{d}_k \qquad \text{where } \mathbf{d}_k = \frac{-\mathbf{Q}_k\mathbf{G}_v^t}{\sqrt{\mathbf{G}_v\mathbf{Q}_k\mathbf{G}_v^t}}$$

Next compute $\mathbf{Y}_k$ as the point on the hyperplane $\mathbf{G}_v\mathbf{x} = g_v$ obtained by moving from $\mathbf{x}_k$ in the direction $\mathbf{d}_k$. Hence $\mathbf{Y}_k = \mathbf{x}_k + \lambda_k\mathbf{d}_k$ for some step length λ_k. Show that $\lambda_k = (g_v - \mathbf{G}_v\mathbf{x}_k)/(\mathbf{G}_v\mathbf{d}_k)$ and that $\lambda_k \geq 0$. Also show that if $\lambda_k \geq 1$, then $S = \phi$ and the procedure can terminate.

8.12 (*Khachian's Main Step* 2) Assuming that $\lambda_k < 1$ in Exercise 8.11, let $\mathbf{x}_{k+1}$ be the point on the line segment $[\mathbf{Y}_k, \mathbf{y}_k]$ that divides this in the ratio $1 : q$. Hence

$$\mathbf{x}_{k+1} = \left(\frac{q}{q+1}\right)\mathbf{Y}_k + \left(\frac{1}{q+1}\right)\mathbf{y}_k = \mathbf{x}_k + \frac{(1+\lambda_k q)}{(1+q)}\mathbf{d}_k$$

Define the new ellipsoid $E_{k+1} = \{\mathbf{x} : (\mathbf{x} - \mathbf{x}_{k+1})^t\mathbf{Q}_{k+1}^{-1}(\mathbf{x} - \mathbf{x}_{k+1}) \leq 1\}$ with center at $\mathbf{x}_{k+1}$ and with $\mathbf{Q}_{k+1}$ given by the expression

$$\mathbf{Q}_{k+1} = \left[\frac{q^2(1-\lambda_k^2)}{(q^2-1)}\right]\left[\mathbf{Q}_k - \frac{2(1+q\lambda_k)}{(q+1)(\lambda_k+1)}\mathbf{d}_k\mathbf{d}_k^t\right]$$

Show that

a. $\mathbf{Q}_{k+1}$ is symmetric and positive definite (given that so is $\mathbf{Q}_k$).
b. The point $\mathbf{y}_k$ lies on the boundary of E_{k+1}.
c. The hyperplane $\mathbf{G}_v\mathbf{x} = \mathbf{G}_v\mathbf{y}_k$ is tangential to E_{k+1} at $\mathbf{y}_k$.
d. $E_k \cap \{\mathbf{x} : \mathbf{G}_v\mathbf{x} \leq g_v\} \subseteq E_{k+1}$, where $E_k = \{\mathbf{x} : (\mathbf{x} - \mathbf{x}_k)^t\mathbf{Q}_k^{-1}(\mathbf{x} - \mathbf{x}_k) \leq 1\}$.

Inductively show that $E_0 \cap S \subseteq E_k \cap S \subseteq E_k$ for all $k = 1, 2, \ldots,$ that is, each of the ellipsoids $E_1, E_2, \ldots$ contains at least that portion of S that was captured by E_0.

8.13 Let $\mathbf{V}_1$ and $\mathbf{V}_2$ be two vectors such that $\mathbf{V}_1^t\mathbf{V}_2 > 0$. Let θ be a scalar, Using suitable row operations, show that

$$\det\left[\mathbf{I} - \theta\mathbf{V}_1\mathbf{V}_2^t/\mathbf{V}_1^t\mathbf{V}_2\right] = (1 - \theta)$$

8.14 Using the fact that the volume of the ellipsoid E_k equals $V_0\sqrt{\det(\mathbf{Q}_k)}$, where V_0 is the volume of the unit sphere in E^q, use Exercise 8.13 to show that

$$\text{vol}(E_k) < e^{-k/2(1+q)}\,\text{vol}(E_0) < e^{-k/2(1+q)}2^{(q+2L)}$$

8.15 Using Exercises 8.10–8.14, conclude that if $S \neq \phi$, then by $6(q+1)^2L$ iterations, the algorithm will find a solution in S. Hence if no solution is found in $6(q+1)^2L$ iterations, then $S = \phi$. Therefore show that the *overall complexity of the algorithm* is $O(q^4L)$.

8.16 Consider the open system of inequalities $\mathbf{Gx} < \mathbf{g}$ that defines the set

$$S = \{\mathbf{x} = (x_1, x_2) : -x_1 - x_2 < -1, x_1 < 1\}$$

Show that with $L = 2$, the assertion of Exercise 8.10 holds. Use Khachian's algorithm with $L = 2$ to find a solution to this system. Illustrate the progress of the algorithm graphically.

8.17 Consider the open inequality system $x_1 - x_2 > 1$, $2x_1 + 3x_2 > 6$, $x_1 > 0$, $x_2 > 0$. Show that with $L = 2$, the assertion of Exercise 8.10 holds. Use Khachian's algorithm with $L = 2$ in order to find a solution to this system. Illustrate the progress of the algorithm graphically.

8.18 In order to be able to use Khachian's algorithm for solving linear programming problems, one needs to be able to determine if the *closed* system of inequalities $\mathbf{Gx} \leq \mathbf{g}$ has a solution, where $\mathbf{G}$ is $r \times q$, r and $q \geq 2$, and $\mathbf{G}$ and $\mathbf{g}$ have all integer components. Toward this end, define $\bar{S} = \{\mathbf{x} : \mathbf{Gx} \leq \mathbf{g}\}$ and consider the following open inequality system, where $\mathbf{1}$ is a row vector of r ones, and where L is defined in Exercise 8.10.

$$\mathbf{Gx} < \mathbf{g} + (2^{-L})\mathbf{1}^t, \text{ that is, } (2^L)\mathbf{Gx} < (2^L)\mathbf{g} + \mathbf{1}^t$$

Use Farkas' lemma (see Chapter 5) to show that $\bar{S}$ is nonempty if and only if this open inequality system has a solution. Show that the complexity of determining whether or not $\bar{S}$ is empty by Khachian's algorithm is $O(rq^5L)$.

8.19 (*Optimal Rounding Routine*). Suppose that we are given a solution $\bar{\mathbf{x}}$ to the open inequality system $\mathbf{Gx} < \mathbf{g} + (2^{-L})\mathbf{1}^t$, where L is defined as in Exercise 8.10. Suppose that we are to determine from that a solution to the system $\mathbf{Gx} \leq \mathbf{g}$. Of course, if $\mathbf{G}\bar{\mathbf{x}} \leq \mathbf{g}$, then we are done. Otherwise define $s_i(\mathbf{x}) = g_i - \mathbf{G}_i\mathbf{x}$ as the slack in the ith constraint $\mathbf{G}_i\mathbf{x} \leq g_i$ corresponding to a given solution $\mathbf{x}$. Note that we have $s_i(\bar{\mathbf{x}}) < 0$ for some $i \in \{1, \ldots, r\}$, but that $s_i(\bar{\mathbf{x}}) > -2^{-L}$ for each $i = 1, \ldots, r$. (This evidently means that only a small adjustment in $\bar{\mathbf{x}}$ may be required to obtain $s_i(\bar{\mathbf{x}}) \geq 0$ for all $i = 1, \ldots, r$;

hence the name of this routine.) Consider the operations of the following two steps:

Step 1. First, refine $\bar{\mathbf{x}}$ to a solution $\bar{\mathbf{x}}_{\text{new}}$ that is such that (i) $s_i(\bar{\mathbf{x}}_{\text{new}}) \geq \min\{0, s_i(\bar{\mathbf{x}})\}$, and (ii) every row of $\mathbf{G}$ can be written as a linear combination of the rows with index set $V = \{i : s_i(\bar{\mathbf{x}}_{\text{new}}) \leq 0\}$. Show how this can be achieved polynomially. (*Hint:* If $\bar{\mathbf{x}}$ does not satisfy these conditions, select a row $\mathbf{G}_k$ of $\mathbf{G}$, with $k \notin V$ that is not spanned by $\mathbf{G}_i$, $i \in V$, and find a solution $\mathbf{d}(\neq \mathbf{0})$ to the system of equations $\mathbf{G}_i\mathbf{d} = 0$ for $i \in V$, $\mathbf{G}_k\mathbf{d} = 1$. Consider the ray $\mathbf{x} = \bar{\mathbf{x}} + \lambda\mathbf{d}$, $\lambda \geq 0$ and take a suitable step along this ray. Repeat.)

Step 2. Given $\bar{\mathbf{x}}_{\text{new}}$ from Step 1 along with the set V, find a maximal linearly independent set of rows from the set $\mathbf{G}_i$, $i \in V$. Let these rows be $\mathbf{G}_i$, $i \in J \subseteq V$. Then show that any solution to the system $\mathbf{G}_i\mathbf{x} = g_i$, $i \in J$ yields the desired solution to the system $\mathbf{G}\mathbf{x} \leq \mathbf{g}$.

Derive a polynomial complexity bound for this routine.

8.20 This exercise describes an alternative approach to determining a solution to a closed system of inequalities via Khachian's algorithm. Consider the closed system of inequalities $\mathbf{G}\mathbf{x} \leq \mathbf{g}$, where $\mathbf{G}$ is $r \times q$. Suppose that we have a solution $\bar{\mathbf{x}}$ satisfying $\mathbf{G}_i\bar{\mathbf{x}} < g_i$ for $i = 1, \ldots, p$, where $p < r$. Examine the open system of inequalities $\mathbf{G}_i\mathbf{x} < g_i$ for $i = 1, \ldots, p+1$, and suppose that this system is empty. Show that $\mathbf{G}_{p+1}x = g_{p+1}$ for *all* feasible solutions to $\mathbf{G}\mathbf{x} \leq \mathbf{g}$. By solving for some variable in this equation in terms of the others, show how the system $\mathbf{G}\mathbf{x} \leq \mathbf{g}$ can be reduced to an equivalent one with one fewer variable and constraint. How can you repeatedly use this process to polynomially determine using Khachian's algorithm whether or not a closed system $\mathbf{G}\mathbf{x} \leq \mathbf{g}$ (with integer data) has a feasible solution? What is the complexity of this procedure?

8.21 Consider the solution of the linear program in canonical form: Minimize $\mathbf{cx}$, subject to $\mathbf{Ax} \geq \mathbf{b}$, $\mathbf{x} \geq \mathbf{0}$, where $\mathbf{A}$ is $m \times n$. Show that solving this problem is equivalent to determining a solution to a certain closed system of inequalities. How would you decide if the given linear program is infeasible or unbounded if the inequality system has no solution. Using Exercise 8.18, show that the overall complexity of using Khachian's algorithm is $O[(m+n)^6L]$, where L is given by Equation (8.1).

8.22 Consider the open inequality system in E^2 given by $x_1 + 0x_2 < 1 + 1/\theta$ and $-x_1 + 0x_2 < -1$, where $\theta \geq 5/\sqrt{2}$. Let $\mathbf{x}_0 = (1 + (\sqrt{2}/5), 0)'$ and define $E_0 = \{\mathbf{x} : (\mathbf{x} - \mathbf{x}_0)'\mathbf{Q}_0^{-1}(\mathbf{x} - \mathbf{x}_0) \leq 1\}$, where $\mathbf{Q}_0 = 2\mathbf{I}$ so that E_0 is a circle with center at $\mathbf{x}_0$ and with radius $\sqrt{2}$. (Hence this corresponds to "L" $= \frac{1}{2}$). Using this as a starting solution, apply Khachian's algorithm (Exercises 8.10–8.15) using $\lambda_k \equiv 0$. (This corresponds to Khachian's original algorithm.)

a. Give the geometric interpretation of using the formulas for $\mathbf{x}_{k+1}$ and $\mathbf{Q}_{k+1}$ of Exercise 8.12 with $\lambda_k = 0$. Illustrate with a sketch similar to Figure 8.8.

b. Derive closed form expressions for $\mathbf{x}_k$ and $\mathbf{Q}_k$ as a function of iteration k.

c. Determine the smallest integer k such that $\mathbf{x}_k$ satisfies the inequality system. Show that by making θ arbitrarily large, we can make the algorithm go through an arbitrarily large number of iterations before terminating. What does this say about the complexity of Khachian's algorithm using real data?

d. Suppose that θ is an integer. Integerize the open inequality system and determine L from Exercise 8.10. What does part (c) say about the practical computational relationship of Khachian's algorithm with the quantity L?

8.23 Show that the matrix $\mathbf{PP}'$ is invertible, where $\mathbf{P}$ is defined as in Equation (8.7), with $\mathbf{A}$ being $m \times n$ of rank m and with $\mathbf{D}_k = \text{diag}\{x_{k1}, \ldots, x_{kn}\}$, given $\mathbf{x}_k > \mathbf{0}$.

8.24 The Karush–Kuhn–Tucker optimality conditions for Problem (8.8) assert that the negative gradient of the objective function should be contained in the cone spanned by the gradients of the binding constraints. Use this fact to show that $\mathbf{y}_{\text{new}}$ defined by Equation (8.9) solves Problem (8.8).

8.25 Consider the problem of minimizing $\|\bar{\mathbf{c}}' - \mathbf{P}'\mathbf{w}\|^2$ over $\mathbf{w}$ unrestricted in sign. Show by setting the gradient of the objective function equal to zero that $\mathbf{w} = (\mathbf{PP}')^{-1}\mathbf{P}\bar{\mathbf{c}}'$ solves this problem.

8.26 a. Using the procedure of Exercise 8.19, provide the details and the complexity analysis of an optimal rounding routine for Karmarkar's algorithm as applied to Problem (8.4) under assumptions (A1) and (A2). In particular, define L as in Equation (8.1). Show that it is sufficient to determine a solution $\mathbf{x}_k$ to Problem (8.4) with objective function value $\mathbf{cx}_k < 2^{-L}$ in order to polynomially round this using Exercise 8.19 to an optimal solution to Problem (8.4) (with a zero objective function value).

b. Illustrate the method of part (a) by providing the details of optimally rounding the solution $\mathbf{x}_{26}$ obtained in Example 8.2.

c. What is the relationship of this procedure with the rounding routine given in Section 8.4?

d. Present a simplex pivot type of scheme for efficiently implementing the optimal rounding routine of Section 8.4. Show that this process has a complexity bound of $O(m^2n)$.

8.27 Use Exercise 8.8 to show that there exists a value of M of magnitude $2^{O(L)}$ such that the artificial variable y_{n+3} in the problem of Equation (8.13) is zero at optimality, given that the problem in Equation (8.12) is feasible.

8.28 Using the development of Sections 8.4 and 8.5, show how one can use Karmarkar's algorithm to determine whether a given linear program is infeasible or whether it is unbounded.

8.29 Consider the potential function $\hat{f}(\mathbf{x}) = \ln[\mathbf{cx}/\mathbf{1x}]$. Show how this transforms in the $\mathbf{y}$-space under the projective transformation (8.5). Can this potential function be used to analyze the complexity of Karmarkar's algorithm? (*Hint:* Examine the relationship between $\hat{f}(\mathbf{x})$ and $\mathbf{cx}$ for Problem (8.4).)

8.30 Determine the value $\bar{\alpha}^*$ of $\bar{\alpha}$ that minimizes the expression $-\bar{\alpha} + \bar{\alpha}^2/2(1-\bar{\alpha})^2$ in Equation (8.22). Why does this not necessarily imply that the value $\bar{\alpha}^*$ is the best value for obtaining a maximum reduction in the potential function $F(\cdot)$? How would you go about finding a better value of $\bar{\alpha}$ from the viewpoint of reducing $F(\cdot)$? (*Hint:* Examine the derivation of the inequalities (8.20) and (8.21).)

8.31 Analyze in detail the complexity of a single iteration of Karmarkar's algorithm for Problem (8.4).

8.32 The Karush–Kuhn–Tucker optimality conditions for Problem (8.24), where the constraints are written as $\leq$ inequalities, require the gradient of the objective function to lie in the cone spanned by the gradients of the binding constraints. Show that at optimality for Problem (8.24), the first constraint is binding and that $\mathbf{y} > \mathbf{0}$. Using the stated optimality conditions, show that at optimality in Problem (8.24) some $1 \leq q \leq (n-1)$ components of $\mathbf{y}$ are equal to a value U and the remaining components are equal to a value V, where U and V satisfy Equation (8.25).

8.33 Provide the details of Karmarkar's algorithm and its complexity for solving Problem (8.4) via Problem (8.27) and the sliding objective method, under the assumption (A1) but treating the optimal objective value as unknown. Include an analysis of the optimal rounding via the purification step.

8.34 Provide the complexity of solving the (general) linear programming problem (8.11) via Karmarkar's algorithm using the transformation of this problem into Problem (8.13) and the details of Exercise 8.33.

8.35 Illustrate the procedure of Exercise 8.33 using Example 8.4 in Section 8.5 with the information that the optimal objective value $v^* \in [-6, 0]$.

8.36 Solve the problem to minimize $2x_1 + 3x_2$, subject to $x_1 + 2x_2 \geq 1$, and $x_1 \geq 0$, $x_2 \geq 0$, using Karmarkar's algorithm.

8.37 Suppose that one uses the techniques of Sections 8.4 and 8.5 to determine a basic solution to the primal–dual formulation (8.14). Does this necessarily yield an optimal primal basis whose complementary dual basis is dual feasible? If not, then under what conditions will this be obtained automatically, and what additional work may be required otherwise? Give the details of a polynomial-time scheme for finding such a basis. Explain the significance of being able to obtain such a basis.

8.38 Consider the following *QR factorization* method for obtaining the least-squares solution $\bar{\mathbf{w}}$ to the problem of minimizing $\|\bar{\mathbf{c}}^t - \mathbf{P}^t\mathbf{w}\|^2$ in order to determine $\mathbf{c}_p$ of Equation (8.9*b*) via $\mathbf{c}_p = \bar{\mathbf{c}}^t - \mathbf{P}^t\bar{\mathbf{w}}$. Given $\mathbf{P}^t$ of full column rank, its *QR* decomposition yields $\mathbf{P}^t = \mathbf{QR}$, where $\mathbf{Q}$ is an $n \times n$ orthogonal matrix (that is, $\mathbf{QQ}^t = \mathbf{I}$) and $\mathbf{R}$ is a $n \times (m+1)$ matrix of the form $\begin{bmatrix} \mathbf{R}_1 \\ \mathbf{0} \end{bmatrix}$. Here $\mathbf{R}_1$ is $(m+1) \times (m+1)$, nonsingular, and upper triangular. (See the Notes and References section for details on performing such a decomposition.) Denote $\mathbf{Q}^t\bar{\mathbf{c}}^t \equiv \begin{bmatrix}\mathbf{c}_1\\ \mathbf{c}_2\end{bmatrix}$, with $\mathbf{c}_1$ being of dimension $(m+1)$ and $\mathbf{c}_2$ being of dimension $(n-m-1)$. Using this decomposition, show that $\bar{\mathbf{w}}$ is given as the solution to the linear, triangular system $\mathbf{R}_1\mathbf{w} = \mathbf{c}_1$. Furthermore, $\mathbf{c}_p = \mathbf{Q}\begin{bmatrix}\mathbf{0}\\ \mathbf{c}_2\end{bmatrix}$.

8.39 Consider the linear assignment problem to minimize $\sum_{i=1}^m \sum_{j=1}^m c_{ij}x_{ij}$ subject to $\sum_{i=1}^m x_{ij} = 1$ for $j = 1,\ldots, m$, $\sum_{j=1}^m x_{ij} = 1$ for $i = 1,\ldots, m$, and $\mathbf{x} \geq \mathbf{0}$. Suggest a suitable transformation to use Karmarkar's algorithm for solving this problem. Provide an algorithmic statement, including a suggested starting solution and any algorithmic specializations. In particular, how would you exploit the fact that the extreme points are integer valued? (*Hint:* With $\mathbf{c}$ integer, once a solution within a unit of optimality is found, it can be optimally rounded by purification.)

8.40 Consider the linear transportation problem to minimize $\sum_{i=1}^{NS}\sum_{j=1}^{ND} c_{ij}x_{ij}$ subject to $\sum_{j=1}^{ND} x_{ij} \leq s_i$ for $i = 1,\ldots, NS$, $\sum_{i=1}^{NS} x_{ij} = d_j$ for $j = 1,\ldots, ND$, and $\mathbf{x} \geq \mathbf{0}$, where NS is the number of sources with respective supplies s_i, $i = 1,\ldots, NS$, and where ND is the number of sinks with respective demands d_j, $j = 1,\ldots, ND$. (Assume all integer data and that $\sum_{i=1}^{NS} s_i \geq \sum_{j=1}^{ND} d_j$.) Repeat the analysis of Exercise 8.39 for this problem.

8.41 Consider the linear programming problem to minimize $\mathbf{cx}$ subject to $\mathbf{Ax} = \mathbf{b}$ and $\mathbf{x} \geq \mathbf{0}$, where $\mathbf{A}$ is $m \times n$ of rank m, the feasible region is bounded, and where the data is all integer. Suppose that the optimal objective value v^* and some (rational) vector $\bar{\mathbf{x}} > \mathbf{0}$ feasible to this problem are known.

a. Show that by using the transformation

$$\mathbf{y} = \frac{\mathbf{D}^{-1}\mathbf{x}}{\mathbf{1D}^{-1}\mathbf{x} + 1}, \qquad y_0 = \frac{1}{\mathbf{1D}^{-1}\mathbf{x} + 1}$$

and by replacing the resulting fractional objective by the objective function $\mathbf{cDy} - v^* y_0$, one obtains an equivalent linear program of the form (8.4).

b. Argue the polynomiality of Karmarkar's algorithm with respect to the original problem using this approach.

c. Is the equivalence of part (a) true if the feasible region of the original problem is unbounded?

d. Consider the following linear programming problem.

$$\begin{aligned}
\text{Minimize}\quad & -x_1 - 2x_2 \\
\text{Subject to}\quad & x_1 - x_2 + x_3 = 1 \\
& x_1 + 2x_2 + x_4 = 2 \\
& x_1, \ x_2, \ x_3, x_4 \geq 0
\end{aligned}$$

Using $\bar{\mathbf{x}} = (0.5, 0.5, 1, 0.5)^t$ and $v^* = -2$, show that this problem is equivalent to that of Example 8.2.

8.42 Consider the linear programming problem

$$\text{Minimize}\ \{\mathbf{cx} : \mathbf{Ax} = \mathbf{b}, \mathbf{x} \geq \mathbf{0}\} \tag{8.28}$$

where $\mathbf{A}$ is $m \times n$ of rank $m < n$ and where the data is all integer. Without loss of generality, assume that the feasible region is bounded and that a feasible solution $\mathbf{x}_0 > \mathbf{0}$ is known. (Show how one may use artificial variables and bounds, if necessary, for this purpose.) Construct the so-called *barrier function problem* from the given linear program (8.28), using a barrier parameter $\mu > 0$.

$$\text{Minimize}\left\{\beta(\mathbf{x}) \equiv \mathbf{cx} - \mu \sum_{j=1}^{n} \ln(x_j) : \mathbf{Ax} = \mathbf{b}\right\} \tag{8.29}$$

Observe that the nonnegativity constraints in Problem (8.28) have been incorporated into the objective function in Problem (8.29) along with a "barrier term" that approaches ∞ if any x_j value approaches zero (from the positive side). It is well known from nonlinear programming (see the Notes and References section) that given any desired level of accuracy $\epsilon > 0$, there exists a $\mu > 0$ small enough such that the corresponding optimal solution to Problem (8.29) is positive and has an objective value in Problem (8.28) within ϵ of optimality. Consider the following *Affine Scaling Variant of Karmarkar's Algorithm*.

Initialization Let $\mathbf{x}_0$ be the given positive feasible solution, set $k = 0$, $q = 0.97$, and select $\mu = \min\{2^{-2L-1}/n,\ 10^{-12}\}$, where $L = \lceil 1 + \Sigma_i \Sigma_j \log(1 + |a_{ij}|)\rceil$.

Step 1 (Dual estimates, projected gradient, and direction of motion, based on a projected gradient step in the $\mathbf{y}$-space, where $\mathbf{y} = \mathbf{D}_k^{-1}\mathbf{x}$. The latter transformation gives the algorithm its name.) Given $\mathbf{x}_k$, define $\mathbf{D}_k = \operatorname{diag}\{x_{k1}, \ldots, x_{kn}\}$, $\bar{\mathbf{c}}_k = \mathbf{cD}_k$, and $\mathbf{P}_k = \mathbf{AD}_k$. Compute the following:

Dual estimate $\mathbf{w}_k = (\mathbf{P}_k\mathbf{P}_k^t)^{-1}\mathbf{P}_k[\bar{\mathbf{c}}_k^t - \mu\mathbf{1}^t]$

Reduced cost vector $\mathbf{R}_k = \mathbf{c}^t - \mu\mathbf{D}_k^{-1}\mathbf{1}^t - \mathbf{A}^t\mathbf{w}_k$

Projected gradient vector $\mathbf{c}_{pk} = [\mathbf{I} - \mathbf{P}_k^t(\mathbf{P}_k\mathbf{P}_k^t)^{-1}\mathbf{P}_k][\bar{\mathbf{c}}_k^t - \mu\mathbf{1}^t]$

Direction of motion $\mathbf{d}_k = -\mathbf{D}_k\mathbf{c}_{pk}$

Step 2 (Termination Test).

If $\mathbf{w}_k$ is dual feasible in Problem (8.28) and $\mathbf{cx}_k - \mathbf{b}^t\mathbf{w}_k < 2^{-2L}$, then purify $\mathbf{x}_k$ to an extreme point of Problem (8.28) and terminate. Otherwise, continue.

Step 3 (New iterate).

Set $\mathbf{x}_{k+1} = \mathbf{x}_k + \lambda_k\mathbf{d}_k$, where the step length λ_k is given by

$$\lambda_k = \text{Minimum}\left\{q\lambda_{\max},\ \frac{(1-q)^2}{\mu}\right\},$$

$$\text{where } \frac{1}{\lambda_{\max}} \equiv \text{Maximum}\left\{-\frac{d_{kj}}{x_{kj}} : j = 1, \ldots, n\right\}$$

Increment k by one and return to Step 1.

a. Show that the algorithm generates a sequence of positive iterates $\mathbf{x}_k$, $k = 0, 1, 2, \ldots$.

b. Show that the algorithm either stops finitely with an optimal solution $\mathbf{x}_k$ to Problem (8.29) in case $\|\mathbf{c}_{pk}\| = 0$ for some k, or else, ignoring the termination criterion, an infinite sequence of primal–dual iterates $(\mathbf{x}_k, \mathbf{w}_k)$ is generated that is such that the sequence $\{\|\mathbf{c}_{pk}\|\} \to 0$. Moreover, for any convergent subsequence $\{\mathbf{x}_k, \mathbf{w}_k\}_K$ indexed by K, with limit point $(\overline{\mathbf{x}}, \overline{\mathbf{w}})$, show that $\overline{\mathbf{x}} > \mathbf{0}$ and that $\overline{\mathbf{x}}$ solves Problem (8.29).

c. Show that $\overline{\mathbf{w}}$ obtained in part (b) is dual feasible to Problem (8.28). Moreover, show that the duality gap between $\overline{\mathbf{x}}$ and $\overline{\mathbf{w}}$ in Problem (8.28) is $\mathbf{c}\overline{\mathbf{x}} - \mathbf{b}'\overline{\mathbf{w}} = n\mu$.

d. Show that the reduced cost vector $\mathbf{R}_k$ approaches zero as $k \to \infty$, and that the dual iterate $\mathbf{w}_k$ becomes dual feasible in Problem (8.28) once $\mathbf{R}_k$ is sufficiently close to zero. Hence, with μ defined as previously show that the algorithm terminates finitely at Step 2.

e. Present a purification process for obtaining a pair of optimal primal and dual extreme point solutions to Problem (8.28) after termination at Step 2.

f. Comment on the use of $\mu = 0$. (This was the original affine scaling variant.)

8.43 Suppose that the system $\mathbf{Ax} = \mathbf{b}$, $\mathbf{x} \geq \mathbf{0}$ has a positive feasible solution. Consider the following Phase I problem, where $\mathbf{x}_0 > \mathbf{0}$ is arbitrarily chosen and λ is an artificial variable.

$$\text{P1:} \quad \text{Minimize } \{\lambda : \mathbf{Ax} - (2-\lambda)\mathbf{b} - (\lambda - 1)\mathbf{Ax}_0 = 0, \mathbf{x} \geq \mathbf{0}, \lambda \geq 0\}$$

Note that $(\mathbf{x}, \lambda) = (\mathbf{x}_0, 2) > \mathbf{0}$ is a feasible, positive solution. Show that P1 has an optimal value less than one. Also show that for any feasible solution $(\overline{\mathbf{x}}, \overline{\lambda})$ to P1, where $\overline{\lambda} \leq 1$, the solution $\hat{\mathbf{x}} = [(1-\overline{\lambda})\mathbf{x}_0 + \overline{\mathbf{x}}]/(2-\overline{\lambda})$ is a positive, feasible solution for the given linear system. Using Exercises 8.41 and 8.42, suggest a two-phase procedure for Karmarkar's algorithm, as well as for its affine scaling variant.

8.44 Consider the linear programming problem given in Exercise 8.41d. Starting with the solution $\mathbf{x}_0 = (0.5, 0.5, 1, 0.5)'$, apply the affine scaling algorithm of Exercise 8.42 to solve this problem. Show the progress of the iterates with respect to feasibility and objective values in the problem (8.28) and its dual, and purify the resulting solution to a pair of primal and dual extreme points of Problem (8.28).

8.45 Solve the problem of Exercise 8.36 using the affine scaling algorithm of Exercise 8.42, starting with the solution $(x_1, x_2, x_3, x_4) = (1, 1, 1, 1)$, where x_3 is the slack in the single constraint and x_4 is a suitable artificial variable.

8.46 Show that the dual iterate $\mathbf{w}_k$ for the affine scaling algorithm of Exercise 8.42 is the solution to a certain least-squares problem. Motivate this latter problem.

8.47 Consider the linear programming problem

$$\text{Minimize } \{\mathbf{cx} : \mathbf{Ax} = \mathbf{b}, -1 \leq x_j \leq 1 \quad \text{for } j = 1, \ldots, n\} \quad (8.30)$$

where $\mathbf{A}$ is $m \times n$ of rank m and where the data is all integer. Show how a problem with general lower and upper bounds l_j and u_j on the variables, where $-\infty < l_j \leq u_j < \infty$, can be put into this form. Assume (using artificial variables if necessary) that there exists some known feasible solution $\mathbf{x}_0$ to Problem (8.30) satisfying $-\mathbf{1}^t < \mathbf{x}_0 < \mathbf{1}^t$. Analogous to Exercise 8.42, construct the following barrier function problem for Problem (8.30) that uses a barrier term to keep the variables strictly within their bounds:

$$\text{Minimize}\left\{\mathbf{cx} - \mu \sum_{j=1}^{n} \left[\log(1 + x_j) + \log(1 - x_j)\right] : \mathbf{Ax} = \mathbf{b}\right\} \quad (8.31)$$

Here $\mu > 0$ is of magnitude $2^{-O(L)}$, and is assumed to be sufficiently small so that a solution near enough to an optimum to Problem (8.31) can be purified to an optimal solution to Problem (8.30). Consider the following *Bounded Variables Affine Scaling Algorithm.*

Initialization Let $-\mathbf{1}^t < \mathbf{x}_0 < \mathbf{1}^t$ be the initial feasible solution. Set the iteration counter $k = 0$, and select $q \in [0.97, 0.99]$ and $\mu = 2^{-O(L)}$.

Step 1 (Dual estimates, projected gradient, and direction of motion, based on a projected gradient step in the **y**-space, where $\mathbf{y} = \mathbf{D}_k^{-1}\mathbf{x}$).

Given $\mathbf{x}_k$, define $\mathbf{D}_k = \text{diag}\{D_{k1}, \ldots, D_{kn}\}$, where $D_{kj} = \min\{1 + x_{kj}, 1 - x_{kj}\}$ for $j = 1, \ldots, n$, and define $\delta(x_k) = [\delta_1(x_{k1}), \ldots, \delta_n(x_{kn})]^t$, where $\delta_j(x_{kj}) = (-2D_{kj}x_{kj})/(1 - x_{kj}^2)$ for $j = 1, \ldots, n$. (Note that $-1 < \delta_j(x_{kj}) < 1$ for $j = 1, \ldots, n$.) Denote $\bar{\mathbf{c}}_k = \mathbf{cD}_k$, $\mathbf{P}_k = \mathbf{AD}_k$, and compute the following:

Dual estimate $\mathbf{w}_k = (\mathbf{P}_k\mathbf{P}_k^t)^{-1}\mathbf{P}_k[\bar{\mathbf{c}}_k^t - \mu\delta(\mathbf{x}_k)]$

Reduced cost vector $\mathbf{R}_k = \mathbf{c}^t - \mu\mathbf{D}_k^{-1}\delta(\mathbf{x}_k) - \mathbf{A}^t\mathbf{w}^k$

Projected gradient vector $\mathbf{c}_{pk} = \mathbf{D}_k\mathbf{R}_k \equiv [\mathbf{I} - \mathbf{P}_k^t(\mathbf{P}_k\mathbf{P}_k^t)^{-1}\mathbf{P}_k][\bar{\mathbf{c}}_k^t - \mu\delta(\mathbf{x}_k)]$

Direction of motion $\mathbf{d}_k = -\mathbf{D}_k\mathbf{c}_{pk}$

Step 2 (Termination Test).

If $\|\mathbf{c}_{pk}\| = 0$, then stop with $\mathbf{x}_k$ as optimal to Problem (8.31). Purify $\mathbf{x}_k$ to an optimal vertex of Problem (8.30). (Practically, one would stop if $\|\mathbf{c}_{pk}\|$ is sufficiently small.)

Step 3 (New Iterate).

Set $\mathbf{x}_{k+1} = \mathbf{x}_k + \lambda_k\mathbf{d}_k$, where the step length λ_k is given by

$$\lambda_k = \text{Minimum}\left\{q\lambda_{\max}, (1 - q^2)^2/4\mu\right\}$$

where

$$\frac{1}{\lambda_{\max}} = \underset{j=1,\ldots,n}{\text{Maximum}}\left[\text{Max}\left\{\frac{d_{kj}}{(1 - x_{kj})}, -\frac{d_{kj}}{(1 + x_{kj})}\right\}\right] > 0$$

Increment k by one and return to Step 1.

a. Show that the bounded variables affine scaling algorithm either stops finitely with an optimal solution $\mathbf{x}_k$ to Problem (8.31) in case $\|\mathbf{c}_{pk}\| = 0$, or else, an infinite sequence of primal and dual iterates $\{\mathbf{x}_k, \mathbf{w}_k\}$ is generated that is such that the sequence $\{\|\mathbf{c}_{pk}\|\} \to 0$. Moreover, for any convergent subsequence $\{\mathbf{x}_k, \mathbf{w}_k\}_K$ indexed by K with limit point $(\bar{\mathbf{x}}, \bar{\mathbf{w}})$, we have $-\mathbf{1}^t < \bar{\mathbf{x}} < \mathbf{1}^t$ and that $\bar{\mathbf{x}}$ solves Problem (8.31).

b. Show that the dual iterates $\mathbf{w}_k$ correspond to a dual feasible solution to Problem (8.30) for all k. Obtain an expression for the duality gap between the primal and dual solutions $\mathbf{x}_k$ and $\mathbf{w}_k$ for Problem (8.30). Suggest a termination criterion based on this duality gap that will permit a purification to a pair of optimal primal and dual extreme point solutions for Problem (8.30).

8.48 Solve the following problem using the bounded variables affine scaling algorithm, after transforming it to the form required by Exercise 8.47.

$$\begin{array}{ll}\text{Minimize} & -x_1 \\ \text{Subject to} & 2x_1 + x_2 + x_3 = 2 \\ & x_1 - x_2 + x_4 = 3 \\ & -1 \le x_1 \le 2, \ -3 \le x_2 \le -1, \ 0 \le x_3 \le 7, \ 0 \le x_4 \le 3\end{array}$$

Use the solution $\mathbf{x} = (0, -2, 4, 1)$ as the starting solution.

8.49 For the bounded variables affine scaling algorithm of Exercise 8.47, consider the following alternative affine transformation to be used in Step 1 of the algorithm.

$$\mathbf{x} = \mathbf{D}_k \mathbf{y} \qquad \text{where } \mathbf{D}_k = \text{diag}\left\{\left(1 - x_{kj}^2\right), \ j = 1, \ldots, n\right\}$$

Using the same barrier function problem (8.31) state an affine scaling algorithm analogous to the one in Exercise 8.47 and establish a similar convergence result.

8.50 Illustrate the algorithm of Exercise 8.49 using the example of Exercise 8.48.

8.51 Consider the problem of estimating the parameters of a linear regression model $\mathbf{b} = \mathbf{Ax} - \mathbf{E}$, where $\mathbf{b} = (b_1, \ldots, b_n)^t$ is a vector of n observations, $\mathbf{A}$ is an $n \times m$ matrix with each row corresponding to values of some m independent variables, $\mathbf{x}$ is a vector of m parameters to be estimated, and $\mathbf{E}$ is a vector of n random errors that are independently and identically distributed according to some distribution. We are interested in finding out a set of values for the parameters $\mathbf{x}$ that minimize $\sum_{i=1}^{n} |E_i|$.

a. Formulate this problem as a linear programming problem.

b. Write the dual to this problem and show how the bounded variables affine scaling algorithm of Exercise 8.47 can be used to solve this problem.

8.52 Write a computer program for using Karmarkar's algorithm for solving the linear program of the form Minimize $\{\mathbf{cx} : \mathbf{Ax} = \mathbf{b}, \mathbf{x} \ge \mathbf{0}\}$, where $\mathbf{A}$ is $m \times n$ of rank m.

8.53 Write a computer program for using the affine scaling algorithm of Exercise 8.42 to solve the linear program: Minimize $\{\mathbf{cx} : \mathbf{Ax} = \mathbf{b}, \mathbf{l} \le \mathbf{x} \le \mathbf{u}\}$ for the case when $l_i = 0$, $u_i = \infty$ for $i = 1, \ldots, n$, where $\mathbf{A}$ is $m \times n$. Repeat using the algorithm of Exercise 8.47 to solve the linear program for the case $-\infty < l_i < u_i < \infty$ for $i = 1, \ldots, n$.

NOTES AND REFERENCES

1. The concept of "good" or polynomially bounded algorithms was independently proposed by Edmonds [1965], and Cobham [1965]. See Cook [1971], Karp [1972], Garey and Johnson [1979], and Papadimitriou and Steiglitz [1982] for further reading on this subject.
2. For the complexity analysis of the simplex method under various pivoting rules, see Klee and Minty [1972], Jeroslow [1973], Avis and Chvatal [1978], and Goldfarb and Sit [1979]. (Exercise 8.6 uses Chvatal's [1983] approach.)
3. Around 1957, Hirsch conjectured (see Dantzig [1963]) that for bounded polyhedral sets with f facets in n dimensions, any two vertices of the polytope are connected by a simplex path of length at most $(f - n)$. This is referred to as the *Hirsch conjecture*. Klee and Walkup [1967] showed the necessity of boundedness of the polyhedral set and established this conjecture under the restriction $f - n \leq 5$. (Klee [1965] showed this conjecture to be true for $n \leq 3$.) The *monotonic bounded Hirsch conjecture* asserts that there exists a simplex path between any two vertices of length no more than $(f - n)$, which additionally has nonincreasing objective function values. Todd [1980] has shown this conjecture to be false for $n \gtrsim 4$. For a related result on the zero-one set partitioning problem, see Balas and Padberg [1970].
4. Average-case probabilistic complexity analyses of the simplex method appear in the novel work of Borgwardt [1982a, b] and Smale [1983a, b]. This independent work has been coordinated and extended by Haimovich [1983]. Also, see Todd [1989] for a recent survey.
5. The Soviet mathematician L. G. Khachian [1979] was the first to introduce linear programming problems in the Class P of problems. Proofs of the results appear in Khachian [1981] and Gacs and Lovasz [1981]. The latter paper contains the ideas of Exercises 8.12, and the optimal rounding scheme of Exercise 8.19. Khachian's ellipsoid algorithm derives its ideas from earlier work by other Soviet mathematicians including Shor [1970a, b] and Yudin and Nemirovski [1976]. Analysis of this algorithm using finite precision arithmetic appears in Khachian [1980] and Grotschel, Lovasz, and Schrijver [1981]. The "deep cut" variation discussed in Exercises 8.10–8.15 was first presented in Shor and Gershovich [1979]. The unbounded complexity of Khachian's algorithm with real, noninteger data was demonstrated by Traub and Wozniakowski [1982] (see Exercise 8.22). A polynomial variant of Khachian's algorithm that uses simplices in lieu of ellipsoids appears in Yaminitsky and Levin [1982]. An excellent survey appears in Bland, Goldfarb, and Todd [1981]. Also see the survey paper of Shor [1983] and the ellipsoid algorithm development in Murty [1983] and Schrijver [1986] for further details.
6. Using a polynomial-time linear programming subroutine, Tardos [1985, 1986] shows that the linear program to minimize $\mathbf{cx}$ subject to $\mathbf{Ax} \leq \mathbf{b}$ with integer data can be solved in a number of elementary arithmetic operations bounded

by a polynomial in the size of A alone. Hence network structured problems, for example, admit genuinely or strongly polynomial algorithms. Megiddo [1983] also derives a strongly polynomial algorithm for linear programs with at most two nonzero coefficients in the objective function and in each of the constraints.

7. The projective polynomial-time algorithm for linear programming was proposed by N. Karmarkar [1984, 1985] at AT & T Bell Laboratories, U.S.A. Lawler [1985] and Padberg [1986] discuss the use of larger step sizes, and Goldfarb and Mehrotra [1985] and Anstreicher [1986a] validate techniques using approximate projection operations. An alternative to using the sliding objective function/bisection scheme is presented in Todd and Burrell [1986] and Anstreicher [1985]. Charnes, Song, and Wolfe [1984] exhibit a worst-case linear convergence rate for Karmarkar's algorithm. However, Iri and Imai [1986] develop a modification that is quadratically convergent. Implementation suggestions and transformations for solving general linear programs by Karmarkar's method appear in Tomlin [1985] and Shanno and Marsten [1985], in particular. The two-phase approach of Exercise 8.43 is from Kojima [1986]. For factorizations in solving the least-squares problem accurately, see Dongarra *et al.* [1979], Heath [1982], George and Heath [1980], and Todd and Burrell [1986]. Relationships of Karmarkar's algorithm to projected Newton Barrier methods have been shown by Gill, Murray, Saunders, Tomlin, and Wright [1985], and to the ellipsoid algorithm have been demonstrated by Todd [1987] and Ye [1987a, b]. For computational experience using linear assignment problems, see Aronson, Barr, Helgason, Kennington, Loh, and Zaki [1985]. The "purification" scheme first found its elements in Charnes, Kortanek, and Raike [1965]. Also, see Kortanek and Jishan [1988]. Sherali, Skarpness, and Kim [1988] show how to use such a scheme for optimally rounding to a basic feasible solution. Ye [1987b] also discusses how to obtain an optimal basis via Karmarkar's algorithm, and Ye and Kojima [1987] discuss how to recover dual solutions. Variants and discussions related to Karmarkar's projective algorithm appear in Anstreicher [1986b, c], Asic, Kovacevic, and Nikolic [1986], Blair [1986], Blum [1985], Chang and Murty [1987], de Ghellinck and Vial [1986], Gay [1987], Gonzaga [1985, 1987b], Kojima [1986], Lustig [1985], Megiddo [1985, 1986], Murty [1985], Nazareth [1986], Rinaldi [1986], Ye [1985], and Ye and Chiu [1985], among others. (See Note 8 for the affine scaling and the path-following variants.) For exploiting special structures using a variant of Karmarkar's projective algorithm, see Todd [1987]. Monma [1987], Todd [1989], and Zimmerman [1986] provide recent surveys.

8. The *affine scaling variant* of Karmarkar's algorithm finds its origin in Dikin [1967, 1974], but was more recently proposed independently by Barnes [1986], Vanderbei, Meketon, and Freedman [1986] and Cavalier and Soyster [1985], following Karmarkar's algorithm. Barnes [1986], Vanderbei *et al.* [1986], Kortanek and Shi [1987], and Sherali [1987] prove convergence under various conditions. Sherali *et al.* [1988] have presented an assumption-free convergence analysis of a perturbation of this algorithm. (Exercises 8.42, 8.47, and

8.49 are derived from this paper.) For exploiting special structures using this variant, see Chandru and Kochar [1986]. Adler, Resende, and Veiga [1986] and Monma and Morton [1987] have also independently developed dual affine variants of Karmarkar's algorithm. Megiddo and Shub [1987] have exhibited a worst-case exponential behavior for the affine scaling algorithm with infinitesimal steps (continuous trajectory) on the Klee–Minty [1972] polytope by starting the algorithm close enough to a vertex. By adopting an additional step that centers the iterate within the polytope, Barnes [1987] has derived a polynomial-time variant of the affine scaling algorithm. This centering amounts to increasing the denominator of Karmarkar's multiplicative potential function, while the usual gradient based step is aimed at reducing the numerator of this potential function. Hence we have polynomial-time behavior. This approach falls under the category of *path-following algorithms*. These algorithms have been motivated by the work of Megiddo [1986b], and include the $O(n^3L)$ algorithms of Ben Daya and Shetty [1988], Gonzaga [1987a], Monteiro and Adler [1987], Renegar [1988], Vaidya [1987], and Ye and Todd [1987]. In particular, McShane, Monma, and Shanno [1988] provide implementation details and computations using the primal-dual path-following method of Monteiro and Adler [1987]. The work of Gonzaga [1987a, b] also motivates a *bidirectional* search in the **y**-space based on the projection of the vector **1**, in addition to that of the vector $-\mathbf{cD}_k$, onto the null space of $\mathbf{P}_k = \mathbf{AD}_k$, in the context of the affine scaling method. Gonzaga [1988] also shows how to derive a polynomial-time affine scaling variant through a potential function minimization process. For a polynomial-time primal-dual affine scaling variant of Karmarkar's algorithm, see Monteiro, Adler, and Resende [1988].

9. AT & T has released a commercial software based on Karmarkar's algorithm, under the trademark of KORBX. This software implements the primal affine scaling, the dual affine scaling, and the primal–dual path following variants of Karmarkar's algorithm.

NINE: MINIMAL COST NETWORK FLOWS

In this chapter we present the specialization of the simplex algorithm to network structured linear programming problems. This specialization, known as the *network simplex algorithm*, performs the simplex operations directly on the graph or the network itself. We also discuss appropriate data structures that facilitate the implementation of such a graph-theoretic procedure on the computer. The overall efficiency with which such a procedure operates enables one to solve problems 200–300 times faster than with a standard simplex approach which ignores any inherent special structures other than sparsity. Hence very large problems can be solved with a reasonable amount of effort. Note that the network simplex algorithm has been shown to exhibit worst-case exponential behavior on certain classes of transportation problems. Moreover, there exist special (strongly) polynomial-time algorithms for network flow problems based on an iterative scaling method or on an implementation of a dual simplex method, as well as Khachian's and Karmarkar's polynomial algorithms. However, the algorithms discussed in this chapter and in Chapter 11 are by far the most effective in practice. Furthermore, this chapter lays the foundation for studying the polynomial-time dual-simplex algorithms, as well as more advanced topics dealing with generalized networks, extracting network or generalized network structures or substructures from given linear programming problems, and solving network flow problems with side constraints and/or side variables. We refer the reader to the Notes and References section for reading material on these topics.

We begin our discussion by introducing network structured linear programs and some associated concepts from graph theory that are relevant to our study. Next we expose an important property possessed by the coefficient matrix that imparts this problem its special structure, and we characterize a basis matrix in graph theoretic terms. This leads us into developing the network simplex algorithm for the nonnegatively constrained as well as for the bounded variables problem. Thereafter we examine data structures for representing the basis matrix that enable us to implement simplex types of algorithms on the computer. This discussion includes terminology and concepts that are important in making the reader sufficiently literate to study much of the modern literature on network flow programming.

9.1 THE MINIMAL COST NETWORK FLOW PROBLEM

Consider a *directed network* or a *digraph* G, consisting of a finite set of *nodes* (vertices or points) $\mathscr{N} = \{1, 2, \ldots, m\}$ and a set of *directed arcs* (links, branches, edges, or lines) $\mathscr{A} = \{(i, j), (k, l), \ldots, (s, t)\}$ joining pairs of nodes in $\mathscr{N}$. Arc (i, j) is said to be *incident* at nodes i and j and is directed from node i to node j. We shall assume that the network has m nodes and n arcs. Figure 9.1 presents a network with four nodes and seven arcs.

With each node i in G we associate a number b_i that is the available supply of an item (if $b_i > 0$) or the required demand for the item (if $b_i < 0$). Nodes with $b_i > 0$ are sometimes called *sources*, and nodes with $b_i < 0$ are sometimes called *sinks*. If $b_i = 0$, then none of the item is available at node i and none is required; in this case node i is sometimes called an *intermediate* (*or transshipment*) node. Associated with each arc (i, j) we let x_{ij} be the amount of flow on the arc (we assume $x_{ij} \geq 0$), and we let c_{ij} be the unit shipping cost along the arc.

We shall assume that the total supply equals the total demand within the network, that is, $\Sigma_{i=1}^{m} b_i = 0$. If this is not the case, that is, $\Sigma_{i=1}^{m} b_i > 0$, then add a dummy demand node, $m + 1$, with $b_{m+1} = -\Sigma_{i=1}^{m} b_i$ and arcs with zero cost from each supply node to the new node.

The minimal cost network flow problem may be stated as follows. Ship the available supply through the network to satisfy demand at minimal cost. Mathematically, this problem becomes (where summations are taken over existing arcs)

$$\text{Minimize} \quad \sum_{i=1}^{m} \sum_{j=1}^{m} c_{ij} x_{ij}$$

$$\text{Subject to} \quad \sum_{j=1}^{m} x_{ij} - \sum_{k=1}^{m} x_{ki} = b_i \qquad i = 1, \ldots, m \tag{9.1}$$

$$x_{ij} \geq 0 \qquad i, j = 1, \ldots, m$$

Constraints (9.1) are called the *flow conservation* or *nodal balance* or *Kirchhoff*

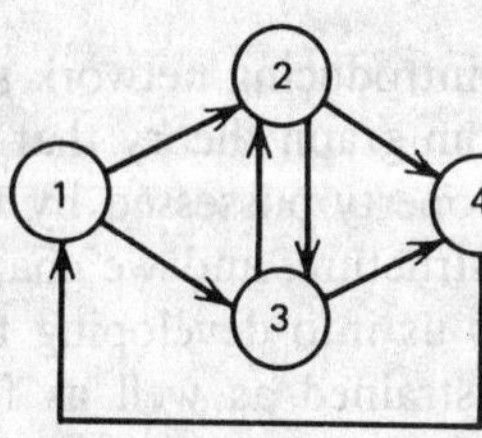

Figure 9.1. Example of a network.

equations and indicate that the flow may be neither created nor destroyed in the network. In the conservation equations, $\Sigma_{j=1}^{m} x_{ij}$ represents the total flow out of node i while $\Sigma_{k=1}^{m} x_{ki}$ indicates the total flow into node i. These equations require that the net flow out of node i, $\Sigma_{j=1}^{m} x_{ij} - \Sigma_{k=1}^{m} x_{ki}$, should equal b_i. If $b_i < 0$, then there should be more flow into i than out of i. This problem is also said to be *uncapacitated*. The problem, which includes upper bounds on arcs, is said to be *capacitated* and is considered subsequently.

The minimal cost flow problem might arise in a logistics network where people and materials are being moved between various points in the world. It may be associated with the movement of locomotives between points in a railroad network to satisfy power for trains at least travel cost. Minimal cost network flow problems occur in the design and analysis of communication systems, oil pipeline systems, tanker scheduling problems, and a variety of other areas.

Clearly, the minimal cost flow problem is a linear program and can be solved in any one of several ways. One way is to apply the ordinary primal simplex algorithm to the problem. What we seek in this chapter is a simplification of the simplex method so that it can be applied directly on the network without the need of a simplex tableau.

9.2 SOME BASIC DEFINITIONS AND TERMINOLOGY FROM GRAPH THEORY

Consider the *digraph* $G(\mathscr{N}, \mathscr{A})$ introduced earlier with node set $\mathscr{N}$ and arc set $\mathscr{A}$. This graph is said to be *proper* if the cardinalities of $\mathscr{N}$ and $\mathscr{A}$ satisfy $|\mathscr{N}| \geq 2$ and $|\mathscr{A}| \geq 1$. Two nodes in the graph are said to be *adjacent* if they are directly connected by some arc. For the directed arc $(i, j) \in \mathscr{A}$, node i is called its *from-node* and node j is called its *to-node*. For any node i, the set of nodes j for which there exists an arc (i, j) is called the *forward-star* of i. Similarly, for any node i, the set of nodes j for which there exists an arc (j, i) is called the *reverse-star* of i.

Throughout our study, we will be concerned with graphs that have directed arcs only, that is, with digraphs. In contrast an arc that has no orientation and that permits flow in either direction is called an *undirected arc*. A graph that has all undirected arcs is called an *undirected graph*. If a graph has both directed and undirected arcs, then it is called a *mixed graph*. Note that in the context of the minimum cost network flow problem, if there exists an undirected arc (i, j) with a cost $c_{ij} \geq 0$, then this can be equivalently replaced by two oppositely oriented directed arcs (i, j) and (j, i), each with the same unit cost c_{ij}. However, this construction is not valid if $c_{ij} < 0$. In such a case, if the problem is feasible, then we can take any feasible solution to the original problem and superimpose on it an infinite circulation of flow from i to j along (i, j) and back from j to i along (j, i) in the transformed network. In this manner, we can maintain feasibility in Problem (9.1) and drive the objective function to $-\infty$, while in reality such a circulation of flow results in a zero additional flow and cost in the original problem.

Continuing, a *path* (from node i_0 to i_p) is a sequence of arcs $P = \{(i_0, i_1), (i_1, i_2), \ldots, (i_{p-1}, i_p)\}$ in which the initial node of each arc is the same as the terminal node of the preceding arc in the sequence and $i_0, \ldots, i_p$ are all distinct nodes. Thus each arc in the path is directed "toward" i_p and "away from" i_0. A *chain* is a similar structure to a path except that not all arcs are necessarily directed toward node i_p. Figure 9.2*a* illustrates a path, and Figure 9.2*b* presents a chain. A *circuit* is a path from some node i_0 to i_p plus the arc (i_p, i_0). Thus a circuit is a *closed path*. Similarly, a *cycle* is a closed chain. Figures 9.2*c* and 9.2*d* depict circuits and cycles. Every path is a chain but not vice versa. Every circuit is a cycle but not conversely.

These definitions refer to what are known as *simple* paths, chains, circuits or cycles. We will assume that these definitions apply unless otherwise specified. A *nonsimple* path, for example, can permit nodes in the set $i_0, i_1, \ldots, i_p$ to repeat and hence may include circuits. For example, the set of arcs $\{(1,2), (2,3), (3,4), (4,2), (2,3), (3,5), (5,6), (6,7), (7,5), (5,8)\}$ describes a nonsimple path from node 1 to 8 with the sequence of nodes $i_0, i_1, \ldots, i_p$ visited being $i_0 = 1, 2, 3, 4, 2, 3, 5, 6, 7, 5$, and $8 = i_p$.

Throughout this chapter we will assume that the underlying graph G for the minimum cost network flow problem is *connected*, that is, there exists a chain between every pair of nodes in G. This property is also referred to as yielding a *weakly connected* graph. A *strongly connected* graph is one where there is a (directed) path from each node to every other node. A *complete* graph is one where each node is connected by an arc to every other node. Hence a complete

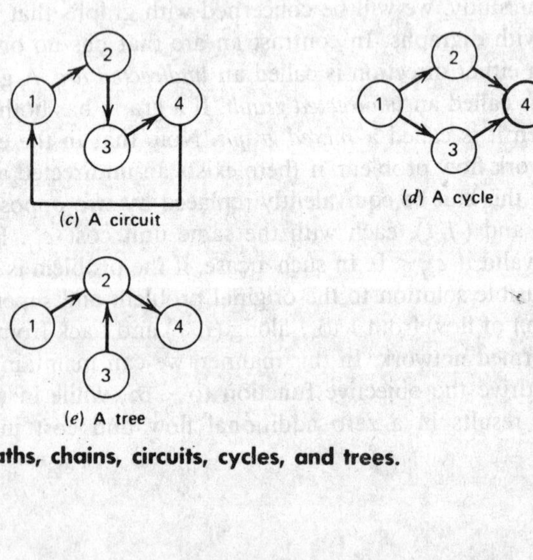

Figure 9.2. Paths, chains, circuits, cycles, and trees.

digraph on m nodes has $m(m-1)$ arcs. A *subgraph* $G'(\mathcal{N}', \mathcal{A}')$ of a graph $G(\mathcal{N}, \mathcal{A})$ is one that satisfies $\mathcal{N}' \subseteq \mathcal{N}$ and $\mathcal{A}' \subseteq \mathcal{A}$, with the understanding that if $(i, j) \in \mathcal{A}'$, then both i and j are in $\mathcal{N}'$. If $G' \neq G$, then G' is said to be a *proper-subgraph* of G. If $\mathcal{N}' = \mathcal{N}$, then G' is said to be a *spanning* subgraph of G, that is, it "spans" all the nodes of G. A subgraph $G'(\mathcal{N}', \mathcal{A}')$ of $G(\mathcal{N}, \mathcal{A})$ is said to be *induced by the node set* $\mathcal{N}'$ if $\mathcal{A}'$ includes all the arcs in $\mathcal{A}$ that have both their associated nodes present in $\mathcal{N}'$. For example, referring to Figure 9.1, the subgraph induced by the node set $\mathcal{N}' = \{1, 2, 4\}$ is the graph with nodes 1, 2, and 4, and with the arc set $\mathcal{A}' = \{(1,2), (2,4), (4,1)\}$. A *component* of a graph G is a subgraph that is connected and that is not a proper-subgraph of another connected subgraph. Hence the components of a graph are *maximal connected subgraphs*—they are "separable pieces" of a graph. For example, if we delete arcs $(1, 2)$, $(1, 3)$, and $(4, 1)$ in Figure 9.1, the resulting graph has two components (which ones?).

A *tree* is a connected graph with no cycles. A *spanning tree*, defined with respect to some underlying graph G is a tree that includes every node of the graph, that is, it is a spanning, connected subgraph with no cycles. Figure 9.2e illustrates a spanning tree of the graph in Figure 9.1. Figures 9.2a and 9.2b are also examples of spanning trees. Figures 9.2c and 9.2d are not trees. Note that a graph with no cycles has a tree as each of its components (why?). Such a graph is called a *forest*. Similarly, a *spanning forest* defined with respect to a graph G is a spanning subgraph of G with no cycles. The *rank* or *degree* of a node is the number of arcs incident at it. The *in-degree* of a node i is the number of arcs that have i as their to-node, and the *out-degree* of i is the number of arcs that have i as their from-node. Hence (in-degree of i) + (out-degree of i) = (degree of i). For example, in Figure 9.1, the in-degree of node 1 is 1 and its out-degree is 2; its degree is 3. A node with degree 0 or 1 is an *end node*. (Sometimes the end node of a tree is referred to as a *leaf node*.) For example, nodes 1 and 4 are end(leaf) nodes of the tree in Figure 9.2a.

We will see shortly that trees play a very central role in network flow problems because they help us to characterize bases of the coefficient matrix in Problem (9.1). Let us examine some additional facts concerning tree graphs.

Property 1

Let T be a proper tree graph with $m(\geq 2)$ nodes, and let $(i, j) \in T$. Then *disconnecting* (i, j) from T, that is, removing the arc (i, j) from T but leaving the nodes i and j in T, decomposes T into two trees T_1 and T_2. This follows since disconnecting (i, j) must result in $r \geq 2$ components, each component being a tree, or else by putting back (i, j), we would have a cycle in T, a contradiction. Since putting back (i, j) can connect no more than two components by definition and since T is connected, we must have $r = 2$.

Property 2

A proper tree graph has at least two end nodes. This is clearly true for a tree with only two nodes. By induction, suppose that this is true for a tree with $(m-1)$

nodes. Let us show that it is true for a tree with m nodes, where $m \geq 3$. Select some $(i, j) \in T$ and disconnect it from T, creating two trees T_1 and T_2, each with no more than $(m - 1)$ nodes. If T_1 or T_2 has only one node, then by the induction hypothesis, the total number of end nodes in T_1 and T_2 is at least 3 (why?). Moreover, by putting back (i, j), we can lose at most one of these end nodes. Hence T has at least two nodes. Otherwise, if both T_1 and T_2 have at least two nodes, then they have at least four end nodes in total, and putting back (i, j) can result in a loss of at most two end nodes. This again shows that T has at least two end nodes.

Property 3

A tree with m nodes has $(m - 1)$ arcs. This is clearly true for $m = 1$ or 2. By induction, assume that this property holds for a tree with $(m - 1)$ nodes and consider a tree with m nodes, $m \geq 3$. By Property 2, an end node i exists. Disconnect the (unique) arc incident at the end node and obtain two trees T_1 and T_2 (by Property 1), where $T_1 \equiv \{i\}$. Hence T_1 has zero arcs, and T_2 has $(m - 1)$ nodes. By the induction hypothesis, it has $(m - 2)$ arcs. Therefore T has $(m - 2) + 1 = (m - 1)$ arcs.

These fundamental properties help establish several equivalent characterizations of a tree graph.

Equivalent Characterizations of a Tree Graph T:

(a) T is connected and has no cycles.
(b) T is connected and has $(m - 1)$ arcs.
(c) T has $(m - 1)$ arcs and has no cycles.
(d) T is connected, but disconnecting any arc from T results in two components.
(e) T has no cycles, but adding any new arc to T results in a graph with exactly one cycle. (Such a graph is called a *one-tree*.)
(f) T has a unique chain connecting each pair of its nodes.

Observe that the statements (a), (b), and (c) assert that any two of the three characteristics of being connected, having no cycles, and having $(m - 1)$ arcs implies the third. Furthermore we will see later that statements (d) and (f) play a role in computing and updating dual variables. Statement (e) pertains to the representation of a nonbasic variable column in terms of the basic variable columns.

The equivalence of these characterizations is readily established. For example, consider statements (a) and (b). Note that (a) implies (b) by Property 3. (b) also implies (a) since if T is connected, has $(m - 1)$ arcs, and contains a cycle, then one can delete an arc from the cycle and not destroy connectedness (why?). Using this method, one can obtain a connected graph with no cycles, but one that has fewer than $(m - 1)$ arcs—a contradiction to Property 3. Hence (a) and (b) are equivalent. Similarly, (a) and (c) are equivalent since (a) implies (c) by Property 3, and (c) implies (a) because if it did not, then we would have a graph

T with $(m - 1)$ arcs and with no cycles that is not connected. Then by adding an arc between any two components, we can reduce the number of components by one without creating any cycle (why?). Continuing in this fashion, we can obtain a connected graph with no cycles, but one that has more than $(m - 1)$ arcs—a contradiction to Property 3. Hence (a) and (c) are equivalent. The equivalence of (a) with the other characterizations is easily shown in a similar manner using these properties. This is left to the reader in Exercise 9.1.

9.3 PROPERTIES OF THE A MATRIX

Consider the coefficient matrix **A** associated with the constraint set (9.1). The matrix **A** has one row for each node of the network and one column for each arc. Each column of **A** contains exactly two nonzero coefficients: a "+1" and a "−1." The column associated with arc (i, j) contains a "+1" in row i, a "−1" in row j, and zeros elsewhere. Thus the columns of **A** are given by

$$\mathbf{a}_{ij} = \mathbf{e}_i - \mathbf{e}_j$$

where $\mathbf{e}_i$ and $\mathbf{e}_j$ are unit vectors in E^m, with 1's in the ith and jth positions respectively. The **A** matrix is called the *node-arc incidence matrix* for the graph. The **A** matrix for the network of Figure 9.1 is

$$\begin{array}{c} \\ \mathbf{A} = \end{array} \begin{array}{c} \begin{array}{ccccccc} (1,2) & (1,3) & (2,3) & (2,4) & (3,2) & (3,4) & (4,1) \end{array} \\ \left[\begin{array}{ccccccc} 1 & 1 & 0 & 0 & 0 & 0 & -1 \\ -1 & 0 & 1 & 1 & -1 & 0 & 0 \\ 0 & -1 & -1 & 0 & 1 & 1 & 0 \\ 0 & 0 & 0 & -1 & 0 & -1 & 1 \end{array}\right] \end{array} \begin{array}{c} \\ 1 \\ 2 \\ 3 \\ 4 \end{array}$$

Rank of the A Matrix

Assume that **A** is a node-arc incidence matrix of a connected digraph. Clearly the **A** matrix does not have full rank since the sum of its rows is the zero vector. To show that **A** has rank $m - 1$ we need only select an $(m - 1) \times (m - 1)$ submatrix from **A** that is nonsingular.

Let T be any spanning tree in the network G. By deleting arcs involved in a cycle, we can show constructively that such a tree exists. From the earlier discussion, T consists of the m nodes of G together with $m - 1$ arcs of G that do not form a cycle. Consider the $m \times (m - 1)$ submatrix $\mathbf{A}_T$ of **A** associated with the nodes and arcs in T. Since $m \geq 2$, T has at least one end k. In such a case the kth row of $\mathbf{A}_T$ contains a single nonzero entry. Permute the rows and columns of $\mathbf{A}_T$ so that this nonzero entry is in the first row and first column. Then $\mathbf{A}_T$ becomes

$$\mathbf{A}_T = \left[\begin{array}{c|c} \pm 1 & \mathbf{0} \\ \hline \mathbf{p} & \mathbf{A}_{T'} \end{array}\right]$$

Delete the first row and column of $\mathbf{A}_T$ and consider the matrix $\mathbf{A}_{T'}$, which is $(m - 1) \times (m - 2)$. Correspondingly obtain the graph T' from T by removing

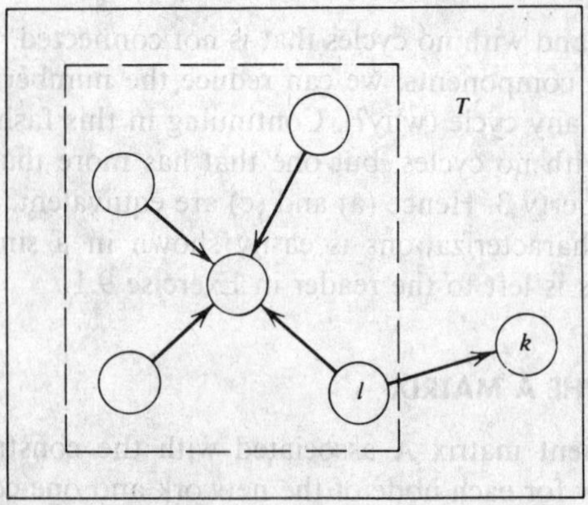

Figure 9.3. Reduction of T to T'.

node k and the incident arc (see Figure 9.3). Here T' is also a tree. It must contain at least one end, say, node l. Permuting the rows and columns of $\mathbf{A}_{T'}$ so that the single nonzero entry in row l is in the first row and column, we may write $\mathbf{A}_T$ as

$$\mathbf{A}_T = \left[\begin{array}{c|c|c} \pm 1 & 0 & \mathbf{0} \\ \hline p_1 & \pm 1 & \mathbf{0} \\ \hline \mathbf{p}_2 & \mathbf{q} & \mathbf{A}_{T''} \end{array}\right]$$

We can continue in this manner exactly $m-1$ times, after which all $m-1$ columns of $\mathbf{A}_T$ are fixed. Deleting the remaining bottom row of $\mathbf{A}_T$, we have an $(m-1)\times(m-1)$ matrix $\mathbf{B}_T$ that is lower triangular with nonzero diagonal elements and therefore nonsingular. Thus the rank of $\mathbf{A}$ is $m-1$.

If we select columns (1, 3), (2, 3), and (3, 4) from the node-arc incidence matrix for the network of Figure 9.1, we get the following lower triangular matrix after using the foregoing permutation process, with (end) nodes selected in the order 1, 2, and 3, and finally discarding row 4:

$$\mathbf{B}_T = \begin{array}{c} \\ 1 \\ 2 \\ 3 \end{array} \begin{array}{c} \begin{array}{ccc} (1,3) & (2,3) & (3,4) \end{array} \\ \left[\begin{array}{ccc} 1 & 0 & 0 \\ 0 & 1 & 0 \\ -1 & -1 & 1 \end{array}\right] \end{array}$$

The corresponding spanning tree is given as follows.

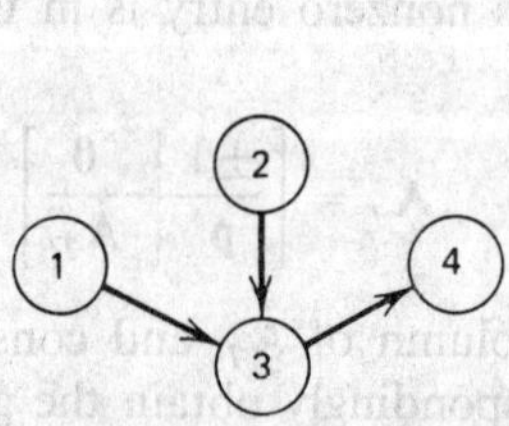

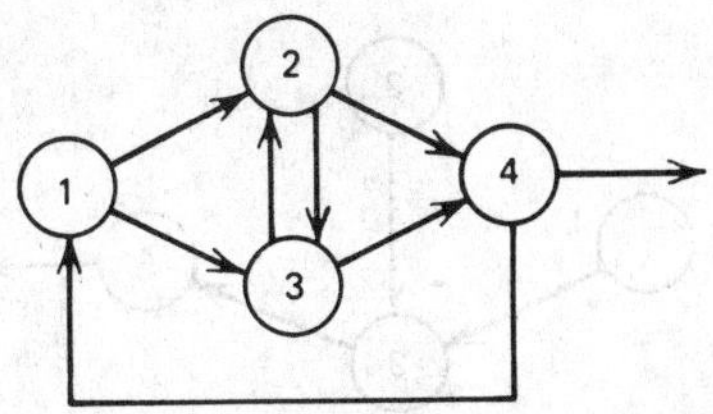

Figure 9.4. **A generalized graph *G*.**

The Artificial Variable

Recall that the simplex method always starts with a full rank constraint matrix. We demonstrated earlier that the rank of **A** is $m - 1$. Therefore an artificial variable is required so that the rank of the new matrix is m. Introducing an artificial variable corresponding to node m (any other node will do) leads to the constraint matrix $(\mathbf{A}, \mathbf{e}_m)$. We need not penalize the artificial variable since it must be zero in any feasible solution to the artificial problem (why?).

Any basic solution must contain m linearly independent columns, and hence the artificial variable must appear in every basic solution. If we liberalize our definition of an arc, then the new column can be viewed as an arc beginning at node m and terminating in space (see Figure 9.4). This one-ended arc is called a *root arc*. The associated node (m) is called a *root node*.

Characterization of a Basic Matrix

We determined the rank of **A** by examining the submatrix associated with any spanning tree. This also demonstrates that a spanning tree together with a single artificial variable corresponds to a basis for the **A** matrix. A graph of this type is called a *rooted spanning tree* and is illustrated in Figure 9.5. Note that the associated basis matrix **B** obtained by adding the root arc and the root node as the last column and row respectively to $\mathbf{B}_T$ is lower triangular because $\mathbf{B}_T$ is lower triangular and the root arc column is $\mathbf{e}_m$. This is shown here for the previous example.

$$\mathbf{B} = \begin{array}{c} 1 \\ 2 \\ 3 \\ 4 \end{array} \overset{\begin{array}{cccc} (1,3) & (2,3) & (3,4) & \text{root arc} \end{array}}{\begin{bmatrix} 1 & 0 & 0 & 0 \\ 0 & 1 & 0 & 0 \\ -1 & -1 & 1 & 0 \\ 0 & 0 & -1 & 1 \end{bmatrix}}$$

A rooted spanning tree corresponds to a (lower triangular) basis for $(\mathbf{A}, \mathbf{e}_m)$. The converse of this is also true. Namely, any basis for $(\mathbf{A}, \mathbf{e}_m)$ is a node-arc incidence matrix of a rooted spanning tree. To see this, let **B** be an $m \times m$ basis matrix of $(\mathbf{A}, \mathbf{e}_m)$, and let G_B be the graph whose node-arc incidence matrix is **B**. Since the artificial column must be a part of every basis and since **B** is $m \times m$, G_B is a *rooted spanning subgraph* of G, that is, it is a spanning subgraph plus the root arc. Besides the root arc, it has $(m-1)$ arcs. Furthermore it must be

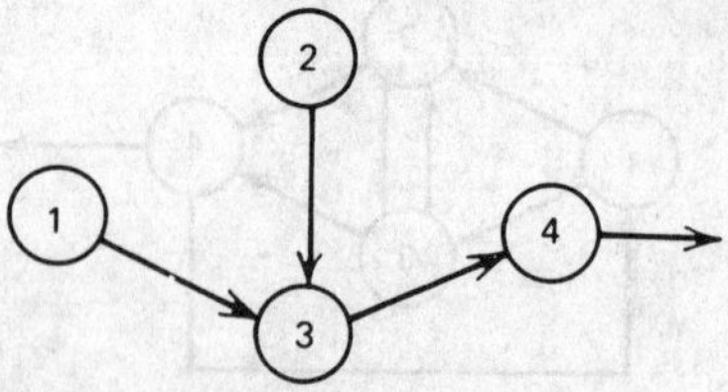

Figure 9.5. A basis subgraph is a rooted spanning tree.

connected or else it contains a component with no root arc. However the rows of **B** corresponding to the nodes in such a component must add to zero since each basic arc with nonzero entries in these rows has a $+1$ and a -1 in these rows and zeros elsewhere. Hence G_B has m nodes, one root arc, it is connected, and has $(m-1)$ arcs. By the equivalent characterization (b) of a tree in the foregoing section, G_B is a rooted spanning tree.

Although the previous argument implies that G_B cannot contain a cycle, it is instructive to see this property more directly. Suppose that G_B is a rooted spanning subgraph of G, and by contradiction, suppose that G_B contains a cycle. (See Figure 9.6.) Select some arc (i, j) in the cycle and orient the cycle in the direction of that arc. Then, for each column of **B** associated with an arc in the cycle, assign a coefficient of $+1$ if the arc is in the direction of orientation of the cycle and -1 otherwise. Applying these coefficients to the respective columns in **B**, we find that the weighted sum of these columns is the zero vector. In Figure 9.6 we have

$$\mathbf{a}_{ij} - \mathbf{a}_{kj} - \mathbf{a}_{lk} + \mathbf{a}_{lp} + \mathbf{a}_{pq} + \cdots = (\mathbf{e}_i - \mathbf{e}_j) - (\mathbf{e}_k - \mathbf{e}_j) - (\mathbf{e}_l - \mathbf{e}_k) + (\mathbf{e}_l - \mathbf{e}_p) + (\mathbf{e}_p - \mathbf{e}_q) + \cdots = \mathbf{0}$$

Thus the columns of **B** could not have been linearly independent. This contradiction shows that G_B contains no cycles. But G_B contains m nodes and $m-1$ arcs. Hence G_B is a tree (why?).

To summarize, we have shown that each basis consists of the root together with a spanning tree (see Figure 9.5), and conversely. Thus we have the following theorem.

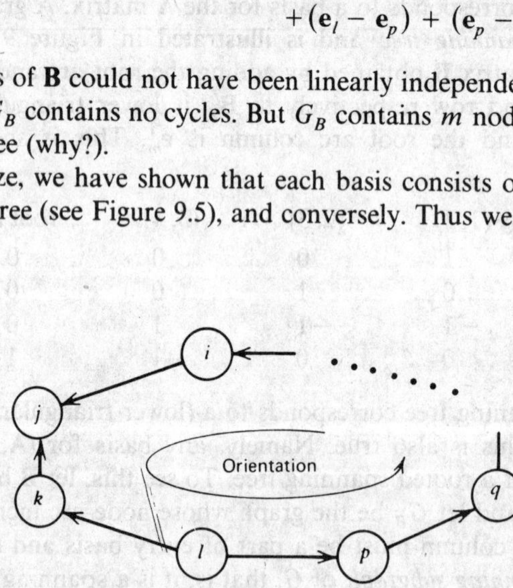

Figure 9.6. Illustration of a linearly dependent set of columns.

***Theorem* 9.1**

Consider a minimum cost network flow problem defined on a connected digraph G with one root arc. Then $\mathbf{B}$ is a basis matrix for this problem if and only if it is the node-arc incidence matrix of a rooted spanning tree of G.

If we have a minimum cost network flow problem defined on a graph G with more than one component, each component having at least one root arc, or if the problem is defined on a graph G with only one component but more than one root arc, then a basis is characterized in general by a *rooted spanning forest* (see Exercise 9.36). However, as shown in Section 9.7, one can easily transform such a problem equivalently to a form where G is a connected digraph with a single root arc. Hence Theorem 9.1 characterizes a basis matrix without loss of generality.

Triangularity, Integrality, and Total Unimodularity

For a minimum cost network flow program defined on a digraph G with one root arc, we have seen that every basis $\mathbf{B}$ corresponds to a rooted spanning tree. Moreover, this basis has elements ± 1 or 0 and can be put in lower triangular form with $+1$ or -1 on the diagonals. This fact implies two very useful features. First, the systems of equations $\mathbf{B}\mathbf{x}_B = \mathbf{b}$ and $\mathbf{w}\mathbf{B} = \mathbf{c}_B$ that determine the values of the basic variables $\mathbf{x}_B$ and the dual variables (simplex multipliers) $\mathbf{w}$ are efficiently solved by a simple forward or backward substitution process. We will see shortly how these equations can be solved directly on the network itself. Second, if the b_i's are all integers, then since $\det \mathbf{B} = \pm 1$ and $\mathbf{B}$ has all integer components, we see by Cramer's rule or by directly solving the triangular system that the solution to $\mathbf{B}\mathbf{x}_B = \mathbf{b}$ is all integer. Hence every extreme point is integer valued. Therefore if the problem requires an integer optimal solution, then we can solve it by simply determining an optimal extreme point solution to the linear programming network flow problem obtained by ignoring the integrality restrictions.

There is a richer property that the matrix $\mathbf{A}$ satisfies, namely, that of total unimodularity. A matrix $\mathbf{A}$ is said to be *totally unimodular* if every square submatrix of $\mathbf{A}$ has determinant $+1$, -1, or 0. In the case of the node-arc incidence matrix $\mathbf{A}$, since all entries are ± 1 or 0, every 1×1 submatrix has determinant ± 1 or 0. Hence by induction, suppose that this property is true for every square submatrix of size $(k-1) \times (k-1)$, and let $\mathbf{A}_k$ be any $k \times k$ submatrix of $\mathbf{A}$, where $k \geq 2$. We must show that $\det \mathbf{A}_k = \pm 1$ or 0. Note that each column of $\mathbf{A}_k$ has either all zeros or only a single nonzero entry that is $+1$ or -1, or it has exactly two nonzero entries, namely a $+1$ and a -1. If any column of $\mathbf{A}_k$ is zero, then $\det \mathbf{A}_k = 0$. If any column of $\mathbf{A}_k$ has a single nonzero entry, then expanding the determinant of $\mathbf{A}_k$ by the minors of that column, we get $\det \mathbf{A}_k = \pm \det \mathbf{A}_{k-1}$, where $\mathbf{A}_{k-1}$ is a square submatrix of size $(k-1) \times (k-1)$. By the induction hypothesis, $\det \mathbf{A}_{k-1} = \pm 1$ or 0 and hence $\det \mathbf{A}_k = \pm 1$ or 0. Otherwise *every* column of $\mathbf{A}_k$ must have a $+1$ and a -1. In this case, since the rows of $\mathbf{A}_k$ add up to the zero vector, we have that $\det \mathbf{A}_k = 0$. Hence $\mathbf{A}$ is totally unimodular.

Note that this proof holds even if we delete rows or columns from **A**, or more importantly, if we add ± unit vector columns to **A**. These columns may correspond to slack or surplus or artificial variable columns. The resulting coefficient matrix in such cases continues to be totally unimodular. Consequently, $\det \mathbf{B} = \pm 1$ for all bases **B**, and every extreme point is integer valued for integer right-hand-sides **b**. For any basis **B**, the inverse matrix $\mathbf{B}^{-1}$ is also comprised of elements that are ± 1 or 0 (why?). Furthermore the updated column $\mathbf{y}_{ij}$ in any canonical representation (simplex tableau) of a basic solution is comprised of ± 1 or 0's. To see this, note that $\mathbf{y}_{ij}$ is given by the system $\mathbf{B}\mathbf{y}_{ij} = \mathbf{a}_{ij}$. Hence the kth element y_{ijk} of $\mathbf{y}_{ij}$ is given by Cramer's rule as

$$y_{ijk} = \frac{\det \mathbf{B}_k}{\det \mathbf{B}}$$

where $\mathbf{B}_k$ is obtained from **B** by replacing its kth column with $\mathbf{a}_{ij}$. Hence $\mathbf{B}_k$ is square submatrix of **A**, and so $\det \mathbf{B}_k = \pm 1$ or 0. Since $\det \mathbf{B} = \pm 1$, it follows that $y_{ijk} = \pm 1$ or 0. Hence any column $\mathbf{a}_{ij}$ can be obtained by the simple addition and subtraction of basic variable columns. (We will also see this fact constructively later.)

To summarize, we have shown that the matrix $\mathbf{A}^0$ comprised of the node-arc incidence matrix of a graph along with ± unit vector columns is totally unimodular. Hence for any basis **B** of $\mathbf{A}^0$, **B** and $\mathbf{B}^{-1}$ have elements that are ± 1 or 0, $\det \mathbf{B} = \det \mathbf{B}^{-1} = \pm 1$, the extreme points of $\{\mathbf{x}\colon \mathbf{A}^0\mathbf{x} = \mathbf{b},\ \mathbf{x} \geq \mathbf{0}\}$ are all integer valued for any integer vector **b**, and $\mathbf{B}^{-1}\mathbf{A}^0$ has all ± 1 or 0 elements.

9.4 REPRESENTATION OF A NONBASIC VECTOR IN TERMS OF THE BASIC VECTORS

Consider the basis subgraph G_B corresponding to a rooted spanning tree; select any nonbasic arc (p, q). Since G_B is a tree, we know that there is a unique chain between nodes p and q. This chain is called a *basis equivalent chain*, and together with the nonbasic arc (p, q), constitutes a cycle. (See Figure 9.7.) Assigning the cycle an orientation consistent with (p, q), we have

$$\begin{aligned}\mathbf{a}_{pq} - \mathbf{a}_{pj} + \mathbf{a}_{kj} + \cdots + \mathbf{a}_{qt} \\ = (\mathbf{e}_p - \mathbf{e}_q) - (\mathbf{e}_p - \mathbf{e}_j) + (\mathbf{e}_k - \mathbf{e}_j) + \cdots + (\mathbf{e}_q - \mathbf{e}_t) = \mathbf{0}\end{aligned}$$

or

$$\mathbf{a}_{pq} = \mathbf{a}_{pj} - \mathbf{a}_{kj} + \cdots - \mathbf{a}_{qt}$$

This development leads to the following simple procedure for representing any nonbasic column in terms of the basic columns. First the unique cycle formed by joining the nonbasic arc to the basis subgraph is determined. The cycle is then given an orientation consistent with the nonbasic variable. A basic column in the cycle along its orientation receives a coefficient of -1 in the representation, and a basic column in the cycle opposite to its orientation receives a coefficient of $+1$ in the representation. Other basic columns receive zero coefficients.

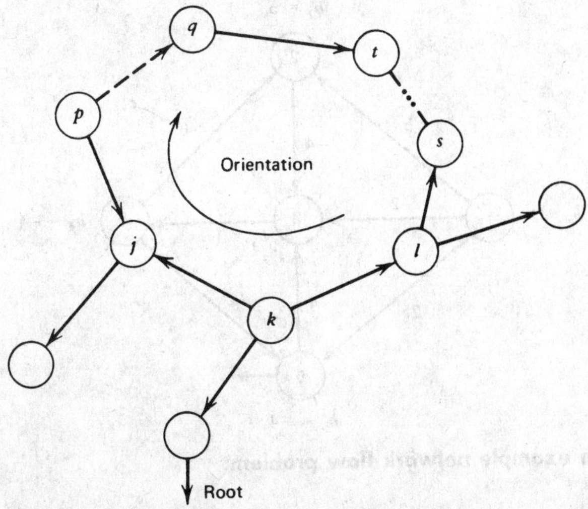

Figure 9.7. Cycle formed by adding a nonbasic arc to the basis tree.

As an example consider the subgraph of Figure 9.5, which is a basis for the network of Figure 9.4. Suppose that we seek the representation of the nonbasic arc $(1, 2)$. Using the foregoing rule, we get

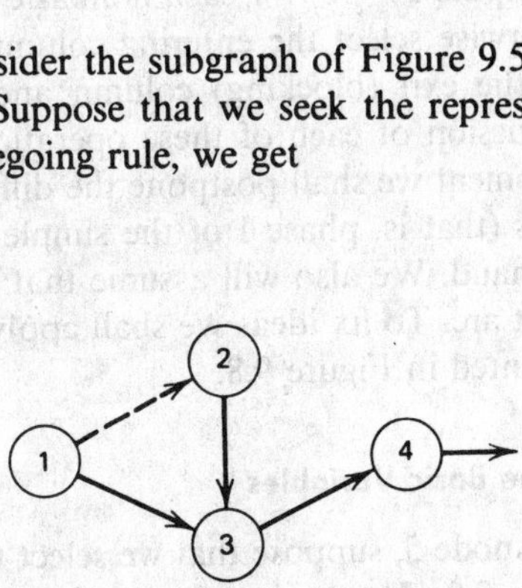

$$\begin{aligned}\mathbf{a}_{12} &= \mathbf{a}_{13} - \mathbf{a}_{23} \\ &= (\mathbf{e}_1 - \mathbf{e}_3) - (\mathbf{e}_2 - \mathbf{e}_3) \\ &= \mathbf{e}_1 - \mathbf{e}_2\end{aligned}$$

Note that the coefficients in the representation of the nonbasic column $\mathbf{a}_{pq}$ in terms of the basic columns give rise to the vector $\mathbf{y}_{pq}$ [that is, the entries in the simplex tableau under the (p, q) column]. Since the artificial column never appears in the representation for any other column and since the artificial variable always remains basic at value zero, we may select any value for the associated cost coefficient, say, $c_a = 0$.

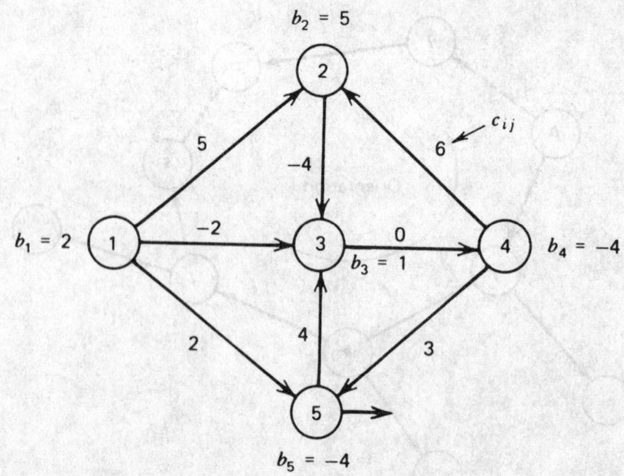

Figure 9.8. **An example network flow problem.**

9.5 THE SIMPLEX METHOD FOR NETWORK FLOW PROBLEMS

The general steps of the simplex method are as follows. First find a starting basic feasible solution. Next compute $z_j - c_j$ for each nonbasic variable x_j. If optimality is achieved, stop; otherwise select the entering column. If unboundedness is not achieved, determine the exit (blocking) column and pivot. The following paragraphs present a discussion of each of these operations applied to network flow problems. For the moment we shall postpone the difficulties associated with identifying a feasible basis (that is, phase I of the simplex method) and assume that a feasible basis is at hand. We also will assume that the problem is defined on a digraph with one root arc. To fix ideas we shall apply each of the foregoing steps to the problem presented in Figure 9.8.

Computing the Values of the Basic Variables

Adding the artificial arc to node 5, suppose that we select the feasible basis given by the subgraph in Figure 9.9. The basic system of equations $\mathbf{Bx}_B = \mathbf{b}$ to be solved is

$$\begin{bmatrix} 1 & 0 & 0 & 0 & 0 \\ 0 & 1 & 0 & 0 & 0 \\ 0 & -1 & 1 & 0 & 0 \\ 0 & 0 & -1 & 1 & 0 \\ -1 & 0 & 0 & -1 & 1 \end{bmatrix} \begin{bmatrix} x_{15} \\ x_{23} \\ x_{34} \\ x_{45} \\ x_5 \end{bmatrix} = \begin{bmatrix} 2 \\ 5 \\ 1 \\ -4 \\ -4 \end{bmatrix}$$

where x_5 is the artificial variable associated with node 5.

Taking advantage of the lower triangular structure of the basis matrix, we may iteratively solve for the basic variables. From the top equation, $x_{15} = 2$. From the second equation, $x_{23} = 5$. From the third equation, $x_{34} = 1 + x_{23} = 6$. Next, $x_{45} = -4 + x_{34} = 2$. Finally, $x_5 = -4 + x_{15} + x_{45} = 0$ and thus the basis is feasible. These same computations can be made directly on the graph in Figure 9.9 as follows.

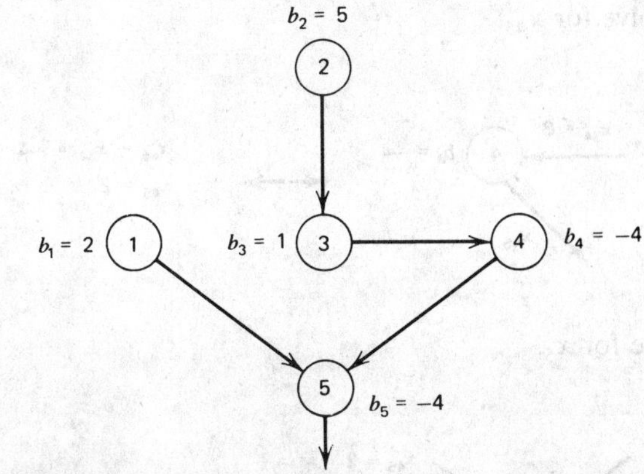

Figure 9.9. A basic subgraph.

Examining node 1 in Figure 9.9, we see that it is an *end* of the basic tree, that is, a node with only one basic arc incident to it. Hence the corresponding basic equation contains only one variable, and the value of that variable can be readily obtained. In this case arc $(1, 5)$ points out of node 1 and thus x_{15} has a $+1$ in row 1. Thus $x_{15} = b_1$ or $x_{15} = 2$.

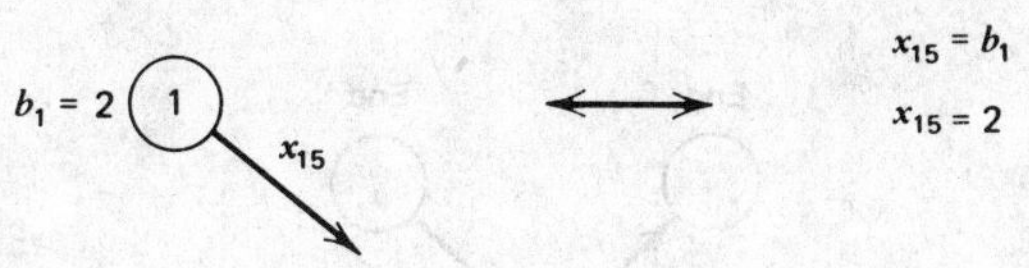

Examining node 2, we see that it is an end and hence x_{23} can be computed similarly:

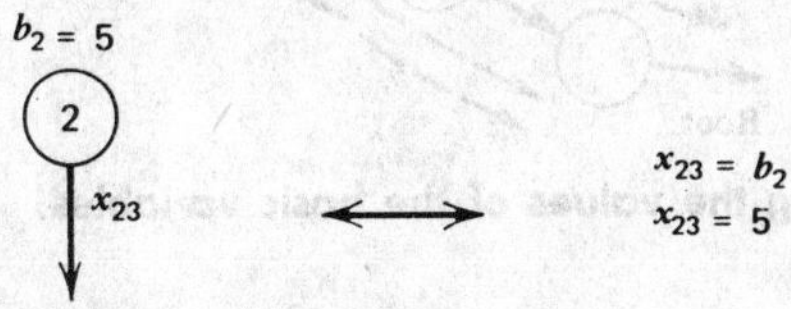

Next notice that node 3 has all of its incident arc variables assigned values except one. Thus we can use the conservation equation for node 3 to solve for the remaining variable.

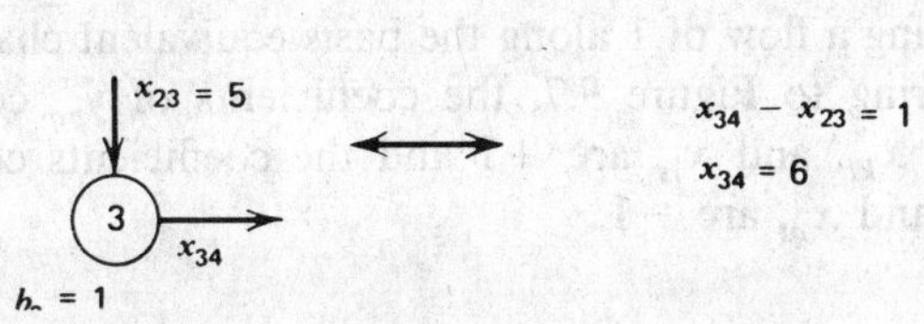

We can now solve for x_{45}:

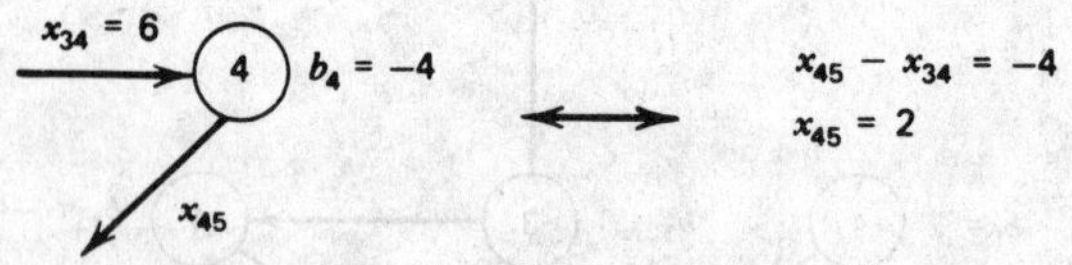

Finally we solve for x_5.

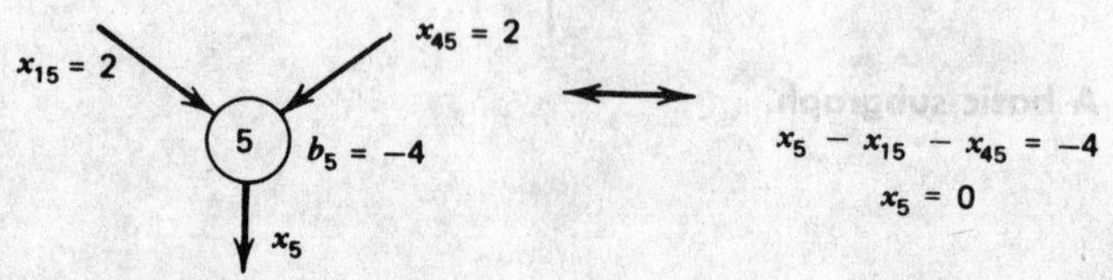

The process of obtaining the basic solution proceeds from the ends of the tree toward the root (see Figure 9.10). As we shall see later, the process of obtaining the dual variables is just reversed.

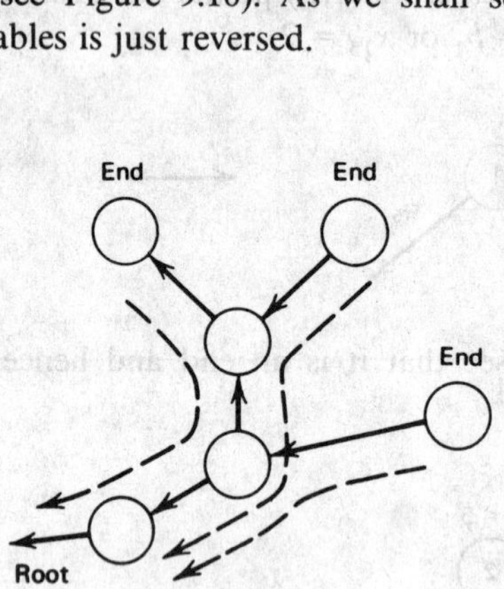

Figure 9.10. Computing the values of the basic variables.

Let us provide an insight with respect to the computation of $\mathbf{y}_{pq}$ via the system $\mathbf{B}\mathbf{y}_{pq} = \mathbf{a}_{pq}$, in Section 9.4. Note that this is similar to the system "$\mathbf{B}\mathbf{x}_B = \mathbf{b}$," where $\mathbf{a}_{pq}$ plays the role of the vector "$\mathbf{b}$." Since $\mathbf{a}_{pq} = \mathbf{e}_p - \mathbf{e}_q$, the situation is analogous to having a supply of 1 at node p and a demand of 1 at node q. All other nodes are transshipment nodes. Hence the solution to $\mathbf{B}\mathbf{y}_{pq} = \mathbf{a}_{pq}$ corresponds to sending a flow of 1 along the basis equivalent chain from p to q. Consequently, referring to Figure 9.7, the coefficients of $\mathbf{y}_{pq}$ corresponding to basic variables x_{pj}, x_{kl}, and x_{ls} are $+1$ and the coefficients corresponding to basic variables x_{kj} and x_{qt} are -1.

Computing Dual Variables w and $z_{ij} - c_{ij}$

Given a basic subgraph, we need to compute $z_{ij} - c_{ij}$ for each nonbasic variable x_{ij} and either stop or proceed by introducing a nonbasic variable with a positive $z_{ij} - c_{ij}$. Toward this end, we may compute the dual or simplex multiplier vector $\mathbf{w}$ through the system $\mathbf{wB} = \mathbf{c}_B$, and then determine $z_{ij} - c_{ij}$ through the expression $z_{ij} - c_{ij} = \mathbf{wa}_{ij} - c_{ij}$. For the basis subgraph of Figure 9.9 we have

$$[w_1, w_2, w_3, w_4, w_5]\begin{bmatrix} 1 & 0 & 0 & 0 & 0 \\ 0 & 1 & 0 & 0 & 0 \\ 0 & -1 & 1 & 0 & 0 \\ 0 & 0 & -1 & 1 & 0 \\ -1 & 0 & 0 & -1 & 1 \end{bmatrix} = [2, -4, 0, 3, 0]$$

Note that the rows of the lower triangular basis matrix are also permuted as 1, 2, 3, 4, and 5.

Using the last $\mathbf{w}$ equation, the one associated with the root, we get $w_5 = 0$. We may now proceed away from the root in the following fashion,

$$w_4 - w_5 = c_{45} \Rightarrow w_4 = 3 + 0 = 3$$

$$w_3 - w_4 = c_{34} \Rightarrow w_3 = 0 + 3 = 3$$

$$w_2 - w_3 = c_{23} \Rightarrow w_2 = -4 + 3 = -1$$

$$w_1 - w_5 = c_{15} \Rightarrow w_1 = 2 + 0 = 2$$

We start with the dual variable for the root node at zero value, then proceed away from the root toward the ends of the tree using the relationship that $w_i - w_j = c_{ij}$ along the basic arcs in the tree.

While the process of computing primal variables consisted of working from the ends of the basis tree inward toward the root (see Figure 9.10), the process of computing dual variables consists of working from the root of the basis tree outward toward the ends (see Figure 9.11).

To compute $z_{ij} - c_{ij}$ for the nonbasic arc (i, j), we apply the definition

$$\begin{aligned} z_{ij} - c_{ij} &= \mathbf{wa}_{ij} - c_{ij} \\ &= \mathbf{w}(\mathbf{e}_i - \mathbf{e}_j) - c_{ij} \\ &= w_i - w_j - c_{ij} \end{aligned}$$

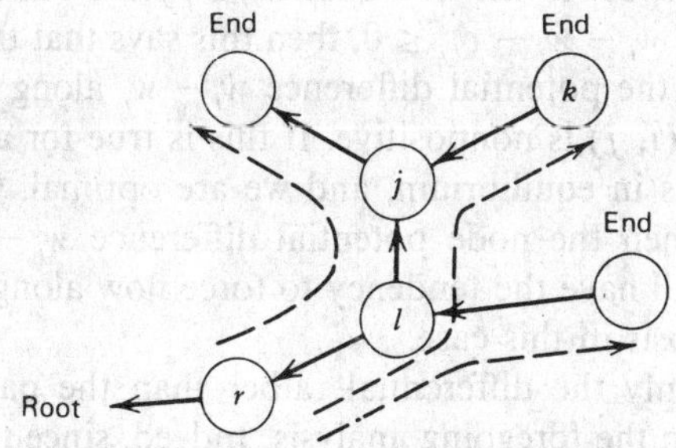

Figure 9.11. Computing the values of the dual variables.

Thus the $z_{ij} - c_{ij}$ can be conveniently computed on the network. Notice also that by requiring that $w_i - w_j = c_{ij}$ along basic arcs, we are actually requiring that $z_{ij} - c_{ij} = 0$ for basic variables.

Using the values of the dual variables obtained earlier, we summarize the value of $z_{ij} - c_{ij}$ for each nonbasic variable x_{ij}.

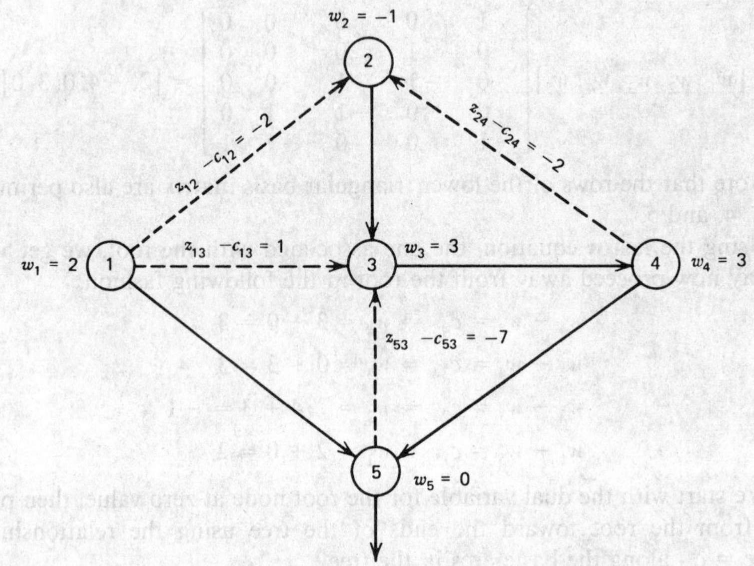

Let us provide some interpretations for the dual variables w_i and for the reduced costs $c_{ij} - z_{ij}$. The dual variables w_i are sometimes referred to as *node potentials* because of the following analogy. Consider a basis tree and suppose that the flow sustained by this tree has resulted in certain pressure heads or "node potentials" at the nodes. Imagine that the c_{ij} values are reverse pressures or forces along the arcs and that the root node is at a zero potential. The equations $w_i - w_j = c_{ij}$ for the basic arcs assert that the net force $w_i - w_j$ along (i, j) due to the node potential difference equals the back pressure c_{ij}. Hence the system $\mathbf{wB} = \mathbf{c}_B$ determines, as if it were, the node potentials $\mathbf{w}$ at equilibrium or steady-state, with respect to the flow sustained by the basis tree. For a nonbasic variable arc (i, j), if $w_i - w_j - c_{ij} \leq 0$, then this says that the back pressure c_{ij} is at least as much as the potential difference $w_i - w_j$ along (i, j), and so the net flow pressure along (i, j) is nonpositive. If this is true for all nonbasic arcs, then the entire network is in equilibrium, and we are optimal. On the other hand, if $w_i - w_j - c_{ij} > 0$, then the node potential difference $w_i - w_j$ exceeds the back pressure c_{ij}. This will have the tendency to force flow along (i, j). Hence we can enter x_{ij} into the basis in this case.

Observe that only the differential rather than the particular values of the $\mathbf{w}$-variables matter in the foregoing analysis. Indeed, since the cost c_a on the root arc can be fixed at any arbitrary value, if $\overline{\mathbf{w}}$ are the dual variables calculated with $c_a = 0$, then with an arbitrary cost value c_a, the $\mathbf{w}$ values are given by $w_i = \overline{w}_i + c_a$ for all $i = 1, \ldots, m$. This is evident from the nature of the system $\mathbf{wB} = \mathbf{c}_B$ (why?). Hence all the dual variables can be scaled up or down by a constant if desired.

The node potentials have an interesting marginal cost economic interpretation as well. Recall that $w_i = \partial z / \partial b_i$ in the usual sense, that is, w_i is the increase in cost if b_i is increased (marginally) by a unit and if all other right-hand-sides and nonbasic variables are held at their current values. But increasing b_i by a unit amounts to putting an extra unit of supply at node i, which must consequently leak out through the root arc via the unique chain joining node i to the root node because of the conservation of flow equations. Hence w_i is the cost of a unit flow along this chain. (Note that some arcs in this chain may have a reverse orientation, and so the flow on that arc would decrease by a unit.) For example, for node 2 in Figure 9.9. $w_2 = c_{23} + c_{34} + c_{45} = -1$, and for node k in Figure 9.11 $w_k = c_{kj} - c_{lj} + c_{lr}$.

Let us now define $\pi_i = -w_i$ for $i = 1, \ldots, m$, so that π_i is interpreted as the increase in cost if an extra unit of demand is created at node i (given that flow can occur only on the arcs in the basis tree). Hence if each node is imagined to be a market, π_i plays the role of the (marginal) cost of supplying an extra unit of the product at market i. (By using c_a as previously, we can make all $\pi_i \geq 0$ if so desired.) For any basic arc (i, j), the equation $w_i - w_j = c_{ij}$ implies that $\pi_j = \pi_i + c_{ij}$, that is, we break even if we purchase a unit at market i, transport it along the arc (i, j), and sell it at market j. Similarly, if $w_i - w_j - c_{ij} \leq 0$ for any nonbasic arc (i, j), then $\pi_j \leq \pi_i + c_{ij}$; therefore the foregoing venture is not profitable. If this is true for all nonbasic arcs, then the market system is in equilibrium—the current set of basic flows have generated a set of prices that hold the flows in steady state. On the other hand, if $w_i - w_j - c_{ij} > 0$ for a nonbasic arc (i, j), then $\pi_j > \pi_i + c_{ij}$, and an entrepreneur would find it profitable to purchase in market i and sell at market j by transporting the product over the arc (i, j). For example, for the nonbasic arc (1, 3), $z_{13} - c_{13} = w_1 - w_3 - c_{13} = 1$ and so, it is profitable to pay $\pi_1 = -2$ per unit at node 1, pay the transportation cost of $c_{13} = -2$ per unit, and receive $\pi_3 = -3$ per unit at node 3. The net gain is $-(-2) - (-2) + (-3) = 1$ per unit.

Note also that $z_{ij} - c_{ij} = \mathbf{c}_B \mathbf{y}_{ij} - c_{ij}$. From Section 9.4, we see that y_{ij} has a $+1$ corresponding to basic arcs that point in a direction reverse to the orientation of flow change in the cycle created by introducing (i, j) in the basis, a -1 corresponding to basic arcs in the cycle that have the same orientation as the flow change, and zero otherwise. Hence $z_{ij} - c_{ij}$ is the *negative* of the cost of sending a unit along the cycle in the same orientation as (i, j). For example, $z_{13} - c_{13} = -c_{34} - c_{45} + c_{15} - c_{13} = 0 - 3 + 2 - (-2) = 1$. Therefore the objective function will fall by a unit for every unit of flow sent around this cycle in the orientation of (i, j). Although this gives another method of computing $z_{ij} - c_{ij}$, it is less efficient than the earlier method of computing $z_{ij} - c_{ij}$ through the dual variables $\mathbf{w}$ (why?). Hence it is not used in practice.

Determining the Exiting Column and Pivoting

When we applied the cycle method to compute $z_{ij} - c_{ij}$ for a nonbasic arc, we essentially identified the pivot process. In the foregoing example $z_{13} - c_{13} > 0$ and so x_{13} is a candidate to enter the basis. What we must do is proceed to increase x_{13}, adjust the basic variables to maintain feasibility with respect to the right-hand-side, and determine the first basic variable to reach zero. This blocking basic variable is the exiting variable and leaves the basis.

Consider the basic tree together with arc (1, 3). If we increase x_{13} by Δ, then to provide balance we must increase x_{34} by Δ, increase x_{45} by Δ, and finally decrease x_{15} by Δ.

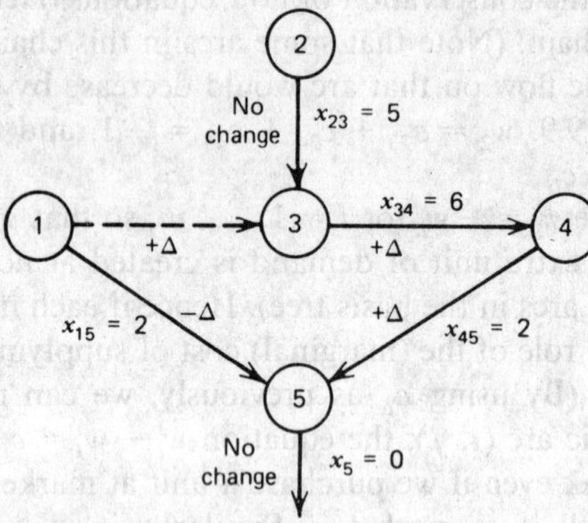

This process of adjustment can be thought of as sending an additional amount of flow Δ around the unique cycle created when the nonbasic arc is added to the basic tree. Sending flow against the direction of an arc corresponds to decreasing flow on the arc.

As x_{13} increases by Δ, the only basic variable to decrease is x_{15} and its new value is $x_{15} = 2 - \Delta$. Thus the critical value of Δ is equal to 2, at which instant x_{15} drops to zero and leaves the basis. All of the other basic variables are adjusted appropriately in value and the new basic solution is given as follows.

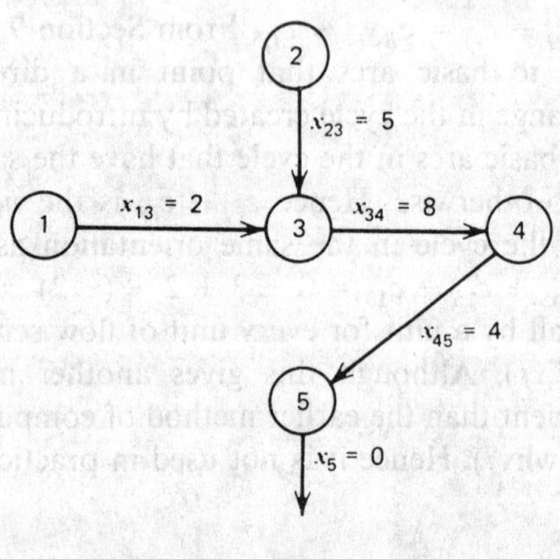

We leave it as an exercise to the reader to show that determining the exiting variable and adjusting the values of the basic variables accordingly as described earlier is equivalent to performing the usual minimum ratio test and pivoting.

We have seen earlier how the basis tree and the flows are updated in a pivot operation. To complete the discussion of an *iteration* of the network simplex method, let us show how the dual variables may be *updated* rather than recomputed from scratch with respect to the new basis tree. Suppose that x_{pq} enters the basis, and x_{uv} is the exiting variable. Disconnect the leaving arc (u, v) from the current basis tree. By Property 1 of tree graphs, this decomposes the tree into two trees T_1 and T_2, with the root node r, say, belonging to tree T_1. Hence the new dual variable values for the nodes in T_1 will remain the same as before, since the chains connecting these nodes to the root node remained unchanged. For tree T_2, let us consider two cases. First suppose that $q \in T_2$. Node q is then called the *leading node* with respect to the entering arc (p, q). Note that for each $(i, j) \in T_2$, we currently have $w_i - w_j = c_{ij}$. If we change all the w_i's in T_2 by a constant, we will still satisfy $w_i - w_j = c_{ij}$ for all $(i, j) \in T_2$. Consequently, once we know the new value $w_{q(\text{new})}$ of the dual variable associated with node q, we can compute the new dual value $w_{i(\text{new})}$ for each node i in T_2 as $w_i + (w_{q(\text{new})} - w_q)$, since $(w_{q(\text{new})} - w_q)$ is the amount by which w_q has increased. However, denoting $\delta_{pq} = z_{pq} - c_{pq} > 0$, we have $\delta_{pq} = w_p - w_q - c_{pq}$, while $w_p - w_{q(\text{new})} = c_{pq}$ (why?). Hence $w_{q(\text{new})} - w_q = \delta_{pq}$, and all the dual variables in T_2 simply *increase* by δ_{pq}.

On the other hand, if p is the leading node, that is, $p \in T_2$, then we have $w_{p(\text{new})} - w_q = c_{pq}$, and $w_{p(\text{new})} - w_p = -\delta_{pq}$. Hence, in this case, all the duals of T_1 will remain the same as previously, but the duals of T_2 will *fall* by δ_{pq}. For example, in the foregoing pivot operation, $(p, q) = (1, 3)$ enters and $(k, l) = (1, 5)$ leaves the basis. Disconnecting (1, 5), we find that the tree T_1 contains the nodes 2, 3, 4, and 5, but T_2 contains node 1 alone. Since $p \in T_2$, the dual of node 1 falls by $\delta_{pq} = 1$ to the value $w_{1(\text{new})} = 1$ and the duals of the other nodes remain the same as before.

Summary of the Network Simplex Algorithm

INITIALIZATION STEP

Find an initial basic feasible solution represented by a rooted spanning tree, with r as the root node. (Section 9.7 discusses how to achieve this, using artificial variables if necessary.) Compute the basic flows $\mathbf{x}_B$ and the dual variables $\mathbf{w}$ associated with this basis tree.

MAIN STEP

Compute $\delta_{pq} = z_{pq} - c_{pq} = w_p - w_q - c_{pq} = \text{Maximum } \{z_{ij} - c_{ij} = w_i - w_j - c_{ij} : (i, j) \text{ is nonbasic}\}$. If $z_{pq} - c_{pq} \leq 0$, then stop; the current solution is optimal. (If any artificial variables are positive at optimality, the problem is infeasible.) Otherwise introduce (p, q) into the basis tree and determine the unique cycle formed by tracing the basis equivalent chain connecting p and q. Find the maximum flow $\Delta \geq 0$ that can be sent along this cycle in the orientation of the entering arc (p, q). If $\Delta \to \infty$, then stop; the problem is unbounded. (If the big-M method is being used and some artificial variable is positive, the original problem is infeasible; otherwise it is unbounded.) If Δ is finite, determine an exiting or blocking variable arc (u, v). Update flows by appropriately adjusting the flows in the cycle by Δ. Disconnect (u, v) from the basis tree and decompose it into two trees T_1 and T_2, where the root node $r \in T_1$. Update the basis tree by adding arc (p, q) to it. Update the dual variables by respectively increasing or decreasing the dual values of nodes in T_2 by δ_{pq}, according to whether q or p is the (leading) node in T_2. Repeat the main step.

9.6 AN EXAMPLE OF THE NETWORK SIMPLEX METHOD

Example 9.1

Consider the network of Figure 9.12. In Figure 9.13 we give the complete solution of this minimal cost network flow problem. The value $\delta_{pq} = z_{pq} - c_{pq}$ for the entering variable is circled. The exiting variable is denoted by * in the figure.

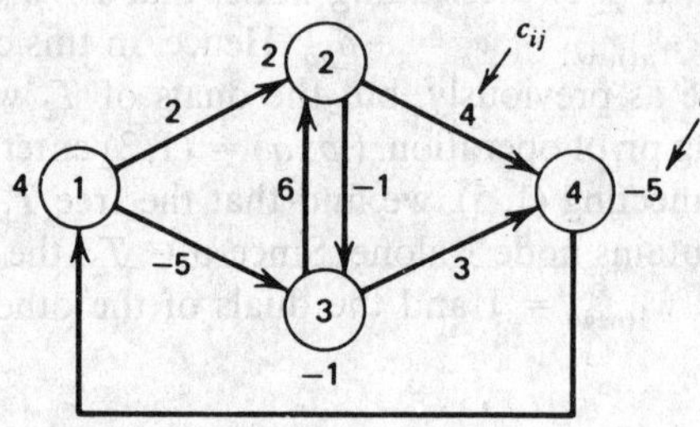

Figure 9.12. **An example network.**

9.7 FINDING AN INITIAL BASIC FEASIBLE SOLUTION

We have assumed thus far that we are given a minimum cost network flow problem defined on a connected digraph with a single root arc and that we have a starting basic feasible solution represented by a rooted spanning tree to commence the network simplex algorithm. We now give a method for generally attaining this situation.

Suppose that by adding any necessary slack or surplus variables and by employing the usual transformations on variables (including the replacement of undirected arcs that have nonnegative costs by two oppositely oriented directed arcs), we have put the given problem in the form Minimize $\{\mathbf{cx} : \mathbf{A}^0\mathbf{x} = \mathbf{b}, \mathbf{x} \geq \mathbf{0}\}$, where $\mathbf{A}^0$ is $m \times n$ and is composed of columns that are either $\pm$ unit vector

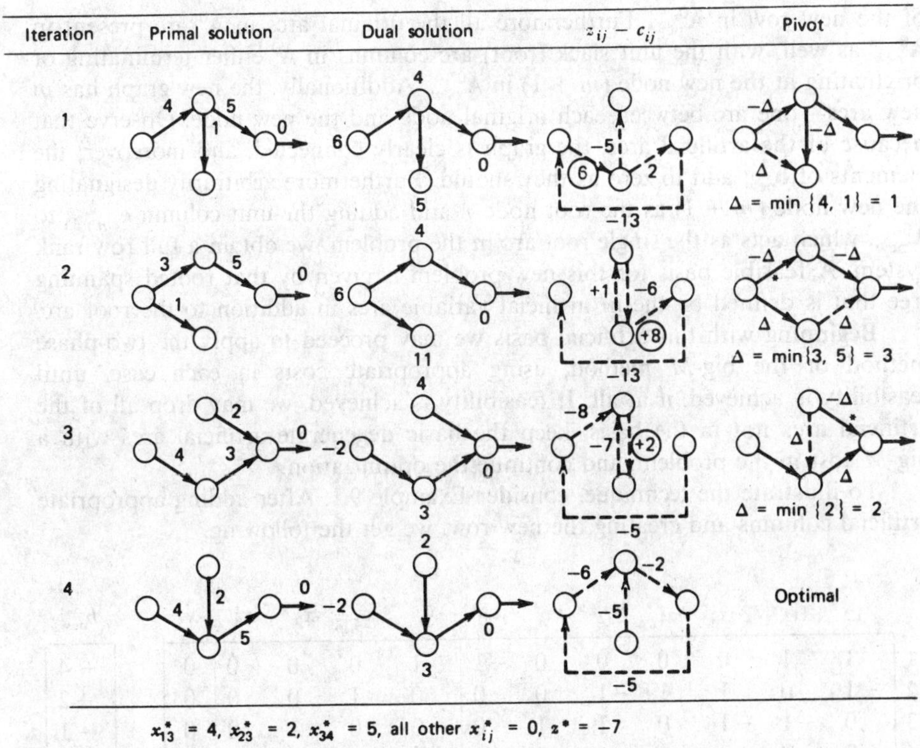

Figure 9.13. The solution to Example 9.1.

columns or that have a $+1$ in some row, a -1 in some other row, and zeros elsewhere. Hence we have a network structured problem. Although we can possibly use some of the arcs (columns) in $\mathbf{A}^0$ itself as part of a starting basis, let us present the general use of an all artificial start.

Suppose that we add an artificial column for *every* row of $\mathbf{A}^0$, the ith artificial column being $\pm\mathbf{e}_i$ depending on the sign of b_i (that is, $+\mathbf{e}_i$ if $b_i \geq 0$; $-\mathbf{e}_i$ otherwise). Also, let us add a redundant row given by the negative of the sum of the rows of the "extended" $\mathbf{A}^0$ matrix. The problem constraints then become as follows, where $\mathbf{x}_a$ is a vector of artificial variables:

$$\left[\begin{array}{c|cccc} \mathbf{A}^0 & \pm 1 & & & \\ & & \pm 1 & & \\ & & & \ddots & \\ & & & & \pm 1 \\ \hline -\mathbf{1A}^0 & \mp 1 & \mp 1 & \cdots & \mp 1 \end{array}\right] \begin{pmatrix} \mathbf{x} \\ \mathbf{x}_a \end{pmatrix} = \left[\begin{array}{c} \mathbf{b} \\ \hline \mathbf{1b} \end{array}\right]$$

Let the foregoing system be denoted by $\mathbf{A}^0_{\text{new}}\mathbf{x}' = \mathbf{b}_{\text{new}}$. Observe that each column in $\mathbf{A}^0_{\text{new}}$ has exactly one $+1$, one -1, and zeros elsewhere. Hence it may be viewed as a graph. This new graph has one additional node $(m + 1)$ because of the new row in $\mathbf{A}^0_{\text{new}}$. Furthermore all the original arcs in $\mathbf{A}^0$ are present in $\mathbf{A}^0_{\text{new}}$ as well, with the unit slack (root) arc columns in $\mathbf{A}^0$ either terminating or originating at the new node $(m + 1)$ in $\mathbf{A}^0_{\text{new}}$. Additionally, the new graph has m new arcs—one arc between each original node and the new node. Observe that because of the artificial arcs, the graph is clearly connected, and moreover, the elements of $\mathbf{b}_{\text{new}}$ add to zero as they should. Furthermore arbitrarily designating the new node $(m + 1)$ as the root node r and adding the unit column $\mathbf{e}_{(m+1)}$ to $\mathbf{A}^0_{\text{new}}$, which acts as the single root arc in the problem, we obtain a full row rank system. A feasible basis for this new problem is given by that rooted spanning tree that is defined by the m artificial variable arcs in addition to the root arc.

Beginning with this artificial basis we may proceed to apply the two-phase method or the big-M method, using appropriate costs in each case, until feasibility is achieved, if at all. If feasibility is achieved, we may drop all of the artificial arcs not in the basis, keep the basic degenerate artificial arcs with a big-M cost in the problem, and continue the optimization.

To illustrate the technique, consider Example 9.1. After adding appropriate artificial columns and creating the new row, we get the following.

	x_{12}	x_{13}	x_{23}	x_{24}	x_{32}	x_{34}	x_{41}	x_1	x_2	x_3	x_4	x_5	b_{new}
1	1	1	0	0	0	0	−1	1	0	0	0	0	+4
2	−1	0	1	1	−1	0	0	0	1	0	0	0	+2
3	0	−1	−1	0	1	1	0	0	0	−1	0	0	−1
4	0	0	0	−1	0	−1	1	0	0	0	−1	0	−5
5	0	0	0	0	0	0	0	−1	−1	1	1	1	0

Selecting the phase I method, the artificial variables x_1, x_2, x_3, and x_4 will have cost coefficients of 1 while all other variables have zero cost coefficients. This leads to the associated network flow problem of Figure 9.14 where the cost coefficient of the root x_5 is zero. With this feasible basis at hand we proceed to solve the phase I problem, using the procedures developed in this chapter.

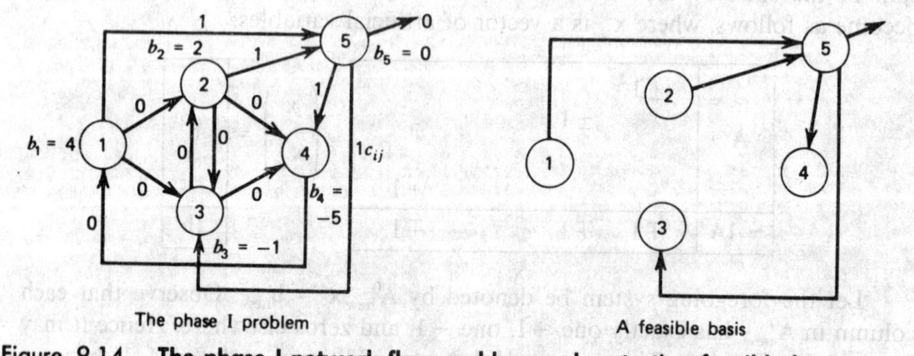

Figure 9.14. The phase I network flow problem and a starting feasible basis.

Example 9.2

As another example, consider the following problem:

$$\begin{array}{lrcrcrcrcrcr}
\text{Minimize} & x_{12} & + & 2x_{13} & + & 3x_{23} & - & 2x_{24} & + & 4x_{34} & & \\
\text{Subject to} & +x_{12} & + & x_{13} & & & & & & & \leq & 6 \\
 & -x_{12} & & & + & x_{23} & + & x_{24} & & & \leq & 4 \\
 & & - & x_{13} & - & x_{23} & & & + & x_{34} & \leq & -2 \\
 & & & & & & - & x_{24} & - & x_{34} & \leq & -5 \\
 & x_{12}, & & x_{13}, & & x_{23}, & & x_{24}, & & x_{34} & \geq & 0
\end{array}$$

This problem states that no more than six and four units of supply are available at nodes 1 and 2 respectively. At least two and five units of demand also must be absorbed at nodes 3 and 4 respectively, with additional units being permissible if this is profitable. We now add slack variables s_1, s_2, s_3, and s_4 with zero costs and add appropriate artificial variables x_1, x_2, x_3, and x_4 with big-M costs to the problem. Creating a new row equal to the negative of the sum of the other rows and including the artificial root arc variable x_5 with a zero cost yields the following system of equations.

	x_{12}	x_{13}	x_{23}	x_{24}	x_{34}	s_1	s_2	s_3	s_4	x_1	x_2	x_3	x_4	x_5	$\mathbf{b}_{\text{new}}$
1	1	1	0	0	0	1	0	0	0	1	0	0	0	0	6
2	−1	0	1	1	0	0	1	0	0	0	1	0	0	0	4
3	0	−1	−1	0	1	0	0	1	0	0	0	−1	0	0	−2
4	0	0	0	−1	−1	0	0	0	1	0	0	0	−1	0	−5
5	0	0	0	0	0	−1	−1	−1	−1	−1	−1	1	1	1	−3

We now ask the reader to draw the graph corresponding to this system, to construct the starting artificial basis tree, and to optimize using the big-M method (see Exercise 9.11).

Note also that we could have used the slacks s_1 and s_2 in lieu of the artificial variables x_1 and x_2 respectively in the starting basis. This use of an *all* artificial start is only for illustrative purpose.

9.8 NETWORK FLOWS WITH LOWER AND UPPER BOUNDS

It is simple and straightforward to make the transition from the ordinary simplex method for network flow problems to the bounded variables simplex method for network flow problems. We briefly review the essential changes required to effect such an extension to this method.

Getting Started and Computing the Flows

Via standard transformations, including the method of Section 9.7, we can assume without loss of generality that the given problem is in the form

$$\text{Minimize } \{\mathbf{cx} : \mathbf{A}^0\mathbf{x} = \mathbf{b}, \mathbf{l} \leq \mathbf{x} \leq \mathbf{u}\}$$

where $\mathbf{A}^0$ is an $(m+1) \times n$ node-arc incidence matrix of a connected digraph, including m appropriate artificial arcs and a single root arc, $\Sigma b_i = 0$, and where $-\infty < l_j < u_j \leq \infty$ for all $j = 1, \ldots, m$. Using the phase I or the big-M method, all the original arcs in the problem may be set at one of their (finite) bounds for the initial basis. This uniquely determines the flow values of the m artificial variables, and uniquely determines the root arc flow value as zero (why?). (Note that the artificial variables should be added with the appropriate signs so that they are all nonnegative in this solution.) Hence we have a starting basic feasible solution. For the remaining iterations, we need not worry about the conservation of flow equations since we will automatically enforce these equations by modifying the flow in cycles. This process also will automatically update the arc flows. However, if we are given a basic partition and we need to compute the flows, this may be done by adjusting the right-hand-side values to $\mathbf{b} - \mathbf{N}_1\mathbf{l}_{N1} - \mathbf{N}_2\mathbf{u}_{N2}$ (in the notation of Section 5.2) and solving the system $\mathbf{B}\mathbf{x}_B = \mathbf{b} - \mathbf{N}_1\mathbf{l}_{N1} - \mathbf{N}_2\mathbf{u}_{N2}$ (as in Section 9.5).

Computing the Dual Variables and the $z_{ij} - c_{ij}$'s

Lower and upper bounds have no effect on the computation of the dual variables and on the computation of the $z_{ij} - c_{ij}$'s. Note, however, that in the presence of lower and upper bounds the optimality criteria are

$$x_{ij} = u_{ij} \Rightarrow z_{ij} - c_{ij} \geq 0$$

and

$$x_{ij} = l_{ij} \Rightarrow z_{ij} - c_{ij} \leq 0$$

These are easy to check and we can readily determine whether some nonbasic variable x_{ij} should be increased or decreased if optimality is not achieved.

Determining the Exiting Column and Pivoting

Once the entering column is selected, it is again an easy task to select the exiting column and to pivot. We add the entering nonbasic arc, regardless of whether the variable is increasing or decreasing, to the basis tree and determine the unique cycle formed. Then, if the entering variable is increasing, we send an amount Δ around the cycle in the direction of the entering variable. If the entering variable is decreasing, we send an amount Δ around the cycle against the direction of the entering variable. Figure 9.15 illustrates these two possibilities. To compute the critical value of Δ we check those basic variables increasing as well as those decreasing and the possibility that x_{ij} may reach its other bound. If the last possibility occurs, x_{ij} remains nonbasic (at its other bound) and all basic variables along the cycle are adjusted accordingly. Otherwise, the nonbasic variable enters and some basic variable exists at one or other of its bounds, and all variables along the cycle are adjusted accordingly. If the basis tree remains the same, the duals are unchanged. Otherwise the duals are updated exactly as before.

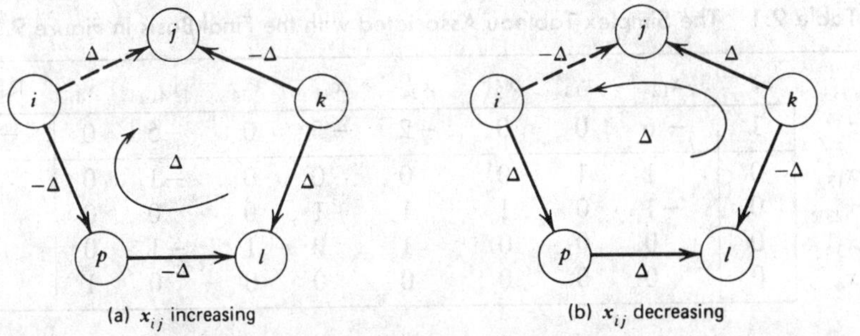

Figure 9.15. Two cases for entering are: (*a*) x_{ij} increasing. (*b*) x_{ij} decreasing.

An Example of the Network Simplex Method in the Presence of Lower and Upper Bounds

Consider the network of Figure 9.16. We present, in Figure 9.17, the complete solution to that network. We have identified a starting basic feasible solution, thus omitting phase I. The notation "$|\xrightarrow{k}$" represents a nonbasic arc at value k. The exiting variable is noted by *.

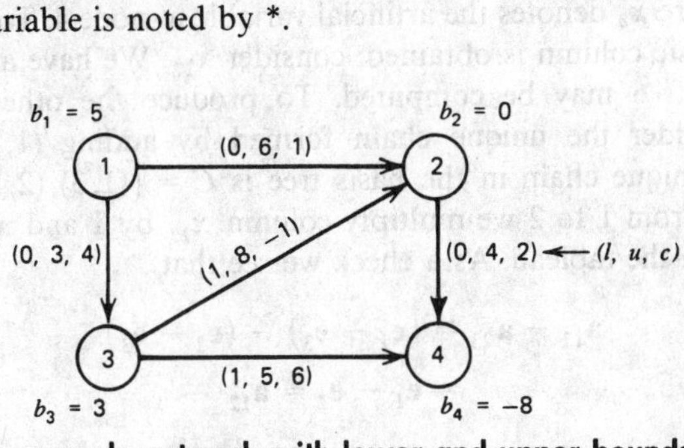

Figure 9.16. An example network with lower and upper bounds.

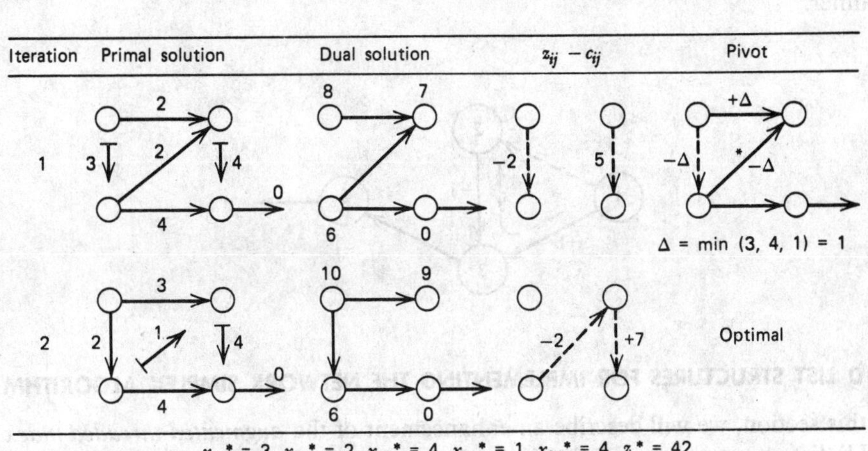

Figure 9.17. The solution to the example in Figure 9.16.

Table 9.1 The Simplex Tableau Associated with the Final Basis in Figure 9.13

	z	x_{12}	x_{13}	x_{23}	x_{24}	x_{32}	x_{34}	x_{41}	x_4	RHS
z	1	−6	0	0	−2	−5	0	−5	0	−7
x_{13}	0	1	1	0	0	0	0	−1	0	4
x_{23}	0	−1	0	1	1	−1	0	0	0	2
x_{34}	0	0	0	0	1	0	1	−1	0	5
x_4	0	0	0	0	0	0	0	0	1	0

9.9 THE SIMPLEX TABLEAU ASSOCIATED WITH A NETWORK FLOW PROBLEM

In Section 9.4 we showed how to construct the column $\mathbf{y}_{ij}$ for any nonbasic arc (i, j). Elsewhere, we have seen how to obtain the values of the basic variables and the $z_{ij} - c_{ij}$'s. Thus it is possible to construct the entire updated tableau by examining the basis subgraph at the corresponding iteration.

As an example, consider the final (optimal) basis in Figure 9.13 for the network flow problem of Example 9.1. The simplex tableau for this basis is given in Table 9.1, where x_4 denotes the artificial variable at node 4. To illustrate how a particular nonbasic column is obtained, consider x_{12}. We have already indicated how $z_{12} - c_{12} = -6$ may be computed. To produce the other entries in the column we consider the unique chain formed by adding (1, 2) to the basic subgraph. The unique chain in the basis tree is $C = \{(1, 3), (2, 3)\}$. To reorient this into a path from 1 to 2 we multiply column x_{13} by 1 and x_{23} by -1; thus the coefficients in the tableau. As a check we see that

$$\begin{aligned}\mathbf{a}_{13} - \mathbf{a}_{23} &= (\mathbf{e}_1 - \mathbf{e}_3) - (\mathbf{e}_2 - \mathbf{e}_3)\\ &= \mathbf{e}_1 - \mathbf{e}_2 = \mathbf{a}_{12}\end{aligned}$$

as required. The other columns of the simplex tableau are obtained in a similar manner.

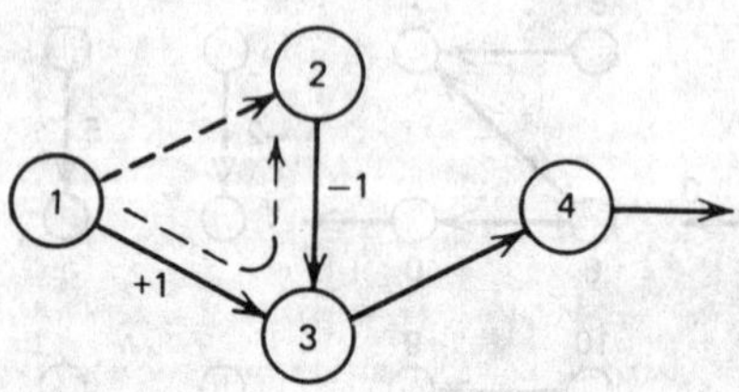

9.10 LIST STRUCTURES FOR IMPLEMENTING THE NETWORK SIMPLEX ALGORITHM

In this section, we will describe an enhancement of the *augmented threaded index* (ATI) list structures for efficiently implementing the network simplex algorithm on the computer. Although simpler labeling techniques are possible, the method we describe is very efficient. We will also introduce terminology and concepts that are now standard in network programming literature.

The main construct in any list structure scheme lies in the representation of a rooted spanning tree. The ATI method prescribes three fundamental node-length arrays that not only represent the layout of the tree, but also facilitate the efficient execution of a simplex iteration, and the updating of the primal and dual solutions and of the list structures. Additionally, three other node-length arrays obtainable from these three fundamental lists may be explicitly maintained in order to enhance the updating of the list structures. These are described later along with the simplex algorithm and the updating operations.

List Structures: Definitions

Imagine that a rooted spanning tree T with m nodes is given, where r is the root node, and that this tree is laid out in the convenient form depicted in Figure 9.18. The list structures described here enable the computer to "see" this tree in the form shown in the figure. The first list is the *predecessor* index list, denoted $p(i)$, $i = 1, \ldots, m$. For any node i, $p(i)$ uniquely gives the node connected to node i on the chain from i to the root node. For this reason, $p(i)$ is also referred to as

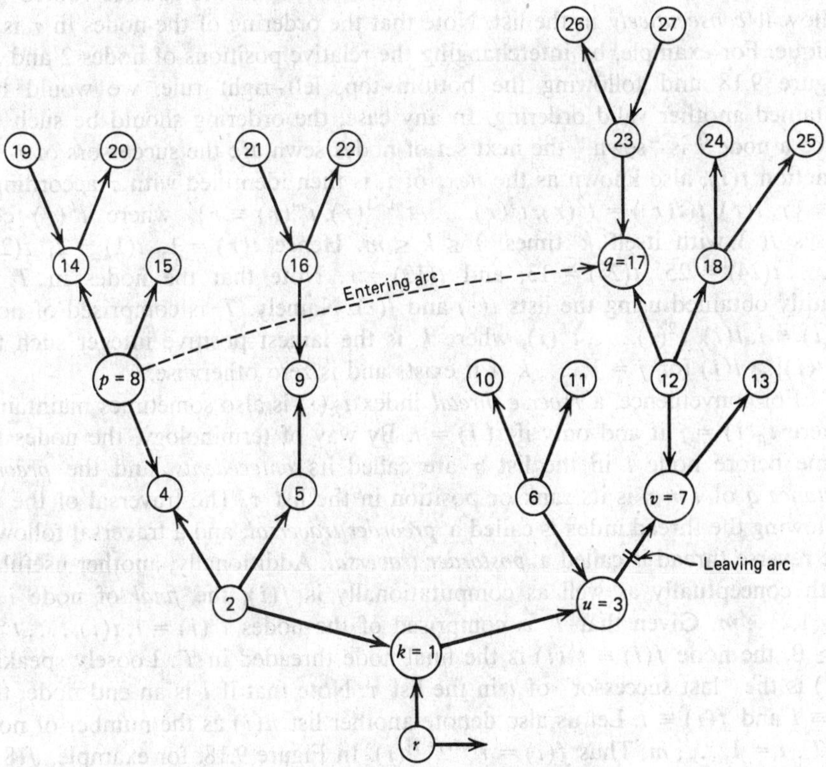

Figure 9.18. List structures for network flow programming.

the "*down*" of i. In Figure 9.18, for example, $p(1) = r$, $p(9) = 5$, $p(17) = 12$, and $p(27) = 23$. Thus $p(i)$ precedes i or is the "parent" of i in the "growth" of the tree. By convention, let $p(r) \equiv 0$.

Define the *level* of node i as the number of arcs in the (unique) chain connecting node i to the root node. Denote this by $l(i)$, $i = 1,\ldots, m$. Hence $l(1) = 1$, $l(10) = 4$, $l(17) = 5$, and $l(i) = 6$ for nodes $i = 19, 20, \ldots, 25$, that is, these nodes are all at "level" 6 with respect to the tree in Figure 9.18.

Define the *thread index* $t(i)$ for $i = 1,\ldots, m$. Imagine the process of sewing the *nodes* of the tree in Figure 9.18 with a needle and a thread; proceed from bottom to top, from left to right, and finally return to node r after sewing all the nodes together. The nodes might then be sewn in the order of the following list $\tau = \{r, 1, 2, 4, 8, 14, 19, 20, 15, 5, 9, 16, 21, 22, 3, 6, 10, 11, 7, 12, 17, 23, 26, 27, 18, 24, 25, 13, r\}$. Formally, let us define the *subtree rooted at node i* to be that tree T_i that is obtained by disconnecting the arc connecting i and $p(i)$ and that contains the node i. For the root node r, the tree T_r is the entire tree T. For example, T_8 is obtained in Figure 9.18 by disconnecting the arc (8, 4) and is the subgraph of T that is induced by the nodes 8, 14, 15, 19, and 20. The nodes in T_i (other than node i) are called the *successors* of i, since node i lies on the chains joining each of these nodes to the root node.The *immediate successors* of i have node i as their predecessor. We now can define τ as an ordering of the nodes such that for any node i, the list of its successors, that is, the nodes in the subtree rooted at i, follow it *consecutively* in the list. Note that the ordering of the nodes in τ is not unique. For example, by interchanging the relative positions of nodes 2 and 3 in Figure 9.18 and following the bottom–top, left–right rule, we would have obtained another valid ordering. In any case, the ordering should be such that once a node i is "sewn," the next set of nodes sewn are the successors of i. The function $t(i)$, also known as the *next* of i, is then identified with τ according to $\tau \equiv \{r, t(r), t(t(r)) = t^2(r), t^3(r), \ldots, t^{m-1}(r), t^m(r) \equiv r\}$, where $t^k(r)$ composes $t(\cdot)$ with itself k times, $1 \le k \le m$. Hence $t(r) = 1$, $t(1) = 2$, $t(2) = 4, \ldots, t(24) = 25$, $t(25) = 13$, and $t(13) = r$. Note that the nodes in T_i are readily obtained using the lists $t(\cdot)$ and $l(\cdot)$. Namely, T_i is comprised of nodes $t^0(i) \equiv i, t(i), t^2(i), \ldots, t^k(i)$, where k is the largest positive integer such that $l[t^j(i)] > l(i)$ for $j = 1,\ldots, k$ if it exists and is zero otherwise.

For convenience, a *reverse thread* index $t_R(\cdot)$ is also sometimes maintained, where $t_R(i) = j$ if and only if $t(j) = i$. By way of terminology, the nodes that come before node i in the list τ are called its *antecedents*, and the *preorder distance* q of $i \neq r$ is its rank or position in the list τ. The traversal of the tree following the thread index is called a *preorder traversal*, and a traversal following the reverse thread is called a *postorder traversal*. Additionally, another useful list both conceptually as well as computationally is $f(i)$, the *final* of node i for $i = 1,\ldots, m$. Given that T_i is comprised of the nodes $t^0(i) = i, t(i), \ldots, t^k(i)$, $k \ge 0$, the node $f(i) \equiv t^k(i)$ is the final node threaded in T_i. Loosely speaking, $f(i)$ is the "last successor" of i in the list τ. Note that if i is an end node, then $T_i \equiv i$ and $f(i) \equiv i$. Let us also denote another list $n(i)$ as the number of nodes in T_i, $i = 1,\ldots, m$. Thus $f(i) = t^{[n(i)-1]}(i)$. In Figure 9.18, for example, $f(8) = 15$, $n(8) = 5$, $f(9) = 22$, $n(9) = 4$, $f(7) = 13$, $n(7) = 10$, $f(21) = 21$, $n(21) = 1$, $f(r) = 13$, and $n(r) = 28$.

Initialization

The foregoing lists are readily constructed for the all artificial basis in which the tree is comprised of arcs connecting each node and the root node. For this tree, we have $p(i) = r$ for all $i \neq r$ and $p(r) \equiv 0$, $l(i) = 1$ for all $i \neq r$, and $l(r) = 0$, $t(r) = 1$, $t(1) = 2$, $t(2) = 3, \ldots, t(m-1) = r$ (assuming that node "m" is the root r). Also $f(i) = i$ and $n(i) = 1$ for all $i \neq r$, while $f(r) = m - 1$ and $n(r) = m$.

On the other hand, if a certain advanced basis is available, then these lists may be initialized as follows. (Use the example of Figure 9.18 while reading this procedure.) Pick some node as the root node r. Let $S(r)$ be the immediate successors of r determined by finding the basic arcs incident at r. Then $p(i) = r$ and $l(i) = 1$ for all $i \in S(r)$. Pick some node $j \in S(r)$, put $t(r) = j$, and remove j from $S(r)$. Similarly, construct $S(j)$ as the set of all nodes other than node r that are adjacent to node j for some basic arc. Assume that $S(j) \neq \phi$. Then $p(i) = j$ and $l(i) = l(j) + 1$ for all $i \in S(j)$. Pick some node $k \in S(j)$, put $t(j) = k$, and remove k from $S(j)$. Continue in this manner until some node e is reached for which $S(e) = \phi$. Hence e is an end node of the basis tree T. In this case, follow the predecessor index to come down the tree until a node q is reached for which $S(q) \neq \phi$. In this event, pick some $l \in S(q)$, put $t(e) = l$, and remove l from $S(q)$. While coming down the tree, each node i that is crossed over downwards, including node e itself, gets node e as its final successor node $f(i)$. Also, the number of new nodes labeled between threading i and giving node i its $f(\cdot)$ value yields $n(i)$. The process now can be continued by constructing $S(l)$ and proceeding "up" the tree again. The procedure stops when following the chain down from some end e, it reaches node r and finds $S(r) = \phi$, from where in particular, among other labels, it sets $f(r) = e$ and $t(e) = r$.

Simplex Operations

Observe that each arc in the problem is uniquely identified with some node via its predecessor index. That is, for each node $i \neq r$, there is an associated, uniquely identified arc connecting i and $p(i)$. This arc may be $[i, p(i)]$ or $[p(i), i]$. Hence a node-length array ORIENT$(\cdot)$ may be carried to record the orientation of the basic arcs, having a $+1$ in one case and a -1 in the other case. For example, for node 20, the associated arc is $[p(20), 20] \equiv (14, 20)$ and ORIENT(20) $= -1$, while for node 8, the associated arc is $[8, p(8)] = (8, 4)$ and ORIENT(8) $= 1$. Consequently, the basic arc flow vector can be stored in a node-length array FLOW$(\cdot)$, where FLOW(i) is the flow on the arc connecting i and $p(i)$. Let DUAL(i) represent the dual variable associated with node i. The arrays DUAL$(\cdot)$ and FLOW$(\cdot)$ are readily initialized via a preorder traversal and a postorder traversal, respectively. For example, in Figure 9.18, the dual values may be determined in the order of the nodes $r, 1, 2, 4, 8, 14, 19, 20, 15, \ldots$ as they appear in τ, based on the arc connecting each node i and its predecessor $p(i)$, since the dual of $p(i)$ is already computed for each i in this order. Similarly, the flows may be computed by following the nodes i in the reverse thread order $13, 25, 24, 18, 27, 26, 23, 17, 12, \ldots,$ and determining the flow on the arc connecting i and $p(i)$. This is based on the nodal balance equation for node i in this order, since the flow on at most one incident arc is unknown in each case.

Suppose that by following the usual pricing scheme, we determine an entering arc (p, q). The basis equivalent chain may be traced by coming down from p and q following the predecessor index list $p(\cdot)$ until a first intersection point k is reached. Note that node k may be p or q. Hence if $l(p) \geq l(q)$, say, then one can come down from p until one reaches the level of q, from where it is known whether or not $k \equiv q$. If not, then one can come down simultaneously one level at a time along the two chains until they intersect. While doing this, a minimum ratio test can be conducted simultaneously (how?), so that the leaving arc is known once the basis equivalent chain has been traced. Let the leaving arc connect nodes u and v, where $u = p(v)$ (see Figure 9.18). Note that the actual leaving arc may be (u, v) or (v, u). When the leaving arc is disconnected from T, the tree not containing the root is the subtree T_v. Let us also *assume* that q is the leading node in T_v—the entering arc may actually be (p, q) or (q, p).

Updating Operations

The flows are updated as usual on the basis equivalent chain by tracing down from p and q to k after determining the flow change Δ. The predecessor index $p(\cdot)$ changes only on the $q \to v$ chain, also known as the *stem*, simply reversing or inverting on this chain with $p(q) = p$. Hence in Figure 9.18, we get $p(v) = 12$, $p(12) = 17$, and $p(17) = 8$, with $p(i)$ as before otherwise. Accordingly, the tree arcs that are associated with the nodes on the stem $q \to v$ through the predecessor index change in this reversal process, and so the ORIENT($\cdot$) and FLOW($\cdot$) vectors should be appropriately updated. For example, FLOW(12) now corresponds to the flow on arc (12, 17) and ORIENT(12) = 1. The dual values in $T - T_v$ remain the same, while the dual values in T_v all increase by $\delta_{pq} = z_{pq} - c_{pq}$ if (p, q) is the entering arc and decrease by $\delta_{qp} = z_{qp} - c_{qp}$ if (q, p) is the entering arc. Hence the dual values can be updated by following the thread index from v to $f(v)$. (Notice how $n(v)$ helps in this respect.)

Finally we need to update $t(\cdot)$, $l(\cdot)$, $f(\cdot)$, and $n(\cdot)$. This is done by a *grafting process*. Notice that T_v is, as if it were, rooted at node v, while $T'_r \equiv T - T_v$ is rooted at r. Suppose that we reroot T_v at q to get T_q, that is, obtain the list structures for the new T_q, and similarly, obtain the list structures for T'_r. Then we can simply graft T_q onto T'_r by connecting p and q to obtain the new tree. In this process, let $x = f(p)$ and $y = t(x)$ in T'_r, and let $z = f(q)$ in T_q. In Figure 9.18, $x = 15$, $y = 5$, and, say, $z = 13$. Then because T_q is a part of T_p in the new tree, we set $t(x) = q$ and $t(z) = y$. Figure 9.19*a* illustrates this grafting process. Also, the levels of the nodes in T_q after grafting will all increase by one plus the level of p in T'_r. Furthermore all nodes that had the final of node p as their last successor in T'_r will now have $z = f(q)$ as their last successor. Letting $\gamma = p(t[f(p)])$ in T'_r, these nodes are on the chain from p to γ, not including γ if $\gamma \neq 0$, and on the chain from p to r otherwise. In Figure 9.18, we have $\gamma = p[t(15)] = p(5) = 2$, and so after grafting, we will have $f(8) = f(4) = 13$. Similarly, the $n(\cdot)$ values for each node on the chain from p to r in T'_r

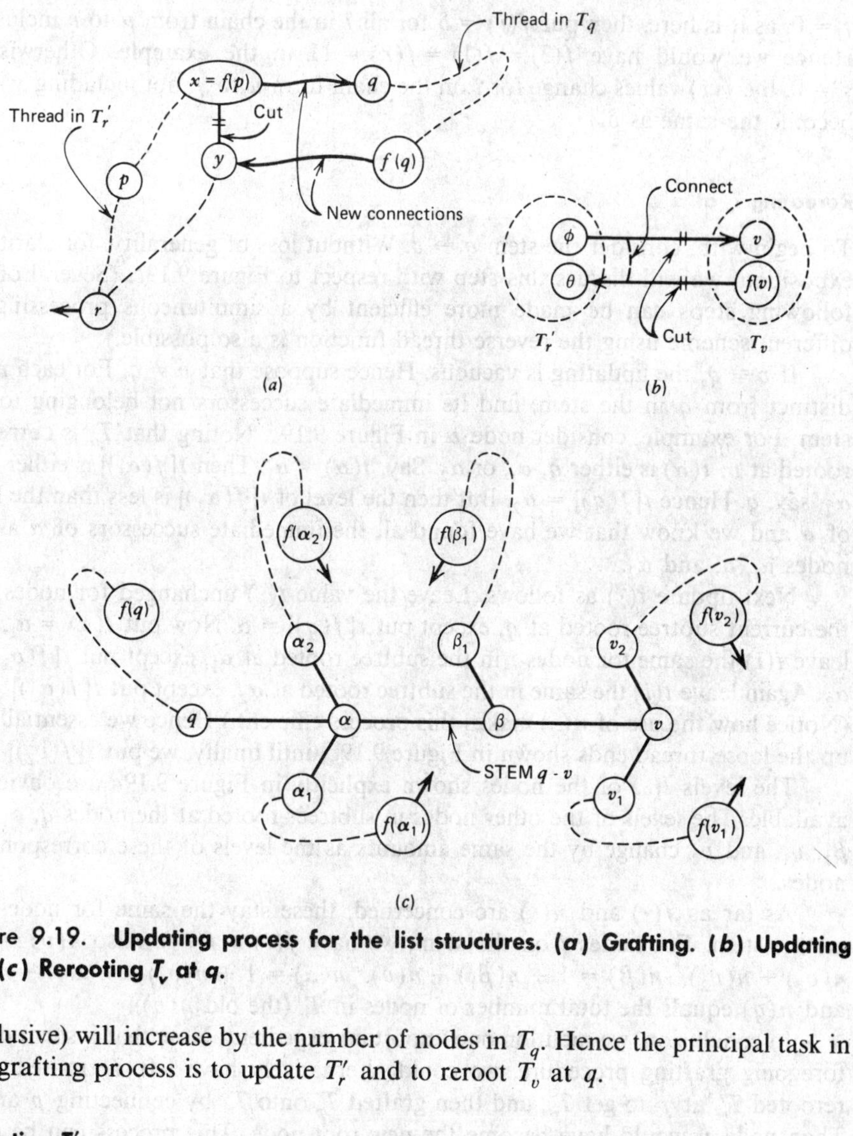

Figure 9.19. Updating process for the list structures. (*a*) Grafting. (*b*) Updating T_r'. (*c*) Rerooting T_v at *q*.

(inclusive) will increase by the number of nodes in T_q. Hence the principal task in the grafting process is to update T_r' and to reroot T_v at q.

Updating T_r'

The $p(\cdot)$ and $l(\cdot)$ values remain the same. For the thread index, since the arc (u, v) is disconnected, the thread entering T_v and the thread leaving T_v to reenter T_r' need to be cut and reconnected as shown in Figure 9.19*b*. Let ϕ and θ be such that $\phi = t_R(v)$ (that is, $t(\phi) = v$) and $\theta = t[f(v)]$. Then we put $t(\phi) = \theta$ in T_r'. The $n(\cdot)$ values of the nodes on the chain from u to r (inclusive) will fall by $n(v)$. For updating $f(\cdot)$ in T_r', note that all the nodes that have $f(v)$ as their final node will get the new $f(u)$ as their final node. With respect to the previous indices, find $\gamma = p[t(f(v))]$, set $\delta = f(u)$ if $f(u) \notin T_v$, and let $\delta = t_R(v)$ otherwise. In Figure 9.18, $f(v) = 13$, $t(13) = r$, $p(r) = 0$, and $\delta = t_R(v) = 11$. If

$\gamma = 0$, as it is here, then put $f(i) = \delta$ for all i in the chain from u to r inclusive. Hence we would have $f(3) = f(1) = f(r) = 11$ in the example. Otherwise if $\gamma \neq 0$, the $f(i)$ values change for i on the chain form u to γ, not including γ, and become the same as δ.

Rerooting T_v at q

To begin with, consider the stem $q \to v$. Without loss of generality, for clarity in exposition, we will discuss this step with respect to Figure 9.19*c*. (Several of the following steps can be made more efficient by a simultaneous processing. A different scheme using the reverse thread function is also possible.)

If $v \equiv q$, the updating is vacuous. Hence suppose that $v \neq q$. For each node distinct from q in the stem, find its immediate successors not belonging to the stem. For example, consider node α in Figure 9.19*c*. Noting that T_v is currently rooted at v, $t(\alpha)$ is either q, α_1, or α_2. Say, $t(\alpha) = \alpha_1$. Then $t[f(\alpha_1)]$ is either q or α_2, say, q. Hence $t[f(q)] = \alpha_2$. But then the level of $t[f(\alpha_2)]$ is less than the level of α and we know that we have found all the immediate successors of α as the nodes α_1, q, and α_2.

Next update $t(\cdot)$ as follows. Leave the value $t(\cdot)$ unchanged for nodes i in the current subtree rooted at q, except put $t[f(q)] = \alpha$. Now put $t(\alpha) = \alpha_1$, and leave $t(i)$ the same for nodes i in the subtree rooted at α_1, except put $t[f(\alpha_1)] = \alpha_2$. Again leave $t(i)$ the same in the subtree rooted at α_2, except put $t[f(\alpha_2)] = \beta$. (Notice how the use of $n(\cdot)$ makes this process efficient.) Hence we essentially tie up the loose thread ends shown in Figure 9.19*c* until finally, we put $t[f(v_2)] = q$.

The levels $l(\cdot)$ of the nodes shown explicitly in Figure 9.19*c* are obviously available. The levels of the other nodes in subtrees rooted at the nodes q, α_1, α_2, β_1, v_1, and v_2 change by the same amounts as the levels of these corresponding nodes.

As far as $f(\cdot)$ and $n(\cdot)$ are concerned, these stay the same for nodes not on the stem. For nodes i on the stem, we have $f(i) \equiv f(v_2)$. Also $n(v) = 1 + n(v_1) + n(v_2)$, $n(\beta) = 1 + n(\beta_1) + n(v)$, $n(\alpha) = 1 + n(\alpha_1) + n(\alpha_2) + n(\beta)$, and $n(q)$ equals the total number of nodes in T_v (the old $n(v)$).

In conclusion, we mention one important expedient. Note that instead of the foregoing grafting procedure, we could alternatively have left T_v rooted at v, rerooted T_r' at p to get T_p, and then grafted T_p onto T_v by connecting p and q. Then node v would have become the new root node. This process can be done exactly as previously. The tree that is rerooted at p or q is called the *upper tree*, while the other is called the *lower tree*. The decision as to which tree to make the upper tree depends on the ease with which the rerooting of the "upper" tree can be effected. Since this effort is directly related to the length of the stem, the tree with the shorter stem $q - v$ or $p - r$ may be chosen as the upper tree.

9.11 DEGENERACY, CYCLING, AND STALLING

In the absence of degeneracy, the network simplex algorithm converges in a finite number of iterations as in Chapter 3. (By degeneracy, we mean a basic arc other than the root arc with a zero flow value; the root arc flow is zero for *all feasible solutions*.) However, simple examples (see Exercise 9.39) have been constructed that show that the network simplex algorithm can cycle, that is, infinitely loop through a sequence of degenerate pivots. Hence cycling prevention rules need to be employed. It turns out that a special lexicographic type of rule for the network simplex algorithm is not only easy to implement with no significant overheads, but is *computationally* beneficial for highly degenerate problems as, for example, the linear assignment problem.

The prescribed cycling prevention rule is based on maintaining what are known as strongly feasible bases. A *strongly feasible* (*basis*) *tree* (*SFT*) for Problem (9.1) (including the root arc) is a feasible rooted spanning tree in which all degenerate arcs (if any) are *pointing toward the root*. A directed arc (i, j) is said to be "pointing toward the root" if $j = p(i)$, that is, if j lies on the chain from i to the root node r. It is instructive to identify the connection between strongly feasible and "lexicographically positive" bases of Chapter 4. The ith row of $\mathbf{B}^{-1}$ is given as the solution $\mathbf{R}_i$ to the system $\mathbf{R}_i\mathbf{B} = \mathbf{e}_i^t$. Observe that this is similar to the system "$\mathbf{wB} = \mathbf{c}_B$." Hence $\mathbf{R}_i$ may be found by giving the ith basic arc a cost of 1, the other arcs a cost of zero, and computing the dual variables. If the ith arc is a root arc, then $\mathbf{R}_i$ has all ones (why?). If the ith arc is not a root arc, then from Section 9.5 the component of $\mathbf{R}_i$ corresponding to node j is zero if the ith arc does not lie on the chain from j to the root. Otherwise it is $+1$ if the ith arc is pointing toward the root and is -1 if the ith arc is pointing away from the root. In particular, for a basis corresponding to a strongly feasible tree, the row of $\mathbf{B}^{-1}$ corresponding to a degenerate arc has all components equal to 0 or 1. In fact, the foregoing argument shows that a basis $\mathbf{B}$ has lexicographic positive rows in $[\mathbf{B}^{-1}\mathbf{b}, \mathbf{B}^{-1}]$ if and only if it corresponds to a strongly feasible tree (why?). Hence from Chapter 4, we are guaranteed finiteness if we maintain strongly feasible trees.

Obtaining an Initial Strongly Feasible Basis

If an all artificial starting basis is being used, then by simply defining the artificial arc connected to a transshipment node as being directed toward the (dummy $(m + 1)st$) root node, we will have a strongly feasible starting basis. On the other hand, if some advanced start basis is given, then consider the spanning graph formed by the arcs with positive flows and let some node be designated as the root node. This graph is a spanning forest. If it is connected, then it is a strongly feasible tree. Otherwise, one at a time, connect a component tree not containing the root node to the one containing the root node using any arc in the problem that is pointing from a node in the former tree to one in the latter tree. Such an arc must of course be degenerate. Notice that if at any stage there are no arcs pointing from the set $X = \{i\text{: node } i \text{ is not as yet connected to the root}\}$ to the complement set $\overline{X}$, then the problem is separable over the node sets X and $\overline{X}$. This follows because we must have $\sum_{i \in X} b_i = \sum_{i \in \overline{X}} b_i = 0$, and the flow on any arc from a node in $\overline{X}$ to a node in X must be zero for all feasible solutions (why?). Hence we may solve these separable problems independently. Note that we already have a strongly feasible tree for the node set $\bar{X}$. Alternatively, we can connect any node in X to the root node via a degenerate artificial arc and continue this process.

Maintaining Strongly Feasible Trees

Given a strongly feasible basis tree at the beginning of an iteration and given an entering arc (p, q), the following simple rule for selecting a leaving variable arc (u, v) ensures that we maintain a strongly feasible tree. Trace the basis equivalent chain by tracing the unique chains from p and q to the root node r until they intersect at some common node k. Figure 9.20*a* depicts the situation generically. Note that k can coincide with p and/or r or with q and/or r. Define the cycle $C(e)$ created as having the orientation of the flow along (p, q) and commencing at node k, proceeding first along the $k \to p$ chain, then including arc (p, q), and finally proceeding along the $q \to k$ chain. Determine the exiting variable in $C(e)$, breaking ties (if any) by selecting the *last* blocking arc in $C(e)$. For example, in Figure 9.20*b*, when (4, 7) enters, we have $p = 4$, $q = 7$, and $k = 5$, and of the blocking arcs (4, 3) and (6, 7), we select (6, 7) as the leaving variable arc.

It is easy to see following Figure 9.20*a* that this rule maintains strongly feasible bases. Note that any degenerate arc not involved in the cycle $C(e)$ is still pointing toward the root after the pivot, irrespective of the choice for the leaving variable (why?). Observe that the pivot is degenerate if and only if there exists a

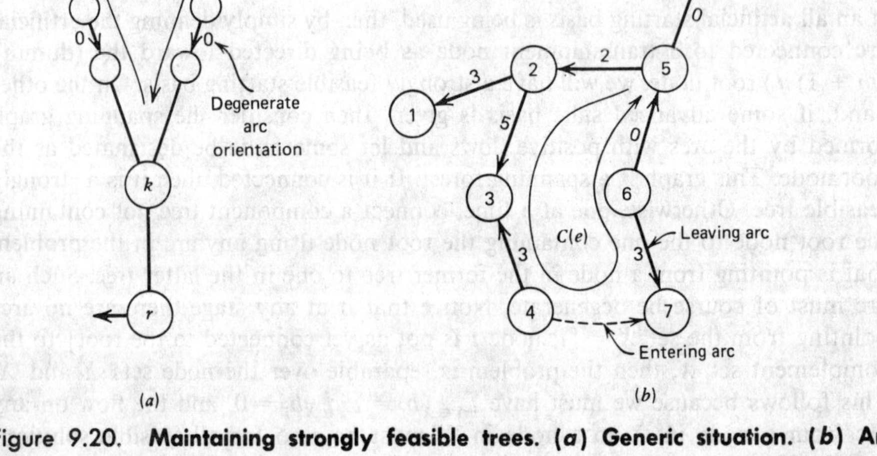

Figure 9.20. Maintaining strongly feasible trees. (*a*) Generic situation. (*b*) An example.

degenerate arc on the chain $k \to p$. In such a case, since the last degenerate arc on the chain $k \to p$ leaves the basis, the other degenerate arcs on $k \to p$ are still pointing toward the root, as are arcs (p, q) and any degenerate arcs that may have existed on the chain $q \to k$. If the pivot is nondegenerate, then there are no degenerate arcs on the chain $k \to p$. Moreover, any old degenerate arc on the chain $q \to k$ will now have a positive flow. Hence we only need to examine the orientation of any new degenerate arcs created, which must correspond to alternate blocking arcs. Since the last blocking arc in $C(e)$ has been selected to leave the basis, the other blocking arcs are pointing toward the root after the pivot, and so we maintain a strongly feasible tree.

In Chapter 4, it was proven that the foregoing algorithm will converge finitely. However, a more direct argument based on the network simplex algorithm is instructive. Note that it is sufficient to show that the algorithm cannot cycle (why?). Consider a degenerate pivot in which (p, q) enters with $\delta_{pq} = z_{pq} - c_{pq} > 0$. The leaving variable arc (u, v) must lie on the chain $k \to p$. Hence p is the leading node in the subtree T_1 that is obtained when (u, v) is disconnected from the current tree and that does not contain the root node. Consequently, the dual values of the nodes in T_1 fall by δ_{pq}, while the dual values of the nodes in $T - T_1$ remain the same. Therefore, in a sequence of degenerate pivots, the sum of the dual variables, for example, provides a strictly (decreasing) monotone function that guarantees that a basis tree cannot repeat. This establishes the finite convergence of the network simplex algorithm. Note that as in Chapter 4 the vector $(\mathbf{c}_B\mathbf{B}^{-1}\mathbf{b}, \mathbf{c}_B\mathbf{B}^{-1})$ is lexicographically decreasing over all iterations (why?).

This cycling prevention rule is readily extended to the bounded variables network simplex algorithm. A strongly feasible basis tree may be defined here as one in which the basic arcs that are degenerate at their lower bounds are pointing toward the root and those basic and degenerate at their upper bounds are pointing away from the root. Then the same leaving variable rule (which includes the case of the entering arc blocking itself) maintains strong feasibility and guarantees finite convergence. We leave the details to the reader in Exercise 9.51.

Stalling and its Prevention

Although the foregoing technique guarantees finite convergence, it is still entirely possible that the algorithm may *stall* at an extreme point, that is, go through a number of consecutive degenerate pivots that, although finite, is exponential in m and n. For example, if $\mathbf{b} \equiv \mathbf{0}$ in Problem (9.1), then *all* possible bases represent the same extreme point, namely the origin, and there may be an exponential number of such bases that one may enumerate before terminating with $z_{ij} - c_{ij} \leq 0$ for all nonbasic variables. In fact, there exist examples exhibiting this phenomenon (see the Notes and References section). In any case, it seems desirable to devise a rule that would guarantee that the number of consecutive *degenerate* pivots is bounded above by a polynomial in m and n.

As seen in Chapter 4, a key to preventing stalling is to keep the length as well as the number of stages small. Recall that a *stage* is a sequence of degenerate pivots in which no nonbasic variable remains enterable throughout and no strict subset of this sequence has the same property. The length of a stage may be controlled by adopting an appropriate "affirmative action" rule for selecting an entering variable. For example, the Least Recently Considered (LRC) variable choice rule discussed in Chapter 4 guarantees that the number of pivots in a stage is no more than n. It turns out that *independent* of the rule for choosing an entering variable, if we maintain strongly feasible basis trees as previously, then the number of stages in a sequence of consecutive degenerate pivots is no more than the number of degenerate basic variables at the current extreme point. Hence the number of stages in a sequence of degenerate pivots is bounded above by m, so that the number of consecutive degenerate pivots using strongly feasible bases in conjunction with the LRC entering rule, for example, is bounded above by the polynomial mn. This underscores the relevance of maintaining strongly feasible bases.

To establish the above bound on the number of stages, let $T_0, T_1, \ldots, T_M$ be the basis trees in a sequence of consecutive degenerate pivots such that either optimality is recognized at T_M or the pivot executed on T_M is nondegenerate. Notice that the nondegenerate basic arcs are common in all of the trees $T_0, T_1, \ldots, T_M$ (why?). Consider the spanning graph formed by these nondegenerate basic arcs. Let this graph have (tree) components $G_1, G_2, \ldots, G_P$, where $P \leq m$. In each of the trees $T_0, \ldots, T_M$, these components $G_1, \ldots, G_P$ are connected to each other via some $P - 1$ degenerate arcs. In particular, let T_M be as shown in Figure 9.21, with the root node $r \in G_1$. Let $L \leq P - 1$ be the maximum number of degenerate arcs in the chain from any node to the root node

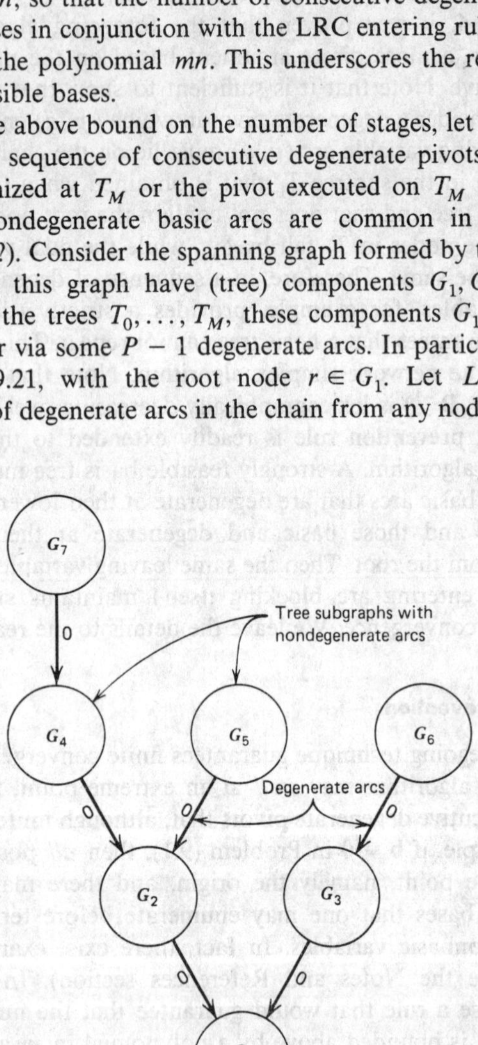

Figure 9.21. Final tree T_M in a sequence of degenerate pivots.

in T_M. Observe that $P = 7$ and $L = 3$ in Figure 9.21. We will show in fact that the number of stages in the sequence $T_0, \ldots, T_M$ is no more than L and hence no more than $m - 1$.

Toward this end, let us make certain observations. First, since $T_0, \ldots, T_M$ are strongly feasible and the pivots in this sequence are degenerate, the dual variable of each node is nonincreasing in this sequence and the sum is strictly decreasing. Second, since the leaving arc is degenerate in each pivot, the dual variables of the nodes in each subgraph G_i either remain the same or all strictly decrease by the same amount for $i = 1, \ldots, P$ in each pivot. Third, denoting $\bar{\mathbf{w}}$ as the dual values in T_M, we have that $w_i = \bar{w}_i$ for all $i \in G_1$ in each of the trees $T_0, \ldots, T_M$, since the arcs in G_1 are basic in all these trees. Fourth, from the first point, as long as the dual values of the nodes in any subgraph are not equal to their values in $\bar{\mathbf{w}}$, they are strictly greater than these values.

By contradiction, suppose that the procedure goes through more than L stages. Consider a subgraph G_k at the first "level" in T_M. By the "level" of a subgraph we will mean the number of degenerate arcs in a chain in T_M connecting any node in this subgraph to the root node. Suppose that for some tree T_j, $j \in \{0, \ldots, M-1\}$, the dual values w_i of nodes $i \in G_k$ are strictly greater than $\bar{w}_i$. In particular, let (p, q) be the degenerate arc connecting G_k to G_1 in T_M. Then in the tree T_j, we have $w_p - w_q - c_{pq} = w_p - \bar{w}_q - c_{pq} > \bar{w}_p - \bar{w}_q - c_{pq} = 0$ (why?). As long as the dual values in G_k are not equal to $\bar{\mathbf{w}}$, arc (p, q) remains enterable and a stage persists. By the time the first stage is over within the sequence $T_0, \ldots, T_M$, we will have $w_i = \bar{w}_i$ for all $i \in G_k$. This equality holds in every tree after this stage.

Inductively assume that the procedure has gone through some $l \geq 1$ stages and the dual values of nodes in subgraphs at levels l or lower are all at their corresponding values in $\bar{\mathbf{w}}$. Consider a subgraph G_k at level $(l + 1)$ and let (p, q) be the first degenerate arc encountered in the chain from any node in G_k to the root node. If $w_i > \bar{w}_i$ for $i \in G_k$ in any tree following stage l, then as before in this tree, we have $w_p - w_q - c_{pq} = w_p - \bar{w}_q - c_{pq} > \bar{w}_p - \bar{w}_q - c_{pq} = 0$ so that arc (p, q) is enterable. As long as the dual values in G_k are not fixed at their values in $\bar{\mathbf{w}}$, (p, q) remains enterable and stage $(l + 1)$ persists. When stage $(l + 1)$ is completed within the sequence $T_0, \ldots, T_M$, the dual values of nodes in subgraphs at level $(l + 1)$ are all at their values in $\bar{\mathbf{w}}$. Consequently, the number of stages in the sequence $T_0, \ldots, T_M$ cannot be more than L. This establishes the required result.

Another Useful Entering Variable Choice Rule that Prevents Stalling

Besides the partial pricing rule for selecting an entering variable described in Chapter 3 and the LRC rule described in this chapter, there is another competitive entering variable selection rule that is well suited to how data is customarily stored. Suppose that the arcs in the problems are stored in a one-dimensional array with arcs having node 1 as their from-node appearing first, then arcs having node 2 as their from-node, and so on. Consider the following rule. Suppose that an arc having node j at its from-node has just entered. (Assume $j = 1$ to initialize.) Then the next entering variable is chosen as the *best* enterable arc with node j again as its from-node, and if none exists, then the *best* enterable arc with node $j+1$ as its from-node, and so on with node $j+2,\ldots,m,1,2,\ldots,$ and node $j-1$ as its from-node. Then, we show that the maximum number of consecutive degenerate pivots using strongly feasible basis trees is no more than mL, which is less than m^2.

Note that the use of strongly feasible basis trees ensures that the number of stages in a sequence of degenerate pivots is no more than L, independent of the rule for choosing an entering variable. To show that the length of a stage is no more than m, consider a tree T_j in the sequence $T_0,\ldots,T_{M-1}$ that leads to a degenerate pivot. Let $\mathbf{w}$ be the dual values in T_j. Suppose that arc (p,q) enters, say, where $p \in G_k$. Referring to Figure 9.21 (assuming this represents tree T_j and *not* T_M), note that the pivot will be degenerate if and only if q does not lie in G_k or in its "successors," that is, if and only if the basis equivalent chain includes the degenerate arc (u, v) that is the first degenerate arc encountered by tracing from any node in G_k to the root node. Moreover, by the rule for breaking ties for the leaving variable (see Figure 9.20*a*), the arc (u, v) must be the leaving variable arc. Consequently, if (p,q) enters degenerately, then the dual values of nodes in G_k and its successors reduce by $\delta_{pq} = z_{pq} - c_{pq} > 0$, while those of the other nodes remain the same. Let these revised dual variables in T_{j+1} be called $\mathbf{w}'$. Since the set of nodes in G_k or in its successors have been unaffected by the pivot, for any node l not in this set we have $w_p' - w_l' - c_{pl} = (w_p - \delta_{pq}) - w_l - c_{pl} = (w_p - w_l - c_{pl}) - \delta_{pq} \leq 0$ since δ_{pq} was the most positive $z_{pt} - c_{pt}$ value among all arcs of the type (p, t). If some arc having node p as its from-node is now enterable, the resulting pivot must be nondegenerate. Consequently, if the next pivot is degenerate, the from-node cannot be node p, which means that in a sequence of degenerate pivots the length of a stage is $\leq m$. This establishes the required bound. In Exercise 9.52 we ask the reader to consider another variation of this rule.

9.12 GENERALIZED NETWORK PROBLEMS

Consider a linear programming problem of the form

$$\begin{aligned}
\text{Minimize} \quad & \mathbf{cx} \\
\text{Subject to} \quad & \mathbf{Ax} = \mathbf{b} \\
& \mathbf{x} \geq \mathbf{0}
\end{aligned}$$

where the matrix $\mathbf{A}$ is $m \times n$ of rank m and where each column $\mathbf{a}_j$ of $\mathbf{A}$ has at most two nonzero entries. Then one may conceptualize the constraints $\mathbf{Ax} = \mathbf{b}$ as a set of generalized nodal flow balance equations on a graph $G(\mathscr{N}, \mathscr{A})$ that is constructed as follows. Let $\mathscr{N}$ be the set of m nodes $\{1,\ldots,m\}$, with node i being associated with the ith row of $\mathbf{A}$. For each column $\mathbf{a}_j$ of $\mathbf{A}$ with two nonzero entries, occurring, say, in rows p and q, construct an undirected link l_j between nodes p and q. For each column $\mathbf{a}_j$ of $\mathbf{A}$ with a single nonzero entry, occurring, say, in row p, construct an undirected link l_j incident at node p. Let $\mathscr{A}$ denote the set of these links, and let x_j denote the flow associated with link l_j. Additionally, for each node p, examine the links that are incident at p and associate with each such link l_j at the point of incidence at p a multiplier equal to the coefficient in column $\mathbf{a}_j$ and row p. Then the nodal balance equations assert that at each node p, the sum of the weighted flows on the incident links, where the weights equal the corresponding incident multipliers, must equal b_p. Because of this interpretation of modified flows, this type of problem is called a *generalized network* or a *network flow with gains* problem. Problems of this type are used to model a variety of situations involving a modification of flow or the transformation of one type of product to another in a network-based system.

As an illustration, consider the following example problem.

$$\begin{array}{llllllll}
\text{Minimize} & 3x_1 + & x_2 - & x_3 + 2x_4 + & x_5 - & 2x_6 + & x_7 & \\
\text{Subject to} & 2x_1 & & & & + 3x_6 & & = 6 \\
& -x_1 + & 3x_2 & & & & - x_7 & = 0 \\
& & & x_3 + 2x_4 & & - x_6 & & = 8 \\
& & & 3x_3 & + 2x_5 & & + 3x_7 & = 8 \\
& & & x_4 - & x_5 & & & = 2 \\
& x_1, & x_2, & x_3, \quad x_4, & x_5, & x_6, & x_7 & \geq 0
\end{array}$$

The corresponding graph $G(\mathscr{N}, \mathscr{A})$ is shown in Figure 9.22a along with the multipliers associated with the seven links.

Consider the subgraph $G_B(\mathscr{N}, \mathscr{A}_B)$ of $G(\mathscr{N}, \mathscr{A})$ shown in Figure 9.22b, where the quantity b_i is shown against each node i, for $i = 1, \ldots, 5$. Suppose that the flows are permitted to be positive only on the links shown in Figure 9.22b, that is, the flows x_6 and x_7 are restricted to zeros. Then at node 1, we have $2x_1 = 6$ or $x_1 = 3$; at node 2, we must therefore have $3x_2 - x_1 = 0$ or $x_2 = x_1/3 = 1$. Notice how the structure of the rooted tree is used to compute these values, as in the network simplex method. Similarly, treating the second component in Figure 9.22b as a tree for a fixed value of x_3, we must have at node 3 that

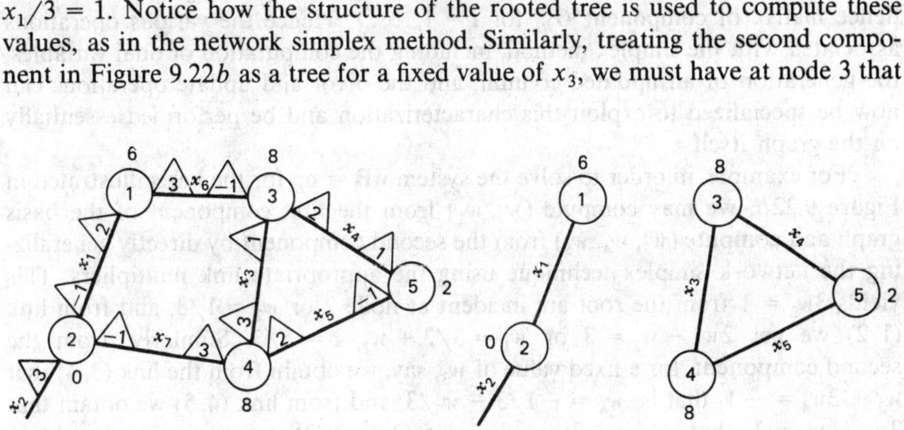

Figure 9.22. Example of a generalized network. (*a*) Network. (*b*) Basis.

$2x_4 = 8 - x_3$ or $x_4 = 4 - (x_3/2)$, we must have at node 4 that $2x_5 = 8 - 3x_3$ or $x_5 = 4 - (3/2)x_3$, and we must have at node 5 that $x_4 - x_5 = 2$. Substituting for x_4 and x_5 in terms of x_3 into this last equation gives $x_3 = 2$, from which we obtain $x_4 = 3$ and $x_5 = 1$. Hence this solution, which is feasible and is uniquely obtained when x_6 and x_7 are fixed at zero, is a *basic* feasible solution (why?). Notice how this directly generalizes the network simplex technique for solving the system $\mathbf{Bx}_B = \mathbf{b}$.

This example illustrates the following general characterization of a basis of **A**. Let **B** be any basis of **A**. Corresponding to **B**, construct a spanning subgraph $G_B(\mathcal{N}, \mathcal{A}_B)$ of $G(\mathcal{N}, \mathcal{A})$ as follows. Let $\mathcal{N}$ be the set of nodes $1, \ldots, m$, and let $\mathcal{A}_B$ be the set of links corresponding to the columns of **B**. Let $G_{Bi}(\mathcal{N}, \mathcal{A}_{Bi})$, $i = 1, \ldots, r$ be the $r \geq 1$ components of G_B. Then, as we show next each component G_{Bi}, $i = 1, \ldots, r$, is either a rooted tree or is a one-tree. Such a graph G_B is called a *pseudorooted spanning forest*. Note that $r = 2$ in the example, and that one component is a rooted tree, while the other component is a one-tree.

To establish the foregoing claim, let $m_i = |\mathcal{N}_i|$ be the number of nodes in component i for $i = 1, \ldots, r$. Note that since G_{Bi} is connected, we have from Property 3 of tree graphs that $|\mathcal{A}_{Bi}| \geq (m_i - 1)$ for $i = 1, \ldots, r$. Suppose that for some i, we have $|\mathcal{A}_{Bi}| = (m_i - 1)$, so that G_{Bi} is a tree graph. Then consider the m_i rows of **B** corresponding to the nodes in $\mathcal{N}_i$. Observe that these rows have nonzero entries only in the $(m_i - 1)$ columns corresponding to the links in $\mathcal{A}_{Bi}$, and so these rows cannot be linearly independent, a contradiction to the fact that **B** is a basis. Consequently, we must have $|\mathcal{A}_{Bi}| \geq m_i$ for $i = 1, \ldots, r$. However, $\Sigma_{i=1}^{r} m_i = m$ and $\Sigma_{i=1}^{m} |\mathcal{A}_{Bi}| = m$, the total number of basic columns. Hence we must have $|\mathcal{A}_{Bi}| = m_i$ for each $i = 1, \ldots, r$. By our equivalent characterizations of tree graphs in Section 9.1, it then follows that each component that is a connected graph on some m_i nodes with m_i links must either be a rooted tree graph or a one-tree graph.

As a result of this characterization, any basis **B** of **A** can be arranged into r diagonal blocks $\mathbf{B}_1, \ldots, \mathbf{B}_r$, with $\mathbf{B}_i$ representing the square (generalized) incidence matrix of component G_{Bi} for $i = 1, \ldots, r$. Hence the various operations associated with the simplex method, including the computation of dual variables, the generation of an updated column, and the pivot and update operations can now be specialized to exploit this characterization and be performed essentially on the graph itself.

For example, in order to solve the system $\mathbf{wB} = \mathbf{c}_B$ for the basis illustrated in Figure 9.22*b*, we may compute (w_1, w_2) from the first component of the basis graph and compute (w_3, w_4, w_5) from the second component by directly generalizing the network simplex technique using the appropriate link multipliers. This yields $3w_2 = 1$ from the root arc incident at node 2 or $w_2 = 1/3$, and from link (1, 2) we get $2w_1 - w_2 = 3$ or $w_1 = 3/2 + w_2/2 = 5/3$. Similarly, from the second component, for a fixed value of w_3, say, we obtain from the link (3, 4) that $w_3 + 3w_4 = -1$, that is, $w_4 = -1/3 - w_3/3$, and from link (4, 5) we obtain that $2w_4 - w_5 = 1$, that is, $w_5 = 2w_4 - 1 = -5/3 - (2/3)w_3$, and finally, we obtain from link (3, 5) that $2w_3 + w_5 = 2$. Substituting for w_5 in this last equation gives $2w_3 - 5/3 - (2/3)w_3 = 2$ or $w_3 = 11/4$, which yields $w_4 = -5/4$, and

$w_5 = -7/2$. Having now determined $\mathbf{w} = (5/3, 1/3, 11/4, -5/4, -7/2)$, we may price the nonbasic variables by computing $z_j - c_j = \mathbf{wa}_j - c_j$. This gives $z_6 - c_6 = 3w_1 - w_3 - c_6 = 5 - 11/4 + 2 = 17/4$, and $z_7 - c_7 = -w_2 + 3w_4 - c_7 = -1/3 - 15/4 - 1 = -61/12$. Hence x_6 is enterable. We may now generate the updated entering column $\mathbf{y}_6$ by solving the system $\mathbf{B}\mathbf{y}_6 = \mathbf{a}_6$ in a manner similar to the solution of the system $\mathbf{B}\mathbf{x}_B = \mathbf{b}$, and then perform the pivot and updating operations. We leave the details of specifying the general rules for this procedure and the completion of the example to the reader in Exercise 9.37.

EXERCISES

9.1 Provide a complete proof for the equivalence of the six characterizations of a tree given in Section 9.2.

9.2 Show that the two root arcs and the chain between them form a dependent set.

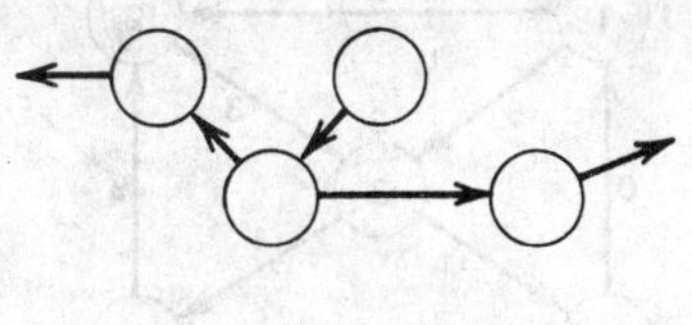

9.3 Solve the following network flow problem using x_{12}, x_{23}, x_{34} as part of a starting basis.

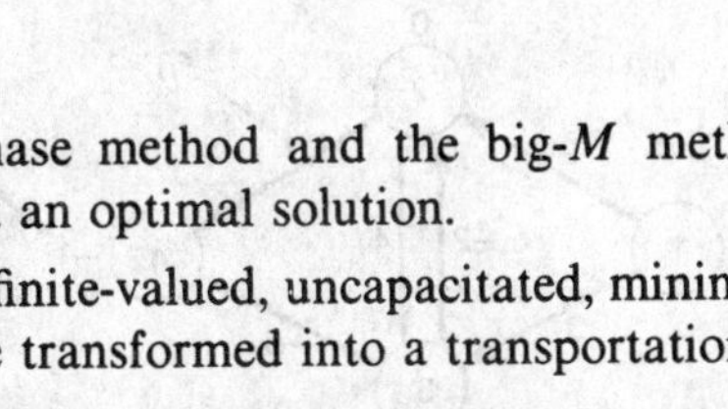

9.4 Apply the two-phase method and the big-M method to the network of Exercise 9.3 to get an optimal solution.

9.5 Indicate how any finite-valued, uncapacitated, minimal cost flow problem of Section 9.1 can be transformed into a transportation problem. Illustrate by

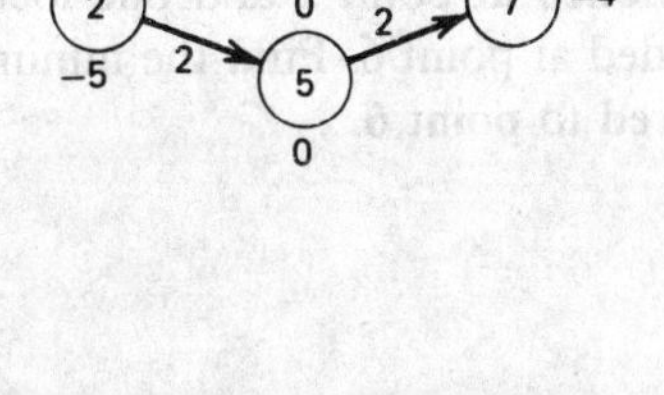

the above network. (*Hint.* Locate the minimal cost path from each supply point to each demand point.) How would you achieve this transformation for capacitated problems? (*Hint.* See Exercise 9.35.)

9.6 Suppose, in the flow problem of Section 9.1, that we locate a path from a supply point to a demand point. Putting as much flow as possible on the path, decreasing the corresponding supply and demand, we repeat the whole process until all supplies are allocated (and demands satisfied).
a. Will the feasible solution obtained be basic?
b. If not, how can the solution obtained be made basic?

9.7 Solve the following network flow problem.

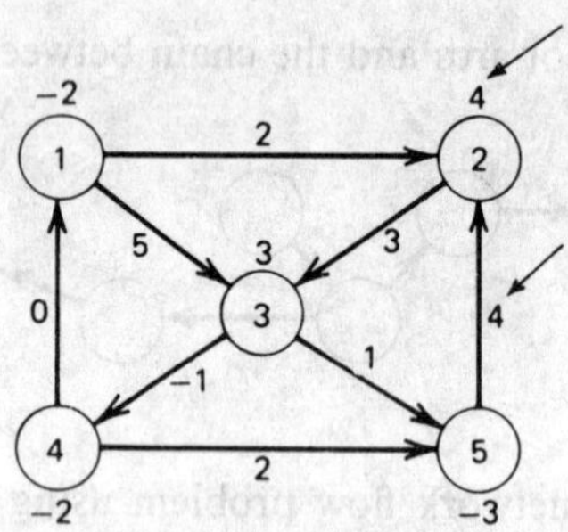

9.8 Solve the following network flow problem.
(*Note.* For this choice of b_i's we have found the shortest path from node 1 to node 4.)

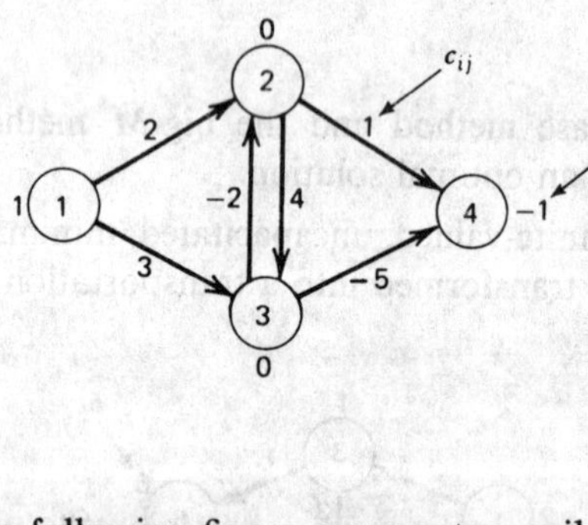

9.9 Suppose that the following figure represents a railroad network. The numbers beside each arc represent the time it takes to traverse the arc. Two locomotives are stationed at point 2 and one locomotive at point 1. Three locomotives are needed at point 6. Find the minimum total time solution to get the power required to point 6.

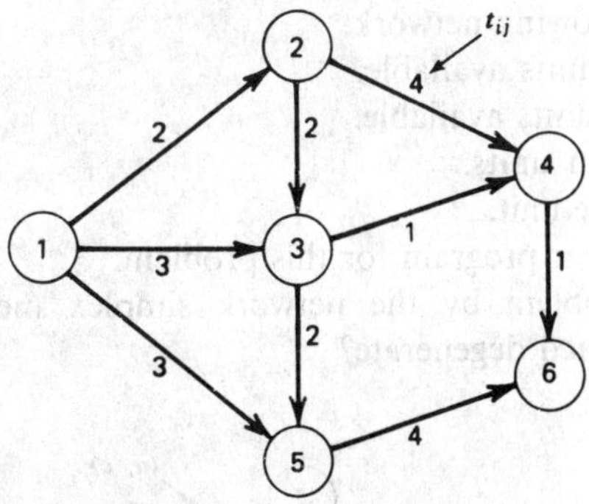

9.10 Solve the following flow problem. (*Note:* $\Sigma b_i \neq 0$.) Does it have alternative optimal solutions?

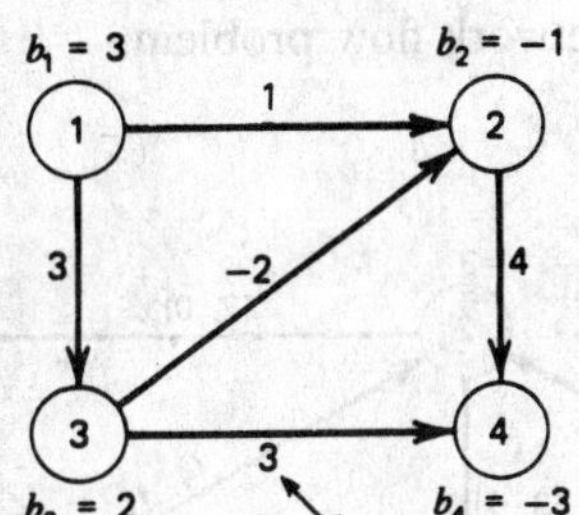

9.11 Draw the underlying graph for the transformed problem of Example 9.2. Optimize using the big-M method. Start with an all artificial variable basis.

9.12 Put the following problem in the standard form (9.1). Construct an all artificial initial basis and optimize using the two-phase or the big-M method.

$$\text{Minimize} \quad 7x_{12} + 2x_{13} + 2x_{23} - 2x_{42} + 6x_{34} + 2x_{35} + 3x_{64} + x_{65} + 4x_{75} + 2x_{76}$$

$$\begin{array}{lrrrrrrrrrrl}
\text{Subject to} & +x_{12} & +x_{13} & & & & & & & & & \leq 5 \\
& -x_{12} & & +x_{23} & -x_{42} & & & & & & & = -2 \\
& & -x_{13} & -x_{23} & & +x_{34} & +x_{35} & & & & & = 3 \\
& & & & +x_{42} & -x_{34} & & -x_{64} & & & & \leq -4 \\
& & & & & & -x_{35} & & -x_{65} & -x_{75} & & \geq -4 \\
& & & & & & & +x_{64} & +x_{65} & & -x_{76} & = 1 \\
& & & & & & & & & +x_{75} & +x_{76} & = 2 \\
& x_{12}, & x_{13}, & x_{23}, & x_{42}, & x_{34}, & x_{35}, & x_{64}, & x_{65}, & x_{75}, & x_{76} & \geq 0
\end{array}$$

9.13 Consider the following network.
Node 1 has five units available.
Node 3 has two units available.
Node 2 needs four units.
Node 4 needs one unit.

a. Set up the linear program for this problem.
b. Solve the problem by the network simplex method. Is the optimal solution obtained degenerate?

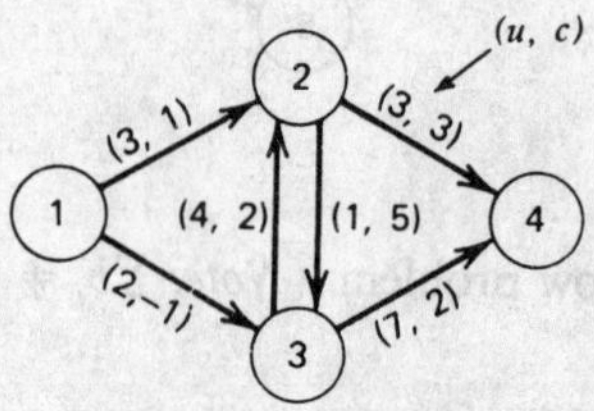

9.14 Solve the following network flow problem.

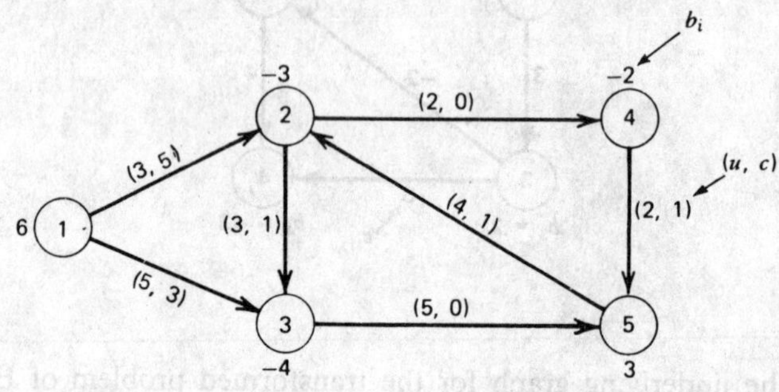

9.15 Starting with x_{12}, x_{24}, and x_{31} as part of a basis where x_{14} is nonbasic at its upper bound and where all other x_{ij}'s are nonbasic at their lower bound, solve the following network flow problem.

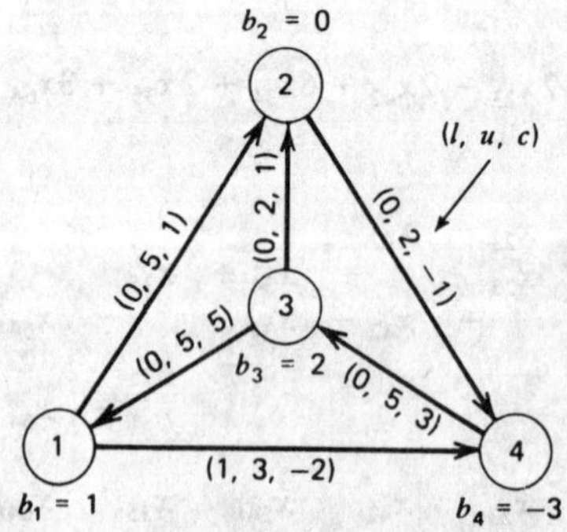

9.16 Apply the two-phase method and the big-M method to Exercise 9.15.

9.17 A company has requirements for a certain type of machine during the next N months of D_i per month, $i = 1, 2, \ldots, N$. Those requirements must be met, although excesses are permitted to any desirable extent. No machines are available at the beginning of month 1. At the beginning of each of the N months, the company may purchase machines. The machines are delivered immediately; that is, there is no lead time. The company may purchase machines that last one month, two months, and so on up to M months. The number of months a machine is usable is called its service life. The cost of a machine depends on the month in which it is bought and on its service life. In particular, a machine bought in the ith month with service life of k months costs c_{ik} dollars. Naturally a machine that lasts longer costs more, so that $c_{ip} < c_{is}$ for $p < s$.

a. Formulate a mathematical statement for this problem, assuming that the objective is to minimize the sum of the monthly costs. Let x_{ik} be the number of machines bought in month i with service life of k months.

b. Formulate the *general* problem as a flow problem. Use $N = 4$ and $M = 3$ to illustrate, summarizing your work on a network. (*Hint.* Consider elementary row operations to obtain a 1 and -1 in every column.)

9.18 Consider the following network flow problem.
In addition:
Node 1 can produce an unlimited supply of units at a cost of two per unit.
Node 2 can produce up to three units at a cost of four per unit.
Node 3 needs two units.
Node 4 needs five units.
Find the optimal shipping policy.

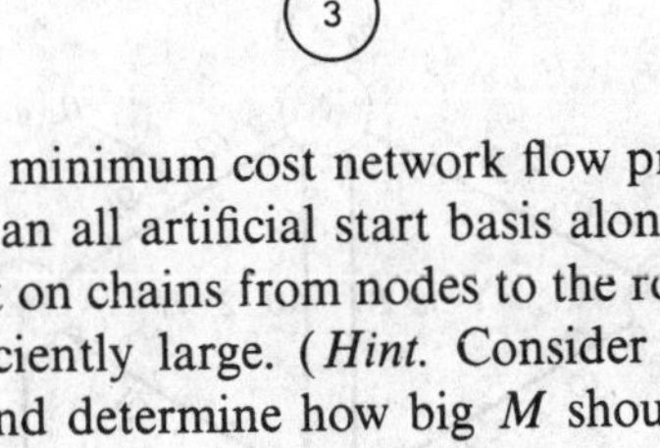

9.19 Consider a feasible minimum cost network flow programming problem (9.1) being solved using an all artificial start basis along with the big-M method. In terms of the cost on chains from nodes to the root node, prescribe a value of M that is sufficiently large. (*Hint.* Consider any optimal basis to the original problem and determine how big M should be so that the artificial nonbasic arcs have $z_{ij} - c_{ij} < 0$.)

9.20 It is often necessary (particularly in the dual simplex algorithm and in integer programs) to generate a row of the simplex tableau. Indicate how this can easily be done for a network flow problem. Illustrate by generating the row associated with x_{12} in the starting simplex tableau for Exercise 9.3.

[*Hint.* Remove some arc (basic in a given row) of the basic tree, thereby separating the nodes into two subsets. Then consider the set of nonbasic arcs going between the two node sets—those in the same direction as the given basic arc and those in the opposite direction.]

9.21 Consider the following network for a minimum cost network flow problem with the net supply (b_i) values shown on the nodes and the arc costs c_{ij} shown on the arcs.

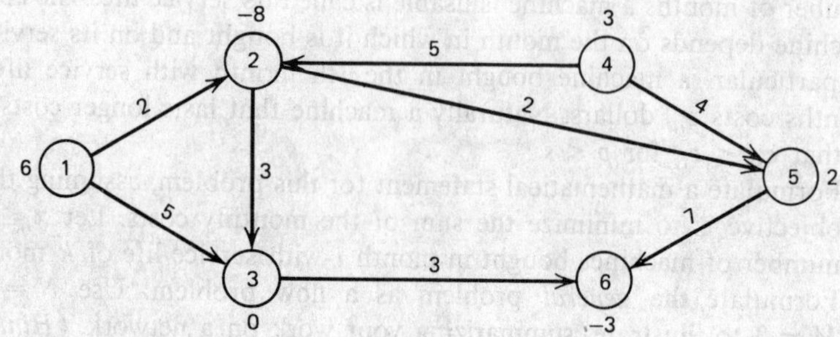

a. Construct a rooted spanning tree with node 2 as the root node and the arcs (1, 2), (2, 3), (2, 5), (3, 6), and (4, 5) along with the root arc as basic. Determine the associated primal and dual solutions for this basis.

b. Show that this basis is dual feasible. On the *graph*, perform a dual simplex iteration in which the dual variable complementary to x_{25} enters the (dual) basis. Show all details.

9.22 Indicate how the lexicographic simplex method may be applied to the network flow problem.

9.23 Develop in detail two algorithms for solving network flow problems that are based on the dual simplex and the primal-dual methods.

9.24 Solve the following network flow problem.

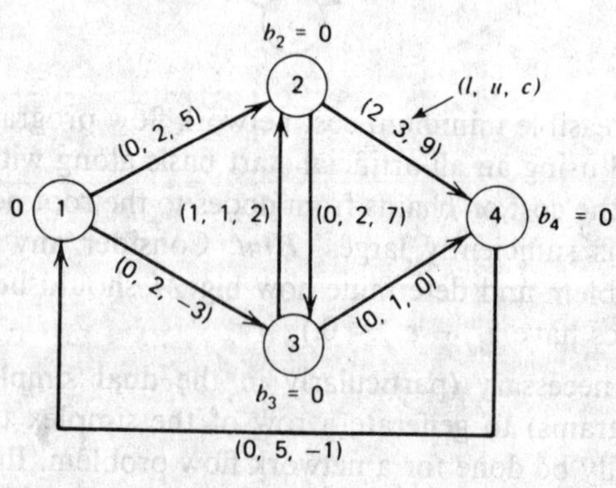

9.25 Indicate how we can generate a row or a column of $\mathbf{B}^{-1}$ associated with a network flow problem. Illustrate by generating the row of $\mathbf{B}^{-1}$ associated with x_{12} in the initial basis for Exercise 9.3.

9.26 According to the rule of Section 9.5, show that the root is the last basic variable to be assigned a value.

9.27 Prove or give a counterexample: In order for a variable to be basic in a given row of the linear program for a network flow problem, the variable must have a nonzero entry (that is, ± 1) in that row of the original constraint matrix; that is, the arc must be incident with the associated node.

9.28 Indicate how we can handle lower and upper capacity constraints on flow through a node i. (*Hint.* Consider splitting node i into two nodes.)

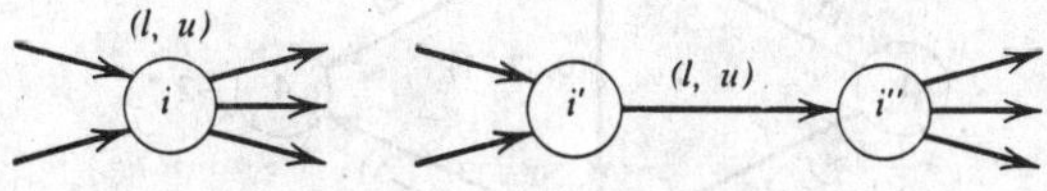

9.29 Consider the following network of cities. Each city must be visited exactly once by some salesman traveling in a nontrivial circuit. Use network flows to indicate the number of salesmen necessary and their routes to minimize the total distance traveled. Assume that an unlimited number of salesmen are available at no cost to be positioned wherever needed. (*Hint.* See Exercise 9.28.) Can we impose the additional constraint that only one salesman be used? (This would be the classical *traveling salesman problem*.)

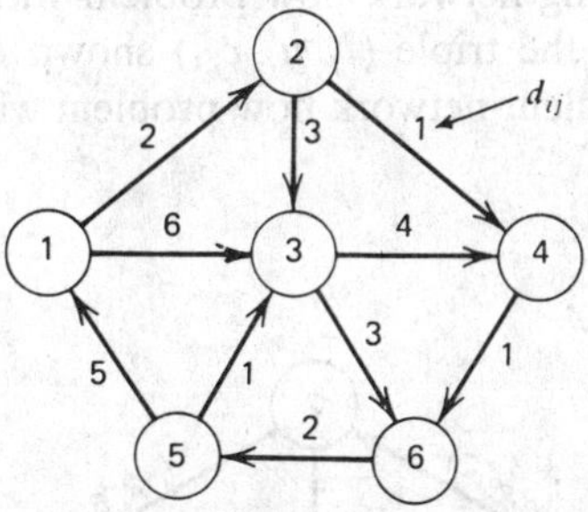

9.30 Suppose that a minimum cost network flow program contains an undirected arc (i, j) with a cost per unit flow of c_{ij}. Show how the corresponding variable x_{ij} appears in the mathematical formulation of the problem *before* using any transformations on this variable.

9.31 How can we handle undirected arcs when $\mathbf{l} = \mathbf{0}$ and $\mathbf{c} \geq \mathbf{0}$ in a network flow problem? (*Hint.* Consider replacing a single undirected arc by two oppo-

sitely directed arcs.) What happens when $c < 0$ or when $l > 0$ for some arcs? Relate the transformation used to solve Exercise 9.30.

9.32 How can we transform a network flow problem with some $l'_{ij} \neq 0$ into one with all $\mathbf{l} = \mathbf{0}$? Illustrate by the following network.

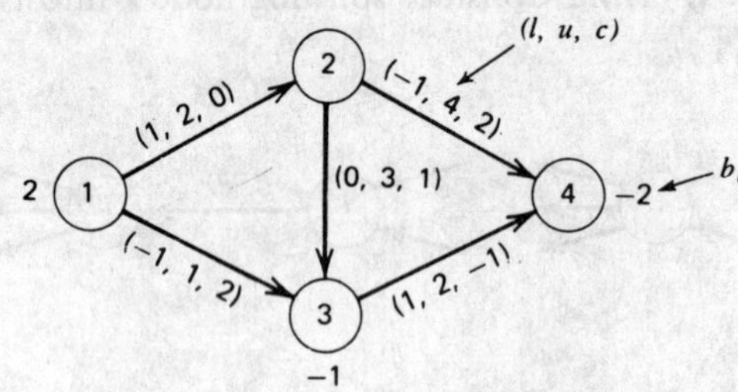

9.33 Show how a network flow problem with $\mathbf{l} = \mathbf{0}$, for which the cost function associated with each arc is piecewise-linear and convex, can be solved by the methods of this chapter. (*Hint.* Consider adding parallel arcs—one for each segment—with proper costs and upper bounds.) Can the methods of this chapter be also used for network problems with piecewise-linear and concave functions?

9.34 Consider the following network flow problem with the b_i values shown on the nodes and with the triple (l_i, u_i, c_{ij}) shown on the arcs. Convert this problem to an equivalent network flow problem with all lower bounds equal to zero.

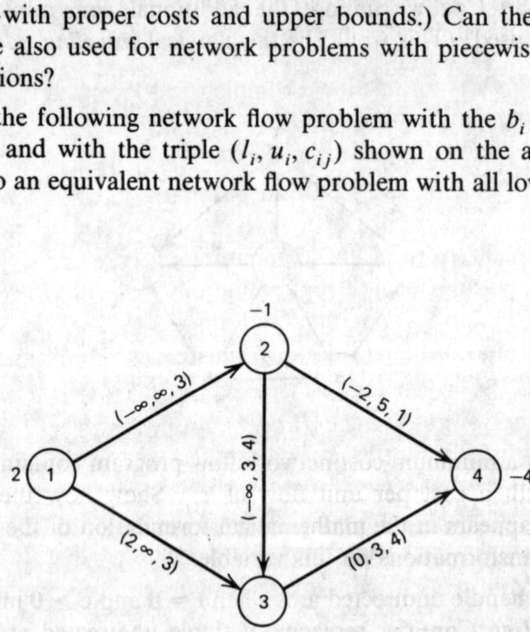

9.35 Show that a network flow problem with zero lower bounds and finite upper bounds can be converted to one with zero lower bounds and no (infinite) upper bounds. [*Hint.* Consider splitting arc (i, j) as shown.]

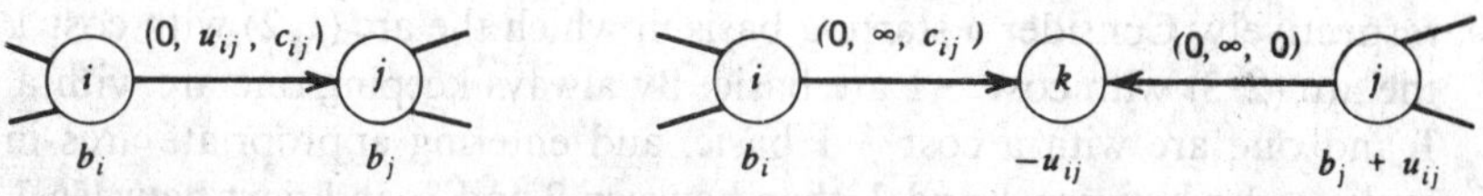

Derive this transformation algebraically by performing appropriate row operations after converting the upper bounding constraints to equalities. Illustrate by converting the following network to one without upper bounds.

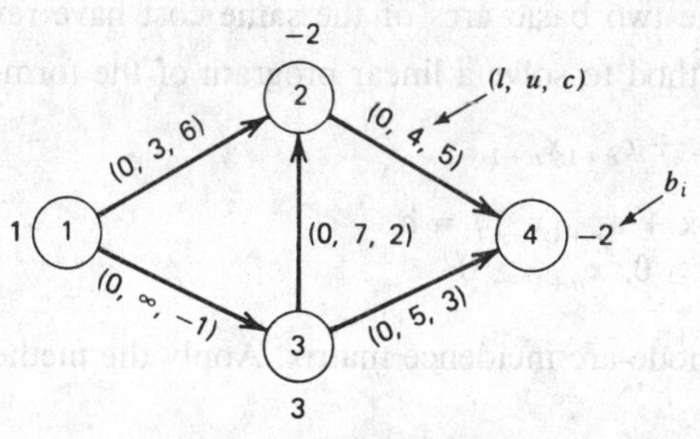

9.36 Prove that when we discard the redundant constraint, a network flow basis is characterized by a *rooted spanning forest* (a collection of trees, each with exactly one root, which spans the node set). Using this characterization, develop the representation of a nonbasic column in terms of the basis. Utilize your results to describe a complete simplex method for network flow problems without the redundant constraint. Apply the method to the network in Exercise 9.3 with the last constraint deleted.

9.37 Consider the generalized network problem introduced in Section 9.12. Given the characterization of a basis as a pseudorooted spanning forest, develop specialized rules for computing the complementary dual basic solution, generating the updated column of an entering variable, and for performing the pivot operation and updating the basis along with the associated primal and dual solutions. Illustrate using the example of Section 9.12.

9.38 a. Provide the values of $p(i), l(i), t(i), t_R(i), f(i), n(i)$ for $i = r, 1, \ldots, 27$ corresponding to the tree in Figure 9.18.

b. If this was an advanced start basis, show how you would construct these lists (*without* using a graphical display) and how you would compute the flow and dual values using these lists.

c. For the pivot shown in Figure 9.18, provide a precise account of how these lists and quantities are updated using T_r' as the lower tree.

d. Repeat part c using T_v as the lower tree.

9.39 Consider a network flow problem with three vertices that each have $b_i = 0$ and with twelve arcs given as follows. Between each pair of nodes i and j, there are two arcs from i to j having costs $+1$ and -1 respectively, and there are also two reverse arcs from j to i having costs $+1$ and -1 respectively. Consider a starting basis in which the arc (1, 2) with cost 1 and the arc (2, 3) with cost -1 are basic. By always keeping one arc with a cost 1 and one arc with a cost -1 basic, and entering appropriate arcs in the cyclic order between 1 and 3, then between 2 and 3, and next between 1 and 2, starting with the entering arc from 1 to 3 with a cost -1, construct a sequence of bases that exhibits the phenomenon of cycling. (*Hint.* Try to construct a cycle of length 12, that is, trees $T_0, T_1, \ldots, T_{11}, T_{12}$, where $T_{12} \equiv T_0$, in which for each $i = 0, 1, \ldots, 5$, the tree T_{i+6} is identical to T_i except that the two basic arcs of the same cost have reverse orientations.)

9.40 Develop a method to solve a linear program of the form

$$\begin{array}{ll} \text{Minimize} & \mathbf{cx} + c_{n+1}x_{n+1} \\ \text{Subject to} & \mathbf{Ax} + \mathbf{a}_{n+1}x_{n+1} = \mathbf{b} \\ & \mathbf{x} \geq \mathbf{0},\ x_{n+1} \geq 0 \end{array}$$

where $\mathbf{A}$ is a node-arc incidence matrix. Apply the method to the following problem.

$$\begin{array}{lrcrcrcrcrcrcl} \text{Minimize} & 2x_1 & + & 3x_2 & + & x_3 & + & 3x_4 & + & 5x_5 & + & 4x_6 & & \\ \text{Subject to} & x_1 & + & x_2 & & & & & & & + & x_6 & = & 2 \\ & -x_1 & & & + & x_3 & + & x_4 & & & - & 2x_6 & = & 3 \\ & & - & x_2 & - & x_3 & & & + & x_5 & + & 3x_6 & = & -1 \\ & & & & & & - & x_4 & - & x_5 & - & x_6 & = & -4 \\ & x_1, & & x_2, & & x_3, & & x_4, & & x_5, & & x_6 & \geq & 0 \end{array}$$

9.41 Can the results of Exercise 9.40 be generalized to the case where the constraint matrix is of the form $(\mathbf{A}, \mathbf{D})$, where $\mathbf{A}$ is a node-arc incidence matrix and $\mathbf{D}$ is any arbitrary matrix? Apply the method to the previous problem with the additional column x_7 with $\mathbf{a}_7 = (2, -5, 2, 0)^t$ and $c_7 = -3$.

9.42 Solve the following transportation problem by the network simplex method of this chapter.

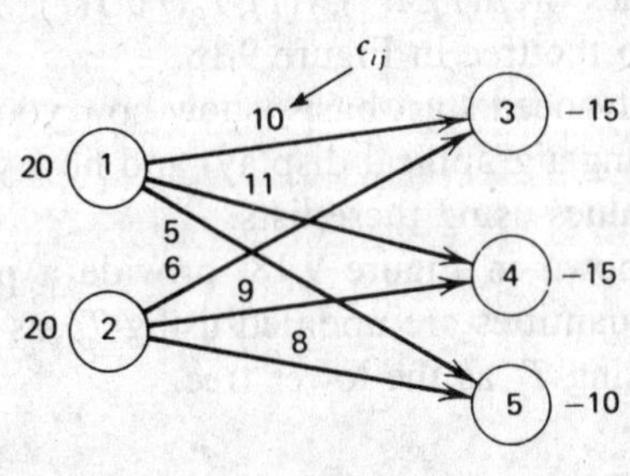

9.43 Consider the following *production-transportation-inventory problem*. A firm produces a single product at two locations in each of two time periods. The unit costs and production limits vary by time period and are given by the following diagram.

PRODUCTION LOCATION	TIME PERIOD 1	TIME PERIOD 2	
1	\$25/6	\$35/2	Unit cost/
2	\$30/10	\$42/9	Production limit

The product will be shipped (instantaneously) to each of two locations to satisfy specified demands over the two periods. These demands are as follows.

CONSUMER LOCATION	TIME PERIOD 1	TIME PERIOD 2
1	3	1
2	5	4

The unit shipping cost varies over time and is given by the following.

PRODUCTION LOCATION	PERIOD 1 CONSUMER LOCATION 1	PERIOD 1 CONSUMER LOCATION 2	PRODUCTION LOCATION	PERIOD 2 CONSUMER LOCATION 1	PERIOD 2 CONSUMER LOCATION 2
1	\$50	\$60	1	\$60	\$80
2	\$40	\$70	2	\$70	\$90

Finally, the product may be held in inventory at the production and consumer locations to satisfy later period needs. The relevant data are given.

PRODUCTION LOCATION 1	PRODUCTION LOCATION 2	CONSUMER LOCATION 1	CONSUMER LOCATION 2	
\$1/2	\$2/3	\$3/1	\$4/3	Unit cost/ Inventory limit

Set up a network flow problem that can be used to solve the problem of minimizing the total cost to satisfy the demand over the two periods. (*Hint.* Create a separate node that represents each location at each time period. Shipping arcs connect nodes for the same time period; inventory arcs oonnect nodes in different time periods.)

9.44 Show that the coefficient matrix for the lower-upper bounded network flow problem is totally unimodular. This matrix is of the form

x	s_1	s_2	RHS
A	**0**	**0**	**b**
I	**−I**	**0**	**l**
I	**0**	**I**	**u**

9.45 a. State the dual problem for the lower-upper bounded network flow problem in Section 9.8.
b. Give the values of all of the optimal dual variables for Figure 9.17 at iteration 2.
c. Verify optimality by computing the dual objective function.

9.46 Consider the following network.

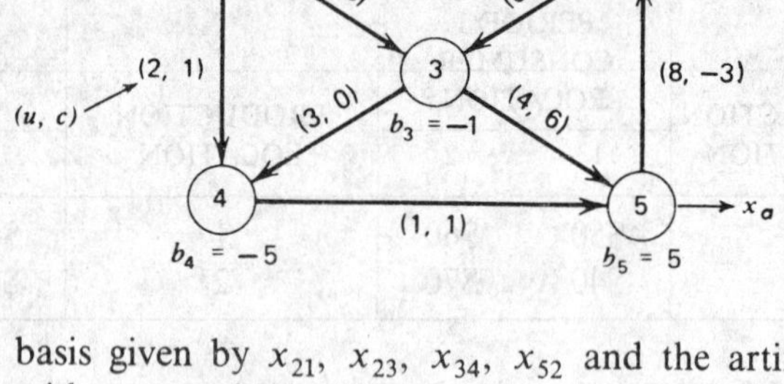

Consider a basis given by x_{21}, x_{23}, x_{34}, x_{52} and the artificial (root arc) variable x_a with x_{14} nonbasic at its upper bound and all other variables nonbasic at their lower bounds, that is, zero.

a. Find the basic solution. Is it feasible?
b. Give the basis in lower triangular form.
c. Give the simplex tableau associated with the specified basic solution.
d. State the dual program.
e. Find the complementary dual solution. Is it feasible?
f. Regardless of costs, perform one pivot graphically to bring x_{14} into the basis.
g. Is the new basis optimal?

9.47 If it is known in advance that a certain arc will be carrying positive flow in any optimal solution to an uncapacitated network flow problem, what simplifications can be made in the solution method?

9.48 Twenty million barrels of oil must be transported from Dhahran in Saudi Arabia to the ports of Rotterdam, Marseilles, and Naples in Europe. The demands of these ports are respectively 4, 12, and 4 million barrels. The following three alternative routes are possible (see accompanying map).

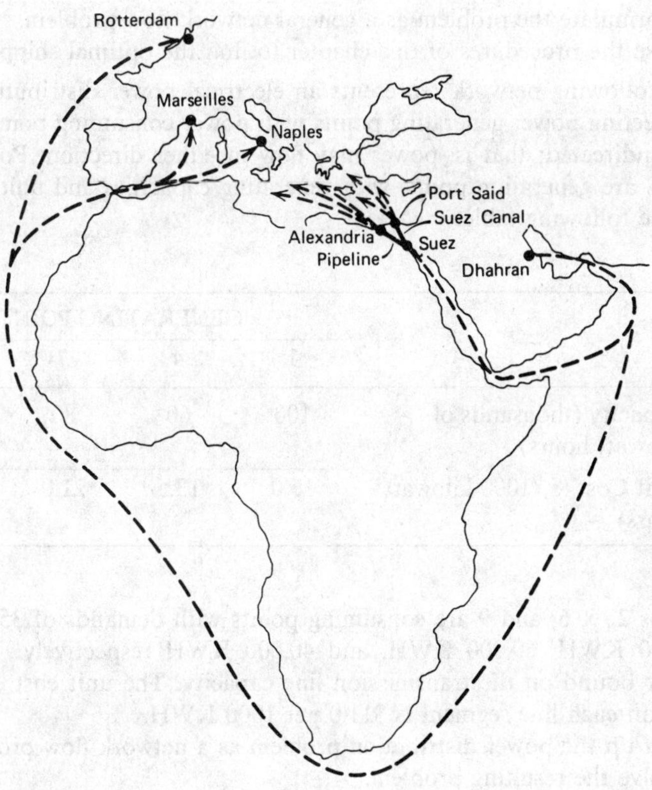

a. From Dhahran, around Africa to Rotterdam, Marseilles, and Naples. The average transportation and handling cost per barrel is \$1.20, \$1.40, and \$1.40 respectively.

b. From Dhahran to the city of Suez, and then through the Suez Canal to Port Said. From Port Said the oil is shipped to Rotterdam, Marseilles, and Naples. The average transportation and handling cost from Dhahran to the city of Suez is \$0.30, and the additional unit cost of transporting through the canal is \$0.20. Finally, the unit transportation costs from Port Said to Rotterdam, Marseilles, and Naples are respectively \$0.25, \$0.20, and \$0.15.

c. From Dhahran to the city of Suez, and then through the proposed pipeline system from Suez to Alexandria. The average transportation cost per barrel through the pipeline system is $0.15, and the unit transportation costs from Alexandria to Rotterdam, Marseilles, and Naples are $0.22, $0.20, and $0.15.

Furthermore, 30 percent of the oil in Dhahran is transported by large tankers that cannot pass through the Suez Canal. Also, the pipeline system from Suez to Alexandria has a capacity of 10 million barrels of oil.

a. Formulate the problem as a general network flow problem.
b. Use the procedures of this chapter to find the optimal shipping pattern.

9.49 The following network represents an electrical power distribution network connecting power generating points with power consuming points. The arcs are undirected; that is, power may flow in either direction. Points 1, 4, 7, and 8 are generation points with generating capacities and unit costs given by the following table.

	GENERATING POINT			
	1	4	7	8
Capacity (thousands of kilowatt hours)	100	60	80	150
Unit Cost ($/1000 kilowatt hours)	15.0	13.5	21.0	23.5

Points 2, 5, 6, and 9 are consuming points with demands of 35,000 KWH, 50,000 KWH, 60,000 KWH, and 40,000 KWH respectively. There is no upper bound on the transmission line capacity. The unit cost of transmission on each line segment is $11.0 per 1000 KWH.

a. Set up the power distribution problem as a network flow problem.
b. Solve the resulting problem.

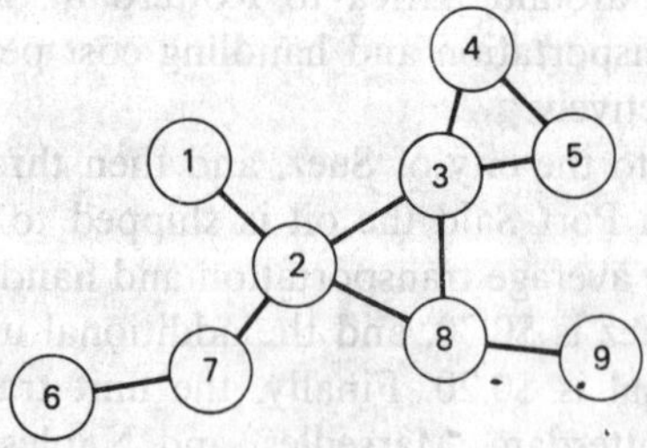

9.50 Formulate Exercise 1.17 as a network flow problem. Solve this problem by the network simplex method.

9.51 Using the definition given in Section 9.11 of a strongly feasible basis tree for bounded variables network flow problems, show how the rule given there maintains strong feasibility and prevents cycling. (*Hint.* See Exercise 9.35!)

9.52 Consider a minimum cost network flow problem in which the nodes are numbered $1, 2, \ldots, m$ and the arcs are stored in the same order as those having node 1 as their from-node, then those having node 2 as their from-node, and so on. Suppose that the following rule for selecting an entering variable is adopted. If an arc having node j as its from-node was last selected, pick the next arc as the best enterable arc having the earliest node in the list $j+1, \ldots, m, 1, 2, \ldots, j$ as its from-node. Using this rule in conjunction with strongly feasible trees, show that the maximum number of consecutive degenerate pivots is bounded above by mL, where L is the largest number of degenerate arcs in any chain from a node to the root node in the final tree T_M of the degenerate sequence. (*Hint.* Follow the original stalling prevention proof of Section 9.11. Note that it is no longer necessarily true that the stage length is less than or equal to m in this case.)

9.53 Consider a directed graph $G(\mathcal{N}, \mathcal{A})$ and suppose that a two-commodity minimum cost network flow problem is defined on this digraph, where $\mathbf{x}$ and $\mathbf{y}$ denote the flow variables for the two respective commodities.

$$\begin{aligned}
\text{Minimize} \quad & \mathbf{cx} + \mathbf{dy} \\
\text{Subject to} \quad & \sum_{j:\,(i,j)\in\mathcal{A}} x_{ij} - \sum_{k:\,(k,i)\in\mathcal{A}} x_{ki} = b_{1i} \qquad \text{for } i = 1, \ldots, m \\
& \sum_{j:\,(i,j)\in\mathcal{A}} y_{ij} - \sum_{k:\,(k,i)\in\mathcal{A}} y_{ki} = b_{2i} \qquad \text{for } i = 1, \ldots, m \\
& \theta_1 \mathbf{x} \le \mathbf{y} \le \theta_2 \mathbf{x} \\
& \mathbf{x}, \mathbf{y} \ge \mathbf{0}
\end{aligned}$$

where $0 < \theta_1 < \theta_2$ are given constants.

a. Add nonnegative slack variable vectors $\mathbf{s}_1$ and $\mathbf{s}_2$ to write $\theta_1\mathbf{x} \le \mathbf{y} \le \theta_2\mathbf{x}$ as $\theta_1\mathbf{x} - \mathbf{y} + \mathbf{s}_1 = 0$ and $\mathbf{y} - \theta_2\mathbf{x} + \mathbf{s}_2 = 0$. Show that the nonnegativity constraints on $\mathbf{x}$ and $\mathbf{y}$ are implied by the other constraints.

b. Treating $\mathbf{x}$ and $\mathbf{y}$ as *unrestricted* above, solve for them in terms of the slack variables and eliminate them from the problem. Demonstrate that the resulting problem is a network flow problem separable in the variables $\mathbf{s}_1$ and $\mathbf{s}_2$.

c. Comment on the algorithmic implications of this strategy.

NOTES AND REFERENCES

1. The simplex method for general network flow problems is a natural extension of the work on the transportation problem and the work of Koopmans [1949] relating linear programming bases for transportation problems and trees in a graph.
2. Computational experience with the network simplex algorithm, as reported by Glover, Karney, and Klingman [1974], Langley and Kennington [1973], Ali *et al.* [1978], and others, indicates that this algorithm compares favorably with

several other procedures for solving network flow problems. However, recent work by Bertsekas [1985] and Bertsekas and Tseng [1988] has resulted in a primal-dual type of algorithm (see Chapter 11) that appears to dominate the primal simplex algorithm. (The codes RELAX-II and RELAXT-II based on this algorithm solve the standard NETGEN benchmark test problems (Klingman *et al.* [1974]) two–four times faster than the best-known primal simplex code RNET written at Rutger's University.

3. Zadeh [1973] has exhibited the worst-case exponential behavior of the primal simplex and the primal-dual network algorithm among others, using a class of modified transportation problems. Edmonds and Karp [1972] were the first to propose a polynomial-time algorithm for network flow problems. Their procedure is based on a scaling technique that solves a sequence of network flow problems with modified (scaled) right-hand-side values, where the solution to each problem is obtained from the previous one via at most m shortest path problems, and where the first problem is a shortest path problem. Tardos [1985] and Orlin [1984] showed how this algorithm can be modified to derive strongly polynomial dual simplex types of algorithms for minimum cost network flow programming problems.
4. The discussion of list structures in Section 9.10 is based on the enhancements to the augmented threaded index (ATI) method of Glover *et al.* [1974a] as discussed in Barr *et al.* [1979]. (See Barr *et al.* [1979] for a discussion on rerooting T_v at q based on reverse threads as an alternative to the technique presented here.) See also Glover and Klingman [1982] for computational comparison studies.
5. The strongly feasible bases of Section 9.11 were introduced by Cunningham [1976]. Its equivalence with lexicographic positivity was exhibited by Orlin [1985]. The instance of cycling in Exercise 9.39 is from Cunningham [1979]. Cunningham [1979] also exhibits the occurrence of stalling and discusses its prevention. Orlin [1985] also has shown that using Dantzig's entering criterion along with the lexicographic cycling prevention rule, the maximum number of consecutive degenerate (primal simplex) pivots for the minimum cost network flow problem is bounded by $O(m^2 n \log m)$.
6. A general technique for converting linear programs to network flow problems wherever possible via elementary row operations and variable-scaling (finding *hidden networks*) is presented in Bixby and Cunningham [1980]. See also Klingman [1977]. Cunningham [1983] shows an instance (see Exercise 9.39) in which the elimination of unrestricted variables can help in such a conversion process. In cases where a complete conversion is not possible, methods for exploiting network substructures are discussed in Glover *et al.* [1978], Glover and Klingman [1981], and Frendewey and Glover [1981]. Also see Patty [1987] for a survey.
7. Exercise 9.37 discusses the extension of the models of this chapter to the *flow with gains* or the *generalized network* models. Jewell [1962] first solved this problem by a primal-dual method. Ellis Johnson [1965], Langley [1973], and others have since treated the problem via the simplex method. For further reading on this subject, see McBride [1981], Elam *et al.* [1979], Glover *et al.* [1974b], Kennington and Helgason [1980], and Bertsekas and Tseng [1985].

TEN: THE TRANSPORTATION AND ASSIGNMENT PROBLEMS

Transportation and assignment problems are a pair of important network structured linear programming problems that arise in several contexts and that have deservedly received a great deal of attention in the literature. Although these problems are special cases of network flow problems, we have seen in the previous chapter that any finite-valued, capacitated or uncapacitated, minimum cost network flow problem can be transformed into an equivalent (uncapacitated) transportation problem (see Exercise 9.5). In fact, although the assignment problem is itself a special case of the transportation problem, any transportation problem and hence any (capacitated) minimum cost network flow problem can be equivalently transformed into an assignment problem, as shown in Section 10.7. However, this is computationally inadvisable. Nonetheless this feature is one reason for the interest generated by this pair of problems.

Algorithmically, the uncapacitated or the capacitated transportation problem may be solved using the techniques of the previous chapter. Some specialization in implementation is possible, but for the most part, the basic technique remains the same. For the sake of exposition, we will present some of these specializations in the context of the *transportation tableau*, which is customarily used to pedagogically present the network simplex algorithm for transportation problems. Although we will rely on the previous chapter for justifications and proofs, and we will exhibit the relationships of the method presented here with the general techniques of the foregoing chapter, the discussion in the present chapter is more or less self-contained.

On the other hand, because of its very special structure, a host of specialized algorithms have been proposed for the assignment problem. The well-known and instructive Hungarian algorithm, which led to the development of a general primal-dual algorithm for linear programming problems, is presented here. We also comment on a specialization of the strongly feasible network simplex algorithm for assignment problems, known as the alternating basis algorithm, which yields a competitive solution procedure. Another algorithm that runs up to seven times faster than the latter algorithm is based on a successive shortest-path procedure. We present this method here, but postpone the actual solution of the shortest-path subproblems to Chapter 12.

10.1 DEFINITION OF THE TRANSPORTATION PROBLEM

Consider m origin points, where origin i has a supply of s_i units of a particular item (commodity). In addition, there are n destination points, where destination j requires d_j units of the commodity. We assume that $s_i, d_j > 0$. Associated with each link (i, j), from origin i to destination j, there is a unit cost c_{ij} for transportation. The problem is to determine a feasible "shipping pattern" from origins to destinations that minimizes the total transportation cost. This problem is known as the *Hitchcock* or the *transportation* problem.

Let x_{ij} be the number of units shipped along link (i, j) from origin i to destination j. Further assume that the problem is *balanced*, that is, the total supply equals the total demand. Hence

$$\sum_{i=1}^{m} s_i = \sum_{j=1}^{n} d_j$$

If the total supply exceeds the total demand, then a dummy destination can be created with demand $d_{n+1} = \Sigma_i s_i - \Sigma_j d_j$, and $c_{i,\,n+1} = 0$ for $i = 1, \ldots, m$. Assuming that the total supply equals the total demand, the linear programming model for the transportation problem becomes as follows.

Minimize

$$c_{11}x_{11} + \cdots + c_{1n}x_{1n} + c_{21}x_{21} + \cdots + c_{2n}x_{2n} + \cdots + c_{m1}x_{m1} + \cdots + c_{mn}x_{mn}$$

Subject to

$$\begin{array}{lllllll}
x_{11} + \cdots + x_{1n} & & & & & & = s_1 \\
 & x_{21} + \cdots & + x_{2n} & & & & = s_2 \\
 & & & \ddots & & & \vdots \\
 & & & & x_{m1} + \cdots & + x_{mn} & = s_m \\
x_{11} & + \; x_{21} & & \cdots & + x_{m1} & & = d_1 \\
\quad \ddots & \quad \ddots & & & \quad \ddots & & \vdots \\
\quad x_{1n} & & + x_{2n} & & \vdots & + x_{mn} & = d_n \\
x_{11}, \ldots, & x_{1n}, \; x_{21}, \ldots, & x_{2n}, \ldots, & & x_{m1}, \ldots, & x_{mn} & \geq 0
\end{array}$$

The transportation problem is graphically illustrated in Figure 10.1.

The underlying graph, which is comprised of the origin and destination nodes O_i, $i = 1, \ldots, m$ and D_j, $j = 1, \ldots, n$ respectively and the connecting links or arcs, is said to be *bipartite*. That is, the nodes are partitioned into two sets such that all the arcs in the network are directed from a node in the first set to a node in the second set. It is *complete* in the sense that all such possible arcs are present. Here we assume that if any connection between an origin–destina-

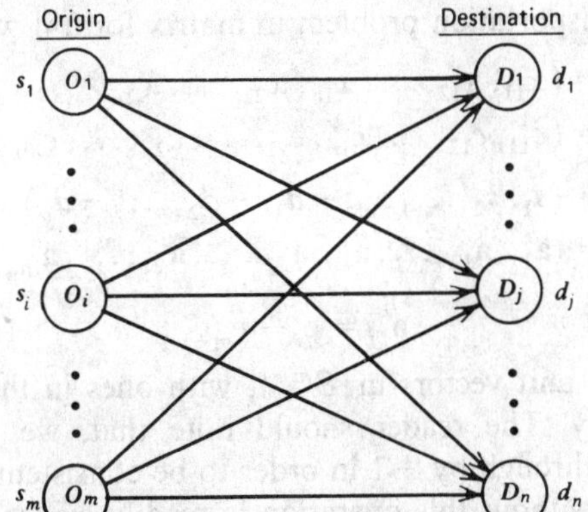

Figure 10.1. Illustration of the graph of a transportation problem.

tion pair (i, j) is prohibited, then the corresponding cost coefficient c_{ij} is large enough so that the variable x_{ij} is essentially an artificial variable in the problem. The problem also may be represented by a *transportation tableau* in which the rows $1, \ldots, m$ represent the origin nodes, the columns $j = 1, \ldots, n$ represent the destination nodes, and the *cell* in row i and column j represents the flow variable x_{ij}. The corresponding cost coefficient c_{ij} is often displayed as shown in cell (i, j).

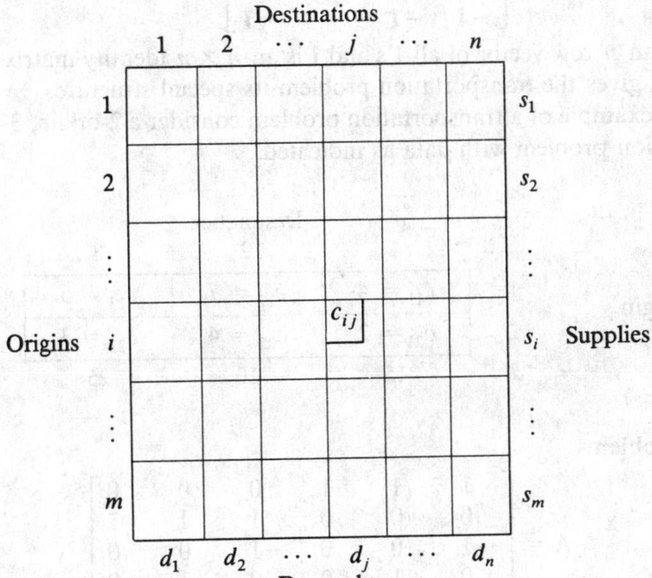

We can cast the transportation problem in matrix form if we let

$$\begin{aligned}
\mathbf{x} &= (x_{11}, x_{12}, \ldots, x_{1n}, x_{21}, \ldots, x_{2n}, \ldots, x_{mn})^t \\
\mathbf{c} &= (c_{11}, c_{12}, \ldots, c_{1n}, c_{21}, \ldots, c_{2n}, \ldots, c_{mn}) \\
\mathbf{b} &= (s_1, s_2, \ldots, s_m, -d_1, -d_2, \ldots, -d_n)^t \\
\mathbf{A} &= (\mathbf{a}_{11}, \mathbf{a}_{12}, \ldots, \mathbf{a}_{1n}, \mathbf{a}_{21}, \ldots, \mathbf{a}_{2n}, \ldots, \mathbf{a}_{mn})
\end{aligned}$$

where

$$\mathbf{a}_{ij} = \mathbf{e}_i - \mathbf{e}_{m+j}$$

and $\mathbf{e}_i$ and $\mathbf{e}_{m+j}$ are unit vectors in E^{m+n}, with ones in the ith and $(m+j)$th positions respectively. The reader should note that, we have multiplied the demand constraints through by -1 in order to be consistent with Chapter 9. An identical derivation without this operation is readily evident. With these definitions the problem takes the following form.

$$\begin{aligned}
\text{Minimize} \quad & \mathbf{cx} \\
\text{Subject to} \quad & \mathbf{Ax} = \mathbf{b} \\
& \mathbf{x} \geq \mathbf{0}
\end{aligned}$$

The **A** matrix, which is the node-arc incidence matrix defined in Chapter 9 and which is of dimension $(m+n) \times mn$ has the following special form.

$$\mathbf{A} = \overbrace{\begin{bmatrix}
\mathbf{1} & \mathbf{0} & \cdots & \mathbf{0} \\
\mathbf{0} & \mathbf{1} & \cdots & \mathbf{0} \\
\vdots & \vdots & & \vdots \\
\mathbf{0} & \mathbf{0} & \cdots & \mathbf{1} \\
-\mathbf{I} & -\mathbf{I} & \cdots & -\mathbf{I}
\end{bmatrix}}^{mn \text{ columns}} m+n \text{ rows}$$

where **1** is an n row vector of all 1's and **I** is an $n \times n$ identity matrix. It is the **A** matrix that gives the transportation problem its special structure.

As an example of a transportation problem consider a 2-origin, 3-destination transportation problem with data as indicated.

		Destination 1	Destination 2	Destination 3	s_i
Origin	1	$c_{11} = 4$	$c_{12} = 7$	$c_{13} = 5$	30
	2	$c_{21} = 2$	$c_{22} = 4$	$c_{23} = 3$	20
	d_j	15	10	25	

For this problem

$$\mathbf{A} = \begin{bmatrix}
1 & 1 & 1 & 0 & 0 & 0 \\
0 & 0 & 0 & 1 & 1 & 1 \\
-1 & 0 & 0 & -1 & 0 & 0 \\
0 & -1 & 0 & 0 & -1 & 0 \\
0 & 0 & -1 & 0 & 0 & -1
\end{bmatrix}$$

Feasibility of the Transportation Problem

Under the assumption that supply equals demand, the transportation problem always has a feasible solution. For example, it is easy to show that

$$x_{ij} = \frac{s_i d_j}{d} \qquad i = 1, \ldots, m, \; j = 1, \ldots, n$$

where $d = \Sigma_i s_i = \Sigma_j d_j$, is a feasible solution. Note that for each feasible vector $\mathbf{x}$, every component x_{ij} is bounded as follows:

$$0 \leq x_{ij} \leq \text{Minimum}\{s_i, d_j\}$$

We know that a bounded linear program with a feasible solution has an optimal solution. Thus it is now only a matter of finding the optimal solution.

10.2 PROPERTIES OF THE A MATRIX

We shall examine some of the properties of the **A** matrix that give the transportation problem its special structure. As we shall see, these properties permit a simple and efficient application of the simplex method for transportation problems.

Rank of the A Matrix

Clearly, rank $(\mathbf{A}) \neq m + n$ since the sum of the rows equals zero. In fact, as seen in Chapter 9, the rank of **A** is $m + n - 1$. Hence we are left with two choices for a basis—we can either delete the last row or any row leaving $m + n - 1$ linearly independent constraints for which a basis exists, or we can add an artificial vector for one constraint. When applying the simplex method, we shall select the latter approach and augment **A** with a new artificial or *root arc* variable x_a with column vector $\mathbf{e}_{m+n}$. Henceforth we denote $\mathbf{A}^0 = (\mathbf{A}, \mathbf{e}_{m+n})$.

Total Unimodularity of the $\mathbf{A}^0$ Matrix

The single most important property that the transportation matrix possesses is the total unimodularity property. In Chapter 9, we showed that the $\mathbf{A}^0$ matrix is *totally unimodular*, that is, the determinant of every square submatrix formed from it has value -1, 0, or $+1$. This means that every basis **B** of $\mathbf{A}^0$ and hence its inverse has determinant ± 1, and that the elements of $\mathbf{B}^{-1}$ are all ± 1 or 0 (why?). Using Cramer's rule (see Section 2.2), we conclude that the updated simplex column $\mathbf{y}_{ij}$ of the variable x_{ij} in a canonical representation with respect to the basis **B**, which is given as a solution to the linear system $\mathbf{B}\mathbf{y}_{ij} = \mathbf{a}_{ij}$, is comprised of elements that are ± 1 or 0. This means that any vector $\mathbf{a}_{ij}$ can be obtained by the simple addition and subtraction of basic vectors. It is instructive to see how one can construct such a representation on the transportation tableau.

In particular, in the representation of the nonbasic vector $\mathbf{a}_{ij} = \mathbf{e}_i - \mathbf{e}_{m+j}$ in terms of basic vectors there must be a basic vector of the form $\mathbf{a}_{ik} = \mathbf{e}_i - \mathbf{e}_{m+k}$

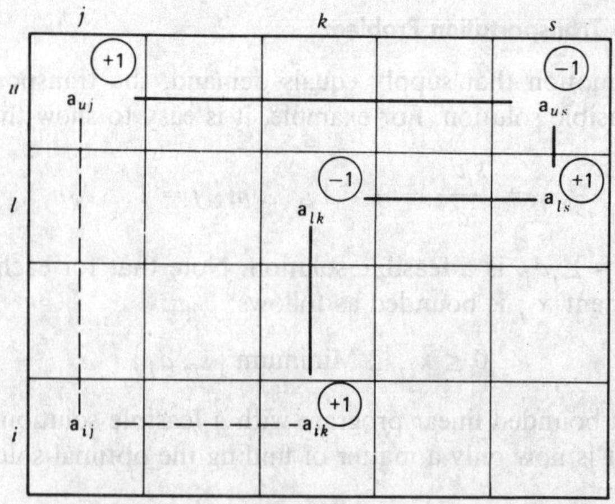

Figure 10.2. **Illustration of the representation of a_{ij} in terms of the basic vectors.**

with a coefficient of +1. Then there must exist a basic vector of the form $\mathbf{a}_{lk} = \mathbf{e}_l - \mathbf{e}_{m+k}$ with a coefficient of −1 in the representation. This process continues until finally there must exist a vector of the form $\mathbf{a}_{uj} = \mathbf{e}_u - \mathbf{e}_{m+j}$ with a coefficient of +1 in the representation. A typical representation of $\mathbf{a}_{ij}$ is

$$\begin{aligned}\mathbf{a}_{ij} &= \mathbf{a}_{ik} - \mathbf{a}_{lk} + \mathbf{a}_{ls} - \mathbf{a}_{us} + \mathbf{a}_{uj} \\ &= (\mathbf{e}_i - \mathbf{e}_{m+k}) - (\mathbf{e}_l - \mathbf{e}_{m+k}) + (\mathbf{e}_l - \mathbf{e}_{m+s}) \\ &\quad - (\mathbf{e}_u - \mathbf{e}_{m+s}) + (\mathbf{e}_u - \mathbf{e}_{m+j}) \\ &= \mathbf{e}_i - \mathbf{e}_{m+j}\end{aligned}$$

Representing the nonbasic vector $\mathbf{a}_{ij}$ in terms of the basic vectors is illustrated on the transportation matrix (tableau) in Figure 10.2. Note that the cell (i, j) together with the cells (i, k), (l, k), (l, s), (u, s), and (u, j) form a *cycle* in the matrix. The cells (i, k), (l, k), (l, s), (u, s), and (u, j) form a *chain* in the matrix between cell (i, k) and cell (u, j). Other basic cells which do not appear in the representation of $\mathbf{a}_{ij}$ are deleted from Figure 10.2. Also, note that the signs of the coefficients alternate throughout the chain.

Characterization of a Basis on a Transportation Tableau

In Chapter 9, we saw that any basis $\mathbf{B}$ of $\mathbf{A}^0$ corresponds to a *rooted spanning tree* and vice versa. In the sense of the transportation tableau, the interpretation of this fact is as follows. The basis is *rooted*, that is, it (necessarily) contains the root arc x_a. The remaining $(m + n - 1)$ basic variables correspond to some cell in the transportation tableau. The basis must be *spanning*, that is, there must be at least one basic cell in each row and in each column of the transportation tableau, or

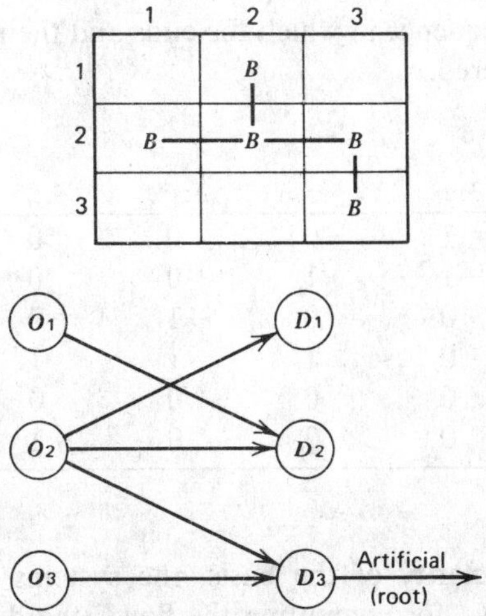

Figure 10.3. Illustration of a basis on the transportation tableau and the transportation graph.

else a row of **B** would have all zero elements (why?), which would disqualify it as a basis. Finally the basis must be a *tree*, that is, the $(m + n - 1)$ basic cells should not contain a cycle, or else we can write one column of **B** in terms of the others.

Figure 10.3 shows a basis on the transportation tableau as well as its graph representation. The basic cells are identified with **B**'s on the transportation tableau and the corresponding arcs are depicted on the graph. Note that there is a unique chain connecting every pair of basic cells shown by the lines in Figure 10.3, that is, there is a unique way of getting from any row or column of the tableau to another row or column while stepping along the basic cells. This corresponds to a unique chain comprised of basic arcs connecting any pair of nodes in the basis graph. For example, to get from origin 1 to destination 3, we step along the basic cells or arcs (1, 2), (2, 2), and (2, 3). This *connectedness* property is a consequence of the fact that the basis has $(m + n - 1)$ basic arcs or cells and has no cycles.

The basis is also lower triangular (after a suitable permutation of its rows and columns). Because the basis graph is a tree with at least two nodes, it has at least two *ends*—at least one of which is not the root node, that is, not a node at which the root arc is incident. Since an *end node* is one at which only one arc is incident, this corresponds to a row or column of the transportation tableau that has only one basic cell. In Figure 10.3 the origin nodes 1 and 3 and the destination node 1 are the ends of the basis tree. Let us now select the end O_1, say, and permute the rows and columns of **B** so that its first row and column correspond to O_1 and x_{12} respectively. Deleting O_1 and its incident arc in the graph, and hence deleting row 1 along with the basic cell (1, 2) in the transportation tableau, the remaining graph or tableau represents another tree with one less node. Repeating this process, we obtain **B** in lower triangular form. The rows and columns identify the sequence in which the ends and their corresponding incident arcs have been considered.

$$
\mathbf{B} = \begin{array}{c} O_1 \\ D_2 \\ D_1 \\ O_2 \\ O_3 \\ D_3 \end{array}
\overset{\begin{array}{cccccc} x_{12} & x_{22} & x_{21} & x_{23} & x_{33} & x_a \end{array}}{\begin{bmatrix} 1 & 0 & 0 & 0 & 0 & 0 \\ -1 & -1 & 0 & 0 & 0 & 0 \\ 0 & 0 & -1 & 0 & 0 & 0 \\ 0 & 1 & 1 & 1 & 0 & 0 \\ 0 & 0 & 0 & 0 & 1 & 0 \\ 0 & 0 & 0 & -1 & -1 & 1 \end{bmatrix}}
$$

Because of the triangularity of the basis, the systems of equations $\mathbf{B}\mathbf{x}_B = \mathbf{b}$, $\mathbf{w}\mathbf{B} = \mathbf{c}_B$, and $\mathbf{B}\mathbf{y}_{ij} = \mathbf{a}_{ij}$ for computing the flows, the dual variable values, and the representation of a nonbasic variable column in terms of the basic variable columns respectively can be solved very simply and efficiently. The solution of these systems of equations on the transportation tableau, along with the accompanying simplex operations are addressed next.

10.3 REPRESENTATION OF A NONBASIC VECTOR IN TERMS OF THE BASIC VECTORS

We have determined that each nonbasic cell together with a subset of the basic cells forms a cycle and that the basic cells in this cycle provide the required representation for the nonbasic cell. Then we found that the set of basic cells formed a spanning tree on the transportation matrix. We further know that there is a unique chain between every pair of cells in the tree; otherwise cycles would be created. All of this suggests that to find the representation of a given nonbasic cell (i, j) we use the chain in the basic tree between some basic cell in row i and some basic cell in column j. This is essentially accurate except that not all basic cells in this chain (in the transportation matrix) are in the representation.

To produce the proper representation for a given nonbasic cell (variable) we simply locate the unique cycle, in the basis graph, containing the arc associated with the particular nonbasic cell. Then all of the basic cells of the transportation matrix associated with the arcs of the cycle in the graph are required for the representation of the nonbasic cell. The process of locating the representation directly on the transportation matrix is essentially the same except that not all basic cells in the unique cycle are used. In this case we use only those cells of the chain for which there is another cell of the chain in the same row and another cell of the chain in the same column, that is, the *corners of the cycle*.

To illustrate, consider Figure 10.4. Suppose that we want to represent $\mathbf{a}_{14}$ in terms of the basic vectors. In Figure 10.5 the unique cycle of the graph is given by (1, 4), (3, 4), (3, 1), (1, 1). Deleting the nonbasic arc (1, 4), we are left with the unique basic chain (3, 4), (3, 1), (1, 1). As we already know, these are assigned

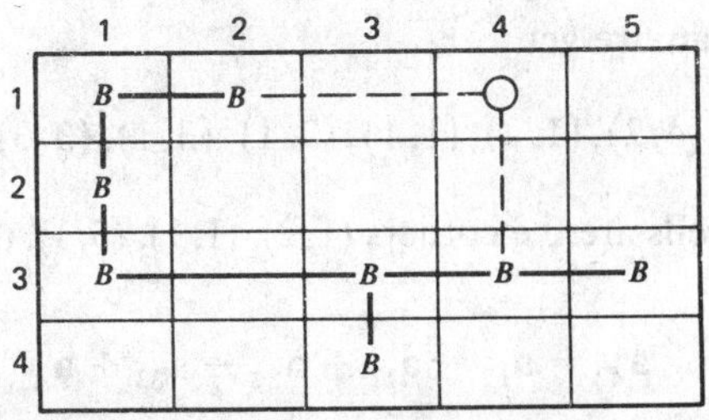

Figure 10.4. **Illustration of finding the representation of a nonbasic cell.**

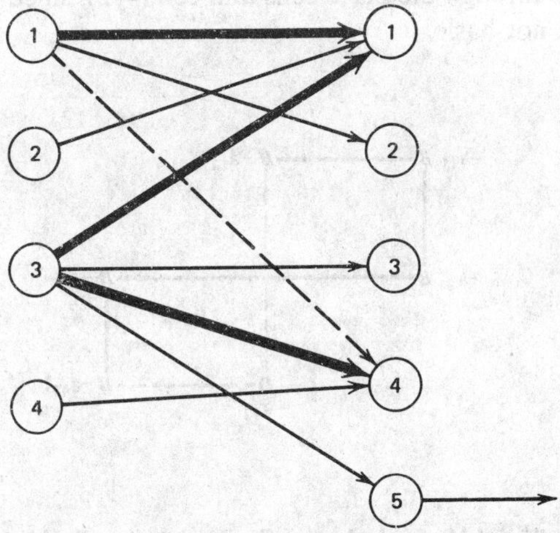

Figure 10.5. **The cycle in the graph associated with the basis of Figure 10.4.**

alternating signs of $+1$ and -1, giving the following representation:

$$\mathbf{a}_{14} = \mathbf{a}_{11} - \mathbf{a}_{31} + \mathbf{a}_{34}$$

which we may verify by

$$\mathbf{e}_1 - \mathbf{e}_{4+4} = (\mathbf{e}_1 - \mathbf{e}_{4+1}) - (\mathbf{e}_3 - \mathbf{e}_{4+1}) + (\mathbf{e}_3 - \mathbf{e}_{4+4})$$

If we had sought the representation from the transportation tableau, we would first obtain the unique cycle (1, 4), (1, 2), (1, 1), (2, 1), (3, 1), (3, 3), (3, 4). Other than the nonbasic cell (1, 4), the corners of the cycle are cells (1, 1), (3, 1), and (3, 4).

The representation is given here.

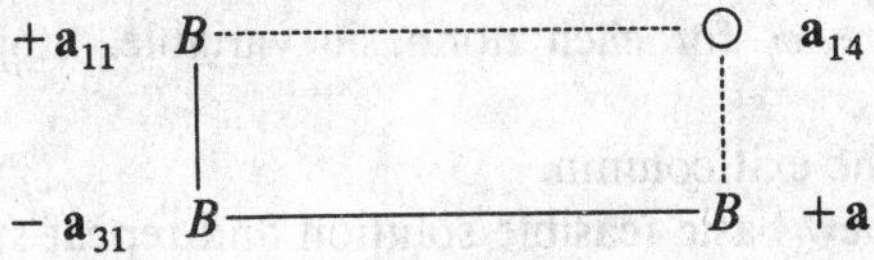

As another example, we shall represent the vector $\mathbf{a}_{42}$ associated with the nonbasic cell (4, 2) in terms of the basic variables. Tracing the cycle C in the transportation tableau, we get

$$C = \{(4,2),(1,2),(1,1),(2,1),(3,1),(3,3),(4,3)\}$$

The required basic cells are the corners (1, 2), (1, 1), (3, 1), (3, 3), and (4, 3). Hence the representation is

$$\mathbf{a}_{42} = \mathbf{a}_{12} - \mathbf{a}_{11} + \mathbf{a}_{31} - \mathbf{a}_{33} + \mathbf{a}_{43}$$

which can easily be verified as correct. Pictorially, the representation is (ignoring other basic points) as follows. Although this appears to be two cycles, it is actually one cycle through the basic cells and cell (4, 2), since the cell (3, 2), where the lines cross, is not basic.

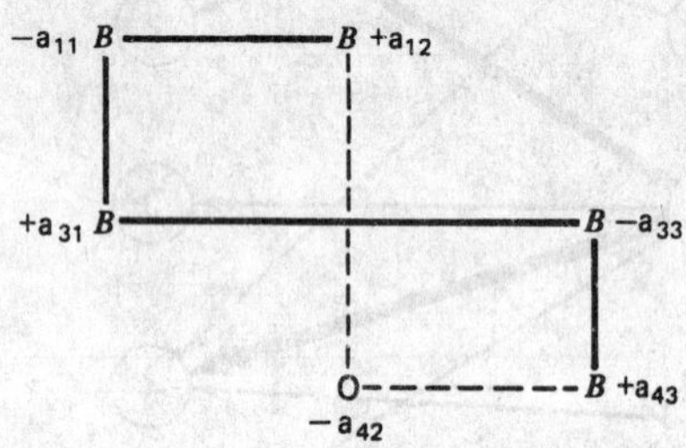

The Role of the Artificial Variable in the Transportation Problem

We note that the representation of a nonbasic vector involves only basic vectors associated with the unique chain through the basic cells. In particular, the artificial vector never becomes involved in any representation, and therefore the artificial variable will always remain zero. This fact will allow us essentially to ignore the artificial variable in the application of the simplex method to transportation problems.

10.4 THE SIMPLEX METHOD FOR TRANSPORTATION PROBLEMS

The general steps in the application of the simplex method to a linear program are as follows.

1. Find a starting basic feasible solution.
2. Compute $z_j - c_j$ for each nonbasic variable. Stop or select the entry column.
3. Determine the exit column.
4. Obtain the new basic feasible solution and repeat step 2.

We shall show how each of these steps can be carried out directly on the transportation tableau.

Finding a Starting Basic Feasible Solution

In Section 10.2 we produced a feasible solution for the transportation problem. However, the solution was not basic. While it would not be a difficult procedure to convert that solution into a basic feasible solution, we shall consider another procedure for obtaining a basic feasible solution. This procedure is called the *northwest corner rule*. During the procedure as a variable x_{ij} is assigned a value, we reduce the corresponding s_i and d_j by that value. Let the reduced values of s_i and d_j be denoted by $\hat{s}_i$ and $\hat{d}_j$ respectively. In particular, to start with, $\hat{s}_i = s_i$ and $\hat{d}_j = d_j$.

Assuming that the total supply equals the total demand, beginning in cell (1, 1) we let

$$x_{11} = \text{Minimum}\left\{\hat{s}_1, \hat{d}_1\right\}$$

and replace $\hat{s}_1$ by $\hat{s}_1 - x_{11}$ and $\hat{d}_1$ by $\hat{d}_1 - x_{11}$. Then, if $\hat{s}_1 > 0$, move to cell (1, 2); let

$$x_{12} = \text{Minimum}\left\{\hat{s}_1, \hat{d}_2\right\}$$

and replace $\hat{s}_1$ by $\hat{s}_1 - x_{12}$ and $\hat{d}_2$ by $\hat{d}_2 - x_{12}$. However, if $\hat{s}_1 = 0$, move to cell (2, 1); let

$$x_{21} = \text{Minimum}\left\{\hat{s}_2, \hat{d}_1\right\}$$

and replace $\hat{s}_2$ by $\hat{s}_2 - x_{21}$ and $\hat{d}_1$ by $\hat{d}_1 - x_{21}$. Note that the case $s_1 = d_1$ produces *degeneracy*, because after computing x_{11} the revised values $\hat{s}_1$ and $\hat{d}_1$ are both zeros. Hence the basic variable $x_{21} = 0$. (Observe that in this case we could have alternatively moved to the cell (1, 2) and made the variable x_{12} basic at value zero.) In the general case, after assigning some variable x_{kl} the value

$$x_{kl} = \text{Minimum}\left\{\hat{s}_k, \hat{d}_l\right\}$$

we replace $\hat{s}_k$ by $\hat{s}_k - x_{kl}$ and $\hat{d}_l$ by $\hat{d}_l - x_{kl}$. If $(k, l) = (m, n)$, we terminate. Otherwise if $\hat{s}_k > 0$, we move to the cell $(k, l + 1)$ which must exist (why?), and compute $x_{k,(l+1)} = \text{Minimum}\{\hat{s}_k, \hat{d}_{l+1}\}$. On the other hand, if $\hat{s}_k = 0$, we move to the cell $(k + 1, l)$ which must exist (why?), and compute $x_{(k+1),l} = \text{Minimum}\{\hat{s}_{k+1}, \hat{d}_l\}$. Note that if $\hat{s}_k = \hat{d}_l$ in the computation of x_{kl}, then in the latter case we will have the revised $\hat{d}_l = 0$ and we would obtain $x_{(k+1),l}$ as a degenerate basic variable. The process of assigning a variable the minimum of the remaining supply or demand, adjusting both, and moving to the right, or down, one cell continues until all supplies and demands are allocated. Figure 10.6 illustrates how the process might work.

The northwest corner rule produces exactly $m + n - 1$ basic, nonnegative x_{ij}'s. Each time an x_{ij} is made basic and assigned some nonnegative value, either a supply or a demand constraint is satisfied. When $m + n - 1$ variables have been assigned values, then $m + n - 1$ of the constraints are satisfied. Noting that one of the constraints of the transportation problem is redundant, then all the constraints are met.

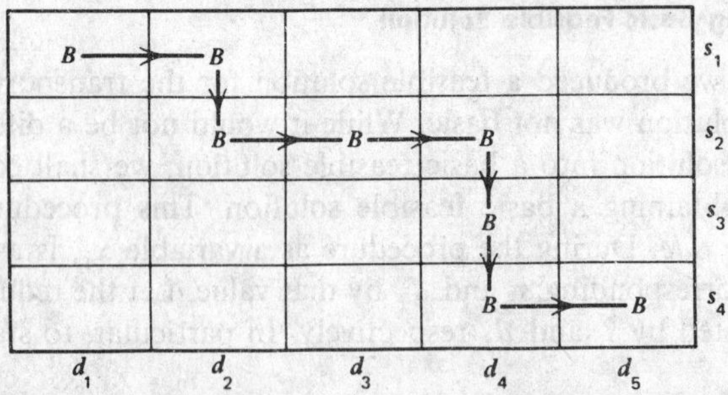

Figure 10.6. Graphical illustration of how the northwest corner rule might allocate values to the variables.

The graphical structure of the basic cells is obviously connected and spanning. To demonstrate that the graph is a spanning tree and therefore a basis, it remains only to show that it contains no cycles. (Actually, this follows from the additional fact that the graph has $m + n - 1$ basic arcs.) Since, at each step, either the row or column index of the basic variable is increased by 1, it is not possible to make a new variable in an earlier row or column basic—the only way to produce cycles. Therefore the northwest corner method produces a basic feasible solution. An alternative procedure, called *Vogel's approximation method*, is described in Exercise 10.29.

To illustrate, consider the transportation tableau of Figure 10.7 where the supplies and demands are indicated. We first let $x_{11} = \text{Minimum}\{\hat{s}_1, \hat{d}_1\} = 15$, decrease $\hat{s}_1$ and $\hat{d}_1$ by $x_{11} = 15$, and move to cell (1, 2) since $\hat{s}_1 > 0$. Next, let $x_{12} = \text{Minimum}\{\hat{s}_1, \hat{d}_2\} = \text{Minimum}\{15, 20\} = 15$, decrease $\hat{s}_1$ and $\hat{d}_2$, and move to cell (2, 2) since $\hat{s}_1 = 0$. This process is continued until all supplies and demands are satisfied. Notice that we do have the required number of basic variables, namely $7 = m + n - 1$. The blank cells are nonbasic and the associated variables have zero values.

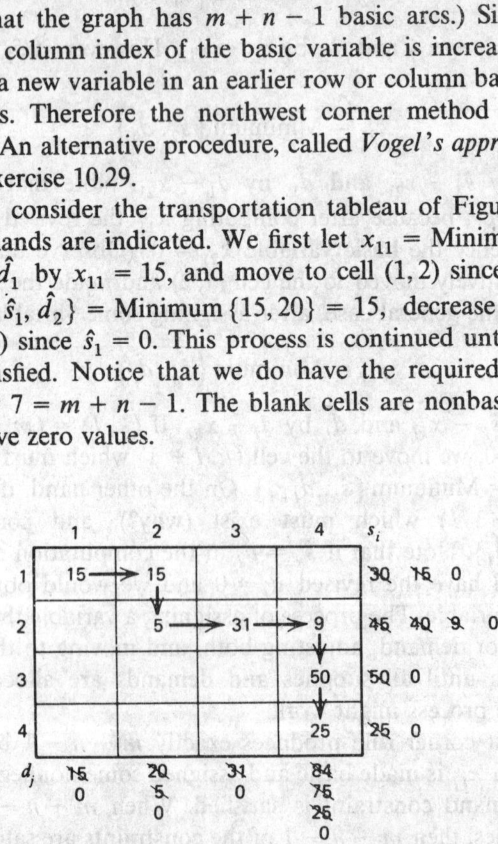

Figure 10.7. Example of the northwest corner rule.

Computing $z_{ij} - c_{ij}$ for Each Nonbasic Cell

Given a basic feasible solution, our next task is to determine whether that solution is optimal or to select an entering variable. Now

$$z_{ij} - c_{ij} = \mathbf{c}_B \mathbf{y}_{ij} - c_{ij}$$

We have shown how to obtain the components of $\mathbf{y}_{ij}$, that is, the coefficients of the representation of $\mathbf{a}_{ij}$ in terms of the basic vectors. Since $\mathbf{y}_{ij}$ consists of 1's, -1's, and 0's, then $\mathbf{c}_B\mathbf{y}_{ij}$ is calculated by simply adding and subtracting the costs of some basic variables. To illustrate, consider Figure 10.2. The $z_{ij} - c_{ij}$ value for the nonbasic variable x_{ij} is given by

$$z_{ij} - c_{ij} = (c_{uj} - c_{us} + c_{ls} - c_{lk} + c_{ik}) - c_{ij}$$

Using the data of the example in Section 10.1 and the basis indicated here, we get the following result.

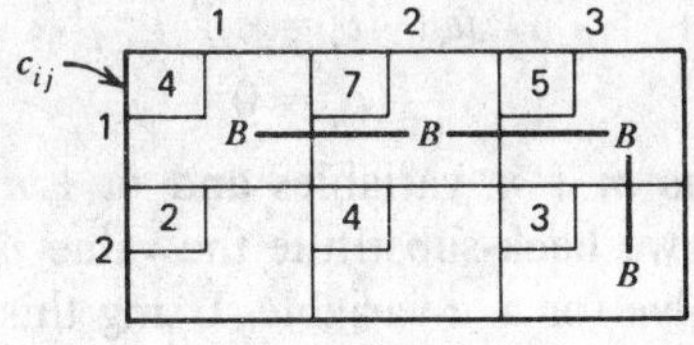

$$z_{21} - c_{21} = 4 - 5 + 3 - 2 = 0$$

$$z_{22} - c_{22} = 7 - 5 + 3 - 4 = 1$$

and x_{22} is a candidate to enter the basis.

The optimality criterion for the transportation problem is given by $z_{ij} - c_{ij} \leq 0$ for each nonbasic x_{ij}. A given cell (k, l) is a candidate to enter the basis if $z_{kl} - c_{kl} > 0$.

The foregoing procedure for calculating $z_{ij} - c_{ij}$ utilizes the form

$$z_{ij} - c_{ij} = \mathbf{c}_B\mathbf{B}^{-1}\mathbf{a}_{ij} - c_{ij} = \mathbf{c}_B\mathbf{y}_{ij} - c_{ij}$$

The vector $\mathbf{y}_{ij}$ is determined by constructing the unique cycle through cell (i, j) and some of the basic cells as discussed earlier. Hence this method is sometimes called the *cycle method*. Note, however, that calculating $z_{ij} - c_{ij}$ can be alternatively performed as follows, which is the preferable method.

$$z_{ij} - c_{ij} = \mathbf{c}_B\mathbf{B}^{-1}\mathbf{a}_{ij} - c_{ij} = \mathbf{w}\mathbf{a}_{ij} - c_{ij}$$

Let w_i, $i = 1, \ldots, m$ be denoted by u_i and w_{m+j}, $j = 1, \ldots, n$ by v_j. Then the dual vector $\mathbf{w}$ is given by

$$\mathbf{w} = (u_1, \ldots, u_m, v_1, \ldots, v_n)$$

Since $\mathbf{a}_{ij}$ has a 1 in the ith and a -1 in the $(m + j)$th position, then $\mathbf{w}\mathbf{a}_{ij} = u_i - v_j$. Hence $z_{ij} - c_{ij} = u_i - v_j - c_{ij}$. This method of calculating $z_{ij} - c_{ij}$ is thus called the *dual variable* or (u_i, v_j) method.

Since the dual vector $\mathbf{w} = \mathbf{c}_B\mathbf{B}^{-1}$, then $\mathbf{w}$ is the solution to the system

$$\mathbf{w}\mathbf{B} = \mathbf{c}_B$$

where

$$\mathbf{B} = (\mathbf{a}_{pq}, \ldots, \mathbf{a}_{st}, \mathbf{e}_{m+n})$$
$$\mathbf{c}_B = (c_{pq}, \ldots, c_{st}, c_a)$$

and $\mathbf{a}_{pq}, \ldots, \mathbf{a}_{st}$ are $m + n - 1$ basic columns, $c_{pq}, \ldots, c_{st}$ are their corresponding cost coefficients, $\mathbf{e}_{m+n}$ is the artificial column, and c_a is its cost coefficient. Since we have previously shown that the value of the artificial variable will never vary from zero, the value of c_a does not matter. For convenience, we shall select $c_a = 0$. Since $\mathbf{B}$ is triangular, we have an easy system to solve. The system

$$(u_1, \ldots, u_m, v_1, \ldots, v_n)(\mathbf{a}_{pq}, \ldots, \mathbf{a}_{st}, \mathbf{e}_{m+n}) = (c_{pq}, \ldots, c_{st}, 0)$$

is equivalent to (since $\mathbf{a}_{ij} = \mathbf{e}_i - \mathbf{e}_{m+j}$):

$$u_p - v_q = c_{pq}$$
$$\vdots$$
$$u_s - v_t = c_{st}$$
$$v_n = 0$$

The foregoing system has $m + n$ variables and $m + n$ equations. Utilizing the concept of triangularity, we back-substitute the value $v_n = 0$ into each equation where v_n appears and solve for a u-variable. Using this newly found u-variable, we back-substitute to find some v, and so on.

As an illustration, consider the example problem of Section 10.1 with the basis as indicated. The last dual variable v_3 receives the value zero from the artificial column.

Use the artificial column $\Rightarrow v_3 = 0$
Use the basic cell (2, 3): $u_2 - v_3 = 3 \Rightarrow u_2 = 3$
Use the basic cell (1, 3): $u_1 - v_3 = 5 \Rightarrow u_1 = 5$
Use the basic cell (1, 2): $u_1 - v_2 = 7 \Rightarrow v_2 = -2$
Use the basic cell (1, 1): $u_1 - v_1 = 4 \Rightarrow v_1 = 1$

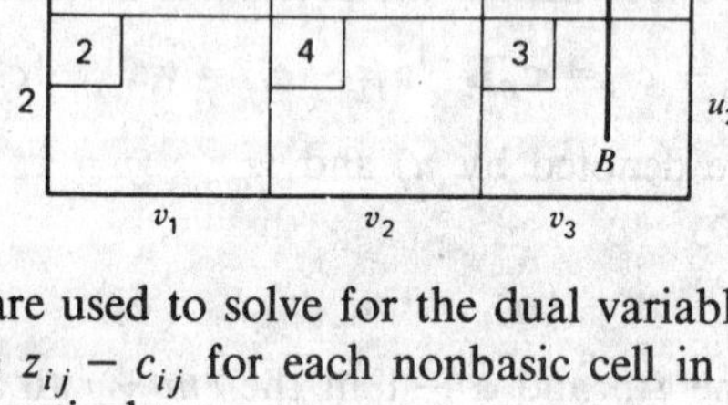

Only the basic cells are used to solve for the dual variables. Given the $\mathbf{w}$ vector, we may compute the $z_{ij} - c_{ij}$ for each nonbasic cell in order to determine the entering column. In particular:

$$z_{21} - c_{21} = u_2 - v_1 - c_{21}$$
$$= 3 - 1 - 2 = 0$$
$$z_{22} - c_{22} = u_2 - v_2 - c_{22}$$
$$= 3 + 2 - 4 = 1$$

and, again, we see that x_{22} is a candidate to enter the basis.

Determination of the Exit Column

Once a column (cell), say, (k, l), has been selected to enter the basis, it is an easy matter to determine the exit column. Recall that the coefficients in the basic representation for that column are the negatives of the rates of change of the corresponding basic variables with a unit increase in the nonbasic (entering) variable. Thus if the entry, in column $\mathbf{y}_{kl}$, corresponding to a basic variable is -1, then the basic variable will increase at the same rate as the nonbasic variable x_{kl} increases. If this entry is $+1$, then the basic variable will decrease at the same rate as the nonbasic variable x_{kl} increases.

Let $\hat{x}_{ij}$ be the value of x_{ij} in the current solution and let Δ be the amount by which the nonbasic variable, x_{kl} increases. Since each component of $\mathbf{y}_{kl}$ is either 1, -1, or 0, then the usual minimum ratio test gives

$$\Delta = \text{Minimum}\left\{\hat{x}_{ij}: \text{basic cell } (i, j) \text{ has a } +1 \text{ in the representation of the nonbasic cell } (k, l)\right\}$$

Given Δ, we proceed to adjust the values of the variables around the cycle by this amount, according to the sign of the coefficient in the representation. For the example of Section 10.1 with the basis indicated here, as x_{22} enters the basis, we get the following result.

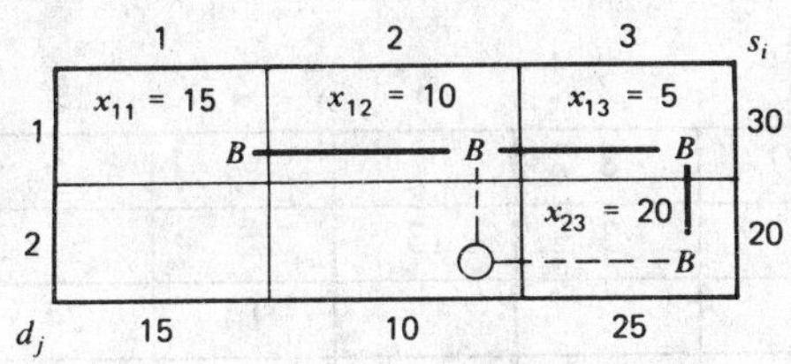

$$\Delta = \text{Minimum}\{x_{12}, x_{23}\} = \text{Minimum}\{10, 20\} = 10$$

The new solution is given by

$$x_{12} = \hat{x}_{12} - \Delta = 10 - 10 = 0 \quad \text{(leaves the basis)}$$

$$x_{13} = \hat{x}_{13} + \Delta = 5 + 10 = 15$$

$$x_{23} = \hat{x}_{23} - \Delta = 20 - 10 = 10$$

$$x_{22} = \Delta = 10$$

$$x_{11} = 15 \quad \text{(unchanged)}$$

The new basis is given by the following.

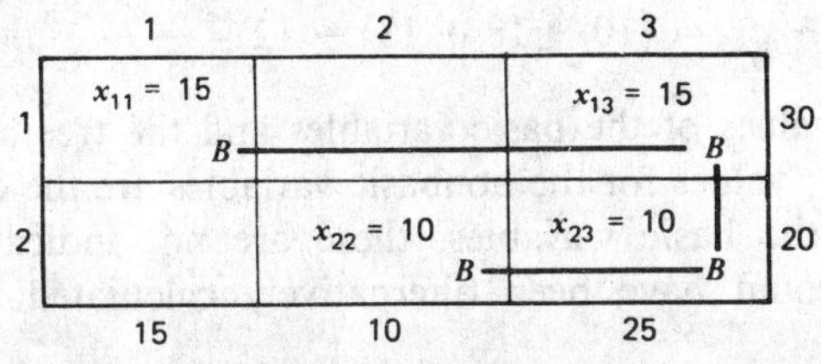

10.5 ILLUSTRATIVE EXAMPLES AND A NOTE ON DEGENERACY

Consider the transportation problem indicated by the data of Figure 10.8. The number in the upper left hand corner of each cell is the cost associated with the particular variable.

The starting basic feasible solution produced by the northwest corner method is illustrated in Figure 10.9.

	1	2	3	4	5	6	s_i
1	10	12	13	8	14	19	18
2	15	18	12	16	19	20	22
3	17	16	13	14	10	18	39
4	19	18	20	21	12	13	14
d_j	10	11	13	20	24	15	

Figure 10.8. Example data for a transportation problem.

	1	2	3	4	5	6
1	10	8				
2		3	13	6		
3				14	24	1
4						14

Figure 10.9. The northwest corner basic solution.

Beginning with (1, 3), we price out each of the nonbasic cells by the cycle method:

$$z_{13} - c_{13} = (12 - 18 + 12) - 13 = -7$$

$$z_{14} - c_{14} = (12 - 18 + 16) - 8 = 2$$

$$z_{15} - c_{15} = (12 - 18 + 16 - 14 + 10) - 14 = -8$$

$$\vdots$$

$$z_{45} - c_{45} = (10 - 18 + 13) - 12 = -7$$

The current values of the basic variables and the tree are shown in Figure 10.10; the $z_{ij} - c_{ij}$ values for the nonbasic variables are the circled values. Since $z_{ij} - c_{ij} = 0$ for the basic variables, these are not indicated. Note that the $z_{ij} - c_{ij}$ values could have been alternatively calculated more efficiently as

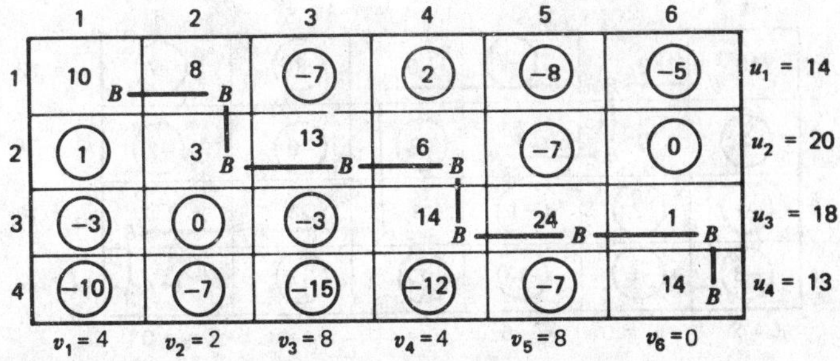

Figure 10.10. The $z_{ij} - c_{ij}$ values for the nonbasic cells.

follows. First solve for the dual variables (their values are summarized in Figure 10.10).

$$\begin{array}{ll}
\text{artificial column} & \Rightarrow\ v_6 = 0 \\
\text{basic cell }(4,6):\ u_4 - v_6 = 13 & \Rightarrow\ u_4 = 13 \\
\text{basic cell }(3,6):\ u_3 - v_6 = 18 & \Rightarrow\ u_3 = 18 \\
\text{basic cell }(3,5):\ u_3 - v_5 = 10 & \Rightarrow\ v_5 = 8 \\
\text{basic cell }(3,4):\ u_3 - v_4 = 14 & \Rightarrow\ v_4 = 4 \\
\text{basic cell }(2,4):\ u_2 - v_4 = 16 & \Rightarrow\ u_2 = 20 \\
\text{basic cell }(2,3):\ u_2 - v_3 = 12 & \Rightarrow\ v_3 = 8 \\
\text{basic cell }(2,2):\ u_2 - v_2 = 18 & \Rightarrow\ v_2 = 2 \\
\text{basic cell }(1,2):\ u_1 - v_2 = 12 & \Rightarrow\ u_1 = 14 \\
\text{basic cell }(1,1):\ u_1 - v_1 = 10 & \Rightarrow\ v_1 = 4
\end{array}$$

Then $z_{ij} - c_{ij} = u_i - v_j - c_{ij}$, for example, $z_{14} - c_{14} = u_1 - v_4 - c_{14} = 14 - 4 - 8 = 2$. Figure 10.10 indicates that the maximal $z_{ij} - c_{ij}$ is $z_{14} - c_{14} = 2$. Therefore x_{14} enters the basis. Referring to Figure 10.10, we see that

$$\mathbf{a}_{14} = \mathbf{a}_{12} - \mathbf{a}_{22} + \mathbf{a}_{24}$$

and the corresponding cycle is as follows.

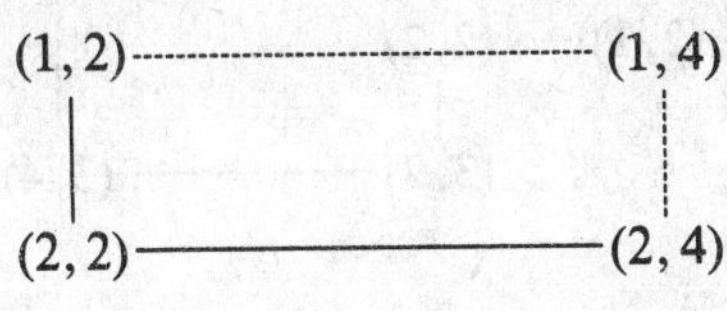

From this we find

$$x_{14} = \text{Minimum}\,\{\hat{x}_{12}, \hat{x}_{24}\} = \text{Minimum}\,\{8,6\} = 6$$

$$x_{12} = 8 - 6 = 2$$

$$x_{22} = 3 + 6 = 9$$

$$x_{24} = 6 - 6 = 0 \qquad \text{(leaves the basis)}$$

The new basis and the values of the basic variables are summarized in Figure 10.11. Using the dual variable method, $z_{ij} - c_{ij}$ is calculated for each nonbasic variable.

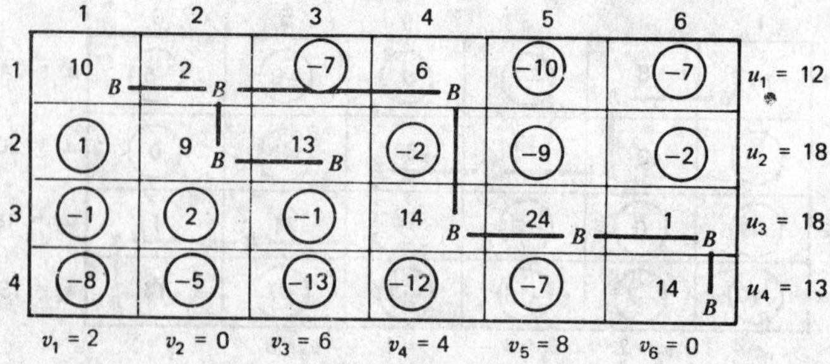

Figure 10.11. Second basic feasible solution.

As indicated by the circled entries in Figure 10.11, cell (3, 2) is the entry cell. The cycle associated with (3, 2) is as follows.

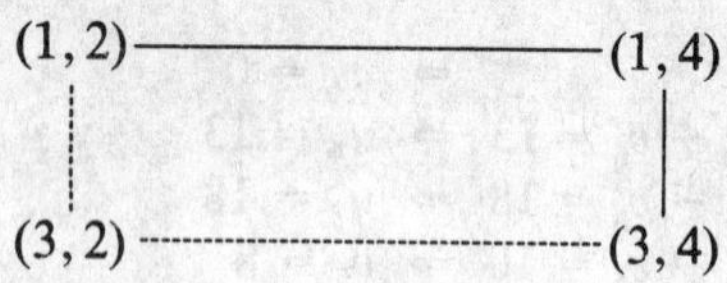

From this we obtain

$$x_{32} = \text{Minimum}\{\hat{x}_{12}, \hat{x}_{34}\} = \text{Minimum}\{2, 14\} = 2$$
$$x_{12} = 2 - 2 = 0 \qquad \text{(leaves the basis)}$$
$$x_{14} = 6 + 2 = 8$$
$$x_{34} = 14 - 2 = 12$$

The new basic feasible solution and the new $z_{ij} - c_{ij}$ values are given in Figure 10.12. Examining the $z_{ij} - c_{ij}$ entries, we find that (2, 1) is the entry cell. The associated cycle is as follows.

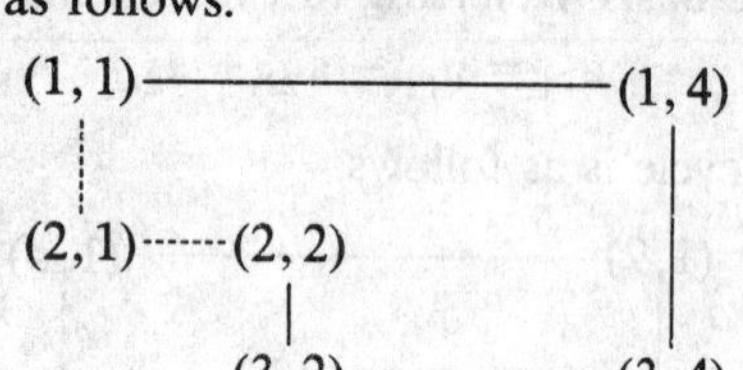

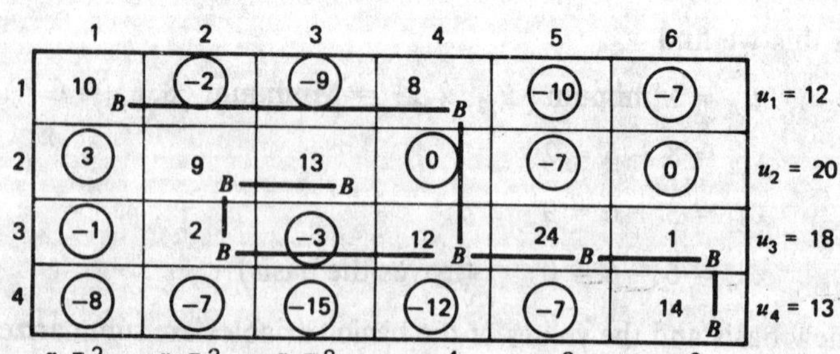

Figure 10.12. Third basic feasible solution.

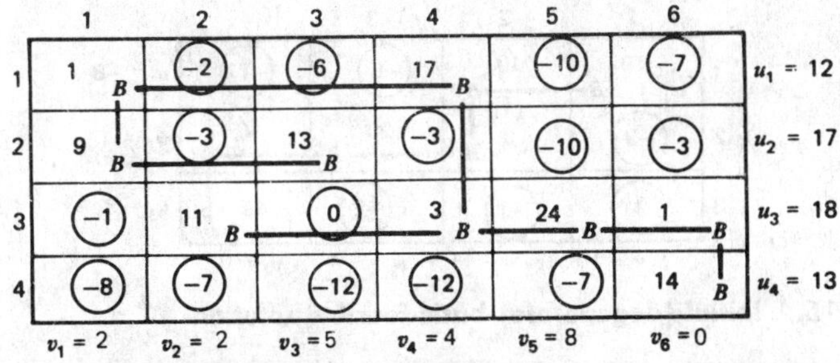

Figure 10.13. **Fourth basic feasible solution.**

From this we obtain

$$x_{21} = \text{Minimum}\,\{\hat{x}_{11}, \hat{x}_{34}, \hat{x}_{22}\} = \text{Minimum}\,\{10, 12, 9\} = 9$$

$$x_{11} = 10 - 9 = 1$$

$$x_{14} = 8 + 9 = 17$$

$$x_{34} = 12 - 9 = 3$$

$$x_{32} = 2 + 9 = 11$$

$$x_{22} = 9 - 9 = 0 \qquad \text{(leaves the basis)}$$

Figure 10.13 presents the new basic feasible solution and the new $z_{ij} - c_{ij}$'s. Since $z_{ij} - c_{ij} \leq 0$ for each nonbasic variable, the indicated solution is optimal.

Example 10.2

Degeneracy

Consider the example of Figure 10.14. Aplying the northwest corner rule to this example we obtain the sequence of calculations:

$$x_{11} = \text{Minimum}\,\{\hat{s}_1, \hat{d}_1\} = \text{Minimum}\,\{20, 10\} = 10$$

$$\hat{s}_1 = 20 - 10 = 10, \ \hat{d}_1 = 10 - 10 = 0$$

$$x_{12} = \text{Minimum}\,\{\hat{s}_1, \hat{d}_2\} = \text{Minimum}\,\{10, 10\} = 10$$

$$\hat{s}_1 = 10 - 10 = 0, \ \hat{d}_2 = 10 - 10 = 0$$

	1	2	3	4	s_i
1	2	3	4	9	20
2	14	12	5	1	30
3	12	15	9	3	40
d_j	10	10	20	50	

Figure 10.14. **An example of degeneracy.**

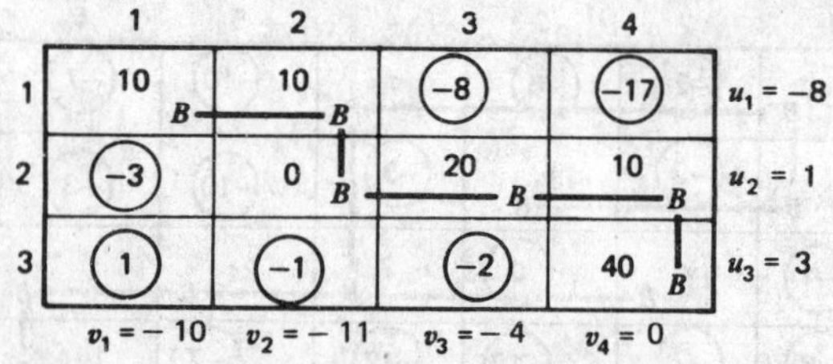

Figure 10.15. **Initial (degenerate) basic feasible solution.**

At this point we may move to either (2, 2), or to (1, 3). Suppose that we move to (2, 2).

$$x_{22} = \text{Minimum}\left\{\hat{s}_2, \hat{d}_2\right\} = \text{Minimum}\{30, 0\} = 0$$
$$\hat{s}_2 = 30 - 0 = 30, \ \hat{d}_2 = 0 - 0 = 0$$
$$x_{23} = \text{Minimum}\left\{\hat{s}_2, \hat{d}_3\right\} = \text{Minimum}\{30, 20\} = 20$$
$$\hat{s}_2 = 30 - 20 = 10, \ \hat{d}_3 = 20 - 20 = 0$$
$$x_{24} = \text{Minimum}\left\{\hat{s}_2, \hat{d}_4\right\} = \text{Minimum}\{10, 50\} = 10$$
$$\hat{s}_2 = 10 - 10 = 0, \ \hat{d}_4 = 50 - 10 = 40$$
$$x_{34} = \text{Minimum}\left\{\hat{s}_3, \hat{d}_4\right\} = \text{Minimum}\{40, 40\} = 40$$
$$\hat{s}_3 = 40 - 40 = 0, \ \hat{d}_4 = 40 - 40 = 0$$

All other x_{ij}'s are nonbasic and are assigned value zero. The initial basic feasible solution is given in Figure 10.15. As required there are $m + n - 1 = 3 + 4 - 1 = 6$ basic variables forming a connected tree. Note, however, that the basic feasible solution is degenerate since the basic variable $x_{22} = 0$. For each nonbasic variable we calculate $z_{ij} - c_{ij}$ by the dual variables method. The values are depicted in circles in Figure 10.15 for the nonbasic variables. Since $z_{31} - c_{31} = 1$, then x_{31} enters the basis. The corresponding cycle is as follows.

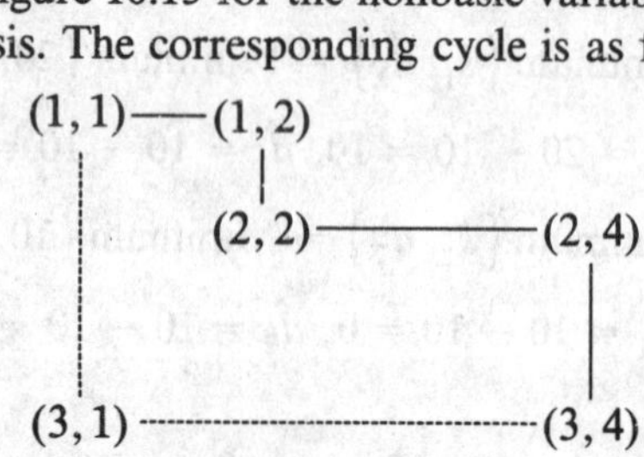

$$x_{31} = \text{Minimum}\{\hat{x}_{11}, \hat{x}_{22}, \hat{x}_{34}\} = \text{Minimum}\{10, 0, 40\} = 0$$
$$x_{11} = 10 - 0 = 10$$
$$x_{21} = 10 + 0 = 10$$
$$x_{22} = 0 - 0 = 0 \quad \text{(leaves the basis)}$$
$$x_{24} = 10 + 0 = 10$$
$$x_{34} = 40 - 0 = 40$$

Note that x_{22} leaves the basis and x_{31} enters the basis at zero level. We have the same extreme point but a different basis. The new basis and the new $z_{ij} - c_{ij}$'s for the nonbasic variables are shown in Figure 10.16. Since $z_{ij} - c_{ij} \leq 0$ for each nonbasic variable, the current solution is optimal.

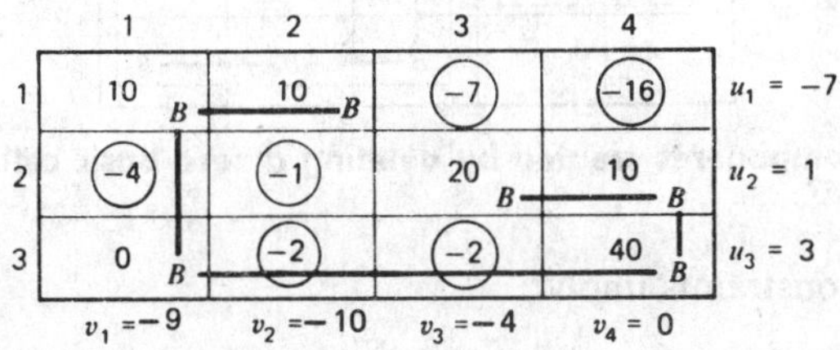

Figure 10.16. Optimal basic feasible solution.

Notice that in this example we have

$$20 = s_1 = d_1 + d_2 = 10 + 10$$

or, in other words, a subset of the supplies equals a subset of the demands. We shall show that this is always true when degeneracy occurs in the transportation problem.

A Necessary Condition for Degeneracy in the Transportation Problem

Suppose that at some iteration of the transportation algorithm we obtain a degenerate basic feasible solution, as that shown in Figure 10.17. Deleting one of the degenerate cells separates the connected tree into several components as shown in Figure 10.18. Sum the supply constraints over the variables in one of the components, say, C_1, to obtain

$$\sum_{C_1} x_{ij} = \sum_{C_1} s_i$$

Summing the demand constraints over the variables in the same component, we get

$$\sum_{C_1} x_{ij} = \sum_{C_1} d_j$$

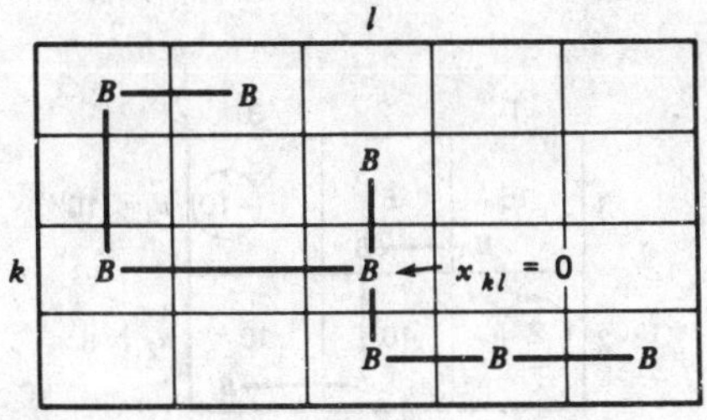

Figure 10.17. A degenerate basis.

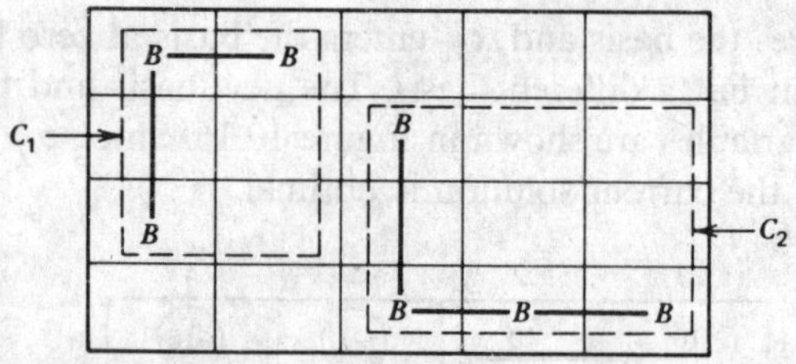

Figure 10.18. **Components created by deleting a zero basic cell.**

Together, these constraints imply

$$\sum_{C_1} s_i = \sum_{C_1} d_j$$

Thus a necessary condition for the presence of degeneracy is that a proper subset of the rows and columns have the total supply equal to the total demand. Therefore there is no degeneracy (and cycling) if no such proper subset exists (see Exercise 10.17).

10.6 THE SIMPLEX TABLEAU ASSOCIATED WITH A TRANSPORTATION TABLEAU

We have all of the information available to construct the simplex tableau associated with a transportation tableau if we so desire. In Section 10.3 a method was described for calculating the updated column vector $\mathbf{y}_{ij}$. In Section 10.4 we saw how to calculate $z_{ij} - c_{ij}$ for a nonbasic variable x_{ij}. This information together with the basic solution provide all the necessary entries in the simplex tableau.

As an example, consider the transportation problem given by the following data.

c_{ij}	1	2	3	s_i
1	10	11	20	20
2	6	9	8	20
d_j	15	15	10	

Figure 10.19 presents the initial transportation tableau and the associated sim-

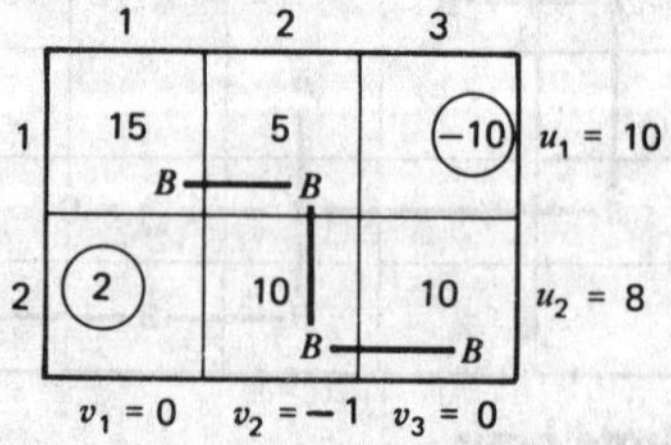

Figure 10.19. **An initial transportation tableau and the associated simplex tableau.**

plex tableau (including the artificial variable). Examining either tableau, we see that x_{21} enters the basis and x_{22} leaves. We ask the reader to verify the entries in the simplex tableau by generating the $z_{ij} - c_{ij}$'s and the $\mathbf{y}_{ij}$'s from the transportation tableau.

	z	x_{11}	x_{12}	x_{13}	x_{21}	x_{22}	x_{23}	x_a	RHS
z	1	0	0	-10	2	0	0	0	375
x_{11}	0	1	0	0	1	0	0	0	15
x_{12}	0	0	1	1	-1	0	0	0	5
x_{22}	0	0	0	-1	1	1	0	0	10
x_{23}	0	0	0	1	0	0	1	0	10
x_a	0	0	0	0	0	0	0	1	0

10.7 THE ASSIGNMENT PROBLEM: (KUHN'S) HUNGARIAN ALGORITHM

An important special case of the transportation problem is the case where $m = n$, each $s_i = 1$, and each $d_j = 1$. This special case is called the *assignment problem*. As an example, suppose we have m individuals and m jobs. If individual i is assigned to job j, the cost incurred will be c_{ij}. We wish to find the minimal cost assignment or a one-to-one *matching* of individuals to jobs. In each basic feasible solution $x_{ij} = 1$ means that individual i is assigned to job j, and $x_{ij} = 0$ indicates that individual i is not assigned to job j.

Since the assignment problem is a special case of the transportation problem, we could apply the transportation procedure developed in this chapter. Note, however, as will be discussed in more detail, that the constraints of the assignment problem admit exactly m positive variables at each basic feasible solution. The number of basic variables is $2m - 1$. Thus if the transportation algorithm is used we would have $m - 1$ basic variables at zero level, leading to a highly degenerate problem. A partial resolution of this difficulty is addressed in Section 10.8. In this section we shall exploit the special structure of the assignment problem to get a competitive primal-dual type of algorithm.

A mathematical model for the assignment problem is given by the following.

$$\text{Minimize} \quad \sum_{i=1}^{m}\sum_{j=1}^{m} c_{ij}x_{ij}$$

$$\begin{aligned} \text{Subject to} \quad & \sum_{j=1}^{m} x_{ij} = 1 && i = 1,\ldots, m \\ & -\sum_{i=1}^{m} x_{ij} = -1 && j = 1,\ldots, m \\ & x_{ij} = 0 \text{ or } 1 && i, j = 1,\ldots, m \end{aligned}$$

In matrix form, the assignment problem becomes as follows.

$$\begin{array}{ll} \text{Minimize} & \mathbf{cx} \\ \text{Subject to} & \mathbf{Ax} = \mathbf{b} \\ & x_{ij} = 0 \text{ or } 1 \qquad i, j = 1, \ldots, m \end{array}$$

where $\mathbf{x} = (x_{11}, \ldots, x_{1m}, \ldots, x_{m1}, \ldots, x_{mm})'$, $\mathbf{A}$ is a $2m \times m^2$ matrix whose (i, j) column is $\mathbf{a}_{ij} = \mathbf{e}_i - \mathbf{e}_{m+j}$ for $i = 1, 2, \ldots, m$ and $j = 1, 2, \ldots, m$, and $b = (\mathbf{1}, -\mathbf{1})'$, where $\mathbf{1}$ is a vector of m ones. Thus we see that $\mathbf{A}$ is the same constraint matrix as that for the transportation problem. The underlying graph is bipartite, and so the assignment problem is sometimes referred to as a "minimum weighted matching problem on a bipartite graph," with weights c_{ij}. Applying the total unimodularity property of $\mathbf{A}$, we know that an optimal basic feasible solution to the assignment problem with the constraint $x_{ij} = 0$ or 1 replaced by $x_{ij} \geq 0$ will be all integer. Furthermore, as a result of the constraints no x_{ij} can exceed 1. Hence all x_{ij} will be either zero or 1 in an optimal solution to the linear program. This permits us to replace the constraint $x_{ij} = 0$ or 1 by the constraint $x_{ij} \geq 0$. Thus we obtain the following.

$$\begin{array}{ll} \text{Minimize} & \mathbf{cx} \\ \text{Subject to} & \mathbf{Ax} = \mathbf{b} \\ & \mathbf{x} \geq \mathbf{0} \end{array}$$

It is interesting to note that any transportation problem can be equivalently transformed into (its special case) an assignment problem. This is readily, although inefficiently, accomplished by making s_i copies of origin O_i, each with a unit supply for $i = 1, \ldots, m$, and similarly, by making d_j copies of destination D_j, each with a unit demand for $j = 1, \ldots, n$. The problem hence becomes one of matching the $d = \Sigma_i s_i = \Sigma_j d_j$ units of supply and demand, where the cost of matching a unit of supply from a copy of O_i to a unit of demand at a copy of D_j is c_{ij}.

Observe that the assignment polytope has $m!$ extreme points. Viewing the assignment constraints as matching m "origins" O_i, $i = 1, \ldots, m$ to m destinations D_j, $j = 1, \ldots, m$, there are clearly $m!$ possible matchings, each corresponding to some permutation of the destinations being matched with the origins $O_1, \ldots, O_m$. For each matching, choosing the corresponding x_{ij} variables that equal 1 as basic variables, we obtain m basic variables that are unity. Figure 10.20a depicts the situation for some matching with $m = 4$. To complete the basis, we need to select $(m - 1)$ (degenerate) variables or arcs so that the resulting graph is connected and hence a tree. (Note that a connected assignment subgraph with $(2m - 1)$ arcs cannot contain a cycle.) Figure 10.20b shows one choice of degenerate basic arcs. Since every feasible integer solution is a matching

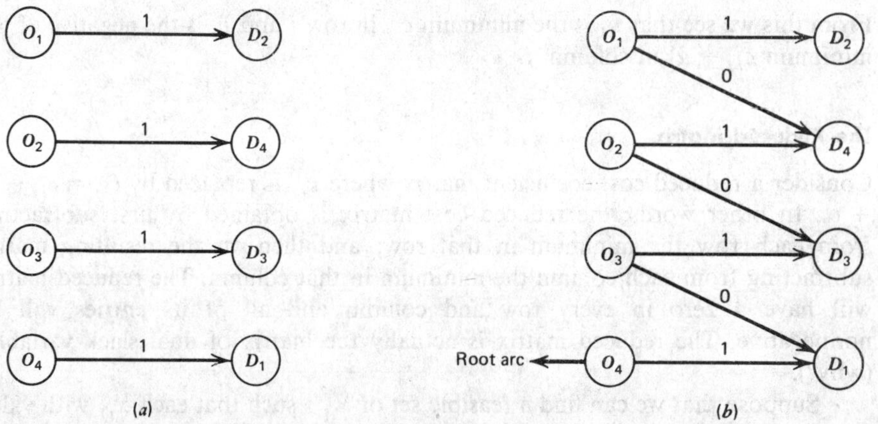

Figure 10.20. Matchings and basic feasible solutions.

and every matching is an extreme point of the assignment polytope, the assignment polytope has $m!$ vertices. Moreover, every basic feasible solution has $(m-1)$ degenerate arcs as noted earlier. In fact, it can be shown that every vertex has $(2^{m-1})(m^{m-2})$ bases representing it. Furthermore, given any pair of extreme points $\mathbf{x} \neq \mathbf{y}$, either $\mathbf{x}$ and $\mathbf{y}$ are adjacent vertices or there exists a third extreme point $\mathbf{z}$ such that $\mathbf{x}$ and $\mathbf{z}$ are adjacent and so are $\mathbf{y}$ and $\mathbf{z}$. (See Exercises 10.35 and 10.36.)

The Dual Problem

The dual of the assignment problem with the nonnegativity restrictions replacing the $0-1$ constraints becomes as follows.

$$\text{Maximize} \quad \sum_{i=1}^{m} u_i - \sum_{j=1}^{m} v_j$$

$$\begin{array}{llll} \text{Subject to} & u_i - v_j \leq c_{ij} & & i, j = 1, \ldots, m \\ & u_i, v_j \quad \text{unrestricted} & & i, j = 1, 2, \ldots, m \end{array}$$

The complementary slackness conditions are given by

$$(c_{ij} - u_i + v_j)x_{ij} = 0 \qquad i, j = 1, \ldots, m$$

Thus, if we can find a set of feasible u's, v's, and x's that satisfy complementary slackness, those u's, v's, and x's will be optimal.

A feasible dual solution is given by

$$\hat{u}_i = \underset{1 \leq j \leq m}{\text{Minimum}} \{c_{ij}\} \qquad i = 1, \ldots, m$$

$$\hat{v}_j = -\underset{1 \leq i \leq m}{\text{Minimum}} \{c_{ij} - \hat{u}_i\} \qquad j = 1, \ldots, m$$

From this we see that $\hat{u}_i$ is the minimum c_{ij} in row i and $\hat{v}_j$ is the negative of the minimum $c_{ij} - \hat{u}_i$ in column j.

The Reduced Matrix

Consider a reduced cost coefficient matrix where c_{ij} is replaced by $\hat{c}_{ij} = c_{ij} - u_i + v_j$. In other words, the reduced cost matrix is obtained by first subtracting from each row the minimum in that row; and then on the resulting matrix subtracting from each column the minimum in that column. The reduced matrix will have a zero in every row and column and all of its entries will be nonnegative. The reduced matrix is actually the matrix of dual slack variables (why?).

Suppose that we can find a feasible set of x_{ij}'s such that each x_{ij} with value 1 is associated with a zero cell of the reduced matrix. Then by complementary slackness we conclude that we have the optimal solution. What, then, constitutes a set of feasible x_{ij}'s? Reviewing the constraints of the assignment problem, it is clear that we must have exactly one x_{ij} in each row equal to 1 and exactly one x_{ij} in each column equal to 1. Thus in a feasible solution there will be exactly m of the x_{ij}'s equal to 1, the rest being zero.

Let us illustrate the foregoing ideas with an example. Consider the following cost coefficient matrix for an assignment problem.

	1	2	3	4	Row minimum
1	3	2	5	4	2
2	0	1	2	3	0
3	4	1	−1	3	−1
4	2	5	3	4	2

Subtracting the row minimum from each element in the row, we get the following tableau.

	1	2	3	4
1	1	0	3	2
2	0	1	2	3
3	5	2	0	4
4	0	3	1	2
Column minimum	0	0	0	2

Subtracting the column minimum in the new matrix from each element in the column, we get the reduced matrix as follows.

	1	2	3	4
1	1	[0]	3	0
2	[0]	1	2	1
3	5	2	[0]	2
4	0	3	1	[0]

$(= \hat{c}_{ij})$

Now if we let $x_{12}^* = x_{21}^* = x_{33}^* = x_{44}^* = 1$ and if we let all other x_{ij}^*'s be zero, then we have a feasible solution with positive x_{ij}'s associated with zero cells of the reduced matrix, thus producing an optimal solution.

It is not always so easy to find an optimal solution. Take, for example, the following cost matrix.

	1	2	3
1	2	5	7
2	4	2	1
3	2	6	5

$(= c_{ij})$

The reduced matrix is given by:

	1	2	3
1	0	2	5
2	3	0	0
3	0	3	3

$(= \hat{c}_{ij})$

Here it is not possible to set three of the x_{ij}'s equal to 1 such that all three positive x_{ij}'s occur in zero cells and no two positive x_{ij}'s occur in the same row or column.

A Partial Solution

Notice that for the reduced matrix, the maximum number of x_{ij}'s, from among the zero cells, which can be set equal to 1 without any two positive x_{ij}'s occurring in the same row or column is 2. For example, we might let $x_{11} = x_{22} = 1$, $x_{11} = x_{23} = 1$, $x_{31} = x_{22} = 1$, or $x_{31} = x_{23} = 1$. In this case the maximum number of cells with zero $\hat{c}_{ij}$ such that no two cells occupy the same row or column is 2. The corresponding cells are called *independent*. Notice also that if one were to draw a set of lines through the rows and columns to *cover* the zeros so that there is at least one line through each zero, the minimum number of such lines for this matrix is 2, a line through column 1 and a line through row 2.

	1	2	3
1	0	2	5
2	3	0	0
3	0	3	3

We see in this example that the maximum number of independent zero cells and the minimum number of lines required to cover the zeros are equal. This result, which is true in general, is given by the following theorem. We shall not prove this theorem here. (In Chapter 12, Exercise 12.18 asks the reader to show that this theorem is a special case of the maximal flow–minimal cut theorem for networks. At that time we also suggest a method for systematically finding the required number of lines. Also see Exercise 10.37 and Exercise 10.41.)

***Theorem* 10.1**

The maximum number of independent zero cells in a reduced assignment matrix is equal to the minimum number of lines to cover all zeros in the matrix.

Modifying the Reduced Matrix

Suppose that we have not yet obtained the optimal solution, that is, cannot find a feasible set of positive x_{ij}'s from among the zero cells of the reduced matrix. Consider the covered matrix, the reduced matrix with zeros covered by the fewest number of lines. Let k be the number of lines required. Also let $S_r = \{i_1, i_2, \ldots\}$ be the set of uncovered rows and $S_c = \{j_1, j_2, \ldots\}$ be the set of uncovered columns. Define $\bar{S}_r = M - S_r$ and $\bar{S}_c = M - S_c$, where $M = \{1, 2, \ldots, m\}$. Finally, let p be the number of rows in S_r and q the number of columns in S_c. Then $k = (m - p) + (m - q)$.

Let c_0 be the minimum uncovered element, that is,

$$c_0 = \underset{\substack{i \in S_r \\ j \in S_c}}{\text{Minimum}} \{\hat{c}_{ij}\} > 0$$

It can be easily demonstrated that a new dual feasible solution is given by

$$\begin{aligned}
\bar{u}_i &= \hat{u}_i + c_0 \quad & i \in S_r \\
\bar{u}_i &= \hat{u}_i & i \in \bar{S}_r \\
\bar{v}_j &= \hat{v}_j & j \in S_c \\
\bar{v}_j &= \hat{v}_j + c_0 & j \in \bar{S}_c
\end{aligned}$$

In the reduced matrix with elements $c_{ij} - u_i + v_j$, this is equivalent to subtracting c_0 from each uncovered row and adding c_0 to each covered column. Another way to view this is that c_0 is subtracted from each uncovered element and added to each twice-covered element. The new reduced cost coefficient matrix has nonnegative elements and a zero in every row and column (why?).

For the previous 3×3 matrix we have $c_0 = \text{Minimum}\{2, 5, 3, 3\} = 2$ and the new reduced cost matrix is given by

	1	2	3
1	0	0	3
2	5	0	0
3	0	1	1

Notice that now a feasible set of x_{ij}'s exists with positive x_{ij}'s associated with zero cells (zero dual slack variables).

Note that primal feasibility is attained, dual feasibility is maintained (since the entries in the reduced cost matrix are nonnegative), and finally complementary slackness holds (since $x_{ij} = 1$ only if the corresponding dual slack is zero). Thus the Karush–Kuhn–Tucker conditions hold and the optimal solution, $x_{12}^* = x_{23}^* = x_{31}^* = 1$ (all other x_{ij}^*'s set equal to 0) is at hand.

Summary of the Hungarian Algorithm

The algorithm developed in this section may be summarized as follows.

INITIALIZATION STEP

For each row of the cost matrix, subtract the minimum element in the row from each element in the row. For each column of the resulting matrix, subtract the minimum element in the column from each element in the column. The result is a reduced matrix.

MAIN STEP

1. Draw the minimum number of lines through the rows and columns to cover all zeros in the reduced matrix. If the minimum number of lines is m, then an optimal solution is available. Otherwise go to step 2.

2. Select the minimum uncovered element. Subtract this element from each uncovered element and add it to each twice-covered element. Return to step 1.

An Example

Consider the following cost matrix.

	1	2	3	4	5
1	2	3	5	1	4
2	−1	1	3	6	2
3	−2	4	3	5	0
4	1	3	4	1	4
5	7	1	2	1	2

The reduced matrix is as follows.

	1	2	3	4	5
1	1	2	3	0	2
2	0	2	3	7	2
3	0	6	4	7	1
4	0	2	2	0	2
5	6	0	0	0	0

Here the minimum number of lines to cover all zeros is 3. The minimum uncovered element is 1. Subtract this from each uncovered element and add it to each twice-covered element.

	1	2	3	4	5
1	1	1	2	0	1
2	0	1	2	7	1
3	0	5	3	7	0
4	0	1	1	0	1
5	7	0	0	1	0

Again, we do not have an optimal solution at hand. The minimum uncovered element is 1. Subtract 1 from each uncovered element and add it to each twice-covered element. This leads to the following matrix.

	1	2	3	4	5
1	1	[0]	1	0	1
2	[0]	0	1	7	1
3	0	4	2	7	[0]
4	0	0	0	[0]	1
5	8	0	[0]	2	1

In this matrix an optimal solution is given by $x_{12}^* = x_{21}^* = x_{35}^* = x_{44}^* = x_{53}^* = 1$ and all other x_{ij}^*'s set equal to zero.

In Exercise 10.39 we ask the reader to show that the Hungarian method for assignment problems is precisely the primal-dual algorithm applied to the assignment problem.

Finite Convergence of the Assignment Algorithm

Whenever we cannot find a feasible set of x_{ij}'s from among the zero cells of the reduced matrix, we repeat the process of drawing lines and adjusting the matrix. We can only do this a finite number of times before an optimal solution is found. Clearly, an optimal solution can be found if all reduced costs become zero. To show finiteness we note that the reduced costs are always nonnegative and that

$$\begin{aligned}\sum_i \sum_j \hat{c}_{ij} - \sum_i \sum_j \bar{c}_{ij} &= \sum_{(S_r, S_c)} (\hat{c}_{ij} - \bar{c}_{ij}) + \sum_{(S_r, \bar{S}_c)} (\hat{c}_{ij} - \bar{c}_{ij}) \\ &\quad + \sum_{(\bar{S}_r, S_c)} (\hat{c}_{ij} - \bar{c}_{ij}) + \sum_{(\bar{S}_r, \bar{S}_c)} (\hat{c}_{ij} - \bar{c}_{ij}) \\ &= \sum_{(S_r, S_c)} c_0 + \sum_{(S_r, \bar{S}_c)} 0 + \sum_{(\bar{S}_r, S_c)} 0 + \sum_{(\bar{S}_r, \bar{S}_c)} (-c_0) \\ &= pqc_0 - (m-p)(m-q)c_0 \\ &= m(p+q-m)c_0\end{aligned}$$

But $p + q$ is the number of uncovered rows and columns, so that

$$p + q - m = (2m - k) - m = m - k$$

where k is the number of covered rows and columns. By Theorem 1, k is also the maximum number of independent zero cells. In particular, $k < m$ because otherwise we would have had an optimal solution at the last iteration. Therefore

$$\sum_i \sum_j (\hat{c}_{ij} - \bar{c}_{ij}) = m(m-k)c_0$$

is a positive integer provided that the original cost matrix consists of integers. Since the entries in the reduced cost matrix are always nonnegative by construction, and since the sum of the entries is reduced by a positive integer at each iteration, the algorithm stops in a finite number of steps. At termination we have an optimal solution since the Karush–Kuhn–Tucker conditions hold.

10.8 ALTERNATING BASIS ALGORITHM FOR ASSIGNMENT PROBLEMS

We have seen that every basic feasible solution to the assignment problem has degeneracy of order $(m-1)$. In fact, it has been empirically observed that typically 90 percent of the pivots used to solve the assignment problem by the usual primal simplex algorithm are degenerate when $m \geq 500$. This should not be surprising since we know that every nonoptimal vertex is either adjacent to an optimal vertex or is adjacent to a neighbor of this optimal vertex. Hence it stands to reason that if one devises a primal simplex algorithm that circumvents a sizeable portion of the degenerate pivots, then a significant savings in computational effort will result. The alternating basis algorithm is a step in this direction. This algorithm is a special case of the strongly feasible network simplex algorithm discussed in Chapter 9, in which one additionally designates an origin node as the root node. For the assignment problem, a strongly feasible basis that has an origin node as a root node is called an *alternating path basis*. Since the algorithm encounters only strongly feasible bases, that is, bases in which all degenerate arcs point toward the root, it automatically circumvents several bases representing the extreme points of the assignment polytope. As a result, this procedure yields about a 25 percent savings in pivots over the usual network simplex algorithm.

Figure 10.20*b* shows an alternating path basis that is a chain graph. Such a basis can always be used to initialize the algorithm, given any choice of a starting one-to-one matching. (Exercise 10.46 indicates why this may be an advantageous starting point in the absence of any other information.) Figure 10.21*a* shows a typical alternating path basis. Figure 10.21*b* shows a basis representing this same extreme point that is *not* an alternating path basis (why?).

Note that the name "alternating path" arises from the fact that the arcs with flows of one and zero alternate on the chain from any node to the root node. These arcs are respectively called *one-arcs* and *zero-arcs*. Observe, however, that not any feasible basis with an origin node as a root node and with one-arcs and zero-arcs alternating is an alternating path basis. For example, consider the basis obtained (with $m = 5$) in Figure 10.21*b* by removing O_3, D_3, O_7, and D_7 along with their incident arcs. In fact, it can be easily shown (see Exercise 10.48) that a feasible basis is an alternating path basis if and only if all one-arcs are pointing away from the root. This holds if and only if an origin node is a root node with one incident basic arc (other than the root arc) and all other origin nodes have

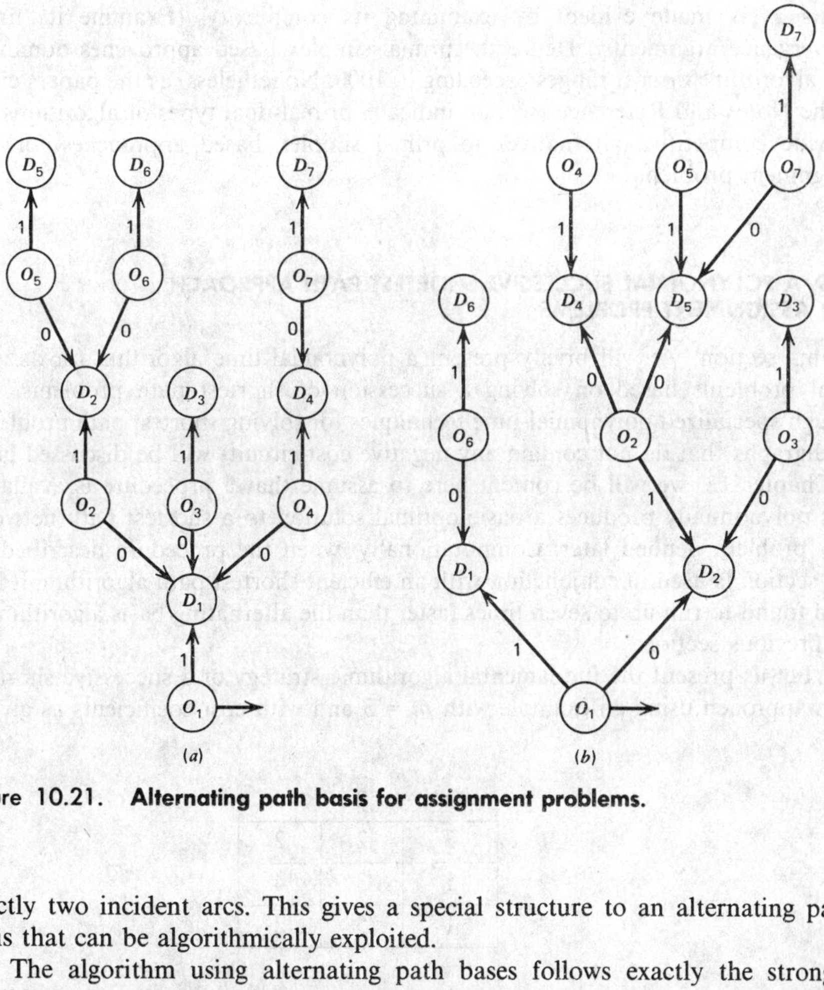

Figure 10.21. Alternating path basis for assignment problems.

exactly two incident arcs. This gives a special structure to an alternating path basis that can be algorithmically exploited.

The algorithm using alternating path bases follows exactly the strongly feasible tree network simplex procedure of Chapter 9. However, some specializations are possible. For example, observe that a pivot with entering arc (O_i, D_j) is nondegenerate if and only if D_j is a successor of O_i, that is, O_i lies on the chain from D_j to the root node. Whether the pivot is degenerate or not, the leaving arc, by the strongly feasible tree rule, is the one incident at O_i in the basis equivalent chain or in the cycle that is formed by adding (O_i, D_j) to the basis graph. By noting that each origin node has a unique immediate successor that is the destination node assigned to it, the tree may be stored by considering each such origin–destination pair as a single "married" unit and thereby reducing the storage and updating requirements by roughly one-half. We ask the reader to explore these specializations in Exercise 10.49.

Computationally, the primal-dual algorithm of the previous section dominates the alternating basis algorithm for totally dense problems when the cost parameters are in the range 0–100. However, unlike the primal simplex based approaches, the primal-dual algorithm is quite sensitive to ranges in cost coefficients as is made evident by examining its complexity. (Examine its finite convergence arguments.) Hence the primal simplex based approaches dominate this algorithm for cost ranges exceeding 0–1000. Nonetheless, as the papers cited in the Notes and References section indicate, primal-dual types of algorithms do provide competitive alternatives to primal simplex based approaches for the assignment problem.

10.9 A POLYNOMIAL SUCCESSIVE SHORTEST PATH APPROACH FOR ASSIGNMENT PROBLEMS

In this section, we will briefly present a polynomial-time algorithm for assignment problems based on solving a succession of shortest path problems. Although specialized polynomial-time techniques for solving shortest path problems on digraphs that do not contain any negative cost circuits will be discussed later in Chapter 12, we will be content here to assume that a procedure is available that polynomially produces a basic optimal solution to a shortest path network flow problem defined later. Computationally, when the procedure described in this section is used in conjunction with an efficient shortest path algorithm, it has been found to run up to seven times faster than the alternating basis algorithm of the previous section.

Let us present the fundamental algorithmic strategy of a successive shortest path approach using an example with $m = 3$ and with cost coefficients as given.

	1	2	3
1	3	2	2
2	1	3	4
3	1	5	2

Suppose that at any intermediate step of the procedure, we have some partial solution in which some origin–destination pairs are *tentatively* matched. Let the sets UO and UD contain the *unassigned* or *active* origin and destination nodes respectively, and let $\overline{UO}$ and $\overline{UD}$ be their respective complements. To begin with, $\overline{UO}$ and $\overline{UD}$ are both empty. Let $\bar{\mathbf{x}}$ denote the present partial solution. Hence $\bar{\mathbf{x}} \equiv \mathbf{0}$ at the initialization step, and whenever an origin–destination assignment is made, the corresponding component of $\bar{\mathbf{x}}$ is made unity. Also assume that currently $|UO| = |UD| \geq 1$ so that $\bar{\mathbf{x}}$ is infeasible to the assignment problem.

Given a partial solution $\bar{\mathbf{x}}$, we first execute the *shortest path step*. For this purpose, let us construct a special network from the assignment graph as follows. For convenience, let the origin node O_i be indexed as i for $i = 1, \ldots, m$ and let the destination node D_j be indexed as $m + j$ for $j = 1, \ldots, m$. Create an additional dummy root node 0 and connect this to each origin node in UO using a zero cost arc. For each O_i, D_j pair for which $\bar{x}_{ij}$ is currently unity, create the reverse arc $(m + j, i)$ and let the cost on this arc be $-c_{ij}$. All the other (O_i, D_j) arcs in the assignment graph for which $\bar{x}_{ij} = 0$ are present as the usual arcs $(i, m + j)$ in this new graph with the same cost c_{ij}. Next place a supply of $|UD| \equiv |UO|$ at node 0, and place a demand of one unit at each of the destination nodes $m + j$ for $D_j \in UD$. Call the resulting network flow problem the *shortest-path problem* $SP(\bar{\mathbf{x}})$. Note that $SP(\bar{\mathbf{x}})$ is feasible since we are considering a complete bipartite assignment network problem. (For a problem on a noncomplete bipartite graph, if $SP(\bar{\mathbf{x}})$ is infeasible, then so is the original assignment problem—see Exercise 10.50.) Hence if $SP(\bar{\mathbf{x}})$ is not unbounded, that is, it does not contain any circuit with a total negative cost, then it has an optimal solution. This is clearly so for the starting problem $SP(\bar{\mathbf{x}})$ with $\bar{\mathbf{x}} = \mathbf{0}$, since the graph for this problem has all the original assignment graph arcs, and has a supply of m at the root node 0 and a demand of one at each of the destination nodes $m + j$, $j = 1, \ldots, m$. This is depicted in Figure 10.22*a*, where the supplies and demands are shown against each node and the costs are shown on the respective arcs. In fact, an optimal basic feasible solution here has basic arcs $(0, i)$, $i = 1, \ldots, m$ and $(k_j, m + j)$, $j = 1, \ldots, m$, where $k_j \in \operatorname{argmin}\{c_{ij} : i = 1, \ldots, m\}$. Figure 10.22*b* shows an optimal basis for the example problem, where the optimal dual variables are shown against the nodes. Later we show that the primal feasible problem $SP(\bar{\mathbf{x}})$ is always dual feasible as well and hence always has an optimal solution. In this case, efficient polynomial-time algorithms described in Chapter 12 may be used to obtain a basic optimal solution.

The next step of the algorithm is the *flow augmentation step*. The basis tree obtained at the previous step gives a shortest path from node 0 through some unassigned origin node(s), to all the unassigned demand nodes. Let us pick any such demand node $m + j$ and trace the chain (reverse *path*) to the root node 0. This chain must pass through some unassigned origin node i, say. On this path from i to $m + j$, complement the partial assignment values. Note that since the origin and destination nodes alternate on this path as well as the zero and one partial assignment values, the origin and destination nodes in $\overline{UO}$ or $\overline{UD}$ on this path remain in these respective sets. Additionally, we have $i \in \overline{UO}$ and $(m + j) \in \overline{UD}$, so that the cardinality of UO (and hence UD) falls by one. If there is another chain (reverse path) from another unassigned destination node to the root node 0 that goes through some other origin node in the updated UO set, then this step is repeated with respect to this path. This is carried out sequentially until the final updated graph contains no such path. Note that any two paths considered sequentially at this step must be disjoint since the basis tree has no cycles. If UD (and hence UO) is empty at the end of this step, then as shown later the current assignment $\bar{\mathbf{x}}$ is optimal. Otherwise we return to the shortest path step and repeat the procedure. Since $|UD|$ drops by at least one at each iteration, we can loop through these two steps at most m times, which yields an $O(m^4)$ polynomial-time algorithm (assuming that the complexity of the shortest path algorithm is $O(m^3)$).

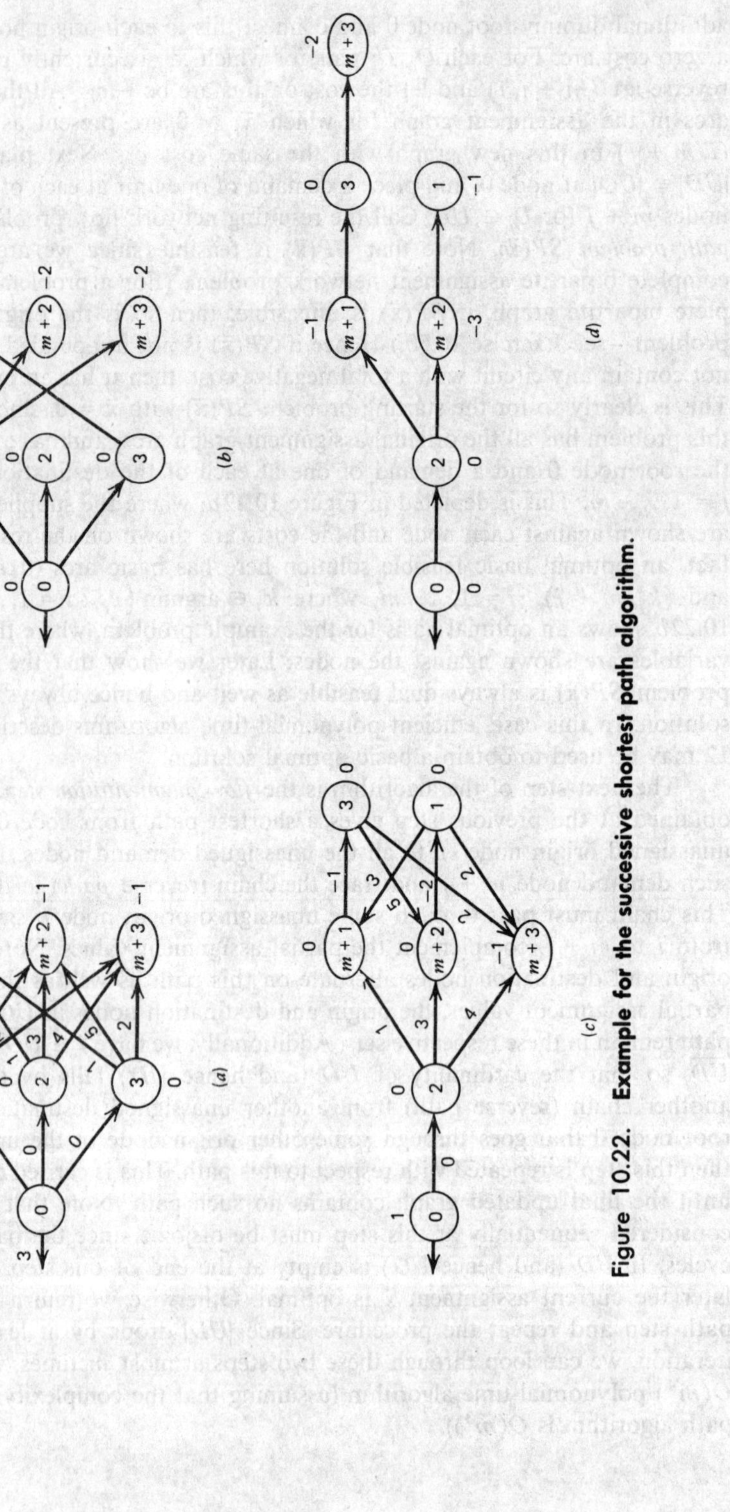

Figure 10.22. Example for the successive shortest path algorithm.

For the example problem, we can trace a path $\{0, 3, m+1\}$ to destination node 1 in Figure 10.22*b* at the flow augmentation step. This gives $\bar{x}_{31} = 1$, $UO = \{1, 2\}$, and $UD = \{2, 3\}$. A second path to destination 2 can be traced as $\{0, 1, m+2\}$, which yields $\bar{x}_{12} = 1$, $UO = \{2\}$, and $UD = \{3\}$. No additional paths can be traced and so currently we have the partial assignment solution $\bar{x}_{12} = \bar{x}_{31} = 1$ and $\bar{x}_{ij} = 0$ otherwise.

Repeating the shortest path step, the new $SP(\bar{\mathbf{x}})$ graph obtained is shown in Figure 10.22*c*. An optimal basis for this problem is shown in Figure 10.22*d*. At the flow augmentation step, we trace the path $\{0, 2, m+1, 3, m+3\}$ from node 0 to the unassigned destination node $m+3$. Note that presently $\bar{x}_{21} = 0$, $\bar{x}_{31} = 1$ and $\bar{x}_{33} = 0$ alternate on this path. Complementing these values, we get $\bar{x}_{21} = 1$, $\bar{x}_{31} = 0$, and $\bar{x}_{33} = 1$. Hence the resulting assignment solution is $\bar{x}_{12} = \bar{x}_{21} = \bar{x}_{33} = 1$, and $\bar{x}_{ij} = 0$ otherwise, with $UO = UD = \varnothing$. We terminate claiming this solution to be optimal.

To justify this algorithm, consider an optimal basis tree $T(\bar{\mathbf{x}})$ for the shortest path problem $SP(\bar{\mathbf{x}})$ at some iteration. Let $\bar{\mathbf{w}} = (\bar{w}_1, \ldots, \bar{w}_{m+m})$ be the associated set of optimal dual multipliers. Note that if $\bar{x}_{ij} = 0$, then arc $(i, m+j)$ is present in the problem $SP(\bar{\mathbf{x}})$ and we have $\bar{w}_i - \bar{w}_{m+j} \le c_{ij}$. On the other hand, if $\bar{x}_{ij} = 1$, then arc $(m+j, i)$ is present in the problem $SP(\bar{\mathbf{x}})$ with a cost $-c_{ij}$, and so we have $\bar{w}_{m+j} - \bar{w}_i \le -c_{ij}$ or $\bar{w}_i - \bar{w}_{m+j} \ge c_{ij}$. Consequently, $\bar{\mathbf{w}}$ is dual feasible to the assignment problem: Minimize$\{\mathbf{cx} : \mathbf{Ax} = \mathbf{b}, \mathbf{0} \le \mathbf{x} \le \mathbf{1}\}$ and is complementary slack with respect to $\bar{\mathbf{x}}$ (why?). After revising $\bar{\mathbf{x}}$ to get $\bar{\mathbf{x}}_{\text{new}}$, say, following the flow augmentation step, observe that the only $\bar{x}_{ij}$ values that change correspond to basic arcs in $T(\bar{\mathbf{x}})$. Since $\bar{w}_i - \bar{w}_{m+j} = c_{ij}$ for these arcs, we must have that $\bar{\mathbf{x}}_{\text{new}}$ is also complementary slack with respect to $\bar{\mathbf{w}}$. Consequently, if $\bar{\mathbf{x}}_{\text{new}}$ is a feasible assignment solution, that is, the revised $UD = \varnothing$, then it is optimal. Otherwise we construct the problem $SP(\bar{\mathbf{x}}_{\text{new}})$. But for $SP(\bar{\mathbf{x}}_{\text{new}})$, the solution $\bar{\mathbf{w}}$ is still dual *feasible* since the arcs that get reversed in orientation and in the sign on the cost coefficients in going from $SP(\bar{\mathbf{x}})$ to $SP(\bar{\mathbf{x}}_{\text{new}})$ are basic in $T(\bar{\mathbf{x}})$ and hence have $\bar{w}_i - \bar{w}_{m+j} = c_{ij}$. Therefore $SP(\bar{\mathbf{x}}_{\text{new}})$ is primal and dual feasible; therefore it has an optimal solution. This process can now continue until after at most m iterations, we achieve optimality.

10.10 THE TRANSSHIPMENT PROBLEM

In the transportation problem studied in this chapter we have assumed that each point is either an origin, where goods are available, or a destination, where goods are required. Suppose that, in addition, there are intermediate points where goods are neither available nor required, but where goods can be transshipped. The problem of finding the shipping pattern with the least cost is called the *transshipment problem* and is illustrated in Figure 10.23.

This problem is a particular type of network flow problem that can be solved directly using the network simplex algorithm of Chapter 9, for example. It is also possible to convert this problem into a transportation problem by tracing shortest paths between each origin–destination pair. (See Exercise 9.5.)

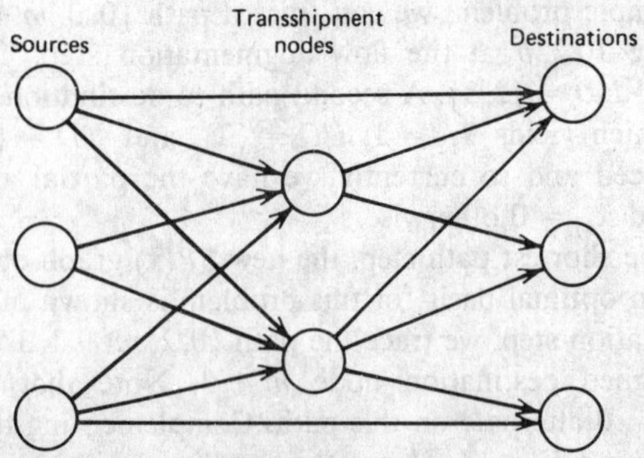

Figure 10.23. Example of a transshipment problem.

Another procedure for converting the model into a transportation problem is to add buffer stocks at certain nodes as discussed in Exercise 10.52.

EXERCISES

10.1 Describe in detail how any finite-valued, capacitated minimum cost network flow programming problem can be transformed into an equivalent assignment problem.

10.2 Solve the following transportation problem.

		Destination 1	2	3	s_i
Origin	1	3	4	1	2
	2	2	7	5	5
	d_j	1	3	3	

c_{ij} matrix

10.3 On Tuesday the GT Railroad Company will have four locomotives at IE Junction, one locomotive at Centerville, and two locomotives at Wayover City. Student trains each requiring one locomotive will be at A-Station, Fine Place, Goodville, and Somewhere Street. The local map gives the following distances.

	A-STATION	FINE PLACE	GOODVILLE	SOMEWHERE STREET
IE Junction	13	35	42	9
Centerville	6	61	18	30
Wayover City	15	10	5	9

How should they assign power (locomotives) so that the total distance traveled is minimized?

10.4 Referring to the transportation problem defined in Section 10.1, state whether the following are true or false and provide a brief justification.

a. If the total supply of a proper subset of the origin nodes equals the total demand of a proper subset of the destination nodes, then there exists a degenerate basic feasible solution.
b. The entries in any updated simplex tableau column add to unity.
c. Given an optimal basic feasible solution, if some of the cost coefficients of the basic variables are decreased, the solution remains optimal.

10.5 The following is a transportation tableau.

(Each cell: c_{ij} in the upper-left box; x_{ij} below.)

	1	2	3	4	s_i
1	c_{ij} = 9; x_{ij} = 4	7; 14	12	8	18
2	15	12	12; 4	15	4
3	8; 2	9	6; 4	12	6
4	14	12	11; 7	12; 5	12
	6	14	15	5	

a. Is the solution basic?
b. Show that the solution is optimal.
c. Does this problem have alternative optimal solutions?
d. Give the original linear programming problem and its dual.
e. Derive the optimal solution to the dual problem.
f. Give the optimal simplex tableau associated with the foregoing transportation tableau.
g. Suppose that c_{43} is increased from 11 to 13. Is the solution still optimal? If not, find the new optimal solution.

10.6 The following tableau depicts a feasible solution to a transportation problem. Is this solution basic? If not, then starting with this solution, construct a basic feasible solution. Compare the costs of the two solutions. Do the columns of variables x_{11}, x_{12}, x_{23}, x_{32}, x_{33}, x_{34}, and x_a yield a basis?

c_{ij} / x_{ij}	1	2	3	4	s_i
1	5 / 15	6	2	4 / 20	35
2	3	6 / 10	8 / 22	2 / 10	42
3	4 / 5	1	9 / 18	10	23
	20	10	40	30	

10.7 Consider the following transportation problem.

	1	2	3	s_i	
1	2	5	− 1	1	c_{ij} matrix
2	1	3	4	2	
d_j	1	1	1		

a. Construct a basic feasible solution by the northwest corner method. Show the basic tree.
b. Find the optimal solution.

10.8 Consider Exercise 10.5.

a. Add 10 to each c_{ij}. Applying the cycle method, is the tableau still optimal?
b. Multiply each c_{ij} by 10. Applying the cycle method, is the tableau still optimal?
c. Repeat parts (a) and (b) for a general constant k.
d. What can we conclude from part (c) for transportation problems?

10.9 Consider the following transportation problem.

		Destination				
		1	2	3	s_i	
Origin	1	2	1	1	3	c_{ij} matrix
	2	1	4	4	5	
	d_j	1	3	4		

a. Prove by duality theory that $(x_{11}, x_{12}, x_{13}, x_{21}, x_{22}, x_{23}) = (0, 3, 0, 1, 0, 4)$ is an optimal solution.
b. Interpret, economically, the dual variables for the solution in part (a).

10.10 Consider the following transportation problem.

	1	2	3	4	s_i	
1	6	2	−1	0	5	
2	4	2	2	3	25	c_{ij} matrix
3	3	1	2	1	25	
d_j	10	10	20	15		

a. Give the northwest corner starting basic solution.
b. Find an optimal solution.
c. What is the effect on the optimal solution if c_{11} is changed to 0?

10.11 Consider the data of Exercise 10.10. Apply the method of Section 10.1 to obtain a feasible solution. Convert the feasible solution into a basic feasible solution. Show the basic tree.

10.12 Devise a method for applying the lexicographic simplex method directly on the transportation tableau. Show how the northwest corner rule can be used to obtain a starting basic feasible solution with lexicographic positive rows in $(\mathbf{B}^{-1}\mathbf{b}, \mathbf{B}^{-1})$.

10.13 An airline company can buy gasoline from three suppliers. The suppliers have available $2K$, $6K$, and $6K$ gallons respectively. The company needs gasoline at three locations with each location requiring $5K$, $3K$, and $2K$ gallons' respectively. The per/K gallon quoted price for gas delivered to each location is as follows.

		Location 1	Location 2	Location 3
Suppliers	1	2	1	1
	2	4	2	5
	3	1	8	9

How can the company buy the gasoline to minimize the total cost?

10.14 Show that if the cost coefficient for the artificial variable is increased by an amount θ, then all of the u_i's and v_j's will change by the same amount θ. Show that the cost coefficient of the artificial variable does not matter in the computation of $z_{ij} - c_{ij}$.

10.15 Consider a transportation problem with m sources and n sinks with respective supplies and demands s_i, $i = 1, \ldots, m$ and d_j, $j = 1, \ldots, n$. Assume that only a subset $\mathscr{A}$ of the possible arcs (i, j) are present in the problem with given cost coefficients c_{ij}, each source and sink node having at least one incident arc. Create a dummy source node $(m+1)$ with supply equal to $\Sigma_{j=1}^{n} d_j$, and create a dummy sink node $(n+1)$ with demand equal to $\Sigma_{i=1}^{m} s_i$. Also construct slack arcs $(i, n+1)$ for $i = 1, \ldots, m+1$ with zero cost coefficients, and construct artificial arcs $(m+1, j)$ $j = 1, \ldots, n$ with cost coefficients equal to M.

a. Show that the following value of M is sufficient to ensure that the total flow on the artificial arcs is at its minimum feasible value at optimality.

$$M = 1 + \left\{\text{sum of the } Q \text{ largest } c_{ij} \text{ values, } (i, j) \in \mathscr{A}\right\}$$

$$- \left\{\text{sum of the } (Q-1) \text{ smallest } c_{ij} \text{ values, } (i, j) \in \mathscr{A}\right\}$$

where $Q = \text{minimum}\{m, n, \lceil |\mathscr{A}|/2 \rceil\}$, and where $\lceil \cdot \rceil$ denotes the rounded-up integer value.

b. Examining the role played by M in your proof, comment on how alternative bounds on M may be derived.

10.16 Show how a transportation problem with rational supplies, demands, and cost coefficients can be scaled into an equivalent transportation problem with integer data.

10.17 Show that if we define

$$\hat{s}_i = s_i + \epsilon \qquad i = 1, \ldots, m$$

$$\hat{d}_j = d_j \qquad j = 1, \ldots, n-1$$

$$\hat{d}_n = d_n + m\epsilon$$

then by a proper choice of ϵ we can totally avoid degeneracy in the transportation problem (and maintain an equivalent problem).

10.18 An automobile manufacturer has assembly plants located in the Northwest, Midwest, and Southeast. The cars are assembled and sent to major markets in the Southwest, West, East, and Northeast. The appropriate distance matrix, availabilities, and demands are given by the following chart.

	SOUTHWEST	EAST	WEST	NORTHEAST	s_i
NORTHWEST	1000	8000	1800	2000	2,000,000
MIDWEST	400	700	900	1400	1,300,000
SOUTHEAST	800	1200	900	1100	1,600,000
d_j	1,000,000	1,500,000	1,200,000	700,000	

a. Assuming that cost is proportional to distance, find the optimal shipment pattern.

b. Assuming that cost is proportional to the square of distance, find the optimal shipment pattern.

10.19 Solve the following assignment problem by the transportation method.

PERSON \ JOB	1	2	3	s_i
1	2	1	0	1
2	1	6	4	1
3	1	5	6	1
d_j	1	1	1	

c_{ij} matrix

10.20 Consider the following problem.

$$\text{Minimize} \quad \sum_{i=1}^{m} \sum_{j=1}^{n} c_{ij}x_{ij}$$

$$\begin{aligned} \text{Subject to} \quad & \sum_{j=1}^{n} x_{ij} = s_i && i = 1, 2, \ldots, m \\ & \sum_{i=1}^{m} p_{ij}x_{ij} = d_j && j = 1, 2, \ldots, n \\ & x_{ij} \geq 0 && i = 1, \ldots, m \\ & && j = 1, \ldots, n \end{aligned}$$

where $p_{ij} > 0$ for all i, j. Extend the transportation algorithm of this chapter to handle the foregoing problem (which is sometimes referred to as the *generalized transportation problem*).

10.21 Formulate the problem of Exercise 1.15 as a generalized transportation model, and use the procedure of Exercise 10.20 to find the optimal solution.

10.22 Devise a procedure for identifying the unique cycle of basic cells associated with an entering nonbasic cell. The procedure must specify the cells with coefficient 1, -1, and 0 in the representation of the nonbasic cell. Such a procedure must be developed when the transportation algorithm is coded on a digital computer.

10.23 Show how sensitivity analysis of the c_{ij}, s_i, and d_j may be carried out on a transportation tableau.

10.24 Give the dual of a transportation problem. Is it possible to readily specify a feasible solution to the dual problem? If so, starting with this solution, devise a method for applying the primal-dual method directly to a transportation tableau.

10.25 Prove or give a counterexample: For a variable to be basic in a particular row of the linear program for a transportation problem, the variable must have a nonzero entry in that row.

10.26 A company has contracted for five jobs. These jobs can be performed in six of its manufacturing plants. Because of the size of the jobs, it is not feasible to assign more than one job to a particular manufacturing facility. Also, the second job cannot be assigned to the third manufacturing plant. The cost estimates, in thousands of dollars, of performing the jobs in the different manufacturing plants, are summarized.

	PLANT					
JOB	1	2	3	4	5	6
1	50	55	42	57	48	52
2	66	70	—	68	75	63
3	81	78	72	80	85	78
4	40	42	38	45	46	42
5	62	55	58	60	56	65

a. Formulate the problem of assigning the jobs to the plants so that the total cost is minimized.
b. Solve the problem by the transportation algorithm.
c. Solve the problem by the Hungarian assignment algorithm.
d. Apply the primal-dual algorithm to this problem. Make all possible simplifications.

10.27 Consider the transportation problem corresponding to the following tableau.

	1	2	3	4	s_i	
1	4	3	6	5	20	
2	7	10	5	6	30	c_{ij} matrix
3	8	9	7	7	50	
d_j	15	35	20	30		

a. Solve the problem by the transportation algorithm.
b. Suppose that c_{24} is replaced by 4. Without resolving the problem, find the new optimal solution.
c. How large should c_{12} be made before some optimality condition of the solution in part (a) is violated?

10.28 Consider the following data for a transportation problem:

	1	2	3	4	s_i
1	14	56	48	27	71
2	82	35	21	81	47
3	99	31	71	63	93
d_j	71	35	45	60	

c_{ij} matrix

a. Indicate a starting basic solution by the northwest corner rule. Give the basic tree.
b. Find the optimal solution.
c. Give the simplex tableau associated with the basic feasible solution in part (a).

10.29 The following is *Vogel's approximation method* for obtaining a reasonably good starting basic feasible solution.

Step 0. Begin with all cells unallocated.

Step 1. In the problem at hand, compute for each row and each column the difference between the lowest and next lowest cost cell in the row or column.

Step 2. Among those rows and columns at hand, select the one with maximum difference.

Step 3. Allocate as much as possible to the x_{ij} with the lowest cost cell in the selected row or column. Decrease the corresponding supply and demand. Drop the row or column whose supply or demand is zero.

Step 4. Make any allocations where only one unallocated cell remains in a row or column. After reducing the corresponding supply and demands and dropping the row or column, repeat Step 4 as necessary.

Step 5. Stop if no rows and columns remain. Otherwise return to Step 1 with the reduced problem.

a. Apply Vogel's method to the data of Exercise 10.2.
b. Apply Vogel's method to the data of Exercise 10.10.
c. Apply Vogel's method to the data of Exercise 10.19.

10.30 Show that Vogel's approximation method leads to a basic feasible solution, after including any required zero cells.

10.31 The following is the *matrix minimum method* for obtaining a reasonably good starting feasible solution.

Step 0. Begin with all cells unallocated.

Step 1. Identify the lowest-cost unallocated cell in the matrix and allocate as much as possible to this cell.

Step 2. Reduce the corresponding supply and demand, dropping the one going to zero, and repeat Step 1 until all supplies and demands are allocated.

a. Show that the procedure produces a basic feasible solution.

b. Apply the procedure to Exercises 10.2 and 10.10.

10.32 Show how the dual simplex method can be applied to transportation problems directly on the transportation tableau.

10.33 Consider the following *capacitated transportation problem*.

$$\text{Minimize} \quad \sum_{i=1}^{m} \sum_{j=1}^{n} c_{ij} x_{ij}$$

$$\begin{aligned}
\text{Subject to} \quad & \sum_{j=1}^{n} x_{ij} = s_i && i = 1, 2, \ldots, m \\
& \sum_{i=1}^{m} x_{ij} = d_j && j = 1, 2, \ldots, n \\
& 0 \le x_{ij} \le u_{ij} && i = 1, \ldots, m \\
& && j = 1, \ldots, n
\end{aligned}$$

Specialize the bounded variables simplex method of Chapter 5 (or Chapter 9) to solve the foregoing transportation problem on the *transportation tableau*. Describe all details: finding a starting basic feasible solution, computing the dual variables and $z_{ij} - c_{ij}$, the entering cell, and the leaving cell.

10.34 Consider Exercise 1.16.

a. Formulate the problem.

b. Solve the problem by the capacitated transportation algorithm that you developed in Exercise 10.33.

c. Suppose that the third manufacturing company lost one of its other contracts so that 700 tons are available to the furniture company. What is the optimal solution?

10.35 Show that every vertex of the assignment polytope of Section 10.7 has $(2^{m-1})(m^{m-2})$ bases representing it.

10.36 Referring to the assignment polytope of Section 10.7, show that two extreme points $\mathbf{x} \neq \mathbf{y}$ are adjacent if and only if $P(\mathbf{x}) \cup P(\mathbf{y})$ contains exactly one cycle, where $P(\mathbf{x})$, for a given vertex $\mathbf{x}$, is the subgraph of the assignment graph that contains only the positive flow arcs (see Figure 10.20*a* for example). Show that given any pair of extreme points $\mathbf{x} \neq \mathbf{y}$ of the assignment polytope, either $\mathbf{x}$ and $\mathbf{y}$ are adjacent or there exists an extreme point $\mathbf{z}$ such that $\mathbf{x}$ and $\mathbf{z}$ as well as $\mathbf{y}$ and $\mathbf{z}$ are adjacent. (*Hint.* Recall that $\mathbf{x}$ and $\mathbf{y}$ are adjacent if and only if they have bases representing them such that the union of the respective spanning tree graphs contains exactly one cycle.)

10.37 Use the theorems of duality to prove that the minimum number of lines to cover all zeros in a reduced assignment matrix equals the maximum number of independent zero cells.
(*Hint.* Consider the problem:

$$\text{Maximize} \quad \sum_i \sum_j x_{ij}$$

$$\text{Subject to} \quad \sum_{\hat{c}_{ij}=0} \mathbf{a}_{ij} x_{ij} \leq \mathbf{1}$$

$$x_{ij} \geq 0$$

where $\mathbf{a}_{ij} = \mathbf{e}_i + \mathbf{e}_{m+j}$ and the sum is taken only over zero cells of the reduced matrix. Take the dual of this linear program and examine its properties.)

10.38 Show that each new dual solution in the assignment procedure specified in Section 10.7 is feasible.

10.39 Compare the Hungarian method for the assignment problem with the primal-dual method applied to the assignment problem.

10.40 Apply the Hungarian method to the following assignment problem.

Cost matrix

	1	2	3	4	5
1	2	6	4	−1	3
2	1	5	4	4	6
3	0	2	5	1	5
4	4	1	3	2	5
5	6	2	5	2	5

10.41 Describe a procedure suitable for computer coding that will find, directly on the reduced matrix, the maximum number of independent zero cells in the reduced matrix (or equivalently the minimum number of lines to cover all zeros). The reader may wish to study Exercise 12.18.

10.42 Sally, Susan, and Sandra will go on a date with Bob, Bill, and Ben. Sally likes Bill twice as much as Bob and three times as much as Ben. Susan likes Bill three times as much as Bob and five times as much as Ben (Ben is a loser!). Sandra likes Bob about as much as Bill but likes them both about five times as much as Ben. How should the couples pair up so that in the aggregate the girls are as happy as possible? If one girl is willing to stay home, which one should it be? Which boy will lose out? (You guessed it!)

10.43 A carpenter, plumber, and engineer are available to perform certain tasks. Each person can perform only one task in the allotted time. There are four tasks available, three of which must be done. The inefficiency matrix for person i assigned to task j is as follows.

	SOLDERING	FRAMING	DRAFTING	WIRING
Carpenter	2	6	4	4
Plumber	3	4	4	3
Engineer	2	5	6	5

Which person should be assigned to which job? (*Hint.* Create a dummy person.) Which job will go unfinished? Now suppose that each person can perform up to two tasks, and all tasks must be done. What should they do?

10.44 Given the optimal reduced matrix for an assignment problem, show how to construct the basic tree for the associated linear program. Demonstrate by the following reduced matrix.

2	3	0	4
0	1	4	0
0	0	2	4
1	0	2	1

10.45 Show how one can construct the simplex tableau associated with the optimal assignment matrix.

10.46 Show that given any basis representing an extreme point of the assignment polytope, the maximum number of potential nondegenerate pivots on this basis graph is $m(m-1)/2$, and the minimum number is $(m-1)$. Show that the maximum number is achieved if and only if the basis tree is a chain graph (see Figure 10.20b) and the minimum number is achieved on a basis in which all the degenerate arcs originate at some node or all terminate at some node. Can both types of bases be alternating path bases?

10.47 Is it true that a feasible basis for an assignment problem in which some origin node is the root node, and in which the one-arcs and the zero-arcs alternate on the chain from any node to the root node is an alternating path basis?

10.48 a. Show that a feasible basis for an assignment problem is an alternating path basis if and only if all one-arcs are pointing away from the root, which is also equivalent to having an origin node as the root node with a single arc incident at the root node and with all other origin nodes having exactly two incident arcs.

b. Give equivalent definitions of "alternating path bases" by designating a demand node as the root node.

10.49 Using the alternating path basis algorithm for solving assignment problems, provide details for all the specializations you can make in the algorithmic operations, including any specializations in the list structures of Chapter 9.

10.50 Consider an assignment problem for which the underlying bipartite graph is not complete. Suppose that we construct the shortest path problem $SP(\bar{\mathbf{x}})$ for some partial solution $\bar{\mathbf{x}}$ as in the successive shortest path algorithm of Section 10.9. Show that if $SP(\bar{\mathbf{x}})$ is infeasible, then so is the original assignment problem. (*Hint.* Show that based on a feasible assignment, if one exists, one can construct a (reverse) path from any unassigned destination node to the root node 0 in $SP(\bar{\mathbf{x}})$.)

10.51 Solve the problem in Exercise 10.19 using the successive shortest path algorithm of Section 10.9. At each iteration, give the basic primal and dual optimal solutions to the shortest path problem $SP(\bar{\mathbf{x}})$. Demonstrate that this dual solution is feasible to the original assignment problem as well as to the updated shortest path problem $SP(\mathbf{x}_{\text{new}})$, and that it is complementary slack with respect to both $\bar{\mathbf{x}}$ and $\bar{\mathbf{x}}_{\text{new}}$.

10.52 Given a transshipment problem, the following procedure is suggested to convert it into a transportation problem. First the nodes are classified into the following mutually exclusive categories.

1. Pure source: a node that only ships.
2. Pure sink: a node that only receives.
3. Transshipment node: a node that may ship and receive.

A transportation tableau is constructed as follows. The origins are the pure sources and the transshipment nodes. The availability at each transshipment node i is replaced by $s_i + B$, where s_i is the maximum of zero and the net out of node i, and B is a buffer stock to be specified later. The destinations are the pure sinks and the transshipment nodes. The requirement at a transshipment node i is $d_i + B$, where d_i is the maximum of 0 and the net into node i. If there is no direct link from node i to node j, then c_{ij} is equal to M, where M is a large positive number. Also $c_{jj} = 0$ for transshipment nodes. Finally, B is a large positive number, say

$$B = \sum_i s_i$$

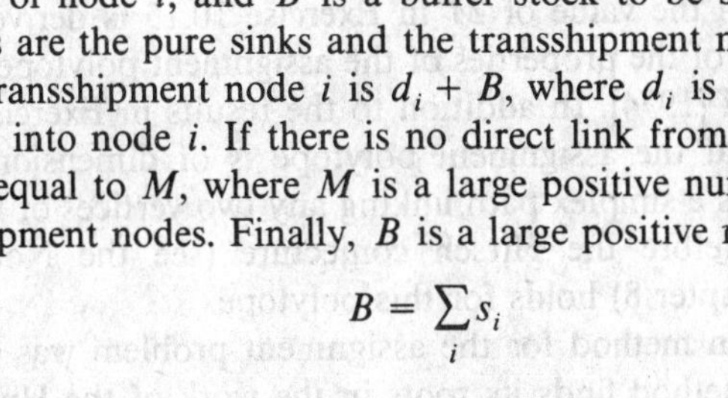

a. Using the foregoing instructions, form the transportation tableau corresponding to the above transshipment problem, where the availability at nodes 1, 2, and 3 are respectively 10, 20, and 15, and the requirement at nodes 5, 6, and 7 are respectively 10, 25, and 10.

b. Solve the problem using the transportation algorithm. Interpret the solution. What is the interpretation of the buffer B?

c. Show that the procedure outlined in this exercise is valid in general. (*Hint.* On the original flow conservation constraints, use appropriate substitutions of the type where a new variable equals B minus the sum of some variables.)

d. Convert the problem into a transportation problem using the least cost method discussed in Section 10.10. Apply the transportation algorithm and interpret your solution.

10.53 If it is known in advance that a certain variable will be positive in any optimal solution to a transportation problem, what simplifications can result in the solution method?

NOTES AND REFERENCES

1. Hitchcock [1941] is credited with the first formulation and discussion of a transportation model. Dantzig [1951c] adapted his simplex method to solve transportation problems. Charnes and Cooper [1954] developed an intuitive presentation of Dantzig's procedure through what is called the "stepping stone" method. In this chapter we label this the "cycle method."
2. Koopmans [1949] was the first to note the relationship between basic solutions in the transportation problem and the tree structure of a graph. Other good discussions are provided by Dantzig [1963a] and Ellis Johnson [1965].
3. The bound on the value of M in Exercise 10.15 is derived in Sherali [1987].
4. A good study of the properties of the assignment polytope appears in Balinski and Russakoff [1974]. In addition to the results in Exercises 10.35 and 10.36, they show that the assignment polytope is of dimension $(m - 1)^2$, has m^2 facets, and has a simplex path linking any two vertices of length no more than $2m - 1$. Therefore the Hirsch conjecture (see the Notes and References section of Chapter 8) holds for this polytope.
5. The Hungarian method for the assignment problem was developed by Kuhn [1955]. This method finds its roots in the work of the Hungarian mathematician Egerváry [1931], hence its name. Kuhn's paper led to the general primal-dual method for linear programs in the following year.
6. McGinnis [1983] presents a computational comparison between the Hungarian algorithm and the primal simplex alternating path basis algorithm. For improvements and variations in the Hungarian algorithm, see Bertsekas [1981], Carpenito and Toth [1980], Redlack and Huang [1987], and Jonker and Volgenant [1986]. Balinski [1984, 1985] provides an excellent study of the dual polyhedra of transportation and assignment problems, and exhibits why dual based simplex approaches may be more suitable for these problems than primal procedures.

7. The alternating path basis algorithm was developed by Barr *et al.* [1977]. This algorithm turns out to be a special case of the strongly feasible tree algorithm of Cunningham [1976] that was independently developed for general network flow problems. The results in Exercises 10.48 and 10.49 are from the former paper. The result in Exercise 10.46 is from Bazaraa and Sherali [1982].
8. The successive shortest path algorithm for assignment problems discussed in Section 10.9 is from Glover, Glover, and Klingman [1986]. The threshold shortest path subroutine is used to derive a computationally very efficient algorithm. For another successive shortest path algorithm, see Engquist [1982]. Other polynomial algorithms for the assignment problem appear in Akgul [1986], Balinski [1985] (see Goldfarb [1985] for an improvement of this "signature algorithm"), Balinski [1986], and Hung [1983]. Also see Orlin [1985] for some relevant discussions.
9. The reader interested in a further study of transportation problems involving aggregation and disaggregation methods for large-scale problems is referred to Balas [1965], Zipkin [1980], and Zipkin and Raimer [1983].

ELEVEN: THE OUT-OF-KILTER ALGORITHM

In Chapter 9 we presented a network simplex method for solving minimal cost network flow problems. In this chapter we present another method for solving minimal cost network flow problems, called the out-of-kilter algorithm. This algorithm is similar to the primal-dual algorithm in that it begins with dual feasibility but not necessarily primal feasibility and iterates between primal and dual problems until optimality is achieved. However, it differs from the primal-dual algorithm (as strictly interpreted) in that the out-of-kilter algorithm does not always maintain complementary slackness. In fact, the principal thrust is to attain complementary slackness. A version of the algorithm can be designed in which one maintains primal and dual feasibility (not necessarily basic solutions) and strives to achieve complementary slackness. Computationally, the state-of-the-art primal simplex codes run two–three times faster than traditional out-of-kilter codes. However, recent advances in primal-dual methods that use the basic *ingredients* of the out-of-kilter algorithm, although in a manner different from this algorithm, have resulted in algorithms that run two–four times faster than the best primal simplex code with the speed-up factor increasing by an order of magnitude for large problems. (See the Notes and References section.) From this viewpoint the *concepts* of the various algorithmic steps presented in this chapter are important.

11.1 THE OUT-OF-KILTER FORMULATION OF A MINIMAL COST NETWORK FLOW PROBLEM

For convenience of presentation, the form of the minimal cost flow problem that we shall work with is

$$\begin{aligned}
\text{Minimize} \quad & \sum_{i=1}^{m}\sum_{j=1}^{m} c_{ij}x_{ij} \\
\text{Subject to} \quad & \sum_{j=1}^{m} x_{ij} - \sum_{k=1}^{m} x_{ki} = 0 \qquad i = 1,\ldots,m \\
& x_{ij} \geq l_{ij} \qquad i, j = 1,\ldots,m \\
& x_{ij} \leq u_{ij} \qquad i, j = 1,\ldots,m
\end{aligned} \tag{11.1}$$

where it is understood that the sums and inequalities are taken over existing arcs only. We call a *conserving flow* any flow (choice of the x_{ij}'s) satisfying constraints (11.1). A conserving flow that satisfies the remaining constraints $l_{ij} \le x_{ij} \le u_{ij}$ is a *feasible flow* (solution). We shall assume that c_{ij}, l_{ij}, and u_{ij} are integers and that $0 \le l_{ij} \le u_{ij} < \infty$.

The foregoing formulation is completely equivalent to the formulation of the minimal cost network flow problem presented in Chapter 9. This is readily seen by noting that a minimum cost network flow problem can be first of all put in the form (11.1) with a general (integer) right-hand-side vector **b** instead of a zero vector in the flow conservation constraints and with $0 \le l_{ij} < u_{ij} < \infty$ for all (i, j). Here we assume that the variables x_{ij} that are unbounded from above are artificially bounded by using $u_{ij} = M$, where M is sufficiently large. (See Exercise 11.6.) By replacing each right-hand-side magnitude $|b_i|$ with a variable y_i, where y_i is bound restricted as $|b_i| \le y_i \le |b_i|$, it is readily seen how one can transform it into the form (11.1). Hence in particular, note that we may possibly have some $l_{ij} = u_{ij}$ in Problem (11.1). We emphasize again that the homogeneous form in Problem (11.1) and the assumption that $l_{ij} \ge 0$ for all (i, j) are only for the sake of convenience. The same algorithm is applicable with a general right-hand-side vector and with $-\infty < l_{ij} \le u_{ij} < \infty$, given a starting solution that is flow-conserving (perhaps to an artificial network). We ask the reader to explore this in Exercises 11.13, 11.14, and 11.15. In Exercise 11.6 we ask the reader to provide the details of this transformation between the two formulations.

Since all right-hand-side values of the flow conservation equations (11.1) are zero, we conclude that the flow in the network does not have a starting point or an ending point, but *circulates* continuously throughout the network. Thus all conserving flow in the network will be along circuits (directed cycles). For this reason, the representation (11.1) is known as a *circulatory network flow problem*.

The Dual of the Circulatory Network Flow Problem and Its Properties

If we associate a dual variable w_i with each node conservation equation (11.1), a dual variable h_{ij} with the constraint $x_{ij} \le u_{ij}$ (which is treated as $-x_{ij} \ge -u_{ij}$ for the purpose of taking the dual), and a dual variable v_{ij} with the constraint $x_{ij} \ge l_{ij}$, the dual of the out-of-kilter formulation for the minimal cost network flow problem is given by

$$\begin{aligned}
\text{Maximize} \quad & \sum_{i=1}^{m}\sum_{j=1}^{m} l_{ij}v_{ij} - \sum_{i=1}^{m}\sum_{j=1}^{m} u_{ij}h_{ij} & \\
\text{Subject to} \quad & w_i - w_j + v_{ij} - h_{ij} = c_{ij} & i, j = 1,\ldots, m \\
& h_{ij}, v_{ij} \ge 0 & i, j = 1,\ldots, m \\
& w_i \quad \text{unrestricted} & i = 1,\ldots, m
\end{aligned}$$

where the summations and the constraints are taken over existing arcs. The dual problem has a very interesting structure. Suppose that we select any set of w_i's (we shall assume throughout the development that the w_i's are integers). Then the dual constraint for arc (i, j) becomes

$$v_{ij} - h_{ij} = c_{ij} - w_i + w_j, \qquad h_{ij} \geq 0, \qquad v_{ij} \geq 0$$

and is satisfied by

$$v_{ij} = \text{Maximum}\{0, c_{ij} - w_i + w_j\}$$
$$h_{ij} = \text{Maximum}\{0, -(c_{ij} - w_i + w_j)\}$$

Thus the dual problem always possesses a feasible solution given any set of w_i's. In fact, the choices of v_{ij} and h_{ij} yield the optimal values of v_{ij} and h_{ij} for a fixed set of w_i's (why?).

The Complementary Slackness Conditions

The complementary slackness conditions for optimality of the out-of-kilter formulation are (review the Karush–Kuhn–Tucker optimality conditions) the following:

$$(x_{ij} - l_{ij})v_{ij} = 0 \qquad i, j = 1, 2, \ldots, m \tag{11.2}$$

$$(u_{ij} - x_{ij})h_{ij} = 0 \qquad i, j = 1, 2, \ldots, m \tag{11.3}$$

Define $z_{ij} - c_{ij} \equiv w_i - w_j - c_{ij}$. Then by the definition of v_{ij} and h_{ij} we get

$$v_{ij} = \text{Maximum}\{0, -(z_{ij} - c_{ij})\} \tag{11.4}$$

$$h_{ij} = \text{Maximum}\{0, z_{ij} - c_{ij}\} \tag{11.5}$$

Note that $z_{ij} - c_{ij}$ would be the familiar coefficient of x_{ij} in the objective function row of the lower-upper bounded simplex tableau if we had a basic solution to the primal problem. However, we need not have one here, and no such implication of a basis is being made here by this notation.

Given a set of w_i's we can compute $z_{ij} - c_{ij} = w_i - w_j - c_{ij}$. Noting Equations (11.4) and (11.5), the complementary slackness conditions (11.2) and (11.3) hold only if

$$z_{ij} - c_{ij} < 0 \Rightarrow v_{ij} > 0 \Rightarrow x_{ij} = l_{ij} \qquad i, j = 1, 2, \ldots, m$$

$$z_{ij} - c_{ij} > 0 \Rightarrow h_{ij} > 0 \Rightarrow x_{ij} = u_{ij} \qquad i, j = 1, 2, \ldots, m$$

Conversely, if $x_{ij} = l_{ij}$ for all (i, j) such that $z_{ij} - c_{ij} < 0$, and if $x_{ij} = u_{ij}$ for all (i, j) such that $z_{ij} - c_{ij} > 0$, then defining v_{ij} and h_{ij} as in Equations (11.4) and (11.5) respectively, we have complementary slackness holding. Hence we obtain the following key result that embodies the optimality conditions of Problem (11.1) as well as the principal thrust of the out-of-kilter algorithm.

***Theorem* 11.1**

Let $\mathbf{x}$ be any conserving flow, and let $\mathbf{w} = (w_1, \ldots, w_m)$ be any integer vector. Then $\mathbf{x}$ and $\mathbf{w}$ are respectively primal and dual optimal solutions to Problem

(11.1) if and only if for all (i, j)

$$z_{ij} - c_{ij} < 0 \text{ implies } x_{ij} = l_{ij}$$
$$z_{ij} - c_{ij} > 0 \text{ implies } x_{ij} = u_{ij}$$
$$z_{ij} - c_{ij} = 0 \text{ implies } l_{ij} \le x_{ij} \le u_{ij}$$

where $(z_{ij} - c_{ij}) \equiv w_i - w_j - c_{ij}$ for all (i, j).

The problem then is to search over values of w_i's and conserving x_{ij}'s until the three conditions of Theorem 11.1 are satisfied.

Consider Figure 11.1*a*. Selecting a set of starting w_i's, say, each $w_i = 0$, and a conserving flow, say, each $x_{ij} = 0$, we can check for optimality. Figure 11.1*b*, gives $z_{ij} - c_{ij}$, x_{ij}, and w_i for the network of Figure 11.1*a*. In Figure 11.1*b* we see that $z_{12} - c_{12} = -2$ and $x_{12} = 0\ (= l_{12})$ and thus arc $(1, 2)$ is said to be *in-kilter*, that is, *well*. On the other hand, $z_{23} - c_{23} = 3$ and $x_{23} = 0\ (< u_{23})$ and hence arc $(2, 3)$ is said to be *out-of-kilter*, that is, *sick*. Thus the name *out-of-kilter*.

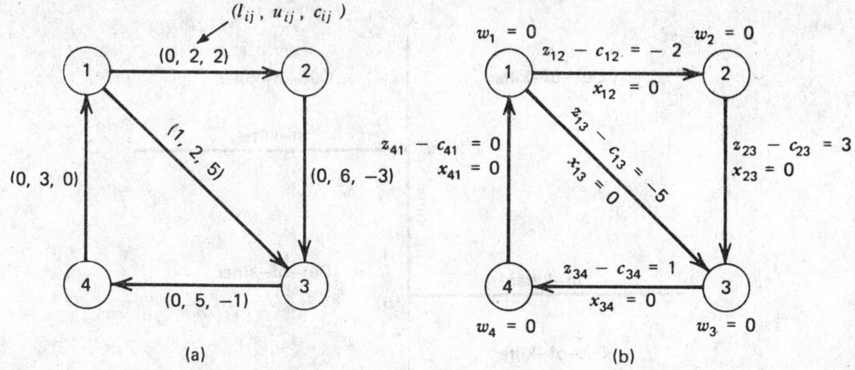

Figure 11.1. An example network: (*a*) The network. (*b*) w_i, $z_{ij} - c_{ij}$, x_{ij}.

To bring arc $(2, 3)$ into kilter we must either increase x_{23} or decrease $z_{23} - c_{23}$ by changing the w_i's. This is exactly what the out-of-kilter algorithm attempts to do. During the primal phase of the out-of-kilter algorithm we shall be changing the x_{ij}'s in an attempt to bring arcs into kilter. During the dual phase we change the w_i's in an attempt to reach an in-kilter state.

The Kilter States and Kilter Numbers for an Arc

The in-kilter and out-of-kilter states for each arc in a network are given in Figure 11.2. Note that an arc is in kilter if $l_{ij} \le x_{ij} \le u_{ij}$ and the conditions (of Theorem 11.1) hold. As we change the flow on arc (i, j), the arc moves up and down a particular column in Figure 11.2 depending on whether x_{ij} increased or decreased. As we change the w_i's the arc moves back and forth along a row. Figure 11.2*b* gives a graphical depiction of the kilter states of an arc. Each of the cells in the matrix in Figure 11.2*a* corresponds to a particular subregion in Figure 11.2*b*.

	$z_{ij} - c_{ij} < 0$	$z_{ij} - c_{ij} = 0$	$z_{ij} - c_{ij} > 0$
$x_{ij} > u_{ij}$	Out–of–kilter	Out–of–kilter	Out–of–kilter
$x_{ij} = u_{ij}$	Out–of–kilter	In–kilter	In–kilter
$l_{ij} < x_{ij} < u_{ij}$	Out–of–kilter	In–kilter	Out–of–kilter
$x_{ij} = l_{ij}$	In–kilter	In–kilter	Out–of–kilter
$x_{ij} < l_{ij}$	Out–of–kilter	Out–of–kilter	Out–of–kilter

(a)

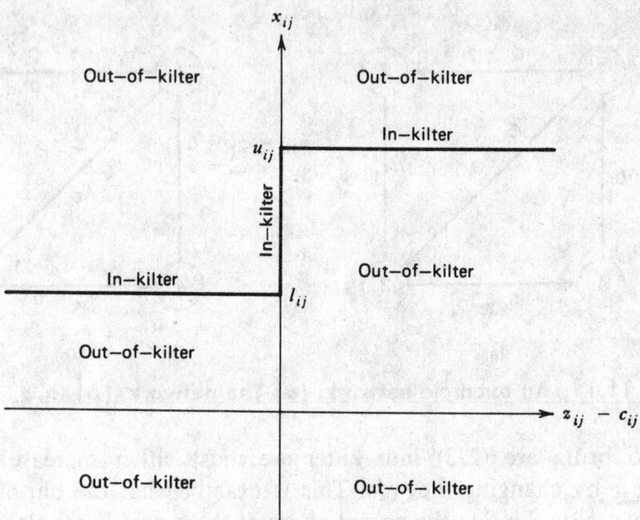

(b)

Figure 11.2. The possible kilter states for an arc.

In order to assure that the algorithm will converge, we need some measure of the "distance" from optimality. If we can construct an algorithm that periodically (at finite intervals) reduces the distance from optimality by an integer, then the algorithm will eventually converge.

There are many different measures of distance for the out-of-kilter method. We present in Figure 11.3 one measure of distance that we call the *kilter number* K_{ij} for an arc (i, j). The kilter number is defined here to be the *minimal change of flow on the arc that is needed to bring it into kilter*. The kilter number of an arc

	$z_{ij} - c_{ij} < 0$	$z_{ij} - c_{ij} = 0$	$z_{ij} - c_{ij} > 0$
$x_{ij} > u_{ij}$	$\lvert x_{ij} - l_{ij} \rvert$	$\lvert x_{ij} - u_{ij} \rvert$	$\lvert x_{ij} - u_{ij} \rvert$
$x_{ij} = u_{ij}$	$\lvert x_{ij} - l_{ij} \rvert$	0	0
$l_{ij} < x_{ij} < u_{ij}$	$\lvert x_{ij} - l_{ij} \rvert$	0	$\lvert x_{ij} - u_{ij} \rvert$
$x_{ij} = l_{ij}$	0	0	$\lvert x_{ij} - u_{ij} \rvert$
$x_{ij} < l_{ij}$	$\lvert x_{ij} - l_{ij} \rvert$	$\lvert x_{ij} - l_{ij} \rvert$	$\lvert x_{ij} - u_{ij} \rvert$

(a)

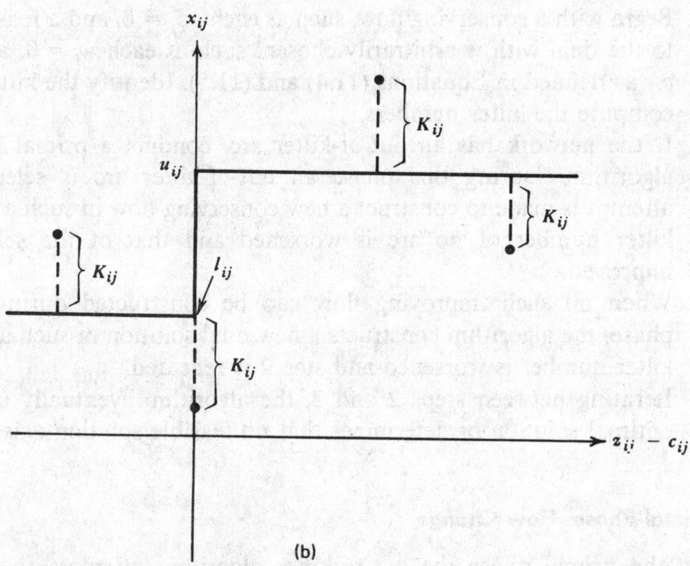

(b)

Figure 11.3. The kilter numbers K_{ij}'s.

is illustrated graphically in Figure 11.3*b*. Notice that since all terms involve absolute values, the kilter number for an arc is nonnegative. Also notice that if the arc is in-kilter, the associated kilter number is zero, and if the arc is out-of-kilter, the associated kilter number is strictly positive. Note that if $z_{ij} - c_{ij} < 0$, then arc (i, j) is in-kilter only if the flow is equal to l_{ij} and hence the kilter number $|x_{ij} - l_{ij}|$ indicates how far the current flow x_{ij} is from the ideal case l_{ij}. Similarly, if $z_{ij} - c_{ij} > 0$, then the kilter number $|x_{ij} - u_{ij}|$ gives the distance from the ideal flow of u_{ij}. Finally, if $z_{ij} - c_{ij} = 0$, then the arc is in-kilter if $l_{ij} \leq x_{ij} \leq u_{ij}$. In particular, if $x_{ij} > u_{ij}$, then the arc is brought in-kilter if the flow decreases by $|x_{ij} - u_{ij}|$, and if $x_{ij} < l_{ij}$, then the arc is brought in-kilter if the flow increases by $|x_{ij} - l_{ij}|$, and hence the entries in Figure 11.3 under the column $z_{ij} - c_{ij} = 0$.

One method of assuring finite convergence of the out-of-kilter algorithm is to guarantee the following.

1. The kilter number of any arc never increases.
2. At finite intervals the kilter number of some arc is reduced (by an integer).

This is exactly what we shall be able to achieve.

11.2 STRATEGY OF THE OUT-OF-KILTER ALGORITHM

As indicated before, the out-of-kilter algorithm may be generally viewed as a primal-dual type of algorithm. In this respect the generic steps of the algorithm are:

1. Begin with a conserving flow, such as each $x_{ij} = 0$, and a feasible solution to the dual with $\mathbf{w}$ arbitrarily chosen, such as each $w_i = 0$, and with h_{ij}, v_{ij} as defined in Equations (11.4) and (11.5). Identify the kilter states and compute the kilter numbers.
2. If the network has an out-of-kilter arc, conduct a primal phase of the algorithm. During this phase an out-of-kilter arc is selected and an attempt is made to construct a new conserving flow in such a way that the kilter number of no arc is worsened and that of the selected arc is improved.
3. When no such improving flow can be constructed during the primal phase, the algorithm constructs a new dual solution in such a way that no kilter number is worsened and step 2 is repeated.
4. Iterating between steps 2 and 3, the algorithm eventually constructs an optimal solution or determines that no feasible solution exists.

The Primal Phase: Flow Change

During the primal phase the out-of-kilter algorithm attempts to decrease the kilter number on an out-of-kilter arc by changing the conserving flows in such a way that the kilter number on any other arc is not worsened. Examining Figure 11.3, we see that the flows must be changed in such a way that the corresponding kilter states move closer to the in-kilter states. For example, for the out-of-kilter state $x_{ij} > u_{ij}$ and $z_{ij} - c_{ij} < 0$, we can decrease x_{ij} by as much as $|x_{ij} - l_{ij}|$ before the arc comes into kilter. If we decrease x_{ij} beyond this, the arc will pass the in-kilter state (we do not want this to happen). Also, we do not permit any increase in this x_{ij}. A similar analysis of the other kilter states produces the results in Figure 11.4*a*.

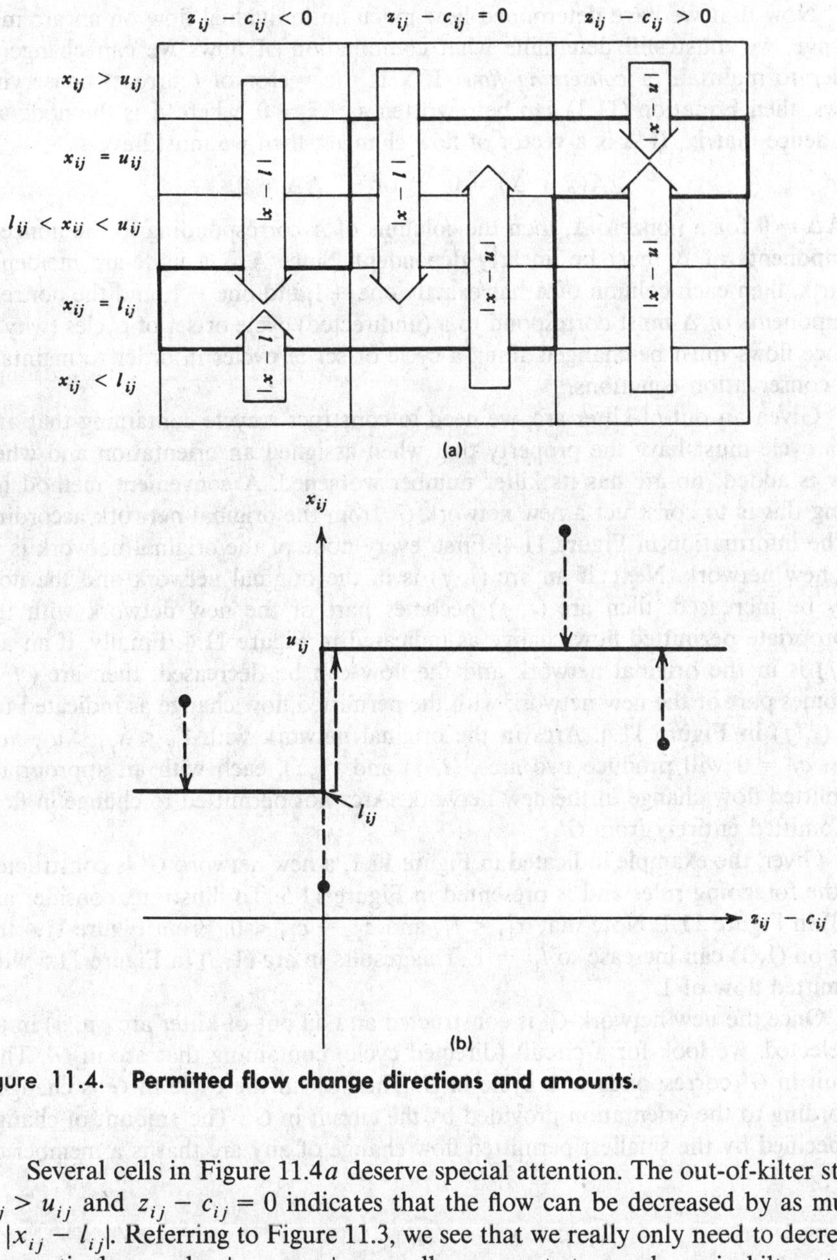

Figure 11.4. Permitted flow change directions and amounts.

Several cells in Figure 11.4a deserve special attention. The out-of-kilter state $x_{ij} > u_{ij}$ and $z_{ij} - c_{ij} = 0$ indicates that the flow can be decreased by as much as $|x_{ij} - l_{ij}|$. Referring to Figure 11.3, we see that we really only need to decrease the particular x_{ij} by $|x_{ij} - u_{ij}|$, a smaller amount, to reach an in-kilter state. However, as can be seen in Figure 11.3, we may continue to decrease x_{ij} by an amount up to $|x_{ij} - l_{ij}|$ and the arc will still remain in-kilter. It is often desirable to do this to aid other arcs in reaching in-kilter states. Also, an arc in the in-kilter state $l_{ij} < x_{ij} < u_{ij}$ and $z_{ij} - c_{ij} = 0$ may have its flow either increased or decreased, while still maintaining its in-kilter status. Again we may illustrate the permitted flow changes graphically. This is done in Figure 11.4b.

Now that we have determined how much an individual flow on an arc may change, we must still determine what combination of flows we can change in order to maintain a *conserving flow*. If $\bar{\mathbf{x}}$ is the vector of (current) conserving flows, then Equation (11.1) can be rewritten as $\mathbf{A}\bar{\mathbf{x}} = \mathbf{0}$, where $\mathbf{A}$ is the node-arc incidence matrix. If Δ is a vector of flow changes, then we must have

$$\mathbf{A}(\bar{\mathbf{x}} + \Delta) = \mathbf{0} \qquad \text{or} \qquad \mathbf{A}\Delta = \mathbf{0}$$

If $\mathbf{A}\Delta = \mathbf{0}$ for a nonzero Δ, then the columns of $\mathbf{A}$ corresponding to the nonzero components of Δ must be linearly dependent. Since $\mathbf{A}$ is a node-arc incidence matrix, then each column of $\mathbf{A}$ has exactly one $+1$ and one -1, and the nonzero components of Δ must correspond to a (undirected) cycle or set of cycles (why?). Hence flows must be changed along a cycle or set of cycles in order to maintain the conservation equations.

Given an out-of-kilter arc, we need to construct a cycle containing that arc. This cycle must have the property that when assigned an orientation and when flow is added, no arc has its kilter number worsened. A convenient method for doing this is to construct a new network G' from the original network according to the information in Figure 11.4. First, every node of the original network is in the new network. Next, if an arc (i, j) is in the original network and the flow may be increased, then arc (i, j) becomes part of the new network with the appropriate permitted flow change as indicated in Figure 11.4. Finally, if an arc (i, j) is in the original network and the flow can be decreased, then arc (j, i) becomes part of the new network with the permitted flow change as indicated for arc (i, j) in Figure 11.4. Arcs in the original network with $l_{ij} < x_{ij} < u_{ij}$ and $z_{ij} - c_{ij} = 0$ will produce two arcs, (i, j) and (j, i), each with an appropriate permitted flow change in the new network. Arcs not permitted to change in flow are omitted entirely from G'.

Given the example indicated in Figure 11.1, a new network G' is constructed by the foregoing rules and is presented in Figure 11.5. To illustrate, consider arc $(1, 3)$ in Figure 11.1. Note that $x_{13} < l_{13}$ and $z_{13} - c_{13} < 0$. From Figure 11.4 the flow on $(1, 3)$ can increase to $l_{13} = 1$. This results in arc $(1, 3)$ in Figure 11.5 with permitted flow of 1.

Once the new network G' is constructed and an out-of-kilter arc (p, q) in G' is selected, we look for a circuit (directed cycle) containing that arc in G'. This circuit in G' corresponds to a cycle in G. The flow in the cycle in G is changed according to the orientation provided by the circuit in G'. The amount of change is specified by the smallest permitted flow change of any arc that is a member of

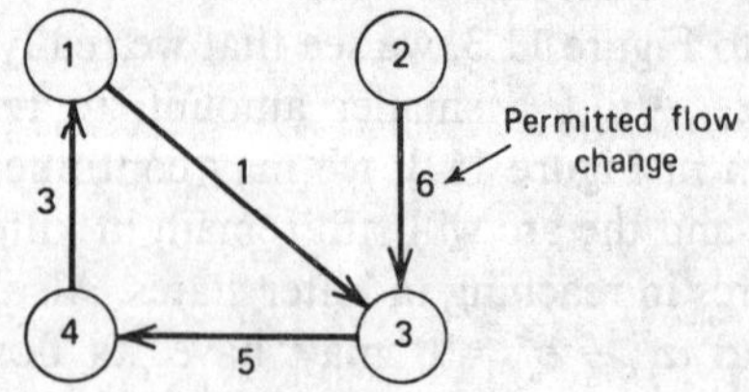

Figure 11.5. The modified network G' for Figure 11.1.

the circuit in G'. If no circuit containing the selected out-of-kilter arc exists in G', then we proceed to the dual phase of the algorithm.

We remark here that the construction of G' presented is only for pedagogical purposes. Given an actual out-of-kilter arc (p, q) in G', we wish to determine if there exists a circuit in G' that involves this arc. Note that this arc may be (p, q) or (q, p) in G, depending on whether an increase or a decrease in the flow on this arc is required in G. Hence in G' we need to find a (directed) path from q to p. This may be done by constructing a tree T that begins with the node q and at each stage scans for a pair of nodes i and j with $i \in T$ and $j \notin T$ such that there is an arc (i, j) in G that can permit a flow increase or such that there is an arc (j, i) in G that can permit a flow decrease. Arcs of this type are called *label eligible arcs*. If such an arc exists, it is used to include node j in the tree T and the process repeats. If node p gets included in T at any step, then we will have found a circuit in G' oriented along (p, q) that permits a positive flow change. This situation is called a *breakthrough*. If no additional nodes can be added to T at some step and $p \notin T$, then there does not exist any circuit in G' involving (p, q) (why?). This situation is called a *nonbreakthrough*. In this event, we proceed to the dual phase with the nodes in T forming a set X and the remaining nodes in $\bar{X}$. A labeling scheme that constructs such a tree T is described at the end of this chapter. For now, let us continue to work with the network G'.

As an illustration of the primal phase, consider the modified network G' of Figure 11.5. We select an out-of-kilter arc, say, (1, 3). From Figure 11.5, we see, that a circuit exists containing arc (1, 3), namely $C = (\{1, 3), (3, 4), (4, 1)\}$. Hence we can change the flow around the associated cycle in G, increasing flows on arcs with the orientation of the circuit in G' and decreasing flows on arcs against the orientation of the circuit in G', and obtain an improved (in the kilter number sense) solution. The amount of permitted change in flow is $\Delta =$ Minimum$\{1, 5, 3\} = 1$. The new solution and associated modified network is given in Figure 11.6*a*. Arcs (2, 3) and (3, 4) are still out-of-kilter in G. Selecting one of the associated arcs in G' (see Figure 11.6*b*), say, (2, 3), we attempt to find a circuit in G' containing the selected arc. Because no such circuit exists, we must pass to the dual phase of the out-of-kilter algorithm. The tree T in this case consists of nodes 3, 4, and 1, and arcs (3, 4) and (4, 1) from G'.

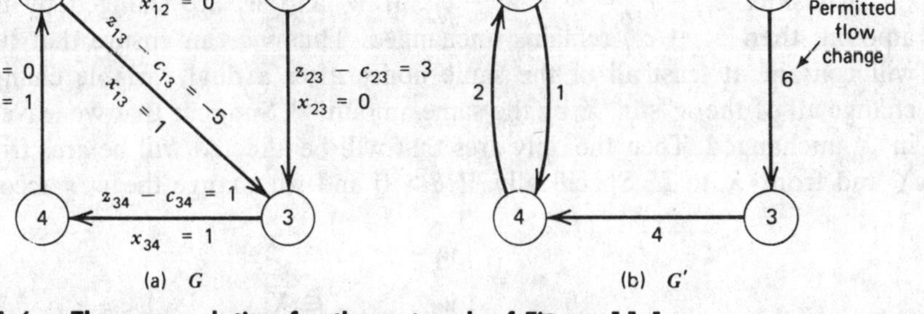

Figure 11.6. The new solution for the network of Figure 11.1.

It is convenient (but not necessary) for the various proofs of convergence to work on the same out-of-kilter arc (p, q) until it comes in-kilter. We shall assume throughout our discussion of the algorithm that this is done.

The Dual Phase: Dual Variable Change

When it is no longer possible to construct a circuit in G' containing a specific out-of-kilter arc, then we must change the $z_{ij} - c_{ij}$'s so that no kilter number is worsened and either the out-of-kilter arc is brought into kilter or some new arcs are introduced into G' that would eventually allow us to find a circuit containing the out-of-kilter arc under consideration.

Since $z_{ij} - c_{ij} = w_i - w_j - c_{ij}$, we must change the w_i's in order to change the $z_{ij} - c_{ij}$'s. Let (p, q) be an out-of-kilter arc in G', and let X be the set of nodes in G' that can be reached from node q along some *path* in G'. Hence X is the set of nodes in the tree T constructed earlier. Let $\bar{X} \equiv \mathcal{N} - X$, where $\mathcal{N} = \{1, \ldots, m\}$. Note that neither X nor $\bar{X}$ is empty since $q \in X$ and $p \in \bar{X}$ when we pass to the dual phase. For $(p, q) = (2, 3)$ in Figure 11.6 we have $X = \{3, 4, 1\}$ and $\bar{X} = \{2\}$. In Figure 11.7 we illustrate the sets X and $\bar{X}$.

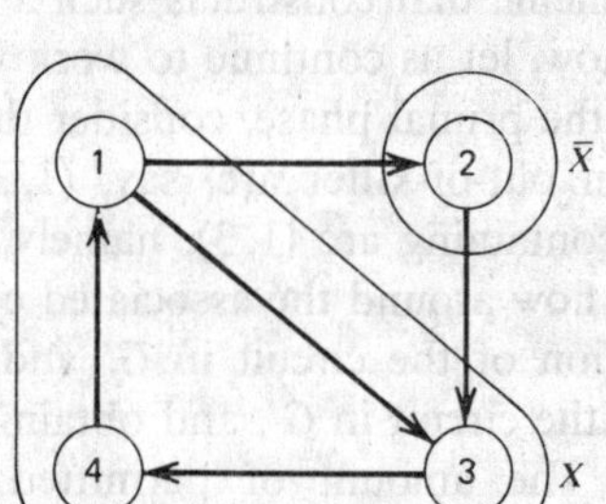

Figure 11.7. X and $\bar{X}$ in G for $(p, q) = (2, 3)$ in Figure 11.6.

We would like to change the w_i's so that no kilter number is worsened and the set X gets larger periodically. If another node comes into X at finite intervals, then eventually p will come into X and a circuit is created in G'. We have implicitly assumed that X will not get smaller. To ensure that this will not happen we should change the w_i's so that all arcs with both ends in X are retained in the modified graph.

Consider $z_{ij} - c_{ij} = w_i - w_j - c_{ij}$. If w_i and w_j are changed by the same amount, then $z_{ij} - c_{ij}$ remains unchanged. Thus we can ensure that the set X will contain at least all of the same nodes after a dual variable change if we change all of the w_i's in X by the same amount θ. Suppose that we leave the w_i's in $\bar{X}$ unchanged. Then the only arcs that will be affected will be arcs from X to $\bar{X}$ and from $\bar{X}$ to X. Specifically, if $\theta > 0$ and we change the w_i's according to

$$w_i = \begin{cases} w_i + \theta & i \in X \\ w_i & i \in \bar{X} \end{cases}$$

then the revised $(z_{ij} - c_{ij})$ values are given by

$$(z_{ij} - c_{ij})' = z_{ij} - c_{ij} \quad \text{if} \quad i \in X, \ j \in X$$
$$\text{or} \quad i \in \overline{X}, \ j \in \overline{X}$$

However, if $i \in X$ and $j \in \overline{X}$, we get

$$\begin{aligned}(z_{ij} - c_{ij})' &= (w_i + \theta) - w_j - c_{ij} \\ &= (z_{ij} - c_{ij}) + \theta\end{aligned}$$

Also for $i \in \overline{X}$ and $j \in X$ we get

$$\begin{aligned}(z_{ij} - c_{ij})' &= w_i - (w_j + \theta) - c_{ij} \\ &= (z_{ij} - c_{ij}) - \theta\end{aligned}$$

Thus arcs from X to $\overline{X}$ will have their $z_{ij} - c_{ij}$'s increased by θ and those arcs from $\overline{X}$ to X will have their $z_{ij} - c_{ij}$'s decreased by θ. We must determine θ so that the kilter number of no arc is worsened and the kilter state of some arc is changed. First we must identify the arcs that can be in the set $(X, \overline{X})$ and in the set $(\overline{X}, X)$. (The notation (X, Y) represents the set $S = \{(x, y) : x \in X, y \in Y\}$). The set of arcs in G going between X and $\overline{X}$ are said to constitute the *cut-set* with respect to X and $\overline{X}$.

Examining Figure 11.4, we see that the cut-set $(X, \overline{X})$ cannot contain an arc associated with the kilter state $x_{ij} < l_{ij}$ and $z_{ij} - c_{ij} < 0$, since such an arc (i, j) in G would become an arc in G' with the result that if i can be reached (along a path) from q, then j can be reached from q and thus $j \in X$ (a contradiction). Examining the remaining kilter states, we find that the only candidates for membership in $(X, \overline{X})$ are those identified in Figure 11.8. Recall that arcs from X to $\overline{X}$ in G have their $z_{ij} - c_{ij}$'s increased. Thus these arcs change kilter states in a left-to-right fashion as indicated in Figure 11.8*a*. Examining an arc from X to $\overline{X}$ in G that is in the kilter state $x_{ij} > u_{ij}$ and $z_{ij} - c_{ij} < 0$, we see from Figure 11.3 that as θ increases, K_{ij} decreases from $K_{ij} = |x_{ij} - l_{ij}|$ to $K_{ij} = |x_{ij} - u_{ij}|$ and thereafter remains constant. Thus, for such an arc, we can increase θ as much as we like and the arc's kilter number will never increase. Hence such an arc gives rise to an upper limit on θ of ∞ as indicated in Figure 11.8*a*. Any arc from X to $\overline{X}$ in G that is in the kilter state $x_{ij} = u_{ij}$ and $z_{ij} - c_{ij} < 0$ will have its kilter number first decrease and then remain unchanged as θ increases (why?). Thus, again ∞ is an upper limit on the permitted change in θ for such an arc to ensure that no kilter number will worsen. However, examining an arc from X to $\overline{X}$ in G in the kilter state $l_{ij} < x_{ij} < u_{ij}$ and $z_{ij} - c_{ij} < 0$, we see that the associated kilter number K_{ij} first decreases (to zero), then starts to increase. In order to eliminate the potential increase in K_{ij} for the arc we must place a limit of $|z_{ij} - c_{ij}|$ on θ. Similarly, we must place a limit of $|z_{ij} - c_{ij}|$ on θ for arcs in the state $x_{ij} = l_{ij}$ and $z_{ij} - c_{ij} < 0$. This analysis justifies the entries in Figure 11.8. Each of the permissible kilter states for arcs in $(X, \overline{X})$ is graphically portrayed in Figure 11.8*b*.

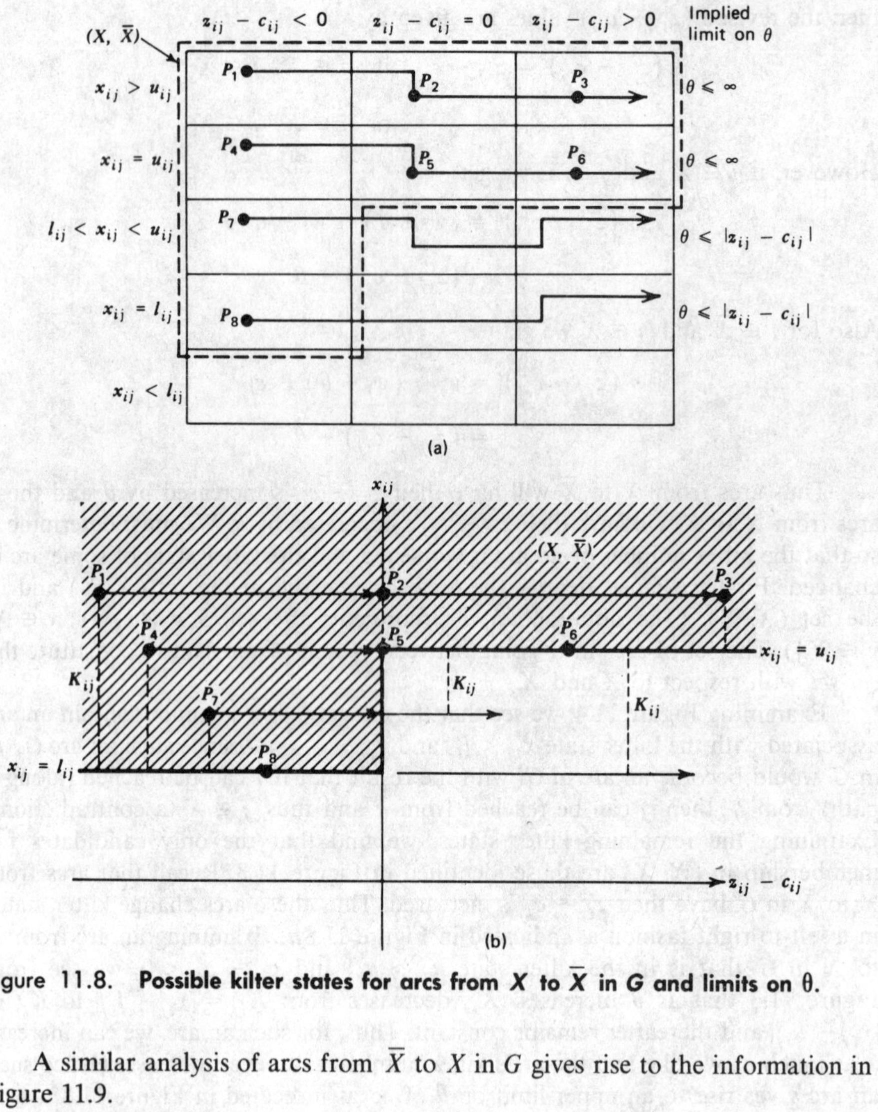

Figure 11.8. Possible kilter states for arcs from X to $\overline{X}$ in G and limits on θ.

A similar analysis of arcs from $\overline{X}$ to X in G gives rise to the information in Figure 11.9.

Insofar as worsening of kilter numbers is concerned, Figures 11.8 and 11.9 indicate that we need only compute θ based on arcs from X to $\overline{X}$ with $x_{ij} < u_{ij}$ and arcs from $\overline{X}$ to X with $x_{ij} > l_{ij}$. However, if we proceed to define a method of computing θ based only on these considerations, difficulties would arise in interpreting the meaning of the value $\theta = \infty$. Matters are greatly simplified if, instead of strict inequalities on flow (that is, $x_{ij} < u_{ij}$ and $x_{ij} > l_{ij}$), we admit weak inequalities on flow (that is, $x_{ij} \leq u_{ij}$ and $x_{ij} \geq l_{ij}$). The reason for this deviation from intuition will become apparent when we proceed to establish convergence of the algorithm.

The previous discussion concerning limits on θ based on kilter number considerations and on (yet to be established) convergence properties leads to the following formal procedure for computing θ.

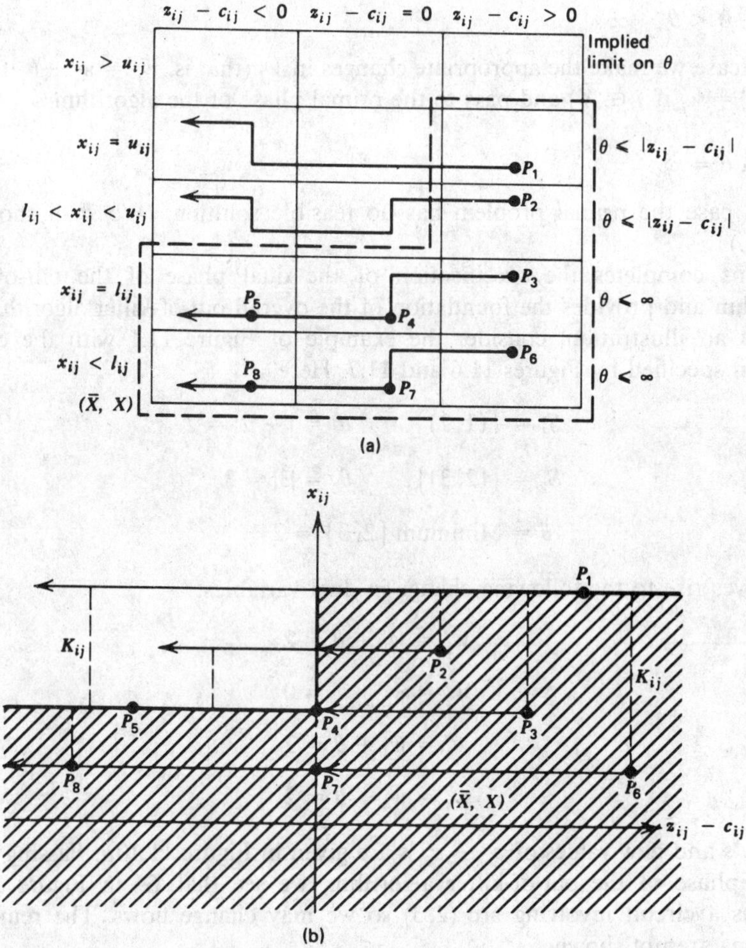

Figure 11.9. Possible kilter states for arcs from $\bar{X}$ to X in G and limits on θ.

In G define S_1 and S_2 by

$$S_1 \equiv \left\{(i, j) : i \in X,\ j \in \bar{X},\ z_{ij} - c_{ij} < 0,\ x_{ij} \leq u_{ij}\right\}$$

and

$$S_2 \equiv \left\{(i, j) : i \in \bar{X},\ j \in X,\ z_{ij} - c_{ij} > 0,\ x_{ij} \geq l_{ij}\right\}$$

Let

$$\theta_1 = \underset{(i,j) \in S_1}{\text{Minimum}} \left\{|z_{ij} - c_{ij}|\right\}$$

$$\theta_2 = \underset{(i,j) \in S_2}{\text{Minimum}} \left\{|z_{ij} - c_{ij}|\right\}$$

$$\theta = \text{Minimum}\left\{\theta_1, \theta_2\right\}$$

where $\theta_i \equiv \infty$ if S_i is empty. Thus θ is strictly positive (why?). Also, θ is either a positive integer or ∞ (why?). These two possibilities are briefly discussed.

Case 1: $0 < \theta < \infty$.

In this case we make the appropriate changes in w_i (that is, $w_i' = w_i + \theta$ if $i \in X$ and $w_i' = w_i$ if $i \in \overline{X}$) and pass to the primal phase of the algorithm.

Case 2: $\theta = \infty$.

In this case the primal problem has no feasible solution. (We shall show this shortly.)

This completes the specification of the dual phase of the out-of-kilter algorithm and provides the foundation of the overall out-of-kilter algorithm.

As an illustration, consider the example of Figure 11.1 with the current solution specified by Figures 11.6 and 11.7. Here

$$S_1 = \{(1,2)\}, \qquad \theta_1 = |-2| = 2$$

$$S_2 = \{(2,3)\}, \qquad \theta_2 = |3| = 3$$

$$\theta = \text{Minimum}\{2,3\} = 2$$

This gives rise to the following change in dual variables:

$$w_1' = w_1 + \theta = 2$$

$$w_2' = w_2 \qquad = 0$$

$$w_3' = w_3 + \theta = 2$$

$$w_4' = w_4 + \theta = 2$$

The x_{ij}'s and new values of $z_{ij} - c_{ij}$'s are given in Figure 11.10a. Passing to the primal phase of the out-of-kilter algorithm, we see that G' in Figure 11.10b contains a circuit involving arc (2, 3) so we may change flows. The remaining iterations are not shown.

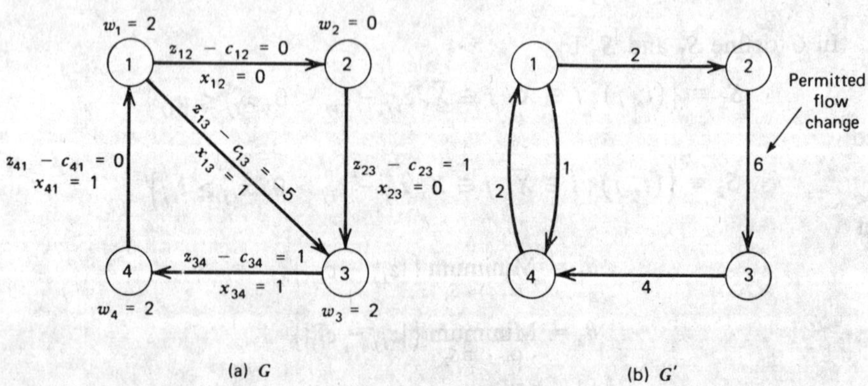

Figure 11.10. The new solution obtained from Figure 11.6 after the first dual variable change.

There is really no need to work directly with the dual variables themselves since we may transform the $z_{ij} - c_{ij}$'s directly by

$$(z_{ij} - c_{ij})' = \begin{cases} (z_{ij} - c_{ij}) & \text{if } i \in X,\ j \in X \text{ or } i \in \overline{X},\ j \in \overline{X} \\ (z_{ij} - c_{ij}) + \theta & \text{if } i \in X,\ j \in \overline{X} \\ (z_{ij} - c_{ij}) - \theta & \text{if } i \in \overline{X},\ j \in X \end{cases}$$

In Exercise 11.23 we ask the reader to show how the dual variables can be recovered anytime we need them. Note that $(z_{ij} - c_{ij})'$ is an integer (why?).

As an example of infeasibility, consider the example network of Figure 11.11*a*. Selecting a set of x_{ij}'s and w_i's, we find in Figure 11.11*b* that arc (2, 1) is out-of-kilter. Setting up G' in Figure 11.11*c*, we find no circuit containing the arc (2, 1). In this case $X = \{1\}$ and $\overline{X} = \{2\}$. Here $S_1 = \varnothing$ (the empty set) and $S_2 = \varnothing$ and thus $\theta = \infty$. It is clear by examining u_{12} and l_{21} that no feasible solution exists.

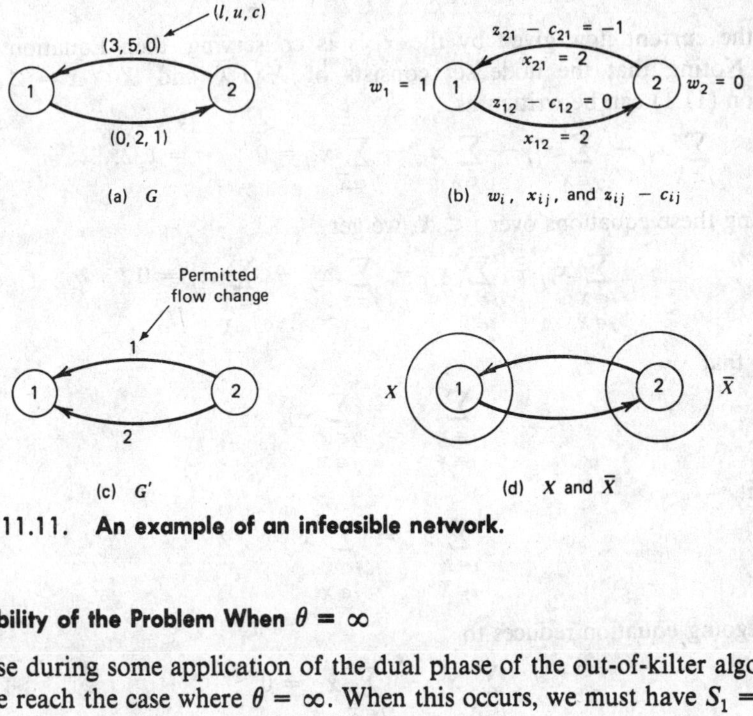

Figure 11.11. An example of an infeasible network.

Infeasibility of the Problem When $\theta = \infty$

Suppose during some application of the dual phase of the out-of-kilter algorithm that we reach the case where $\theta = \infty$. When this occurs, we must have $S_1 = S_2 = \varnothing$. Since $S_1 = \varnothing$, then by reviewing the definition of S_1 we conclude that $i \in X$ and $j \in X$ imply one of the following cases:

1. $z_{ij} - c_{ij} < 0$ and $x_{ij} > u_{ij}$
2. $z_{ij} - c_{ij} = 0$
3. $z_{ij} - c_{ij} > 0$

From Figure 11.8 and since $i \in X$ and $j \in \overline{X}$, possibility (2) or (3) can hold only if $x_{ij} \geq u_{ij}$. Hence $S_1 = \varnothing$ holds only if $x_{ij} \geq u_{ij}$ for $i \in X$ and $j \in \overline{X}$. Similarly, $S_2 = \varnothing$ holds only if $i \in \overline{X}$ and $j \in X$ implies that $x_{ij} \leq l_{ij}$. Hence $S_1 = S_2 = \varnothing$ implies

$$x_{ij} \geq u_{ij} \quad \text{if} \quad i \in X,\ j \in \overline{X} \tag{11.6}$$

and

$$x_{ij} \leq l_{ij} \quad \text{if} \quad i \in \overline{X},\ j \in X \tag{11.7}$$

In particular, consider the out-of-kilter arc (p, q) in G'. If (p, q) is in G, then by inequality (11.7) $x_{pq} \leq l_{pq}$. Suppose that $x_{pq} = l_{pq}$. Since (p, q) is out-of-kilter, then $z_{pq} - c_{pq} > 0$, violating the assumption that $S_2 = \varnothing$. Thus, $x_{pq} < l_{pq}$. If, on the other hand, (q, p) is in G then by a similar argument we may show that $x_{qp} > u_{qp}$. Thus, at least one of the inequalities (11.6) or (11.7) is strict. Summing these two inequalities we get

$$\sum_{\substack{i \in X \\ j \in \overline{X}}} x_{ij} - \sum_{\substack{i \in \overline{X} \\ j \in X}} x_{ij} > \sum_{\substack{i \in X \\ j \in \overline{X}}} u_{ij} - \sum_{\substack{i \in \overline{X} \\ j \in X}} l_{ij} \tag{11.8}$$

Since the current flow given by the x_{ij}'s is conserving, then Equation (11.1) holds. Noting that the node set consists of $X \cup \overline{X}$ and $X \cap \overline{X} = \varnothing$, then Equation (11.1) can be written as

$$\sum_{j \in X} x_{ij} + \sum_{j \in \overline{X}} x_{ij} - \sum_{j \in X} x_{ji} - \sum_{j \in \overline{X}} x_{ji} = 0 \qquad i = 1, 2, \ldots, m$$

Summing these equations over $i \in X$, we get

$$\sum_{\substack{i \in X \\ j \in X}} x_{ij} + \sum_{\substack{i \in X \\ j \in \overline{X}}} x_{ij} - \sum_{\substack{j \in X \\ i \in X}} x_{ji} - \sum_{\substack{i \in X \\ j \in \overline{X}}} x_{ji} = 0$$

Noting that

$$\sum_{\substack{i \in X \\ j \in X}} x_{ij} = \sum_{\substack{j \in X \\ i \in X}} x_{ji}$$

and that

$$\sum_{\substack{i \in X \\ j \in \overline{X}}} x_{ji} = \sum_{\substack{i \in \overline{X} \\ j \in X}} x_{ij}$$

the foregoing equation reduces to

$$\sum_{\substack{i \in X \\ j \in \overline{X}}} x_{ij} - \sum_{\substack{i \in \overline{X} \\ j \in X}} x_{ij} = 0 \tag{11.9}$$

Substituting in inequality (11.8), we get

$$0 > \sum_{\substack{i \in X \\ j \in \overline{X}}} u_{ij} - \sum_{\substack{i \in \overline{X} \\ j \in X}} l_{ij} \tag{11.10}$$

Suppose by contradiction that there is a feasible flow represented by $\hat{x}_{ij}$ for $i, j = 1, 2, \ldots, m$. Therefore $u_{ij} \geq \hat{x}_{ij}$ and $-l_{ij} \geq -\hat{x}_{ij}$ and so inequality (11.10) gives

$$0 > \sum_{\substack{i \in X \\ j \in \overline{X}}} u_{ij} - \sum_{\substack{i \in \overline{X} \\ j \in X}} l_{ij} \geq \sum_{\substack{i \in X \\ j \in \overline{X}}} \hat{x}_{ij} - \sum_{\substack{i \in \overline{X} \\ j \in X}} \hat{x}_{ij} \tag{11.11}$$

But since the $\hat{x}_{ij}$'s represent a feasible flow, they must be conserving. In a fashion similar to Equation (11.9) it is clear that the right-hand-side of inequality (11.11) is equal to zero. Therefore inequality (11.11) implies that $0 > 0$, which is impossible. This contradiction shows that if $\theta = \infty$, there could be no feasible flow.

Note that if we had defined S_1 and S_2 by strict inequalities on x (namely, $x_{ij} < u_{ij}$ and $x_{ij} > l_{ij}$ respectively), we could not have produced the strict inequality needed in (11.8).

Convergence of the Out-of Kilter Algorithm

For the purpose of the following finite convergence argument we make the assumption that the vectors $\mathbf{l}$, $\mathbf{u}$, and $\mathbf{c}$ are integer valued.

In developing a finite convergence argument for the out-of-kilter algorithm, there are several properties of the algorithm that should be noted. First, every time a circuit is constructed in G' containing an out-of-kilter arc, the kilter number of that arc and of the total network is reduced by an integer (why?). We can construct only a finite number of circuits containing out-of-kilter arcs before an optimal solution is obtained (why?). Second, after each dual variable change, the kilter state of each arc in G that has both ends in X remains unchanged. Hence, if (p, q) is not in kilter, then after a dual variable change, each node in X before the change is in X after the change. Two possibilities exist. One possibility is that a new node k may be brought into X by virtue of an arc being added in G' from some node in X to node k. Each time this occurs the set X grows by at least one node. This can occur at most a finite number of times before node p becomes a member of X and a circuit is created containing (p, q). Thus, if the algorithm is not finite, it must be the case that an infinite number of dual variable changes take place without the set X increasing or $\theta = \infty$. We shall show that this cannot occur.

Suppose that after a dual variable change no new node becomes a member of X; that is, X does not increase. Then upon passing to the next dual phase we have the same sets X and $\overline{X}$ and the same x_{ij}'s. In addition, each arc from X to $\overline{X}$ has had its $z_{ij} - c_{ij}$ increased and each arc from $\overline{X}$ to X has had its $z_{ij} - c_{ij}$ decreased. Thus, after the dual variable change, the new sets S_1' and S_2' satisfy

$$S_1' \subseteq S_1 \qquad \text{and} \qquad S_2' \subseteq S_2$$

(why?). Further, by the choice of the (finite) value of θ, at least one arc has been dropped from either S_1 or S_2. Thus at least one of the foregoing inclusions is proper. Now S_1 and S_2 may decrease at most a finite number of times before $S_1 \cup S_2 = \varnothing$ and $\theta = \infty$, in which case the algorithm stops.

This completes a finiteness argument for the out-of-kilter algorithm. We now summarize the algorithm and present an example.

11.3 SUMMARY OF THE OUT-OF-KILTER ALGORITHM

The complete algorithm consists of three phases: the initiation phase, the primal phase, and the dual phase.

Initiation Phase

Begin with a conserving (integer) flow, say, each $x_{ij} = 0$, and an initial set of, (integer) dual variables, say, each $w_i = 0$. Compute $z_{ij} - c_{ij} = w_i - w_j - c_{ij}$.

Primal Phase

Determine the kilter state and the kilter number for each arc. If all arcs are in kilter, stop; the optimal solution is obtained. Otherwise select or continue with a previously selected out-of-kilter arc. From the network G construct a new network G' according to Figure 11.4. For each arc (i, j) in G that is in one of the kilter states that permit a flow increase, place an arc (i, j) in G' with a permitted flow increase as indicated in Figure 11.4. For any arc (i, j) in G that is in one of the kilter states that permit a flow decrease, place an arc (j, i) in G' with the permitted flow indicated in Figure 11.4. For those arcs in G that are members of states that permit no flow change, place no arc in G'. In G', attempt to construct a circuit containing the selected out-of-kilter arc (p, q). If such a circuit is available, we have a breakthrough. Determine a flow change Δ equal to the minimum of the permitted flow changes on arcs of the circuit. Change the flow on each arc of the associated cycle in G by the amount Δ using the orientation specified by the circuit as the direction of increase. In particular, let $x'_{ij} = x_{ij} + \Delta$ if (i, j) was a member of the circuit in G'; let $x'_{ij} = x_{ij} - \Delta$ if (j, i) was a member of the circuit in G'; let $x'_{ij} = x_{ij}$ otherwise. Repeat the primal phase. If no circuit containing arc (p, q) is available in G', we have a nonbreakthrough, and we pass to the dual phase.

Dual Phase

Determine the set of nodes X that can be reached from node q along a path in G'. (This is available via the tree T constructed as in Section 11.2.) Let $\bar{X} = \mathcal{N} - X$. In G, define S_1 and S_2 by

$$S_1 = \left\{(i, j) : i \in X,\ j \in \bar{X},\ z_{ij} - c_{ij} < 0,\ x_{ij} \leq u_{ij}\right\}$$

$$S_2 = \left\{(i, j) : i \in \bar{X},\ j \in X,\ z_{ij} - c_{ij} > 0,\ x_{ij} \geq l_{ij}\right\}$$

Let

$$\theta = \underset{(i,\,j)\in S_1 \cup S_2}{\text{Minimum}} \left\{ |z_{ij} - c_{ij}|, \infty \right\}$$

If $\theta = \infty$, stop; no feasible solution exists. Otherwise, change the w_i's and the $z_{ij} - c_{ij}$'s according to

$$w_i' = \begin{cases} w_i + \theta & \text{if } i \in X \\ w_i & \text{if } i \in \overline{X} \end{cases}$$

$$(z_{ij} - c_{ij})' = \begin{cases} (z_{ij} - c_{ij}) & \text{if } (i, j) \in (X, X) \cup (\overline{X}, \overline{X}) \\ (z_{ij} - c_{ij}) + \theta & \text{if } (i, j) \in (X, \overline{X}) \\ (z_{ij} - c_{ij}) - \theta & \text{if } (i, j) \in (\overline{X}, X) \end{cases}$$

and pass to the primal phase.

11.4 AN EXAMPLE OF THE OUT-OF-KILTER ALGORITHM

Consider the network given in Figure 11.12. Initiating the out-of-kilter algorithm with each $x_{ij} = 0$ and each $w_i = 0$, we get the sequence of primal and dual phases given in Figure 11.13.

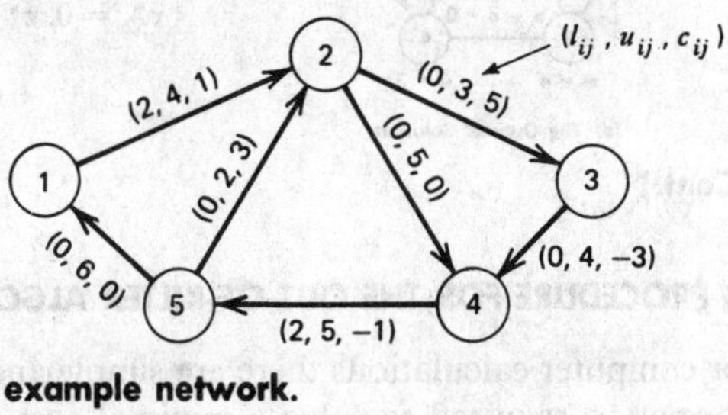

Figure 11.12. An example network.

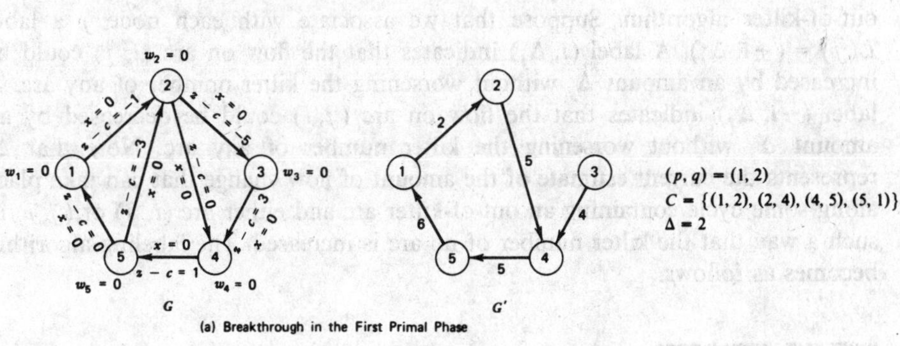

(a) Breakthrough in the First Primal Phase

Figure 11.13. The out-of-kilter method solution for Figure 11.12. (Continued on next page.)

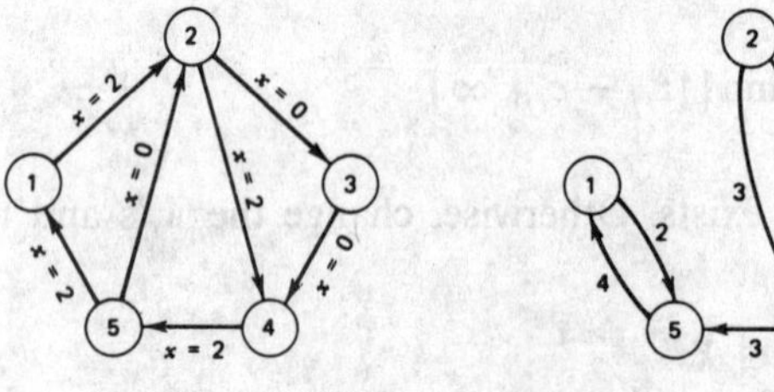

$(p, q) = (3, 4)$
$X = \{1, 2, 4, 5\}, \bar{X} = \{3\}$
$S_1 = \{(2, 3)\}, S_2 = \{(3, 4)\}$
$\theta = 3$

(b) Nonbreakthrough and the First Dual Phase

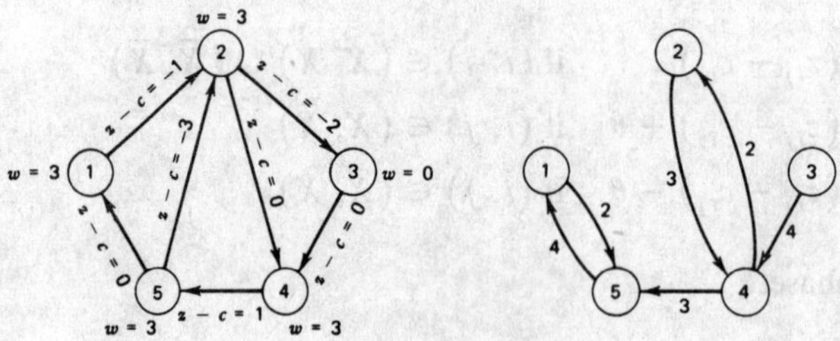

(3, 4) is in - kilter
$(p, q) = (4, 5)$
$X = \{1, 5\}, \bar{X} = \{2, 3, 4\}$
$S_1 = \{(1, 2), (5, 2)\}$
$S_2 = \{(4, 5)\}$
$\theta = 1$

(c) Nonbreakthrough and the Second Dual Phase

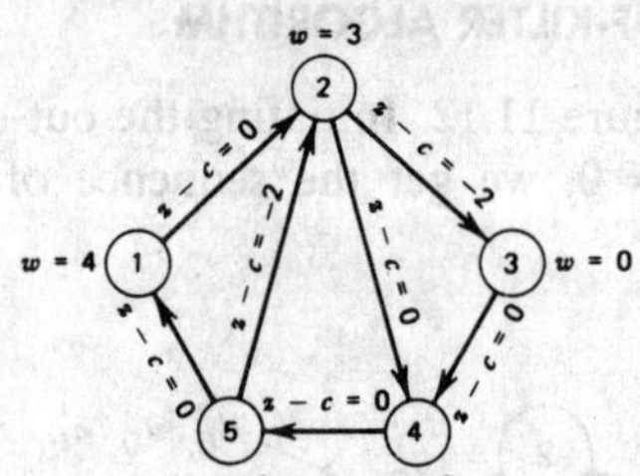

$x_{12}^* = 2, x_{23}^* = 0$
$x_{24}^* = 2, x_{34}^* = 0$
$x_{45}^* = 2, x_{51}^* = 2$
$x_{52}^* = 0, z^* = 0$

(d) The Optimal Solution

Figure 11.13. **(Cont.)**

11.5 A LABELING PROCEDURE FOR THE OUT-OF-KILTER ALGORITHM

Either for hand or computer calculations there are simple and convenient ways to maintain the information required to solve a minimal cost flow problem by the out-of-kilter algorithm. Suppose that we associate with each node j a label $L(j) = (\pm i, \Delta_j)$. A label (i, Δ_j) indicates that the flow on arc (i, j) could be increased by an amount Δ_j without worsening the kilter number of any arc. A label $(-i, \Delta_j)$ indicates that the flow on arc (j, i) could be decreased by an amount Δ_j without worsening the kilter number of any arc. Note that Δ_j represents the current estimate of the amount of flow change that can take place along some cycle containing an out-of-kilter arc and either arc (i, j) or (j, i) in such a way that the kilter number of no arc is increased. The labeling algorithm becomes as follows.

INITIALIZATION STEP

Select a conserving flow, for example, each $x_{ij} = 0$, and a set of dual variables, such as each $w_i = 0$.

Main Step

1. If all arcs are in kilter according to Figure 11.2, stop; the optimal solution is obtained. Otherwise, select (or continue with a previously selected) out-of-kilter arc, say (s, t). Erase all labels. If (s, t) is in one of the states where a flow increase, Δ_{st} is required according to Figure 11.4, then set $q = t$, $p = s$, and $L(q) = (+p, \Delta_{st})$. Otherwise, if (s, t) is in one of the states where a flow decrease is required according to Figure 11.4, then set $q = s$, $p = t$, and $L(q) = (-p, \Delta_{st})$.
2. If node i has a label, node j has no label, and flow may be increased by an amount Δ_{ij} along arc (i, j) according to Figure 11.4, then assign node j the label $L(j) = (+i, \Delta_j)$ where $\Delta_j = \text{Minimum}\{\Delta_i, \Delta_{ij}\}$. If node i has a label, node j has no label, and flow may be decreased by an amount Δ_{ji} along arc (j, i) according to Figure 11.4, then give node j the label $L(j) = (-i, \Delta_j)$ where $\Delta_j = \text{Minimum}(\Delta_i, \Delta_{ji}\}$. Repeat step 2 until either node p is labeled or until no more nodes can be labeled. If node p is labeled, go to step 3 (*breakthrough* has occurred); otherwise, go to step 4 (*nonbreakthrough* has occurred and the labeled nodes are in the tree T).
3. Let $\Delta = \Delta_p$. Change flow along the identified cycle as follows. Begin at node p. If the first entry in $L(p)$ is $+k$, then add Δ to x_{kp}. Otherwise, if the first entry in $L(p)$ is $-k$, then subtract Δ from x_{pk}. Backtrack to node k and repeat the process until node p is reached again in the backtrack process. Go to step 1.
4. Let X be the set of labeled nodes and let $\overline{X} = \mathcal{N} - X$. Define $S_1 = \{(i, j) : i \in X,\ j \in \overline{X},\ z_{ij} - c_{ij} < 0,\ x_{ij} \leq u_{ij}\}$ and $S_2 = \{(i, j) : i \in \overline{X},\ j \in X,\ z_{ij} - c_{ij} > 0,\ x_{ij} \geq l_{ij}\}$. Let $\theta = \text{Minimum}\{|z_{ij} - c_{ij}|,\ \infty : (i, j) \in S_1 \cup S_2\}$. If $\theta = \infty$, stop; no feasible solution exists. Otherwise, let

$$w_i' = \begin{cases} w_i + \theta & \text{if } i \in X \\ w_i & \text{if } i \in \overline{X} \end{cases}$$

and

$$(z_{ij} - c_{ij})' = \begin{cases} (z_{ij} - c_{ij}) & \text{if } (i, j) \in (X, X) \cup (\overline{X}, \overline{X}) \\ (z_{ij} - c_{ij}) + \theta & \text{if } (i, j) \in (X, \overline{X}) \\ (z_{ij} - c_{ij}) - \theta & \text{if } (i, j) \in (\overline{X}, X) \end{cases}$$

Go to Step 1.

An Example of the Labeling Algorithm

We shall illustrate the labeling method for the out-of-kilter algorithm by performing the first two iterations represented in Figure 11.13a and b. From Figure 11.13a we find that arc (1, 2) is an out-of-kilter arc whose flow must be increased.

The sequence of operations of the labeling algorithm are as follows:

1. $s, t = (1,2)$, $q = 2$, $p = 1$, $L(2) = (+1,2)$
2. $L(4) = (+2,2)$
3. $L(5) = (+4,2)$
4. $L(1) = (+5,2)$
5. Breakthrough: $\Delta = 2$
6. $L_1(1) = +5 \Rightarrow x'_{51} = x_{51} + \Delta = 2$
7. $L_1(5) = +4 \Rightarrow x'_{45} = x_{45} + \Delta = 2$
8. $L_1(4) = +2 \Rightarrow x'_{24} = x_{24} + \Delta = 2$
9. $L_1(2) = +1 \Rightarrow x'_{12} = x_{12} + \Delta = 2$
10. Erase all labels, $(s, t) = (3,4)$, $q = 4$, $p = 3$, $L(4) = (+3,4)$
11. $L(5) = (+4,3)$
12. $L(1) = (+5,3)$
13. $L(2) = (-4,2)$
14. Nonbreakthrough: $X = \{1,2,4,5\}$, $\overline{X} = \{3\}$, $\theta = 3$
15. $w_1 = w_2 = w_4 = w_5 = 3$, $w_3 = 0$

Since arc $(3,4)$ is now in-kilter, we select another out-of-kilter arc, erase all labels, and continue.

11.6 INSIGHT INTO CHANGES IN PRIMAL AND DUAL FUNCTION VALUES

It is instructive to see how the primal (penalty) and dual objective function values change in the primal and dual phases of the out-of-kilter algorithm. Not only does this provide an alternative convergence argument in which one need not necessarily work on the same out-of-kilter arc until it comes in kilter, but it also provides an insight into other acceptable ways of modifying primal and dual solutions.

First consider the primal phase. Define for each arc (i, j) in G a function $P_{ij}(x_{ij})$ that measures the infeasibility of a conserving flow $\mathbf{x}$ as follows:

$$P_{ij}(x_{ij}) = \text{Maximum}\{0, l_{ij} - x_{ij}\} + \text{Maximum}\{0, x_{ij} - u_{ij}\}$$

Note that $P_{ij}(x_{ij}) = 0$ if and only if $l_{ij} \le x_{ij} \le u_{ij}$ and is positive otherwise. Construct a primal *penalty function* $f(\mathbf{x})$ defined for any conserving flow $\mathbf{x}$ as follows:

$$f(\mathbf{x}) = \mathbf{cx} + M \sum_{(i,j)} P_{ij}(x_{ij}) \qquad (11.12a)$$

where M is sufficiently large (for example, $M > 2\Sigma_{(i,j)}|c_{ij}|u_{ij}$, assuming that $0 \le l_{ij} \le u_{ij} < \infty$ and that the algorithm is initialized with $x_{ij} = 0$ for all (i, j)). Observe that the function $f(\mathbf{x})$ composes the original objective function with a penalty term and is similar to the big-M objective function. For $\mathbf{x}$ restricted to be flow-conserving, we have $\mathbf{Ax} = \mathbf{0}$ so that $\mathbf{wAx} = 0$. Subtracting $\mathbf{wAx} = 0$ from the expression in Equation (11.12a) and noting that $(\mathbf{c} - \mathbf{wA})$ has components $(c_{ij} - z_{ij})$, we can equivalently rewrite this function as

$$f(\mathbf{x}) = \sum_{(i,j)} (c_{ij} - z_{ij})x_{ij} + M \sum_{(i,j)} P_{ij}(x_{ij}) \qquad (11.12b)$$

for any conserving flow **x**. Whenever we have a breakthrough in the primal phase, we modify flow in a cycle in G such that no kilter number worsens, and the kilter number of at least one arc strictly reduces. In particular, no $P_{ij}(x_{ij})$ term increases (why?). If any such term decreases (by an integer), then since M is large enough, so does the value of $f(\cdot)$. On the other hand, if no $P_{ij}(x_{ij})$ term decreases in the cycle, then we must have $l_{ij} \leq x_{ij} \leq u_{ij}$ for all arcs (i, j) in this cycle (why?). Thus the penalty term in Equation (11.12) remains constant during the flow change. However, from Figure 11.4 and the foregoing fact, if any arc (i, j) in the cycle satisfies $z_{ij} - c_{ij} < 0$, then we must have $l_{ij} < x_{ij} \leq u_{ij}$ and x_{ij} must be decreasing in the breakthrough. Similarly, if $z_{ij} - c_{ij} > 0$ for any arc (i, j) in the cycle, then we must have $l_{ij} \leq x_{ij} < u_{ij}$ and x_{ij} must be increasing in the breakthrough. In either event, the first term in Equation (11.12*b*) strictly falls. Since we cannot have $z_{ij} - c_{ij} = 0$ for all arcs (i, j) in the cycle in this case (why?), we again obtain a strict decrease in $f(\cdot)$. Therefore $f(\cdot)$ falls by an integer at every primal breakthrough.

Next consider the dual phase. Note from Equations (11.4) and (11.5) that $h_{ij} - v_{ij} - (z_{ij} - c_{ij}) = 0$ for all (i, j). Using the current primal flow **x**, we can equivalently rewrite the dual objective function as follows:

$$\sum_{(i,j)} x_{ij}(c_{ij} - z_{ij}) + \sum_{(i,j)} (l_{ij} - x_{ij})v_{ij} + \sum_{(i,j)} (x_{ij} - u_{ij})h_{ij} \qquad (11.13)$$

During the dual phase, assuming $\theta < \infty$, we note that $z_{ij} - c_{ij}$ increases by θ for all $(i, j) \in (X, \bar{X})$, and decreases by θ for all $(i, j) \in (\bar{X}, X)$. Hence from Equation (11.9), the first term in Equation (11.13) remains constant (why?). Next consider an arc $(i, j) \in (X, \bar{X})$. Referring to Equations (11.4) and (11.5) and Figure 11.8, suppose that $\bar{c}_{ij} \equiv z_{ij} - c_{ij} < 0$. Since $\bar{c}_{ij}$ increases by θ, v_{ij} decreases by θ and h_{ij} remains zero in case $l_{ij} \leq x_{ij} \leq u_{ij}$, and v_{ij} decreases by $\min\{\theta, |\bar{c}_{ij}|\}$ and h_{ij} increases by $\max\{0, \theta - |\bar{c}_{ij}|\}$ in case $x_{ij} > u_{ij}$. Therefore, the corresponding term $(l_{ij} - x_{ij})v_{ij} + (x_{ij} - u_{ij})h_{ij}$ in Equation (11.13) remains unchanged if and only if $x_{ij} = l_{ij}$, that is, arc (i, j) is in kilter, and increases by a positive integer otherwise. Furthermore if $z_{ij} - c_{ij} \geq 0$, then h_{ij} increases by θ, v_{ij} remains zero, and we have $x_{ij} \geq u_{ij}$. Consequently, the corresponding term in Equation (11.13) remains unchanged if and only if $x_{ij} = u_{ij}$, that is, arc (i, j) is in kilter, and increases by a positive integer otherwise. Similarly, consider $(i, j) \in (\bar{X}, X)$. If $\bar{c}_{ij} \equiv z_{ij} - c_{ij} > 0$, then h_{ij} falls by θ and v_{ij} remains zero in case $l_{ij} \leq x_{ij} \leq u_{ij}$, and h_{ij} falls by $\min\{\theta, \bar{c}_{ij}\}$ and v_{ij} increases by $\max\{0, \theta - \bar{c}_{ij}\}$ in case $x_{ij} < l_{ij}$. On the other hand, if $z_{ij} - c_{ij} \leq 0$, then v_{ij} increases by θ and h_{ij} remains zero, while we have $x_{ij} \leq l_{ij}$. Again, the corresponding term in Equation (11.13) remains unchanged if and only if (i, j) is in kilter and increases by a positive integer otherwise. Since the out-of-kilter arc (p, q) is either in $(X, \bar{X})$ or $(\bar{X}, X)$ in G, we obtain a strict increase (by an integer) in the dual objective value during an iteration of the dual phase.

Since the primal penalty function $f(\cdot)$ in Equation (11.12) falls by an integer in a primal breakthrough and since the dual objective function (11.13) increases by an integer in every dual phase, the difference given by the primal minus the dual function values (call it the *duality gap*) falls by an integer every iteration. Hence if $S_1 = \varnothing$ and $S_2 = \varnothing$ is not realized, the duality gap must become nonpositive finitely. In this event, if $P_{ij}(x_{ij}) = 0$ for all (i, j), that is, we have feasibility, then the current solution $(\mathbf{x}, \mathbf{w})$ is primal-dual optimal. On the other hand, if some $P_{ij}(x_{ij})$ is positive, then since it is an integer, we have from Equation (11.12a) that the dual value is $\geq \mathbf{cx} + M > \Sigma_{(i,j)}|c_{ij}|u_{ij}$. This exceeds any possible feasible value realizable in Equation (11.1). Hence the dual is unbounded and the primal is infeasible, and so we may terminate.

For the illustrative example of Section 11.4, we compute $M > 2(4 + 6 + 5 + 15 + 12) = 84$, say, $M = 85$. For the starting solution $(\bar{\mathbf{x}}, \bar{\mathbf{w}}) = (\mathbf{0}, \mathbf{0})$, we have $f(\mathbf{0}) = M(2 + 2) = 4M = 340$ and the dual objective value is $2 - 5 - 12 = -15$ from Equations (11.12) and (11.13) respectively. After the first breakthrough, the primal penalty function value becomes $2(1) + 2(-1) + M(0) = 0$. Note that feasibility is achieved and will be maintained by the algorithm henceforth (why?). After the first dual phase, the dual objective value becomes $1(2 - 5) = -3$, which is an increase of 12 units from -15. After the second (consecutive) dual phase, the dual objective value also becomes zero. Thus we achieve optimality.

Observe that by keeping a track of the duality gap, we need not work with the same out-of-kilter arc until it comes in kilter in order to guarantee finite convergence. Furthermore, any algorithmic scheme that ensures that the duality gap decreases from its previous lowest value (not necessarily monotonically) by an integer at finite intervals, is guaranteed to converge finitely.

EXERCISES

11.1 Solve the following problem by the out-of-kilter algorithm.

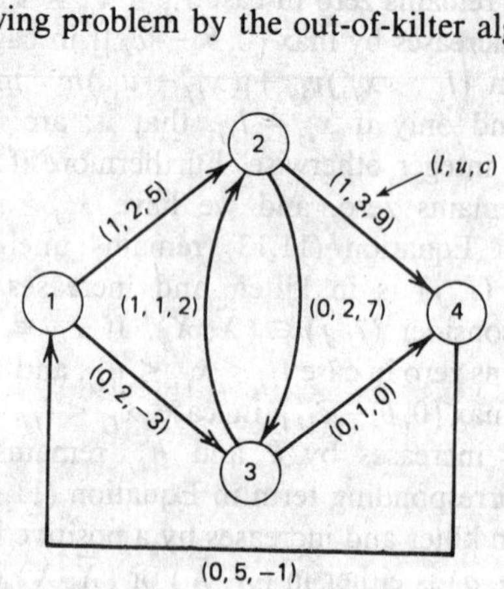

11.2 Consider the following network flow problem.

a. Give the kilter state of each arc.

b. Solve the problem by the out-of-kilter algorithm.

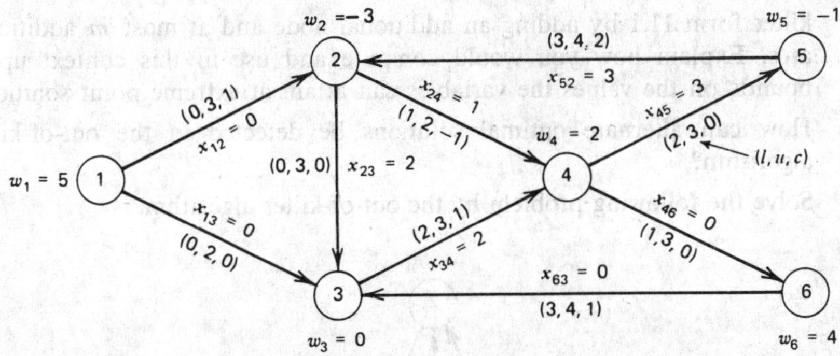

11.3 Solve the following problem by the out-of-kilter algorithm.

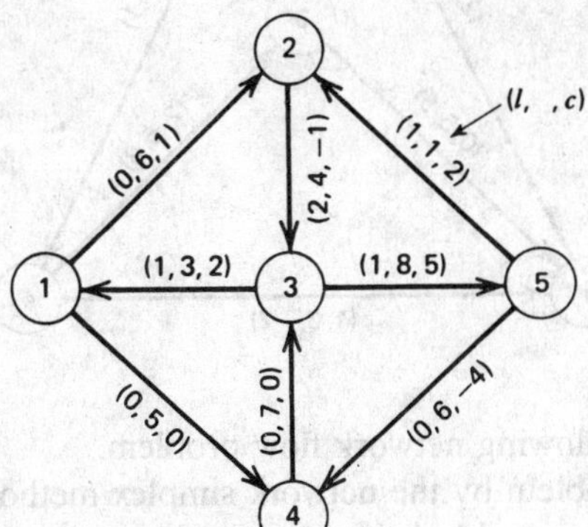

11.4 Explain in detail the data and list structures you would use in order to efficiently implement the out-of-kilter algorithm. Compare your storage requirement with that for a primal (network) simplex implementation.

11.5 Show how the out-of-kilter algorithm detects infeasibility in the following problem.

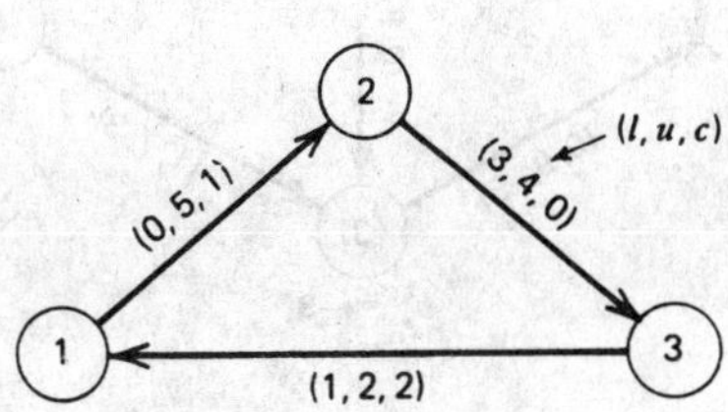

11.6 Show by manipulating the constraint equations mathematically that any minimal cost network flow problem can be transformed into the out-of-kilter form 11.1 by adding an additional node and at most m additional arcs. Explain how you would compute and use in this context upper bounds on the values the variables can attain at extreme point solutions.

11.7 How can alternate optimal solutions be detected in the out-of-kilter algorithm?

11.8 Solve the following problem by the out-of-kilter algorithm.

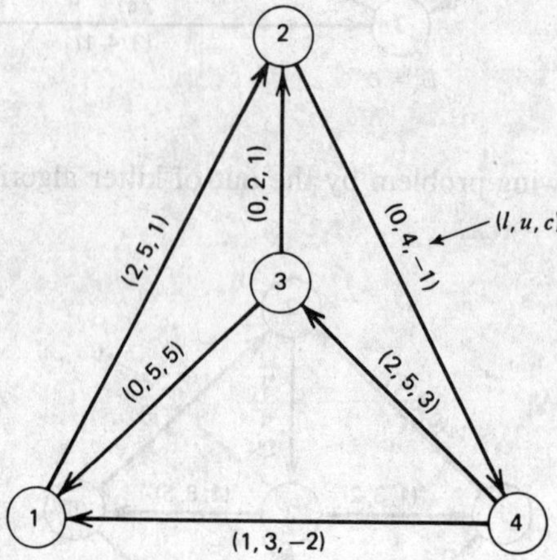

11.9 Consider the following network flow problem.

a. Solve the problem by the network simplex method of Chapter 9.
b. Transform the problem into a circulation form and solve it by the out-of-kilter algorithm.

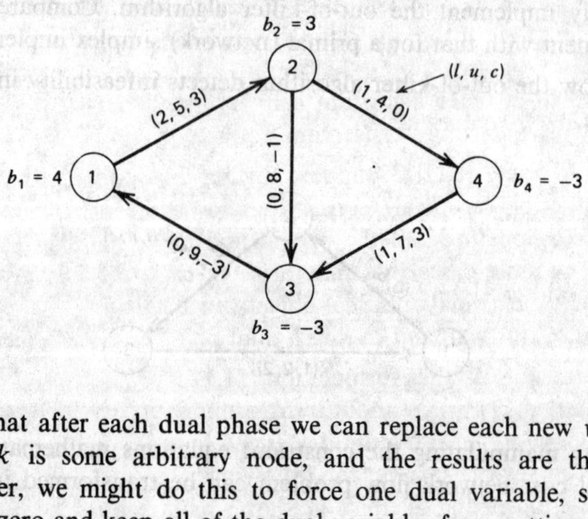

11.10 Show that after each dual phase we can replace each new w_i by $w_i - w_k$, where k is some arbitrary node, and the results are the same. In a computer, we might do this to force one dual variable, such as w_k, to remain zero and keep all of the dual variables from getting too large.

11.11 In the primal phase of the out-of-kilter algorithm, suppose one needs to find a (directed) path from q to p (in G'), given an out-of-kilter arc (p, q). Consider the construction of the usual flow change tree T described in Sections 11.2 and 11.5. Recall that starting with $T = \{q\}$, each node that is included in T receives a label equal to the (positive) flow it can receive from q. Suppose that at each step, we add to T that node which among all candidates that can be connected to nodes already in T can receive the largest flow change label. Show that if $p \in T$, then this procedure will have found the maximum flow change path from q to p in G'.

11.12 Consider the following network flow problem with flows as indicated.

a. Ignoring the fact that the b_i's are not zero, apply the out-of-kilter algorithm directly to the foregoing network with the starting x_{ij}'s as given.

b. Solve by the network simplex method of Chapter 9.

c. Are the solutions of part (a) and (b) the same? Discuss!

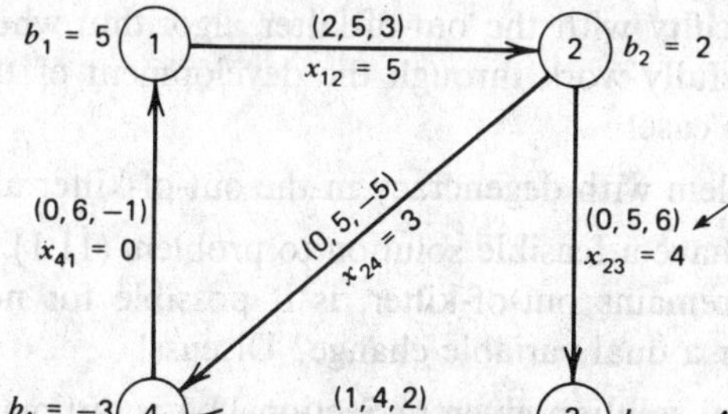

.1.13 Consider the minimal cost network flow problem: Minimize $\mathbf{cx}$ subject to $\mathbf{Ax} = \mathbf{b}$, $\mathbf{l} \le \mathbf{x} \le \mathbf{u}$, where $\mathbf{A}$ is a node-arc incidence matrix. Define a "conserving flow" to be any $\mathbf{x}$ satisfying $\mathbf{Ax} = \mathbf{b}$ (the conservation equations). Show that without transforming the network to the out-of-kilter form, the out-of-kilter algorithm can be applied directly on the original network with a starting "conserving flow" to solve the problem.

11.14 Referring to Exercise 11.13, suppose further that we initialize the algorithm with a (possibly artificial) conserving flow that is also feasible to the bounds. Develop in detail the primal and dual phases of the algorithm. State the possible cases that may arise in this specialization. Also show directly that the primal and dual objective values strictly improve during a (primal) breakthrough and during a dual phase respectively.

11.15 Consider the out-of-kilter formulation (11.1). Assume that $-\infty < l_{ij} \le u_{ij} < \infty$ for all arcs (i, j). Without transforming to $l_{ij} \ge 0$ for all (i, j), show that the out-of-kilter algorithm developed in this chapter is directly applicable. In this respect, answer the following.

a. Illustrate by solving Exercise 11.8 after replacing $(\mathbf{l}, \mathbf{u}, \mathbf{c})$ for arc (3,1) by $(-1, 5, 5)$.

b. Develop a value for M that can be used in the primal penalty function (11.12) in this case.

c. Given a network flow problem of the type in Exercise 11.13 and given any conserving flow $\bar{\mathbf{x}}$, show how you can transform this problem into the form of the present exercise.

11.16 Interpret the dual of the out-of-kilter formulation (11.1) from a marginal cost viewpoint.

11.17 Considering the out-of-kilter problem, show that a feasible solution exists if and only if for every choice of X and $\bar{X} = \mathcal{N} - X$ we have

$$\sum_{\substack{i \in X \\ j \in \bar{X}}} l_{ij} \leq \sum_{\substack{i \in \bar{X} \\ j \in X}} u_{ij}$$

(*Hint*. Review the section on the case where $\theta = \infty$.)

11.18 Demonstrate directly that the dual objective function value is unbounded as $\theta \to \infty$ when $S_1 = \emptyset$ and $S_2 = \emptyset$ during the dual phase of the out-of-kilter algorithm.

11.19 Is there any difficulty with the out-of-kilter algorithm when $l_{ij} = u_{ij}$ for some (i, j)? Carefully work through the development of the out-of-kilter algorithm for this case!

11.20 Is there any problem with degeneracy in the out-of-kilter algorithm?

11.21 Suppose that we have a feasible solution to problem (11.1). Assuming that the selected arc remains out-of-kilter, is it possible for no new node to come into X after a dual variable change? Discuss!

11.22 Show that the dual solution given in Section 11.1 is optimal for a fixed set of w_i's.

11.23 Suppose that we work only with the $z_{ij} - c_{ij}$'s after the initial dual solution and never bother to change the w_i's. Show how the w_i's can be recovered anytime we want them. (*Hint*. The w_i's are not unique. Set any one $w_i = 0$.)

11.24 In the out-of-kilter algorithm, show that if no cycle exists in the subset of arcs in G with $x_{ij} \neq l_{ij}$ and $x_{ij} \neq u_{ij}$, then the current solution corresponds to a basic solution of the associated linear program. Indicate how the out-of-kilter algorithm can be initiated with a basic solution if one is

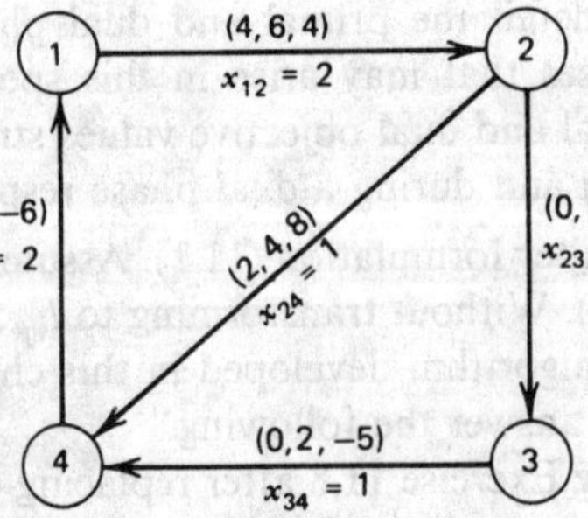

not readily available. Illustrate by the above network with the indicated conserving flow. (*Hint*. Start with a conserving flow. If a cycle exists among arcs where $x_{ij} \neq l_{ij}$ and $x_{ij} \neq u_{ij}$, consider modifying the flow around the cycle in a direction that improves the penalty function value in Equation (11.12).)

11.25 Using the results of Exercise 11.24, if the out-of-kilter algorithm is initiated with a basic solution to the linear program, show how a basic solution can be maintained thereafter. (*Hint*. Let $E = \{(i, j) : x_{ij} \neq l_{ij} \text{ and } x_{ij} \neq u_{ij}\}$. Start with only appropriate arcs associated with E as members of G'. Whenever a circuit exists, change flows. Eliminate any residual cycles in E as in Exercise 11.24. Otherwise, after developing X, add an appropriate arc to G', which is not a member of E and which enlarges X; then work with E as much as possible again. If no circuit still exists in G', add another arc that is not in E but does not enlarge X. Continue as often as necessary. If no such arc not in E exists that enlarges X, then pass to the dual phase. This is an example of *block pivoting*.)

11.26 Provide an interpretation of the primal network simplex algorithm for bounded variables in terms of an appropriately restricted execution of primal and dual phases of the out-of-kilter algorithm. In particular, specify the flow change $\delta \geq 0$ along with the appropriate circuit in the "primal phase," the sets X, $\bar{X}$, and the value θ for the "dual phase".

11.27 Extend the out-of-kilter algorithm to handle rational values of c_{ij}, l_{ij}, and u_{ij} directly.

11.28 If during the primal phase we permit some kilter numbers to increase as long as the sum of all the kilter numbers decreases, will the out-of-kilter algorithm work? How could this be made operational?

11.29 Suppose that we are given a network with m nodes and n arcs, all lower bounds equal to zero, positive upper bounds, and no costs involved. Show how the out-of-kilter algorithm can be used to find the maximum amount of flow from node 1 to node m. (*Hint*. Consider adding an arc from node m to node 1 with $l_{m1} = 0$, $u_{m1} = \infty$, $c_{m1} = -1$ with all other c_{ij}'s set at zero.)

11.30 Find the maximum flow in the following network from node 1 to node 4 using the out-of-kilter algorithm. (*Hint*. Refer to Exercise 11.29.)

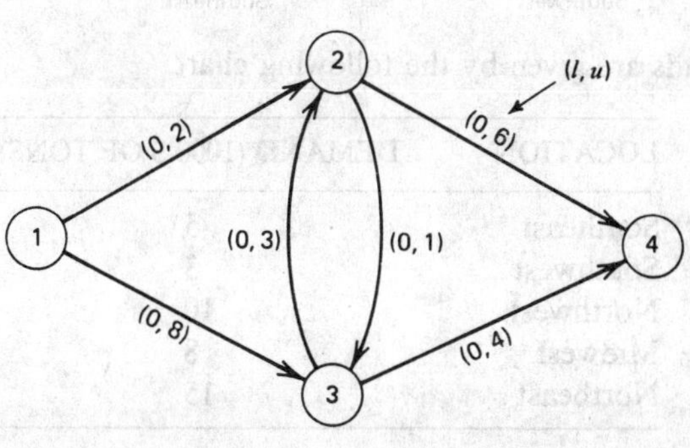

11.31 Suppose that we are given a network of m nodes and n arcs with a cost c_{ij} for each arc and no lower or upper bounds involved. Assume that there are no negative total cost (directed) circuits. Show how the out-of-kilter algorithm can be used to find the shortest (least) cost path from node 1 to node m via the following construction. Add an arc from node m to node 1 of $l_{m1} = u_{m1} = 1$ and $c_{m1} = 0$. Set the lower and upper bounds of all other arcs at 0 and 1 respectively. Can this scheme be used to find the shortest simple path from node 1 to node m in the presence of negative cost circuits?

11.32 Develop an expression for the rate in change in dual objective value as the value of the dual variables of nodes in some subset X of $\mathscr{N}$ are (marginally) increased. If this rate is positive, how would you determine the best possible value by which to increase the dual variables of nodes in X? How could you use this information algorithmically?

11.33 Consider a general linear program of the form: Minimize $\mathbf{cx}$ subject to $\mathbf{Ax} = \mathbf{b}$, $\mathbf{l} \le \mathbf{x} \le \mathbf{u}$. Suppose that we begin with a solution $\mathbf{x}$ that satisfies $\mathbf{Ax} = \mathbf{b}$. Develop primal and dual phases of a linear programming algorithm, based on the out-of-kilter algorithm, for solving this general linear program.

11.34 Coal is being hauled out of Kentucky bound for locations in the Southeast, Southwest, Midwest, Northwest, and Northeast. The network of routes is given.

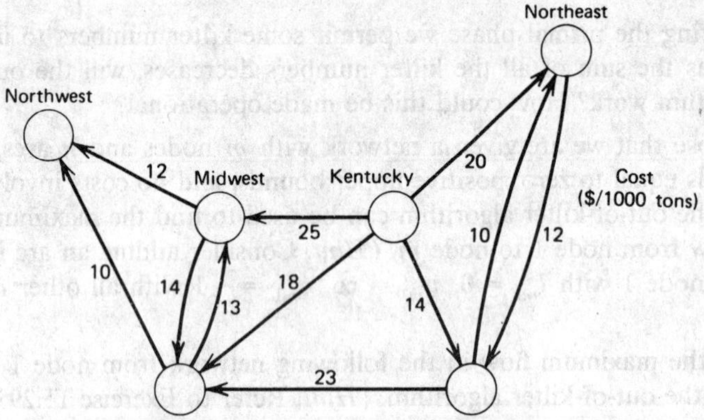

The demands are given by the following chart.

LOCATION	DEMAND (1000'S OF TONS)
Southeast	5
Southwest	3
Northwest	10
Midwest	8
Northeast	15

Kentucky has a supply of 65,000 tons per week. Except for nonnegativity there is an upper limit on flow of 17,000 tons on each arc. Ignoring the return route for coal cars, use the out-of-kilter algorithm to find the least cost distribution system for coal.

11.35 Show how a transportation problem and an assignment problem can be solved by the out-of-kilter algorithm.

11.36 The "Plenty of Water" Company wishes to deliver water for irrigation to three oases: the sin oasis, the devil's oasis, and the pleasure oasis. The company has two stations A and B in the vicinity of these oases. Because of other commitments, at most 600 kilo-tons and 200 kilo-tons can be delivered by the two stations to the oases. Station A is connected with the sin oasis by a 10 kilometer pipeline system and with the devil's oasis by a 15 kilometer pipeline system. Similarly station B is connected with the pleasure oasis by a 15 kilometer pipeline system and with the devil's oasis by a 5 kilometer pipeline system. Furthermore, the pleasure oasis and the devil's oasis are connected by a road allowing the transportation of water by trucks. Suppose that the sin oasis, the devil's oasis, and the pleasure oasis require 200, 350, and 150 kilo-tons of water. Further suppose that the transportation cost from station A is $0.01 per kilo-ton per kilometer, and the transportation cost from station B is $0.012 per kilo-ton per kilometer. Finally suppose that the transportation cost between the pleasure oasis and the devil's oasis is $0.15 per kilo-ton.

a. Formulate the problem so that the out-of-kilter algorithm can be used.
b. Solve the problem by the out-of-kilter algorithm.
c. Suppose that a road is built joining the sin oasis and the devil's oasis with a shipping cost of $0.10 per kilo-ton. Would this affect the optimal solution? If so, find the new optimal.

11.37 Suppose that the air freight charge per ton between locations is given by the following table (except where no direct air freight service is available).

LOCATION	1	2	3	4	5	6	7
1	—	12	27	—	45	35	15
2	12	—	10	25	32	—	22
3	27	10	—	28	50	28	10
4	—	25	28	—	16	20	32
5	45	32	50	16	—	26	35
6	36	—	28	20	26	—	20
7	15	22	10	32	35	20	—

A certain corporation must ship a certain perishable commodity from locations 1, 2, and 3 to locations 4, 5, 6, and 7. A total of 30, 50, and 20 tons of this commodity are to be sent from locations 1, 2, and 3 respectively. A total of 15, 30, 25, and 30 tons are to be sent to locations 4, 5, 6, and 7 respectively. Shipments can be sent through intermediate locations at a cost equal to the sum of the costs for each of the legs of the journey. The problem is to determine the shipping plan that minimizes the total freight cost. Formulate the problem and solve it by the out-of-kilter algorithm.

11.38 A manufacturer must produce a certain product in sufficient quantity to meet contracted sales in the next four months. The production facilities available for this product are limited, but by different amounts in the respective months. The unit cost of production also varies according to the facilities and personnel available. The product can be produced in one month and then held for sale in a later month, but at an estimated storage cost of \$1 per unit per month. No storage cost is incurred for goods sold in the same month in which they are produced. There is presently no inventory of this product, and none is desired at the end of the four months. Pertinent data are given.

MONTH	CONTRACTED SALES	MAXIMUM PRODUCTION	UNIT COST OF PRODUCTION	UNIT STORAGE COST PER MONTH
1	20	40	14	1
2	30	50	16	1
3	50	30	15	1
4	40	50	17	1

Formulate the production problem as a network problem and solve it by the out-of-kilter algorithm.

11.39 Water is to be transported through a network of pipelines from the big dam to the low valley for irrigation. A network is shown where arcs represent pipelines and the number on each arc represents the maximum permitted rate of water flow in kilo-tons per hour. It is desired to determine the maximum rate of flow from the big dam to the low valley.

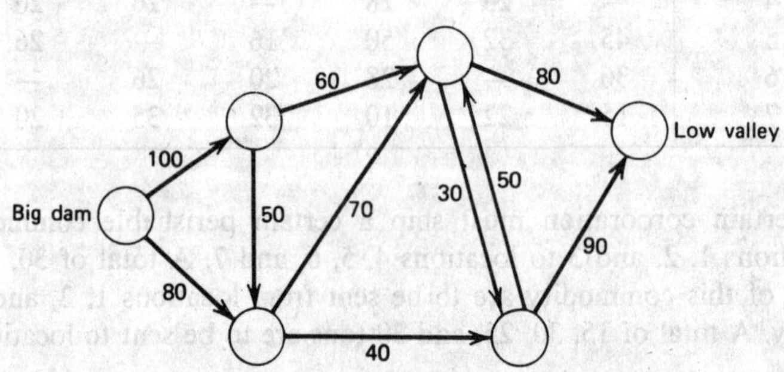

a. Formulate the problem so that it can be solved by the out-of-kilter algorithm.
b. Solve the problem by the out-of-kilter algorithm.
c. Through the use of a more powerful pumping system the maximum rate of flow on any arc can be increased by a maximum of 10 kilo-tons of water per hour. If the rate is to be increased on only one pipeline, which one would you recommend and why?

11.40 Let c_{ij} be the length associated with arc (i, j) in a given network with no (directed) circuits. It is desired to find the path with shortest distance and that with the maximum distance between any two given nodes. Formulate the two problems so that the out-of-kilter algorithm can be used. Make all possible simplifications in the application of the out-of-kilter algorithm for these two problems. What is the significance of assuming that the network has no circuits? (*Hint*. See Exercise 11.31.)

11.41 An assembly consists of three parts A, B, and C. These parts go through the following operations in order: forging, drilling, grinding, painting, and assembling. The duration of these operations in days is summarized.

	DURATION OF OPERATION			
PART	FORGING	DRILLING	GRINDING	PAINTING
A	1.0	0.5	1.0	0.3
B	2.0	0.3	0.5	0.3
C	3.0	1.0	—	0.4

Upon painting, parts A and B are assembled in two days and then A, B, and C are assembled in one day. It is desired to find the least time required for the assembly (this problem is called the *critical path problem*).
a. Formulate the problem as a network problem.
b. Solve the problem by any method you wish.
c. Solve the problem by the out-of-kilter algorithm.
d. Solve the problem by the simplified procedure you obtained in Exercise 11.40.
e. Because of the shortage of forging machines, suppose that at most two parts can go through forging at any particular time. What is the effect of this restriction on the total processing duration?

NOTES AND REFERENCES

1. Fulkerson [1961a] developed the out-of-kilter algorithm for network flow problems. For a slightly different development of the out-of-kilter algorithm, see Ford and Fulkerson [1962], and for a specialization, see Kennington and Helgason [1980].

2. The presentation of the out-of-kilter algorithm in this chapter follows that of Clasen [1968], especially the division of states according to values of flows x_{ij} and reduced costs $z_{ij} - c_{ij}$.
3. The spirit of the out-of-kilter algorithm can be extended to a procedure for general linear programs. This has been done by Jewell [1967]. The corresponding steps in the general case require the solution to linear programming subproblems instead of finding cycles or changing dual variables in a simple way.
4. Barr *et al.* [1974] provide a streamlined implementation scheme for the out-of-kilter algorithm along with computational results. Another implementation scheme is described in Singh [1986]. Computational experience and comparisons are also provided by Glover and Klingman [1978] and Hatch [1975].
5. Bertsekas and Tseng [1988] describe a primal-dual network flow algorithm that operates two–three times faster on the NETGEN benchmark test problems compared with RNET, the incumbent primal simplex code developed at Rutger's University by Professor Grigoriadis and Professor Hsu. This algorithm maintains a flow feasible to the bounding constraints, and complementary slack with respect to the current dual (feasible) solution. It strives to satisfy the flow conservation constraints via flow augmentations, dual variable changes (accompanied with flow modifications), and ascents in the dual objective value via appropriate cut-sets. The principal computational advantage comes from the quick dual ascent steps using cut-sets based on single nodes as opposed to determining steepest ascent cut-sets. (See Exercise 11.32.) A computer code description (of the versions RELAX-II and RELAXT-II) appears in Bertsekas and Tseng [1986].

TWELVE: MAXIMAL FLOW, SHORTEST PATH, MULTICOMMODITY FLOW, AND NETWORK SYNTHESIS PROBLEMS

Two special and important network flow problems are the maximal flow problem and the shortest path problem. Both of these problems can be solved by either the network simplex method of Chapter 9 or the out-of-kilter algorithm of Chapter 11. However, their frequent occurrence in practice and the specialized, more efficient procedures for handling these two problems provide a strong case for considering them separately.

We also include in this chapter an introduction to the class of network flow problems called multicommodity network flows. In Chapters 9, 10, and 11 we have considered network flow problems in which it was not necessary to distinguish among the units flowing in the networks. There was essentially a single commodity or type of unit. There are network flow problems in which different types of units must be treated. In these instances supplies and demands are by commodity type, and the distinction among the commodities must be maintained. We shall examine this multicommodity flow problem, consider the difficulty in dealing with it, and present a decomposition-based procedure for solving it.

Another topic that we introduce is network synthesis. The network flow problems considered thus far are *analysis* problems. A network is *given*, and some analysis is performed on it. On the other hand, the network synthesis problem requires one to *construct* an optimal network (in some defined sense) that satisfies certain specifications. Several of the tools developed in this book, for example, including network analysis and decomposition methods, are used to study such problems.

12.1 THE MAXIMAL FLOW PROBLEM

We present here Ford and Fulkerson's *labeling method* for solving maximal flow problems. This method is simply a specialization of the out-of-kilter algorithm applied to the maximal flow problem. Specialized primal simplex implementations have been found to be significantly faster than this method (with a speed-up factor of 1.5–5) and require only one-third of the amount of storage. However, this method does provide useful insights, and as with the general out-of-kilter algorithm, improved primal-dual implementations may make it more competitive.

Consider a network with m nodes and n arcs through which a single commodity will flow. We associate with each arc (i, j) a lower bound on flow of $l_{ij} = 0$ and an upper bound on flow of u_{ij}. We shall assume throughout the development that the u_{ij}'s (*arc capacities*) are finite integers. There are no costs involved in the maximal flow problem. In such a network, we wish to find the maximum amount of flow from node 1 to node m.

Let f represent the amount of flow in the network from node 1 to node m. Then the maximal flow problem may be stated as follows:

$$\begin{array}{ll} \text{Maximize} & f \\ \text{Subject to} & \displaystyle\sum_{j=1}^{m} x_{ij} - \sum_{k=1}^{m} x_{ki} = \begin{cases} f & \text{if } i = 1 \\ 0 & \text{if } i \neq 1 \text{ or } m \\ -f & \text{if } i = m \end{cases} \\ & x_{ij} \leq u_{ij} \qquad i, j = 1, 2, \ldots, m \\ & x_{ij} \geq 0 \qquad i, j = 1, 2, \ldots, m \end{array}$$

where the sums and inequalities are taken over existing arcs in the network. This is called the node-arc formulation for the maximal flow problem since the constraint matrix is a node-arc incidence matrix. (See Exercise 12.20 for another formulation.) Noting that f is a variable and denoting the node-arc incidence matrix by $\mathbf{A}$, we can write the maximal flow problem in matrix form as

$$\begin{array}{ll} \text{Maximize} & f \\ \text{Subject to} & (\mathbf{e}_m - \mathbf{e}_1)f + \mathbf{A}\mathbf{x} = \mathbf{0} \\ & \mathbf{x} \leq \mathbf{u} \\ & \mathbf{x} \geq \mathbf{0} \end{array}$$

Since the activity vector for f is $(\mathbf{e}_m - \mathbf{e}_1)$, the difference of two unit vectors, we may view f as a flow variable on an arc from node m to node 1. This provides the direct formulation of the maximal flow problem in (out-of-kilter) circulatory form (with zero right-hand-side values for the flow conservations). Recalling that the out-of-kilter problem dealt with minimization, we assign a cost coefficient of zero to every flow variable except $x_{m1} \equiv f$, which receives -1.

Arc $(m, 1)$ is sometimes called the *return arc*. Figure 12.1 presents an example of the maximal flow problem and its equivalent out-of-kilter network flow problem. In Figure 12.1 the lower bound $l_{m1} = l_{41} = 0$ is derived from the fact that all $x_{ij} = 0$ and $x_{m1} = 0$ is a feasible solution to the maximal flow problem. Thus the maximal value of x_{m1} will never be less than zero.

Before continuing with the development of the maximal flow problem, we introduce the useful and important concept of cut-sets.

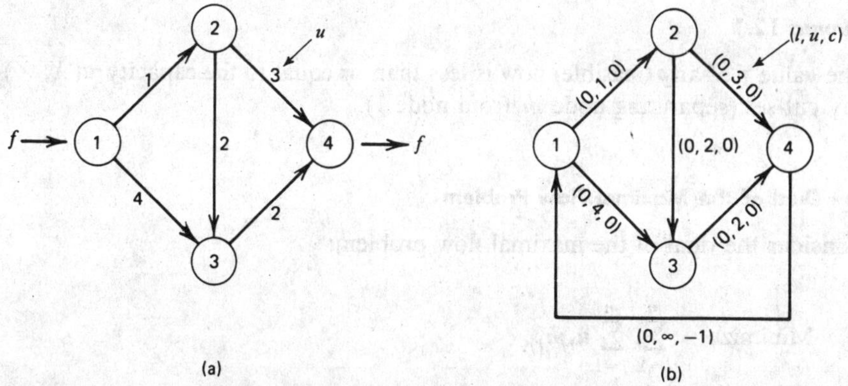

Figure 12.1. An example of a maximal flow problem: (*a*) Maximal flow problem. (*b*) Out-of-kilter equivalent problem.

Cut-Set (Separating Node *m* from Node 1)

Let X be any set of nodes in the network such that X contains node 1 but not node m. Let $\overline{X} = \mathscr{N} - X$. Then $(X, \overline{X}) \equiv \{(i, j) : i \in X,\ j \in \overline{X}\}$ is called a *cut-set* separating node m from node 1.

Capacity of a Cut-Set

Let $(X, \overline{X})$ be any cut-set in a network G. Then $u(X, \overline{X}) \equiv \Sigma_{(i,j) \in (X, \overline{X})} u_{ij}$ is called the capacity of the cut-set. In Figure 12.1a there are several cut-sets separating node 4 from node 1 in G. They are

$$X = \{1\}, \quad \overline{X} = \{2,3,4\} \quad (X, \overline{X}) = \{(1,2),(1,3)\}, \quad u(X, \overline{X}) = 5$$

$$X = \{1,2\}, \quad \overline{X} = \{3,4\}, \quad (X, \overline{X}) = \{(1,3),(2,3),(2,4)\}, \quad u(X, \overline{X}) = 9$$

$$X = \{1,3\}, \quad \overline{X} = \{2,4\}, \quad (X, \overline{X}) = \{1,2),(3,4)\}, \quad u(X, \overline{X}) = 3$$

$$X = \{1,2,3\}, \quad \overline{X} = \{4\}, \quad (X, \overline{X}) = \{(2,4),(3,4)\}, \quad u(X, \overline{X}) = 5$$

Let $(X, \overline{X})$ be any cut-set separating node m from node 1 in G. Summing the flow conservation equations of the maximal flow problem over nodes in X, flow variables with both ends in X cancel and we get

$$\sum_{\substack{i \in X \\ j \in \overline{X}}} x_{ij} - \sum_{\substack{i \in \overline{X} \\ j \in X}} x_{ij} = f \tag{12.1}$$

Using $x_{ij} \geq 0$ and $x_{ij} \leq u_{ij}$, we get

$$\sum_{\substack{i \in X \\ j \in \overline{X}}} u_{ij} - 0 \geq f \tag{12.2}$$

This leads to the following.

***Lemma* 12.1**

The value f of any (feasible) flow is less than or equal to the capacity $u(X, \overline{X})$ of any cut-set (separating node m from node 1).

The Dual of the Maximal Flow Problem

Consider the dual of the maximal flow problem:

$$\text{Minimize} \quad \sum_{i=1}^{m} \sum_{j=1}^{m} u_{ij} h_{ij}$$

$$\begin{aligned} \text{Subject to} \quad w_m - w_1 &= 1 & & \\ w_i - w_j + h_{ij} &\geq 0 & \quad i, j &= 1, 2, \ldots, m \\ h_{ij} &\geq 0 & \quad i, j &= 1, 2, \ldots, m \end{aligned}$$

where $\mathbf{w}$ corresponds to the conservation equations and $\mathbf{h}$ corresponds to $\mathbf{x} \leq \mathbf{u}$. Note that the first dual constraint is assoicated with the flow f whose column is $\mathbf{e}_m - \mathbf{e}_1$. A typical column $\mathbf{a}_{ij}$ of the node-arc incidence matrix $\mathbf{A}$ has $+1$ at the ith position and -1 at the jth position, which leads to the dual constraints $w_i - w_j + h_{ij} \geq 0$.

Let $(X, \overline{X})$ be any cut-set and consider the foregoing dual problem. If we let

$$w_i = \begin{cases} 0 & \text{if} \quad i \in X \\ 1 & \text{if} \quad i \in \overline{X} \end{cases}$$

$$h_{ij} = \begin{cases} 1 & \text{if} \quad (i, j) \in (X, \overline{X}) \\ 0 & \text{otherwise} \end{cases}$$

then this particular choice of $\mathbf{w}$ and $\mathbf{h}$ provides a feasible solution to the dual problem (why?) whose dual objective is equal to the capacity of the cut-set.

Thus, Lemma 12.1 also follows from the duality theorem, which states that any feasible solution to a minimization problem has an objective value greater than or equal to that of the associated maximization problem. As the reader may suspect, we shall show that the capacity of the minimal cut-set (the one with minimal capacity) is equal to the value of the maximal flow. We shall prove this constructively. We ask the reader to prove this result in Exercise 12.3 using the earlier duality relationship.

An Algorithm for the Maximal Flow Problem

From Lemma 12.1 if we are able to find a flow and a cut-set such that $u(X, \overline{X}) = f$, we will have the maximal flow (and the minimal cut-set). We shall do this constructively by simply specializing the out-of-kilter algorithm to this problem.

Suppose that we start with any feasible (integer) flow in G, say, each $x_{ij} = 0$. From G construct G' as follows.

1. If arc (i, j) is in G and $x_{ij} < u_{ij}$, then place arc (i, j) in G' with permitted flow change value $\Delta_{ij} = u_{ij} - x_{ij}$.
2. If arc (i, j) is in G and $x_{ij} > 0$, then place arc (j, i) in G' with permitted flow change value $\Delta_{ji} = x_{ij}$.

Now, in G' two possibilities exist:

Case 1

A *path* P exists, in G', from node 1 to node m.

Case 2

No path exists, in G', from node 1 to node m.

In case 1 we may construct a new feasible flow with greater objective value (that is, flow out of node 1). Let Δ be equal to the minimum permitted flow change on the path P from node 1 to node m in G', that is, $\Delta = \text{Minimum } \{\Delta_{ij} : (i, j) \text{ is in the path}\}$. Now Δ is a positive integer (why?). Consider the associated chain P' (undirected path) in G. Construct a new flow as follows. Add Δ to flows on arcs of the associated chain in G with the direction of the path in G', subtract Δ from flows on arcs of the associated chain in G against the direction of the path in G', and leave all other arc flows unchanged. The new flow is feasible (why?). The value of the new flow is $f' = f + \Delta$ (why?).

Assuming that the capacities are finite, case 1 can occur only a finite number of times before case 2 occurs (why?). When case 2 occurs, let X be the set of nodes in G' that can be reached along some path in G' from node 1. Let $\overline{X} = \mathcal{N} - X$ and note that node m belongs to $\overline{X}$ (why?). Consider the arcs in G between X and $\overline{X}$. First, every arc (i, j) in G from X to $\overline{X}$ must have $x_{ij} = u_{ij}$; otherwise there would be an arc (i, j) in G' and j would be a member of X (a contradiction). Second, every arc (i, j) in G from $\overline{X}$ to X must have $x_{ij} = 0$; otherwise there would be an arc (j, i) in G' and i would be a member of X (a contradiction). Substituting in Equation (12.1), we get

$$\sum_{\substack{i \in X \\ j \in \overline{X}}} u_{ij} = f \quad \text{or} \quad u(X, \overline{X}) = f.$$

Thus we must have the maximal flow by noting inequality (12.2). Hence we have constructively proved the following.

***Theorem* 12.1** (Maximal Flow–Minimal Cut Theorem)

The value of the maximal flow in G is equal to the capacity of the minimal cut-set in G.

Summary of the Maximal Flow Algorithm

The constructive proof of the maximal flow–minimal cut theorem leads to the following maximal flow algorithm.

INITIALIZATION STEP

Select a set of feasible (integer) flows, say each $x_{ij} = 0$.

MAIN STEP

From G construct G' as follows.

1. All of the nodes in G are in G'.
2. If $x_{ij} < u_{ij}$ in G, place (i, j) in G' with the permitted flow change on (i, j) of $\Delta_{ij} = u_{ij} - x_{ij}$.
3. If $x_{ij} > 0$ in G, place (j, i) in G' with the permitted flow change on (j, i) of $\Delta_{ji} = x_{ij}$.

Note that arc (i, j) in G will give rise to two arcs in G' if $0 < x_{ij} < u_{ij}$. Attempt to locate a path P in G' from node 1 to node m. If no such path exists, stop; the optimal solution is obtained. Otherwise, let Δ be the minimum permitted flow change on P in G'. Add Δ to flows on arcs of the associated chain in G with the direction of the path in G', subtract Δ from flows on arcs of the associated chain in G against the direction of the path in G', and leave all other arc flows unchanged. Repeat the main step.

Locating a path is called *breakthrough*, whereas finding no path is called *nonbreakthrough*.

An Example of the Maximal Flow Problem

Consider the network of Figure 12.1*a*. Figure 12.2 presents the complete solution to the maximal flow problem for this network.

Basic Solutions in the Maximal Flow Algorithm

In Chapter 9 we characterized basic solutions to a network flow problem. Recall that a basic solution to a network flow problem consists of a set of nonbasic variables at one of their lower or upper bounds plus a set of variables that form a rooted spanning tree. Thus we may conclude that if the set $E = \{(i, j): 0 < x_{ij} < u_{ij}\}$ does not contain a cycle, then we have a basic feasible solution at each iteration of the maximal flow algorithm (why?).

To identify a basis after each flow change in the maximal flow algorithm we take all of the variables in the set E plus an additional number of variables at one of their bounds to form a spanning tree. This set together with the artificial variable (located at node m) forms a rooted spanning tree (and the nonbasic variables are at their bounds). Note that since f, the flow in the network, is a

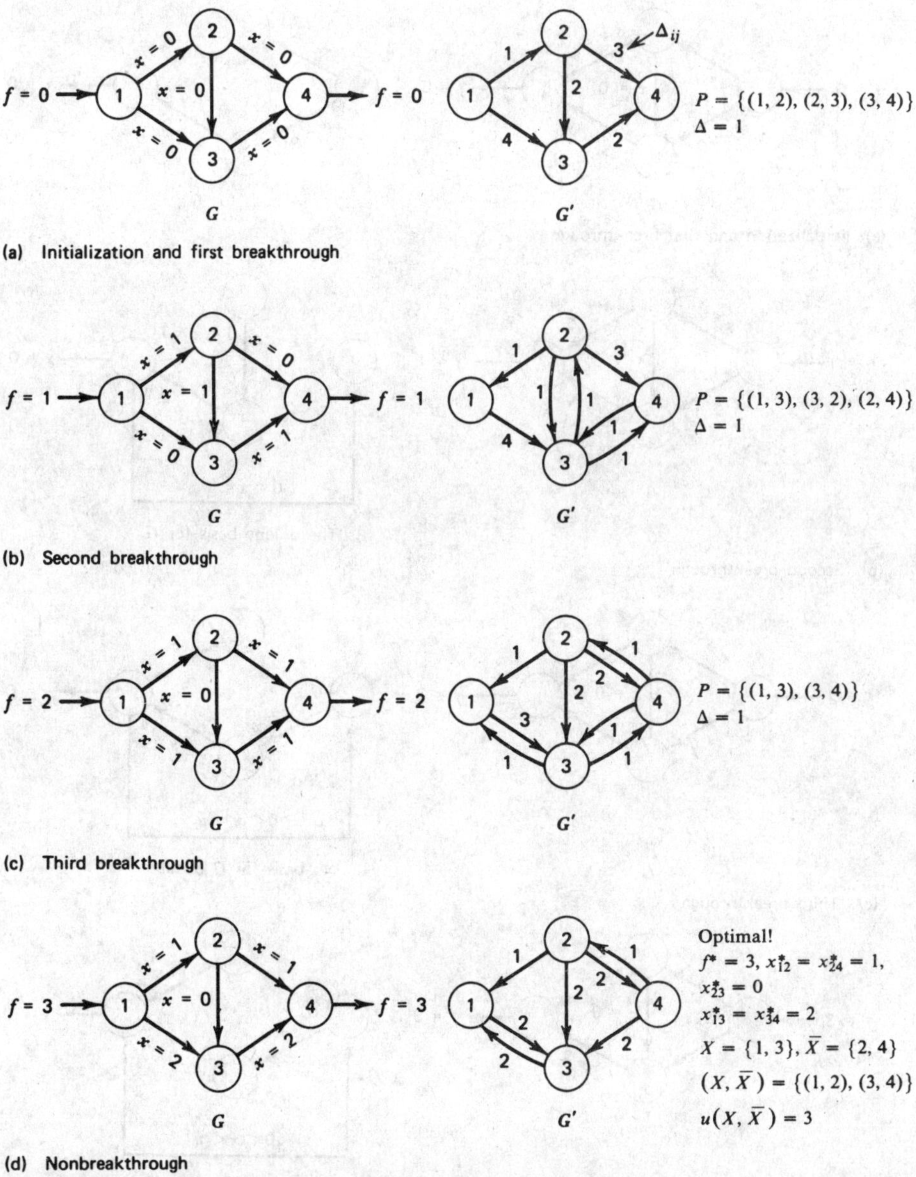

Figure 12.2. The solution for the network of Figure 12.1*a*.

variable, it must be taken onto the left-hand-side of the constraint system and becomes an arc from node *m* to node 1 (as discussed previously).

In Figure 12.3 we present bases corresponding to the solutions at each iteration of the example in Figure 12.2. Notice that the bases in Figure 12.3*b* and *d* are unique. Also, notice that in Figure 12.3*c* no basis is possible that corresponds to the maximal flow algorithm solution at that point. In Exercise

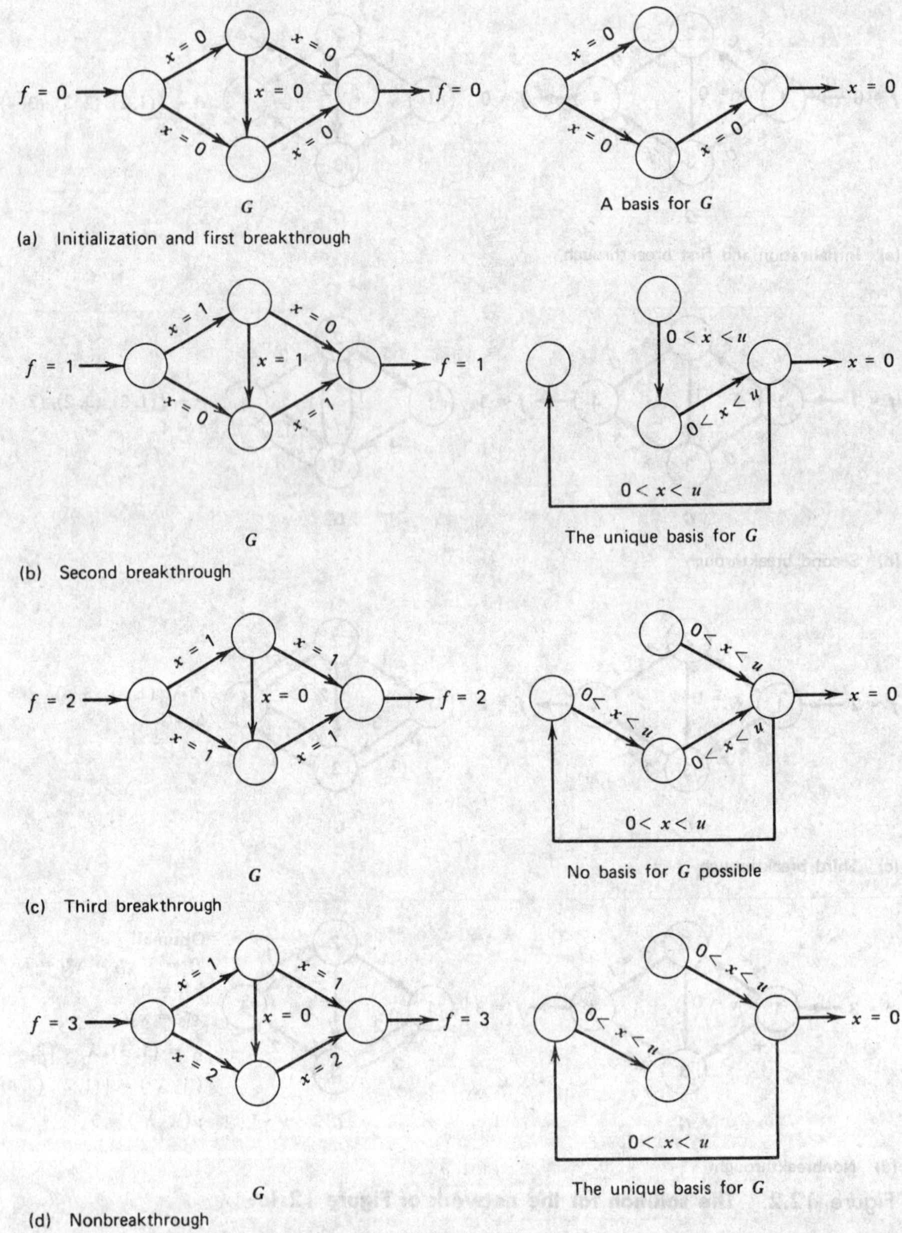

Figure 12.3. Comparison between solutions in the maximal flow algorithm and bases (when possible) for Figure 12.2.

12.12 we suggest a procedure for finding paths in G' in such a way that we shall always have a basic solution available. Finally, notice that the bases presented in Figures 12.3*a* and *b* are not *adjacent*. To obtain the basis in Figure 12.3*b* we have replaced two basic variables in the basis of Figure 12.3*a* by two nonbasic variables. This is an example of *block pivoting* discussed in Chapter 3.

A Labeling Algorithm for the Maximal Flow Problem

Either for hand or computer calculations, there are simple and convenient ways to maintain the information required to solve a maximal flow problem. As in the out-of-kilter method, we shall present a (tree construction) labeling algorithm that does not require the creation of the network G'. Suppose that we associate with each node j a label $L(j) = (\pm i, \Delta_j)$ containing two pieces of information. The second entry, Δ_j, in $L(j)$ indicates the amount of flow that can be sent to node j from node 1 through the current network with given flows without violating the capacity constraints $0 \le x_{ij} \le u_{ij}$. The first entry, $\pm i$, in $L(j)$ indicates the previous node in the chain along which flow can be changed. If the first entry in $L(j)$ is $+i$, then flow will be added to arc (i, j); otherwise, if the first entry is $-i$, then flow will be subtracted from arc (j, i). The labeling algorithm becomes as follows.

INITIALIZATION STEP

Set $x_{ij} = 0$ for $i, j = 1, \ldots, m$.

MAIN STEP

1. Erase any labels and set $L(1) = (-, \infty)$.
2. If node i has a label, node j has no label, and $x_{ij} < u_{ij}$, then set $L(j) = (+i, \Delta_j)$, where $\Delta_j = \text{Minimum}\{\Delta_i, u_{ij} - x_{ij}\}$. If node i has a label, node j has no label, and $x_{ji} > 0$, then set $L(j) = (-i, \Delta_j)$, where $\Delta_j = \text{Minimum}\{\Delta_i, x_{ji}\}$. Repeat step 2 until either node m is labeled or until no more nodes can be labeled. Go to step 3.
3. If node m is not labeled, stop; the optimal solution is obtained. Otherwise, if node m is labeled, then change flows in the network as follows. Set $\Delta = \Delta_m$. Begin at node m and consider the first entry of $L(m)$. If the first entry is $+k$, then add Δ to x_{km}. If the first entry of $L(m)$ is $-k$, then subtract Δ from x_{mk}. Backtrack to node k and repeat the process until node 1 is reached. Return to step 1.

When the algorithm stops, let X be the set of labeled nodes and $\bar{X} = \mathcal{N} - X$. The set $(X, \bar{X})$ is the minimal cut-set.

An Example of the Labeling Algorithm

Suppose that we apply the labeling algorithm to the maximal flow problem of Figure 12.1*a* to produce the first two iterations of the maximal flow algorithm represented in Figure 12.2*a* and Figure 12.2*b*. We begin with each $x_{ij} = 0$.

The sequence of labeling operations are as follows:

1. $L(1) = (-, \infty)$
2. $L(2) = (+1, 1)$
3. $L(3) = (+2, 1)$
4. $L(4) = (+3, 1)$
5. Breakthrough: $\Delta = 1$
6. $L_1(4) = +3 \Rightarrow x_{34} = 0 + \Delta = 1$
7. $L_1(3) = +2 \Rightarrow x_{23} = 0 + \Delta = 1$
8. $L_1(2) = +1 \Rightarrow x_{12} = 0 + \Delta = 1$
9. Erase all labels, $L(1) = (-, \infty)$
10. $L(3) = (+1, 4)$
11. $L(2) = (-3, 1)$
12. $L(4) = (+2, 1)$
13. Breakthrough: $\Delta = 1$
14. $L_1(4) = +2 \Rightarrow x_{24} = 0 + \Delta = 1$
15. $L_1(2) = -3 \Rightarrow x_{23} = 1 - \Delta = 0$
16. $L_1(3) = +1 \Rightarrow x_{13} = 0 + \Delta = 1$

Having completed the change of flows, we erase all labels and continue. The foregoing sequence of labels is not unique. We selected it because it illustrated the method completely. (See Exercise 12.10.)

12.2 THE SHORTEST PATH PROBLEM

Suppose that we are given a network G with m nodes, n arcs, and a cost c_{ij} associated with each arc (i, j) in G. The shortest path problem is: Find the shortest (least costly) path from node 1 to node m in G. The cost of the path is the sum of the costs on the arcs in the path.

The Mathematical Formulation of a Shortest Path Problem

We may think of the shortest path problem in a network flow context if we set up a network in which we wish to send a single unit of flow from node 1 to node m at minimal cost. Thus $b_1 = 1$, $b_m = -1$, and $b_i = 0$ for $i \neq 1$ or m. The mathematical formulation becomes:

$$\text{Minimize} \quad \sum_{i=1}^{m} \sum_{j=1}^{m} c_{ij} x_{ij}$$

$$\text{Subject to} \quad \sum_{j=1}^{m} x_{ij} - \sum_{k=1}^{m} x_{ki} = \begin{cases} 1 & \text{if} \quad i = 1 \\ 0 & \text{if} \quad i \neq 1 \text{ or } m \\ -1 & \text{if} \quad i = m \end{cases}$$

$$x_{ij} = 0 \text{ or } 1 \qquad i, j = 1, 2, \ldots, m$$

where the sums and the $0-1$ requirements are taken over existing arcs in G. The constraints $x_{ij} = 0$ or 1 indicate that each arc is either in the path or not.

Ignoring the $0-1$ constraints, we again find the familiar flow conservation equations. From Chapter 9 we know that the node-arc incidence matrix associated with the flow conservation equations is totally unimodular. Thus if we replace $x_{ij} = 0$ or 1 by $x_{ij} \geq 0$, and if an optimal solution exists, then the simplex method would still obtain an integer solution where the value of each variable is zero or one. (This can be shown by selecting any basis and applying Cramer's rule.) Thus we may solve the integer program as the following linear program.

$$\text{Minimize} \quad \sum_{i=1}^{m} \sum_{j=1}^{m} c_{ij} x_{ij}$$

$$\text{Subject to} \quad \sum_{j=1}^{m} x_{ij} - \sum_{k=1}^{m} x_{ki} = \begin{cases} 1 & \text{if} \quad i = 1 \\ 0 & \text{if} \quad i \neq 1 \text{ or } m \\ -1 & \text{if} \quad i = m \end{cases}$$

$$x_{ij} \geq 0 \qquad i, j = 1, 2, \ldots, m$$

Since the shortest path problem is a minimal cost network flow problem, we can solve it by one of the methods described in Chapter 9 or 11. However, we shall soon see that more efficient methods exist for this problem.

Consider the dual of the shortest path problem:

$$\text{Maximize} \quad w_1 - w_m$$

$$\begin{aligned} \text{Subject to} \quad w_i - w_j &\leq c_{ij} & i, j &= 1, 2, \ldots, m \\ w_i &\quad \text{unrestricted} & i &= 1, 2, \ldots, m \end{aligned}$$

It will be more convenient to make the substitution $w_i' \equiv -w_i$. As we shall shortly see $w_i' - w_1'$ is the shortest distance from node 1 to node i at optimality. Hence we can get the shortest distance from node 1 to all nodes of the network.

A Shortest Path Procedure When All Costs Are Nonnegative

Consider the case when all $c_{ij} \geq 0$. In this case a very simple and efficient procedure, known as Dijkstra's algorithm, exists for finding the shortest path (from node 1 to node m). This method also automatically yields the shortest path from node 1 to all of the other nodes as well.

INITIALIZATION STEP

Set $w_1' = 0$ and let $X = \{1\}$.

MAIN STEP

Let $\overline{X} = \mathcal{N} - X$ and consider the arcs in the set $(X, \overline{X}) = \{(i, j) : i \in X, J \in \overline{X}\}$. Let

$$w'_p + c_{pq} = \underset{(i,j)\in(X,\overline{X})}{\text{Minimum}} \{w'_i + c_{ij}\}$$

Set $w'_q = w'_p + c_{pq}$ and place node q in X. Repeat the main step exactly $m - 1$ times (including the first time) and then stop; the optimal solution is at hand.

Validation of the Algorithm

We now prove that the algorithm produces an optimal solution. Assume, inductively, that each w'_i for $i \in X$ represents the cost of the shortest path from node 1 to node i. This is certainly true for $i = 1$ (why?). Consider the algorithm at some point when a new node q is about to be added to X. Suppose that

$$w'_p + c_{pq} = \underset{(i,j)\in(X,\overline{X})}{\text{Minimum}} \{w'_i + c_{ij}\} \tag{12.3}$$

We shall show that the shortest path from node 1 to node q has length $w'_q = w'_p + c_{pq}$ and can be constructed iteratively as the shortest path from node 1 to node p plus the arc (p, q). Let P be any path from node 1 to node q. It suffices to show that the length of P is $\geq w'_q$. Since node 1 is in X and node q is currently in $\overline{X}$, then P must contain an arc (i, j) where $i \in X$ and $j \in \overline{X}$ (i and j could be p and q respectively). The length of the path P is thus equal to the sum of the following.

1. The length from node 1 to node i.
2. The length of arc (i, j), that is, c_{ij}.
3. The length from j to q.

By the induction hypothesis the length from node 1 to node i is greater than or equal to w'_i. Since the costs of all arcs are assumed nonnegative, then the length in part 3 is ≥ 0. Therefore the length of P is greater than or equal to $w'_i + c_{ij}$. In view of Equation (12.3) and since $w'_q = w'_p + c_{pq}$, it is clear that the length of P is $\geq w'_q$. This completes the induction argument and the algorithm is verified.

An Example of the Shortest Path Problem with Nonnegative Costs

Consider the network of Figure 12.4. It is desired to find the shortest path from node 1 to all other nodes. Figure 12.5 presents the complete solution for this example. The darkened arcs are those used in the selection of the node to be added to X at each iteration. These arcs can be used to trace the shortest path from node 1 to any given node i (how?). As the reader may suspect, it is no accident that the darkened arcs form a tree!

Indeed, each time a new node is added to the set X, it is connected to the nodes in X with an accompanying arc that clearly does not create a cycle.

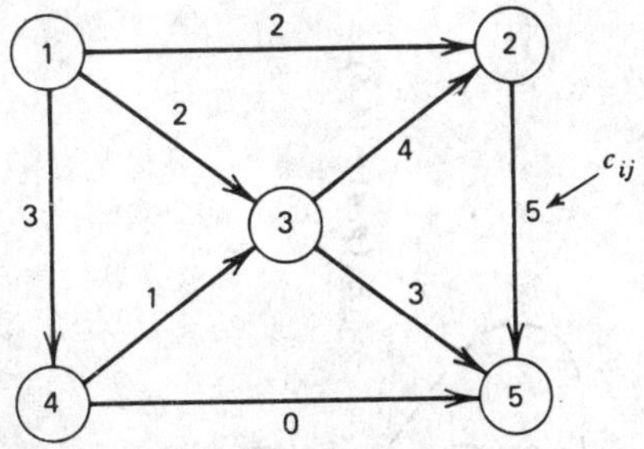

Figure 12.4. An example of a shortest path problem.

(Notice how the predecessor labels of Chapter 9 can be used here with node 1 as the root node.) The tree thus generated is known as the *shortest path tree*.

Complexity of the Algorithm and a Primal Simplex Interpretation

As a rough complexity analysis of the foregoing algorithm, observe that the number of elementary computations in each of the $(m - 1)$ iterations of the algorithm are bounded above by a polynomial of order $O(m^2)$. Thus, the algorithm has a complexity bound of $O(m^3)$. One should *update* the calculations of the path lengths to the nodes rather than recompute them from scratch every iteration. That is, whenever a new node is added to X, its forward star may be scanned to possibly update any of the distance labels w_i' for the nodes in $\bar{X}$. The node with the smallest label w_i' then can be transferred to X, and the current distance calculations for the nodes in $\bar{X}$ should be retained instead of being erased. In this manner, each iteration is of complexity $O(m)$, and the algorithm is of complexity $O(m^2)$. Observe also by the foregoing inductive validation of the algorithm that at each (nonnull) step of the algorithm, the procedure determines a shortest path from node 1 to the newly labeled node i. In fact, if at any iteration $(\leq m - 1)$ of the algorithm, it turns out that the arc set $(X, \bar{X})$ is empty, then the unlabeled nodes in $\bar{X}$ are all obviously unreachable from node 1 and the nodes in X have their shortest paths from node 1 identified by the current shortest path tree. If all nodes are reachable from node 1, the algorithm has found the shortest paths from node 1 to all the nodes at the end of iteration $(m - 1)$.

The foregoing algorithm is actually an implementation of the primal simplex method. To see the relationship, assume for convenience that all nodes are reachable from node 1. Consider a minimum cost network flow problem in which a supply of $(m - 1)$ is placed at node 1 and each of the other nodes has a demand of one unit. Suppose that we start with an all artificial basis, designate node 1 as the root node, and construct artificial arcs having big-M cost coefficients to directly connect node 1 with each of the nodes $i = 2, \ldots, m$. Then it is easy to verify that the first simplex iteration using the rule of entering the variable with the most positive $z_{ij} - c_{ij}$ value is identical to the first iteration of Dijkstra's algorithm. In fact, this is true for all the iterations. To see this, suppose

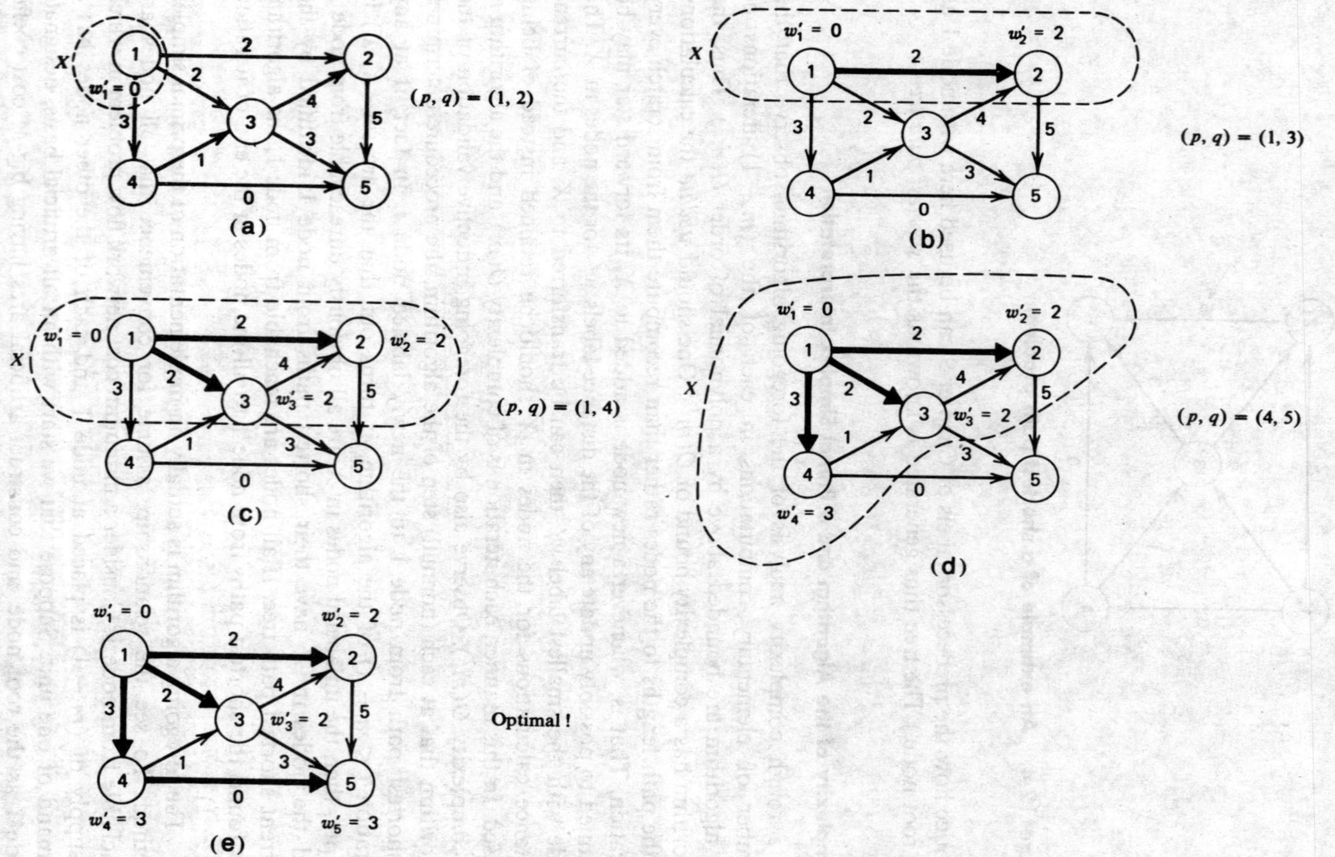

Figure 12.5. **Solution of the example of Figure 12.4.**

that at some stage, we have a set of nodes still connected to node 1 via the artificial arcs, while the remaining nodes are connected to node 1 by (directed) paths in the current shortest path (sub)tree. Let X be a set that contains the latter nodes, and let $\overline{X}$ contain the former nodes. Note that the dual variables w_i for nodes $i \in \overline{X}$ are all $-M$, while the dual variables w_i for the nodes $i \in X$ are equal to the negative of the total cost on the arcs in the respective paths connecting the nodes to node 1 in the current basis tree. Hence we currently have $w_i = -w_i'$ as given by Dijkstra's algorithm. Note that the maximum $z_{ij} - c_{ij}$ value is realized for an arc (i, j) with $i \in X$ and $j \in \overline{X}$ (why?). Therefore, the value

$$\underset{(i,j)}{\text{Max}} \{z_{ij} - c_{ij}\} = \text{Max}\left\{w_i - w_j - c_{ij} : i \in X,\ j \in \overline{X}\right\}$$

$$= -\underset{(i,j)}{\text{Min}}\left\{-M + w_i' + c_{ij} : i \in X,\ j \in \overline{X}\right\}$$

is realized by the same arc (p, q) as in Dijkstra's algorithm (why?). When the arc (p, q) enters the basis, the only reverse arc in the resulting cycle is the artificial arc that therefore leaves the basis. Moreover, the subtree of the resulting basis tree induced by the nodes in $X \cup \{q\}$ is identical to the current shortest path tree constructed by Dijkstra's algorithm. Hence this algorithm coincides with a primal simplex implementation and produces an optimal basis tree for the above-mentioned network flow problem at termination. This is verified by the foregoing validation of Dijkstra's algorithm (why?).

A Shortest Path Procedure for Arbitrary Costs

The shortest path algorithm described earlier in this section does not generalize to the case when the costs are allowed to be negative. Figure 12.6 illustrates this where the previous algorithm would select node 3 to enter X with $w_3' = w_1' + c_{13} = 2$ as the value of the shortest path from node 1 to node 3. However, it would be better to first travel to node 2, incurring a higher cost, and then go on to node 3 for a saving.

There still is a fast and efficient method for the shortest path problem with negative costs. We shall assume, however, that the sum of the costs on arcs of any

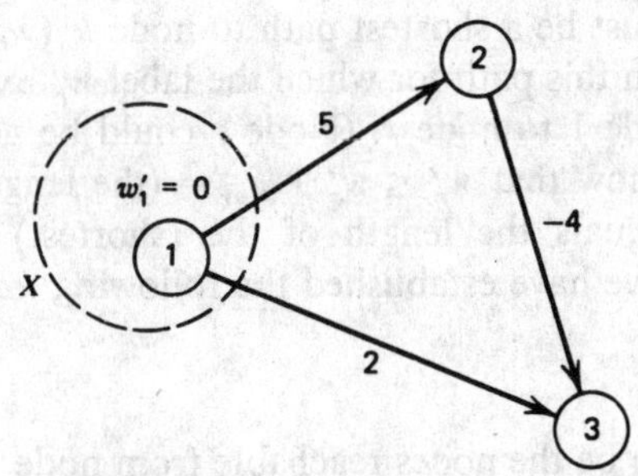

Figure 12.6. An example where the nonnegative cost algorithm will not work.

circuit in G is nonnegative. Without this assumption, a "traveler" would proceed directly to the circuit in G and traverse it an infinite number of times with the cost decreasing after each time around the circuit. Note that placing an upper bound of unity on the arcs does not help solve this problem! (See Exercise 12.26.) Indeed, the problem of finding the shortest *simple* path on networks with mixed-sign costs is NP-complete (see the computational complexity references of Chapter 8).

The algorithm works with the dual of the shortest path problem. Recall that the dual problem with the substitution $w_i' = -w_i$ for $i = 1, 2, \ldots, m$ is given by the following.

$$\begin{array}{lll} \text{Maximize} & w_m' - w_1' & \\ & & \\ \text{Subject to} & w_j' - w_i' \leq c_{ij} & i, j = 1, 2, \ldots, m \\ & w_i' \quad \text{unrestricted} & i = 1, 2, \ldots, m \end{array}$$

Since the objective and the constraints involve only differences in variables we may set one variable to any value, say $w_1' = 0$ (why?).

In the algorithm for negative costs we shall begin with a choice of $\mathbf{w}'$ that is "superoptimal" with respect to the dual objective but which may violate one or more of the dual constraints. We shall show that by iteratively modifying $\mathbf{w}'$ to satisfy the constraints, one at a time, we shall be able to terminate in a finite number of steps with the optimal solution.

Considering the set of nodes reachable from node 1 and assuming these nodes to be $i = 1, \ldots, m$, the fundamental task is to determine a set of labels w_i' equal to the length of some path from node 1 to node i for $i = 1, \ldots, m$ such that

$$w_j' = \underset{i:(i,j) \text{ exists}}{\text{Minimum}} \{ w_i' + c_{ij} \} \quad \text{for all } j = 1, \ldots, m \tag{12.4}$$

If w_i', $i = 1, \ldots, m$ are the lengths of the shortest (simple) paths, then Equation (12.4) must be satisfied in the absence of negative cost circuits or else we would be able to obtain a strictly shorter (simple) path to some node j for which Equation (12.4) is violated (how?). Conversely, if Equation (12.4) is satisfied, then the labels w_i', $i = 1, \ldots, m$ must be the lengths of the shortest paths. If this is not true for some node j, say, then consider a shortest path $\{1, p, q, r, \ldots, j\}$ from node 1 to node j. Note that for each node k in this path, the path from node 1 to node k in this path must be a shortest path to node k (why?). Hence let node r, say, be the first node in this path for which the label w_r' exceeds the length of the shortest path from node 1 to node r. (Node r could be node j itself.) But from Equation (12.4), we know that $w_r' \leq w_q' + c_{qr} =$ (the length of the shortest path to q) $+ c_{qr}$, which equals the length of the (shortest) path from 1 to r, a contradiction. Hence we have established the following *key result*.

***Theorem* 12.2**

Let nodes $i = 1, \ldots, m$ be the nodes reachable from node 1 in a network with no negative cost (directed) circuit reachable from node 1. Suppose that w_i' is the length of some path from node 1 to node i for $i = 1, \ldots, m$. Then w_i' is the length of the shortest path from node 1 to node i for $i = 1, \ldots, m$ if and only if Equation (12.4) holds.

Consequently, our task is to ensure that Equation (12.4) holds. The astute reader will have observed the relationship between Theorem 12.2 and the (Karush–Kuhn–Tucker) primal-dual feasibility and complementary slackness conditions for the shortest path problem. For the case of nonnegative arc costs, we were able to satisfy Equation (12.4) by a *label setting* procedure in which the first time a node was selected, it received a label equal to the length of its shortest path from node 1. In view of the example in Figure 12.6, such labels may need to be revised in the presence of costs with mixed-signs. Hence the following modification of Dijkstra's algorithm is known as a *label correcting* algorithm.

The algorithm for finding the shortest path distances from node 1 to all the nodes proceeds as follows, assuming that G has no negative cost (directed) circuits. Recall that for a node p, the *forward star* of p is the set of nodes j such that (p, j) is an arc in the network. We also follow the convention that a node not reachable from node 1 has a shortest path distance of ∞ and that $\infty + c_{ij} = \infty$ for $-\infty < c_{ij} < \infty$.

INITIALIZATION STEP

Set $w_1' = 0$ and $w_i' = \infty$ for $i \neq 1$. Let the *scan eligible list* $SE = \{1\}$.

MAIN STEP

If the scan eligible list $SE = \varnothing$, stop, the labels w_i' are the shortest path distances from node 1 to nodes $i = 1, \ldots, m$. (The shortest paths are traceable via the predecessor labels or the "darkened" arcs.) Otherwise select a node $p \in SE$ and scan the forward star of p. For all arcs (p, q) corresponding to this set for which $w_q' > w_p' + c_{pq}$, set $w_q' = w_p' + c_{pq}$, let p be the predecessor of node q, and replace SE by $SE \cup \{q\}$. (At the end of the step, the "darkened" edges correspond to the arcs connecting each node i that has $w_i' < \infty$ to its predecessor node.) Remove node p from SE and repeat the main step.

An Example of the Shortest Path Algorithm for Arbitrary Costs

Consider the network of Figure 12.7 where we wish to find the shortest path from node 1 to all the nodes. In Figure 12.8 we present the complete solution of the example by the previous algorithm. There is no required order in which the arcs must be considered for the algorithm to converge. When the algorithm stops, Figure 12.8*d* gives the shortest paths and their values w_i'. Arcs along the shortest path have $w_j' - w_i' = c_{ij}$.

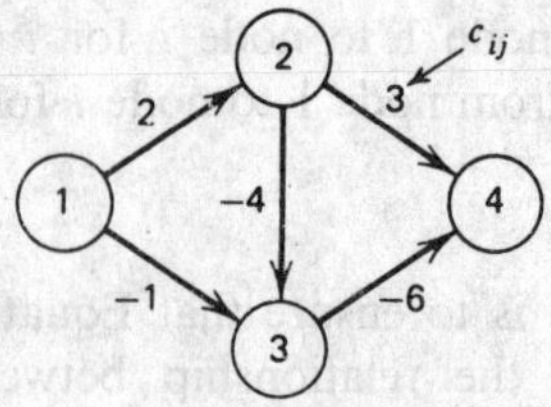

Figure 12.7. **A shortest path example with negative costs.**

Verification of the Algorithm for Arbitrary Costs

We shall first show that when $w_i' < \infty$, it represents the cost of some path (not necessarily simple) from node 1 to node i. We note that in computing the cost of a (not necessarily simple) path, we must count the cost of an arc as many times as the arc appears in the path. Thus for a nonsimple path $P = \{(1, 3), (3, 4), (4, 5), (5, 3), (3, 4), (4, 6)\}$ the associated cost would be $c_{13} + c_{34} + c_{45} + c_{53} + c_{34} + c_{46}$. As we shall see, nonsimple paths play a part in the algorithm only when there is a negative circuit in G.

We shall demonstrate that w_i' $(< \infty)$ represents the cost of a path from node 1 to node i at every iteration of the procedure.

Now, since we start with $w_1' = 0$ and $w_i' = \infty$ for $i \neq 1$, the result is true at the first iteration. This is so because $w_1' = 0$ represents the cost of the (empty) path, which contains no arcs, from node 1 to node 1.

Assume that the result is true at iteration t; that is, that w_i' $(< \infty)$ represents the cost of some (not necessarily simple) path from node 1 to node i. Consider iteration $t + 1$. Suppose that for $i \in SE$, we find that $w_i' + c_{ij} < w_j'$ in which case we set $w_j' = w_i' + c_{ij}$. By assumption there exists a path P_i from node 1 to node i at iteration t with cost w_i'. Consider the path $P_j = P_i \cup \{(i, j)\}$. This path has cost $w_i' + c_{ij} = w_j'$ and the result is true at iteration $t + 1$. Thus, if $w_k' < \infty$, then there exists a (not necessarily simple) path P from node 1 to node k along which $\Sigma_{(i, j) \in P}\, c_{ij} = w_k'$.

If there are no negative circuits, then the cost of any nonsimple path is greater than or equal to the cost of the corresponding simple path after deleting the circuits (why?). Hence, if there are no negative circuits, w_i' is bounded from below by the cost of the shortest simple path and thus by a finite integer (why?). Finally, treating ∞ as a large integer, since $0 \leq |SE| \leq m$ and since the two-tuple $\{\Sigma_i w_i', |SE|\}$ lexicographically decreases by a positive integer at each iteration (why?), the shortest path algorithm will stop in a finite number of steps if there are no negative circuits. Moreover, at termination if any $w_i = \infty$, then it is readily verified that node i is unreachable from node 1 and vice versa.

Let us show that at termination, the labels w_i' are the shortest path distances. From Theorem 12.2, it is sufficient to show that condition (12.4) holds for the nodes reachable from node 1. Suppose on the contrary that for some node q, we have $w_q' > \underset{i}{\text{Minimum}}\, \{w_i' + c_{iq}\} \equiv w_p' + c_{pq}$, say. (We know that $w_q' = w_r' + c_{rq} < \infty$ for some r.) However, after $w_p' < \infty$ was set for the final time in the algorithm and was later selected from SE for the final time, we should have

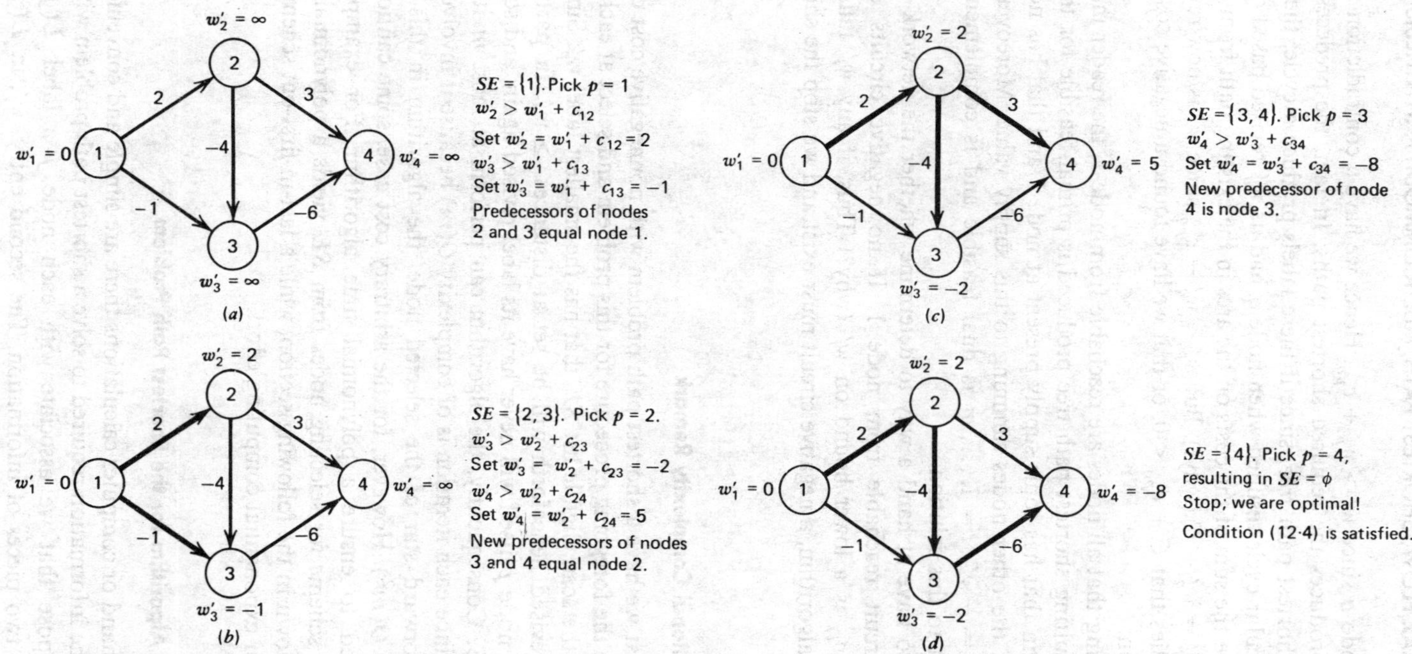

Figure 12.8. The solution for the network of Figure 12.7.

relabeled node q since $w'_q > w'_p + c_{pq}$. Hence we have a contradiction and so the algorithm produces the required shortest paths. In fact, the predecessor labels produce a shortest path *tree*, since if these labels produce a cycle, then suppose that a particular cycle resulted when node q had its w'_q revised based on node p. Letting C be the sum of the costs on the arcs in the existing path from q to p, we have that $w'_p = w'_q + C$ (why?). But $w'_p < w'_q - c_{pq}$ since w'_q is now revised. This in turn implies that $C + c_{pq} < 0$, or that we have found a negative cost circuit, a contradiction.

Assuming that all nodes are reachable from node 1, the reader may observe that the resulting shortest path tree produced is primal feasible for the shortest path problem that has some supply present at node 1 and that has nonnegative demands at the other nodes summing to this supply value. Moreover, the dual solution $w_i = -w'_i$, $i = 1, \ldots, m$ is dual feasible and is complementary slack with respect to this solution.

We also have at hand a way to determine whether the network contains a negative circuit reachable from node 1. If no negative circuits exist, then $c_0 = \Sigma_{c_{ij}<0} c_{ij}$ is a lower bound on w'_i (why?) Thus, if any w'_i falls below c_0 during the algorithm, a negative circuit must exist and we stop the shortest path algorithm.

A Computational Complexity Remark

Suppose that we have a shortest path problem with nonnegative cost coefficients. We will use the foregoing procedure for this problem and select at each stage that node from the scan eligible list SE that has the smallest label w'_p. Since $c_{ij} \geq 0$, all labels revised henceforth will be set at distances $\geq w'_p$. In particular, this means that node p itself will never have its label revised again, and so will never re-enter SE. Consequently, the algorithm can perform at most m iterations in this case. Since each iteration is of complexity $O(m)$ because it involves scanning only the forward star of the selected node, the algorithm in this case is of complexity $O(m^2)$. However, for the arbitrary cost case, some caution needs to be exercised to ensure a polynomial time algorithm. For example, a first-in-first-out scheme for selecting nodes from SE yields a polynomial-time algorithm as shown in the following section, while a last-in-first-out scheme is known to admit an exponential computing effort.

A Labeling Algorithm for the Shortest Path Problem

Either for hand or computer calculations there are simple and convenient ways to maintain the information required to solve a shortest path problem with arbitrary costs. Suppose that we associate with each node j a label $L(j) = (i, w'_j)$ containing two pieces of information. The second entry, w'_j, in $L(j)$ indicates the cost (length) of the current "best" path from node 1 to node j. The first entry, i, in $L(j)$ gives the predecessor node, that is, the node just prior to node j in the path. Let $c_0 = \Sigma_{c_{ij}<0} c_{ij}$. The labeling algorithm becomes as follows.

INITIALIZATION STEP

Set $L(1) = (-,0)$, $L(i) = (-,\infty)$ for $i = 2,\ldots, m$, and $SE = \{1\}$.

MAIN STEP

If $SE = \phi$, stop; the required shortest paths are available. The second label gives the shortest path distances and the paths are traceable using the first (predecessor) labels. Otherwise select $p \in SE$. For all arcs (p, q) with q in the forward star of p such that $w_q' > w_p' + c_{pq}$, set $L(q) = \{p, w_q' = w_p' + c_{pq}\}$, and replace SE by $SE \cup \{q\}$. If $w_q' < c_0$, stop; there is a negative cost circuit in G. Otherwise remove p from SE and repeat the main step.

An Example of the Labeling Algorithm

Suppose that we use the labeling algorithm to solve the shortest path problem of Figure 12.7. First, $c_0 = -1 - 4 - 6 = -11$.

The sequence of operations of the labeling algorithms are as follows.

1. $L(1) = (-,0)$, $L(2) = (-,\infty)$, $L(3) = (-,\infty)$, $L(4) = (-,\infty)$, $SE = \{1\}$.
2. $p = 1$, $L(2) = (1,2)$, $L(3) = (1,-1)$, $SE = \{2,3\}$.
3. $p = 2$, $L(3) = (2,-2)$, $L(4) = (2,5)$, $SE = \{3,4\}$.
4. $p = 3$, $L(4) = (3,-8)$, $SE = \{4\}$.
5. $p = 4$, $SE = \varnothing$.

STOP; $L(2) = (1,2)$, $L(3) = (2,-2)$, and $L(4) = (3,-8)$. The shortest paths to nodes 2, 3, and 4 are of lengths 2, -2, and -8 respectively, and are respectively determined as $\{(1,2)\}$, $\{(1,2),(2,3)\}$ and $\{(1,2),(2,3),(3,4)\}$ by backtracking using the first labels in $L(\cdot)$.

Identifying Negative Circuits With the Shortest Path Algorithm

We have already indicated that if $w_k' < c_0$ at some point in the shortest path algorithm, then a negative circuit reachable from node 1 exists in G. To find such a negative circuit, begin at node k and apply the following procedure.

INITIALIZATION STEP

Let $p = k$, $C = \{k\}$.

MAIN STEP

Let $L_1(p)$ be the first entry in $L(p)$.

1. If $L_1(p) > 0$, let $l = L_1(p)$, and replace $L_1(p)$ by $-L_1(p)$, and C by $C \cup \{l\}$. Set $p = l$ and repeat the main step.

2. If $L_1(p) < 0$, then stop. A negative circuit has been found, and may be traced by starting at the last element in C and proceeding backwards through C until this element repeats.

Note that the original node k may not be part of the circuit.

12.3 POLYNOMIAL SHORTEST PATH ALGORITHMS FOR NETWORKS WITH ARBITRARY COSTS

We will now slightly modify the algorithm for determining the shortest paths from node 1 to all the other nodes in a network having arbitrary cost coefficients in order to derive an efficient polynomial-time (label correcting) algorithm. The particular algorithm we discuss is known as the *partitioned shortest path algorithm*, because it effectively partitions the scan eligible list SE into two sets, namely NOW and NEXT. As implied by the names of these sets, the nodes in the list NOW are scanned first. The nodes in NOW postpone the consideration of any new scan eligible nodes into the list NEXT until NOW is exhausted. At this point, if NEXT is nonempty, its contents are transferred to NOW and the procedure repeats. This continues until $SE \equiv \text{NOW} \cup \text{NEXT}$ is empty. Notice that the first-in-first-out scheme for selecting a node from SE is a particular case of this procedure. The algorithm is specified next.

Partitioned Shortest Path (PSP) Algorithm

INITIALIZATION

Set $w_1' = 0$, $w_i' = \infty$ for $i \neq 1$, NOW $= \{1\}$, NEXT $= \varnothing$, and the iteration counter $k = 1$. Let c_0 be the sum of the negative costs in G.

STEP 1

If NOW $= \varnothing$, proceed to step 2. Otherwise select a node $p \in$ NOW. For each arc (p, q) with q in the forward star of p such that $w_q' > w_p' + c_{pq}$, set $w_q' = w_p' + c_{pq}$ and let p be the predecessor of node q. If $w_q' < c_0$ for any such node, stop; there exists a negative cost circuit in G. If node q is not currently scan eligible, that is, it is in neither NOW nor NEXT, add it to NEXT. Remove node p from NOW and repeat step 1.

STEP 2

If NEXT $= \varnothing$, then stop; the labels w_i' give the shortest path distances. The shortest paths are traceable by backtracking using the predecessor labels. Otherwise put NOW $=$ NEXT and NEXT $= \varnothing$ (perhaps by simply renaming the lists), increment k by one, and return to step 1.

Example of the PSP Algorithm

To illustrate, consider the example of Figure 12.7. As before, let $L(\cdot, \cdot)$ be the predecessor and path length label as defined in the previous section. We begin at iteration $k = 1$ with NOW = {1}, NEXT = ∅, $L(1) = (-, 0)$, and $L(i) = (-, \infty)$ for $i \neq 1$. The operations proceed as follows.

1. $k = 1$ with NOW = {1}, NEXT = ∅:
   $p = 1$, $L(2) = (1, 2)$, $L(3) = (1, -1)$, NOW = ∅, NEXT = {2, 3}.
2. $k = 2$ with NOW = {2, 3}, NEXT = ∅:
   $p = 2$, $L(3) = (2, -2)$, $L(4) = (2, 5)$, NOW = {3}, NEXT = {4}.
   $p = 3$, $L(4) = (3, -8)$, NOW = ∅, NEXT = {4}.
3. $k = 3$ with NOW = {4}, NEXT = ∅:
   $p = 4$, NOW = ∅, NEXT = ∅.

The algorithm terminates with $L(2) = (1, 2)$, $L(3) = (2, -2)$, and $L(4) = (3, -8)$. These labels give the shortest paths from node 1 to nodes 2, 3, and 4 as before.

Complexity Analysis of the PSP Algorithm

First of all note that the algorithm terminates finitely, and produces the shortest paths from node 1 to all the nodes reachable from node 1, provided that G does not contain any negative cost circuit (reachable from node 1). This follows from the previous section by simply noting that SE = NOW ∪ NEXT and that the algorithm only specifies a (partial) order in which the nodes in SE must be considered. However, it is this partitioning of SE that guarantees a polynomial-time algorithm. As we now verify, the algorithm performs at most $(m - 1)$ iterations. Since each iteration scans no arc of the problem more than once, it is of complexity $O(|\mathscr{A}|)$, where $|\mathscr{A}|$ is the number of arcs. Hence the algorithm is of complexity $O(m|\mathscr{A}|)$ which translates to $O(m^3)$ for dense networks.

The key property of the algorithm is as follows. Consider the current (shortest path) partial tree that exists at the beginning of step 2 at each iteration k in the algorithm. Then the *level* of each node in NEXT in this tree is at least k, where the level of a node is the number of arcs in the chain from this node to node 1 in the tree. This is clearly true for $k = 1$. Inductively, assume that this is true for iteration k and consider iteration $(k + 1)$. Note that at the beginning of iteration $(k + 1)$, the list NOW coincides with the list NEXT at step 2 of the previous iteration. Hence by the induction hypothesis, each of the nodes in this list has a level of at least k in the tree. A node that has its label revised in this iteration is either already in NOW or enters NEXT, but with its level in the tree being one more than the level of the node that revises its label. Hence the levels of the nodes in NEXT at the end of step 1 (or the beginning of step 2) of iteration $(k + 1)$ are at least $(k + 1)$, thus establishing the required result.

Because of this key property, since a shortest path tree is constructed by the algorithm in the absence of negative cost circuits in G, the shortest paths from node 1 to the nodes that are in NEXT at the beginning of step 2 of each iteration k involve at least k arcs. This follows since the shortest path to any node is verified or established in the iteration following the one in which it shows up in NEXT at step 2 for the last time. This means that the algorithm cannot possibly perform more than $(m - 1)$ iterations (why?). In fact, some node in NEXT at step 2 must have its shortest path already determined at each iteration. This follows because the current list NEXT becomes the list NOW for the next iteration, and since all nodes relabeled from this point onward must have a node in this list on the chain to node 1, we cannot possibly have all the nodes in this list repeating or else we would have a cycle. For this example, at the end of iteration $k = 1$, node $2 \in$ NEXT has its shortest path determined, and at the end of iteration $k = 2$, node $4 \in$ NEXT has its shortest path determined. Node 3 shows up in NEXT for the last time at (the end of) iteration $k = 1$ and has its shortest path determined during iteration $k = 2$. The levels of nodes 2, 3, and 4 in the final shortest path tree are 1, 2, and 3 respectively.

Threshold Partitioned Shortest Path Algorithm

In concluding this section, we remark that a computationally efficient variant of the PSP algorithm, particularly for nonnegative cost coefficients, has been designed to operate as follows. At the end of every iteration k, a *threshold value* t is computed. Instead of transferring the entire list NEXT into NOW, only those nodes $i \in$ NEXT are placed in NOW that have $w_i' \leq t$. The value t is tuned heuristically based on the network structure, but of course always satisfies $t \geq \text{Minimum}\{w_i' : i \in \text{NEXT}\}$. Additionally, at step 1 of the PSP algorithm, whenever a node q has its label updated and q is not a member of NOW, it is checked whether the revised w_q' exceeds the current threshold value t or not. If $w_q' > t$, then node q is added to NEXT as before. Otherwise node q is added to a separate list NOW′ (unless it is already there) and it is removed from NEXT if it happens to be there. Then when NOW becomes empty, the contents of NOW′ are transferred to NOW and the process repeats by scanning the nodes in NOW as before, until both NOW and NOW′ are empty. At this point, the list NEXT is processed as earlier. If NEXT $= \varnothing$ also, then the method terminates.

Observe that when all costs are nonnegative, the nodes with the smallest w_i' labels in the list NOW′ or NEXT which are transferred into NOW have their shortest path distances already determined, and so will not show up in any subsequent set. Hence there can be no more than $(m - 1)$ *total* transfers from NOW′ or NEXT into NOW. Since the effort involved between each such transfer is of complexity $O(|\mathscr{A}|)$, the algorithm is of complexity $O(m|\mathscr{A}|)$. Computationally, any new nodes added to NOW′ or NEXT are appended to the bottom of these lists. It is also suggested that nodes from NOW be selected in a last-in-first-out order. Empirically, such an algorithm has been found to dominate all existing procedures on several types of networks with nonnegative costs (see the Notes and References section).

12.4 MULTICOMMODITY FLOWS

In all of the flow problems we have considered to this point, it has not been necessary to distinguish among the units flowing in the network. This class of network flow problems is called single-commodity flow problems. There is also a class of network flow problems called multicommodity flow problems in which it is necessary to distinguish among the flows in the network.

The most natural example of multicommodity flows occurs in rush hour traffic in any metropolitan city. If the area is divided into zones, then there are a number of people in zone i who must travel to work in zone j. There are also a number of people who must travel from zone j to work in zone i. Where people are located corresponds to supply ($b > 0$) and where they wish to go corresponds to demand ($b < 0$). If we treat the problem as a single-commodity flow problem, a minimal cost flow procedure (network simplex or out-of-kilter) would use the supply of people in a given zone to satisfy the demand in the same zone. This is an unacceptable solution. In this problem and ones like it, we must distinguish between the different types of flow and be careful to retain their identity and flow pattern throughout the optimization procedure. That is, we must essentially have a different flow vector and set of conservation equations for each commodity.

As we shall see, multicommodity flow problems do not enjoy the same special properties as single-commodity flow problems. As an example, consider the network of Figure 12.9. Suppose that there are three commodities that flow through the network. The source for commodity 1 is node 1, and the sink for commodity 1 is node 3. That is, commodity 1 must originate only at node 1 and terminate only at node 3. Similarly, let the source and sink for commodity 2 be nodes 2 and 1 respectively. Finally, the source and sink for commodity 3 are nodes 3 and 2 respectively. With the restriction that the sum of all commodities flowing on an arc should not exceed the arc capacity $u_{ij} = 1$, what is the maximal sum of commodity flows, $f_1 + f_2 + f_3$, possible in the network?

Finding the maximal flow for the three-commodity problem of Figure 12.9 is relatively simple since there is only one path that each commodity can take on its

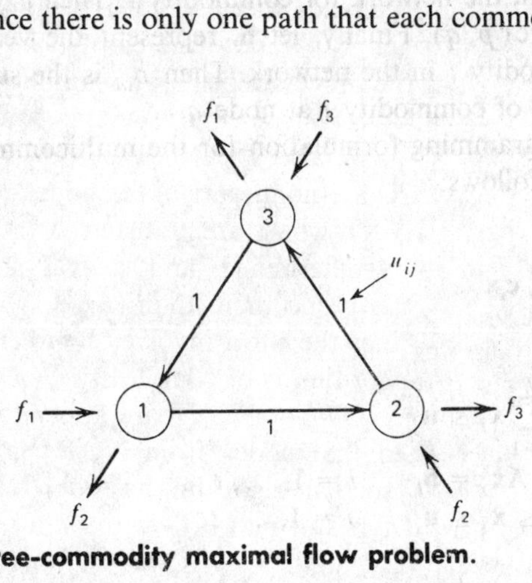

Figure 12.9. **A three-commodity maximal flow problem.**

way from its source to its sink. The paths for commodity 1, 2, and 3 respectively are

$$P_1 = \{(1,2),(2,3)\}$$
$$P_2 = \{(2,3),(3,1)\}$$
$$P_3 = \{(3,1),(1,2)\}$$

If we place a single unit of flow on any one of the paths, then the other paths are completely blocked (that is, must have zero flow) and thus the total flow would be 1. However, there is a better solution available if we do not require integer flows. Suppose that we place $\frac{1}{2}$ unit of flow of commodity 1 on P_1, $\frac{1}{2}$ unit of flow of commodity 2 on P_2, and $\frac{1}{2}$ unit of flow of commodity 3 on P_3. In this case none of the arc capacities are violated and the total flow of all commodities is $\frac{3}{2}$. From this we see that multicommodity flow problems do not necessarily provide integer flows.

Even though multicommodity flow problems do not have as "nice" a structure as single-commodity flow problems, they still are linear programs (if we ignore integrality of the variables). As we shall soon see, multicommodity flow problems do have some special structure that permits the application of decomposition techniques.

The Multicommodity Minimal Cost Flow Problem

Suppose that we are given a network G with m nodes and n arcs in which there will flow t different commodities. Let $\mathbf{u}_i$ represent the vector of upper limits on flow for commodity i in the arcs of the network. Then u_{ipq} is the upper limit on flow of commodity i in arc (p, q). Also, let $\mathbf{u}$ represent the vector of upper limits on the sum of all commodities flowing in the arcs of the network. Then u_{pq} is the upper limit on the sum of all commodity flows in arc (p, q). Let $\mathbf{c}_i$ represent the vector of arc costs in the network for commodity i. Then c_{ipq} is the unit cost of commodity i on arc (p, q). Finally, let $\mathbf{b}_i$ represent the vector of supplies (or demands) of commodity i in the network. Then b_{iq} is the supply (if $b_{iq} > 0$) or demand (if $b_{iq} < 0$) of commodity i at node q.

The linear programming formulation for the multicommodity minimal cost flow problem is as follows:

$$\text{Minimize} \quad \sum_{i=1}^{t} \mathbf{c}_i \mathbf{x}_i$$

$$\begin{aligned} \text{Subject to} \quad & \sum_{i=1}^{t} \mathbf{x}_i \leq \mathbf{u} & \\ & \mathbf{A}\mathbf{x}_i = \mathbf{b}_i & i = 1, \ldots, t \\ & \mathbf{0} \leq \mathbf{x}_i \leq \mathbf{u}_i & i = 1, \ldots, t \end{aligned}$$

where $\mathbf{x}_i$ is the vector of flows of commodity i in the network and $\mathbf{A}$ is the node-arc incidence matrix of the graph. The foregoing formulation is called the *node-arc* formulation for the multicommodity flow problem since it uses the node-arc incidence matrix.

The multicommodity minimal cost flow problem possesses the block diagonal structure discussed in Section 7.5 of the decomposition chapter. Thus we may apply the block diagonal decomposition technique to the foregoing problem. The multicommodity minimal cost flow problem has $(t + 1)n$ variables and $n + mt$ constraints (including the slack variables for the coupling constraints and ignoring the nonnegativity and upper bound constraints $\mathbf{0} \leq \mathbf{x}_i \leq \mathbf{u}_i$). Thus, even for moderate-sized problems, the constraint matrix will be large. For example, suppose that we have a problem with 100 nodes, 250 arcs, and 10 commodities. The problem will have 2750 variables and 1250 constraints.

Consider the application of the decomposition algorithm to the minimal cost multicommodity flow problem. Let $X_i = \{\mathbf{x}_i : \mathbf{A}\mathbf{x}_i = \mathbf{b},\ \mathbf{0} \leq \mathbf{x}_i \leq \mathbf{u}_i\}$. Assume that each component of $\mathbf{u}_i$ is finite so that X_i is bounded (see Exercise 12.52 for a relaxation of this assumption). Then any $\mathbf{x}_i$ can be expressed as a convex combination of the extreme points of X_i as follows:

$$\mathbf{x}_i = \sum_{j=1}^{k_i} \lambda_{ij}\mathbf{x}_{ij}$$

where

$$\sum_{j=1}^{k_i} \lambda_{ij} = 1$$

$$\lambda_{ij} \geq 0 \qquad j = 1, \ldots, k_i$$

and $\mathbf{x}_{i1}, \mathbf{x}_{i2}, \ldots, \mathbf{x}_{ik_i}$ are the extreme points of X_i. Substituting for $\mathbf{x}_i$ in the multicommodity minimal cost flow problem and denoting the vector of slacks by $\mathbf{s}$, we get the following.

$$\begin{array}{lll}
\text{Minimize} & \displaystyle\sum_{i=1}^{t} \sum_{j=1}^{k_i} (\mathbf{c}_i\mathbf{x}_{ij})\lambda_{ij} & \\
\text{Subject to} & \displaystyle\sum_{i=1}^{t} \sum_{j=1}^{k_i} \mathbf{x}_{ij}\lambda_{ij} + \mathbf{s} = \mathbf{u} & \\
& \displaystyle\sum_{j=1}^{k_i} \lambda_{ij} = 1 & i = 1, \ldots, t \\
& \lambda_{ij} \geq 0 & j = 1, \ldots, k_i \\
& & i = 1, \ldots, t \\
& \mathbf{s} \geq \mathbf{0} &
\end{array}$$

Suppose that we have a basic feasible solution to the multicommodity minimal cost flow problem in terms of the λ_{ij}'s and let $(\mathbf{w}, \boldsymbol{\alpha})$ be the vector of dual variables corresponding to the basic feasible solution ($\mathbf{w}$ has n components and $\boldsymbol{\alpha}$

has t components). Then dual feasibility is given by the following two conditions:

(i) $w_{pq} \leq 0$ corresponding to each s_{pq}, and
(ii) $\mathbf{w}\mathbf{x}_{ij} + \alpha_i - \mathbf{c}_i\mathbf{x}_{ij} \leq 0$ corresponding to each λ_{ij}.

If any of these conditions is violated, the corresponding variable (s_{pq} or λ_{ij}) is a candidate to enter the master basis. Here s_{pq} is a candidate to enter the basis if $w_{pq} > 0$. For a given commodity i, a nonbasic variable among the λ_{ij}'s could enter the basis if the optimal objective of the following subproblem is positive (why?).

$$\begin{array}{ll} \text{Maximize} & (\mathbf{w} - \mathbf{c}_i)\mathbf{x}_i + \alpha_i \\ \text{Subject to} & \mathbf{A}\mathbf{x}_i = \mathbf{b}_i \\ & \mathbf{0} \leq \mathbf{x}_i \leq \mathbf{u}_i \end{array}$$

But, since $\mathbf{A}$ is a node-arc incidence matrix, this is simply a single-commodity flow problem. Thus it may be solved by one of the efficient techniques for solving single-commodity network flow problems (network simplex or out-of-kilter).

Summary of the Decomposition Algorithm Applied to the Multicommodity Minimal Cost Flow Problem

We now specialize the decomposition algorithm of Chapter 7 to the multicommodity minimal cost flow problem.

INITIALIZATION STEP

Begin with a basic feasible solution to the master problem. Store $\mathbf{B}^{-1}$, $\bar{\mathbf{b}} = \mathbf{B}^{-1}\begin{pmatrix}\mathbf{u}\\ \mathbf{1}\end{pmatrix}$, and $(\mathbf{w}, \boldsymbol{\alpha}) = \hat{\mathbf{c}}_B\mathbf{B}^{-1}$, where $\hat{c}_{ij} = \mathbf{c}_i\mathbf{x}_{ij}$ for the λ_{ij} variables. (The two-phase or the big-M method may be required.)

MAIN STEP

1. Let $(\mathbf{w}, \boldsymbol{\alpha})$ be the vector of dual variables corresponding to the current basic feasible solution to the master problem. If any $w_{pq} > 0$, then the corresponding s_{pq} is a candidate to enter the master basis. If $w_{pq} \leq 0$ for each arc, consider the following ith subproblem.

$$\begin{array}{ll} \text{Maximize} & (\mathbf{w} - \mathbf{c}_i)\mathbf{x}_i + \alpha_i \\ \text{Subject to} & \mathbf{A}\mathbf{x}_i = \mathbf{b}_i \\ & \mathbf{0} \leq \mathbf{x}_i \leq \mathbf{u}_i \end{array}$$

   This is a single-commodity flow problem. If the solution $\mathbf{x}_{ik}$ to this problem has $z_{ik} - c_{ik} = (\mathbf{w} - \mathbf{c}_i)\mathbf{x}_{ik} + \alpha_i > 0$, then λ_{ik} is a candidate to enter the master basis.

2. If there is no candidate to enter the master basis, then stop; the optimal solution is at hand. Otherwise, select a candidate variable, update its column accordingly, $\mathbf{B}^{-1}\begin{pmatrix}\mathbf{e}_{pq}\\ \mathbf{0}\end{pmatrix}$ for s_{pq} and $\mathbf{B}^{-1}\begin{pmatrix}\mathbf{x}_{ik}\\ \mathbf{e}_i\end{pmatrix}$ for λ_{ik}, and pivot. [Note that $\mathbf{e}_{pq}$ is a unit vector with the 1 in the row associated with arc(p, q).] This updates the basis inverse, the dual variables, and the right-hand-side. Return to step 1.

An Example of the Multicommodity Minimal Cost Flow Algorithm

Consider the two-commodity minimal cost flow problem whose data are given in Figure 12.10.

The constraint matrix and the right-hand-side are displayed in Figure 12.11 (the lower and upper bound constraints $\mathbf{0} \le \mathbf{x}_1 \le \mathbf{u}_1$ and $\mathbf{0} \le \mathbf{x}_2 \le \mathbf{u}_2$ are not displayed). Notice the structure of the coupling constraints and the special structured block diagonal constraints. Also note that x_1 and x_2 represent the artificial variables for the two commodities.

INITIALIZATION

To avoid the two-phase or the big-M methods, suppose that we begin with the following feasible solutions.

$$\mathbf{x}_{11} = \begin{bmatrix} x_{112} \\ x_{123} \\ x_{134} \\ x_{141} \\ x_{142} \end{bmatrix} = \begin{bmatrix} 2 \\ 3 \\ 2 \\ 0 \\ 0 \end{bmatrix} \quad \text{and} \quad \mathbf{x}_{21} = \begin{bmatrix} x_{212} \\ x_{223} \\ x_{234} \\ x_{241} \\ x_{242} \end{bmatrix} = \begin{bmatrix} 0 \\ 0 \\ 3 \\ 0 \\ 3 \end{bmatrix}$$

Note that the master basis (in the space of the slack variables and the λ_{ij}'s) consists of all the slacks, λ_{11} and λ_{21}. The basis and its inverse are

$$\begin{array}{c} \\ \mathbf{B} = \end{array} \begin{array}{c} \begin{matrix} s_{12} & s_{23} & s_{34} & s_{41} & s_{42} & \lambda_{11} & \lambda_{21} \end{matrix} \\ \begin{bmatrix} 1 & 0 & 0 & 0 & 0 & 2 & 0 \\ 0 & 1 & 0 & 0 & 0 & 3 & 0 \\ 0 & 0 & 1 & 0 & 0 & 2 & 3 \\ 0 & 0 & 0 & 1 & 0 & 0 & 0 \\ 0 & 0 & 0 & 0 & 1 & 0 & 3 \\ 0 & 0 & 0 & 0 & 0 & 1 & 0 \\ 0 & 0 & 0 & 0 & 0 & 0 & 1 \end{bmatrix} \end{array}$$

$$\mathbf{B}^{-1} = \begin{bmatrix} 1 & 0 & 0 & 0 & 0 & -2 & 0 \\ 0 & 1 & 0 & 0 & 0 & -3 & 0 \\ 0 & 0 & 1 & 0 & 0 & -2 & -3 \\ 0 & 0 & 0 & 1 & 0 & 0 & 0 \\ 0 & 0 & 0 & 0 & 1 & 0 & -3 \\ 0 & 0 & 0 & 0 & 0 & 1 & 0 \\ 0 & 0 & 0 & 0 & 0 & 0 & 1 \end{bmatrix}$$

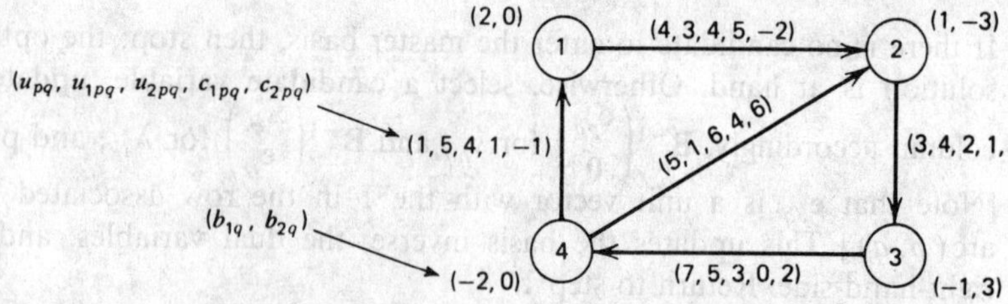

Figure 12.10. A two-commodity minimal cost flow problem.

Here $\mathbf{c}_1\mathbf{x}_{11} = 13$ and $\mathbf{c}_2\mathbf{x}_{21} = 24$. Denoting $\begin{bmatrix}\mathbf{u}\\ \mathbf{1}\end{bmatrix}$ by $\hat{\mathbf{b}}$, we have

$$(\mathbf{w}, \boldsymbol{\alpha}) = \hat{\mathbf{c}}_B\mathbf{B}^{-1} = (0,0,0,0,0,13,24)\mathbf{B}^{-1} = (0,0,0,0,0,13,24)$$

$$\mathbf{x}_B = \mathbf{B}^{-1}\hat{\mathbf{b}} = \mathbf{B}^{-1}\begin{bmatrix}\mathbf{u}\\ 1\\ 1\end{bmatrix} = \mathbf{B}^{-1}\begin{bmatrix}4\\3\\7\\1\\5\\1\\1\end{bmatrix} = \begin{bmatrix}2\\0\\2\\1\\2\\1\\1\end{bmatrix}$$

$$z = \hat{\mathbf{c}}_B\mathbf{B}^{-1}\hat{\mathbf{b}} = 37$$

Setting up the revised simplex array

$(\mathbf{w}, \boldsymbol{\alpha})$	$\hat{\mathbf{c}}_B\mathbf{B}^{-1}\hat{\mathbf{b}}$
$\mathbf{B}^{-1}$	$\mathbf{B}^{-1}\hat{\mathbf{b}}$

for the master problem, we get the following.

	w_{12}	w_{23}	w_{34}	w_{41}	w_{42}	α_1	α_2	RHS
z	0	0	0	0	0	13	24	37
s_{12}	1	0	0	0	0	−2	0	2
s_{23}	0	1	0	0	0	−3	0	0
s_{34}	0	0	1	0	0	−2	−3	2
s_{41}	0	0	0	1	0	0	0	1
s_{42}	0	0	0	0	1	0	−3	2
λ_{11}	0	0	0	0	0	1	0	1
λ_{12}	0	0	0	0	0	0	1	1

Iteration 1

First, all $w_{pq} \leq 0$. Next we check whether a candidate from either subproblem (or commodity) is eligible to enter the master basis.

	FIRST COMMODITY VARIABLES						SECOND COMMODITY VARIABLES						SLACK VARIABLES					
	x_{112}	x_{123}	x_{134}	x_{141}	x_{142}	x_1	x_{212}	x_{223}	x_{234}	x_{241}	x_{242}	x_2	s_{12}	s_{23}	s_{34}	s_{41}	s_{42}	RHS
	−5	−1	0	−1	−4	0	2	0	−2	1	−6	0	0	0	0	0	0	0
Coupling Constraints	1	0	0	0	0	0	1	0	0	0	0	0	1	0	0	0	0	4
	0	1	0	0	0	0	0	1	0	0	0	0	0	1	0	0	0	3
	0	0	1	0	0	0	0	0	1	0	0	0	0	0	1	0	0	7
	0	0	0	1	0	0	0	0	0	1	0	0	0	0	0	1	0	1
	0	0	0	0	1	0	0	0	0	0	1	0	0	0	0	0	1	5
Node-arc incidence matrix for subproblem 1	1	0	0	−1	0	0	0	0	0	0	0	0	0	0	0	0	0	2
	−1	1	0	0	−1	0	0	0	0	0	0	0	0	0	0	0	0	1
	0	−1	1	0	0	0	0	0	0	0	0	0	0	0	0	0	0	−1
	0	0	−1	1	1	1	0	0	0	0	0	0	0	0	0	0	0	−2
Node-arc incidence matrix for subproblem 2	0	0	0	0	0	0	1	0	0	−1	0	0	0	0	0	0	0	0
	0	0	0	0	0	0	−1	1	0	0	−1	0	0	0	0	0	0	−3
	0	0	0	0	0	0	0	−1	1	0	0	0	0	0	0	0	0	3
	0	0	0	0	0	0	0	0	−1	1	1	1	0	0	0	0	0	0

Figure 12.11. The constraint matrix for the two-commodity problem of Figure 12.10.

SUBPROBLEM 1

$$\mathbf{w} - \mathbf{c}_1 = \mathbf{0} - \mathbf{c}_1 = (-5, -1, 0, -1, -4)$$

Subproblem 1 is the single-commodity flow problem defined in Figure 12.12. The optimal (maximal cost) solution is $\mathbf{x}_{12} = (2, 3, 2, 0, 0)'$ and the value of the subproblem 1 objective is

$$z_{12} - c_{12} = (\mathbf{w} - \mathbf{c}_1)\mathbf{x}_{12} + \alpha_1 = -13 + 13 = 0$$

Thus there is no candidate from subproblem 1.

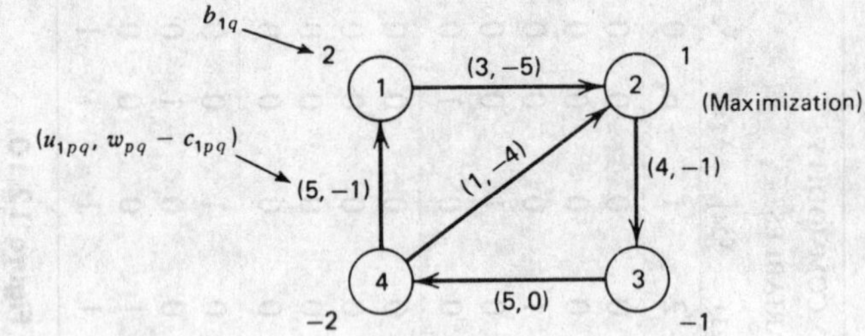

Figure 12.12. Subproblem 1 at the first iteration.

SUBPROBLEM 2

$$\mathbf{w} - \mathbf{c}_2 = \mathbf{0} - \mathbf{c}_2 = (2, 0, -2, 1, -6)$$

Subproblem 2 is the single-commodity flow problem defined in Figure 12.13. The optimal (maximal cost) solution is $\mathbf{x}_{22} = (3, 0, 3, 3, 0)'$ and

$$z_{22} - c_{22} = (\mathbf{w} - \mathbf{c}_2)\mathbf{x}_{22} + \alpha_2 = 3 + 24 = 27$$

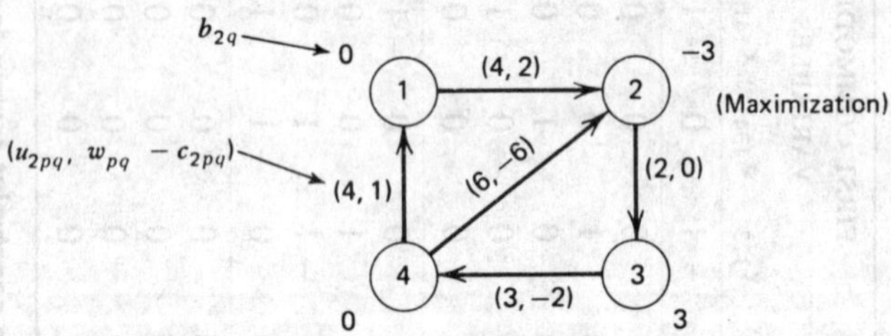

Figure 12.13. Subproblem 2 at the first iteration.

Thus λ_{22} is a candidate to enter the basis. The updated column for λ_{22} (exclusive of $z_{22} - c_{22}$) is

$$\mathbf{B}^{-1}\begin{bmatrix} \mathbf{x}_{22} \\ 0 \\ 1 \end{bmatrix} = (3, 0, 0, 3, -3, 0, 1)'$$

The pivoting process is as follows.

	w_{12}	w_{23}	w_{34}	w_{41}	w_{42}	α_1	α_2	RHS		λ_{22}
z	0	0	0	0	0	13	24	37		27
s_{12}	1	0	0	0	0	-2	0	2		3
s_{23}	0	1	0	0	0	-3	0	0		0
s_{34}	0	0	1	0	0	-2	-3	2		0
s_{41}	0	0	0	1	0	0	0	1		③
s_{42}	0	0	0	0	1	0	-3	2		-3
λ_{11}	0	0	0	0	0	1	0	1		0
λ_{21}	0	0	0	0	0	0	1	1		1

	w_{12}	w_{23}	w_{34}	w_{41}	w_{42}	α_1	α_2	RHS
z	0	0	0	-9	0	13	24	28
s_{12}	1	0	0	-1	0	-2	0	1
s_{23}	0	1	0	0	0	-3	0	0
s_{34}	0	0	1	0	0	-2	-3	2
λ_{22}	0	0	0	$\frac{1}{3}$	0	0	0	$\frac{1}{3}$
s_{42}	0	0	0	1	1	0	-3	3
λ_{11}	0	0	0	0	0	1	0	1
λ_{21}	0	0	0	$-\frac{1}{3}$	0	0	1	$\frac{2}{3}$

Iteration 2

Again all $w_{pq} \leq 0$ (so no s_{pq} is a candidate to enter the master basis).

SUBPROBLEM 1

$$(\mathbf{w} - \mathbf{c}_1) = (-5, -1, 0, -10, -4)$$

Subproblem 1 is the single-commodity flow problem defined in Figure 12.14. The optimal solution is $\mathbf{x}_{13} = (2, 3, 2, 0, 0)^t$ with

$$z_{13} - c_{13} = (\mathbf{w} - \mathbf{c}_1)\mathbf{x}_{13} + \alpha_1 = -13 + 13 = 0$$

Thus there is no candidate from subproblem 1.

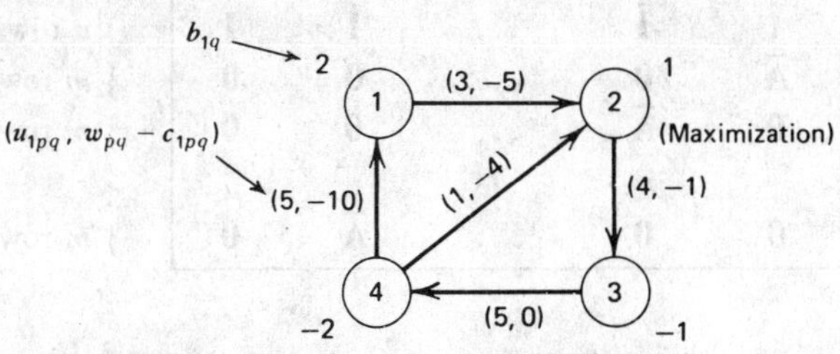

Figure 12.14. Subproblem 1 at the second iteration.

SUBPROBLEM 2

$$(\mathbf{w} - \mathbf{c}_2) = (2, 0, -2, -8, -6)$$

Subproblem 2 is the single-commodity flow problem defined in Figure 12.15. An optimal solution is $\mathbf{x}_{23} = (3, 0, 3, 3, 0)^t$ with

$$z_{23} - c_{23} = (\mathbf{w} - \mathbf{c}_2)\mathbf{x}_{23} + \alpha_2 = -24 + 24 = 0$$

Thus there is no candidate from subproblem 2.

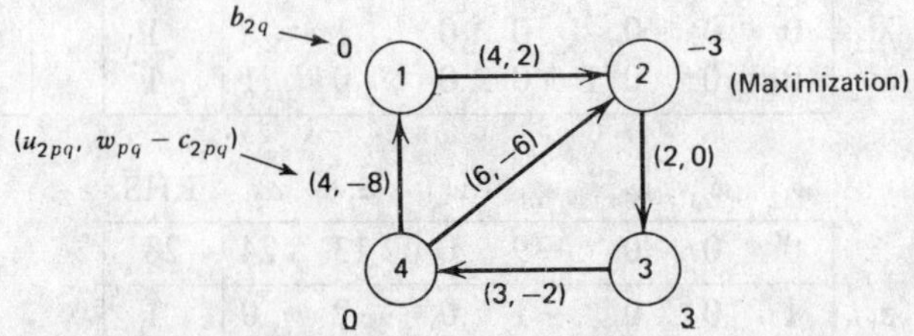

Figure 12.15. **Subproblem 2 at the second iteration.**

Therefore we already have the optimal solution as follows:

$$\begin{aligned} z^* &= 28 \\ \mathbf{x}_1^* &= \lambda_{11}\mathbf{x}_{11} = (2, 3, 2, 0, 0)^t \\ \mathbf{x}_2^* &= \lambda_{21}\mathbf{x}_{21} + \lambda_{22}\mathbf{x}_{22} \\ &= \tfrac{2}{3}(0, 0, 3, 0, 3)^t + \tfrac{1}{3}(3, 0, 3, 3, 0)^t \\ &= (1, 0, 3, 1, 2)^t \end{aligned}$$

12.5 CHARACTERIZATION OF A BASIS FOR THE MULTICOMMODITY MINIMAL COST FLOW PROBLEM

Suppose that we proceed to apply the simplex method directly to the multicommodity minimal cost flow problem. We first note from Chapter 9 that the system $\mathbf{A}\mathbf{x}_i = \mathbf{b}_i$ has rank $m - 1$ so that it is necessary to add an artificial variable for each commodity. Adding this artificial column vector, the overall constraint matrix is given by

$$\begin{array}{ccccc|l} \mathbf{x}_1 & \mathbf{x}_2 & \cdots & \mathbf{x}_t & \mathbf{x} & \\ \hline \bar{\mathbf{I}} & \bar{\mathbf{I}} & \cdots & \bar{\mathbf{I}} & \mathbf{I} & \} \; n \text{ rows} \\ \bar{\mathbf{A}} & \mathbf{0} & \cdots & \mathbf{0} & \mathbf{0} & \} \; m \text{ rows} \\ \mathbf{0} & \bar{\mathbf{A}} & \cdots & \mathbf{0} & \mathbf{0} & \} \; m \text{ rows} \\ & & \cdots & & & \vdots \\ \mathbf{0} & \mathbf{0} & \cdots & \bar{\mathbf{A}} & \mathbf{0} & \} \; m \text{ rows} \end{array}$$

where $\bar{\mathbf{A}} = [\mathbf{A}, \mathbf{e}_m]$ and $\bar{\mathbf{I}} = [\mathbf{I}, \mathbf{0}]$. Selecting a basis submatrix from this matrix, we get

$$\hat{\mathbf{B}} = \begin{array}{|ccccc|} \mathbf{E}_1 & \mathbf{E}_2 & \cdots & \mathbf{E}_t & \mathbf{E} \\ \overline{\mathbf{A}}_1 & \mathbf{0} & \cdots & \mathbf{0} & \mathbf{0} \\ \mathbf{0} & \overline{\mathbf{A}}_2 & \cdots & \mathbf{0} & \mathbf{0} \\ \vdots & \vdots & & \vdots & \vdots \\ \mathbf{0} & \mathbf{0} & \cdots & \overline{\mathbf{A}}_t & \mathbf{0} \end{array}$$

where $\mathbf{E}_i$ and $\mathbf{E}$ are matrices formed by taking selected columns of $\bar{\mathbf{I}}$ and $\mathbf{I}$ respectively. The row location of the 1 for a particular column of $\mathbf{E}_i$ identifies the arc used in $\overline{\mathbf{A}}_i$.

Since $\hat{\mathbf{B}}$ is a basis matrix, each $\overline{\mathbf{A}}_i$ must contain a submatrix that spans E^m. Therefore each $\overline{\mathbf{A}}_i$ contains an $m \times m$ basis (why?). Partition $\overline{\mathbf{A}}_i$ into $[\mathbf{B}_i|\mathbf{D}_i]$ where $\mathbf{B}_i$ is a basis matrix for $\mathbf{A}\mathbf{x}_i = \mathbf{b}_i$. Note that $\mathbf{B}_i$ must contain the artificial column (why?). From Chapter 9, since $\mathbf{B}_i$ is a basis for a set of single-commodity flow conservation constraints, $\mathbf{B}_i$ must correspond to a *rooted spanning tree* in G with the artificial variable as the root. Similarly partition $\mathbf{E}_i$ into $[\mathbf{E}_i'|\mathbf{E}_i'']$. Substituting into $\hat{\mathbf{B}}$ and rearranging the columns, we get

$$\hat{\mathbf{B}} = \begin{array}{|ccccccccc|} \mathbf{E}_1' & \mathbf{E}_2' & \cdots & \mathbf{E}_t' & \mathbf{E}_1'' & \mathbf{E}_2'' & \cdots & \mathbf{E}_t'' & \mathbf{E} \\ \mathbf{B}_1 & \mathbf{0} & \cdots & \mathbf{0} & \mathbf{D}_1 & \mathbf{0} & \cdots & \mathbf{0} & \mathbf{0} \\ \mathbf{0} & \mathbf{B}_2 & \cdots & \mathbf{0} & \mathbf{0} & \mathbf{D}_2 & \cdots & \mathbf{0} & \mathbf{0} \\ \vdots & \vdots & & \vdots & \vdots & \vdots & & \vdots & \vdots \\ \mathbf{0} & \mathbf{0} & \cdots & \mathbf{B}_t & \mathbf{0} & \mathbf{0} & \cdots & \mathbf{D}_t & \mathbf{0} \end{array}$$

In other words, the basis matrix $\hat{\mathbf{B}}$ has the following general structure:

$$\hat{\mathbf{B}} = \begin{bmatrix} \mathbf{E}' & \mathbf{E}'' & \mathbf{E} \\ \mathbf{B} & \mathbf{D} & \mathbf{0} \end{bmatrix}$$

Denoting the right-hand-side $\begin{bmatrix} \mathbf{u} \\ \mathbf{b} \end{bmatrix}$ by $\hat{\mathbf{b}}$ (where $\mathbf{b}$ is a column vector consisting of $\mathbf{b}_1, \mathbf{b}_2, \ldots,$ and $\mathbf{b}_t$), the basic system $\hat{\mathbf{B}}\mathbf{x}_B = \hat{\mathbf{b}}$ reduces to

$$\begin{bmatrix} \mathbf{E}' & \mathbf{E}'' & \mathbf{E} \\ \mathbf{B} & \mathbf{D} & \mathbf{0} \end{bmatrix} \begin{bmatrix} \mathbf{x}_F \\ \mathbf{x}_D \\ \mathbf{s}_B \end{bmatrix} = \begin{bmatrix} \mathbf{u} \\ \mathbf{b} \end{bmatrix}$$

where $\mathbf{x}_B$ is decomposed into $\begin{bmatrix} \mathbf{x}_F \\ \mathbf{x}_D \\ \mathbf{s}_B \end{bmatrix}$. This system is not easy to solve. However by utilizing the following change of variables, the structure in the system can be exploited as we shall outline shortly:

$$\begin{bmatrix} \mathbf{x}_F \\ \mathbf{x}_D \\ \mathbf{s}_B \end{bmatrix} = \begin{bmatrix} \mathbf{I} & -\mathbf{B}^{-1}\mathbf{D} & \mathbf{0} \\ \mathbf{0} & \mathbf{I} & \mathbf{0} \\ \mathbf{0} & \mathbf{0} & \mathbf{I} \end{bmatrix} \begin{bmatrix} \mathbf{x}_F' \\ \mathbf{x}_D' \\ \mathbf{s}_B' \end{bmatrix} \tag{12.5}$$

This is a nonsingular transformation and thus we have an equivalent system with which to work. On substituting for $\mathbf{x}_B$ in $\hat{\mathbf{B}}\mathbf{x}_B = \hat{\mathbf{b}}$, we get

$$\begin{bmatrix} \mathbf{E}' & \mathbf{E}'' & \mathbf{E} \\ \mathbf{B} & \mathbf{D} & \mathbf{0} \end{bmatrix} \begin{bmatrix} \mathbf{I} & -\mathbf{B}^{-1}\mathbf{D} & \mathbf{0} \\ \mathbf{0} & \mathbf{I} & \mathbf{0} \\ \mathbf{0} & \mathbf{0} & \mathbf{I} \end{bmatrix} \begin{bmatrix} \mathbf{x}'_F \\ \mathbf{x}'_D \\ \mathbf{s}'_B \end{bmatrix} = \begin{bmatrix} \mathbf{u} \\ \mathbf{b} \end{bmatrix}$$

$$\begin{bmatrix} \mathbf{E}' & \mathbf{E}'' - \mathbf{E}'\mathbf{B}^{-1}\mathbf{D} & \mathbf{E} \\ \mathbf{B} & \mathbf{0} & \mathbf{0} \end{bmatrix} \begin{bmatrix} \mathbf{x}'_F \\ \mathbf{x}'_D \\ \mathbf{s}'_B \end{bmatrix} = \begin{bmatrix} \mathbf{u} \\ \mathbf{b} \end{bmatrix}$$

Now the second set of equations $\mathbf{B}\mathbf{x}'_F = \mathbf{b}$ is easy to solve since it corresponds to a series of rooted spanning trees, one for each commodity.

Consider the first set of equations in the transformed system after having solved for $\mathbf{x}'_F$:

$$[\mathbf{E}'' - \mathbf{E}'\mathbf{B}^{-1}\mathbf{D}, \mathbf{E}] \begin{bmatrix} \mathbf{x}'_D \\ \mathbf{s}'_B \end{bmatrix} = \mathbf{u} - \mathbf{E}'\mathbf{x}'_F$$

The solution is

$$\begin{bmatrix} \mathbf{x}'_D \\ \mathbf{s}'_B \end{bmatrix} = [\mathbf{E}'' - \mathbf{E}'\mathbf{B}^{-1}\mathbf{D}, \mathbf{E}]^{-1}(\mathbf{u} - \mathbf{E}'\mathbf{x}'_F)$$

and requires the inversion of the matrix $[\mathbf{E}'' - \mathbf{E}'\mathbf{B}^{-1}\mathbf{D}, \mathbf{E}]$. While the matrix is not as easy to invert as $\mathbf{B}$, it is easy to form and understand. Here $\mathbf{E}$ is a matrix of unit columns corresponding to the slack variables in the basis. Now let us investigate the matrix $\mathbf{E}'' - \mathbf{E}'\mathbf{B}^{-1}\mathbf{D}$.

A typical column of $\mathbf{E}'' - \mathbf{E}'\mathbf{B}^{-1}\mathbf{D}$ corresponding to commodity k is given by $\mathbf{e}_{ij} - \mathbf{E}'_k\mathbf{B}_k^{-1}\mathbf{a}_{ij}$, where $\mathbf{e}_{ij}$ is a unit vector in E^n with a 1 in the row corresponding to arc (i, j) and $\mathbf{a}_{ij}$ is a vector in E^m with a 1 in row i and a -1 in row j. From Chapter 9 recall that $\mathbf{y}_{ij} = \mathbf{B}_k^{-1}\mathbf{a}_{ij}$ corresponds to a chain from node i to node j through the basis tree. Note that the coefficients of $\mathbf{y}_{ij}$ actually reorient the chain into a path. Then $-\mathbf{B}_k^{-1}\mathbf{a}_{ij}$ corresponds to a chain from j to i in the rooted spanning tree of commodity k. Each coefficient in $-\mathbf{B}_k^{-1}\mathbf{a}_{ij}$ corresponds to a basic variable in $\mathbf{B}_k$. Now $\mathbf{E}'_k$ is an $n \times m$ matrix with its columns being unit vectors in E^n identifying the basic variables in $\mathbf{B}_k$. Thus $-\mathbf{E}'_k\mathbf{B}_k^{-1}\mathbf{a}_{ij}$ simply expands the m vector $-\mathbf{B}_k^{-1}\mathbf{a}_{ij}$ to an n vector by assigning zero coefficients corresponding to all nonbasic arcs of commodity k. Hence $-\mathbf{E}'_k\mathbf{B}_k^{-1}\mathbf{a}_{ij}$ is an n vector corresponding to the chain from node j to node i in the rooted spanning tree of commodity k. Finally $\mathbf{e}_{ij} - \mathbf{E}'_k\mathbf{B}_k^{-1}\mathbf{a}_{ij}$ corresponds to the unique cycle formed when the arc (i, j) is added to the basis tree (and the basis arcs are properly oriented). Thus, knowing $\mathbf{B}$, it is easy to form $\mathbf{E}'' - \mathbf{E}'\mathbf{B}^{-1}\mathbf{D}$.

The important conclusion is the following.

***Theorem* 12.4**

A transformed basis matrix for the multicommodity minimal cost flow problem corresponds to a rooted spanning tree for each commodity plus a set of cycles and slack arcs.

Once $[\mathbf{E}'' - \mathbf{E}'\mathbf{B}^{-1}\mathbf{D}, \mathbf{E}]$ is formed as described earlier, we can solve for $\mathbf{x}'_D$ and $\mathbf{s}'_B$. With the vector

$$\begin{bmatrix} \mathbf{x}'_F \\ \mathbf{x}'_D \\ \mathbf{s}'_B \end{bmatrix}$$

now known, we can solve for the basic variables $\mathbf{x}_F$, $\mathbf{x}_D$, and $\mathbf{s}_B$ from Equation (12.5). In Exercise 12.60 we ask the reader to develop a systematic procedure for computing the dual variables, updating the column of the entering variable, and the basis inverse. This coupled with the foregoing procedure for computing the basic variables, represents a direct application of the simplex method for solving multicommodity flow problems.

An Example of a Basis Matrix for the Multicommodity Minimal Cost Flow Problem

Consider the multicommodity minimal cost flow problem of Figure 12.10 without the upper bound constraints on the individual commodities. Recall that the constraint matrix is shown in Figure 12.11.

Suppose that we select the basis submatrix (the reader is asked to verify that this is a basis submatrix) indicated in Figure 12.16.

x_{112}	x_{123}	x_{134}	x_{141}	x_1	x_{212}	x_{234}	x_{241}	x_{242}	x_2	s_{12}	s_{34}	s_{42}
1	0	0	0	0	1	0	0	0	0	1	0	0
0	1	0	0	0	0	0	0	0	0	0	0	0
0	0	1	0	0	0	1	0	0	0	0	1	0
0	0	0	1	0	0	0	1	0	0	0	0	0
0	0	0	0	0	0	0	0	1	0	0	0	1
1	0	0	−1	0	0	0	0	0	0	0	0	0
−1	1	0	0	0	0	0	0	0	0	0	0	0
0	−1	1	0	0	0	0	0	0	0	0	0	0
0	0	−1	1	1	0	0	0	0	0	0	0	0
0	0	0	0	0	1	0	−1	0	0	0	0	0
0	0	0	0	0	−1	0	0	−1	0	0	0	0
0	0	0	0	0	0	1	0	0	0	0	0	0
0	0	0	0	0	0	−1	1	1	1	0	0	0

Figure 12.16. A basis submatrix.

Applying the transformation of Equation (12.5), we get the matrix

$$\begin{bmatrix} \mathbf{E}'_1 & \mathbf{E}'_2 & \mathbf{E}''_1 - \mathbf{E}'_1\mathbf{B}_1^{-1}\mathbf{D}_1 & \mathbf{E}''_2 - \mathbf{E}'_2\mathbf{B}_2^{-1}\mathbf{D}_2 & \mathbf{E} \\ \mathbf{B}_1 & \mathbf{0} & \mathbf{0} & \mathbf{0} & \mathbf{0} \\ \mathbf{0} & \mathbf{B}_2 & \mathbf{0} & \mathbf{0} & \mathbf{0} \end{bmatrix} \tag{12.6}$$

Here $\mathbf{B}_1$ consists of x_{112}, x_{123}, x_{134}, and x_1 whereas $\mathbf{B}_2$ consists of x_{234}, x_{241}, x_{242}, and x_2. These two rooted spanning trees are illustrated in Figure 12.17. In

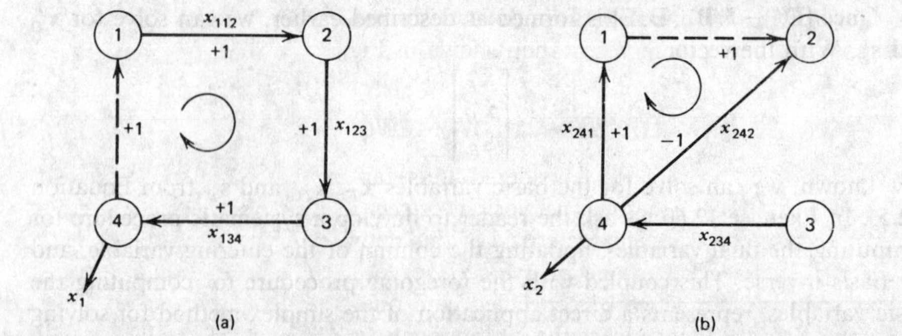

Figure 12.17. Graphical illustration of the basis matrix: (*a*) Commodity 1 subbasis. (*b*) Commodity 2 subbasis.

addition $\mathbf{D}_1$ and $\mathbf{D}_2$ are represented by x_{141} and x_{212} and correspond to the cycles of Figure 12.17.

Examining Figure 12.17, we see that

$$\mathbf{E}_1'' - \mathbf{E}_1'\mathbf{B}_1^{-1}\mathbf{D}_1 = \begin{bmatrix} 0-(-1) \\ 0-(-1) \\ 0-(-1) \\ 1-0 \\ 0-0 \end{bmatrix} \begin{matrix} (1,2) \\ (2,3) \\ (3,4) \\ (4,1) \\ (4,2) \end{matrix}$$

$$\mathbf{E}_2'' - \mathbf{E}_2'\mathbf{B}_2^{-1}\mathbf{D}_2 = \begin{bmatrix} 1-0 \\ 0-0 \\ 0-0 \\ 0-(-1) \\ 0-(+1) \end{bmatrix} \begin{matrix} (1,2) \\ (2,3) \\ (3,4) \\ (4,1) \\ (4,2) \end{matrix}$$

x'_{112}	x'_{123}	x'_{134}	x'_1	x'_{234}	x'_{241}	x'_{242}	x'_2	x'_{141}	x'_{212}	s_{12}	s_{34}	s_{42}
1	0	0	0	0	0	0	0	1	1	1	0	0
0	1	0	0	0	0	0	0	1	0	0	0	0
0	0	1	0	1	0	0	0	1	0	0	1	0
0	0	0	0	0	1	0	0	1	1	0	0	0
0	0	0	0	0	0	1	0	0	−1	0	0	1
1	0	0	0	0	0	0	0	0	0	0	0	0
−1	1	0	0	0	0	0	0	0	0	0	0	0
0	−1	1	0	0	0	0	0	0	0	0	0	0
0	0	−1	1	0	0	0	0	0	0	0	0	0
0	0	0	0	0	−1	0	0	0	0	0	0	0
0	0	0	0	0	0	−1	0	0	0	0	0	0
0	0	0	0	1	0	0	0	0	0	0	0	0
0	0	0	0	−1	1	1	1	0	0	0	0	0

Figure 12.18. The transformed basis submatrix.

Substituting this information into the transformed basis submatrix of matrix (12.6), we get the basis representation shown in Figure 12.18.

12.6 SYNTHESIS OF MULTITERMINAL FLOW NETWORKS

The network flow problems considered thus far have been network *analysis* problems, in which a certain network is *given* and one then is required to determine some specified characteristic of this network. In contrast, the problem studied in this section is one of *synthesis* or *design*. Here a certain required characteristic of the network is specified, and one has to synthesize or design a least cost network that satisfies the requirements.

More specifically, the particular problem analyzed here is as follows:* Suppose that we are to construct some *undirected* network on a specified set of nodes $1, 2, \ldots, m$. This network can be designed by constructing links between node pairs i and j in some arc set $\mathscr{A}$. This set may or may not be totally dense, but we assume that if all the potential arcs in $\mathscr{A}$ are constructed, then the resulting network will be connected. In addition to selecting the arcs from $\mathscr{A}$, we have the freedom of designing the capacities of these arcs. If arc (i, j) is constructed with capacity u_{ij}, then a corresponding cost of $c_{ij} \geq 0$ per unit capacity will be incurred.

The constructed network must independently sustain a certain maximal flow requirement $r_{ij} \geq 0$ between each pair of nodes or terminals i and j (hence the name of this problem). Note that the requirements r_{ij} are specified between each pair of nodes i and j and not only for $(i, j) \in \mathscr{A}$. The problem may be mathematically stated as follows, where f_{ij} denotes the maximum flow between node pair i and j and is of course a function of the constructed network.

$$\text{Minimize}\left\{\sum_{(i,j)\in\mathscr{A}} c_{ij}u_{ij} : f_{ij} \geq r_{ij} \text{ for each } i = 1, \ldots, m-1,\ j = i+1, \ldots, m,\right.$$
$$\left.\text{and} \quad u_{ij} \geq 0 \text{ for } (i, j) \in \mathscr{A}\right\}$$

Before proceeding to solve this problem, there are two manipulations that we need to perform. First we do not need all the $m(m-1)/2$ constraints in this problem. Intuitively, it should be evident that if the maximal flow requirements are satisfied for certain key node pairs, then the other flow requirements ought to be also automatically satisfied. In fact, these key node pairs turn out to be associated with a certain tree T_r known as the *dominant requirement tree*. Only the particular node pairs associated with the $(m-1)$ arcs in T_r need to have their flow requirement constraints explicitly represented. The second manipulation involves the representation of the constraints themselves. The quantities f_{ij} are functions of the variables u_{ij}. This translation must be made in order for us to be able to solve the problem mathematically. Note that this is readily done using the max flow–min cut theorem. Namely, for any node pair (i, j), $f_{ij} \geq r_{ij}$

---

*This problem arises in the design of communication networks.

if and only if the capacity of *each* cut-set separating nodes i and j exceeds r_{ij} (why?). Let $k = 1, \ldots, K_{ij}$ index the possible cut-sets separating nodes i and j. For each such $k \in K_{ij}$, define a column vector $\mathbf{a}_{ijk}$ having $|\mathscr{A}|$ components, one corresponding to each arc in $\mathscr{A}$, with the component corresponding to arc (p, q) being 1 if this arc belongs to the cut-set and zero otherwise. Then for any node pair (i, j), we may replace $f_{ij} \geq r_{ij}$ by the constraints

$$(\mathbf{a}_{ijk})^t \mathbf{u} \geq r_{ij} \qquad \text{for } k = 1, \ldots, K_{ij}$$

where $\mathbf{u}$ is the vector $(u_{ij}, (i, j) \in \mathscr{A})$. Before we consider the first manipulation, let us introduce an example problem.

An Example for the Synthesis Problem

Consider the potential network shown in Figure 12.19*a*, where $m = 5$, $|\mathscr{A}| = 6$ with the permissible arcs as shown, and the corresponding costs c_{ij} per unit capacity are as indicated. Additionally, the requirements r_{ij} are as follows

$$r_{12} = 3, \quad r_{13} = 5, \quad r_{14} = 3, \quad r_{15} = 2, \quad r_{23} = 6,$$
$$r_{24} = 7, \quad r_{25} = 5, \quad r_{34} = 8, \quad r_{35} = 1, \quad r_{45} = 12$$

Dominant Requirement Tree T_r

To define the dominant requirement tree T_r, construct a complete graph G_r on the node set $1, \ldots, m$, with the (undirected) arc between each node pair i and j having a weight equal to the requirement r_{ij}. Then, T_r is defined as the *maximum spanning tree* for G_r, that is, it is a spanning tree for G_r that has the *maximum total weight.*

The process of constructing T_r involves a simple one-pass procedure and may be done as follows. Suppose that the "weights" r_{ij} are sorted from largest to smallest in a list L. (A complete sort is not necessary.) Select arcs for inclusion in T_r by proceeding in the order of list L. Skip any arc that results in a cycle with respect to the arcs already selected. Stop when $(m - 1)$ arcs have been selected. For the example problem, we would select and skip arcs as follows: select (4, 5), select (3, 4), select (2, 4), skip (2, 3) (why?), skip (2, 5), select (1, 3), and stop. The resulting tree T_r is shown in Figure 12.19*b*.

The fact that this "greedy" procedure produces a maximal spanning tree T_r may be verified as follows. Suppose that the tree T^* is some maximal spanning tree. We will show that the total weight of T^* and the constructed tree T_r are the same. Toward this end, proceed down the list L and find the *first* arc (p, q) that belongs to T_r but is not in T^*. Add (p, q) to T^* and examine the (unique) cycle formed by (p, q) and the chain C connecting p and q in T^*. Note that $r_{ij} \geq r_{pq}$ for all $(i, j) \in C$ since T^* is maximal. If any $(i, j) \in C$ satisfies $r_{ij} > r_{pq}$, then $(i, j) \in T_r$. If not, then because (i, j) appears before (p, q) in L and it was not selected for inclusion in T_r, it must have formed a cycle with the arcs selected for T_r before (i, j). But all such arcs are also in T^* by the definition of (p, q). Hence we have a contradiction (why?). But all the arcs in C cannot also be in T_r since

$(p, q) \in T_r$. Hence there exists an arc $(s, t) \in C$, $(s, t) \notin T_r$ such that $r_{st} = r_{pq}$ (why?). Now add (p, q) to T^* and remove (s, t) from T^*. The resulting tree T^*_{new} is still maximal. However the number of arcs in T_r that are not in T^*_{new} is one less than the same number with respect to the previous tree T^*. Continuing in this fashion, we will ultimately obtain a maximal spanning tree T^* that coincides with T_r.

The claim with respect to T_r is that if we satisfy $f_{ij} \geq r_{ij}$ for all $(i, j) \in T_r$, then we will automatically satisfy $f_{ij} \geq r_{ij}$ for all $i = 1, \ldots, m-1$, $j = i + 1, \ldots, m$. To see this, suppose that for a constructed network, we have satisfied $f_{ij} \geq r_{ij}$ for all $(i, j) \in T_r$, and consider any $(p, q) \notin T_r$. Examine the chain C connecting nodes p and q in T_r. Then we have $r_{pq} \leq \text{Min}\{r_{ij} : (i, j) \in C\}$ since T_r is maximal. But $f_{ij} \geq r_{ij}$ for all $(i, j) \in C$ since C is a chain in T_r. Hence $r_{pq} \leq \text{Min}\{f_{ij} : (i, j) \in C\}$. In the constructed network, f_{pq} equals the capacity $u(X, \overline{X})$ of some cut-set $(X, \overline{X})$ separating p and q. However, $(X, \overline{X})$ is also a cut-set for some node pair s and t with $(s, t) \in C$ (why?). (The node s may be p or the node t may be q.) Consequently, using the max flow–min cut theorem, we have that

$$f_{pq} = u(X, \overline{X}) \geq f_{st} \geq \text{Minimum}\{f_{ij} : (i, j) \in C\} \geq r_{pq}$$

This establishes the required result.

Column-Generation Algorithm for the Synthesis Problem

Using the foregoing two manipulations and denoting the vector of cost coefficients $(c_{ij}, (i, j) \in \mathscr{A})$ as $\mathbf{c}$, we can formulate the multiterminal network design problem as follows:

$$\begin{aligned} \text{P:} \quad & \text{Minimize} \quad \mathbf{cu} \\ & \text{Subject to} \quad (\mathbf{a}_{ijk})^t \mathbf{u} \geq r_{ij} \quad \text{for } k = 1, \ldots, K_{ij}, \quad \text{for each } (i, j) \in T_r \\ & \qquad\qquad\quad \mathbf{u} \geq \mathbf{0} \end{aligned}$$

Denoting by $\mathbf{w} = (w_{ijk}, k = 1, \ldots, K_{ij}, (i, j) \in T_r)$ the dual variables associated with the constraints, we may write the dual to this problem as follows:

$$\text{D:} \quad \text{Maximize} \quad \sum_{(i,j) \in T_r} \sum_{k=1}^{K_{ij}} r_{ij} w_{ijk}$$

$$\text{Subject to} \quad \sum_{(i,j) \in T_r} \sum_{k=1}^{K_{ij}} (\mathbf{a}_{ijk}) w_{ijk} \leq \mathbf{c} \qquad (12.7)$$

$$\mathbf{w} \geq \mathbf{0}$$

Observe that similar to the Dantzig–Wolfe master problem of Chapter 7, this problem has several columns that are not all worthwhile enumerating explicitly.

Hence we adopt a column generation procedure in the context of a revised simplex algorithm. Since $\mathbf{c} \geq \mathbf{0}$, we commence with all slacks as basic in the starting basic feasible solution. Note that the corresponding complementary dual variables $\bar{\mathbf{u}}$ associated with the constraints (12.7) are all zeros. At any stage, given a basic feasible solution to D with a simplex multiplier vector $\bar{\mathbf{u}}$, we need to price out the nonbasic variables. For the variable w_{ijk}, the reduced cost "$c_{ijk} - z_{ijk}$" is given by $r_{ij} - (\mathbf{a}_{ijk})'\bar{\mathbf{u}}_{ij}$. Since we are interested in finding the variable with the largest reduced cost, we examine the problem

$$\underset{k=1,\ldots,K_{ij}}{\text{Maximize}}\left\{r_{ij} - (\mathbf{a}_{ijk})'\bar{\mathbf{u}}_{ij}\right\} = r_{ij} - \underset{k=1,\ldots,K_{ij}}{\text{Minimum}}\,(\mathbf{a}_{ijk})'\bar{\mathbf{u}}_{ij} \tag{12.8}$$

for each $(i, j) \in T_r$. Observe that this problem essentially seeks the minimum capacity cut-set separating i and j with respect to the capacity vector $\bar{\mathbf{u}}$ (why?). Hence this is equivalent to solving the maximal flow problem between nodes i and j on the current network using capacities $\bar{\mathbf{u}}$ and determining the corresponding quantity f_{ij}. If $f_{ij} \geq r_{ij}$, that is, if the associated minimal cut gives a nonpositive value in Equation (12.8) for all $(i, j) \in T_r$, then we are optimal. Otherwise we may select the variable w_{ijk} which yields the most positive reduced cost in Equation (12.8) to enter the basis, pivot, and repeat.

Summary of the Column-Generating Algorithm

INITIALIZATION

Construct a revised-simplex tableau for Problem D associated with the starting basic feasible solution having all the slack variables s, say, as basic. Hence $\mathbf{B}^{-1} = \mathbf{I}$ is the basis inverse of size $|\mathscr{A}| \times |\mathscr{A}|$, the right-hand-side is $\mathbf{c}$, and the dual variable vector $\bar{\mathbf{u}}$ is zero.

MAIN STEP

If any $\bar{u}_{ij} < 0$, then the corresponding slack s_{ij} is a candidate to enter the basis. Generate its updated simplex column and pivot it into the basis. Otherwise if $\bar{u}_{ij} \geq 0$ for all (i, j), examine the (constructed) network with capacity vector $\bar{\mathbf{u}}$ and find the maximal flow f_{ij} between each pair of nodes $(i, j) \in T_r$. If $f_{ij} \geq r_{ij}$ for all $(i, j) \in T_r$, then stop; $\bar{\mathbf{u}}$ is an optimal set of capacities. Otherwise select the most violated constraint, say for $(p, q) \in T_r$, and using the associated minimal cut k, generate the column $\mathbf{a}_{pqk}$ for the variable w_{pqk}. From Equation (12.8), the value "$z_{pqk} - c_{pqk}$" $= f_{pq} - r_{pq} < 0$. Update the column $\mathbf{a}_{pqk}$ (by premultiplying with $\mathbf{B}^{-1}$), and perform the usual simplex pivot. Repeat the main step.

Note that since this is a simplex based procedure, it is finitely convergent provided some appropriate cycling prevention rule such as the lexicographic method is employed (see Chapter 5).

Illustrative Example

For the example of Figure 12.19, observe that the Problem D in Equation (12.7) is of the following form:

Maximize

$$\sum_k 5w_{13k} + \sum_k 7w_{24k} + \sum_k 8w_{34k} + \sum_k 12w_{45k}$$

Subject to

$$\sum_k (\mathbf{a}_{13k})w_{13k} + \sum_k (\mathbf{a}_{24k})w_{24k} + \sum_k (\mathbf{a}_{34k})w_{34k} + \sum_k (\mathbf{a}_{45k})w_{45k} \le \begin{bmatrix} c_{12} \\ c_{13} \\ c_{23} \\ c_{24} \\ c_{35} \\ c_{45} \end{bmatrix} \equiv \begin{bmatrix} 2 \\ 3 \\ 2 \\ 4 \\ 3 \\ 2 \end{bmatrix}$$

$$\mathbf{w} \ge \mathbf{0}$$

Iteration 1

$\mathbf{B}^{-1} = \mathbf{I}$, $\bar{\mathbf{u}} = \mathbf{0}$, $\mathbf{B}^{-1}\mathbf{c} = \mathbf{c}$. The maximal flows $f_{ij} = 0$ for all $(i, j) \in T_r$, and so the constraint $f_{45} \ge 12$ is most violated. Arbitrarily select the minimal cut-set $(X, \bar{X})$ with $X = \{4\}$, $\bar{X} = \{1, 2, 3, 5\}$. The corresponding column $\mathbf{a}_{45k} \equiv \mathbf{a}_{451}$ is $(0, 0, 0, 1, 0, 1)'$, where the 1's are in the positions corresponding to arcs (2, 4) and (4, 5). The pivot results in s_{45} leaving the basis. The updated revised simplex tableau is as follows:

			RHS		w_{451}
	0		0		-12
s_{12}			2		0
s_{13}			3		0
s_{23}		**I**	2		0
s_{24}			4		1
s_{35}			3		0
s_{45}			2		①

							RHS		w_{341}	
	0	0	0	0	0	12	24		-8	$z_{341} - c_{341}$
s_{12}	1	0	0	0	0	0	2		1	
s_{13}	0	1	0	0	0	0	3		1	
s_{23}	0	0	1	0	0	0	2		0	
s_{24}	0	0	0	1	0	-1	2		①	$\mathbf{B}^{-1}\mathbf{a}_{341}$
s_{35}	0	0	0	0	1	0	3		1	
w_{451}	0	0	0	0	0	1	2		0	

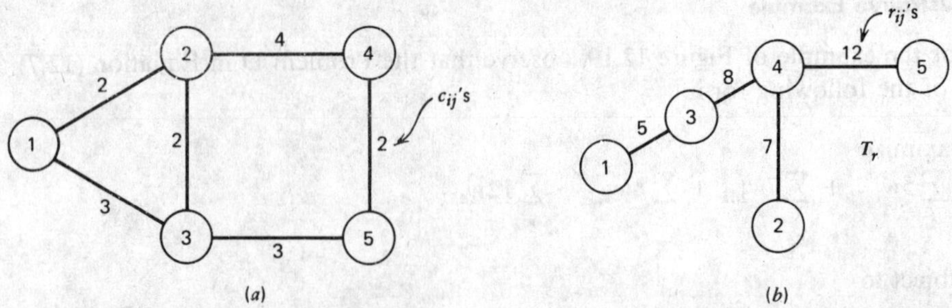

Figure 12.19. Potential network and T_r for the example problem: (a) Network. (b) Tree T_r.

Iteration 2

Now $\bar{u}_{45} = 12$, and $\bar{u}_{ij} = 0$ otherwise. We find that $f_{45} = 12$ and $f_{ij} = 0$ otherwise for $(i, j) \in T_r$. Hence $f_{34} = 0 < 8$ is most violated. A minimal cut $(X, \bar{X})$ has $X = \{2, 3\}$ and $\bar{X} = \{1, 4, 5\}$, for example. The corresponding column for w_{341} is $\mathbf{a}_{341} = (1, 1, 0, 1, 1, 0)'$ and its updated version is shown with the foregoing tableau. The simplex pivot results in s_{24} leaving the basis (using the lexicographic rule to break ties). This results in the following revised simplex tableau.

							RHS
	0	0	0	8	0	4	40
s_{12}	1	0	0	−1	0	1	0
s_{13}	0	1	0	−1	0	1	1
s_{23}	0	0	1	0	0	0	2
w_{341}	0	0	0	1	0	−1	2
s_{35}	0	0	0	−1	1	1	1
w_{451}	0	0	0	0	0	1	2

Note that presently $\bar{u}_{24} = 8$, $\bar{u}_{45} = 4$ and $\bar{u}_{ij} = 0$ otherwise. This gives $f_{13} = 0$, $f_{24} = 8$, $f_{34} = 0$, and $f_{45} = 4$ as the maximal flow values. Hence the constraints $f_{34} \geq 8$ and $f_{45} \geq 12$ are both most violated. One may now continue by generating the appropriate column of w_{342} or w_{452}. (Note that a sequential pivot of more than one improving column per iteration is also permissible.) We ask the reader to complete the solution of this example in Exercise 12.62.

An Efficient Algorithm for a Special Case

We conclude this section by presenting a particularly elegant algorithm for the special case of the foregoing synthesis problem in which all possible links are permissible, that is, $\mathscr{A} \equiv \{(i, j), i = 1, \ldots, m-1, j = i+1, \ldots, m\}$ and $c_{ij} \equiv 1$ for all $(i, j) \in \mathscr{A}$. Hence one is to design a network with the *minimal total capacity* that will sustain the flow requirements. The procedure for this case is presented using the example of Figure 12.19 as an illustration.

STEP 1

First construct the dominant requirement tree T_r. This is illustrated in Figure 12.19*b*.

STEP 2

Decompose T_r into a "sum" of *uniform requirement trees*. Namely, construct a tree T_{r1} that is identical to T_r but with all arcs having the same weight as the smallest r_{ij} value for $(i, j) \in T_r$. Let this weight be δ. Subtract δ from all the r_{ij} values for $(i, j) \in T_r$ and drop the arcs in T_r whose weight has reduced to zero. This results in a forest of two or more trees. Repeat with each proper tree in the

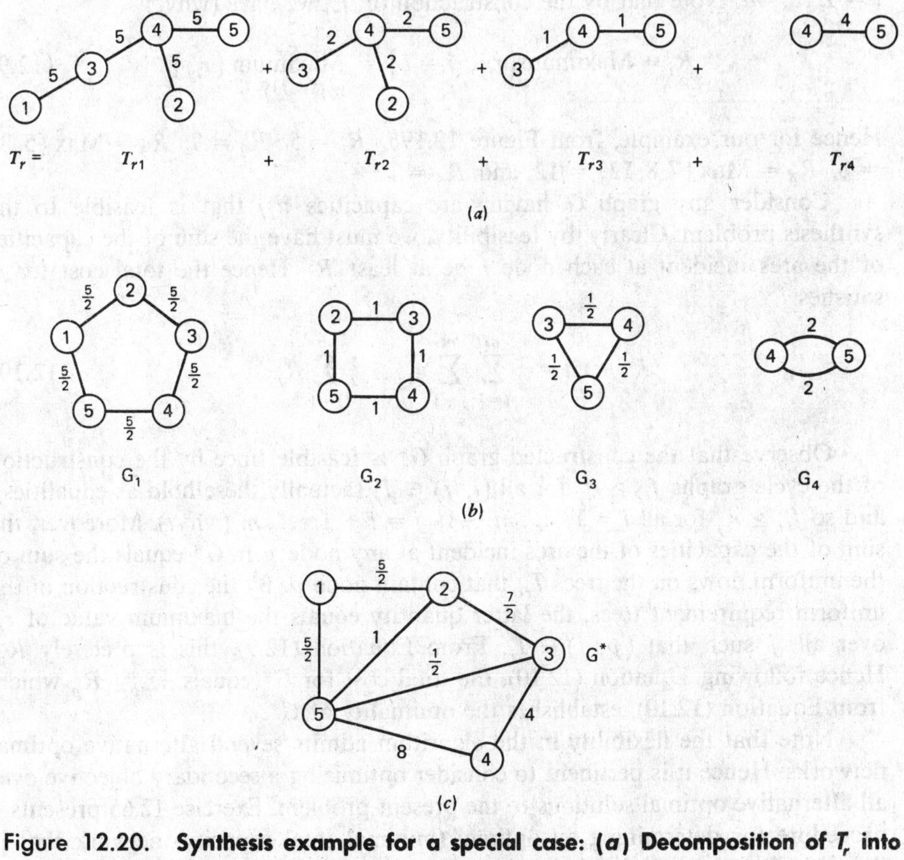

Figure 12.20. Synthesis example for a special case: (*a*) Decomposition of T_r into uniform requirement trees. (*b*) Cycle graphs G_i corresponding to T_{ri}, $i = 1, \ldots, K = 4$. (*c*) Optimal graph G^*.

resulting forest, creating the uniform requirement trees $T_{r1}, T_{r2}, T_{r3}, \ldots, T_{rk}$ in the process.

For the tree T_r of Figure 12.19*b*, the uniform requirement trees are depicted in Figure 12.20*a*, where the (uniform) weights in each tree are shown against the arcs. Here we obtain $K = 4$ uniform requirement trees.

STEP 3

For each uniform requirement tree T_{ri}, $i = 1, \ldots, K$, design a graph G_i that is an arbitrarily ordered undirected cycle on the nodes in T_{ri}, with each of the arcs in the cycle having a capacity $u(G_i)$ equal to one-half the uniform requirement weight of T_{ri}. Then an optimal network G^* is simply a superposition (sum) of the graphs G_i, $i = 1, \ldots, K$.

For the example, Figure 12.20*b* shows the cycle graphs G_1, G_2, G_3, and G_4. Figure 12.20*c* gives their sum G^*.

In order to show that G^* is indeed optimal, define $R_i = \text{Max}\{r_{ij}, j \neq i\}$ for $i = 1, \ldots, m$. Note that by the construction of T_r, we have (why?)

$$R_i \equiv \text{Maximum}\{r_{ij}, j \neq i\} = \underset{j:(i,j) \in T_r}{\text{Maximum}}\{r_{ij}\} \tag{12.9}$$

Hence for our example, from Figure 12.19*b*, $R_1 = 5$, $R_2 = 7$, $R_3 = \text{Max}\{5, 8\} = 8$, $R_4 = \text{Max}\{7, 8, 12\} = 12$, and $R_5 = 12$.

Consider any graph G having arc capacities u_{ij} that is feasible to the synthesis problem. Clearly, by feasibility, we must have the sum of the capacities of the arcs incident at each node i be at least R_i. Hence the total cost for G satisfies

$$\mathbf{cu} = \tfrac{1}{2} \sum_{i=1}^{m} \sum_{j=1}^{m} u_{ij} \geq \tfrac{1}{2} \sum_{i=1}^{m} R_i \tag{12.10}$$

Observe that the constructed graph G^* is feasible since by the construction of the cycle graphs $f_{ij} \geq r_{ij}$ for all $(i, j) \in T_r$ (actually these hold as equalities), and so $f_{ij} \geq r_{ij}$ for all $i = 1, \ldots, m-1$, $j = i+1, \ldots, m$ (why?). Moreover, the sum of the capacities of the arcs incident at any node p in G^* equals the sum of the uniform flows on the trees T_{ri} that contain node p. By the construction of the uniform requirement trees, the latter quantity equals the maximum value of r_{pj} over all j such that $(p, j) \in T_r$. From Equation (12.9), this is precisely R_p. Hence following Equation (12.10), the total cost for G^* equals $\frac{1}{2}\sum_{p=1}^{m} R_p$ which, from Equation (12.10), establishes the optimality of G^*.

Note that the flexibility in the algorithm admits several alternative optimal networks. Hence it is pertinent to consider optimizing a secondary objective over all alternative optimal solutions to the present problem. Exercise 12.65 presents a procedure for determining an optimal (minimal total capacity) network that is flow dominant, that is, it admits a maximal flow between *any* node pair that is at least as great as the value obtainable via *any other* optimal network.

EXERCISES

12.1 Find the maximal flow from node 1 to node 7 in the following network. Identify the associated minimal cut-set.

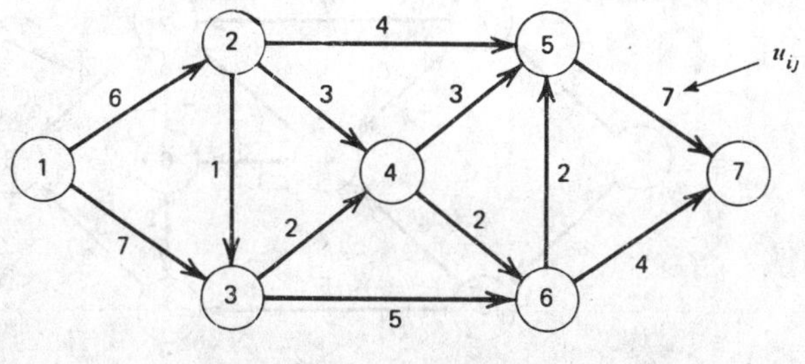

12.2 Discuss the economic meaning of the dual variables in the maximal flow problem. Consider both the w_i's and the h_{ij}'s.

12.3 Show that any basic feasible solution to the dual of the maximal flow problem (with node 1 taken as the root node) corresponds to a certain cut-set with the dual objective value equal to the capacity of the cut-set. Use this along with Lemma 12.1 to establish the maximal flow–minimal cut theorem.

12.4 Consider the production process shown that indicates the various paths that a product can take on its way to assembly through a plant. The number in each box represents the upper limit on items per hour that can be processed at the station.

a. What is the maximal number of parts per hour that the plant can handle?

b. Which operations should you try to improve?

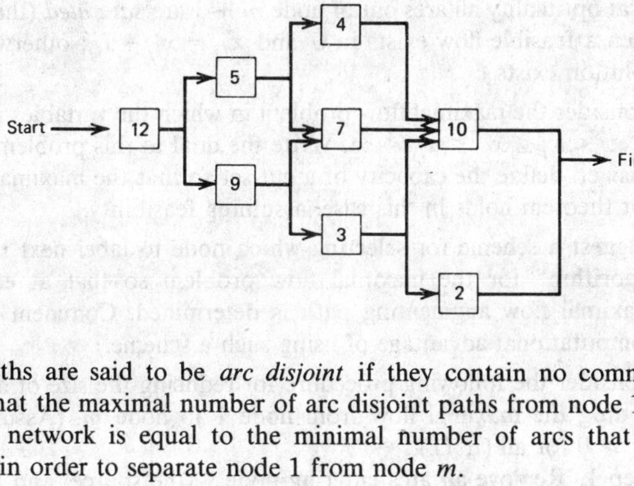

2.5 Two paths are said to be *arc disjoint* if they contain no common arcs. Prove that the maximal number of arc disjoint paths from node 1 to node m in a network is equal to the minimal number of arcs that must be deleted in order to separate node 1 from node m.

12.6 Find the maximal flow from node 1 to node 8 in the following network. Identify the associated minimal cut-set.

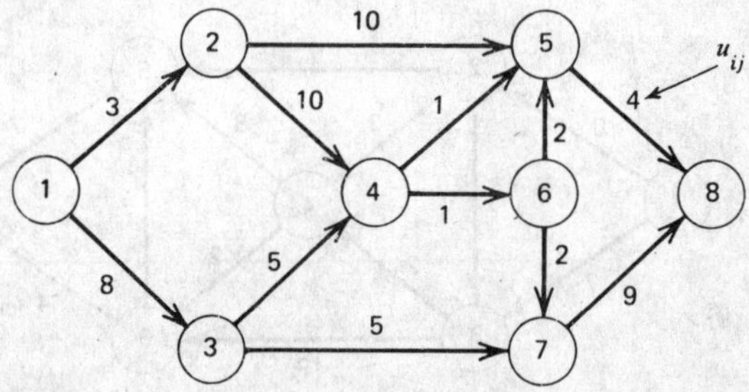

12.7 In a command and control communications network a commander is located at one node and his subordinate at another node. Associated with each link in the network is an effort u_{ij} required to eliminate that link from the network.

a. Present a mathematical model that could be used to find the minimal effort to block all communications from the commander to his subordinate.

b. Indicate how the problem can be solved by a special algorithm.

12.8 Indicate how lower bounds on flow can be handled in the maximal flow algorithm. (*Hint*. Apply a phase I procedure as follows. From G construct G' by: (1) All nodes in G are in G'. (2) In addition G' contains two new nodes $m + 1$ and $m + 2$. (3) All arcs in G are in G'. (4) $l'_{ij} = 0$, $u'_{ij} = u_{ij} - l_{ij}$. (5) If $l_{ij} > 0$, then place arc $(i, m + 2)$ in G' with $u'_{i, m+2} = l_{ij}$ and $l'_{i, m+2} = 0$; place arc $(m + 1, j)$ in G' with $u'_{m+1, j} = l_{ij}$ and $l'_{m+1, j} = 0$. (6) Solve for the maximal flow from node $m + 1$ to node $m + 2$ in G'. (7) If at optimality all arcs out of node $m + 1$ are *saturated* (that is, $x'_{ij} = u'_{ij}$), then a feasible flow exists in G and $x_{ij} = x'_{ij} + l_{ij}$; otherwise, no feasible solution exists.)

12.9 Consider the maximal flow problem in which the variables are restricted as $-\infty < l_{ij} \leq x_{ij} \leq u_{ij} < \infty$. Write the dual to this problem. Based on this answer, define the capacity of a cut-set so that the maximal flow–minimal cut theorem holds in this case, assuming feasibility.

12.10 Suggest a scheme for selecting which node to label next in the "labeling algorithm" for the maximal flow problem so that at each iteration a maximal flow augmenting path is determined. Comment on the relative computational advantage of using such a scheme.

12.11 Consider the following procedure for reducing the size of a network while finding the maximal flow from node 1 to node m. (Assume $l_{ij} = 0$ and $u_{ij} > 0$ for all (i, j).)

Step 1. Remove all arcs entering node 1 (the source) and leaving node m (the sink).

Step 2. Discard any node that has no arcs incident with it.

Step 3. Discard any node, except node 1, that only has arcs leaving it. Also discard these arcs.

Step 4. Discard any node, except node m, that only has arcs entering it. Also discard these arcs.

Repeat steps 2, 3, and 4 until no change results. If node 1 or m is discarded, stop; the maximum flow is zero. Otherwise use the maximum flow algorithm.

a. Show that the maximal flow in the resulting network is the same as in the original network.

b. Is it true that there is a path from node 1 to every node in the resulting network (assuming that node 1 has not been discarded)? If not, then what additional operation is required to ensure this?

12.12 Show that we have a basic feasible solution to the maximal flow problem if there exist no cycles in the set $E = \{(i, j): 0 < x_{ij} < u_{ij}\}$. In this case show how the basic variables and the simplex tableau can be obtained at any iteration of the maximal flow algorithm. Furthermore, show that the following procedure will maintain basic feasible solutions in the maximal flow algorithm.

Step 1. At each iteration, begin with E as defined earlier.

Step 2. Try to find a path from node 1 to node m in G' associated only with arcs in E.

Step 3. If no path is available from step 2, then add one arc in G' not associated with arcs in E to E if its permits the labeling of a new node. With this new arc in E, return to step 2.

12.13 Exhibit in detail how the labeling algorithm for the maximal flow problem is equivalent to an application of the out-of-kilter algorithm in which several primal phases are performed and the first dual phase adjusts the dual multipliers and terminates. Is this always the case when using the out-of-kilter algorithm for any general problem with only one initial out-of-kilter arc?

12.14 What simplifications would result if the network simplex method (of Chapter 9) is used to solve the maximal flow problem? Give all details.

12.15 Develop a dual simplex method for the maximal flow problem.

12.16 How can node capacities be handled in the maximal flow algorithm?

12.17 Modify the maximal flow algorithm to handle undirected arcs.

12.18 Consider the problem of finding the minimum number of lines to cover all zeros in the assignment algorithm (refer to Section 10.7). Show that the maximal flow algorithm can be used to provide the result. (*Hint*. Given the reduced assignment matrix, construct a maximal flow network G as follows. Let nodes $1, \ldots, n$ represent the n rows of the assignment matrix and nodes $n + 1, \ldots, 2n$ represent the columns of the assignment matrix. If the ijth entry in the reduced matrix is zero, draw an arc from node i to node $n + j$ with $u_{i,n+j} = \infty$. Add two additional nodes $2n + 1$ and

$2n+2$. Add an arc $(2n+1, i)$ with $u_{2n+1,i} = 1$ for $i = 1, \ldots, n$ and an arc $(n+i, 2n+2)$ with $u_{n+i,2n+2} = 1$ for $i = 1, \ldots, n$. Solve for the maximal flow from node $2n+1$ to node $2n+2$ in G. The value of the maximal flow is equal to the minimum number of lines to cover all zeros in the reduced matrix. To find which lines to use, consider the sets X and $\overline{X}$ when the maximal flow algorithm stops. If $(2n+1, i)$ is in $(X, \overline{X})$, draw a line through row i. If $(n+i, 2n+2)$ is in $(X, \overline{X})$ draw a line through column i. It still must be shown that this procedure works.)

12.19 Apply the procedure of the previous problem to find the minimum number of lines to cover all zeros in the following reduced assignment matrix.

	1	2	3	4
1	0	0	0	1
2	2	5	3	0
3	1	0	2	0
4	2	0	4	1

12.20 In this chapter we have provided the node-arc formulation for the maximal flow problem. Consider an *arc-path* formulation as follows. Let $j = 1, 2, \ldots, t$ be an enumeration of all of the paths from node 1 to node m in the network. Number the arcs from 1 to n and let

$$p_{ij} = \begin{cases} 1 & \text{if arc } i \text{ is in path } j \\ 0 & \text{otherwise} \end{cases}$$

The arc-path formulation for the maximal flow problem is

$$\text{Maximize} \quad \sum_{j=1}^{t} x_j$$

$$\text{Subject to} \quad \sum_{j=1}^{t} p_{ij} x_j \le u_i \qquad i = 1, \ldots, n$$

$$x_j \ge 0 \qquad j = 1, 2, \ldots, t$$

where x_j represents the flow on path j.

a. Give the complete arc-path formulation for the maximal flow problem of Figure 12.1.

b. Solve the linear program of part (a).

12.21 Consider the arc-path formulation for the maximal flow problem as given in Exercise 12.20. Suppose that we do not enumerate any paths to begin with but, decide to apply the revised simplex method with all slack variables in the starting basic feasible solution. At any iteration of the revised simplex method let $\mathbf{w}$ be the dual vector.

a. What is the simplex entry criterion for (i) a slack variable and (ii) a path variable?

b. Show that there is an easy method to test the simplex entry criterion for path variables using the shortest path algorithm.
c. If you always first enter slacks until no more slacks are eligible to enter, show that you may use the shortest path algorithm for nonnegative costs to test the entry criterion for path variables.
d. Describe the complete steps of the revised simplex method thus obtained.
e. Apply the revised simplex method developed in this exercise to the maximal flow problem in Figure 12.1.

12.22 a. Give the dual of the arc-path formulation for the maximal flow problem as stated in Exercise 12.20.
b. If we add the restriction that the dual variables must be 0 or 1, what interpretation can you give the dual problem?
c. Interpret the dual solution obtained in part (e) of Exercise 12.21.

12.23 Is the constraint matrix for the arc-path formulation of a maximal flow problem always unimodular? Prove or give a counterexample.

12.24 Find the shortest path from node 1 to all nodes of the following network using Dijkstra's algorithm as well as the PSP algorithm. Identify the shortest path tree obtained.

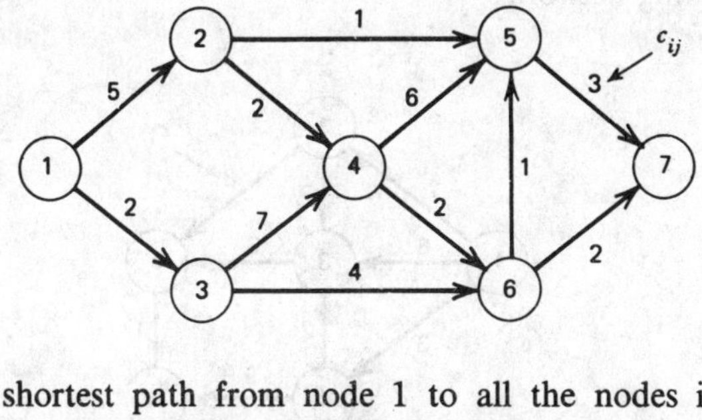

12.25 Find the shortest path from node 1 to all the nodes in the following network. Use the algorithms of both Sections 12.2 and 12.3. Identify the shortest path tree obtained.

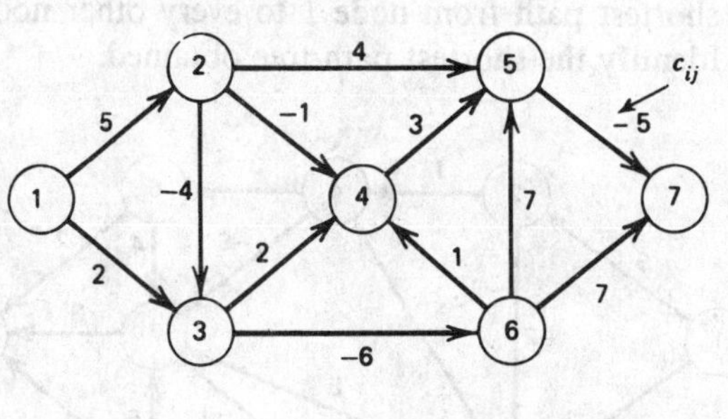

12.26 Suppose that you are to find the shortest simple path from node 1 to node m in a network that may contain negative cost circuits. Further suppose that you have obtained an optimal solution to the minimum cost flow problem on this network after designating $b_1 = 1$, $b_m = -1$, and $b_i = 0$ otherwise, and setting bounds $0 \le x_{ij} \le 1$ on all the arcs (i, j) in the problem. How would you identify the simple path from 1 to m that has a flow of one unit in this solution? Why is this not necessarily the shortest simple path from 1 to m? Provide a mathematical formulation of the problem of determining a shortest simple path.

12.27 a. Show how the shortest path algorithm can be used to find the longest path from node 1 to node m in a network.
b. When finding the longest path, what assumption must be made?
c. Use the results of part (a) to devise direct algorithms for the longest path problem.

12.28 Show that at optimality of the shortest path problem $w_i - w_m$ represents a lower bound on the cost of the shortest path from node i to node m for each i.

12.29 In the shortest path algorithm for negative costs, show that at optimality there always exists a path from node 1 to any node k along which $w_j' = w_i' + c_{ij}$ provided that $w_k' < \infty$. Also show that if $w_k' = \infty$, then no path exists from node 1 to node k.

12.30 Find both the shortest path and the longest path from node 1 to node 6 in the following network.

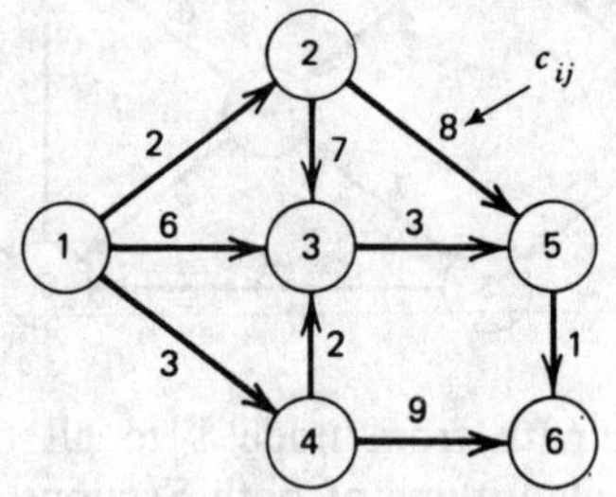

12.31 Find the shortest path from node 1 to every other node in the following network. Identify the shortest path tree obtained.

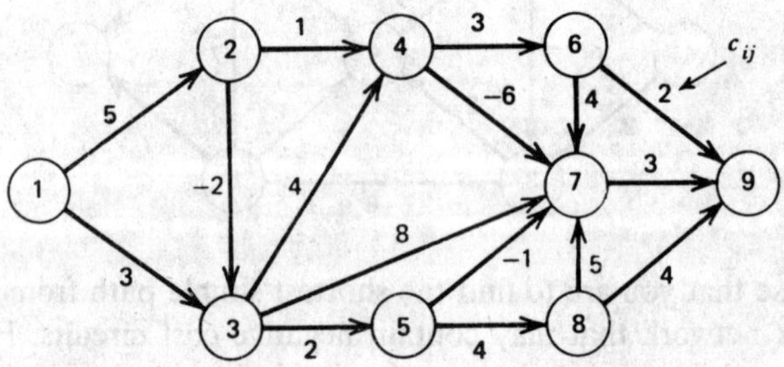

12.32 Find the shortest path from every node to node 6 in the following network. Identify the shortest path tree obtained. (*Hint*. Apply the shortest path algorithm with nonnegative costs in reverse.)

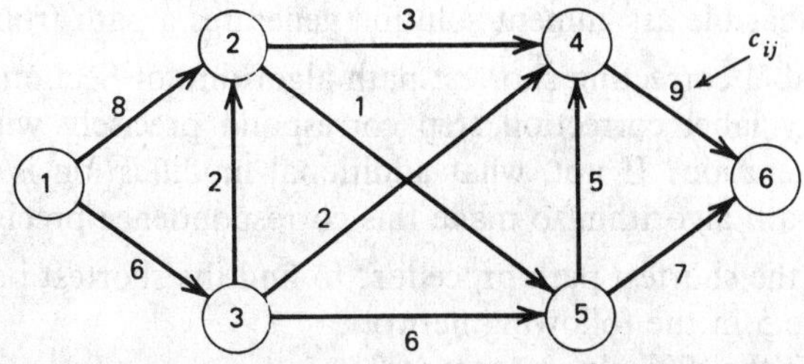

12.33 Modify the shortest path algorithm for negative costs to find the shortest path from every node to node m. Illustrate by finding the shortest path from every node to node 7 in the following network.

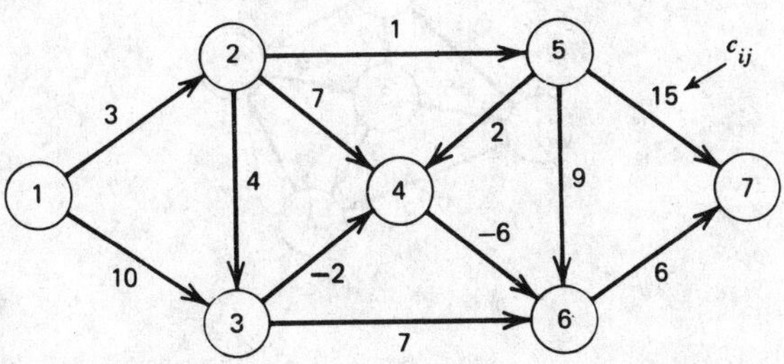

12.34 a. Modify the shortest path algorithm for nonnegative costs to handle an undirected network.

b. Apply the procedure of part (a) to find the shortest path from node 1 to node 5 in the following network.

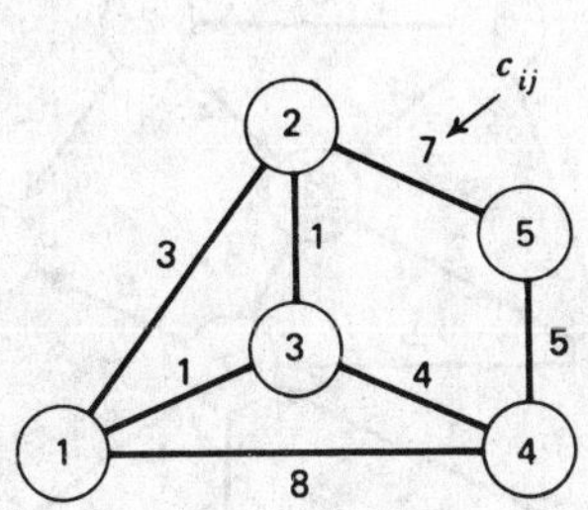

12.35 Can the shortest path algorithm for negative costs be modified to handle undirected arcs? (*Hint*. Consider one of the arcs with negative cost.)

12.36 Consider the problem of determining the shortest path from node 1 to node m in a directed network having no negative cost circuits. How can you formulate and solve this as an assignment problem? (*Hint*: Think of a path in terms of assigning to each node the next node to visit. This ensures that any feasible assignment solution generates a path from 1 to m.)

12.37 For the label correcting shortest path algorithm of Sections 12.2 and 12.3, does every label correction step correspond precisely with some primal simplex iteration? If not, what additional modification is required to the shortest path algorithm to make this correspondence precise?

12.38 a. Apply the shortest path procedure to find the shortest path from node 1 to node 5 in the following network.
b. What is the difficulty in part (a)?
c. Solve the problem by the network simplex method of Chapter 9. Compare with the result in (a).

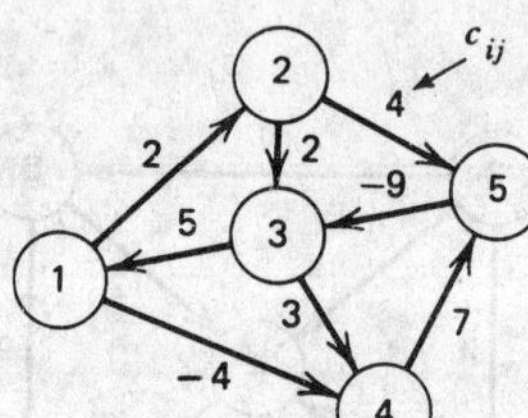

12.39 Bob, Ed, and Stu are in a car pool. They each live at the points 1, 2, and 7 respectively in the following network. They agree to meet at point 10 every morning at a certain time and proceed from there to their work in a single car. The numbers on the arcs represent the travel times in minutes.

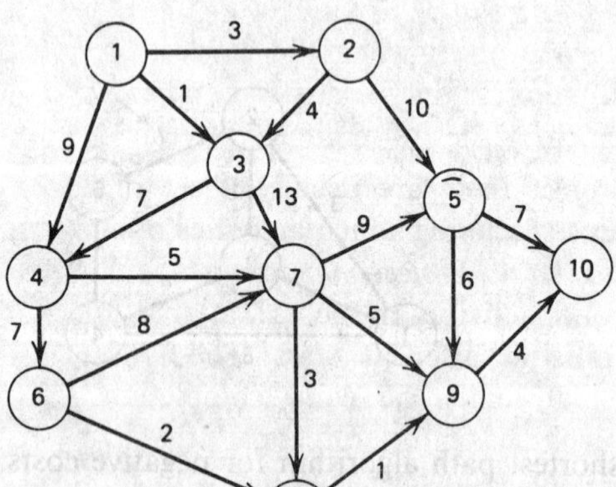

a. What is the fastest route for each man to the meeting point?
b. How early (counting back from the meeting time) should each man leave? Would anyone be agreeable to give another a ride to point 10?

12.40 A single machine is needed to perform a specified function for the next four years, after which the function and machine will no longer be needed. The purchase price of a machine varies over the next four years according to the following table.

YEAR	NOW	ONE YEAR FROM NOW	TWO YEARS FROM NOW	THREE YEARS FROM NOW
Purchase price	$25,000	$33,000	$38,000	$47,000

The salvage value of a machine depends only on its length of service and is given by the following table.

LENGTH OF SERVICE	1 YEAR	2 YEARS	3 YEARS	4 YEARS
salvage value	$17,000	$6,000	$3,000	$1,000

The annual operating cost varies with length of service, as follows.

LENGTH OF SERVICE	NEW	1 YEAR	2 YEARS	3 YEARS
Annual operating cost	$3,000	$5,000	$8,000	$18,000

Construct a network in which the shortest path will yield an optimal policy of purchasing, operating, and salvaging machines over the next four years if management wishes to minimize the total cost.

12.41 In any project there are usually a set of activities that constitute the relationships specifying which activities must be completed before a given activity can start. *Project management* is concerned with the scheduling and control of activities in such a way that the project can be completed as soon as possible after its start. The *critical path* is that sequence of activities that limits the early completion time of the project. (It is generally activities on the critical path that project managers watch closely.)

Consider the following activities with indicated completion times and precedence relationships.

ACTIVITY	COMPLETION TIME (DAYS)	PREDECESSORS
A	3	—
B	6	—
C	2	D, E
D	6	A, B
E	8	B, F
F	5	A
G	3	B, C, F

Find the critical path for this project and the associated project time. [*Hint*. Draw an arc with its own beginning and ending nodes for each activity with the arc cost equal to the completion time of the activity. If activity Q must precede activity R, then draw an arc from the ending node of activity Q to the beginning node of activity R with zero cost on the arc. Provide a starting node to proceed all activities and a finishing node to succeed all activities. In the network thus obtained the longest path (why not shortest path?) will be the critical path.]

12.42 a. How can the shortest path algorithm be used to obtain a starting (not necessarily feasible) solution when the out-of-kilter algorithm is applied to a minimal cost network flow problem with nonzero right-hand-side values?

b. Is there any advantage to doing this?

12.43 Since Dijkstra's shortest path algorithm for nonnegative costs is extremely efficient, it would be highly desirable to be able to convert a network with negative costs to an equivalent network with nonnegative costs. Consider the following procedure for accomplishing this in a network G with m nodes.

INITIALIZATION STEP

Let $t = 1$ (t is the iteration counter).

MAIN STEP

1. Let $i = 1$.
2. Let $\bar{c}_i = \underset{j}{\text{Minimum}}\, c_{ij}$. If $\bar{c}_i < 0$, replace c_{ij} by $c_{ij} - \bar{c}_i$ for all j and replace c_{ki} by $c_{ki} + \bar{c}_i$ for all k.
3. If $i < m$ replace i by $i + 1$ and go to step 2. Otherwise go to step 4.
4. If all $c_{ij} \geq 0$, step, the equivalent network is obtained. Otherwise, if $t < m + 1$ replace t by $t + 1$ and go to step 1; if $t = m + 1$, stop, there is a negative circuit in G.

Upon completing this procedure if all $c_{ij} \geq 0$ we may apply Dijkstra's shortest path algorithm to the equivalent network. Note that although, the proper path will be found, its length must be adjusted.

a. Show that the method works.
b. Apply the method to the networks in Exercises 12.25 and 12.38.
c. Show that if $c_{ij} < 0$ at iteration $m + 1$, then there is a negative circuit in G that includes node j. Is it possible to develop a labeling procedure to find the negative circuit?
d. What is the complexity of this procedure? Comment on its relative advantage.

12.44 Consider a network with upper bounds and costs (all lower bounds are zero). Suppose that we wish to find, among all maximal flows, from node 1 to node m, that maximum flow that minimizes the total cost. This is sometimes called the *minimal cost–maximal flow* problem.

a. Give a linear programming formulation for the minimal cost–maximal flow problem from node 1 to node m in a network.
b. Show how the out-of-kilter method can be used to solve this problem.
c. Apply parts (a) and (b) to the following network to obtain the minimal cost–maximal flow from node 1 to node 4.

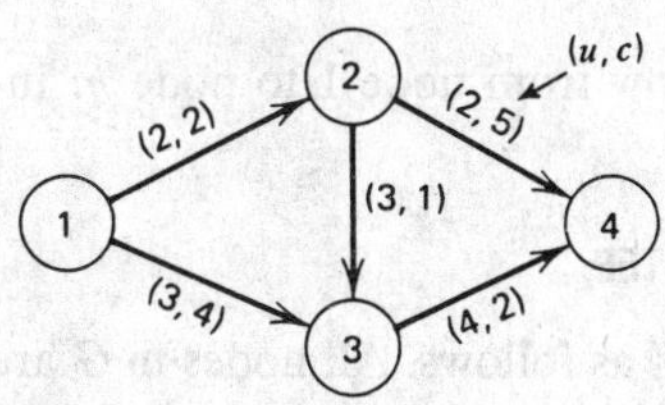

12.45 Consider the following procedure, from Busacker and Gowen, for finding the minimal cost–maximal flow from node 1 to node m in a network with nonnegative costs and all $l'_{ij} = 0$.

INITIALIZATION STEP

Let all $x_{ij} = 0$.

SHORTEST PATH STEP

From G construct G' as follows. All nodes in G are in G'. If $x_{ij} < u_{ij}$ in G, place (i, j) in G' with $\Delta_{ij} = u_{ij} - x_{ij}$ and $c'_{ij} = c_{ij}$. If $x_{ij} > 0$ in G, place (j, i) in G' with $\Delta_{ji} = x_{ij}$ and $c'_{ji} = -c_{ij}$. Find the shortest path from node 1 to node m in G'. If no path exists, stop; the optimal solution is at hand. Otherwise pass to the flow change step.

FLOW CHANGE STEP

Let $\Delta = \text{Minimum }\{\Delta_{ij}: (i, j) \text{ is in the shortest path}\}$. Adjust flows along the associated chain in G by Δ, increasing flows on arcs with the orientation of the path and decreasing flows on arcs against the orientation of the path. Pass to the shortest path step.

a. Apply the algorithm to the example network of the previous problem.
b. Prove that the algorithm converges to the optimal solution in a finite number of steps. It is necessary to show that (i) negative circuits never occur in G', (ii) after a finite number of flow changes no path from node 1 to node m will exist, and (iii) on termination the optimal solution is obtained. (*Hint*. Consider the flow in the network as a parameter and show that after each flow change we have the minimal cost solution for that amount of flow.)
c. What difficulties would occur when we admit negative costs?

12.46 Consider the following algorithm, from Klein, for finding the minimal cost-maximal flow from node 1 to node m in a network with arbitrary costs and all $l_{ij} = 0$.

INITIALIZATION STEP

Find the maximal flow from node 1 to node m in G.

NEGATIVE CIRCUIT STEP

From G construct G' as follows. All nodes in G are in G'. If $x_{ij} < u_{ij}$ in G, place (i, j) in G' with $\Delta_{ij} = u_{ij} - x_{ij}$ and $c'_{ij} = c_{ij}$. If $x_{ij} > 0$ in G, then place (j, i) in G' with $\Delta_{ji} = x_{ij}$ and $c'_{ji} = -c_{ij}$. Use the shortest path algorithm or the method of Exercise 12.43 to find a negative circuit in G'. If no negative circuit exists, stop; the optimal solution is at hand. Otherwise pass to the flow change step.

FLOW CHANGE STEP

Let $\Delta = \text{Minimum }\{\Delta_{ij}: (i, j) \text{ is the negative circuit}\}$. Adjust flows along the associated cycle in G by Δ, increasing flows on arcs with the orientation of the circuit and decreasing flows on arcs against the orientation of the circuit. Pass to the negative circuit step.

a. Apply the algorithm to the network of Exercise 12.44.
b. Prove that the algorithm converges to the optimal solution in a finite number of steps.

12.47 a. Give the linear programming formulation for the two-commodity maximal flow problem shown (with no individual commodity upper bounds).
b. Find the two commodity maximal flow in the network.

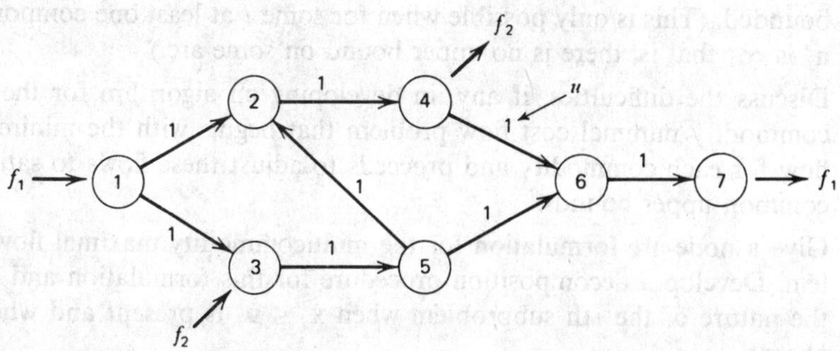

12.48 How can lower bounds be handled in the multicommodity minimal cost flow problem?

12.49 Apply the decomposition algorithm to the following three commodity minimal cost flow problem.

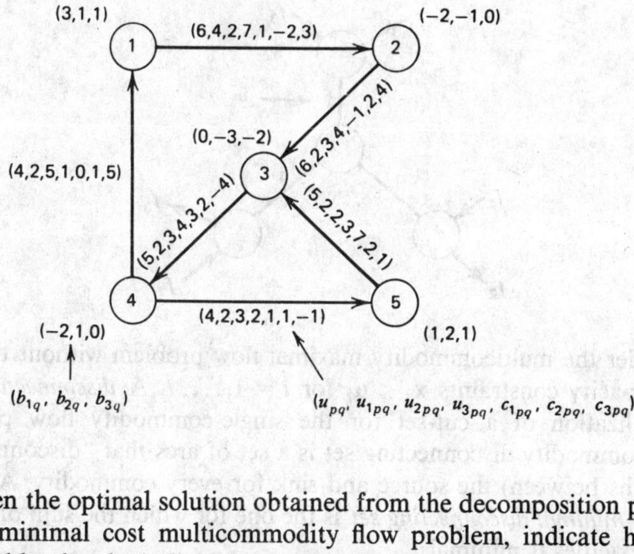

12.50 Given the optimal solution obtained from the decomposition procedure for the minimal cost multicommodity flow problem, indicate how the dual variables for the individual commodity constraints ($\mathbf{Ax}_i = \mathbf{b}_i$ and $\mathbf{x}_i \leq \mathbf{u}_i$) can be recovered. Apply the procedure to the example problem in Section 12.4.

12.51 Discuss the economic meaning of the dual variables for the various constraints in the multicommodity minimal cost flow problem (namely, $\Sigma_i \mathbf{x}_i \leq \mathbf{u}$, $\mathbf{Ax}_i = \mathbf{b}_i$, and $\mathbf{x}_i \leq \mathbf{u}_i$).

12.52 Modify the decomposition algorithm for the minimal cost multicommodity flow problem when the set $X_i = \{\mathbf{x}_i : \mathbf{Ax}_i = \mathbf{b}_i,\ \mathbf{0} \leq \mathbf{x}_i \leq \mathbf{u}_i\}$ is not bounded. (This is only possible when for some i at least one component of $\mathbf{u}_i$ is ∞; that is, there is no upper bound on some arc.)

12.53 Discuss the difficulties, if any, in developing an algorithm for the multicommodity minimal cost flow problem that begins with the minimal cost flow for each commodity and proceeds to adjust these flows to satisfy the common upper bounds.

12.54 Give a node-arc formulation for the multicommodity maximal flow problem. Develop a decomposition procedure for this formulation and discuss the nature of the ith subproblem when $\mathbf{x}_i \leq \mathbf{u}_i$ is present and when it is absent.

12.55 Develop an arc-path formulation for the multicommodity maximal flow problem without the presence of the constraints $\mathbf{x}_i \leq \mathbf{u}_i$ for $i = 1, \ldots, t$. Develop a decomposition procedure for this formulation. (*Hint*. Consider the formulation given in Exercise 12.20.)

12.56 How can undirected arcs be handled in the multicommodity maximal flow problem? Illustrate on the following three-commodity network.

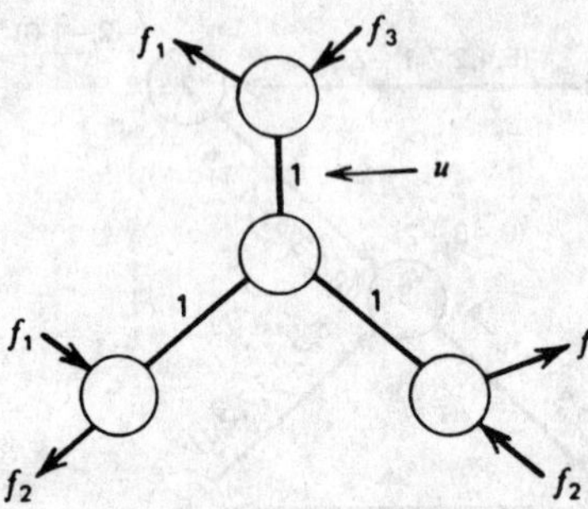

12.57 Consider the multicommodity maximal flow problem without the individual capacity constraints $\mathbf{x}_i \leq \mathbf{u}_i$ for $i = 1, \ldots, t$. A *disconnecting set* is a generalization of a cut-set for the single-commodity flow problem. A multicommodity disconnecting set is a set of arcs that "disconnects" (cuts all paths between) the source and sink for every commodity. A *multicommodity minimal disconnecting set* is the one for which the sum of (common) arc capacities is minimal.

a. Give a mathematical formulation for the minimal disconnecting set problem. (*Hint*. Take the dual of the arc-path formulation for the maximal flow problem and require the dual variables to be zero or 1. Give the interpretation of this dual problem.)
b. Show that the capacity of the multicommodity minimal disconnecting set is greater than or equal to the value of the multicommodity maximal flow. (*Hint*. Apply duality theorems to the formulation in part (a).)
c. Give a minimal disconnecting set for the network of Figure 12.9 and to the network of Exercise 12.47.

d. Compare the capacity of the minimal disconnecting set and the value of the maximal flow for both problems of part (c).

12.58 Show that a multicommodity minimal disconnecting set is the union of single-commodity cut-sets. Is a multicommodity minimal disconnecting set necessarily the union of single commodity *minimal* cut-sets?

12.59 Consider a metropolitan city with the area divided into four zones and a highway network connecting the zones. Let the following matrix, called the *origin-destination* matrix, specify the travel requirements from each (row) zone to every other (column) zone.

	1	2	3	4	5
1	0	10	7	8	5
2	2	0	3	4	4
3	6	2	0	1	5
4	2	4	7	0	5
5	1	1	3	4	0

Travel times and arc (upper) capacities are given by the following.

ARC	(1,2)	(2,4)	(2,5)	(3,2)	(4,3)	(5,1)	(5,4)
Travel time (min)	15	35	15	20	10	15	10
Capacity	43	38	37	27	35	20	10

Find the minimal time traffic assignment in the network.

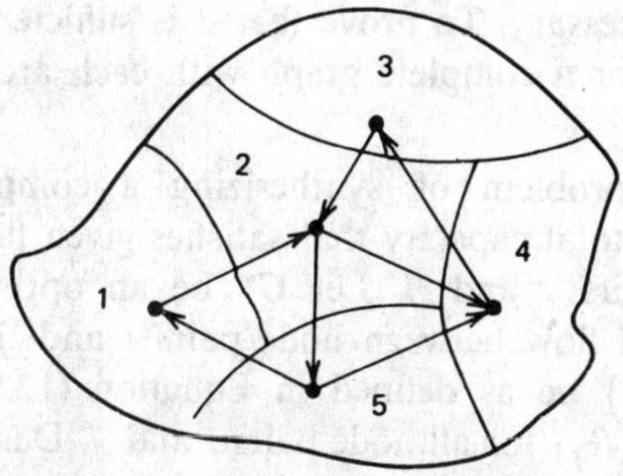

12.60 In the multicommodity minimal cost flow problem suppose that we have the capability of inverting the matrix $[\mathbf{E}'' - \mathbf{E}'\mathbf{B}^{-1}\mathbf{D}, \mathbf{E}]$. Show how the original primal and dual variables can be obtained. Use this information to develop a simplex procedure for solving the multicommodity flow problem directly on the graph. Give all details and illustrate by the problem of Figure 12.10.

12.61 Resolve the multicommodity minimal cost flow problem of Figure 12.10 with $u_{234} = 4$ instead of 3. Is it reasonable to expect this solution in practice?

12.62 Determine an optimal solution to the network synthesis example in Section 12.6.

12.63 Consider the problem of synthesizing a network having the following structure, where the costs per unit capacity are given on the arcs.

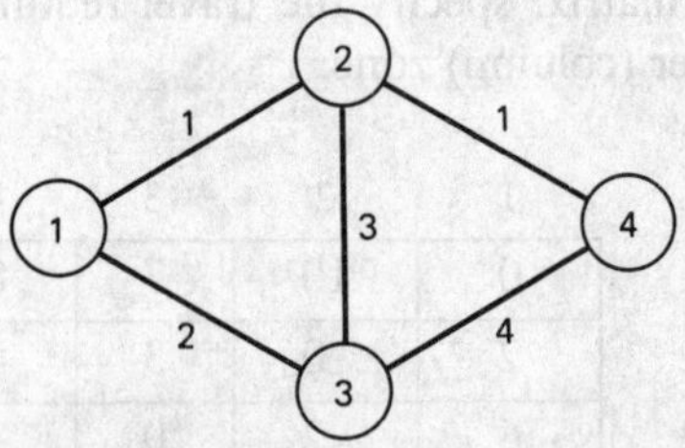

a. Determine a set of optimal capacities for these arcs in order to be able to sustain the following flow requirements: $r_{12} = 5$, $r_{13} = 3$, $r_{14} = 2$, $r_{23} = 4$, $r_{24} = 2$, and $r_{34} = 7$.
b. Repeat, assuming that arc (1, 4) can also be constructed and that $c_{ij} = 1$ for all (i, j).

12.64 Let F be a symmetric nonnegative matrix of size $m \times m$ with elements f_{ij}, where $f_{ii} = \infty$ for $i = 1, \ldots, m$. Then F is said to be *max flow realizable* if there exists some undirected graph on m nodes for which f_{ij} is the maximal flow between nodes i and j. Show that F is max flow realizable if and only if $f_{ij} \geq \text{Minimum } \{f_{ik}, f_{kj}\}$ for all i, j, k. Furthermore show that if F is realizable, then it is realizable by a tree graph. What does this say about the maximum number of distinct max flow values between pairs of nodes in any undirected graph? (*Hint*: Section 12.6 proves that this condition is necessary. To prove that it is sufficient, construct a maximal spanning tree for a complete graph with each arc (i, j) having a weight f_{ij}.)

12.65 Consider the problem of synthesizing a complete undirected graph with minimum total capacity that satisfies given flow requirements r_{ij} between node pairs i and j. Let G^* be an optimal graph, and let f_{ij}^* be the maximal flow between node pairs i and j in G^*. Now let $R_i = \text{Max}\{r_{ij},\ j \neq i\}$ be as defined in Equation (12.9), and compute $r_{ij}^* = \text{Minimum } \{R_i, R_j\}$ for all node pairs i and j. Denote by G^{**} the optimal graph obtained by assuming that the requirements are r_{ij}^* in lieu of r_{ij} for all (i, j), and let f_{ij}^{**} be the maximal flow between node pairs i and j in G^{**}. Show that G^{**} is also optimal for the problem with requirements r_{ij}, and that $f_{ij}^{**} \geq f_{ij}^*$ for all node pairs i and j. Interpret the significance of G^{**}. Construct G^{**} for the example of Section 12.6 and for the example of Exercise 12.63. (*Hint*: First show that (a) $f_{ij}^* \leq \text{Minimum } \{R_i, R_j\}$ for all

(i, j), and (b) $R_i^* \equiv \text{Max}\{r_{ij}^*,\ j \neq i\} = R_i$ for all i. Use these results to establish the required results.)

NOTES AND REFERENCES

1. Ford and Fulkerson [1956] were the first to develop the maximal flow algorithm for networks. Dantzig and Fulkerson [1956] provided a proof of the maximal flow–minimal cut theorem. Glover *et al.* [1984a, b] studied specialized primal simplex algorithms for the maximal flow problem, and have shown them to be significantly (1.5–5 times) faster than the presented labeling techniques, while requiring one-third the amount of storage. They have also shown that this specialized scheme is roughly 1.8 times faster than a version of the incumbent general primal simplex code RNET, even when it is tuned for maximal flow problems. For determining the maximal flow between all pairs of nodes in an undirected graph via an equivalent tree network, see Gomory and Hu [1961].
2. The first algorithms for shortest path problems were developed by Bellman]1958], Dijkstra [1959], Dantzig [1960], Whiting and Hiller [1960], and Floyd [1962]. A good comparison of these procedures appears in Dreyfus [1969]. More recent and significantly more efficient methods have been developed by Glover *et al.* [1958a, b]. Microcomputer implementations of these algorithms are described by Klingman and Scheider [1986]. A comprehensive computational study appears in Glover [1984] and Hung and Divoky [1988], where it is shown that threshold based partitioned shortest path algorithms dominate other procedures. A good exposition on the complexity behavior of some shortest path implementations is given by Shier and Witzgall [1981]. Also see Fredman and Tarjan [1984] for improved implementations using Fibonacci Heaps. For the use of decomposition techniques in solving large-scale shortest path problems on special networks, see Hu [1968], Shier [1973], Jarvis and Tufekci [1981], and Tufekci [1983]. For a more detailed exposition on the relationship between shortest path algorithms and the simplex method, see Akgul [1986b], Dial *et al.* [1979], and Zedah [1979]. The procedure of Exercise 12.43 was developed by Bazaraa and Langley [1974], and is based on ideas presented by Nemhauser [1972].
3. Ford and Fulkerson [1958b] first proposed a column generation procedure for the multicommodity maximal flow problem. This was the forerunner to the Dantzig–Wolfe decomposition procedure for general linear programs. Hartmand and Lasdon [1972] proposed a procedure based on the simplex-method for solving multicommodity flow problems. For the various exercises on multicommodity maximal flow problems, see Robacker [1952], Ford and Fulkerson [1958b], Bellmore *et al.* [1970], Grigoriadis and White [1972], Hartman and Lasdon [1972], and Jarvis and Keith [1974]. A good discussion and survey on multicommodity flow problems appears in Kennington and Halgason [1980].
4. The Network Synthesis problem described in Section 12.6 is from Gomory and Hu [1962]. Also see Ford and Fulkerson [1962].

BIBLIOGRAPHY

Abadie, J. On the Decomposition Principle, ORC Report 63-20, Operations Research Center, University of California, Berkeley, Calif., August 1963.

Abadie, J., and A. C. Williams. "Dual and Parametric Methods in Decomposition," in R. Graves and P. Wolfe (eds.), *Recent Advances in Mathematical Programming*, McGraw-Hill Book Co., New York, 1963.

Adler, I., M. G. C. Resende, and G. Veiga. "An Implementation of Karmarkar's Algorithm for Linear Programming," Operations Research Center, Report 86-8, University of California at Berkeley, May 1986.

Advani, Suresh. "A Linear Programming Approach to Air-Cleaner Design," *Operations Research*, **22** (2), pp. 295–297, March–April, 1974.

Aggarwal, S. P. "A Simplex Technique for a Particular Convex Programming Problem," *Canadian Operational Research Society Journal*, **4** (2), pp. 82–88, July 1966.

Akgul, M. "An Algorithmic Proof of the Polyhedral Decomposition Theorem," *Naval Research Logistics Quarterly*, **35**, pp. 463–472, 1988.

Akgul, M. "A Note on Shadow Prices in Linear Programming," *Journal of the Operational Research Society*, **35(5)** pp. 425–431, 1984.

Akgul, M. "Topics in Relaxation and Ellipsoidal Methods," Ph.D. Dissertation, Department of Combinatorics and Optimization, University of Waterloo, Waterloo, Ont., Canada, 1981.

Akgul, M. "A Genuinely Polynomial Primal Simplex Algorithm for the Assignment Problem," Department of Computer Science and Operations Research, North Carolina State University, Raleigh, NC 27695-8206, 1986a.

Akgul, M. "Shortest Paths and the Simplex Method," Working Paper, Department of Computer Science, North Carolina State University, Raleigh, NC 27695-7913, 1986b.

Albers, D. J. and C. Reid. "An Interview with George B. Dantzig: The Father of Linear Programming," *The College Mathematics Journal*, **17** (4), pp. 292–314, September 1986.

Ali, A., R. Helgason, J. Kennington, and H. Lall. "Primal-Simplex Network Codes: State-of-the-Art Implementation Technology," *Networks*, **8**, pp. 315–339, 1978.

Ali, A. R. Helgason, J. Kennington, and H. Lall. "Computational Comparison Among Three Multicommodity Network Flow Algorithms," *Operations Research*, **28** (4), pp. 995–1000, 1980.

Anstreicher, K. M. "Analysis of a Modified Karmarkar Algorithm for Linear Programming," *Working Paper Series B#84*, *Yale School of Organization and Management*, *Box 1A*, *New Haven*, *CT* 06520, 1985.

Anstreicher, K. M. "A Strengthened Acceptance Criterion for Approximate Projection in Karmarkar's Algorithm," *Operations Research Letters*, **5** (4), pp. 211–214, 1986a.

Anstreicher, K. M. "A Monotonic Projective Algorithm for Fractional Linear Programming," *Algorithmica*, **1**, pp. 483–498, 1986b.

Anstreicher, K. M. "A Combined 'Phase I–Phase II' Projective Algorithm for Linear Programming (To appear in *Mathematical Programming*), Technical Report, Yale School of Organization and Management, New Haven, CT, 1986c.

Arinal, J. C. "Two Algorithms for Hitchcock Problem," *Revue Française de Recherche Opérationnelle* (France), **33**, pp. 359–374, 1964.

Aronofsky, J. S., and A. C. Williams. "The Use of Linear Programming and Mathematical Models in Underground Oil Production," *Management Science*, **8** (4), pp. 394–407, July 1962.

Aronson, J., R. Barr, R. Helgason, J. Kennington, A Loh, and H. Zaki. "The Projective Transformation Algorithm by Karmarkar: A Computational Experiment with Assignment Problems," Technical Report 85-OR-3, Department of Operations Research, Southern Methodist University, and Department of Industrial Engineering, University of Houston, Houston, 1985.

Arrow, K. J., L. Hurwicz, and H. Uzawa (eds.). *Studies in Linear and Nonlinear Programming*, Stanford University Press, Stanford, Calif., 1958.

Asher, D. T. "A Linear Programming Model for the Allocation of *R* and *D* Efforts," *IEEE Transactions on Engineering Management*, *EM-9* (4), pp. 154–157, December 1962.

Asic, M., V. V. Kovacevic-Vujcic, and M. D. Radosavljevic-Nikolic. "Asymptotic Behavior and Numerical Stability of Karmarkar's Method for Linear Programming," University of Belgrade, Belgrade, Yugoslavia, 1986.

Aspvall, B. and R. E. Stone. "Khachiyan's Linear Programming Algorithm," *Journal of Algorithms*, **1**, pp. 1–13, 1980.

Assad, A. A. "Multicommodity Network Flows: A Survey," *Networks*, **8** (1), pp. 37–92, 1978.

Aucamp, D. C. and D. I. Steinberg. "The Computation of Shadow Prices in Linear Programming," *Journal of the Operational Research Society*, **33**, pp. 557–565, 1982.

Au, Tung, and Thomas E. Stelson. *Introduction to Systems Engineering, Deterministic Models*, Addison-Wesley, Reading, Mass. 1969.

Avis, D. and V. Chvatal. "Notes on Bland's Pivoting Rule," *Mathematical Programming Study*, **8**, pp. 24–34, 1978.

Azpeitia, A. G., and D. J. Dickinson. "A Decision Rule in the Simplex Method that Avoids Cycling," *Numerische Mathematik*, **6**, pp. 329–331, 1964.

Balas, E. "The Dual Method for the Generalized Transportation Problem," *Management Science*, **12**, pp. 555–568, 1966.

Balas, E. "An Infeasibility Pricing Decomposition Method for Linear Programs," *Operations Research*, **14**, pp. 847–873, 1966.

Balas, E. "Solution of Large Scale Transportation Problems Through Aggregation," *Operations Research*, **13**, pp. 82–93, 1965.

Balas, E., and P. L. Ivanescu. "On the Generalized Transportation Problem," *Management Science*, **11**, pp. 188–203, 1964.

Balas, E. and M. W. Padberg. "On the Set Covering Problem," Management Sciences Research Report #197, Carnegie Mellon University, 1970.

Balinski, M. L. "Integer Programming: Methods, Uses, Computation," *Management Science*, **12** (3), pp. 253–313, November 1965.

Balinski, M. L. "An Algorithm for Finding all Vertices of Convex Polyhedral Sets," *SIAM Journal on Applied Mathematics IX*, pp. 72–88, 1961.

Balinski, M. L. "Signature Methods for the Assignment Problem," *Operations Research*, **33** (3), pp. 527–536, 1985.

Balinski, M. L. "The Hirsch Conjecture for Dual Transportation Polyhedra," *Mathematics of Operations Research*, 1984.

Balinski, M. L. "A Competitive (Dual) Simplex Method for the Assignment Problem," *Mathematical Programming*, **34** (2), pp. 125–141, 1986.

Balinski, M. L., and R. E. Gomory. "A Mutual Primal-Dual Simplex Method," in R. L. Graves and P. Wolfe (eds.), *Recent Advances in Mathematical Programming*, McGraw-Hill Book Co., New York, 1963.

Balinski, M. L., and R. E. Gomory. "A Primal Method for the Assignment and Transportation Problems," *Management Science*, **10** (3), pp. 578–593, April 1964.

Balinski, M. L. and A. Russakoff. "On the Assignment Polytope," *SIAM Review*, **16** (4), pp. 516–525, 1974.

Barnes, E. R. "A Variation on Karmarkar's Algorithm for Solving Linear Programming Problems," *Mathematical Programming*, **36**, pp. 123–134, 1986.

Barnes, E. R. "A Polynomial Time Version of the Affine Scaling Algorithm," Presented at the Joint National ORSA/TIMS Meeting, St. Louis, October 1987.

Barnes, J. W., and R. M. Crisp. "Linear Programming: A Survey of General Purpose Algorithms," *AIIE Transactions*, **7** (3), September 1975.

Barnett, S. "Stability of the Solution to a Linear Programming Problem," *Operational Research Quarterly*, **13** (3), pp. 219–228, September 1962.

Barnett, S. "A Simple Class of Parametric Linear Programming Problems," *Operations Research*, **16**, (6), pp. 1160–1165, November–December 1968.

Barr, R. S., F. Glover, and D. Klingman. "An Improved Version of the Out-of-Kilter Method and a Comparative Study of Computer Codes," *Mathematical Programming*, **7** (1), pp. 60–86, 1974.

Barr, R. S., F. Glover, and D. Klingman. "The Alternating Basis Algorithm for Assignment Problems," *Mathematical Programming*, **13** (1), pp. 1–13, 1977.

Barr, R. S., F. Glover, and D. Klingman. "The Generalized Alternating Path Algorithm for Transportation Problems," *European Journal of Operational Research*, **2**, pp. 137–144, 1978.

Barr, R. S., F. Glover, and D. Klingman. "Enhancements of Spanning Tree Labelling Procedures for Network Optimization," *INFOR*, **17** (1), pp. 16–34, 1979.

Bartels, R. H., and G. H. Golub. "The Simplex Method of Linear Programming Using LU-Decomposition," *Communications of the ACM*, **12**, pp. 266–268 and 275–278, 1969.

Battersby, Albert. *Network Analysis for Planning and Scheduling*, St. Martin's Press, New York, 1964.

Bazaraa, M. S. An Infeasibility Pricing Algorithm for the Multicommodity Minimum Cost Flow Problem, unpublished paper, School of Industrial and Systems Engineering, Georgia Institute of Technology, Atlanta, Georgia, 1973.

Bazaraa, M. S., and R. W. Langley. "A Dual Shortest Path Algorithm," *SIAM Applied Mathematics*, **26** (3), pp. 496–501, May 1974.

Bazaraa, M. S., and J. J. Goode. "A Survey of Various Tactics for Generating Lagrangian Multipliers in the Context of Lagrangian Duality," *European Journal of Operational Research*, **3**, 322–338, 1979.

Bazaraa, M. S., and J. J. Goode. A New Decomposition Technique, unpublished paper, School of Industrial and Systems Engineering, Georgia Institute of Technology, Atlanta, Georgia, 1975.

Bazaraa, M. S., and H. D. Sherali. "A Versatile Scheme for Ranking the Extreme Points of an Assignment Polytope," *Naval Research Logistics Quarterly*, **28** (4), pp. 545–558, 1981.

Bazaraa, M. S., and H. D. Sherali. "A Property of Degenerate Pivots for Linear Assignment Networks," *Networks*, **12** (4), pp. 469–474, 1982.

Bazaraa, M. S., and C. M. Shetty. *Nonlinear Programming: Theory and Algorithms*, John Wiley and Sons, Inc., New York, NY, 1979.

Bazaraa, M. S. and C. M. Shetty. *Foundations of Optimization*, Springer Verlag, New York. 1976.

Beale, E. M. L. "Cycling in the Dual Simplex Algorithm," *Naval Research Logistics Quarterly*, **2** (4), pp. 269–276, December 1955.

Beale, E. M. L. "An Algorithm for Solving the Transportation Problem When the Shipping Cost Over Each Route is Convex," *Naval Research Logistics Quarterly*, **6** (1), pp. 43–56, March 1959.

Bell, Earl J. "Primal-Dual Decomposition Programming," U.S.G.R.& D.R. Order AD-625 365 from CFSTI, Operations Research Center, University of California, Berkeley, Calif., August 1965.

Bellman, Richard. "On a Routing Problem," *Quarterly Applied Mathematics*, **16** (1), pp. 87–90, April 1958.

Bellmore, M., H. J. Greenberg, and J. J. Jarvis. "Multicommodity Disconnecting Sets," *Management Science*, **16**, pp. B427–B433, 1970.

Bellmore, M., and R. Vemuganti. "On Multicommodity Maximal Dynamic Flows," *Operations Research*, **21**, pp. 10–21, 1973.

Ben Daya, M., and C. M. Shetty. "Polynomial Barrier Function Algorithms for Linear Programming," School of Industrial and Systems Engineering, Georgia Institute of Technology, Atlanta, GA. 30332-0205, 1988.

Benders, J. F. "Partitioning Procedures for Solving Mixed Variables Programming Problems," *Numerische Methamatik*, **4**, *pp*. 238–252, 1962.

Benichoi, M., J. M. Gauthier, G. Hentges, and G. Ribiere. "The Efficient Solution of Large-Scale Linear Programming Problems—Some Algorithmic Techniques and Computational Results," *Mathematical Programming*, **13**, pp. 280–322, December 1977.

Ben-Israel, A., and A. Charnes. "An Explicit Solution of a Special Class of Linear Programming Problems," *Operations Research*, **16** (6), pp. 1160–1175, November–December 1968.

Ben-Israel, Adi, and Philip D. Robers. "A Decomposition Method for Interval Linear Programming," *Management Science*, **16**, (5), pp. 374–387, January 1970.

Bennett, John M., and David R. Green. "An Approach to Some Structured Linear Programming Problems," *Operations Research*, **17** (4), pp. 749–750, July–August 1969.

Berge, C., and A. Ghouila-Houri. *Programming, Games and Transportation Networks*, John Wiley & Sons, New York, 1965.

Bertsekas, D. P. "A New Algorithm for the Assignment Problem," *Mathematical Programming*, **21**, pp. 152–171, 1981.

Bertsekas, D. P. "A Unified Framework for Primal-Dual Methods in Minimum Cost Network Flow Problems," *Math Programming*, **32**, pp. 125–145, 1985.

Bertsekas, D. P. "A Distributed Asynchronous Relaxation Algorithm for the Assignment Problem," *Proceedings 24th IEEE Conference on Decision and Control*, Ft. Lauderdale, Fla, pp. 1703–1704, December 1985.

Bertsekas, D. P. "Distributed Relaxation Methods for Linear Network Flow Problems," *Proceedings of 25th IEEE Conference on Decision and Control*, Athens, Greece, pp. 2101–2106, 1986.

Bertsekas, D. P. "The Auction Algorithm: A Distributed Relaxation Method for the Assignment Problem," Report LIDS-P-1653, March 1987.

Bertsekas, D. P., and D. El Baz. "Distributed Asynchronous Relaxation Methods for Convex Network Flow Problems," LIDS Report P-1417, *SIAM Journal on Control and Optimization*, **25**, pp. 74–85, 1987.

Bertsekas, D. P., and J. Eckstein. "Distributed Asychronous Relaxation Methods for Linear Network Flow Problems," Report LIDS-P-1606, *Proceedings of IFAC '87*, Munich, Germany, Pergamon Press, Oxford, England, July 1987.

Bertsekas, D. P., P. Hosein, and P. Tseng. "Relaxation Methods for Network Flow Problems with Convex Arc Costs," *SIAM Journal on Control and Optimization*, **25**, pp. 1219–1243, 1987.

Bertsekas, D. P., and P. Tseng. "The RELAX Codes for Linear Minimum Cost Network Flow Problems," in *Fortran Codes for Network Optimization*,

Annals of Operations Research (ed. B. Simeone), 1988 (LIDS Report P-1469, MIT, 1986).

Bertsekas, D. P., and P. Tseng. "Relaxation Methods for Minimum Cost Ordinary and Generalized Network Flow Problems," *Operations Research*, **36(1)**, pp. 93–114, 1988.

Best, M. J., and K. Ritter. *Linear Programming Active Set Analysis and Computer Programs*, Prentice-Hall, Inc., Englewood Cliffs, New Jersey, 1985.

Bialy, H. "An Elementary Method for Treating the Case of Degeneracy in Linear Programming," *Unternehmensforschung* (Germany), **10** (2), pp. 118–123, 1966.

Birge, J. R. "A Dantzig–Wolfe Decomposition Variant Equivalent to Basis Factorization," *Mathematical Programming Study 24*, ed. R. W. Cottle, pp. 43–64, 1985.

Birge, J. R., and L. Qi. "Solving Stochastic Linear Programs Via a Variant of Karmarkar's Algorithm," Technical Report 85-12, University of Michigan, College of Engineering, Ann Arbor, Michigan 48109-2117, 1985.

Bitran, G. R., and A. G. Novaes. "Linear Programming with a Fractional Objective Function," *Operations Research*, **21** (1), pp. 22–29, January–February 1973.

Bixby, R. E., and W. H. Cunningham. "Converting Linear Programs to Network Problems," *Mathematics of Operations Research*, **5** (3), pp. 321–357, 1980.

Blair, C. E. "The Iterative Step in the Linear Programming Algorithm of N. Karmarkar," *Algorithmica*, **1** (4), pp. 537–539, 1986.

Bland, R. G. "New Finite Pivoting Rules for the Simplex Method," *Mathematics of Operations Research*, **2**, pp. 103–107, May 1977.

Bland, R. G., D. Goldfarb, and M. J. Todd. "The Ellipsoid Method: A Survey," *Operations Research*, **29** (6), pp. 1039–1091, November/December 1981.

Blum, L. "Towards an Asymptotic Analysis of Karmarkar's Algorithm," Mills College, Oakland, CA 94613, and the Department of Mathematics, University of California, Berkeley, 97420, 1985.

Boldyreff, A. W. "Determination of the Maximal Steady State Flow of Traffic Through a Railroad Network," *Operations Research*, **3** (4), pp. 443–465, November 1955.

Borgwardt, K. H. "Some Distribution Independent Results About the Asymptotic Order of the Average Number of Pivot Steps in the Simplex Method," *Mathematics of Operations Research*, **7** (3), pp. 441–462, August 1982a.

Borgwardt, K. H. "The Average Number of Pivot Steps Required by the Simplex Method is Polynomial," *Zeitschrift fur Operations Research*, **26**, pp. 157–177, 1982b.

Borgwardt, K. H. "A Probabilistic Analysis of the Simplex Method," Habilitation Thesis, Kaiserslautern, October 1984.

Borgwardt, K. H. *The Simplex Method: A Probabilistic Analysis*, Algorithms and Combinatorics 1, Springer-Verlag, p. 268, 1987.

Boulding, K. E., and W. A. Spivey. *Linear Programming and the Theory of the Firm*, The Macmillan Company, New York, 1960.

Bouška, Jiři, and Martin Cerný. "Decomposition Methods in Linear Programming," *Ekonomicko-matematicky obzor* (Czechoslovakia), **1** (4), pp. 337–369, 1965.

Bowman, E. H. "Production Scheduling by the Transportation Method of Linear Programming," *Operations Research*, **4** (1), 1956.

Bowman, E. H. "Assembly-Line Balancing by Linear Programming," *Operations Research*, **8**, pp. 385–389, 1960.

Boyer, D. D. A Modified Simplex Algorithm for Solving the Multicommodity Maximum Flow Problem, Technical Memorandum TM-14930, The George Washington University, Washington, D.C., 1968.

Bradley, G. H. "Survey of Deterministic Networks," *AIIE Transactions*, **7** (3), September 1975.

Bradley, G. H., G. G. Brown, and G. W. Graves. "Design and Implementation of Large-Scale Primal Transshipment Algorithms," *Management Science*, **24** (1), pp. 1–34, 1977.

Bramucci, Franco. "The Inversion Algorithm in Linear Programming," *Metra* (France), **6** (2), pp. 357–381, June 1967.

Briggs, F. E. A. "A Dual Labelling Method for the Hitchcock Problem," *Operations Research*, **10** (4), pp. 507–517, July–August 1962.

Brosius, L. "Comment on a Paper by M. C. Cheng," *Mathematical Programming*, **21**, pp. 229–232, 1981.

Brown, G. W., and T. C. Koopmans. "Computational Suggestions for Maximizing a Linear Function Subject to Linear Inequalities," in T. C. Koopmans (ed.), *Activity Analysis of Production and Allocation*, John Wiley & Sons, New York, pp. 377–380, 1951.

Burdet, C. A. "Generating all the Faces of a Polyhedron," *SIAM Journal on Applied Mathematics XXVI*, pp. 479–489, 1974.

Burkard, R. E., H. W. Hamacher, and J. Tind. "On General Decomposition Schemes in Mathematical Programming," *Mathematical Programming*, **24**, pp. 238–252, 1985.

Burrell, B. P., and M. J. Todd. "The Ellipsoid Method Generates Dual Variables," *Mathematics of Operations Research*, **10**, pp. 688–700, 1985.

Busacker, R. G., and P. J. Gowen. A Procedure for Determining a Family of Minimal-Cost Network Flow Patterns, ORO Technical Report 15, Operations Research Office, John Hopkins University, 1961.

Busacker, R. G., and T. L. Saaty. *Finite Graphs and Networks*: *An Introduction with Applications*, McGraw-Hill Book Co., New York, 1965.

Cabot, A. Victor, Richard L. Francis, and Michael A. Stary. "A Network Flow Solution to a Rectilinear Distance Facility Location Problem," *AIIE Transactions*, **2** (2), pp. 132–141, June 1970.

Cahn, A. S. "The Warehouse Problem," *Bulletin of American Mathematical Society*, **54**, p. 1073, 1948.

Calman, Robert F. *Linear Programming and Cash Management*, *CASH ALPHA*, The M.I.T. Press, Cambridge, Mass., 1968.

Camion, P. "Characterization of Totally Unimodular Matrices," *Proceedings of American Mathematical Society*, **16**, pp. 1068–1073, 1965.

Camm, J. D., P. M. Dearing, and S. K. Tadisina. "The Calhoun Textile Mill Case: An Exercise on the Significance of Linear Programming Model Formulation," *IIE Transactions*, **19** (1), pp. 23–28, March 1987.

Caratheodory, C. "Uber den variabilitatsbereich der koeffizienten von potenzreihen, die gegebene werte nicht annehmen," *Mathematische Annalen*, **64**, pp. 95–115, 1907.

Carpenito, G., and L. Toth. "Algorithm 548—Solution of the Assignment Problem," *ACM Transactions on Mathematical Software*, **6** (1), 1980.

Catchpole, A. R. "The Application of Linear Programming to Integrated Supply Problems in the Oil Industry," *Operational Research Quarterly*, **13** (2), pp. 161–169, June 1962.

Cavalier, T. M., and A. L. Soyster. "Some Computational Experience and a Modification of the Karmarkar Algorithm," Presented at the 12th International Symposium on Mathematical Programming, Massachusetts Institute of Technology, Cambridge, Massachusetts, August 1985.

Cederbaum, I. "Matrices All of Whose Elements and Subdeterminants are 1, −1, or 0," *Journal of Mathematics and Physics*, **36**, pp. 351–361, 1958.

Chadda, S. S. "A Decomposition Principle for Fractional Programming," *Opsearch*, **4** (3), pp. 123–132, 1967.

Chandrascharan, R. "Total Unimodularity of Matrices," *SIAM Journal of Applied Mathematics*, **17**, pp. 1032–1034, 1969.

Chandru, V., and B. S. Kochar. "Exploiting Special Structures Using a Variant of Karmarkar's Algorithm," Research Memorandum 86-10, School of Industrial Engineering, Purdue University, West Lafayeete, Indiana 47907, 1986.

Chandru, V., and B. S. Kochar. "A Class of Algorithms for Linear Programming," Research Memorandum 85-14, School of Industrial Engineering, Purdue University, West Lafayette, Indiana 47907, 1985 (revised June 1986).

Chandy, K. M., and Tachen Lo. "The Capacitated Minimum Spanning Tree," *Networks*, **3** (2), pp. 173–181, 1973.

Chang, S. Y., and K. G. Murty. "The Steepest Descent Gravitational Method for Linear Programming," Technical Report #87-14, Department of Industrial and Operations Engineering, The University of Michigan, Ann Arbor, Michigan 48109-2117, 1987.

Charnes, A. "Optimality and Degeneracy in Linear Programming," *Econometrica*, **20** (2), pp. 160–170, 1952.

Charnes, A., and W. W. Cooper. "The Stepping Stone Method for Explaining Linear Programming Calculations in Transportation Problems," *Management Science*, **1** (1), 1954.

Charnes, A., and W. W. Cooper. *Management Models and Industrial Applications of Linear Programming*, John Wiley & Sons, New York, 1961.

Charnes, A., and W. W. Cooper. "Programming with Linear Fractional Functionals," *Naval Research Logistics Quarterly*, **9**, pp. 181–186, 1962.

Charnes, A., and W. W. Cooper. "On Some Works of Kantorovich, Koopmans, and Others," *Management Science*, **8** (3), pp. 246–263, April 1962.

Charnes, A., W. W. Cooper, and A. Henderson. *An Introduction to Linear Programming*, John Wiley & Sons, New York, 1953.

Charnes, A. W. W. Cooper, and R. Mellon. "Blending Aviation Gasolines: A Study in Programming Interdependent Activities in an Integrated Oil Company," *Econometrica*, **20** (2), 1952.

Charnes, A. W. W. Cooper, and G. L. Thompson. "Some Properties of Redundant Constraints and Extraneous Variables in Direct and Dual Linear Programming Problems," *Operations Research*, **10**, pp. 711–723, 1962.

Charnes, A., and K. O. Kortanek. "An Opposite Sign Algorithm for Purification to an Extreme Point Solution," Office of Naval Research, Memorandum No. 84, Northwestern University, Evanston, 1963.

Charnes, A., K. O. Kortanek, and W. Raike. "Extreme Point Solutions in Mathematical Programming: An Opposite Sign Algorithm, Systems Research Memorandum No. 129, Northwestern University, Evanston, 1965.

Charnes, A., and C. E. Lemke. A Modified Simplex Method for Control of Round-Off Error in Linear Programming, unpublished paper, Carnegie Institute of Technology, 1952.

Charnes, A. and C. E. Lemke. "Minimization of Nonlinear Separable Convex Functionals," *Naval Research Logistics Quarterly*, **1**, (4), pp. 301–312, December 1954.

Charnes, A., and C. E. Lemke. Computational Theory of Linear Programming, I: The Bounded Variables Problem, ONR Research Memorandum 10, Graduate School of Industrial Administration, Carnegie Institute of Technology, Pittsburgh, Pennsylvania, January 1954.

Charnes, A., and W. M. Raike. "One Pass Algorithms for Some Generalized Network Problems," *Operations Research*, **14**, pp. 914–924, 1966.

Charnes, A., T. Song, and M. Wolfe. "An Explicit Solution Sequence and Convergence of Karmarkar's Algorithm," Research Report CCS 501, College of Business Administration 5.202, The University of Texas at Austin, Austin, Texas 78712-1177, 1984.

Cheng, M. C. "New Criteria for the Simplex Algorithm," *Mathematical Programming*, **19**, pp. 230–236, 1980.

Chester, Louie B. Analysis of the Effect of Variance on Linear Programming Problems. U.S.G.R.& D.R. Order AD-611 273 from Clearinghouse, Air Force Institute of Technology, Wright-Patterson AFB, Washington, D.C., August 1964.

Chien, R. T. "Synthesis of a Communication Net," *IBM Journal of Research and Development*, **4**, 1960.

Chisman, James A. "Using Linear Programming to Determine Time Standards," *Journal of Industrial Engineering*, **17** (4), pp. 189–191, April 1966.

Chiu, S. S., and Yinyu Ye. "Recovering the Shadow Price in Projection Methods of Linear Programming," Engineering–Economics Systems Department, Stanford University, Stanford, California 94305, 1985.

Chiu, S. S., and Yinyu Ye. "Simplex Method and Karmarkar's Algorithm: A Unifying Structure," Engineering–Economic Systems Department, Stanford University, Stanford, California 94305, 1985.

Christofides, N. *Graph Theory: An Algorithmic Approach*, Academic Press, New York, 1975.

Chung, An-min. *Linear Programming*, Charles E. Merrill Books, Columbus, Ohio, 1963.

Chvatal, V. *Linear Programming*, W. H. Freeman and Company, New York/ San Francisco, 1980.

Clark, Charles E. "The Optimum Allocation of Resources Among the Activities of a Network," *Journal of Industrial Engineering*, **12** (1), pp. 11–17, January–February 1961.

Clasen, R. J. The Numerical Solution of Network Problems Using Out-of-Kilter Algorithm, RAND Report RM-5456 PR, March 1968.

Clements, R. A. "Linear Programming for Multiple Feed Formulation," *NZOR* (New Zealand), **2** (2), pp. 100–107, July 1974.

Cline, R. E. "Representations for the Generalized Inverse of Matrices Partitioned as **A** = [U,V]," in R. Graves and P. Wolfe (eds.), *Recent Advances in Mathematical Programming*, McGraw-Hill Book Co., New York, 1963.

Cobham, A. "The Intrinsic Computational Difficulty of Functions," *Proceedings of the 1964 International Congress for Logic, Methodology, and Philosophy of Science*, Y. Bar-Hille, (ed.), North-Holland, Amsterdam, pp. 24–30, 1965.

Commoner, F. G. "A Sufficient Condition for a Matrix to be Totally Unimodular," *Networks*, **3**, pp. 351–365, 1973.

Cook, S. A. "The Complexity of Theorem-Proving Procedures," *Proceedings of the Third Annual ACM Symposium on Theory of Computing*, Association for Computing Machinery, New York, pp. 151–158, 1971.

Cooper, L., and J. Kennington. "Non-Extreme Point Solution Strategies for Linear Programs," *Naval Research Logistics Quarterly*, **26** (3), 1979.

Courtillot, M. "On Varying All the Parameters in a Linear Programming Problem and Sequential Solution of a Linear Programming Problem," *Operations Research*, **10** (4), 1962.

Craven, B. D. "A Generalization of the Transportation Method of Linear Programming," *Operational Research Quarterly*, **14** (2), pp. 157–166, June 1963.

Crowder, H. and J. M. Hattingh. "Partially Normalized Pivot Selection," *Mathematical Programming Study Number 4*, December 1975.

Crowder, H., E. L. Johnson, and M. Padberg. "Solving Large-Scale Zero-One Linear Programming Problems," *Operations Research*, **31** (5), pp. 803–834, 1983.

Cunningham, W. H. "A Network Simplex Method," *Mathematical Programming*, **11** (2), pp. 105–116, 1976.

Cunningham, W. H. "Theoretical Properties of the Network Simplex Method," *Mathematics of Operations Research,* **4** (2), pp. 196–208, 1979.

Cunningham, W. H. "A Class of Linear Programs Convertible to Network Problems," *Operations Research*, **31** (2), pp. 387–390, 1983.

Curtis, Floyd H. "Linear Programming the Management of a Forest Property," *Journal of Forestry*, **61** (9), pp. 611–616, September 1962.

Dantzig, G. B. Programming in a Linear Structure, Comptroller, United States Air Force, Washington, D.C., February 1948.

Dantzig, G. B. "Programming of Interdependent Activities, II, Mathematical

Model," in T. C. Koopmans (ed.), *Activity Analysis of Production and Allocation*, John Wiley & Sons, New York, 1951, also published in *Econometrica*, **17** (3–4), pp. 200–211, July–October 1949.

Dantzig, G. B. "Maximization of a Linear Function of Variables Subject to Linear Inequalities," in T. C. Koopmans (ed.), *Activity Analysis of Production and Allocation*, John Wiley & Sons, New York, pp. 339–347, 1951.

Dantzig, G. B. "A Proof of the Equivalence of the Programming Problem and the Game Problem," in T. C. Koopmans (ed.), *Activity Analysis of Production and Allocation*, John Wiley & Sons, New York, pp. 359–373, 1951.

Dantzig, G. B. "Application of the Simplex Method to a Transportation Problem," in T. C. Koopmans (ed.), *Activity of Production and Allocation*, John Wiley & Sons, New York, pp. 359–373, 1951.

Dantzig, G. B. Computational Algorithm of the Revised Simplex Method, RAND Report RM-1266, The Rand Corporation, Santa Monica, Calif. 1953.

Dantzig, G. B. Notes on Linear Programming, Part VII. The Dual Simplex Algorithm, RAND Report RM-1270, The Rand Corporation, Santa Monica, Calif., July 1954.

Dantzig, G. B. Notes on Linear Programming, Part XI, Composite Simplex-Dual Simplex Algorithm-I, Research Memorandum RM-1274, The Rand Corporation, Santa Monica, Calif., April 1954.

Dantzig, G. B. Notes on Linear Programming: Parts VIII, IX, X-Upper Bounds, Secondary Constraints, and Block Triangularity in Linear Programming, Research Memorandum RM-1367, The Rand Corporation, Santa Monica, Calif., October 1954, also published in *Econometrica*, **23** (2), pp. 174–183, April 1955.

Dantzig, G. B. "Discrete Variable Extremum Problems," *Operations Research*, **5** (2), pp. 266–277, April 1957.

Dantzig, G. B. On the Shortest Route Through a Network, RAND Report P-1345, The Rand Corporation, Santa Monica, Calif., April 1958. Also, *Management Science*, **6** (2), pp. 187–190, 1960.

Dantzig, G. B. On the Significance of Solving Linear Programming Problems with Some Integer Variables, RAND Report P-1486, The Rand Corporation, Santa Monica, Calif., September 1958.

Dantzig, G. B. *Linear Programming and Extensions*. Princeton University Press, Princeton, N.J., 1963.

Dantzig, G. B. "Compact Basis Triangularization for the Simplex Method," in R. Graves and P. Wolfe (eds.), *Recent Advances in Mathematical Programming*, McGraw-Hill Book Co., New York, 1963.

Dantzig, G. B. Optimization in Operations Research: Some Examples, U.S.G.R.& D.R. Order AD-618 748, Operations Research Center, University of California, Berkeley, Calif., April 1965.

Dantzig, G. B. All Shortest Routes in a Graph, Technical Report 66-3, Operations Research House, Stanford University, Stanford, Calif., November 1966.

Dantzig, G. B. "Reminiscences About the Origins of Linear Programming," *Operations Research Letters*, **1** (2), pp. 43–48, April 1982.

Dantzig, G. B., W. D. Blanttner, and M. R. Rao. All Shortest Routes from a Fixed Origin in a Graph, Technical Report 66-2, Operations Research House, Stanford University, Stanford, Calif., November 1966.

Dantzig, G. B., L. R. Ford, and D. R. Fulkerson. "A Primal-Dual Algorithm for Linear Programs," in H. W. Kuhn and A. W. Tucker (eds.), *Linear Inequalities and Related Systems*, Annals of Mathematics Study No. 38, Princeton University Press, Princeton, N.J., pp. 171–181, 1956.

Dantzig, G. B., and D. R. Fulkerson. "Minimizing the Number of Tankers to Meet a Fixed Schedule," *Naval Research Logistics Quarterly*, **1** (3), pp. 217–222, September 1954.

Dantzig, G. B., and D. R. Fulkerson. "Computation of Maximal Flows in Networks," *Naval Research Logistics Quarterly*, **2** (4), 1955.

Dantzig, G. B., and D. R. Fulkerson. "On the Max-Flow Min-Cut Theorem of Networks," in H. W. Kuhn and A. W. Tucker (eds.), *Linear Inequalities and Related Systems*, Annals of Mathematics Study No. 38, Princeton University Press, Princeton, N.J., pp. 215–221, 1956.

Dantzig, G. B., D. R. Fulkerson, and S. M. Johnson. "Solution of a Large-Scale Traveling-Salesman Problem," *Operations Research*, **2** (4), pp. 393–410, November 1954.

Dantzig, G. B., D. R. Fulkerson, and S. M. Johnson. On a Linear Programming-Combinatorial Approach to the Traveling Salesman Problem, RAND Report P-1281, The Rand Corporation, Santa Monica, Calif., April 1958.

Dantzig, G. B., R. Harvey, and R. McKnight. Updating the Product Form of the Inverse for the Revised Simplex Method, ORC Report 64-33, Operations Research Center, University of California, Berkeley, Calif., December 1964.

Dantzig, G. B., and D. L. Johnson. "Maximum Payloads Per Unit Time Delivered Through an Air Network," *Operations Research*, **12** (2), March–April 1964.

Dantzig, G. B., S. Johnson, and W. White. A Linear Programming Approach to the Chemical Equilibrium Problem, RAND Report P-1060, The Rand Corporation, Santa Monica, Calif., April 1958.

Dantzig, G. B., and A. Orden. Notes on Linear Programming: Part II-Duality Theorems, Research Memorandum RM-1265, The Rand Corporation, Santa Monica, Calif., October 1953.

Dantzig, G. B., A. Orden, and P. Wolfe. "The Generalized Simplex Method for Minimizing a Linear Form Under Linear Inequality Restraints," *Pacific Journal of Mathematics*, **5** (2), pp. 183–195, June 1955.

Dantzig, G. B., and W. Orchard-Hays. "The Product Form for the Inverse in the Simplex Method," *Mathematical Tables and Aids to Computation*, **8** (46), pp. 64–67, 1954.

Dantzig, G. B., and W. Orchard-Hays. Notes on Linear Programming: Part V—Alternate Algorithm for the Revised Simplex Method Using Product Form for the Inverse, Research Memorandum RM-1268, The RAND Corporation, Santa Monica, Calif., November 1953.

Dantzig, G. B., and R. M. Van Slyke. Generalized Upper Bounded Techniques for Linear Programming I, II, ORC reports 64-17 (1964), 64-18 (1965), Operations Research Center, University of California, Berkeley, Calif.

Dantzig, G. B. and P. Wolfe. "Decomposition Principle for Linear Programs," *Operations Research*, **8** (1), pp. 101–111, January–February 1960.

Dantzig, G. B., and P. Wolfe. "The Decomposition Algorithm for Linear Programs," *Econometrica*, **29** (4), pp. 767–778, October, 1961.

Dantzig, G. B., and P. Wolfe. Linear Programming in a Markov Chain. Notes on Linear Programming and Extensions. Part 59, Research Memorandum RM-2957-PR, The Rand Corporation, Santa Monica, Calif., April 1962.

deGhellinck, G., and J.-P. Vial. "A Polynomial Newton Methods for Linear Programming," *Algorithmica*, **1** (4), pp. 425–454, 1986.

deGhellinck, G., and J. Vial. "An Extension of Karmarkar's Algorithm for Solving a System of Linear Homogeneous Equations on the Simplex," *Mathematical Programming*, **39** (1), pp. 79–92, 1987.

Denardo, Eric V. "On Linear Programming in a Markov Decision Problem," *Management Science*, **16** (5), pp. 281–288, January 1970.

Dennis, J. B. *Mathematical Programming and Electrical Networks*, John Wiley & Sons, New York, 1959.

Dennis, J. E., A. M. Morshedi, and K. Turner. "A Variable-Metric Variant of the Karmarkar Algorithm for Linear Programming," *Mathematical Programming*, **39** (1), pp. 1–30, 1987.

Dent, J. B., and H. Casey. *Linear Programming and Animal Nutrition*, J. B. Lippincott Co., Philadelphia, Pa., 1968.

Dial, R., F. Glover, D. Karney, and D. Klingman. "A Computational Analysis of Alternative Algorithms and Labeling Techniques for Finding Shortest Path Trees," *Networks*, **9**, pp. 215–248, 1979.

Dickson, J. C., and F. P. Frederick. "A Decision Rule for Improved Efficiency in Solving Linear Programming Problems with the Simplex Algorithm," *Communications of the ACM*, **3**, pp. 509–512, 1960.

Dijkstra, E. W. "A Note on Two Problems in Connection with Graphs," *Numerical Mathematics*, **1**, pp. 269–271, 1959.

Dikin, I. I. "Iterative Solution of Problems of Linear and Quadratic Programming," *Soviet Mathematics Doklady*, **8**, pp. 674–675, 1967.

Dikin, I. I. "On the Speed of an Iterative Process," *Upravlyaemye Sistemi*, **12**, pp. 54–60, 1974.

Doig, A. G. "The Minimum Number of Basic Feasible Solutions to a Transportation Problem," *Operational Research Quarterly*, **14** (4), 1963.

Doig, A. G., and A. H. Land. "An Automatic Method of Solving Discrete Programming Problems," *Econometrica*, **28**, pp. 497–520, 1960.

Dongarra, J. J., C. B. Moler, J. R. Bunch, and G. W. Stewart. "LINPACK User's Guide," *SIAM*, Philadelphia, 1979.

Dorfman, R. *Application of Linear Programming to the Theory of the Firm*, University of California Press, Berkeley, Calif., 1951.

Dorfman, R., P. A. Samuelson, and R. M. Solow. *Linear Programming and Economic Analyses*, McGraw-Hill Book Co., New York, 1958.

Dreyfus, S. E. "An Appraisal of Some Shortest Path Algorithms," *Operations Research*, **17** (3), pp. 395–412, 1969.

Duffin, R. J. "Infinite Programs, in H. W. Kuhn and A. W. Tucker (eds.), *Linear Inequalities and Related Systems*, Annals of Mathematics Study No. 38, Princeton University Press, Princeton, N.J., 1956.

Duffin, R. J. The Extremal Length of a Network, Office of Technical Services, Document No. AD-253 665, 1961.

Duffin, R. J. "Dual Programs and Minimum Costs," *Journal of the Society for Industrial and Applied Mathematics*, **10**, pp. 119–123, 1962.

Duffin, R. J. "Convex Analysis Treated by Linear Programming," *Mathematical Programming* (Netherlands), **4** (2), pp. 125–143, April 1973.

Dyer, M. E., and L. G. Proll. "An Algorithm for Determining all Extreme Points of a Convex Polytope," *Mathematical Programming*, **12**, pp. 81–96, 1977.

Eckhardt, U. "Theorems on the Dimension of Convex Sets," *Linear Algebra and its Applications*, **12**, pp. 63–76, 1975.

Edmonds, J. "Paths, Trees, and Flowers," *Canadian Journal of Mathematics*, pp. 449–467, 1965.

Edmonds, J., and R. M. Karp. "Theoretical Improvements in Algorithmic Efficiency for Network Flow Problems," *Journal of the ACM*, **19** (2), pp. 248–264, April 1972.

Egerváry, E. "On Combinatorial Properties of Matrices (1931)," translated by H. W. Kuhn, Paper No. 4, George Washington University Logistics Research Project, 1954.

Eggleston, H. G. *Convexity*, Cambridge University Press, New York, 1958.

Eisemann, K. "The Primal-Dual Method for Bounded Variables," *Operations Research*, **12** (1), January–February 1964.

Elam, J., F. Glover, and D. Klingman. "A Strongly Convergent Primal-Simplex Algorithm for Generalized Networks," *Mathematics of Operations Research*, **4** (1), pp. 39–59, 1979.

Elmaghraby, Salah E. "Sensitivity Analysis of Multiterminal Flow Networks," *Operations Research*, **12** (5), pp. 680–688, September–October 1964.

Elmaghraby, Salah E. "The Theory of Networks and Management Science. Part I," *Management Science*, **17** (1), pp. 1–34, September 1970.

Elmaghraby, Salah E. "The Theory of Networks and Management Science. Part II," *Management Science*, **17** (2), pp. B54–B71, October 1970.

Engquist, M. "A Successive Shortest Path Algorithm for the Assignment Problem," *INFOR*, **20** (4), pp. 370–384, 1982.

Eriksson, J. R. "An Iterative Primal-Dual Algorithm for Linear Programming," Report LITH-MAT-R-1985-10, Institute of Technology, Linkoping University, S-581 83 Linkoping, Sweden, 1985.

Evans, J. R. "A Combinatorial Equivalence Between a Class of Multicommodity Flow Problems and the Capacitated Transportation Problem," *Mathematical Programming*, **10** (3), pp. 401–404, 1976.

Even, S. *Graph Algorithms*, Computer Science Press, Maryland, 1979.

Everett, H. "Generalized Lagrange Multiplier Method for Solving Problems of Optimum Allocation of Resources," *Operations Research*, **11** pp. 399–417, 1963.

Falk, J. E. "Lagrange Multipliers and Nonlinear Programming," *Journal of Mathematical Analysis and Applications*, **19**, pp. 141–159, 1967.

Farbey, B. A., A. H. Lard, and J. D. Murchland. "The Cascade Algorithm for Finding All Shortest Distances in a Directed Graph," *Management Science*, **14** (1), pp. 19–28, September 1967.

Farkas, J. "Uber die Theorie der einfachen Ungleichungen," *Journal für die reine und angewandte Mathematik*, **124**, *pp.* 1–27, 1901.

Fathi, Y., and C. Tovey. "Affirmative Action Algorithms," *Mathematical Programming*, **34**, pp. 292–301, 1986.

Ferguson, Allen R., and G. B. Dantzig. Notes on Linear Programming: Part XVI–The Problem of Routing Aircraft–a Mathematical Solution, Research Memorandum, RM-1369, also RAND Paper P-561, 1954, also *Aeronautical Engineering Review*, **14** (4), pp. 51–55, April 1955.

Ferguson, Allen R., and G. B. Dantzig. "The Allocation of Aircraft to Routes—An Example of Linear Programming Under Uncertain Demand," *Management Science*, **3** (1), pp. 45–73, October 1956.

Fisher, F. P. "Speed Up the Solution to Linear Programming Problems," *Journal of Industrial Engineering*, **12** (6), pp. 412–416, November–December 1961.

Fletcher, R., and M. J. D. Powell. "On the Modification of LDL^T Factorizations," *Math. Comp.*, pp. 1067–1087, 1974.

Flood, M. M. "On the Hitchcock Distribution Problem," *Pacific Journal of Mathematics*, **3** (2), 1953.

Flood, M. M. "Application of Transportation Theory to Scheduling a Military Tanker Fleet," *Operations Research* **2** (2), 1954.

Flood, M. M. "An Alternative Proof of a Theorem of Konig as an Algorithm for the Hitchcock Distribution Problem," in R. Bellman and M. Hall (eds.), *Proceedings of Symposia in Applied Mathematics*, American Mathematical Society, Providence, R.I., pp. 299–307, 1960.

Flood, M. M. "A Transportation Algorithm and Code," *Naval Research Logistics Quarterly*, **8** (3), September 1961.

Florian, M., and P. Robert. "A Direct Search Method to Locate Negative Cycles in a Graph," *Management Science*, **17** (5), pp. 307–310, January 1971.

Floyd, R. W. "Algorithm 97: Shortest Path," *Communications of the ACM*, **5** (6), p. 345, 1962.

Ford, L. R., and D. R. Fulkerson. "Maximal Flow Through a Network," *Canadian Journal of Mathematics*, **8** (3), pp. 399–404, 1956.

Ford, L. R., and D. R. Fulkerson. "A Simple Algorithm for Finding Maximal Network Flows and an Application to the Hitchcock Problem," *Canadian Journal of Mathematics*, **9** (2), pp. 210–218, 1957.

Ford, L. R., and D. R. Fulkerson. "Solving the Transportation Problem," *Management Science*, **3** (1), 1956.

Ford, L. R., and D. R. Fulkerson. "A Primal-Dual Algorithm for the Capacitated Hitchcock Problem," *Naval Research Logistics Quarterly*, **4** (1), 1957.

Ford, L. R., and D. R. Fulkerson. "Constructing Maximal Dynamic Flows from Static Flows," *Operations Research*, **6** (3), 1958.

Ford, L. R., and D. R. Fulkerson. "Suggested Computation of Maximal Multi-Commodity Network Flows," *Management Science*, **5** (1), pp. 97–101, October 1958.

Ford, L. R., and D. R. Fulkerson. *Flows in Networks*, Princeton University Press, Princeton, N.J., 1962.

Forrest, J. J. H., and J. A. Tomlin. "Updating Triangular Factors of the Basis to Maintain Sparsity in the Product Form Simplex Method," *Mathematical Programming*, **2**, pp. 263–278, 1972.

Fox, Bennett. "Finding Minimal Cost-Time Ratio Circuits," *Operations Research*, **17** (3), pp. 546–551, May–June 1969.

Frank, H., and I. T. Frisch. *Communication, Transmission, and Transportation Networks*, Addison-Wesley, Reading, Mass., 1971.

Francis, R. L., and J. A. White. *Facility Layout and Location*, Prentice-Hall, Englewood Cliffs, N.J., 1974.

Frazer, Ronald J. *Applied Linear Programming*, Prentice-Hall, Englewood Cliffs, N.J., 1968.

Fredman, M. L. and R. E. Tarjan, ''Fibonacci Heaps and their Uses in Improved Network Optimization Algorithms,'' *Proceedings of the 25th IEEE Symposium on the Foundations of Computer Science*, 1984.

Frendewey, J., and F. Glover. "Improved Algorithmic and Computational Procedures for Large Scale Embedded Network Linear Programming Problems," ORSA/TIMS Meeting, Houston, October 1981.

Freund, R. M. "Postoptimal Analysis of a Linear Program under Simultaneous Changes in Matrix Coefficients," *Mathematical Programming Study 24*, ed. R. W. Cottle, pp. 1–13, 1985.

Frisch, K. R. "The Logarithmic Potential Method of Convex Programming," Memorandum of May 13, 1955, University Institute of Economics, Oslo, 1955.

Frisch, R. The Multiplex Method for Linear Programming, Memorandum of Social Economic Institute, University of Oslo, Oslo, Norway, 1955.

Fulkerson, D. R. "A Network Flow Feasibility Theorem and Combinatorial Applications," *Canadian Journal of Mathematics*, **11** (3), 1959.

Fulkerson, D. R. "Increasing the Capacity of a Network, the Parametric Budget Problem," *Management Science*, **5** (4), pp. 472–483, July 1959.

Fulkerson, D. R. On the Equivalence of the Capacity-Constrained Transshipment Problem and the Hitchcock Problem, Research Memorandum RM-2480, The Rand Corporation, Santa Monica, Calif., 1960.

Fulkerson, D. R. "An Out-of-Kilter Method for Minimal Cost Flow Problems," *Journal of the Society for Industrial and Applied Mathematics*, **9** (1), 1961.

Fulkerson, D. R. "A Network Flow Computation for Project Cost Curves," *Management Science*, **7** (2), pp. 167–178. January 1961.

Fulkerson, D. R. "Flow Networks and Combinatorial Operations Research," *The American Mathematical Monthly*, **73** (2), February 1966.

Fulkerson, D. R. "Networks, Frames, and Blocking Systems," *Mathematics of the Decision Sciences, Part* 1, Lectures in Applied Mathematics, **11** (eds. G. B. Dantzig and A. F. Veinott), American Mathematical Society, pp. 304–334, 1968.

Fulkerson, D. R., and G. B. Dantzig. "Computations of Maximal Flows in Networks," *Naval Research Logistic Quarterly*, **2** (4), pp. 277–283, December 1955.

Gacs, P., and L. Lovasz. "Khacian's Algorithm for Linear Programming," *Mathematical Programming Study 14*, North-Holland, Amsterdam, The Netherlands, pp. 61–68, January 1981.

Gal, T. "Shadow Prices and Sensitivity Analysis Under Degeneracy," *OR Spektrum*, **8**, pp. 59–71, 1986.

Gale, D. "Neighboring Vertices on a Convex Polyhedron," in H. W. Kuhn and A. W. Tucker (eds.), *Linear Inequalities and Related Systems*, Annals of Mathematics Study No. 38, Princeton University Press, Princeton, N.J., pp. 255–263, 1956.

Gale, D. "A Theorem on Flows in Networks," *Pacific Journal of Mathematics*, **7**, pp. 1073–1082, 1957.

Gale, D. Transient Flows in Networks, Research Memorandum RM-2158, The Rand Corporation, Santa Monica, Calif., 1958.

Gale, D. *The Theory of Linear Economic Models*, McGraw-Hill Book Co., New York, 1960.

Gale, D. On the Number of Faces of a Convex Polytope, Technical Report No. 1, Department of Mathematics, Brown University, 1962.

Gale, D., H. W. Kuhn, and A. W. Tucker. "Linear Programming and the Theory of Games," Chapter 19 of T. C. Koopmans (ed.), *Activity Analysis of Production and Allocation*, Cowles Commission Monograph 13, John Wiley & Sons, New York, 1951.

Garey, M. R., and D. S. Johnson. *Computers and Intractability: A Guide to the Theory of NP-Completeness*, Freeman, San Francisco, 1979.

Garfinkel, R. S., and G. L. Nemhauser. *Integer Programming*, John Wiley & Sons, New York, 1972.

Garvin, W. W., H. W. Crandall, J. B. John, and R. A. Spellman. "Application of Linear Programming in the Oil Industry," *Management Science*, **3** (4), pp. 407–430, July 1957.

Garvin, W. W. *Introduction to Linear Programming*, McGraw-Hill Book Co., New York, 1960.

Gass, S. I. "A First Feasible Solution to the Linear Programming Problem," in H. Antosiewicz (ed.), *Proceedings of the Second Symposium in Linear Programming*, vols. 1 and 2, DCS/Comptroller, Headquarters, U.S. Air Force, Washington, D.C., pp. 495–508, January, 1955.

Gass, S. I. *Linear Programming: Methods and Applications*, 4th ed., McGraw-Hill, New York, 1975.

Gass, S. I. The Dualplex Method for Large-scale Linear Programs, ORC Report 66-15, Operations Research Center, University of California, Berkeley, Calif., June 1966.

Gass, S. I., and T. Saaty. "The Computational Algorithm for the Parametric Objective Function," *Naval Research Logistics Quarterly*, **2** (1–2), 1955.

Gass, S. I., and T. L. Saaty. "Parametric Objective Function. Part II: Generalization," *Operations Research*, **3**, 1955.

Gassner, Betty J. "Cycling in the Transportation Problem," *Naval Research Logistics Quarterly*, **11** (1), March 1964.

Gay, D. M. "Electronic Mail Distribution of Linear Programming Test Problems," *COAL Newsletter*, **13**, pp. 10–12, 1985.

Gay, D. M. "A Variant of Karmarkar's Linear Programming Algorithms for Problems in Standard Form," *Mathematical Programming*, **37**, pp. 81–90, 1987.

Geary, R. C., and M. C. McCarthy. *Elements of Linear Programming, with Economic Applications*, Charles Griffin & Co., London, 1964.

Geoffrion, A. M. "Primal Resource-Directive Approaches for Optimizing Nonlinear Decomposable Systems," *Operations Research*, **18** (3), pp. 375–403, May–June 1970.

Geoffrion, A. M. "Duality in Nonlinear Programming: A Simplified Applications-Oriented Development," *SIAM Review*, **13**, pp. 1–37, 1971.

Geoffrion, A. M. "Generalized Benders Decomposition," *Journal of Optimization Theory and Applications*, **10**, pp. 237–260, 1972.

Geoffrion, A. M. "The Purpose of Mathematical Programming is Insight, Not Numbers," *Interfaces*, **7** (1), pp. 81–92, 1976.

Geoffrion, A. M. "An Introduction to Structural Modelling," *Management Science*, **33** (5), pp. 547–588, 1987.

George, J. A., and M. T. Heath. "Solution of Sparse Linear Least Squares Problems Using Givens Rotations," *Linear Algebra and Its Applications*, **34**, pp. 69–83, 1980.

Gibby, D., F. Glover, D. Klingman, and M. Mead. "A Comparison of Pivot Selection Rules for Primal Simplex Based Network Codes," *Operations Research Letters*, **2** (5), pp. 199–202, 1983.

Gill, P. E., G. H. Golub, W. Murray, and M. A. Saunders. "Methods for Modifying Matrix Factorizations," *Mathematics of Computation*, **28**, pp. 505–535, 1974.

Gill, P. E., and W. Murray. "A Numerical Investigation of Ellipsoid Algorithms for Large-Scale Linear Programming," Technical Report SOL 80-27, Systems Optimization Laboratory, Stanford University, Stanford, Calif., October 1980.

Gill, P. E., W. Murray, M. A. Saunders, J. A. Tomlin, and M. H. Wright. "On Projected Newton Barrier Methods for Linear Programming and an Equivalence to Karmarkar's Projective Method," *Mathematical Programming*, **36**, pp. 183–209, 1986.

Gill, P. E., W. Murray, M. A. Saunders, and M. H. Wright. "A Note on Nonlinear Approaches to Linear Programming," Technical Report SOL 86-7, Department of Operations Research, Stanford University, Stanford, CA 94305, 1986.

Gill, P. E., W. Murray, M. A. Saunders, and M. H. Wright. "Practical Anti-Cycling Procedure for Linear and Nonlinear Programming," Technical Report SOL 88-4, Systems Optimization Laboratory, Department of Operations Research, Stanford University, Stanford, CA 94305, July 1988.

Gilmore, P. C., and R. E. Gomory. "A Linear Programming Approach to the Cutting Stock Problem—Part 1," *Operations Research*, **9**, pp. 849–859, 1961.

Gilmore, P. C., and R. E. Gomory. "A Linear Programming Approach to the Cutting Stock Problem—Part 2," *Operations Research*, **11** (6), pp. 863–887, 1963.

Glassey, C. Roger. "Nested Decomposition and Multi-Stage Linear Programs," *Management Science*, **20** (3), pp. 282–292, November 1973.

Glickman, T., and H. D. Sherali. "Large-Scale Network Distribution of Pooled Empty Freight Cars Over Time, With Limited Substitution and Equitable Benefits," *Transportation Research, B (Methodology)*, **19** (2), pp. 85–94, 1985.

Glicksman, M. A. *Linear Programming and Theory of Games*, John Wiley & Sons, New York, 1963.

Glover, F. "A New Foundation for a Simplified Primal Integer Programming Algorithm," *Operations Research*, **16** (4), pp. 727–740, July–August, 1968.

Glover, F., R. Glover, and D. Klingman. "Threshold Assignment Algorithm," *Mathematical Programming Study 26*, pp. 12–37, 1986.

Glover, F., D. Karney, and D. Klingman. "The Augmented Predecessor Index Method for Locating Stepping-Stone Paths and Assigning Dual Prices in Distributions Problems," *Transportation Science*, **6**, pp. 171–180, 1972.

Glover, F., D. Karney, D. Klingman, and A. Napier. "A Computational Study on Start Procedures, Basis Change Criteria, and Solution Algorithms for Transportation Problems," *Management Science*, **20** (5), pp. 793–813, 1974.

Glover, F., D. Karney, D. Klingman, and R. Russell. "Solving Singly Constrained Transshipment Problems," *Transportation Science*, **12** (4), pp. 277–297, 1978.

Glover, F., and D. Klingman. "Comments on Note by Hatch on Network Algorithms," *Operations Research*, **29** (2), pp. 370–373, 1978.

Glover, F., and D. Klingman. "The Simplex SON Algorithm for LP/Embedded Network Problems," *Mathematical Programming Study*, **15**, pp. 148–176, 1981.

Glover, F., and D. Klingman. "Recent Developments in Computer Implementation Technology for Network Flow Algorithms," *INFOR*, **20** (4), 1982.

Glover, F., D. Klingman, M. Mead, and J. Mote. "A Note on Specialized Versus Unspecialized Methods for Maximum-Flow Problems," *Naval Research Logistics Quarterly*, **31**, pp. 63–65, 1984b.

Glover, F., D. Klingman, J. Mote, and D. Whitman. "A Primal Simplex Variant for the Maximum-Flow Problem," *Naval Research Logistics Quarterly*, **31**, pp. 41–61, 1984a.

Glover, F., D. Klingman, and N. Phillips. "A New Polynomially Bounded Shortest Path Algorithm," *Operations Research*, **33** (1), pp. 65–73, 1985.

Glover, F., D. D. Klingman, N. V. Phillips, and R. F. Schneider. "New Polynomial Shortest Path Algorithms and Their Computational Attributes," *Management Science*, **31** (9), pp. 1106–1128, 1985.

Glover, F., D. Klingman, and J. Stutz. "Augmented Threaded Index Method for Network Optimization," *INFOR*, **12** (3), pp. 293–298, 1974.

Glover, F., D. Klingman, and J. Stutz. "Extensions of the Augmented Predecessor Index Method to Generalized Network Problems," *Transportation Science*, **7** (4), pp. 377–384, 1974.

Glover, F., and D. Klingman. "Basic Dual Feasible Solutions for a Class of Generalized Networks," *Operations Research*, **20**, pp. 126–136, 1972.

Glover, F., D. Klingman, and R. S. Barr. An Improved Version of the Out-of-Kilter Method and a Comparative Study of Computer Codes, Report CS-102, Center for Cybernetic Studies, University of Texas, Austin, Texas, 1972.

Glover, F., D. Karney, and D. Klingman. "Implementation and Computational Comparisons of Primal, Dual and Primal-Dual Computer Codes for Minimum Cost Network Flow Problems," *Networks*, **4** (3), pp. 191–212, 1974.

Golden, B. L., and T. L. Magnanti. "Deterministic Network Optimization: A Bibliography," *Networks*, **7** (2), pp. 149–183, 1977.

Goldfarb, D. "Efficient Dual Simplex Algorithms for the Assignment Problem," *Mathematical Programming*, **33**, pp. 187–203, 1985.

Goldfarb, D., and S. Mehrotra. "A Relaxed Version of Karmarkar's Method," *Mathematical Programming*, **40** (3), pp. 289–316, 1988.

Goldfarb, D., and S. Mehrotra. "A Relaxed Variant of Karmarkar's Algorithm for Linear Programs with Unknown Optimal Objective Value," Report, Department of Industrial Engineering and Operations Research, Columbia University, New York, NY 10027, 1986.

Goldfarb, D., and J. K. Reid. "A Practicable Steepest-Edge Simplex Algorithm," *Mathematical Programming*, **12**, pp. 361–371, 1977.

Goldfarb, D., and W. Y. Sit. "Worst Case Behavior of Steepest Edge Simplex Method," *Discrete Applied Mathematics*, **1**, pp. 277–285, 1979.

Goldman, A. J., and D. Kleinman. "Examples Relating to the Simplex Method," *Operations Research*, **12** (1), pp. 159–161, 1964.

Goldman, A. J. "Resolution and Separation Theorems for Polyhedral Convex Sets," in H. W. Kuhn and A. W. Tucker (eds.), *Linear Inequalities and Related Systems*, Annals of Mathematics Study No. 38, Princeton University Press, Princeton, N.J., pp. 41–51, 1956.

Goldman, A. J., and A. W. Tucker. "Theory of Linear Programming," in H. W. Kuhn and A. W. Tucker (eds.), *Linear Inequalities and Related Systems*, Annals of Mathematics Study No. 38, Princeton University Press, Princeton, N.J., pp. 53–98, 1956.

Goldman, A. J., and A. W. Tucker. "Polyhedral Convex Cones," in H. W. Kuhn and A. W. Tucker (eds.), *Linear Inequalities and Related Systems*, Annals of Mathematics Study No. 38, Princeton University Press, Princeton, N.J., pp. 19–39, 1956.

Golomski, William A. "Linear Programming in Food Blending," Annual Convention Transactions, 17th Annual Convention, American Society for Quality Control, pp. 147–152, 1963.

Golub, G. H., and C. F. Van Loan. *Matrix Computations*, John Hopkins University Press, 1983.

Gomory, R. E. An Algorithm for the Mixed Integer Problem, Research Memorandum RM-2597, The Rand Corporation, Santa Monica, Calif., 1960.

Gomory, R. E. "An Algorithm for Integer Solutions to Linear Programs," in R. L. Graves and P. Wolfe (eds.), *Recent Advances in Mathematical Programming*, McGraw-Hill Book Co., New York, pp. 269–302, 1963.

Gomory, R. E. "All-Integer Integer Programming Algorithm," in J. F. Muth and G. L. Thompson (eds.), *Industrial Scheduling*, Prentice-Hall, Englewood Cliffs, N.J., pp. 193–206, 1963.

Gomory, R. E., and W. J. Baumol. "Integer Programming and Pricing," *Econometrica*, **28** (3), pp. 521–550, 1960.

Gomory, R. E., and T. C. Hu. "Multi-Terminal Network Flows," *SIAM*, **9** (4), 1961.

Gomory, R. E., and T. C. Hu. "An Application of Generalized Linear Programming to Network Flows," *SIAM*, **10** (2), 1962.

Gomory, R. E., and T. C. Hu. "Synthesis of a Communication Network," *SIAM* **12** (2), June 1964.

Goncalves, A. S. "Basic Feasible Solutions and the Dantzig–Wolfe Decomposition Algorithm," *Operations Research Quarterly*, **19** (4), pp. 465–469, December 1968.

Gonzaga, C. C. "A Conical Projection Algorithm for Linear Programming," Memo No. UCB/ERL M85/61, Electronics Research Laboratory, College of Engineering, University of California, Berkeley, CA 94720, 1985.

Gonzaga, C. "An Algorithm for Solving Linear Programming Problems in O(n^3L) Operations," Memo No. UCB/ERL M87/10, Electronics Research Laboratory, College of Engineering, University of California, Berkeley, CA 94720, 1987a.

Gonzaga, C. C. "Search Directions for Interior Linear Programming Methods," Memo #UCB/ERL M87/44, Electronics Research Laboratory, College of Engineering, University of California, Berkeley, CA 94720, 1987b.

Gonzaga, C. D. "Polynomial Affine Algorithms for Linear Programming," Department of Systems Engineering and Computer Science, COPPE-Federal University of Rio de Janeiro, Caixa Postal 6811, 21941 Rio de Janeiro, RJ, Brazil, ES-139/88, 1988.

Gorham, William. "An Application of a Network Flow Model to Personnel Planning," *IEEE Transactions on Engineering Management*, **10** (3), pp. 121–123, September 1963.

Gould, F. J. "Proximate Linear Programming: A Variable Extreme Point Method," *Mathematical Programming*, **3** (3), pp. 326–338, December 1972.

Granot, F., and D. Klingman. "Editorial: Some Recent Advances in the Theory, Computation, and Applications of Network Flow Models," *INFOR*, **20** (4), 1982.

Graves, R. L., and P. Wolfe (eds.). *Recent Advances in Mathematical Programming*, McGraw-Hill Book Co., New York, 1963.

Greenberg, H. J. "An Analysis of Degeneracy," *Naval Research Logistics Quarterly*, **33**, pp. 635–656, 1986.

Greenberg, H. J. "A Functional Description of ANALYZE: A Computer-Assisted Analysis System for Linear Programming Models," *ACM Transactions on Mathematical Software*, **9** (1), pp. 18–56, 1983.

Greenberg, Harold. "Modification of the Primal-Dual Algorithm for Degenerate Problems," *Operations Research*, **16** (6), pp. 1227–1230, November–December 1968.

Grigoriadis, M. D. "A Dual Generalized Upper Bounding Technique," *Management Science*, **17** (5), pp. 269–284, January 1971.

Grigoriadis, M. D., and W. W. White. "A Partitioning Algorithm for the Multicommodity Network Flow Problem," *Mathematical Programming*, **3**, pp. 157–177, 1972.

Grigoriadis, M. D., and W. W. White. "Computational Experience with a Multicommodity Network Flow Algorithm," in R. Cottle and J. Krarup (eds.), *Optimization Methods for Resource Allocation*, English University Press, 1972.

Grinold, R. C. "A Multicommodity Max-Flow Algorithm," *Operations Research*, **16**, pp. 1234–1238, 1968.

Grinold, R. C. "A Note on Multicommodity Max-Flow Algorithm," *Operations Research*, **17**, p. 755, 1969.

Grinold, R. C. "Calculating Maximal Flows in a Network with Positive Gains," working paper WP CP-337, Center for Research in Management Science, University of California, Berkeley, Calif., 1971.

Grinold, R. C. "Steepest Ascent for Large Scale Linear Programs," *SIAM Review*, **14**, pp. 447–464, 1972.

Gross, O. "The Bottleneck Assignment Problem," Paper P-1630, The Rand Corporation, Santa Monica, Calif., March 1959.

Gross, Oliver. "A Linear Program of Prager's. Notes on Linear Programming and Extensions," Part 60, Research Memorandum RM-2993-PR, The Rand Corporation, Santa Monica, Calif., April 1962.

Grotschel, M., L. Lovasz, and A. Schrijver. "The Ellipsoid Method and its Consequences in Combinatorial Optimization," *Combinatorica*, pp. 169–197, 1981.

Grotschel, M., L. Lovasz, and A. Schrijver. *The Ellipsoid Method and Combinatorial Optimization*, Springer, Berlin, 1986.

Gunther, Paul. "Use of Linear Programming in Capital Budgeting," *Operations Research*, **3** (2), pp. 219–224, May 1955.

Gutnik, L. A. "On the Problem of Cycling in Linear Programming," *Dokl. AN SSR* (USSR), **170** (1), pp. 53–56, 1966.

Hadley, G. *Linear Algebra*, Addison-Wesley, Reading, Mass., 1961.

Hadley, G. *Linear Programming*, Addison-Wesley, Reading, Mass., 1962.

Hadley, G., and M. A. Simonnard. A Simplified Two-Phase Technique for the Simplex Method," *Naval Research Logistics Quarterly*, **6** (3), 1959.

Hagelschuer, Paul B. *Theory of Linear Decomposition*, Springer-Verlag, New York, 1971.

Haimovich, M. "The Simplex Method is Very Good!—On the Expected Number of Pivot Steps and Related Properties of Random Linear Programs," Columbia University Press, New York, NY, 1983.

Hakimi, S. L. On Simultaneous Flows in a Communication Network, Document No. AD-267 090, Office of Technical Services, 1961.

Haley, K. B. "The Existence of a Solution to the Multi-Index Problem" *Operational Research Quarterly*, **16** (4), pp. 471–474, December, 1965.

Halmos, P. R., and H. E. Vaughan. "The Marriage Problem," *American Journal of Mathematics*, **72**, (1), pp. 214–215, January 1950.

Harris, Milton Y. "A Mutual Primal-Dual Linear Programming Algorithm," *Naval Research Logistics Quarterly*, **17** (2), pp. 199–206, June 1970.

Harris, P. M. J. "Pivot Selection Methods of the Devex LP Code," *Mathematical Programming Study*, **4**, pp. 30–57, 1975. (Also, see *Mathematical Programming*, 5, pp. 1–28, 1973.)

Hartman, J. K., and L. S. Lasdon. "A Generalized Upper Bounding Algorithm for Multicommodity Network Flow Problems," *Networks* **1**, pp. 333–354, 1972.

Hatch, R. S. "Bench Marks Comparing Transportation Codes Based on Primal-Simplex and Primal-Dual Algorithms," *Operations Research*, **23** (6), pp. 1167–1171, 1975.

Haverly, C. A. "Results of a New Series of Case Runs Using the Karmarkar Algorithm," Haverly Systems, Inc., Denville, New Jersey 07834, 1985a.

Haverly, C. A. "Number of Simplex Iterations for Four Model Structures," Haverly Systems, Inc., Denville, New Jersey 07834, 1985b.

Haverly, C. A. "Studies on Behavior of the Karmarkar Method," Haverly Systems, Inc., Denville, New Jersey 07834, 1985c.

Heady, E. O., and W. Candler. *Linear Programming Methods*, Iowa State College Press, Ames, Iowa, 1958.

Heath, M. "Some Extensions of an Algorithm for Sparse Linear Least Squares Problems," *SIAM Journal on Scientific and Statistical Computing*, **3**, pp. 233–237, 1982.

Held, M., P. Wolfe, and H. D. Crowder. "Validation of Subgradient Optimization," *Mathematical Programming*, **6**, pp. 62–68, 1974.

Helgason, R. V., J. L. Kennington, and H. A. Zaki. "A Parallelization of the Simplex Method," Department of Operations Research, Southern Methodist University, Dallas, Texas 75275, 1987.

Heller, I. "Constraint Matrices of Transportation-Type Problems," *Naval Research Logistics Quarterly*, **4** pp. 73–78, 1957.

Heller, I. "On Linear Programs Equivalent to the Transportation Problem," *SIAM*, **12**, (1), pp. 31–42, March 1964.

Heller, I., and C. B. Tompkins. "An Extension of a Theorem of Dantzig's," in H. W. Kuhn and A. W. Tucker (eds.), *Linear Inequalities and Related Systems*, Annals of Mathematics Study No. 38, Princeton University Press, Princeton, N.J., pp. 247–254, 1956.

Hillier, F. S., and G. J. Lieberman. *Introduction to Operations Research*, 4th ed., Holden-Day, Inc., San Francisco, 1986.

Himmelblau, David M. *Decomposition of Large Scale Problems*, North-Holland, Amsterdam, 1973.

Hitchcock, F. L. "Distribution of a Product from Several Sources to Numerous Localities," *Journal of Mathematical Physics*, **20**, pp. 224–230, 1941.

Ho, J. K., and E. Loute. "An Advanced Implementation of the Dantzig–Wolfe Decomposition Algorithm for Linear Programming," *Mathematical Programming*, **20**, pp. 303–326, 1981.

Ho, J. K., and E. Loute. "Computational Experience with Advanced Implementation of Decomposition Algorithms for Linear Programming," *Mathematical Programming*, **27** (3), pp. 283–290, 1983.

Hoffman, A., N. Mannos, D. Sokolowsky, and N. Wiegmann. "Computational Experience in Solving Linear Programs," *Journal of the Society for Industrial and Applied Mathematics*, **1** (1), 1953.

Hoffman, A. J. "How to Solve a Linear Program," in H. Antosiewicz (ed.), *Proceedings of the Second Symposium in Linear Programming*, vols. 1 and 2, DCS/Comptroller, Headquarters U.S. Air Force, Washington, D.C., pp. 397–424, January, 1955.

Hoffman, A. J. Cycling in the Simplex Algorithm, Report No. 2974, National Bureau of Standards, Washington, D.C., 1953.

Hoffman, A. J., and J. B. Kruskal. "Integral Boundary Points of Convex Polyhedra," in H. W. Kuhn and A. W. Tucker (eds.), *Linear Inequalities and Related Systems*, Annals of Mathematics Study No. 38, Princeton University Press, Princeton, N.J., pp. 233–246, 1956.

Hu, T. C. "The Maximum Capacity Route Problem," *Operations Research*, **9** (6), pp. 898–900, November–December 1961.

Hu, T. C. "Multi-Commodity Network Flows," *Operations Research*, **11** (3), pp. 344–360, May–June 1963.

Hu, T. C. "On the Feasibility of Multicommodity Flows in a Network," *Operations Research*, **12**, pp. 359–360, 1964.

Hu, T. C. Multi-Terminal Shortest Paths, U.S.G.R.& D.R. Order AD-618 757 from Clearinghouse, Operations Research Center, University of California, Berkeley, Calif., April 1965.

Hu, T. C. "Minimum Convex Cost Flows," *Naval Research Logistics Quarterly*, **13** (1), pp. 1–9, March 1966.

Hu, T. C. "Revised Matrix Algorithms for Shortest Paths in a Network," *SIAM*, **15**, (1), pp. 207–218, January 1967.

Hu, T. C. "Laplace Equation and Network Flows," *Operations Research*, **15** (2), pp. 348–356, April 1967.

Hu, T. C. "Decomposition Algorithm for Shortest Paths in a Network," *Operations Research*, **16** (1), pp. 91–102, January–February 1968.

Hu, T. C. *Integer Programming and Network Flows*, Addison-Wesley, Reading, Mass., 1969.

Hung, M. S., and J. J. Divoky. "A Computational Study of Efficient Shortest Path Algorithms," *Computers and Operations Research*, **15** (6), pp. 567–576, 1988.

Hung, M. S., and W. O. Rom. "Solving the Assignment Problem by Relaxation," *Operations Research*, **28** (4), pp. 969–982, 1980.

Hung, M. S. "A Polynomial Simplex Method for the Assignment Problem," *Operations Research*, **31**, pp. 595–600, 1983.

Hwang, C. L., and A. S. Masud. *Multiple Objective Decision Making Methods and Applications: A State of the Art Survey*, Springer, New York, 1979.

Iri, M., and H. Imai. "A Multiplicative Barrier Function Method for Linear Programming," *Algorithmica*, **1** (4), pp. 455–482, 1986.

Iri, M. "A New Method of Solving Transportation Network Problems," *Journal of the Operations Research Society of Japan*, **3**, (1–2), pp. 27–87, October 1960.

Iri, M. "An Extension of the Maximum-Flow Minimum-Cut Theorem to Multicommodity Networks," *Journal of the Operations Research Society of Japan*, **13**, 1971.

Jacobs, W. W. "The Caterer Problem," *Naval Research Logistics Quarterly*, **1** (2), 1954.

Jarvis, J. J. "On the Equivalence Between the Node-Arc and Arc-Chain Formulations for the Multicommodity Maximal Flow Problem," *Naval Research Logistics Quarterly*, **16**, pp. 515–529, 1969.

Jarvis, J. J., and A. M. Jezior. "Maximal Flow with Gains Through a Special Network," *Operations Research*, **20**, pp. 678–688, 1972.

Jarvis, J. J., and P. D. Keith. Multicommodity Flows with Upper and Lower Bounds, Unpublished paper, School of Industrial and Systems Engineering, Georgia Institute of Technology, Atlanta, Ga., 1974.

Jarvis, J. J., and J. B. Tindall. "Minimum Disconnecting Sets in Directed Multi-Commodity Networks," *Naval Research Logistics Quarterly*, **19**, pp. 681–690, 1972.

Jarvis, J. J., and S. Tufekci. "A Decomposition Algorithm for Locating a Shortest Path Between Two Nodes in a Network," *Networks*, **12**, pp. 161–172, 1981.

Jensen, P. A., and J. W. Barnes. *Network Flow Programming*, Wiley, New York, 1980.

Jeroslow, R. G. "The Simplex Algorithm with the Pivot Rule of Maximizing Criterion Improvement," *Discrete Mathematics*, **4**, pp. 367–378, 1973.

Jewell, W. S. Optimal Flow Through Networks, Interim Technical Report No. 8, M.I.T. Project, Fundamental Investigations in Methods of Operations Research, 1958.

Jewell, W. S. "Optimal Flow Through Networks with Gains," *Operations Research*, **10** (4), 1962.

Jewell, W. S. "A Primal-Dual Multicommodity Flow Algorithm," ORC Report 66-24, University of California, Berkeley, Calif., 1966.

Jewell, W. S. "Multi-Commodity Network Solutions," Research Report ORC 66-23, Mathematical Science Division, Operations Research Center, University of California, Berkeley, Calif., September 1966.

Jewell, W. S. "A Complex, Complementary Slackness, Out-of-Kilter Algorithm for Linear Programming," ORC 67-6, University of California, Berkeley, Calif., 1967.

John, F. "Extremum Problems with Inequalities as Subsidiary Conditions," in *Studies and Essays*, Wiley Interscience, New York, pp. 187–204, 1948.

Johnson, E. L. "Programming in Networks and Graphs," Research Report ORC 65-1, University of California, Berkeley, Calif., 1965.

Johnson, E. L. "Networks and Basic Solutions," *Operations Research*, **14**, pp. 619–623, 1966.

Johnson, E. L., M. M. Kostreva, and U. H. Suhl. "Solving 0–1 Integer Programming Problems Arising from Large-Scale Planning Models," *Operations Research*, **33** (4), pp. 803–819, 1985.

Jones, W. G., and C. M. Rope. "Linear Programming Applied to Production Planning," *Operational Research Quarterly*, **15** (4), December 1964.

Jonker, R., and T. Volgenant. "Improving the Hungarian Assignment Algorithm," *Operations Research Letters*, **5** (4), pp. 171–175, 1986.

Kahle, Robert V. "Application of Linear Programming for Industrial Planning," *Proceedings of the American Institute of Industrial Engineers*, 1962.

Kalaba, R. E., and M. L. Juncosa. "Optimal Design and Utilization of Communication Networks," *Management Science*, **3** (1), pp. 33–44, 1956.

Kantorovich, L. *Mathematical Methods in the Organization and Planning of Production*, Publication House of the Leningrad State University, 1939. Translated in *Management Science*, **6**, pp. 366–422, 1958.

Kantorovich, L. "On the Translocation of Masses," *Compt. Rend. Academy of Sciences, U.R.S.S.*, **37**, pp. 199–201, 1942. Translated in *Management Science*, **5** (1), 1958.

Kantorovich, L. V., and M. K. Gavurin. "The Application of Mathematical Methods to Problems of Freight Flow Analysis," *Akademii Nauk SSSR*, 1949.

Kapur, J. N. "Linear Programming in Textile Industry," *Journal of National Productivity Council* (India), **4** (2), pp. 296–302, April–June 1963.

Karlin, S. *Mathematical Methods and Theory in Games, Programming and Economics*, **1–2**, Addison-Wesley, Reading, Mass. 1959.

Karmarkar, N. "A New Polynomial-Time Algorithm for Linear Programming," *Combinatorica*, **4**, pp. 373–395, 1984.

Karmarkar, N. "Some Comments on the Significance of the New Polynomial-Time Algorithm for Linear Programming," AT & T Bell Laboratories, Murray Hill, New Jersey 07974, 1984.

Karmarkar, N. "A New Polynomial-Time Algorithm for Linear Programming," Presented at the ORSA/TIMS Meeting, Boston, November 1985.

Karmarkar, N. "Recent Developments in New Approaches to Linear Programming," Presented at the SIAM Conference on Optimization, Houston, May 1987.

Karp, R. M. "Reducibility Among Combinatorial Problems," in R. E. Miller and J. W. Thatcher (Eds.), *Complexity of Computer Computations*, Plenum, New York, 1972.

Karush, W. "Minima of Functions of Several Variables with Inequalities as Side Constraints," M.S. Thesis, Department of Mathematics, University of Chicago, 1939.

Karush, W. "Duality and Network Flow," Report TM-1042-201-00, System Development Corporation, Document no. AD-402 643, March 1963.

Kelley, J. E. "Parametric Programming and the Primal-Dual Algorithm," *Operations Research*, **7**, 1959.

Kelley, J. E. "The Cutting Plane Method for Solving Convex Programs," *SIAM*, **8** (4), pp. 703–712, December 1960.

Kelley, J. E. "Critical-Path Planning and Scheduling, Mathematical Basis," *Operations Research*, **9** (2), pp. 296–320, May 1961.

Kennington, J. F., and R. V. Helgason. *Algorithms for Network Programming*, Wiley-Interscience, New York, 1980.

Kennington, J. L. "Multicommodity Network Flows: A Survey," Technical Report CP74015, Department of Computer Science and Operations Research, Southern Methodist University, 1974.

Khachian, L. G. "A Polynomial Algorithm in Linear Programming," *Soviet Mathematics Doklady*, **20**, pp. 191–194, 1979.

Khachian, L. G. "Polynomial Algorithms in Linear Programming," *USSR Computational Mathematics and Mathematical Physics*, **20**, pp. 53–72, 1980.

Khachiyan, L. G. "Polynomial Algorithms in Linear Programming," *Zhurnal Vichislitel'noi Matematiki i Matematicheskoi Fiziki* (in Russian), **20** (1), pp. 51–68, 1980. See also Gacs, P., and L. Lovasz, "Khachian's Algorithm for Linear Programming," *Mathematical Programming Study 14*, North-Holland, Amsterdam, The Netherlands, pp. 61–68.

Klee, V. "A Class of Linear Programming Problems Requiring a Large Number of Iterations," *Numerische Mathematik*, **7**, pp. 313–321, 1965a.

Klee, V. "Paths on Polyhedra I," *Journal of the Society of Industrial and Applied Mathematics*, **13**, pp. 946–956, 1965b.

Klee, V., and G. J. Minty. "How Good is the Simplex Algorithm?" in O. Shisha (Ed.), *Inequalities III*, Academic, New York, pp. 159–175, 1972.

Klee, V., and D. W. Walkup. "The *d*-Step Conjecture for Polyhedra of Dimension $d < 6$," *Acta Mathematica*, **117**, pp. 53–78, 1967.

Klee, V. L. "A String Algorithm for Shortest Paths in a Directed Network," *Operations Research*, **12** (3), pp. 428–432, May–June 1964.

Klein, M. "A Primal Method for Minimal Cost Flows," *Management Science*, **14** (3), pp. 205–220, November 1967.

Kleitman, D. J. "An Algorithm for Certain Multi-Commodity Flow Problems," *Networks*, **1**, 75–90, 1971.

Klingman, D. "Finding Equivalent Network Formulations for Constrained Network Problems," *Management Science*, **23** (7), pp. 737–744, 1977.

Klingman, D., A. Napier, and J. Stutz. "NETGEN: A Program for Generating Large Scale Capacitated Assignment, Transportation, and Minimum Cost Flow Network Problems," *Management Science*, **20** (5), pp. 814–821, 1974.

Klingman, D. D., and R. F. Schneider. "Microcomputer-Based Algorithms for Large Scale Shortest Path Problems," *Discrete Applied Mathematics*, **13**, pp. 183–206, 1986.

Klingman, D., and R. Russell. "Solving Constrained Transportation Problems," *Operations Research*, **23**, pp. 91–106, 1975.

Knolmayer, G. "Computational Experiments in the Formulation of Linear Product-Mix and Non-Convex Production-Investment Models," *Computers and Operations Research*, **9** (3), pp. 207–219, 1982.

Kobayashi, Takashi. "On Maximal Flow Problem in a Transportation Network with a Bundle," *Journal of the Operations Research Society of Japan*, **10** (3–4), pp. 69–75, June 1968.

Koch, James V. "A Linear Programming Model of Resource Allocation in a University," *Decision Sciences*, **4** (4), pp. 494–504, October 1973.

Koenigsberg, E. "Some Industrial Applications of Linear Programming," *Operations Research Quarterly*, **12** (2), pp. 105–114, June 1961.

Kojima, M. "Determining Basic Variables of Optimal Solutions in Karmarkar's New LP Algorithm," *Algorithmica*, **1** (4), pp. 499–516, 1986.

Kondor, Y. "Linear Programming of Income Tax Rates," *Israel Journal of Technology* (Israel), **6** (5), pp. 341–354, November–December 1968.

Koopmans, T. C. "Optimum Utilization of the Transportation System," *Econometrica*, **17** (3–4) 1949.

Koopmans, T. C. (ed.). *Activity Analysis of Production and Allocation*, Cowles Commission Monograph 13, John Wiley & Sons, New York, 1951.

Koopmans, T. C., and S. Reiter. "A Model of Transportation," in T. C. Koopmans (ed.), *Activity Analysis of Production and Allocation*, John Wiley & Sons, New York, pp. 222–259, 1951.

Kortanek, K. O., and Z. Jishan. "New Purification Algorithms for Linear Programming," *Naval Research Logistics Quarterly*, **35**, pp. 571–583, 1988.

Kortanek. K. O., and M. Shi. "Convergence Results and Numerical Experiments on a Linear Programming Hybrid Algorithm," *European Journal of Operational Research*, **32** (1), pp. 47–61, 1987.

Kortanek, K. O., and H. M. Strojwas. "An Application of the Charnes–Kortanek–Raike Purification Algorithm to Extreme Points and Extreme Directions," Department of Mathematics, Carnegie-Mellon University, Pittsburgh, PA, 1984.

Kotiah, T. C. T., and D. I. Steinberg. "On the Possibility of Cycling with the Simplex Method," *Operations Research*, **26** (2), pp. 374–376, 1978.

Kozlov, A. and L. W. Black. "Berkeley Obtains New Results with the Karmarkar Algorithm," Progress Report in *SIAM News*, **3** (19), pp. 3 and 20, 1986.

Kruskal, J. B. "On the Shortest Spanning Subtree of a Graph and the Traveling Salesman Problem," *Proceedings of the American Mathematical Society*, **7**, pp. 48–50, 1956.

Kuhn, H. W. "The Hungarian Method for the Assignment Problem," *Naval Research Logistics Quarterly*, **2** (1–2), March–June 1955.

Kuhn, H. W., and R. E. Quandt. "An Experimental Study of the Simplex Method," *Proceedings of the Symposia in Applied Mathematics*, Vol. XV, American Mathematical Society, 1962.

Kuhn, H. W., and A. W. Tucker. "Nonlinear Programming," in J. Neyman (ed.). *Proceedings of the Second Berkeley Symposium on Mathematical Statistics and Probability*, University of California Press, Berkeley, Calif., pp. 481–492, 1950.

Kuhn, H. W. and A. W. Tucker (eds.). Linear Inequalities and Related Systems, *Annals of Mathematics Study* No. 38, Princeton University Press, Princeton, N.J., 1956.

Kunzi, H. P., and K. Kleibohm. "The Triplex Method," *Unternehmensforschung* (Germany), **12** (3), pp. 145–154, 1968.

Kunzi, H. P., and W. Krelle. *Nonlinear Programming*, Blaisdell, Waltham, Mass., 1966.

Lageman, J. J. "A Method for Solving the Transportation Problem," *Naval Research Logistics Quarterly*, **14** (1), March 1967.

Land, A. H., and A. G. Doig. "An Automatic Method for Solving Discrete Programming Problems," *Econometrica*, **28**, pp. 497–520, 1960.

Land, A. H., and S. W. Stairs. "The Extension of the Cascade Algorithm to Large Graphs," *Management Science*, **14** (1), pp. 29–33, September 1967.

Langley, R. W. "Continuous and Integer Generalized Flow Problems," Ph.D. Dissertation, Georgia Institute of Technology, 1973.

Langley, R. W., and J. L. Kennington. "The Transportation Problem: A Primal Approach," Presented to the 43rd ORSA National Meeting, May 1973.

Larionov, B. A. "Abridgement of the Number of Operations in the Solution of the Transportation Problem of Linear Programming," *Trudy Tashkentskogo Instituta Inzhenernogozheleznodorognogo transporta* (USSR), **29**, pp. 211–214, 1964.

Lasdon, L. S. *Optimization Theory for Large Systems*, Macmillan, New York, 1970.

Lasdon, L. S. "Duality and Decomposition in Mathematical Programming," *IEEE Transactions on Systems Science and Cybernetics*, **4** (2), pp. 86–100, 1968.

Lavallee, R. S. "The Application of Linear Programming to the Problem of Scheduling Traffic Signals," *Operations Research* **3** (4), 1955.

Lawler, E. G. *Combinatorial Optimization: Networks and Matroids*, Holt, Rinehart and Winston, New York, 1976.

Lawler, E. G. "Is Karmarkar's Algorithm for Real?," Presented at the EURO VII Congress on Operations Research, Bologna, Italy, 1985.

Lee, S. M. *Goal Programming for Decision Analysis*, Auerbach, Philadelphia, PA, 1972.

Lemke, C. E. "The Dual Method of Solving the Linear Programming Problem," *Naval Research Logistics Quarterly*, **1** (1), 1954.

Lemke, C. E. "The Constrained Gradient Method of Linear Programming," *SIAM* **9** (1), pp. 1–17, March 1961.

Lemke, C. E., and T. J. Powers. A Dual Decomposition Principle, Document No. AD-269 699, 1961.

Leontief, W. W. *The Structure of the American Economy*, 1919–1939, Oxford University Press, New York, 1951.

Liebling, Thomas M. "On the Number of Iterations of the Simplex Method," *Operations Research Verfahren* (Germany), **17**, pp. 248–264, 1973.

Liebling, T. M. "On the Number of Iterations of the Simplex Method," *Methods of Operations Research*, XVII, V Oberwolfach-Tagung uber Operations Research, **13** (19), pp. 248–264, August 1977.

Llewellyn, Robert W. *Linear Programming*, Holt, Rinehart and Winston, New York, 1964.

Lombaers, H. J. M. *Project Planning by Network Analysis*, North-Holland Publishing Co., Amsterdam, 1969.

Lourie, Janice R. "Topology and Computation of the Generalized Transportation Problem," *Management Science*, **11**, (1), September 1964.

Lovasz, L. "A New Linear Programming Algorithm—Better or Worse than the Simplex Method?" *The Mathematical Intelligencer* **2**, 3, pp. 141–146, 1980.

Luenberger, D. G. *Introduction to Linear and Non-Linear Programming*, Second Edition, Addison-Wesley, Reading, Mass., 1984.

Lustig, I. J. "A Practical Approach to Karmarkar's Algorithm," Technical Report Sol 85-5, Department of Operations Research, Stanford University, Stanford, CA 94305, 1985.

Maier, S. F. Maximal Flows Using Spaning Trees, Report 71-14, Operations Research House, Stanford University, Stanford, Calif., 1971.

Malek-Zavarei, M., and J. K. Aggarwal. "Optimal Flow in Networks with Gains and Costs," *Networks*, **1** (4), pp. 355–365, 1972.

Mandl, C. *Applied Network Optimization*, Academic Press, New York, 1979.

Mangasarian, O. L. *Non-Linear Programming*, McGraw-Hill Book Co., New York, 1969.

Manne, A. S. Notes on Parametric Linear Programming, RAND Report P-468, The Rand Corporation, Santa Monica, Calif., 1953.

Manne, A. S. *Scheduling of Petroleum Refinery Operations*, Harvard Economic Studies No. 48, Harvard University Press, Cambridge, Mass., 1956.

Markland, R. E. Analyzing Multi-Commodity Production-Distribution Networks with Transformed Flows, unpublished paper, School of Business Administration, University of Missouri, St. Louis, 1972.

Markowitz, H. M. "The Elimination Form of the Inverse and its Application to Linear Programming," *Management Science*, **3** (3), 1957.

Markowitz, H. M., and A. S. Manne. "On the Solution of Discrete Programming Problems," *Econometrica*, **25** (1), p. 19, January 1957.

Maros, I. G. "A General Phase-I Method in Linear Programming," *European Journal of Operational Research*, **23** (1), 64–77, 1986.

Marshall, Clifford W. *Applied Graph Theory*, Wiley-Interscience, New York, 1971.

Marshall, K. T., and J. W. Suurballe. "A Note on Cycling in the Simplex Method," *Naval Research Logistics Quarterly*, **16**, pp. 121–137, 1969.

Masse, Pierre, and R. Gibrat. "Applications of Linear Programming to Investments in the Electric Power Industry," *Management Science*, **3** (1), pp. 149–166, January 1957.

Mattheiss, T. H. "An Algorithm for Determining Irrelevant Constraints and all Vertices in Systems of Linear Inequalities," *Operations Research*, **1** (1), pp. 247–260, 1973.

Mattheiss, T. H., and D. S. Rubin. "A Survey and Comparison of Methods for Finding all Vertices of Convex Polyhedral Sets," *Mathematics of Operations Research*, **5** (2), pp. 167–185, May 1980.

Mattheiss, T. H., and B. K. Schmidt. "Computational Results on an Algorithm for Finding all Vertices of a Polytope," *Mathematical Programming*, **18**, pp. 308–329, 1980.

Maurras, J. F. "Optimization of the Flow Through Networks with Gains," *Mathematical Programming*, **3**, pp. 135–144, 1972.

Maurras, J. F., K. Truemper, and M. Akgul. "Polynomial Algorithms for a Class of Linear Programs," *Mathematical Programming*, **21**, pp. 121–136, 1981.

Maurras, J. F. "Optimization of the Flow Through Networks with Gains," *Mathematical Programming*, **3**, pp. 135–144, 1972.

Mayeda, W., and M. E. Van Valkenburg. Set of Cut Sets and Optimum Flow, U.S.G.R.& D.R. Order AD-625 200 from CFSTI, Coordinated Science Laboratory, Illinois University, Urbana, Ill., November 1965.

McBride, R. D. "Efficient Solution of Generalized Network Problems," Working Paper, Finance and Business Economics Department, School of Business Administration, University of Southern California, April 1981.

McGinnis, L. F. "Implementation and Testing of a Primal-Dual Algorithm for the Assignment Problem," *Operations Research*, **31** (2), pp. 277–291, 1983.

McIntosh, Peter T. "Initial Solutions to Sets of Related Transportation Problems," *Operational Research Quarterly*, **14** (1), pp. 65–69, March 1963.

McKeown, P. G. "A Vertex Ranking Procedure for Solving the Linear Fixed Charge Problem," *Operations Research*, **23** (6), pp. 1183–1191, 1975.

McMullen, P. "The Maximum Number of Faces of A Convex Polytope," *Mathematika*, **17**, pp. 179–184, 1970.

McShane, K. A., C. L. Monma, and D. Shanno. "An Implementation of a Primal-Dual Interior Point Method for Linear Programming," School of Operations Research and Industrial Engineering, Cornell University, Ithaca, NY 14853.

Megiddo, N. "Is Binary Encoding Appropriate for the Problem-Language Relationship," *Theoretical Computer Science* **19**, pp. 1–5, 1982.

Megiddo, N. "Toward a Genuinely Polynomial Algorithm for Linear Programming," *SIAM Journal of Computing*, **12**, pp. 347–353, 1983.

Megiddo, N. "A Variation on Karmarkar's Algorithm," Preliminary Report, IBM Research Laboratory, San Jose, CA 95193, 1985.

Megiddo, N. "Introduction: New Approaches to Linear Programming," *Algorithmica*, **1** (4), pp. 387–394, 1986a.

Megiddo, N. "Pathways to the Optimal Set in Linear Programming," Research Report RJ 5295, IBM Almaden Research Center, San Jose, CA, 1986b.

Megiddo, N., and M. Shub. "Boundary Behavior of Interior Point Algorithms in Linear Programming," Presented at the Joint National TIMS/ORSA Meeting, New Orleans, May 1987.

Mehrotra, S. "A Self Correcting Version of Karmarkar's Algorithm," Report, Department of Industrial Engineering and Operations Research, Columbia University, New York, NY 10027, 1986.

Metzger, R. W., and R. Schwarzbeck. "A Linear Programming Application to Cupola Charging," *Journal of Industrial Engineering*, **12** (2), pp. 87–93, March–April 1961.

Mifflin, R. "On the Convergence of the Logarithmic Barrier Function Method," *Numerical Methods of Nonlinear Optimization*, Academic Press, New York, pp. 367–369, 1972.

Miller, R. E. "Alternative Optima, Degeneracy and Imputed Values in Linear Programs," *Journal of Regional Science*, **5** (1), pp. 21–39, 1963.

Mills, G. "A Decomposition Algorithm for the Shortest-Route Problem," *Operations Research*, **14** (2), pp. 279–291, March–April 1966.

Minieka, E. "Optimal Flow in a Network with Gains," *INFOR*, **10**, pp. 171–178, 1972.

Minieka, E. "Parametric Network Flows," *Operations Research*, **20** (6), pp. 1162–1170, November–December 1972.

Minieka, E. *Optimization Algorithms for Networks and Graphs*, Marcel Dekker, New York, 1978.

Minkowski, H. *Geometry der Zahlen*, Teubner, Leipzig, 2nd ed., 1910.

Minty, G. J. "On an Algorithm for Solving Some Network-Programming Problems," *Operations Research*, **10** (3), pp. 403–405, May–June 1962.

Mitra, G., M. Tamiz, J. Yadegar, and K. Darby-Dowman. "Experimental Investigation of an Interior Search Algorithm for Linear Programming," Math. Programming Symposium, Boston, 1985.

Monma, C. L. "Recent Breakthroughs in Linear Programming Methods," Bell Communications Research, Morristown, NJ 07960, 1987.

Monma, C., and A. J. Morton. "Computational Experience with a Dual Affine Variant of Karmarkar's Method for Linear Programming," Bell Communications Research, Morristown, NJ 07960, 1987.

Monteiro, R. D. C., and I. Adler. "Interior Path Following Primal-Dual Algorithms-Part I: Linear Programming," Dept. of Industrial Engineering and Operations Research, University of California, Berkeley, CA, 1987.

Monteiro, R. D. C., I. Adler, and M. G. C. Resende. "A Polynomial-Time Primal-Dual Affine Scaling Algorithm for Linear and Convex Quodratic Programming and its Power Series Extension," Technical Report ESRC 88-8, Dept. of Industrial Engineering and Operations Research, University of California, Berkeley, CA 1988.

Moore, E. F. "The Shortest Path Through a Maze," *Proceedings of the International Symposium on the Theory of Switching*, Part II, April 2–5, 1957, The Annals of the Computation Laboratory of Harvard University, vol. 30, Harvard University Press, 1959.

Motzkin, T. S. Beiträge zur Theorie der Linearen Ungleichungen, Ph.D. Dissertation, University of Zurich, 1936.

Motzkin, T. S. "The Multi-index Transportation Problem," *Bulletin of American Mathematical Society*, **58** (4), p. 494, 1952.

Motzkin, T. S. "The Assignment Problem," *Proceedings of the 6th Symposium in Applied Mathematics*, McGraw-Hill Book Co. New York, pp. 109–125, 1956.

Motzkin, T. S., and I. G. Schoenberg. "The Relaxation Method for Linear Inequalities," *Canadian Journal of Mathematics*, **6**, pp. 393–404, 1954.

Mueller, R. K., and L. Cooper. "A Comparison of the Primal Simplex and Primal-dual Algorithms in Linear Programming," *Communications of the ACM*, **18** (11), November 1965.

Mueller-Merbach, H. "An Approximation Method for Finding Good Initial Solutions for Transportation Problems," *Elektronische Datenverarbeitung* (Germany), **4** (6), pp. 255–261, November–December 1962.

Mueller-Merbach, H. "Several Approximation Methods for Solving the Transportation Problem," IBM-Form 78 106, IBM-Fachbibliothek (Germany), November 1963.

Meuller-Merbach, H. "Optimal Acceleration of Projects by Parametric Linear Programming," *Elektronische Datenverarbeitung* (Germany), **9** (1), pp. 33–39, January 1967.

Mueller-Merbach, H. "The Method of Direct Decomposition in Linear Programming," *Ablauf- und Planungsforschung* (Germany), **6** (2), pp. 306–322, April–June 1965.

Mueller-Merbach, H. "An Improved Starting Algorithm for the Ford–Fulkerson Approach to the Transportation Problem," *Management Science*, **13** (1), pp. 97–104, September 1966.

Mulvey, J. M. "Pivot Strategies for Primal-Simplex Network Codes," *Journal of the Association for Computing Machinery*, **25**, pp. 266–270, 1978.

Munkres, James. "Algorithms for the Assignment and Transportation Problems," *SIAM*, **5** (1), pp. 32–38, March 1957.

Murchland, J. D. "A New Method for Finding All Elementary Paths in a Complete Directed Graph," Report LSE-TNT-22, Transport Network Theory Unit, London School of Economics, London, England, October 1965.

Murchland, J. D. "The Once-Through Method of Finding All Shortest Distances in a Graph from a Single Origin," Report LBS-TNT-56, Transport Network Theory Unit, London School of Economics, London, England, August 1967.

Murphy, F. H., and E. A. Stohr. "An Intelligent System for Formulating Linear Programs," *Decision Support Systems*, **2** (1), 1986.

Murtagh, B. A. *Advanced Linear Programming*, McGraw-Hill, New York, 1981.

Murtagh, B. A., and M. A. Saunders. "MINOS-5.0 Users Guide," Report SOL 83-20, Department of Operations Research, Stanford University, California, 1983.

Murty, K. G. "Solving the Fixed Charge Problem by Ranking the Extreme Points," *Operations Research*, **16**, pp. 268–279, 1968.

Murty, K. G. *Linear and Combinatorial Programming*, Wiley, New York, 1976.

Murty, K. G. "Resolution of Degeneracy in the Bounded Variable Primal Simplex Algorithm," Technical Report 78-1, Department of Industrial Engineering and Operations Research, University of Michigan, Ann Arbor, Michigan, 1978.

Murty, K. G. *Linear Programming*, John Wiley & Sons, Inc. New York, 1983.

Murty, K. G. "Faces of a Polyhedron," *Mathematical Programming Study*, pp. 1–13, 1985.

Murty, K. G. "The Gravitational Method for Linear Programming," *Opsearch*, **23**, pp. 206–214, 1986.

Murty, K. G. and Y. Fathi. "A Feasible Direction Method for Linear Programming," *Operations Research Letters*, Vol. **3** (3), 121–127, 1984.

Naniwada, M. "Multicommodity Flows in a Communication Network," *Electronics Communications of Japan*, **52**, pp. 34–40, 1969.

Nazareth, J. L. "Homotopy Techniques in Linear Programming," *Algorithmica*, **1** (4), pp. 529–536, 1986.

Nemhauser, G. L. "A Generalized Permanent Labeling Setting Algorithm for the Shortest Path Between Specified Nodes," *Journal of Mathematical Analysis and Applications*, **38**, pp. 328–334, 1972.

Nicholson, T. A. J. "Finding the Shortest Route Between Two Points in a Network," *The Computer Journal*, **9** (3), pp. 275–280, November 1966.

Nickels, W., W. Rodder, L. Xu, and H.-J. Zimmermann. "Intelligent Gradient Search in Linear Programming," *European Journal of Operational Research*, **22**, pp. 293–303, 1985.

Ohtsuka, Makoto. "Generalization of the Duality Theorem in the Theory of Linear Programming," *Journal of Science of the Hiroshima University*, Ser. A-I (Japan), **30** (1), pp. 31–39, July 1966.

Ohtsuka, M. "Generalized Capacity and Duality Theorems in Linear Programming," *Journal of Science of the Hiroshima University*, Ser. A-I (Japan), **30** (1), pp. 45–56, July 1966.

Onaga, K. "Optimum Flows in General Communications Networks," *Journal of the Franklin Institute*, **283**, pp. 308–327, 1967.

Onaga, K. "A Multicommodity Flow Theorem," *Electronics Communications of Japan*, **53**, pp. 16–22, 1970.

Orchard-Hays, W. "A Composite Simplex Algorithm-II," Research Memorandum RM-1275, The Rand Corporation, Santa Monica, Calif., May 1954.

Orchard-Hays, W. "Background, Development and Extensions of the Revised Simplex Method," Research Memorandum RM-1433, The Rand Corporation, Santa Monica, Calif., 1954.

Orchard-Hays, W. Matrices, "Elimination and the Simplex Method," Report, CEIR, Inc., Bethesda, Md., 1961.

Orden, A. "A Procedure for Handling Degeneracy in the Transportation Problem," mimeograph, DCS/Comptroller, Headquarters U.S. Air Force, Washington, D.C., 1951.

Orden, A. "The Transshipment Problem," *Management Science*, **2** (3), 1956.

Orden, A. "A Step Toward Probabilistic Analysis of Simplex Method Convergence," *Mathematical Programming*, **19** (1), pp. 3–13, July 1980.

Orlin, J. B. "A Polynomial-Time Parametric Simplex Algorithm for the Minimum Cost Network Flow Problem," Sloan W. P. #1484-83, MIT, Cambridge, Massachusetts 02139, 1983.

Orlin, J. B. "Genuinely Polynomial Simplex and Non-Simplex Algorithms for the Minimum Cost Flow Problem," Sloan W. P. #1615-84, MIT, Cambridge, Massachusetts 02139, 1984.

Orlin, J. B. "On the Simplex Algorithm for Network and Generalized Networks," *Mathematical Programming*, **24**, pp. 166–178, 1985.

Osborne, M. R. "Dual Barrier Functions with Superfast Rates of Convergence for the Linear Programming Problem," Report, Department of Statistics, Research School of Social Sciences, Australian National University, 1986.

Padberg, M. "Solution of a Nonlinear Programming Problem Arising in the Projective Method," Preprint, New York University, NY 10003, 1985.

Padberg, M. "A Different Convergence Proof of the Projective Method for Linear Programming," *Operations Research Letters*, **4**, pp. 253–257, 1986.

Paige, C. C., and M. A. Saunders. "An Algorithm for Sparse Linear Equations and Sparse Least-Squares," *ACM Transactions on Mathematical Software*, **8**, pp. 43–71, 1982.

Papadimitriou, C. H., and K. Steiglitz. *Combinatorial Optimization*, *Algorithms and Complexity*, Prentice-Hall, Englewood Cliffs, New Jersey, 1982.

Paranjape, S. R. "The Simplex Method: Two Basic Variables Replacement," *Management Science*, **12** (1), September 1965.

Patty, B. W. "The Networks with Side Constraints Problem," Working Paper 87-04, University of Southern California, Department of ISE, 1987.

Perold, A. F. "A Degeneracy Exploiting LU-Factorization for the Simplex Method," *Mathematical Programming*, **19** (3), pp. 239–254, November 1980.

Phillips, D. T., A. Ravindran, and J. Solberg. *Operations Research*: *Principles and Practice*, Wiley, New York, 1976.

Phillips, D. T., and P. A. Jensen. "The Out-of-Kilter Algorithm," *Industrial Engineering*, **6**, pp. 36–44, 1974.

Picard, J.-C. and M. Queyranne. "Selected Applications of Minimum Cuts in Networks," *INFOR*, **20** (4), 1982.

Pickel, P. F. "Approximate Projections for the Karmarkar Algorithm," Manuscript, Polytechnique Institute of New York, Farmingdale, NY, 1985.

Pierskalla, W. P. "The Multidimensional Assignment Problem," *Operations Research*, **16**, pp. 422–431, 1968.

Pollack, M., and G. Wallace. "Methods for Determining Optimal Traffic Routes in Large Communication Networks," U.S.G.R.R. Document Number AD-434 856, Standord Research Institute, Menlo Park, Calif., June 1962.

Polyak, B. T., and N. V. Tretiyakov. "Concerning an Iterative Method for Linear Programming and its Economic Interpretation," *Economics and Mathematical Methods*, **8**, (in Russian), pp. 740–751, 1972.

Potts, R. B., and R. M. Oliver. *Flows in Transportation Networks*, Academic Press, New York, 1972.

Prager, W. "On the Caterer Problem," *Management Science*, **3** (1), pp. 15–23, October 1956.

Prager, W. "A Generalization of Hitchcock's Transportation Problem," *Journal of Mathematical Physics* (M.I.T.), **36** (2), pp. 99–106, July 1957.

Pshchenichnii, B. M. "The Connection Between Graph Theory and the Transportation Problem," *Dopovidi Akademii Nauk URSR* (USSR), **4**, pp. 427–430, 1963.

Quandt, R. E., and H. W. Kuhn. "On Upper Bounds for the Number of Iterations in Solving Linear Programs," *Operations Research*, **12** (1), 1964.

Ravi, N., and R. E. Wendell. "The Tolerance Approach to Sensitivity Analysis of Matrix Coefficients in Linear Programming—I," Working Paper 562, Graduate School of Business, University of Pittsburgh, October 1984.

Ravi, N., and R. E. Wendell. "The Tolerance Approach to Sensitivity Analysis of Matrix Coefficients in Linear Programming—II," Working Paper 563, Graduate School of Business, University of Pittsburgh, October 1984.

Ravi, N., and R. E. Wendell. "The Tolerance Approach to Sensitivity Analysis of Matrix Coefficients in Linear Programming: General Perturbations," *Journal of the Operational Research Society*, **36** (10), pp. 943–950, 1985.

Ravindran, A., and T. W. Hill. "A Comment on the Use of Simplex Method for Absolute Value Problems," *Management Science*, **19** (5), pp. 581–582, January 1973.

Reban, K. R. "Total Unimodularity and the Transportation Problem: A Generalization," *Linear Algebra and Its Applications*, **8**, pp. 11–24, 1974.

Redlack, A. and C. C. Huang, "The Assignment Problem: A New Look at an Old Problem," TIMS/ORSA Joint National Meeting, WA11.3, New Orleans, May 4–6, 1987.

Reid, J. K. "A Sparsity-Exploiting Variant of the Bartels–Golub Decomposition for Linear Programming Bases," *Mathematical Programming*, **24**, pp. 55–69, 1982.

Renegar, J. "A Polynomial-Time Algorithm Based on Newton's Method for Linear Programming," *Mathematical Programming*, **40**, pp. 59–93, 1988.

Riesco, A., and M. E. Thomas. "A Heuristic Solution Procedure for Linear Programming Problems with Special Structure," *AIIE Transactions*, **1** (2), pp. 157–163, June 1969.

Riley, V., and S. I. Gass. *Linear Programming and Associated Techniques: A Comprehensive Bibliography on Linear, Nonlinear and Dynamic Programming*, Johns Hopkins Press, Baltimore, Md., 1958.

Rinaldi, G. "A Projective Method for Linear Programming with Box-Type Constraints," *Algorithmica*, **1** (4), pp. 517–528, 1986.

Ritter, K. "A Decomposition Method for Linear Programming Problems with Coupling Constraints and Variables," MRC Report No. 739, Mathematics Research Center, U.S. Army, University of Wisconsin, Madison, Wis., April 1967.

Robacker, J. T. "Concerning Multicommodity Networks," Research Memorandum RM-1799, The Rand Corporation, Santa Monica, Calif., 1956.

Robinson, S. M. "A Characterization of Stability in Linear Programming," *Operations Research*, **23** (3), pp. 435–447, May/June 1977.

Rockafellar, R. T. *Network Flows and Monotropic Optimization*, John Wiley & Sons, New York, NY, 1984.

Rockafeller, R. T. *Convex Analysis*, Princeton University Press, Princeton, N.J., 1970.

Rohde, F. V. "Bibliography on Linear Programming," Operations Research, **5** (1), 1957.

Roos, C. "On Karmarkar's Projective Method for Linear Programming," Report 85-23, Department of Mathematics and Informatics, Delft University of Technology, 2600 AL Delft, 1985.

Roos, R. "A Pivoting Rule for the Simplex Method Which is Related to Karmarkar's Potential Function," Department of Mathematics and Informatics, Delft University of Technology, P.O. Box 356, 2600 AJ Delft, The Netherlands, 1985.

Rosen, J. B. "The Gradient Projection Method for Nonlinear Programming: Part I—Linear Constraints," *SIAM*, **8** (1), pp. 181–217, 1960.

Rosen, J. B. "The Gradient Projection Method for Nonlinear Programming: Part II," *SIAM*, **9** (4), 1961.

Rosen, J. B. "Primal Partition Programming for Block Diagonal Matrices," *Numerical Mathematics*, **6**, 250–260, 1964.

Rothfarb, B., N. P. Shein, and I. T. Frisch. "Common Terminal Multicommodity Flow," *Operations Research*, **16**, pp. 202–205, January–February 1968.

Rothfarb, B., and I. T. Frisch. "On the Three-Commodity Flow Problem," *SIAM*, **17**, pp. 46–58, 1969.

Rothchild, B., and A. Whinston. "On Two-Commodity Network Flows," *Operations Research*, **14** (3), pp. 377–388, May–June 1966.

Rothchild, B., and A. Whinston. "Feasibility of Two-Commodity Network Flows," *Operations Research*, **14**, p. 1121–1129, 1966.

Rothchild, B., and A. Whinston. "Maximal Two-Way Flows," *SIAM*, **15**, (5), pp. 1228–1238, September 1967.

Roy, B. "Extremum Paths," *Gestion* (France), **5**, pp. 322–335, May 1966.

Russel, A. H. "Cash Flows in Networks," *Management Science*, **16** (5), pp. 357–373, January 1970.

Rutenberg, D. P. "Generalized Networks," Generalized Upper Bounding and Decomposition of the Convex Simplex Method," *Management Science*, **16** (5), pp. 388–401, January 1970.

Saaty, T. L. "The Number of Vertices of a Polyhedron," *American Mathematical Monthly*, **62**, 1955.

Saaty, T. L. "A Conjecture Concerning the Smallest Bound on the Iterations in Linear Programs," *Operations Research*, **11**, pp. 151–153, 1963.

Saaty, T. L. "Coefficient Perturbation of a Constrained Extremum," *Operations Research*, **7**, May–June 1959.

Saaty, T. L., and S. I. Gass. "The Parametric Objective Function, Part I," *Operations Research*, **2**, 1954.

Saigal, R. "Multicommodity Flows in Directed Networks," ORC Report 67-38, University of California, Berkeley, Calif., 1967.

Saigal, R. "A Constrained Shortest Route Problem," *Operations Research*, **16** (1), pp. 205–209, January–February 1968.

Saigal, R. "On the Modularity of a Matrix," *Linear Algebra and Its Applications*, **5**, pp. 39–48, 1972.

Sakarovitch, M. "The Multi-Commodity Maximum Flow Problem," ORC Report 66-25, University of California, Berkeley, Calif., 1966.

Sakarovitch, M., and R. Saigal. "An Extension of Generalized Upper Bounding Techniques for Structured Linear Programs," *SIAM*, **15** (4), pp. 906–914, July 1967.

Saunders, M. A. "The Complexity of LU Updating in the Simplex Method," *The Complexity of Computational Problem Solving*, R. S. Anderssen and R. P. Brent, eds., Queensland: Queensland University Press, pp. 214–230, 1976a.

Saunders, M. A. "A Fast, Stable Implementation of the Simplex Method Using the Bartels–Golub Updating," in J. R. Bunch and D. J. Rose (eds.), *Sparse Matrix Computations*, Academic Press, New York, NY, pp. 213–226, 1976b.

Sandor, P. E. "Some Problems of Ranging in Linear Programming," *Journal of the Canadian Operational Research Society*, **2** (1), pp. 26–31, June 1964.

Scoins, H. I. "The Compact Representation of a Rooted Tree and the Transportation Problem," Presented at the International Symposium on Mathematical Programming, London, 1964.

Schrage, L. "Implicit Representation of Variable Upper Bounds in Linear Programming," *Mathematical Programming Study*, **4**, pp. 118–132, 1975.

Schrage, L. "Implicit Representation of Generalized Variable Upper Bounds in Linear Programming," *Mathematical Programming*, **14** (1), pp. 11–20, 1978.

Schreck, H. "Experiences with an Implementation of Karmarkar's LP-Algorithm," Mathematisches Institut der Technischen Universitat Munchen, Arcisstrasse 21, Munchen.

Schrijver, A. "The New Linear Programming Method of Karmarkar," *CWI Newsletter*, 8, Centre for Mathematics and Computer Science, P.O. Box 4079, 1009 AB Amsterdam, The Netherlands, 1985.

Schrijver, A. *Theory of Linear and Integer Programming*, Wiley-Interscience, John Wiley and Sons, Inc., 1986.

Seiffart, E. "An Algorithm for Solution of a Parametric Distribution Problem," *Ekonomicko-matematicky obzor* (Czechoslovakia), **2** (3), pp. 263–283, August 1966.

Sengupta, J. K., and T. K. Kumar. "An Application of Sensitivity Analysis to a Linear Programming Problem," *Unternehmensforschung* (Germany), **9** (1), pp. 18–36, 1965.

Seshu, S., and M. Reed. *Linear Graphs and Electrical Networks*, Addison-Wesley, Reading, Mass., 1961.

Sethi, A. P., and G. L. Thompson. "The Pivot and Probe Algorithm for Solving a Linear Program," *Mathematical Programming*, **29** (2), pp. 219–233, 1984.

Shanno, D. F. "A Reduced Gradient Variant of Karmarkar's Algorithm," Working Paper 85-10, Graduate School of Administration, University of California, Davis, CA 95616, 1985a.

Shanno, D. F. "Computing Karmarkar Projections Quickly," *Mathematical Programming*, **4** (1), pp. 61–72, 1988.

Shanno, D. F., and R. E. Marsten. "On Implementing Karmarkar's Method,"

Working Paper 85-1, Graduate School of Administration, University of California at Davis, California, 1985.

Shanno, D. F. and R. L. Weil. "Linear Programming With Absolute Value Functionals," *Management Science*, **16** (5), p. 408, January 1970.

Shapley, L. S. "On Network Flow Functions," Research Memorandum RM-2338, The Rand Corporation, Santa Monica, Calif., 1959.

Shapiro, J. F. "A Note on the Primal-Dual and Out-of-Kilter Algorithms for Network Optimization Problems," *Networks*, **7** (1), pp. 81–88, 1977.

Sherali, H. D. "Equivalent Weights for Lexicographic Multiple Objective Programs: Characterizations and Computations," *European Journal of Operational Research*, **11** (4), pp. 367–379, 1982.

Sherali, H. D. "An Insightful Marginal Cost Analysis for an Electric Utility Capacity Planning Problem," *IIE Transactions*, **17** (4), pp. 378–387, 1985.

Sherali, H. D. "Algorithmic Insights and a Convergence Analysis for a Karmarkar-type Algorithm for Linear Programs," *Naval Research Logistics Quarterly*, **34**, pp. 399–416, 1987.

Sherali, H. D. "A Constructive Proof of the Representation Theorem for Polyhedral Sets Based on Fundamental Definitions," *American Journal of Mathematial and Management Sciences*, 7 (3/4), pp. 253–270, 1987.

Sherali, H. D., "Bounds on Penalties for Dummy Arcs in Transportation Networks," *European Journal of Operational Research*, **36** (3), 1988.

Sherali, H. D., and C. G. Baines. "Advanced Basis or Block Pivoting Methods for Network Structured Linear Programs," *Aligarh Journal of Statistics*, **3** (1), pp. 17–35, 1984.

Sherali, H. D., and S. E. Dickey. "A Extreme Point Ranking Algorithm for the Extreme Point Mathematical Programming Problem," *Computers and Operations Research*, **13** (4), pp. 465–475, 1986.

Sherali, H. D., and C. M. Shetty. "A Primal Simplex Based Solution Procedure for the Retilinear Distance Multifacility Location Problem," *Operational Research Quarterly*, **29** (4), pp. 373–381, 1978.

Sherali, H. D., and B. O. Skarpness, and B. Kim. "An Assumption-Free Convergence Analysis for a Perturbation of the Scaling Algorithm for Linear Programs, with Application to the L_1 Estimation Problem," *Naval Research Logistics Quarterly*, **35**, pp. 473–492, 1988.

Sherali, H. D., and A. L. Soyster. "Preemptive and Nonpreemptive Multi-Objective Programs: Relationships and Counter Examples," *Journal of Optimization Theory and Applications*, **39** (2), pp. 173–186, 1983a.

Sherali, H. D., and A. L. Soyster. "Analysis of Network Structured Models for Electric Utility Capacity Planning and Marginal Cost Pricing Problems," *Energy Models and Studies: Studies in Management Science and Systems Series*, pp. 113–134, edited by Benjamin Lev, North-Holland Publishing Company, 1983b.

Sherali, H. D. A. L. Soyster, and C. G. Baines. "Non-Adjacent Extreme Point Methods for Solving Linear Programs," *Naval Research Logistics Quarterly*, **30** (1), pp. 145–162, 1983.

Sherali, H. D., A. L. Soyster, F. H. Murphy, and S. Sen. "Linear Programming

Based Analysis of Marginal Cost Pricing in Electric Utility Capacity Expansion," *European Journal of Operational Research*, **11**, pp. 349–360, 1982.

Sherali, H. D., A. L. Soyster, F. H. Murphy, and S. Sen. "Intertemporal Allocation of Capital Costs in Electric Utility Capacity Expansion Planning Under Uncertainty," *Management Science*, **30** (1), pp. 1–19, 1984.

Shetty, C. M. "A Solution to the Transportation Problem with Nonlinear Costs," *Operations Research*, **7** (5), pp. 571–580, September–October 1959.

Shetty, C. M. "Solving Linear Programming Problems with Variable Parameters," *Journal of Industrial Engineering*, **10** (6), 1959.

Shetty, C. M. "On Analyses of the Solution to a Linear Programming Problem," *Operational Research Quarterly*, **12** (2), pp. 89–104, June 1961.

Shier, D. R. "A Decomposition Algorithm for Optimality Problems in Tree-Structured Networks," *Discrete Mathematics*, **6**, pp. 175–189, 1973.

Shier, D., and C. Witzgall. "Properties of Labeling Methods for Determining Shortest Path Trees," *Journal of Research National Bureau of Standards*, **86**, pp. 317–330, 1981.

Shor, N. Z. "Utilization of the Operation of Space Dilation in the Minimization of Convex Functions," *Kibernetika*, **6** (1), pp. 6–12, 1970a, (English translation in *Cybernetics*, **6** (1), pp. 7–15, 1970a).

Shor, N. Z. "Convergence Rate of the Gradient Descent Method with Dilatation of the Space," *Kibernetika*, **2**, pp. 80–85, 1970b. (English translation in *Cybernetics*, **6**, pp. 102–108, 1970b.)

Shor, N. Z. "Cut-off Method with Space Extension in Convex Programming Problems," *Kibernetika*, **1**, pp. 94–95, 1977. (English translation in *Cybernetics*, **13**, pp. 94–96, 1977.)

Shor, N. Z. "Generalized Gradient Methods of Nondifferentiable Optimization Employing Space Dilation Operators," In Bachem, A., M. Grotschel, and B. Korte, "Mathematical Programming: The State of the Art," pp. 501–529, Springer-Verlag, Bonn, 1983.

Shor, N. Z., and V. I. Gershovich. "Family of Algorithms for Solving Convex Programming Problems," *Kibernetika*, **15** (4), pp. 62–67, 1979. (English translation in *Cybernetics*, **15** (4), pp. 502–507, 1979.)

Singh, S. "Improved Methods for Storing and Updating Information in the Out-of-Kilter Algorithm," *Journal of the Association for Computing Machinery*, **33** (3), pp. 551–567, 1986.

Simonnard, M. A. "Transportation-Type Problems," Interim Technical Report No. 11, Massachusetts Institute of Technology, Cambridge, Mass., 1959.

Simonnard, M. A. "Structure des bases dans les problèmes de transport," *Rev. Fr. Rech. Op.*, **12**, 1959.

Simonnard, M. A. *Linear Programming*, Prentice-Hall, Englewood Cliffs, N.J., 1966.

Simonnard, M. A., and G. F. Hadley. "Maximum Number of Iterations in the Transportation Problem," *Naval Research Logistics Quarterly*, **6** (2), 1959.

Smale, S. "On the Average Speed of the Simplex Method of Linear Programming," *Mathematical Programming*, **27** (3), pp. 241–262, 1983a.

Smale, S. "The Problem of the Average Speed of the Simplex Method," *Mathe-*

matical Programming, *The State of the Art–Bonn 1982*, (eds.) A. Bachem, M. Grotschel, and B. Korte, Springer, Berlin, pp. 530–539, 1983b.

Smith, C. W. "Maximal Flow at Minimal Cost Through a Special Network with Gains," unpublished thesis, School of Industrial and Systems Engineering, Georgia Institute of Technology, Atlanta, Ga., 1971.

Smith, D. M., and W. Orchard-Hays. "Computational Efficiency in Product Form LP Codes," in R. Graves and P. Wolfe (eds.), *Recent Advances in Mathematical Programming*, McGraw-Hill Book Co., New York, 1963.

Sommerville, D. M. Y. *An Introduction to the Geometry of N-Dimensions*, Dover, New York, 1958.

Spivey, W. A. *Linear Programming*, Macmillan, New York, 1963.

Srinivasan, V., and G. L. Thompson. "Accelerated Algorithms for Labeling and Relabeling of Trees with Applications to Distribution Problems," *Journal of the ACM*, **19**, pp. 712–726, 1972.

Srinivasan, V., and G. L. Thompson. "Benefit-Cost Analysis of Coding Techniques for the Primal Transportation Algorithm," *Journal of the Association for Computing Machinery*, **20**, pp. 194–213, 1973a.

Srinivasan, V., and G. L. Thompson. "Coding Techniques for Transportation Algorithms," *Journal of the Association for Computing Machinery*, **20**, 1973b.

Staffurth, C. (ed.). *Project Cost Control Using Networks*, The Operational Research Society and the Institute of Cost and Works Accountants, London, 1969.

Stevens, B. H. "Linear Programming and Location Rent," *Journal of Regional Science*, **3** (2), pp. 15–26, 1961.

Stoinova-Penkrova, N. "The Simplex Method without Fractions," *Trudove Vissh Ikonomicheshi Institute Sofia* (Bulgaria), **1**, pp. 357–370, 1964.

Stokes, R. W. "A Geometric Theory of Solution of Linear Inequalities," *Transactions of the American Mathematical Society*, **33**, pp. 782–805, 1931.

Strang, G. "Karmarkar's Algorithm in a Nutshell," *SIAM News*, **18**, p. 13, 1985.

Stroup, J. W. "Allocation of Launch Vehicles to Space Missions: A Fixed-Cost Transportation Problem," *Operations Research*, **15** (6), pp. 1157–1163, November–December 1967.

Strum, J. E. *Introduction to Linear Programming*, Holden Day, San Francisco, 1972.

Swanson, L. W. *Linear Programming*, *Basic Theory and Application*, McGraw-Hill Publishing Company, New York, NY, 1980.

Swart, E. R. "How I Implemented the Karmarkar Algorithm in One Evening," APL Quote Quad 15.3, 1985.

Swart, E. R. "A Modified Version of the Karmarkar Algorithm," Department of Mathematics and Statistics, University of Guelph, Guelph, Ontario, Canada, 1985.

Swarup, K. "Duality for Transportation Problem in Fractional Programming," *CCERO* (Belgium), **10** (1), pp. 46–54, March 1968.

Symonds, G. H. *Linear Programming: The Solution of Refinery Problems*, Esso Standard Oil Company, New York, 1955.

Takahashi, I. "Tree Algorithm for Solving Resource Allocation Problems," *Operations Research Society of Japan*, **8**, pp. 172–191, 1966.

Takahashi, I. "Tree Algorithm for Solving Network Transportation Problems," *Operations Research Society of Japan*, **8**, pp. 192–216, 1966.

Tamir, A. "On Totally Unimodular Matrices," *Networks*, **6**, pp. 373–382, 1976.

Tan, S. T. "Contributions to the Decomposition of Linear Programs," *Unternehmensforschung* (Germany), **10**, (3–4), pp. 168–189, 247–268, 1966.

Tang, D. T. "Bipath Networks and Multicommodity Flows," *IEEE Transactions*, **11**, pp. 468–474, 1964.

Tardos, E. "A Strongly Polynomial Minimum Cost Circulation Algorithm," *Combinatorica*, **5**, pp. 247–255, 1985.

Tardos, E. "A Strongly Polynomial Algorithm to Solve Combinatorial Linear Programs," *Operations Research*, **34** (2), pp. 250–256, 1986.

Thompson, G. L., F. M. Tonge, and S. Zionts. "Techniques for Removing Nonbinding Constraints and Extraneous Variables from Linear Programming Problems," *Management Science*, **12** (7), pp. 588–608, March 1966.

Thrall, R. M. "The Mutual Primal-Dual Simplex Algorithm," Report No. 6426-27, University of Michigan Engineering Summer Conferences on Operations Research, Summer 1964.

Todd, M. J. "The Monotonic Bounded Hirsch Conjecture is False for Dimension at Least 4," *Mathematics of Operations Research*, **5** (4), pp. 599–601, November 1980.

Todd, M. J. "Large-Scale Linear Programming: Geometry, Working Bases and Factorizations," *Mathematical Programming*, **26**, pp. 1–20, 1983.

Todd, M. J. "Exploiting Special Structure in Karmarkar's Linear Programming Algorithm," *Mathematical Programming*, **41** (1), pp. 97–114, 1988.

Todd, M. J. "Recent Developments and New Directions in Linear Programming," in *Mathematical Programming: Recent Developments and Applications*, M. Iri and K. Tanabe, eds., KTK Science Publications, Tokyo, Japan, pp. 109–157, 1989.

Todd, M. J., and B. P. Burrell. "An Extension of Karmarkar's Algorithm for Linear Programming Using Dual Variables," *Algorithmica*, **1** (4), pp. 409–424, 1986.

Tomizawa, N. "On Some Techniques Useful for Solution of Transportation Network Problems," *Networks*, **1**, pp. 173–194, 1971.

Tomlin, J. A. "Minimum-Cost Multi-Commodity Network Flows," *Operations Research*, **14** (1), pp. 45–51, February 1966.

Tomlin, J. A. "On Scaling Linear Programming Problems," *Mathematical Programming Study Number 4*, December 1975.

Tomlin, J. A. "An Experimental Approach to Karmarkar's Projective Method for Linear Programming," Ketron, Inc., Mountain View, CA 94040, 1985.

Tompkins, C. B. "Projection Methods in Calculation," in H. A. Antosiewicz (ed.), *Proceedings of the Second Symposium in Linear Programming*, **2**, National Bureau of Standards, 1955.

Tompkins, C. B. "Some Methods of Computational Attack on Programming Problems Other Than the Simplex Method," *Naval Research Logistics Quarterly*, **4** (1), pp. 95–96, March 1957.

Torng, H. C. "Optimization of Discrete Control Systems Through Linear Programming," *Journal of the Franklin Institute*, **278**, (1), pp. 28–44, July 1964.

Traub, J., and H. Wozniakowski, H. "Complexity of Linear Programming," *Operations Research Letters*, **1**, pp. 59–62, 1982.

Truemper, K. "An Efficient Scaling Procedure for Gain Networks," *Networks*, **6** (2), pp. 151–160, 1976.

Truemper, K. "Unimodular Matrices of Flow Problems with Additional Constraints," *Networks*, **7** (4), pp. 343–358, 1977.

Treumper, K. "Algebraic Characterizations of Unimodular Matrices," *SIAM Journal of Applied Mathematics*, **35**, pp. 328–332, 1978.

Tseng, P., and D. P. Bertsekas. "Relaxation Methods for Linear Programs," *Mathematics of Operations Research*, **12**, pp. 569–596, 1987.

Tucker, A. W. "Linear Programming and Theory of Games," *Econometrica*, **18** (2), p. 189, April 1950.

Tucker, A. W. "Linear Inequalities and Convex Polyhedral Sets," in H. A. Antosiewicz (ed.), *Proceedings of the 2nd Symposium in Linear Programming*, **2**, National Bureau of Standards, pp. 569–602, 1955.

Tucker, A. W. "Linear and Nonlinear Programming," *Operations Research*, **5** (2), pp. 244–257, April 1957.

Tucker, A. W. "Dual Systems of Homogeneous Linear Relations," in H. W. Kuhn and A. W. Tucker (eds.), *Linear Inequalities and Related Systems*, Annals of Mathematics Study No. 38, Princeton University, Princeton, N. J., pp. 3–18, 1960.

Tucker, A. W. "On Directed Graphs and Integer Programs," Technical Report, IBM Mathematical Research Project, Princeton University, Princeton, N.J., 1960.

Tucker, A. W. "Solving a Matrix Game by Linear Programming," *IBM Journal of Research and Development*, **4** (5), pp. 507–517, November 1960.

Tucker, A. W. "Combinatorial Theory Underlying Linear Programs," in R. Graves and P. Wolfe (eds.), *Recent Advances in Mathematical Programming*, McGraw-Hill Book Co., New York, 1963.

Tufekci, S. "Decomposition Algorithms for Finding the Shortest Path Between a Source Node and a Sink Node of a Network," *Naval Research Logistics Quarterly*, **30** (3), pp. 387–396, 1983.

Tyndall, W. F. "An Extended Duality Theorem for Continuous Linear Programming Problems," *SIAM*, **15** (5), pp. 1294–1298, September 1967.

Vaidya, P. "An Algorithm for LP Which Requires $O[((M+N)N+(M\cdot N)^{**}(1.5))NL]$ Operations," Presented at the Joint National ORSA/TIMS Meeting, St. Louis, October 1987.

Vajda, S. *The Theory of Games and Linear Programming*, John Wiley & Sons, New York, 1956.

Vajda, S. *Readings in Linear Programming*, John Wiley & Sons, New York, 1958.

Vajda, S. *Mathematical Programming*, Addison-Wesley, Reading, Mass., 1961.

Van de Panne, C., and A. Whinston. "The Simplex and Dual Method for Quadratic Programming," *Operational Research Quarterly*, **15** (4), December 1964.

Van de Panne, C., and A. Whinston. "An Alternative Interpretation of the Primal-Dual Method and Some Related Parametric Methods," U.S.G.R.& D.R. Order AD-624 499, University of Virginia, Charlottesville, Va., August 1965.

Van de Panne, C. *Linear Programming and Related Techniques*, North-Holland/American Elsevier, New York, 1971.

Vanderbei, R. J., M. S. Meketon, and B. A. Freedman. "A Modification of Karmarkar's Linear Programming Algorithm," *Algorithmica*, **1** (4), pp. 395–408, 1986.

Veinott, A. F., and H. M. Wagner. "Optimum Capacity Scheduling, I and II," *Operations Research*, **10**, pp. 518–546, 1962.

Verkhovskii, B. S. "The Existence of a Solution of a Multi-Index Linear Programming Problem," *Doklady Akademii Nauk SSSR* (USSR), **158** (4), pp. 763–766, 1964.

Von Neumann, J. "Uber ein Okonomisches Gleichungssystem und ein Verallgemeinerung des Brouwerschen Fixpunktsatzes," *Ergebnisse eines Mathematischen Kolloguims*, **8**, 1937.

Von Neumann, J. "On a Maximization Problem," manuscript, Institute for Advanced Studies, Princeton, N.J., 1947.

Von Neumann, J. "A Certain Zero-Sum Two-Person Game Equivalent to the Optimal Assignment Problem" in H. W. Kuhn and A. W. Tucker (eds.), *Contributions to the Theory of Games*, **2**, Annals of Mathematics Study No. 28, Princeton University Press, Princeton, N.J., pp. 12–15, 1953.

Von Neumann, J., and O. Morgenstern. *Theory of Games and Economic Behavior*, Princeton University Press, Princeton, N.J., 1944.

Votaw, D. F., and A. Orden. "Personal Assignment Problem," in A. Orden and L. Goldstein (eds.), *Symposium on Linear Inequalities and Programming*, **10**, Planning Research Division, Director of Management Analysis Service, Comptroller, U.S. Air Force, Washington, D.C., pp. 155–163, 1952.

Wagner, H. M. "A Linear Programming Solution to Dynamic Leontief Type Models," Research Memorandum RM-1343, The Rand Corporation, Santa Monica, Calif., 1954.

Wagner, H. M. "A Two-Phase Method for the Simplex Tableau," *Operations Research*, **4** (4), 1956.

Wagner, H. M. "A Comparison of the Original and Revised Simplex Methods," *Operations Research*, **5** (3), 1957.

Wagner, H. M. "A Supplementary Bibliography on Linear Programming," *Operations Research*, **5** (4), 1957.

Wagner, H. M. "On the Distribution of Solutions in Linear Programming Problems," *Journal of the American Statistical Association*, **53**, pp. 161–163, 1958.

Wagner, H. M. "The Dual Simplex Algorithm for Bounded Variables," *Naval Research Logistics Quarterly*, **5** (3), pp. 257–261, September 1958.

Wagner, H. M. "On the Capacitated Hitchcock Problem," Technical Report No. 54, Stanford University, Stanford, Calif., 1958.

Wagner, H. M. "Linear Programming Techniques for Regression Analysis," *Journal of the American Statistical Association*, March 1959.

Walker, W. E. "A Method for Obtaining the Optimal Dual Solution to a Linear Program Using the Dantzig–Wolfe Decomposition," *Operations Research*, **17** (2), pp. 368–370, March–April 1969.

Walkup, D. W. "The Hirsch Conjecture Fails for Triangulated 27-Spheres," *Mathematics of Operations Research*, **3**, pp. 224–230, 1978.

Waugh, F. V. "The Minimum Cost Dairy Feed," *Journal of Farm Economics*, **33** (3), pp. 299–310, August 1951.

Wendell, R. E. "A Preview of the Tolerance Approach to Sensitivity Analysis in Linear Programming," *Discrete Mathematics*, **38**, pp. 121–124, 1982.

Wendell, R. E. "Using Bounds on the Data in Linear Programming: The Tolerance Approach to Sensitivity Analysis," *Mathematical Programming*, **29**, pp. 304–322, 1984.

Wendell, R. E. "The Tolerance Approach to Sensitivity Analysis in Linear Programming," *Management Science*, **31**, pp. 564–578, 1985a.

Wendell, R. E. "Goal Programming Sensitivity Analysis: The Tolerance Approach," *Decision Making with Multiple Objectives*, Springer-Verlag, New York, pp. 300–307, 1985b.

White, D. J. "A Linear Programming Analogue, a Duality Theorem, and a Dynamic Algorithm," *Management Science*, **21** (1), pp. 47–59, September 1974.

White, W. C., M. B. Shapiro, and A. W. Pratt. "Linear Programming Applied to Ultraviolet Absorption Spectroscopy," *Communications of the ACM*, **6** (2), pp. 66–67, February 1963.

Whiting, P., and J. Hillier. "A Method for Finding the Shortest Routing Through a Road Network," *Operational Research Quarterly*, **11**, pp. 37–40, 1960.

Williams, A. C. "A Treatment of Transportation Problems by Decomposition," *SIAM*, **10** (1), pp. 35–48, January–March 1962.

Williams, A. C. "Marginal Values in Linear Programming," *SIAM*, **11** (1), pp. 82–94, March 1963.

Williams, A. C. "Complementary Theorems for Linear Programming," *SIAM Review*, **12** (1), pp. 135–137, January 1970.

Williamson, J. "Determinants Whose Elements are 0 and 1," *American Mathematical Monthly*, **53**, pp. 427–434, 1946.

Wittrock, R. J. "Dual Nested Decomposition of Staircase Linear Programs," *Mathematical Programming Study 24*, ed. R. W. Cottle, pp. 65–86, 1985.

Wolfe, P. "A Technique for Resolving Degeneracy in Linear Programming," *Journal of SIAM*, **11**, pp. 205–211, 1963.

Wolfe, P., and L. Cutler. "Experiments in Linear Programming," *Recent Advances in Mathematical Programming*, R. L. Graves and P. Wolfe (Eds.), McGraw-Hill, New York, pp. 177–200, 1963.

Wolfe, P. "An Extended Composite Algorithm for Linear Programming," Paper P-2373, The Rand Corporation, Santa Monica, Calif., 1961.

Wolfe, P. "The Composite Simplex Algorithm," *SIAM Review*, **7** (1), 1965.

Wolfe, P. "The Product Form of the Simplex Method," U.S.G.R.& D.R. Order AD-612 381, The Rand Corporation, Santa Monica, Calif., 1965.

Wolfe, P., and G. B. Dantzig. "Linear Programming in a Markov Chain," *Operations Research*, **10**, pp. 702–710, 1962.

Wollmer, R. "Removing Arcs from a Network," *Operations Research*, **12** (6), November–December 1964.

Wollmer, R. "Maximizing Flow Through a Network with Node and Arc Capacities," *Transportation Science*, **2** (3), pp. 213–232, August 1968.

Wollmer, R. "The Dantzig–Wolfe Decomposition Principle and Minimum Cost Multicommodity Network Flows," Paper P-4191, The Rand Corporation, Santa Monica, Calif., 1969.

Wollmer, R. "Multicommodity Networks with Resource Constraints: The General Multicommodity Flow Problem," Report TM393-50, Jet Propulsion Laboratory, California Institute of Technology, Pasadena, Calif., 1971.

Woolsey, R. E. D., and H. S. Swanson. *Operations Research for Immediate Application: A Quick and Dirty Manual*, Harper and Row, New York, 1975.

Yadin, D. B., and A. S. Nemirovskii. "Informational Complexity and Efficient Methods for the Solution of Convex Extremal Problems," *Ekonomika i Mathematicheskie Metody*, **12**, pp. 357–369, 1976. (English translation in *Matekon*, **13** (3), pp. 25–45, 1977.)

Yamnitsky, B., and L. A. Levin. "An Old Linear Programming Algorithm Runs in Polynomial Time," in *Proceedings 23rd Annual Symposium on Foundations of Computer Science*, IEEE Computer Society, Long Beach, Calif., pp. 327–328, 1982.

Yaspan, A. "On Finding a Maximal Assignment," *Operations Research*, **14** (4), pp. 646–651, July–August 1966.

Ye, Y. "Barrier Projection and Sliding Current Objective Method for Linear Programming," Presented at 12th Mathematical Programming Symposium, Boston, Engineering Economic Systems Department, Stanford University, Stanford, CA 94305, 1985a.

Ye, Y. "Cutting-Objective and Scaling Methods—A Polynomial Algorithm for Linear Programming," Engineering–Economic Systems Department, Stanford University, Stanford, CA 94305, August 1985b.

Ye, Y. "Recovering Optimal Basis in Karmarkar's Polynomial Algorithm for Linear Programming," Department of Engineering–Economic Systems, Stanford University, Stanford, CA, 1987.

Ye, Y. "Karmarkar's Algorithm and the Ellipsoid Method," Department of Engineering–Economic Systems, Stanford University, Stanford, CA, 1987a.

Ye, Y. and M. Kojima. "Recovering Optimal Dual Solutions in Karmarkar's Polynomial Algorithm for Linear Programming," *Mathematical Programming*, **39** (3), pp. 305–318, 1987.

Ye, Y., and M. J. Todd. "Containing and Shrinking Ellipsoids in the Path-Following Algorithm," Department of Engineering–Economic Systems, Stanford University, Stanford, CA, 1987.

Yoshida, M. "Some Examples Related to the Duality Theorem in Linear Programming," *Journal of Science of the Hiroshima University*, Series A-I (Japan), **30** (1), pp. 41–43, July 1966.

Young, R. D. "A Primal (All-Integer) Integer Programming Algorithm," *Journal of Research*—National Bureau of Standards: B, Mathematics and Mathematical Physics, **69** (3), pp. 213–250, July–September 1965.

Young, R. D. "A Simplified Primal (All Integer) Integer Programming Algorithm," *Operations Research*, **16** (4), pp. 750–782, July–August 1968.

Zadeh, N. "A Bad Network Problem for the Simplex Method and Other Minimum Cost Flow Algorithms," *Mathematical Programming*, **5**, pp. 255–266, 1973.

Zadeh, N. "Near-Equivalence of Network Flow Algorithms," Technical Report No. 26, Department of Operations Research, Stanford University, Stanford, CA, 1979.

Zangwill, W. I. "The Convex Simplex Method," *Management Science*, **14** (3), pp. 221–238, November 1967.

Zangwill, W. I. *Nonlinear Programming: A Unified Approach*, Prentice-Hall, Englewood Cliffs, N.J., 1969.

Zangwill, W. I. "Minimum Convex Cost Flows in Certain Networks," *Management Science*, **14** (7), pp. 429–450, March 1968.

Zeleny, M. *Linear Multiobjective Programming*, Springer, New York, 1974.

Zionts, S. "The Criss-Cross Method for Solving Linear Programming Problems," *Management Science*, **15** (7), pp. 426–445, March 1969.

Zionts, S. *Linear and Integer Programming*, Prentice-Hall, Englewood Cliffs, N.J., 1974.

Zipkin, P. "Bounds for Aggregating Nodes in Network Problems," *Mathematical Programming*, **19**, pp. 155–177, 1980.

Zipkin, P. and K. Raimer. "An Improved Disaggregation Method for Transportation Problems," *Mathematical Programming*, **26** (2), pp. 238–242, 1983.

Zoutendijk, G. *Methods of Feasible Directions*, Elsevier, New York, 1960.

INDEX